Lecture Notes in Computer Science 6365

Vincent Vigneron Vicente Zarzoso
Eric Moreau Rémi Gribonval
Emmanuel Vincent (Eds.)

Latent Variable Analysis and Signal Separation

9th International Conference, LVA/ICA 2010
St. Malo, France, September 27-30, 2010
Proceedings

Springer

Volume Editors

Vincent Vigneron
Université d'Evry Val d'Essone
Dept. of Electrical Engineering
40 rue du Pelvoux, 91020 Courcouronnes, France
E-mail: vvigne@iup.univ-evry.fr

Vicente Zarzoso
Université de Nice-Sophia Antipolis
Laboratoire I3S, Les Algorithmes - Euclide-B
BP 121, 2000 Route des Lucioles, 06903 Sophia Antipolis Cedex, France
E-mail: zarzoso@i3s.unice.fr

Eric Moreau
Université de Toulon
School of Engineering, Dept. of Telecommunications, ISITV
Avenue George Pompidou, BP 56, 83162 La Valette du Var, Cedex, France
E-mail: moreau@univ-tln.fr

Rémi Gribonval
Emmanuel Vincent
INRIA
Equipe-projet METISS, Centre de Recherche INRIA Rennes-Bretagne Atlantique
Campus de Beaulieu, 35042 Rennes cedex, France
E-mail: {remi.gribonval;emmanuel.vincent}@inria.fr

Library of Congress Control Number: 2010934729

CR Subject Classification (1998): C.3, I.4, I.5, I.6, F.2.2, H.5.5

LNCS Sublibrary: SL 1 – Theoretical Computer Science and General Issues

ISSN 0302-9743
ISBN-10 3-642-15994-X Springer Berlin Heidelberg New York
ISBN-13 978-3-642-15994-7 Springer Berlin Heidelberg New York

springer.com

Printed in Germany

Typesetting: Camera-ready by author, data conversion by Scientific Publishing Services, Chennai, India
Printed on acid-free paper 06/3180

Preface

This volume collects the papers presented at the 9th International Conference on Latent Variable Analysis and Signal Separation, LVA/ICA 2010. The conference was organized by INRIA, the French National Institute for Computer Science and Control, and was held in Saint-Malo, France, September 27–30, 2010, at the Palais du Grand Large.

Ten years after the first workshop on *Independent Component Analysis* (ICA) in Aussois, France, the series of ICA conferences has shown the liveliness of the community of theoreticians and practitioners working in this field. While ICA and blind signal separation have become mainstream topics, new approaches have emerged to solve problems involving signal mixtures or various other types of latent variables: semi-blind models, matrix factorization using sparse component analysis, non-negative matrix factorization, probabilistic latent semantic indexing, tensor decompositions, independent vector analysis, independent subspace analysis, and so on. To reflect this evolution towards more general *latent variable analysis* problems in signal processing, the ICA International Steering Committee decided to rename the 9th instance of the conference LVA/ICA.

From more than a hundred submitted papers, 25 were accepted as oral presentations and 53 as poster presentations. The content of this volume follows the conference schedule, resulting in 14 chapters. The papers collected in this volume demonstrate that the research activity in the field continues to range from abstract concepts to the most concrete and applicable questions and considerations. Speech and audio, as well as biomedical applications, continue to carry the mass of the applications considered. Unsurprisingly the concepts of sparsity and non-negativity, as well as tensor decompositions, have become predominant, reflecting the strong activity on these themes in signal and image processing at large. The organizing committee was pleased to invite leading experts in these fields for keynote lectures: Pierre Comon (Université de Nice, France); Stéphane Mallat (Ecole Polytechnique, France); Mark Girolami, (University of Glasgow, UK); and Arie Yeredor (Tel-Aviv University, Israel). A prize (funded by the Fondation Michel Métivier) was awarded to the Best Student Paper during the conference, and the contributions of the five student nominees also reflected the predominance of these themes.

This year's conference provided a forum for the *2010 Signal Separation Evaluation Campaign (SiSEC 2010)*. SiSEC 2010 successfully continued the series of evaluation campaigns initiated during ICA 2007, in London. Compared with previous campaigns, which focused on audio applications, this year's campaign featured the first evaluation dedicated to biomedical applications, a panel session discussion on evaluation, and two invited papers presenting the results of the campaign.

LVA/ICA 2010 also featured a novel *late-breaking / demo* session dedicated to the presentation and discussion of early results and ideas that had not yet been fully formalized and evaluated, of software and data of interest to the community with a focus on open source resources, and of signal separation systems evaluated in SiSEC 2010 but not associated with a full paper. Another novelty was a panel session discussion on the future of the domain (funded by Adobe Systems Incorporated), chaired by Paris Smaragdis (Adobe, USA), featuring Tony Bell (Redwood Center for Theoretical Neuroscience, USA), Andrzej Cichocki (RIKEN, Japan), Michel Fliess (Ecole Polytechnique, France), and Christian Jutten (GIPSA-Lab, France).

The success of LVA/ICA 2010 was the result of the hard work of many people whom we warmly thank here. First, we wish to thank the authors, the Program Committee, and the reviewers without whom this high-quality volume would not exist. The organizers also express their gratitude to the members of the International ICA Steering Committee for their continued support of the conference and their precious advice, as well as to the panel session organizers and panelists, with particular thanks to the SiSEC 2010 organizers, to the local organization committee at INRIA, and to the team from Palais du Grand Large.

Finally, LVA/ICA 2010 was made possible thanks to the financial support of our sponsors. We would like to thank the publisher Springer for its continuous support from the very beginning of the LVA/ICA conferences series, with a special mention to Frank Holzwarth for his responsive technical help during the preparation of these proceedings.

July 2010

Vincent Vigneron
Vicente Zarzoso
Eric Moreau
Rémi Gribonval
Emmanuel Vincent

Organization

Executive Committee

General Chairs	Rémi Gribonval INRIA, France Emmanuel Vincent INRIA, France
Program Chairs	Vincent Vigneron Université d'Evry - Val d'Essonne, France Vicente Zarzoso Université de Nice - Sophia Antipolis, France Eric Moreau Université de Toulon, France
Evaluation Chairs	Shoko Araki NTT CS Labs, Japan Fabian Theis Helmholtz Zentrum München, Germany Guido Nolte Fraunhofer Institute FIRST IDA, Germany
Panel Session Chair	Paris Smaragdis Adobe, USA

Technical Program Committee

Tülay Adali (USA)
Massoud Babaie-Zadeh (Iran)
Salah Bourennane (France)
Marc Castella (France)
Charles Casimiro Cavalcante (Brazil)
Sergio Cruces (Spain)
Yannick Deville (France)
Christian Jutten (France)
Juha Karhunen (Finland)
Elmar Lang (Germany)
Ali Mansour (Australia)
Ali Mohammad Djafari (France)
Mark Plumbley (UK)
Mohammad Bagher Shamsolahi (Iran)
Paris Smaragdis (USA)
Michel Verleysen (Belgium)

Local Organization Committee

Nancy Bertin	Nobutaka Ito	Boris Mailhé
Frédéric Bimbot	Steeve Tessier	Sangnam Nam
Ngoc Duong	Élisabeth Lebret	Alexey Ozerov
Valentin Emiya	Stéphanie Lemaile	Prasad Sudhakar

Overseas Liaisons

North American Liaison	Tülay Adali (USA)
South American Liaison	Charles Casimiro Cavalcante (Brazil)
Asian Liaison	Andrzej Cichocki (Japan)

Best Student Paper Award Jury

Chair Christian Jutten (France)
Andrzej Cichocki (Japan)
Mark Plumbley (UK)
Pierre Comon (France)
Shoji Makino (Japan)

ICA International Steering Committee

Luis Almeida (Portugal)
Shun-Ichi Amari (Japan)
Jean-François Cardoso (France)
Andrzej Cichocki (Japan)
Scott Douglas (USA)
Rémi Gribonval (France)
Simon Haykin (Canada)
Christian Jutten (France)
Te-Won Lee (USA)
Shoji Makino (Japan)
Klaus-Robert Müller (Germany)
Noboru Murata (Japan)
Erkki Oja (Finland)
Mark Plumbley (UK)
Paris Smaragdis (USA)
Fabian Theis (Germany)
Ricardo Vigario (Finland)

Referees

Absil, P.
Achard, S.
Albera, L.
Anderson, M.
Arenas-García, J.
Attux, R.
Barrere, J.
Belouchrani, A.
Berne, O.
Bertin, N.
Blumensath, T.
Borloz, B.
Boufounos, P.
Caiafa, C.
Castells, F.
Chabriel, G.
Chambers, J.
Choi, S.
Cichocki, A.
Clifford, G.
Comon, P.
Correa, N.
Dapena, A.
de Lannoy, G.
De Lathauwer, L.
De Luigi, C.
Diamantaras, K.
Duarte, L.
Dubroca, R.
Duong, N.
Durán-Díaz, I.
El Fadaili, M.
Erdogmus, D.
Fatemizadeh, E.
Fazel, R.
Fernandes, A.
Fernandes, C. A.
Fernandes, C. E.
François, D.
Friedlander, B.
Fuchs, J.-J.
Fyfe, C.
Gini, F.
Girin, L.
Gorriz, J. M.
Gribonval, R.
Hazan, A.
He, Zhaoshui
Hild II, K. E.
Hosseini, S.
Igual-García, J.
Ilin, A.
Jafari, M.
James, C. J.

Kaban, A.
Keck, I. R.
Klapuri, A.
Kofidis, E.
Kohl, F.
Kokkinakis, K.
Koldovsky, Z.
Kuruoglu, E.
Lee, J.
Le Roux, J.
Li, H.
Li, X.L.
Li, Y.-O.
Lopes, R.
Luo, B.
Maleki, A.
Meilinger, M.
Mitianoudis, N.
Montagne, C.
Montalvão, J.
Moradi, M. H.
Moreau, E.
Mørup, M.
Moussaoui, S.
Murillo-Fuentes, J. J.
Mysore, G. J.
Nam, J.
Nandi, A. K.
Nasrabadi, A. M.
Nesbit, A.
Neves, A.
Niazadeh, R.
Nishimori, Y.
Novey, M.
Ozerov, A.
Paris, S.
Pearlmutter, B.
Pérez Iglesias, H. J.
Pesquet-Popescu, B.
Petraglia, M.
Phan, A.H.
Phlypo, R.
Puigt, M.
Puntonet, C. G.
Raiko, T.
Raj, Bhiksha
Richard, G.
Rieta, J. J.
Rivet, B.
Rodriguez, P.
Sameni, R.
Sanei, S.
Sarmiento, A.
Schachtner, R.
Senhadji, L.
Senninger, D.
Setarehdan, K.
Shamsollahi, M.B.
Shashanka, M.
Shimizu, S.
Slaney, M.
Solé-Casals, J.
Soltanian-Zadeh, H.
Souloumiac, A.
Sudhakar, P.
Suyama, R.
Tabus, I.
Tanaka, T.
Theis, F.
Thirion, N.
Tichavsky, P.
Tomazeli, L.
Tomé, A. M.
Tsalaile, T.
Tzanetakis, G.
Vellido, A.
Vergara, L.
Via, J.
Vigario, R.
Vigneron, V.
Vrabie, V.
Wang, Yide
Xerri, B.
Yang, Zhirong
Ylipaavalniemi, J.
Zarzoso, V.
Zayyani, H.
Zhang, K.
Zhong, M.

Sponsoring Institutions

INRIA, the French National Institute for Computer Science and Control;
SMALL, EU-funded FET-Open project (FP7-ICT-225913-SMALL, small-project.eu);
Fondation Michel Métivier;
Adobe Systems Incorporated;
Université de Rennes 1;
Ministère délégué à la Recherche et l'Enseignement Supérieur;
Conseil Régional de Bretagne.

Table of Contents

Speech and Audio Applications

Convolutive Signal Separation

The 2010 Signal Separation Evaluation Campaign (SiSEC2010)

Audio

Theory

Telecom

Tensor Factorizations

Sparsity I

Sparsity; Biomedical Applications

Non-negativity; Image Processing Applications

Tensors; Joint Diagonalization

Sparsity II

Biomedical Applications

Emerging Topics

Blind Source Separation Based on Time-Frequency Sparseness in the Presence of Spatial Aliasing

Benedikt Loesch and Bin Yang

Chair of System Theory and Signal Processing, University of Stuttgart
{benedikt.loesch,bin.yang}@LSS.uni-stuttgart.de

Abstract. In this paper, we propose a novel method for blind source separation (BSS) based on time-frequency sparseness (TF) that can estimate the number of sources and time-frequency masks, even if the spatial aliasing problem exists. Many previous approaches, such as degenerate unmixing estimation technique (DUET) or observation vector clustering (OVC), are limited to microphone arrays of small spatial extent to avoid spatial aliasing. We develop an offline and an online algorithm that can both deal with spatial aliasing by directly comparing observed and model phase differences using a distance metric that incorporates the phase indeterminacy of 2π and considering all frequency bins simultaneously. Separation is achieved using a linear blind beamformer approach, hence musical noise common to binary masking is avoided. Furthermore, the offline algorithm can estimate the number of sources. Both algorithms are evaluated in simulations and real-world scenarios and show good separation performance.

Keywords: Blind source separation, adaptive beamforming, spatial aliasing, time-frequency sparseness.

1 Introduction

The task of convolutive blind source separation is to separate M convolutive mixtures $x_m[i], m = 1, \ldots, M$ into N different source signals. Mathematically, we write the sensor signals $x_m[i]$ as a sum of convolved source signals

$$x_m[i] = \sum_{n=1}^{N} h_{mn}[i] * s_n[i], \quad m = 1 \ldots M \tag{1}$$

Our goal is to find signals $y_n[i]$, $n = 1 \ldots N$ such that, after solving the permutation ambiguity, $y_n[i] \approx s_n[i]$ or a filtered version of $s_n[i]$. In the case of moving sources, the impulse responses $h_{mn}[i]$ are time-varying.

Our algorithms cluster normalized phase vectors $\bar{\mathbf{X}}[k,l]$ in the time-frequency domain, where k is the frequency index and l is the time frame index, respectively. Each cluster with the associated state vector $\mathbf{c}[k, \mathbf{p}_n]$ corresponds to a different source n with the associated location or direction-of-arrival (DOA) parameter vector $\mathbf{p}_n$. Different from DUET, our algorithms can use more than

V. Vigneron et al. (Eds.): LVA/ICA 2010, LNCS 6365, pp. 1–8, 2010.

two microphones. In contrast to DUET and OVC, our algorithms do not suffer from the spatial aliasing problem. After clustering the phase vectors $\bar{\mathbf{X}}[k,l]$, we apply time-frequency masking or the blind beamformer from [1] to separate the sources.

2 Proposed Offline Algorithm

After a short-time Fourier transform (STFT), we can approximate the convolutive mixtures in the time-domain as instantaneous mixtures at each time-frequency (TF) point $[k,l]$:

$$\mathbf{X}[k,l] \approx \sum_{n=1}^{N} \mathbf{H}_n[k] S_n[k,l] \quad (2)$$

$\mathbf{X} = [X_1, \ldots, X_M]^T$ is called an observation vector and $\mathbf{H}_n = [H_{1n}, \ldots, H_{Mn}]^T$ is the vector of frequency responses from source n to all sensors. We assume that the direct path is stronger than the multipath components. This allows us to exploit the DOA information to perform the separation. Note that we are only interested in separation and not in dereverberation. Hence, we do not aim at a complete inversion of the mixing process. As a consequence, we do not require minimum-phase mixing. The proposed algorithm consists of three steps: normalization, clustering, and reconstruction of the separated signals.

2.1 Normalization

From the observation vectors $\mathbf{X}[k,l]$, we derive the normalized phase vectors $\bar{\mathbf{X}}[k,l]$ which contain only the phase differences of the elements of $\mathbf{X}[k,l]$ with respect to a reference microphone J:

$$\bar{\mathbf{X}}[k,l] = \left[e^{j \cdot \arg(X_m[k,l]/X_J[k,l])} \right], \quad m = 1, \cdots, M \quad (3)$$

For a single active source, the phase of the ratio of two elements of $\mathbf{X}[k,l]$ is a linear function of the frequency index k (modulo 2π):

$$\arg\left(X_m[k,l]/X_J[k,l]\right) = 2\pi \Delta f k \tau_m + 2\pi o, \quad o \in \mathbb{Z} \quad (4)$$

where Δf is the frequency bin width and τ_m is the time-difference of arrival (TDOA) of the source with respect to microphone m and J. If there is no spatial aliasing (i.e. $o = 0$), we can cluster the TDOAs at all TF points because of $\tau_m = \frac{1}{2\pi \Delta f k} \arg\left[\frac{X_m[k,l]}{X_J[k,l]}\right]$. However, in the case of spatial aliasing ($o \neq 0$), we can no longer cluster $\frac{1}{2\pi \Delta f k} \arg\left[\frac{X_m[k,l]}{X_J[k,l]}\right]$ directly. Instead we would need to take into account all possible values of o. However, we can avoid this problem by directly comparing the observed phase difference and the model phase difference for multiple microphone pairs using the distance metric

$$\|\bar{\mathbf{X}}[k,l] - \mathbf{c}[k,\mathbf{p}]\|^2 = 2M - 2 \cdot \sum_{m=1}^{M} \cos\left(\arg\left[\frac{X_m[k,l]}{X_J[k,l]}\right] - 2\pi \Delta f k \tau_m(\mathbf{p}) \right) \quad (5)$$

with the state vector $\mathbf{c}[k,\mathbf{p}] = [c_m]_{1\le m\le M} = \left[e^{j2\pi\Delta f k\tau_m(\mathbf{p})}\right]_{1\le m\le M}$. $\mathbf{c}[k,\mathbf{p}]$ contains the expected phase differences for a potential source at $\mathbf{p}$ with respect to microphones $m = 1,\cdots,M$ and J.

The distance metric (5) allows an estimation of the location or DOA parameters $\mathbf{p}_n$ of all sources even if spatial aliasing occurs. This is achieved by considering all frequency bins simultaneously: Due to the spatial aliasing, (5) contains location ambiguities for higher frequencies. However, these ambiguities are removed by summing across all frequency bins. We define $\mathcal{J}(\mathbf{p})$

$$\mathcal{J}_l(\mathbf{p}) = \sum_k \rho(\|\bar{\mathbf{X}}[k,l] - \mathbf{c}[k,\mathbf{p}]\|^2), \quad \mathcal{J}(\mathbf{p}) = \sum_l \mathcal{J}_l(\mathbf{p}), \tag{6}$$

which has maxima for $\mathbf{p} = \mathbf{p}_n$. $\rho(t)$ is a monotoneously decreasing nonlinear function in the range $[0,1]$ that reduces the influence of outliers and increases spatial resolution.

We estimate $\mathbf{p}_n$ by looking for the maxima of $\mathcal{J}(\mathbf{p})$ and then cluster the TF points as described in the next section.

2.2 Source Number Estimation and Clustering

We need to estimate the number of sources $\hat{N}$ and then find clusters $C_1,\ldots,C_{\hat{N}}$ of $\bar{\mathbf{X}}[k,l]$ with centroids $\mathbf{c}[k,\hat{\mathbf{p}}_n]$. Unlike [2–4], we achieve the clustering by a direct search over all possible TDOAs or DOAs $\mathbf{p}$ and do not use iterative approaches such as k-means or expectation-maximization (EM). This has the advantage, that we are guaranteed to find the global optima of the cost function. Inspired from [5], we propose to use $\rho(t) = 1 - \tanh(\alpha\sqrt{t})$ as the nonlinear function $\rho(t)$ in (6). Independently of our research, [6] proposed a similar cost function $\mathcal{J}_l$ for only two microphones and without the summation over time for localization purposes.

Another advantage of the direct search is that we do not need to know the number of sources beforehand as in [2]. Instead, we can count the number of significant and distinct peaks of $\mathcal{J}(\mathbf{p})$: This is done by finding all peaks $\mathbf{p}_n$ of $\mathcal{J}(\mathbf{p})$ with $\mathcal{J}(\mathbf{p}) > t$ and sorting the peaks in descending order $\mathcal{J}(\mathbf{p}_i) > \mathcal{J}(\mathbf{p}_{i+1})$. Then we start from the first peak and accept the next peak if the minimum distance to a previously accepted peak is larger than a certain threshold t_2. The number of estimated sources $\hat{N}$ is then given as the number of accepted peaks.

Since in (6) the peak height is a function of the amount of source activity it might be difficult to count the number of sources if the amount of source activity differs a lot among the sources. One way to solve this problem is to use the max-approach from [6] to estimate the number of sources by replacing $\mathcal{J}(\mathbf{p})$ by $\tilde{\mathcal{J}}(\mathbf{p}) = \max_l \mathcal{J}_l(\mathbf{p})$. This modified function is less sensitive to the amount of source activity because the peak height is proportional to the coherence of the observed phase. So if for each source there is at least one time frame where it is the single active source, $\tilde{\mathcal{J}}(\mathbf{p})$ will yield a large peak for this source. Fig. 1 shows $\mathcal{J}(\mathbf{p})$ and $\tilde{\mathcal{J}}(\mathbf{p})$ for two scenarios with different amounts of source activity. In Fig. 1(b) the max-approach is clearly superior because the contrast beetween true

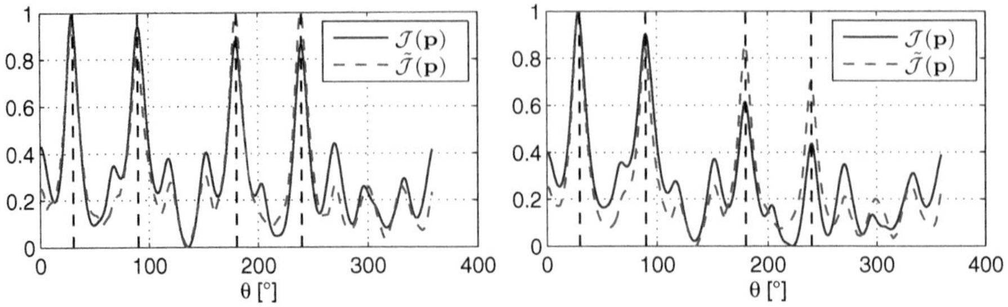

(a) Sources always active

(b) Src 3: start 10 s, Src 4: start 14 s

Fig. 1. Source Number and DOA Estimation for different source acitivity (length 24 s)

peaks and spurious peaks is larger. Furthermore, the max-approach improves TDOA estimation for closely spaced microphones: It selects time frames with high coherence of the observed phase, i.e. a single active source.

The positions/DOAs $\hat{\mathbf{p}}_n, n = 1, \cdots, \hat{N}$ of the sources are given by the relevant peaks of $\mathcal{J}(\mathbf{p})$ or $\tilde{\mathcal{J}}(\mathbf{p})$. For each source, we generate the corresponding state vectors $\mathbf{c}[k, \hat{\mathbf{p}}_n], n = 1, \cdots, \hat{N}$ and assign all TF points to cluster n for which $\|\bar{\mathbf{X}}[k,l] - \mathbf{c}[k, \hat{\mathbf{p}}_n]\|^2$ is minimal.

Comparison with EM Algorithm with GMM Model: [4] uses a related approach for two microphones: They use an EM algorithm with a Gaussian mixture model (GMM) for the phase difference between the two microphones at each TF point. The phase difference of each source is modelled as a mixture of $2K_f + 1$ Gaussians with mean $2\pi k \cdot (1 + \Delta f \mu_q), k = -K_f, \cdots, K_f$, where μ_q is the mean of the q-th component. The observed phase difference for all sources is then described as a mixture of Q such models. Furthermore they use a Dirichlet prior for the mixture weights $\alpha_q, q = 1, \cdots, Q$ to model the sparsity of the source directions, i.e. to represent the phase difference with a small number of Gaussians with large weight α_q. After convergence of the EM algorithm the mean μ_q of the Gaussians with large weight α_q and small variance σ_q^2 reflect the estimated TDOAs of the N sources. Our approach differs in a number of ways:

- We use a direct search instead of an iterative procedure to estimate the source parameters $\mathbf{p}_n$.
- We are guaranteed to find the global optima of the function $\mathcal{J}(\mathbf{p})$, whereas the EM algorithm could converge to local optima.
- We estimate the number of sources by counting the number of significant and distinct peaks instead of checking the weights and variance of the components of a GMM model.
- We do not model the phase difference of each source using $2K_f + 1$ Gaussians. Instead we use the distance metric (5) which incorporates the phase wrapping.

– Our approach is computationally less demanding: We need $N_{\text{grid}} \cdot K \cdot L$ function evaluations, while the EM algorithm requires $N_{\text{iter}} \cdot Q \cdot (2K_f+1) \cdot K \cdot L$ function evaluations. For a typical scenario $N_{\text{grid}} = 180, N_{\text{iter}} = 10, Q = 8, K_f = 5$ and assuming comparable computational cost for each function evaluation, our approach would be about 6 times faster.

The differences between [4] and our approach can be summarized as differences in the model (wrapped Gaussians vs. 2π-periodic distance function) and in the clustering algorithm (iterative EM algorithm vs. direct search and simple clustering).

The authors would like to thank Dr. Shoko Araki for running our proposed algorithm on her dataset from [4]. Using $t = 0.5$ and $t_2 = 5°$, our algorithm estimates the number of sources correctly for all tested cases.

2.3 Reconstruction

To reconstruct the separated signals, we use the blind beamforming approach discussed in [1]. This approach reduces or completely removes musical noise artifacts which are common in binary TF mask based separation. After the beamforming step, additional optional binary TF masks can be used to further suppress the interference. Then we convert the separated signals back to the time domain using an inverse STFT.

3 Online Separation Algorithm

The online algorithm operates on a single frame basis and uses a gradient ascent search for the TDOAs/DOAs of the sources with prespecified maximum number of sources N. Inactive sources result in empty clusters. The cost function $\mathcal{J}_l$ for the l-th STFT frame is

$$\mathcal{J}_l(\mathbf{p}_n) = \sum_k \rho(\|\bar{\mathbf{X}}[k,l] - \mathbf{c}[k,\mathbf{p}_n]\|^2) \tag{7}$$

Its gradient vector with respect to $\mathbf{p}_n$ is

$$\frac{\partial \mathcal{J}_l}{\partial \mathbf{p}_n} = -\sum_k \Re\left\{ j2\pi\Delta f k \cdot \frac{\partial \rho(t)}{\partial t} \cdot \frac{\partial \boldsymbol{\tau}}{\partial \mathbf{p}_n} \cdot \mathbf{c}[k,\mathbf{p}_n] \odot 2(\bar{\mathbf{X}}[k,l] - \mathbf{c}[k,\mathbf{p}_n]) \right\}, \tag{8}$$

where $\boldsymbol{\tau} = [\tau_m]_{1 \le m \le M}$. Similar to [7], we use a time-varying learning rate which is a function of the amount of TF points associated with source n. The separation steps are similar to the offline algorithm. A computationally efficient version of the blind beamforming algorithm can be obtained by using recursive updates of the parameters as in [8].

4 Experimental Evaluation

For the experimental evaluation, we used a sampling frequency of $f_s = 16\,\text{kHz}$, a STFT with frame length 1024 and 75% overlap. SNR was set to 40 dB. We selected six speech signals from the short stories of the CHAINS corpus [9].

4.1 Stationary Sources

First, we evaluate the offline algorithm using the measured impulse responses of room E2A ($T_{60} = 300$ ms) from the RWCP sound scene database [10]. This database contains impulse responses for sources located on a circle with radius 2.02 m with an angular spacing of 20°. The microphone array is shifted by 0.42 m with respect to the center of the circle. We consider a two-microphone array with the spacings $d = 11.3$ cm and $d = 33.9$ cm. All sources have equal power and a length of 5 s. For $d = 11.3$ cm, $\mathcal{J}(\mathbf{p})$ sometimes does not show N distinct maxima. Hence, we use $\tilde{\mathcal{J}}(\mathbf{p})$ for $d = 11.3$ cm since it provides increased resolution by using frames with a single active source for the localization. We have tried different signal combinations and DOA scenarios (A,B,C,D): For scenarios A,B,C we varied the DOAs between 30° ... 150° and tested different angular spacings between the sources. For case D we distributed the sources with maximum angular spacing between 10° ... 170°. The angular spacings between the sources are summarized in Table 1. For each scenario, we considered all signal combinations of $N = 2, 3, 4$ out of the 6 source signals. Table 2 summarizes the performance of the source number estimation and the signal-to-interference (SIR) gain for the different cases. For the source number estimation, we used thresholds $t = 0.46$ for $d = 11.3$ cm and $t = 0.55$ for $d = 33.9$ cm. t_2 was set to 10° for both microphone spacings d. The performance is evaluated by the percentage of estimations for which $\hat{N} = n, n = 1, \cdots, 5$. This is known as the confusion matrix.

Table 1. Considered scenarios

	A	B	C	D
$N = 2$	20°	–	40°	160°
$N = 3$	20°/40°	40°/20°	40°/40°	80°/80°
$N = 4$	20°/40°/40°	40°/20°/40°	40°/40°/40°	60°/40°/60°

Table 2. Source number estimation and SIR gain

		$\hat{N}$ accuracy (%)					SIR gain [dB]			
N	spacing	1	2	3	4	5	A	B	C	D
2	$d = 11.3$ cm	0	97	3	0	0	4.6	–	7.9	15.9
	$d = 33.9$ cm	0	100	0	0	0	6.0	–	9.0	13.0
3	$d = 11.3$ cm	0	1	98	1	0	3.7	5.7	8.7	11.4
	$d = 33.9$ cm	0	0	100	0	0	6.2	8.9	9.4	13.1
4	$d = 11.3$ cm	0	0	2	95	3	2.8	6.0	7.1	6.4
	$d = 33.9$ cm	0	0	0	100	0	6.5	8.9	8.0	8.4

Our offline algorithm estimates the number of sources correctly in almost all cases and shows a good separation performance. A larger microphone spacing achieves a better separation performance for closely spaced sources (case A and B) or for large source numbers. On the short two-source-two-microphone mixtures of the SISEC2010 campaign (`http://irisa.fr/metiss/SiSEC10/short/short_all.html`), separation performance is comparable to other algorithms.

Table 3. Global SIR gain in dB for moving sources

T_{60}	algorithm	source 1	source 2	source 3	mean
100 ms	offline	4.5	18.2	10.1	10.9
	online	25.9	28.1	27.8	27.3
200 ms	offline	5.0	15.9	8.5	9.8
	online	24.6	22.4	24.1	23.7
300 ms	offline	4.9	10.7	6.9	7.5
	online	17.3	16.5	18.0	17.3

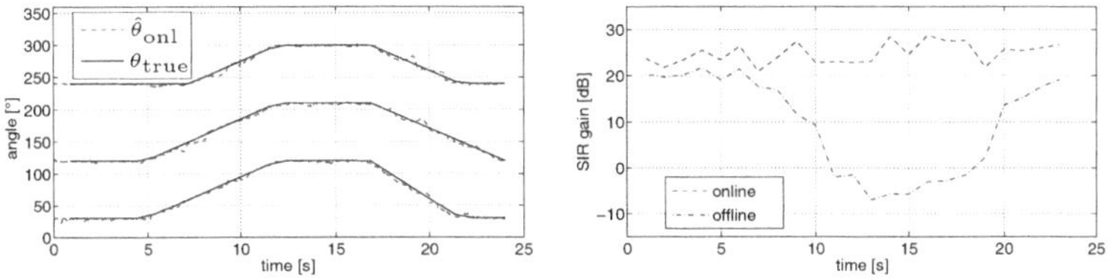

(a) DOA tracking (b) Local SIR gain

Fig. 2. Fixed number of moving sources, $T_{60} = 200\,\text{ms}$

4.2 Moving Sources

In order to have a precise reference for moving source positions, we performed simulations using the MATLAB ISM RoomSim toolbox [11]. The considered room was of size $5\,\text{m} \times 6\,\text{m} \times 2.5\,\text{m}$ and we chose reverberation times of $T_{60} = 100, 200, 300\,\text{ms}$. We used a cross-array (⁘) with $M = 5$ microphones. The microphone spacing was $d = 10\,\text{cm}$, so spatial aliasing occurs above 1700 Hz . We have $N = 3$ sources that move along a circle with radius 1.0 m. Source 1 moves from $\theta_1 = 30°$ to $\theta_1 = 120°$ and back, source 2 moves from $\theta_2 = 120°$ to $\theta_2 = 210°$ and back and source 3 moves from $\theta_3 = 240°$ to $\theta_3 = 300°$. The total simulation time is 24 s. Fig. 2(a) shows the estimated angles $\hat{\theta}_{\text{onl}}[l]$ using our online algorithm as well as the reference angles $\theta_{\text{true}}[l]$. The online-algorithm was initialized with the true angles $\theta_{\text{true}}[0]$. As we see, the online algorithm accurately tracks the sources. During speech pauses, angle estimates are not updated. The separation performance of the online and offline algorithm is summarized in Table 3. As expected, the offline algorithm fails to separate the source signals while our gradient-based online algorithm achieves good results. The reason for the failure of the offline algorithm is that it averages the DOAs of the moving sources. The estimated DOAs are 63°, 119° and 240°. Two of the three estimated DOAs match the initial and final DOAs of the two sources and hence separation quality for these two sources (2 and 3) is acceptable for $T_{60} = 100\,\text{ms}$. However, the separation performance drops significantly when the sources start moving as shown in Fig. 2(b). It shows the local SIR gains which are calculated over non-overlapping segments of 1 second and averaged over the three sources.

5 Conclusion

In this paper we have presented two blind source separation algorithms based on TF sparseness that are able to deal with the spatial aliasing problem by using a distance metric which incorporates phase wrapping (mod 2π) and averaging all frequency bins for the estimation of the location or DOA parameters of the sources. The offline algorithm reliably estimates the source number and achieves the clustering using a direct search. The online algorithm assumes a prespecified maximum number of sources and is able to track moving sources. Both algorithms show good separation performance in midly reverberant environments.

References

1. Cermak, J., Araki, S., Sawada, H., Makino, S.: Blind source separation based on a beamformer array and time frequency binary masking. In: Proc. ICASSP (2007)
2. Araki, S., Sawada, H., Mukay, R., Makino, S.: Underdetermined blind sparse source separation of arbitrarily arranged multiple sensors. Signal Processing 87(8), 1833–1847 (2007)
3. Sawada, H., Araki, S., Mukay, R., Makino, S.: Grouping separated frequency components by estimating propagation model parameters in frequency-domain blind source separation. IEEE Transactions on Audio, Speech and Language Processing 15(5), 1592–1604 (2007)
4. Araki, S., Nakatani, T., Sawada, H., Makino, S.: Stereo source separation and source counting with MAP estimation with Dirichlet prior considering spatial aliasing problem. In: Adali, T., Jutten, C., Romano, J.M.T., Barros, A.K. (eds.) ICA 2009. LNCS, vol. 5441, pp. 742–750. Springer, Heidelberg (2009)
5. Nesta, F., Omologo, M., Svaizer, P.: A novel robust solution to the permutation problem based on a joint multiple TDOA estimation. In: Proc. International Workshop for Acoustic Echo and Noise Control, IWAENC (2008)
6. Chami, Z.E., Guerin, A., Pham, A., Serviere, C.: A phase-based dual microphone method to count and locate audio sources in reverberant rooms. In: Proc. IEEE Workshop on Applications of Signal processing to Audio and Acoustics, WASPAA (2009)
7. Rickard, S., Balan, R., Rosca, J.: Real-time time-frequency based blind source separation. In: Proc. ICA (2001)
8. Loesch, B., Yang, B.: Online blind source separation based on time-frequency sparseness. In: Proc. ICASSP (2009)
9. Cummins, F., Grimaldi, M., Leonard, T., Simko, J.: The CHAINS corpus (characterizing individual speakers) (2006), http://chains.ucd.ie/
10. Real World Computing Partnership, RWCP Sound Scene Database in Real Acoustic Environment (2001), http://tosa.mri.co.jp/sounddb/indexe.htm
11. Lehmann, E.: Image-source method for room impulse response simulation, room acoustics (2008), http://www.watri.org.au/~ericl/ism_code.html

Adaptive Time-Domain Blind Separation of Speech Signals*

Jiří Málek[1], Zbyněk Koldovský[1,2], and Petr Tichavský[2]

[1] Faculty of Mechatronic and Interdisciplinary Studies
Technical University of Liberec, Studentská 2, 461 17 Liberec, Czech Republic
jiri.malek@tul.cz
[2] Institute of Information Theory and Automation, Pod vodárenskou věží 4,
P.O. Box 18, 182 08 Praha 8, Czech Republic

Abstract. We present an adaptive algorithm for blind audio source separation (BASS) of moving sources via Independent Component Analysis (ICA) in time-domain. The method is shown to achieve good separation quality even with a short demixing filter length (L = 30). Our experiments show that the proposed adaptive algorithm can outperform the off-line version of the method (in terms of the average output SIR), even in the case in which the sources do not move, because it is capable of better adaptation to the nonstationarity of the speech.

1 Introduction

The task considered in this paper is the blind separation of d unknown audio sources (BASS) from m recordings, where the unknown mixing process is convolutive and potentially dynamic, e.g., due to moving sources. It is assumed that the system changes slowly and may be considered being static in short time intervals. Therefore, within interval of the length P, the classical convolutive mixing problem is considered, which is described by

$$x_i(n) = \sum_{j=1}^{d} \sum_{\tau=0}^{M_{ij}} h_{ij}(\tau) s_j(n-\tau), \quad i = 1, \ldots, m. \tag{1}$$

Here, $x_1(n), \ldots, x_m(n)$ are the observed signals on microphones, $s_1(n), \ldots, s_d(n)$ are the unknown source signals, and h_{ij} are unknown impulse responses of length M_{ij}. The original sources are assumed to be independent, which allows the basis of the separation to be the Independent Component Analysis (ICA) [1]. For simplicity, we will assume that the number of sources d remains the same throughout the whole recording.

The separation of dynamic mixtures is usually done with block-by-block application of a method intended for stationary mixtures. The method may be

* This work was partly supported by Ministry of Education, Youth and Sports of the Czech Republic through the project 1M0572 and partly by Grant Agency of the Czech Republic through the projects 102/09/1278 and 102/08/0707.

V. Vigneron et al. (Eds.): LVA/ICA 2010, LNCS 6365, pp. 9–16, 2010.

modified more or less to respect the continuity of the (de)mixing process, and the outputting signals are synthesized from the separated signals on blocks. We call such methods *on-line*.

An on-line method applying ICA in the frequency domain was proposed by Mukai et al in [2]. An on-line method working in time-domain based on second-order statistics cost function was proposed by Buchner et al in [3]. Sparseness based on-line algorithm working in frequency domain was presented by Loesch and Yang in [4].

In this paper, we propose an online algorithm that comes from the BASS method from [5]. This original method applies an ICA algorithm to the mixed signals in time-domain to obtain independent components that correspond to randomly filtered versions of the original signals. The components are then grouped into clusters so that components in a cluster correspond to the same source. Finally, components of a cluster are used to reconstruct separated responses (spatial images) of the corresponding source on microphones. In the proposed on-line method, this process is modified so that the ICA and clustering algorithms adapt their internal parameters by performing one iteration in each block only. A new clustering criterion for the similarity of components, which is computationally more effective than the one in [6], is proposed. The speed of adaptivity can be driven by learning parameters and could be made very fast, due to fast convergence of ICA that is based on BGSEP from [7].

The following Section 2 describes all necessary details of the proposed on-line method. Section 3 demonstrates its performance in experiments with real-world recordings and Section 4 concludes the paper.

2 The Proposed Algorithm

The input signals are divided into overlapping blocks of length P, with the shift of T samples such that $R = P/T$ is an integer. The length of overlap of two consecutive blocks is thus $P - T$. The Ith block of the jth input signal will be denoted by

$$x_j^I(n) = x_j((I-1) \cdot T + n), \qquad n = 1, \ldots, T. \tag{2}$$

The uppercase superscript I will be used to denote data and quantities related to the Ith block. A separation procedure described below is successively applied to blocks of input signals and outputs blocks of separated microphone responses (spatial images) of the source signals.

Like the off-line method, the on-line procedure forms delayed copies of the microphone signals, (I) applies a simplified BGSEP algorithm to decompose the data matrix into its independent components and (II) uses a special fuzzy clustering method to group the independent components to form independent subspaces that represent the separated sources. The third step (III) consists in the reconstruction of the separated signals in each block and averaging the signals in the overlapping windows.

2.1 Step I: ICA (Simplified BGSEP Algorithm)

Let $\mathbf{X}^I$ be the data matrix from the Ith block of input signals defined as

$$\mathbf{X}^I = \begin{bmatrix} x_1^I(1) & x_1^I(2) & \ldots\ldots & x_1^I(P) \\ x_1^I(2) & x_1^I(3) & \ldots\ldots & x_1^I(P+1) \\ \vdots & \vdots & \vdots\ \vdots & \vdots \\ x_1^I(L) & x_1^I(L+1) & \ldots\ldots & x_1^I(P+L) \\ x_2^I(1) & x_2^I(2) & \ldots\ldots & x_2^I(P) \\ \vdots & \vdots & \vdots\ \vdots & \vdots \\ \vdots & \vdots & \vdots\ \vdots & \vdots \\ x_m^I(L) & x_m^I(L+1) & \ldots\ldots & x_m^I(P+L) \end{bmatrix}, \tag{3}$$

where L is a free parameter corresponding to the length of the demixing MIMO filter.

The goal of the step is to find a demixing matrix $\mathbf{W}^I$ so that rows of $\mathbf{C}^I = \mathbf{W}^I\mathbf{X}^I$ are as independent as possible, thus, correspond to "independent" components (ICs) of $\mathbf{X}^I$.

The matrix $\mathbf{X}^I$ can be partitioned in a vertical way in M blocks of equal size, $(mL) \times (P/M)$,

$$\mathbf{X}^I = [\mathbf{X}^{I,1}, \ldots, \mathbf{X}^{I,M}]. \tag{4}$$

The simplified BGSEP algorithm estimates $\mathbf{W}^I$ by a joint approximate diagonalization of a set of the covariance matrices

$$\mathbf{R}^{I,k} = \frac{M}{P}\mathbf{X}^{I,k}(\mathbf{X}^{I,k})^T, \qquad k = 1, \ldots, M. \tag{5}$$

For convenience and computation savings we assume that the number of the matrices M is equal to the parameter R that appears in the division of the signal to overlapping blocks. Then, in the transition $\{\mathbf{R}^{I-1,k}\}_{k=1}^M \to \{\mathbf{R}^{I,k}\}_{k=1}^M$, the set of matrices remains unchanged, except for the removed matrix $\mathbf{R}^{I-1,1}$ and the added matrix $\mathbf{R}^{I,M}$.

The diagonalization proceeds by performing one iteration of the WEDGE algorithm - Weighted Exhaustive Diagonalization with Gauss itErations [7], with the weight matrices that are diagonal, optimized for the case when the signals obey the piecewise stationary model. The algorithm uses the estimate of demixing matrix from the previous segment $\mathbf{W}^{I-1}$ to partially diagonalize the matrices in (5)

$$\mathbf{P}^{I,k} = \mathbf{W}^{I-1}\mathbf{R}^{I,k}(\mathbf{W}^{I-1})^T \qquad k = 1, \ldots, M. \tag{6}$$

As in [7], the demixing matrix $\mathbf{W}^I$ is obtained by updating $\mathbf{W}^{I-1}$ as

$$\mathbf{W}^I = (\mathbf{A}^I)^{-1}\mathbf{W}^{I-1} \tag{7}$$

where $\mathbf{A}^I$ has ones on its main diagonal, and the off-diagonal elements are obtained by solving the 2×2 systems

$$\begin{bmatrix} \mathbf{A}_{kl}^I \\ \mathbf{A}_{lk}^I \end{bmatrix} = \beta^I \begin{bmatrix} \mathbf{r}_{ll}^T\mathbf{Z}_{kl}\mathbf{r}_{ll} & \mathbf{r}_{kk}^T\mathbf{Z}_{kl}\mathbf{r}_{ll} \\ \mathbf{r}_{kk}^T\mathbf{Z}_{kl}\mathbf{r}_{ll} & \mathbf{r}_{kk}^T\mathbf{Z}_{kl}\tilde{\mathbf{r}}_{kk} \end{bmatrix}^{-1} \begin{bmatrix} \mathbf{r}_{ll}^T\mathbf{Z}_{kl}\mathbf{r}_{kl} \\ \mathbf{r}_{kk}^T\mathbf{Z}_{kl}\mathbf{r}_{kl} \end{bmatrix}, \tag{8}$$

with

$$\mathbf{r}_{kl} = [(\mathbf{P}^{I,1})_{kl}, \ldots, (\mathbf{P}^{I,M})_{kl}]^T, \tag{9}$$

and

$$\mathbf{Z}_{kl} = \mathrm{diag}\left(\frac{1}{(\mathbf{P}^{I,1})_{kk}(\mathbf{P}^{I,1})_{ll}}, \ldots, \frac{1}{(\mathbf{P}^{I,M})_{kk}(\mathbf{P}^{I,M})_{ll}}\right) \tag{10}$$

for $k, l = 1, \ldots, mL$, $k > l$. The variable β^I in (8) does not exist in the original WEDGE algorithm: it is added here to control the speed of algorithm's convergence. The choice of β^I will be discussed later in Section 2.4.

2.2 Step II: Clustering of Independent Components

Similarity of ICs. Due to the indeterminacy of ICA, the ICs of $\mathbf{X}^I$ are arbitrarily filtered versions of the original signals. To recognize whether two components correspond to the same source, we compute their generalized cross-correlation coefficients known as GCC-PHAT [8]. The coefficients are invariant to the magnitude spectra of the signals and depend on their phase spectra only, which makes them appropriate for the similarity evaluation.

Let $C_i^I(k)$ and $C_j^I(k)$ denote the Fourier transform of the ith and jth component, respectively, $i, j = 1, \ldots, mL$, and k denotes the frequency index. The GCC-PHAT coefficients of the components, denoted by $g_{ij}^I(n)$, are equal to the inverse Fourier transform of

$$G_{ij}^I(k) = \frac{C_i^I(k) \cdot C_j^I(k)^*}{|C_i^I(k)| \cdot |C_j^I(k)|}, \tag{11}$$

where * denotes the complex conjugation. Fast computation of $g_{ij}^I(n)$ can be done by means of the FFT.

If the components correspond exactly to the same source, i.e. without any residual interference, $g_{ij}^I(n)$ is equal to delayed unit impulse function, where the delay cannot be greater than L. Hence, the similarity between the ith and jth component can be measured by $\sum_{n=-L}^{L} |g_{ij}^I(n)|$ and the matrix of mutual similarity $\mathbf{D}^I$ can be computed according to

$$\mathbf{D}_{ij}^I = \sum_{n=-L}^{L} |g_{ij}^I(n)| + \beta_2 \cdot \mathbf{D}_{ij}^{I-1}, \qquad i, j = 1, \ldots, mL, \; i \neq j, \tag{12}$$

where β_2 is a learning parameter, $0 \leq \beta_2 \leq 1$. The diagonal elements of $\mathbf{D}^I$ have no importance for the clustering and are all set to 1.

Clustering Algorithm. For simplicity, we assume that the number of sources d is known and does not change in time. The goal is thus to find d clusters of components according to their mutual similarity given by $\mathbf{D}^I$. We propose to use the Relational Fuzzy C-Means algorithm (RFCM) from [9], which allows tracking of continual changes of the clusters.

The affiliation of a component to a cluster is expressed by a value from [0, 1] where 0 means that the component does not belong to the cluster and vice versa. Let $\boldsymbol{\Lambda}^I_{kj}$ be the kjth element of a $d \times mL$ partition matrix $\boldsymbol{\Lambda}^I$ and represents the affiliation of the jth component to the kth cluster. By definition it holds that $\sum_{k=0}^{d} \boldsymbol{\Lambda}^I_{kj} = 1$.

Now, let $\mathbf{B}$ denotes the dissimilarity matrix whose elements are $\mathbf{B}^I_{ij} = 1/\mathbf{D}^I_{ij}$ for $i \neq j$ and $\mathbf{B}^I_{ii} = 0$. Let $\boldsymbol{\mu}^{I,f}_k$ be a $mL \times 1$ vector defined as $\boldsymbol{\mu}^{I,f}_k = [(\boldsymbol{\Lambda}^I_{k1})^f, \ldots, (\boldsymbol{\Lambda}^I_{k(mL)})^f]^T / \sum_{j=1}^{mL} (\boldsymbol{\Lambda}^I_{kj})^f$ called the prototype of the kth cluster associated with a "fuzzyfication" parameter f, $f > 1$. (We use the experimentally verified value $f = 1.5$). The transition of $\boldsymbol{\Lambda}^{I-1}$ to $\boldsymbol{\Lambda}^I$ is given as one iteration of RFCM as

$$\boldsymbol{\Lambda}^I_{kj} = \Big(\sum_{i=1}^{d} (\mathbf{V}_{kj}/\mathbf{V}_{ij})^{1/(f-1)} \Big)^{-1}, \tag{13}$$

where

$$\mathbf{V}_{kj} = \big(\mathbf{B}^I \boldsymbol{\mu}^{(I-1),f}_k\big)_j - \frac{1}{2} \big(\boldsymbol{\mu}^{(I-1),f}_k\big)^T \mathbf{B}^I \boldsymbol{\mu}^{(I-1),f}_k \tag{14}$$

is the distance of the jth component to the prototype $\boldsymbol{\mu}^{I,f}_k$ (for details see [9]).

2.3 Step III: Reconstruction

The contribution of ICs of the kth cluster to $\mathbf{X}^I$ is given by matrix

$$\widehat{\mathbf{S}}^I_k = (\mathbf{W}^I)^{-1} \mathrm{diag}\big[(\boldsymbol{\Lambda}^I_{k1})^\alpha, \ldots, (\boldsymbol{\Lambda}^I_{k,mL})^\alpha\big]\, \mathbf{C}^I, \tag{15}$$

where α is an adjustable positive parameter. This matrix has analogous structure as $\mathbf{X}^I$ in (3). In the ideal case the rows of $\widehat{\mathbf{S}}^I_k$ contain delayed microphone responses of the kth estimated source only. The response of the kth source at the ith microphone is therefore estimated by summing these rows as

$$\hat{s}^{i,I}_k(n) = \frac{1}{L} \sum_{q=1}^{L} (\widehat{\mathbf{S}}^I_k)_{(i-1)L+q,n+q-1}, \tag{16}$$

where $\big(\widehat{\mathbf{S}}^I_k\big)_{\alpha,\beta}$ is the $\alpha\beta$th element of the matrix $\widehat{\mathbf{S}}^I_k$.

Finally, the overall outputs of the on-line algorithm are synthesized by putting together the estimated blocks of separated signals. The overlapping parts are averaged using a windowing function, for example, the Hann window.

2.4 Implementation Details

The speed of convergence of the ICA can be driven through the parameter β^I in (8). We found it helpful to increase the speed when the clusters of ICs did not seem well separated in the previous block of data. Otherwise, β^I can be close to zero to maintain the continuity. Therefore, we take

$$\beta^I = (1 - \gamma^{I-1})/2. \tag{17}$$

γ^I is the Silhouette index [10] of the hard clustering which is derived from the fuzzy clustering. Let K_k be the set of indices of the components for which $\boldsymbol{\Lambda}_{kj}^I = \max_\ell \boldsymbol{\Lambda}_{\ell j}^I$ (the kth cluster is the closest one to them). The Silhouette index is defined through $\gamma^I = \frac{1}{mL}\sum_{i=1}^{mL} \gamma_i^I$, where

$$\gamma_i^I = \frac{\min_{j \notin K_k}(\mathbf{B}_{ij}^I) - \frac{1}{|K_k - 1|}\sum_{j \in K_k, i \neq j} \mathbf{B}_{ij}^I}{\max\{\min_{j \notin K_k}(\mathbf{B}_{ij}^I), \frac{1}{|K_k - 1|}\sum_{j \in K_k, i \neq j} \mathbf{B}_{ij}^I\}}. \tag{18}$$

The Silhouette index reflects the separateness of clusters as it takes values from [-1,1], where negative values mean poor separateness and vice versa.

The whole algorithm can be initialized so that $\mathbf{W}^0$ is the outcome of the BGSEP algorithm applied to $\mathbf{X}^1$ and the components $\mathbf{W}^0\mathbf{X}^1$ are grouped by the full RFCM algorithm.

3 Experiments

We present two experiments evaluated by means of the BSS_EVAL toolbox [11] using the true sources. The results are presented in the form of three criteria: (i) Signal-to-Interference Ratio (SIR), (ii) Signal-to-Distortion Ratio (SDR), and (iii) Signal-to-Artifact Ratio (SAR).

3.1 Fixed Source Positions

In this experiment, we examine the online algorithm in separating stationary mixtures of speech signals. Positions of the sources and the microphones were fixed. We compare it with the results of the original method from [5]. Hereinafter, the proposed on-line method will be referred to as *on-line T-ABCD*, while the original method will be named *off-line T-ABCD*[1].

To this end, we use data from the publicly available sites of Hiroshi Sawada[2]. The recordings of *four sources using four microphones* are considered. The original signals are utterances of the length 7 s sampled by 8 kHz. The reverberation time of the room is 130 ms. Omnidirectional microphones were used.

The on-line and off-line T-ABCD were both applied with $L = 30$. The other parameters of the on-line method were set to $P = 6144$, $T = 512$, $\beta_2 = 0.95$ and $\alpha = 3$. The separation results are evaluated block-by-block of the same size as in the on-line method. Table 1 summarizes the results averaged over all blocks, separated microphone responses, and sources.

On-line T-ABCD achieves better results in terms of SIR and SDR than the off-line algorithm. It points out to the fact that the on-line method is able to adapt the separating filters throughout the recordings respecting the nonstationarity of sources. On the other hand, the time-invariant separation done by off-line T-ABCD produces less artifacts as indicated by SAR.

[1] The acronym "T-ABCD" comes from the original method as it is does Time-domain Audio source Blind separation based on the Complete Decomposition of the observation space.

[2] `http://www.kecl.ntt.co.jp/icl/signal/sawada/`

Table 1. Results of separation of sources at fixed positions

	SIR[dB]	SDR[dB]	SAR[dB]
on-line T-ABCD	8.43	1.58	4.41
off-line T-ABCD	6.25	1.09	5.38

Table 2. Results of separation of data simulating dynamic conditions

2 cm	SIR[dB]	SDR[dB]	SAR[dB]
o. T-ABCD	10.39	3.87	6.16
Nesta	11.21	4.59	6.54

6 cm	SIR[dB]	SDR[dB]	SAR[dB]
o. T-ABCD	8.77	1.44	4.45
Nesta	7.60	1.37	5.77

3.2 Moving Sources

In this experiment, we consider data given in the task "Determined convolutive mixtures under dynamic conditions" (Audio Signal Separation) in the SiSEC 2010 evaluation campaign organized at this conference[3]. The data simulate dynamic conditions so that *the maximum of two of six speakers* located at fixed positions *around a stereo microphone array* were active at a time. The separating algorithm applied to the data thus needs to adapt to active speakers. The distances of microphones were 2 and 6 cm, and the sampling rate was 16 kHz.

We compare the proposed on-line T-ABCD with the frequency-domain BSS method by Francesco Nesta et al [12,13]. The online method was applied with $L = 30$, $P = 6144$, $T = 512$, $\beta_2 = 0.95$ and $\alpha = 4$. The Nesta's method uses FFT of the length 4096 samples with 75% overlap. As the method works off-line, it was applied independently on disjoint blocks of 1 second where the maximum of two sources were active.

The proposed method appears to be slightly inferior to the frequency-domain method if the distance of the microphones is 2cm, but it achieves better results if the distance is 6cm. We conclude that the on-line T-ABCD seems to outperform the frequency-domain algorithm in cases of larger microphone distances, where the spatial aliasing occurs.

3.3 Computational Demands

The experiments mentioned above were performed on a computer with single core 2.6 Ghz processor with 2 GB RAM. The on-line T-ABCD was implemented in Matlab environment. The computational demands of the algorithm depend on the demixing filter length L. The mixture signals in Section 3.2 were 3 minutes, 29 seconds long, sampled by 16kHz. The on-line T-ABCD separation lasted 14 minutes, 36 seconds ($L = 30$). Although the implementation in Matlab may be considered as rather slow and inefficient, this separation task can be performed in real time when $L = 10$. Mixtures of two sources sampled by 8 kHz can be separated in real-time when $L = 18$.

[3] `http://sisec.wiki.irisa.fr/`

4 Conclusions

We have proposed a method for blind separation of moving audio sources. The algorithm applies the fast converging BGSEP algorithm and fuzzy clustering RFCM algorithm. It is shown that presented time-domain method (with rather short separating filters) is able to achieve results that are comparable to the frequency domain BSS algorithm. The experiment with fixed sources suggests the ability of the proposed method to adapt the separation to the nonstationarity of the data as well.

References

1. Comon, P.: Independent component analysis: a new concept? Signal Processing. 36, 287–314 (1994)
2. Mukai, R., Sawada, H., Araki, S., Makino, S.: Blind Source Separation for Moving Speech Signals Using Blockwise ICA and Residual Crosstalk Subtraction. IEICE Transactions Fundamentals E87-A(8), 1941–1948 (2004)
3. Buchner, H., Aichner, R., Kellermann, W.: A Generalization of Blind Source Separation Algorithms for Convolutive Mixtures Based on Second-Order Statistics. IEEE Trans. on Speech and Audio Proc. 13(1), 120–134 (2005)
4. Loesch, B., Yang, B.: Online blind source separation based on time-frequency sparseness. In: ICASSP 2009, Taipei, Taiwan (2009)
5. Koldovský, Z., Tichavský, P.: Time-domain blind audio source separation using advanced component clustering and reconstruction. In: HSCMA 2008, Trento, Italy, pp. 216–219 (2008)
6. Koldovský, Z., Tichavský, P.: Time-domain blind audio source separation using advanced ICA methods. In: Interspeech 2007, Antwerp, Belgium (2007)
7. Tichavský, P., Yeredor, A.: Fast Approximate Joint Diagonalization Incorporating Weight Matrices. IEEE Transactions of Signal Processing 57(3), 878–891 (2009)
8. Knapp, C.-H., Carter, G.-C.: The Generalized Correlation Method for Estimation of Time Delay. IEEE Transactions on Signal Processing 24(4), 320–327 (1976)
9. Hathaway, R.-J., Bezdek, J.-C., Davenport, J.-W.: On relational data versions of c-means algorithm. Pattern Recognition Letters (17), 607–612 (1996)
10. Rousseeuw, J.P.: Silhouettes: a Graphical Aid to the Interpretation and Validation of Cluster Analysis. Computational and Applied Mathematics 20, 53–65 (1987)
11. Févotte, C., Gribonval, R., Vincent, E.: BSS EVAL toolbox user guide. IRISA, Rennes, France, Tech. Rep. 1706 (2005), `http://www.irisa.fr/metiss/bsseval/`
12. Nesta, F., Svaizer, P., Omologo, M.: A BSS method for short utterances by a recursive solution to the permutation problem. In: SAM 2008, Darmstadt, Germany (2008)
13. Nesta, F., Svaizer, P., Omologo, M.: A novel robust solution to the permutation problem based on a joint multiple TDOA estimation. In: IWAENC 2008, Seattle, USA (2008)

Time-Domain Blind Audio Source Separation Method Producing Separating Filters of Generalized Feedforward Structure*

Zbyněk Koldovský[1,2], Petr Tichavský[2], and Jiří Málek[1]

[1] Institute of Information Technology and Electronics
Technical University of Liberec, Studentská 2, 461 17 Liberec, Czech Republic
zbynek.koldovsky@tul.cz
http://itakura.ite.tul.cz/zbynek

[2] Institute of Information Theory and Automation, Pod vodárenskou věží 4,
P.O. Box 18, 182 08 Praha 8, Czech Republic
p.tichavsky@ieee.org
http://si.utia.cas.cz/Tichavsky.html

Abstract. Time-domain methods for blind separation of audio signals are preferred due to their lower demand for available data and the avoidance of the permutation problem. However, their computational demands increase rapidly with the length of separating filters due to the simultaneous growth of the dimension of an *observation space*. We propose, in this paper, a general framework that allows the time-domain methods to compute separating filters of theoretically infinite length without increasing the dimension. Based on this framework, we derive a generalized version of the time-domain method of Koldovský and Tichavský (2008). For instance, it is demonstrated that its performance might be improved by 4dB of SIR using the Laguerre filter bank.

1 Introduction

Blind Audio Source Separation (BASS) aims at separating unknown audio sources, which are mixed in an acoustical environment according to the convolutive model. The observed mixed signals are

$$x_i(n) = \sum_{j=1}^{d} \sum_{\tau=0}^{M_{ij}-1} h_{ij}(\tau) s_j(n-\tau) = \sum_{j=1}^{d} \{h_{ij} \star s_j\}(n), \quad i = 1, \ldots, m, \tag{1}$$

where $\star$ denotes the convolution, m is the number of microphones, $s_1(n), \ldots, s_d(n)$ are the original sources, and h_{ij} are source-microphone impulse responses each of length M_{ij}. The linear separation consists in finding de-mixing filters that separate original sources in its outputs. Since many methods for finding the filters

* This work was supported by Ministry of Education, Youth and Sports of the Czech Republic through the project 1M0572 and by Grant Agency of the Czech Republic through the project 102/09/1278.

V. Vigneron et al. (Eds.): LVA/ICA 2010, LNCS 6365, pp. 17–24, 2010.

formally assume instantaneous mixtures, i.e., $M_{ij} = 1$ for all i, j, the convolutive model needs to be transformed. This can be done either in the frequency or time domain.

Time-domain approaches, addressed in this paper, consist in decomposing the *observation matrix* defined as [1]

$$\mathbf{X} = \begin{bmatrix} x_1(N_1) & \dots\dots & x_1(N_2) \\ x_1(N_1-1) & \dots\dots & x_1(N_2-1) \\ \vdots & \vdots\ \vdots & \vdots \\ x_1(N_1-L+1) & \dots\dots & x_1(N_2-L+1) \\ x_2(N_1) & \dots\dots & x_2(N_2) \\ x_2(N_1-1) & \dots\dots & x_2(N_2-1) \\ \vdots & \vdots\ \vdots & \vdots \\ x_m(N_1-L+1) & \dots\dots & x_m(N_2-L+1) \end{bmatrix}, \tag{2}$$

where N stands for the number of available samples, and $1 \le N_1 < N_2 \le N$ determine the segment of data used for computations, and L is a free parameter. The decomposition of $\mathbf{X}$ is done by multiplying it by a matrix $\mathbf{W}$. This way FIR filters of the length L whose elements correspond to rows of $\mathbf{W}$ are applied to the mixed signals $x_1(n), \dots, x_m(n)$. This is due to the structure of $\mathbf{X}$ given by (2). The subspace of dimension mL in $\mathbb{R}^{N_2-N_1+1}$ spanned by rows of $\mathbf{X}$ will be called the *observation space.*

It is desired to decompose the observation space into linear subspaces where each of them represents one original signal. It can be done either by some independent subspace analysis (ISA) technique or by an independent component analysis (ICA) method, which is followed by the clustering of the components [2]. Performance of some ISA and ICA methods was studied in [12]. Some other methods utilize block-Sylvester structure of $\mathbf{A} = \mathbf{W}^{-1}$ [1,4]. Computational complexity of all these methods increases most ideally with L^3, which means that L cannot be too large. On the other hand, the frequency response of ordinary rooms is typically several hundreds of taps [3]. Therefore, longer filters would be desired.

Longer separating filters can be obtained by the subband-based separation [3,5]. In this paper, however, we propose to increase the length of the separating filters by changing the definition of the observation space. For a given set of invertible filters $f_{i,\ell}$, $\mathbf{X}$ is defined as

$$\mathbf{X} = \begin{bmatrix} \{f_{1,1} \star x_1\}(N_1) & \dots\dots & \{f_{1,1} \star x_1\}(N_2) \\ \{f_{1,2} \star x_1\}(N_1) & \dots\dots & \{f_{1,2} \star x_1\}(N_2) \\ \vdots & \vdots\ \vdots & \vdots \\ \{f_{1,L} \star x_1\}(N_1) & \dots\dots & \{f_{1,L} \star x_1\}(N_2) \\ \{f_{2,1} \star x_2\}(N_1) & \dots\dots & \{f_{2,1} \star x_2\}(N_2) \\ \vdots & \vdots\ \vdots & \vdots \\ \vdots & \vdots\ \vdots & \vdots \\ \{f_{m,L} \star x_m\}(N_1) & \dots\dots & \{f_{m,L} \star x_m\}(N_2) \end{bmatrix}. \tag{3}$$

Linear combinations of rows of $\mathbf{X}$ defined in this way correspond to outputs of MIMO filters with a generalized feed-forward structure introduced in [8], where the filters $f_{i,\ell}$ are referred to as *eigenmodes*. Note that if $f_{i,\ell}$ realizes backward time-shift by $\ell-1$ samples, i.e. $f_{i,\ell}(n)=\delta(n-\ell+1)$, where $\delta(n)$ stands for the unit impuls function, the construction of $\mathbf{X}$ given by (3) coincides with (2)[1].

The proposed definition (3) extends the class of filters that are applied to signals $x_1(n),\ldots,x_m(n)$ when multiplying $\mathbf{X}$ by $\mathbf{W}$. Time-domain BSS methods searching $\mathbf{W}$ via ICA can thus apply long separating (even IIR) filters without increasing L.

When $\mathbf{X}$ is defined by (2), $\mathbf{A}$ or $\mathbf{W}$ can be assumed to have a special structure (e.g. block-Sylvester) [1,2,4]. In general, the structure does not exist if $\mathbf{X}$ is defined according to (3). It is necessary to apply a separating algorithm that does not rely on the special structure - such as the method from [6,7], referred to as T-ABCD[2]. An extension of T-ABCD working with $\mathbf{X}$ defined through (3) is proposed in the following section. Then, a practical version of T-ABCD using Laguerre eigenmodes is proposed in Section 3, and its performance is demonstrated by Section 4. In Section 5, we present a semi-blind approach to show another potential of the generalized definition of $\mathbf{X}$.

2 Generalized T-ABCD

2.1 The Original Version of T-ABCD

Following the minimal distortion principle, T-ABCD estimates microphone responses of the original signals, $s_k^i(n)=\{h_{ik}\star s_k\}(n)$, $i=1,\ldots,m$, which are signals measured on microphones when the kth source sounds solo. First, we briefly describe the original version of T-ABCD from [6] that proceeds in four main steps.

1. Form the observation matrix $\mathbf{X}$ as in (2).
2. Decompose $\mathbf{X}$ into independent components, i.e., compute the $M\times M$ decomposing matrix $\mathbf{W}$ by an ICA algorithm, $M=mL$.
3. Group the components (rows of) $\mathbf{C}=\mathbf{W}\mathbf{X}$ into clusters so that each cluster contains components that correspond to the same original source.
4. For each cluster, use only components of the cluster to estimate microphone responses of a source corresponding to the cluster.

The details of the fourth step are as follows. For the kth cluster,

$$\widehat{\mathbf{S}_k}=\mathbf{W}^{-1}\mathrm{diag}[\lambda_1^k,\ldots,\lambda_M^k]\,\mathbf{W}\,\mathbf{X}=\mathbf{W}^{-1}\mathrm{diag}[\lambda_1^k,\ldots,\lambda_M^k]\,\mathbf{C},\qquad(4)$$

[1] A further practical generalization is if different number of eigenmodes were considered for a given i, that is $f_{i,\ell}$ for $\ell=1,\ldots,L_i$. For simplicity, we will consider the case $L_1=\cdots=L_m=L$ only.

[2] Time-domain Audio sources Blind separation based on the Complete Decomposition of the observation space.

where $\lambda_1^k, \ldots, \lambda_M^k$ denote positive weights from $[0, 1]$, reflecting degrees of affiliation of components to the kth cluster. Ideally, $\widehat{\mathbf{S}_k}$ is equal to $\mathbf{S}_k$, which is a matrix defined in the same way as $\mathbf{X}$ but consists of the contribution of only the kth source, which is, of the time-shifted copies of the responses $s_k^1(n), \ldots, s_k^m(n)$. Note that since $x_i(n) = s_1^i(n) + \cdots + s_d^i(n)$, it holds that $\mathbf{X} = \mathbf{S}_1 + \cdots + \mathbf{S}_d$.

Taking the structure of $\mathbf{S}_k$ (the same as (2)) into account, the microphone responses are estimated from $\widehat{\mathbf{S}_k}$ as

$$\widehat{s}_k^i(n) = \frac{1}{L}\sum_{\ell=1}^{L} \psi_{k,(i-1)L+\ell}(n + \ell - 1), \tag{5}$$

where $\psi_{k,p}(n)$ is equal to the (p, n)th element of $\widehat{\mathbf{S}}_k$. To clarify, note that $\psi_{k,p}(n)$ provides an estimate of $s_k^i(n - \ell + 1)$ for $p = (i - 1)L + \ell$. See [6] for further details on the method[3].

2.2 Generalization

In the first step of generalized T-ABCD, $\mathbf{X}$ is constructed according to (3). Further steps of the method are the same as described above up to the reconstruction formula given by (5), which is given as follows.

Let $f_{i,\ell}^{-1}$ be the inverse of the filter $f_{i,\ell}$. As $\psi_{k,p}(n)$ defined by the (p, n)th element of $\widehat{\mathbf{S}_k}$, $p = (i - 1)L + \ell$, provides an estimate of $\{f_{i,\ell} \star s_k^i\}(n)$, the microphone responses of the kth separated source are estimated as

$$\widehat{s}_k^i(n) = \frac{1}{L}\sum_{\ell=1}^{L} \{f_{i,\ell}^{-1} \star \psi_{k,(i-1)L+\ell}\}(n). \tag{6}$$

Obviously, (6) coincides with (5) if $f_{i,\ell}(n) = \delta(n - \ell + 1)$.

3 T-ABCD Using Laguerre Filters

In [9,10], Laguerre filters having the feed-forward structure [8] were shown to yield better separation than the ordinary FIR filters, apparently, thanks to increased effective length of their impulse response for certain values of a parameter μ. These filters can be applied within T-ABCD when the eigenmodes $f_{i,\ell}$ in (3) (now we may omit the first index i) are defined through their transfer functions F_ℓ recursively as

$$F_1(z) = 1, \tag{7}$$

$$F_2(z) = \frac{\mu z^{-1}}{1 - (1 - \mu)z^{-1}}, \tag{8}$$

$$F_n(z) = F_{n-1}(z)G(z), \quad n = 3, \ldots, L, \tag{9}$$

[3] Note the missing factor $1/L$ in the formula (9) in [6].

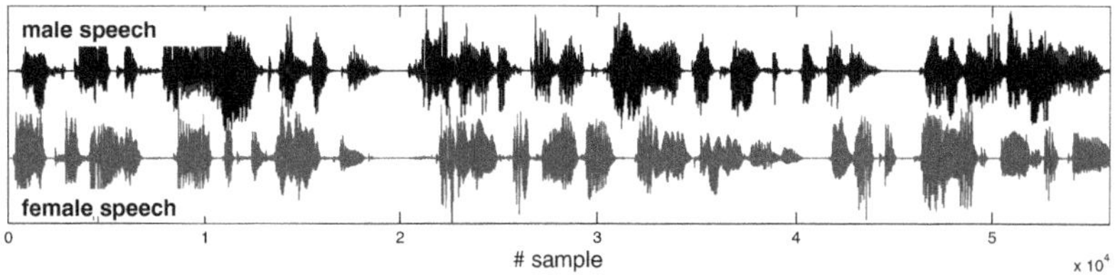

Fig. 1. Original signals used in experiments

where

$$G(z) = \frac{(\mu - 1) + z^{-1}}{1 - (1 - \mu)z^{-1}}, \tag{10}$$

and μ takes values from $(0, 2)$. Note that f_2 is either a low-pass filter (for $0 < \mu < 1$) or a high-pass filter (for $1 < \mu < 2$), and g is an all-pass filter.

The construction of $\mathbf{X}$ through Laguerre eigenmodes embodies (2) as a special case, because for $\mu = 1$, $F_2(z) = G(z) = z^{-1}$, that is $f_2(n) = g(n) = \delta(n - 1)$, consequently, $f_\ell(d) = \delta(n - L + 1)$. This is the only case where the Laguerre filters are FIR of the length L. For $\mu \neq 1$, the filters are IIR.

The effective length of the Laguerre filters denoted by L_* is defined as the minimum length needed to capture 90% of the total energy contained in the impulse response. For the Laguerre filters it approximately holds that [10]

$$L_* = (1 + 0.4|\mu - 1| \log_{10} L)L/\mu. \tag{11}$$

We can see that $L_* > L$ for $\mu < 1$ and vice versa. From here on, we will refer to T-ABCD as the variant proposed in this section as it encompasses the original algorithm when $\mu = 1$.

4 Experiments with Real-World Recordings

The proposed algorithm will be tested in the SiSEC evaluation campaign. The experiments in this paper examine mixtures of Hiroshi Sawada's original signals, which are available on the Internet[4]. The data are a male and a female utterance of the length 7 s recorded at the sampling rate 8kHz; see Fig. 1. For evaluations, we use two standard measures as in [13]: Signal-to-Interference Ratio (SIR) and Signal-to-Distortion Ratio (SDR). The SIR determines the ratio of energies of the desired signal and the interference in the separated signal. The SDR provides a supplementary criterion of SIR that reflects the difference between the desired and the estimated signal in the mean-square sense.

The performance of T-ABCD defined in the previous section was tested by separating Sawada's recordings of the original signals that were recorded in a room with the reverberation time of 130ms using two closely spaced microphones and two loudspeakers placed at a distance of 1.2 m. T-ABCD was applied to

[4] `http://www.kecl.ntt.co.jp/icl/signal/sawada/demo/bss2to4/index.html`

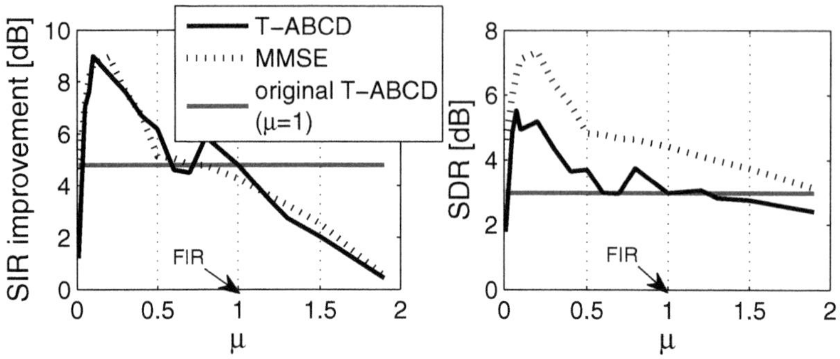

Fig. 2. Results of separation of Sawada's real-world recordings

separate the recordings with $L = 20$ and varying μ. Two seconds of the data were used for computations of separating filters, i.e., $N_1 = 1$ and $N_2 = 16000$. The ICA algorithm applied within T-ABCD is BGSEP from [11] that is based on the approximate joint diagonalization of covariance matrices computed on blocks of $\mathbf{X}$ (we consider blocks of 300 samples). The weighting parameter α for determining weights in (4) was set to 1. A similar setting was used in [6].

For comparison, minimum mean-square error (MMSE) solutions were computed as the best approximations of known responses of signals in the observation space defined by $\mathbf{X}$. It means that the MMSE solutions achieve the best SDR for given L and thus provide an experimental performance bound [10].

Fig. 2 shows resulting values of SIR and SDR averaged over both separated responses of both signals. The potential of Laguerre filters to improve the separation for $\mu < 1$ is demonstrated by the performance of the MMSE separator both in terms of SIR and SDR; similar results were observed in experiments in [10]. T-ABCD improves its performance when μ approaches 0.1 as well, with the optimum at around $\mu = 0.2$. For μ very close to zero ($\mu < 0.1$), the performance usually becomes unstable. Compared to the case $\mu = 1$, where $\mathbf{X}$ coincides with (2) and the separating filters are FIR, the separation is improved by 4dB of SIR and 2dB of SDR. This is achieved at essentially the same computational time (about 1.1 s in Matlab version 7.9 running on a PC, 2.6GHz, 3GB RAM), because the value of μ does not change the dimension of $\mathbf{X}$.

5 Semi-Blind Separation

The goal of this section is to provide another definition example of eigenmodes in (3) that utilizes prior information about the mixing system, otherwise known as the semi-blind approach. Consider the general $m = 2$ and $d = 2$ scenario

$$x_1(n) = \{h_{11} \star s_1\}(n) + \{h_{12} \star s_2\}(n) \tag{12}$$
$$x_2(n) = \{h_{21} \star s_1\}(n) + \{h_{22} \star s_2\}(n). \tag{13}$$

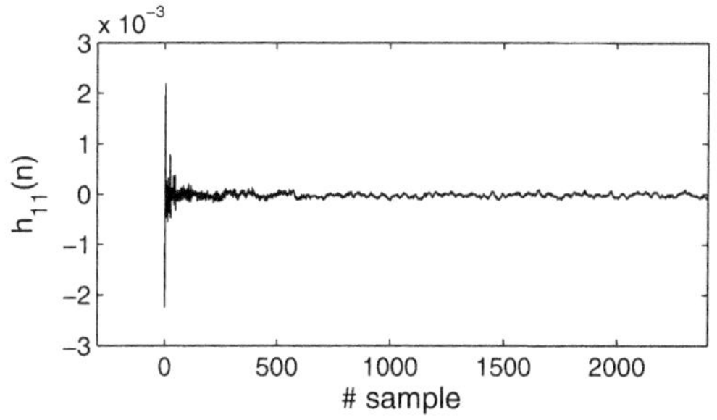

Fig. 3. The microphone-source impulse response $h_{11}(n)$

Almost perfect separation of this mixture can be achieved when taking $L = 2$ and defining $f_{11} = b \star h_{22}$, $f_{12} = -b \star h_{21}$, $f_{21} = -b \star h_{12}$, and $f_{22} = b \star h_{11}$, where $b = (h_{11} \star h_{22} - h_{21} \star h_{12})^{-1}$ assuming that the inversion exists. A trivial verification shows that combinations of signals $\{f_{11} \star x_1\}(n) + \{f_{21} \star x_2\}(n)$ and $\{f_{12} \star x_1\}(n) + \{f_{22} \star x_2\}(n)$ are independent, because they are equal to the original sources s_1 and s_2, respectively. If these combinations were unknown (e.g. when $f_{11}, \ldots, f_{22}$ were known up to a multiple by a constant), we could identify them blindly as independent components of $\mathbf{X}$ that would be defined through (3) with the eigenmodes $f_{11}, \ldots, f_{22}$. The dimension of such $\mathbf{X}$ is only 4, so the computation of ICA is very fast.

Additionally, we can define $f_{11}, \ldots, f_{22}$ with an arbitrary b, e.g., $b(n) = \delta(n)$. Note that b only affects the spectra of independent components of $\mathbf{X}$.

To demonstrate this, we recorded impulse responses of the length 300ms in a lecture room and mixed the original signals from Fig. 1 according to (12)-(13). An example of the recorded impulse response $h_{11}(n)$ is shown in Fig. 3.

The observation matrix $\mathbf{X}$ was constructed as described above with $b(n) = \delta(n)$. BGSEP was applied to $\mathbf{X}$ using only the first second of the recordings ($N_1 = 1$, $N_2 = 8000$) and yielded randomly permuted independent components of $\mathbf{X}$. Signal-to-Interference ratios of two of four components were, respectively, 28.3 dB subject to the male speech and 18.4 dB subject to the female speech, SIRs that represent a highly effective separation.

In comparison, MMSE solutions obtained by optimum FIR filters of the length 20 ($L = 20$ and $\mu = 1$) achieve only 4.8 dB of average SIR subject to the male speech and 6.8 dB subject to the female speech. Although the independent components have different coloration then the original signals (they are close to twice reverberated original signals by the room impulse response), the example reveals the great potential of the general construction of $\mathbf{X}$ in theory. For instance, it is indicative of the possibility to tailor the eigenmodes $f_{i,\ell}$ to room acoustics if the impulse response of the room can be measured with sufficient accuracy.

6 Conclusions

We have proposed a general construction of the observation matrix $\mathbf{X}$ that allows for the application of long separating filters in time-domain BASS methods without increasing the dimension of the observation space. This approach preserves the computational burden as it mostly depends on that dimension. The T-ABCD method was generalized in this way, and its version using Laguerre separating filters was shown to improve the separation with $\mu < 1$, i.e., when the effective length of separating filters L_* is increased compared to ordinary FIR filters with the length L. Future research can be focused on optimizing the choice of the eigenmodes.

References

1. Buchner, H., Aichner, R., Kellermann, W.: A generalization of blind source separation algorithms for convolutive mixtures based on second-order statistics. IEEE Trans. on Speech and Audio Proc. 13(1), 120–134 (2005)
2. Févotte, C., Debiolles, A., Doncarli, C.: Blind separation of FIR convolutive mixtures: application to speech signals. In: 1st ISCA Workshop on Non-Linear Speech Processing (2003)
3. Araki, S., Makino, S., Aichner, R., Nishikawa, T., Saruwatari, H.: Subband-based blind separation for convolutive mixtures of speech. IEICE Trans. Fundamentals E88-A(12), 3593–3603 (2005)
4. Xu, X.-F., Feng, D.-Z., Zheng, W.-X., Zhang, H.: Convolutive blind source separation based on joint block Toeplitzation and block-inner diagonalization. Signal Processing 90(1), 119–133 (2010)
5. Koldovský, Z., Tichavský, P., Málek, J.: Subband blind audio source separation using a time-domain algorithm and tree-structured QMF filter bank. In: Vigneron, V., et al. (eds.) LVA/ICA 2010. LNCS, vol. 6365, pp. 25–32. Springer, Heidelberg (2010)
6. Koldovský, Z., Tichavský, P.: Time-domain blind audio source separation using advanced component clustering and reconstruction. In: HSCMA 2008, Trento, Italy, vol. 2008, pp. 216–219 (2008)
7. Koldovský, Z., Tichavský, P.: Time-Domain blind separation of audio sources on the basis of a complete ICA decomposition of an observation space. Accepted for Publication in IEEE Trans. on Audio, Language, and Speech Processing (April 2010)
8. Principe, J.-C., de Vries, B., de Oliveira, G.: Generalized feedforward structures: a new class of adaptive filters. In: ICASSP 1992, vol. 4, pp. 245–248 (1992)
9. Stanacevic, M., Cohen, M., Cauwenberghs, G.: Blind separation of linear convolutive mixtures using orthogonal filter banks. In: ICA 2001, San Diego, CA (2001)
10. Hild II, K.-E., Erdogmuz, D., Principe, J.-C.: Experimental upper bound for the performance of convolutive source separation methods. IEEE Trans. on Signal Processing 54(2), 627–635 (2006)
11. Tichavský, P., Yeredor, A.: Fast approximate joint diagonalization incorporating weight matrices. IEEE Transactions of Signal Processing 57(3), 878–891 (2009)
12. Koldovský, Z., Tichavský, P.: A comparison of independent component and independent subspace analysis algorithms. In: EUSIPCO 2009, Glasgow, England, pp. 1447–1451 (2009)
13. Schobben, D., Torkkola, K., Smaragdis, P.: Evaluation of blind signal separation methods. In: ICA 1999, Aussois, France, pp. 261–266 (1999)

Subband Blind Audio Source Separation Using a Time-Domain Algorithm and Tree-Structured QMF Filter Bank*

Zbyněk Koldovský[1,2], Petr Tichavský[2], and Jiří Málek[1]

[1] Institute of Information Technology and Electronics
Technical University of Liberec, Studentská 2, 461 17 Liberec, Czech Republic
zbynek.koldovsky@tul.cz
http://itakura.ite.tul.cz/zbynek
[2] Institute of Information Theory and Automation, Pod vodárenskou věží 4,
P.O. Box 18, 182 08 Praha 8, Czech Republic
p.tichavsky@ieee.org
http://si.utia.cas.cz/Tichavsky.html

Abstract. T-ABCD is a time-domain method for blind linear separation of audio sources proposed by Koldovský and Tichavský (2008). The method produces short separating filters (5-40 taps) and works well with signals recorded at the sampling frequency of 8-16 kHz. In this paper, we propose a novel subband-based variant of T-ABCD, in which the input signals are decomposed into subbands using a tree-structured QMF filter bank. T-ABCD is then applied to each subband in parallel, and the separated subbands are re-ordered and synthesized to yield the final separated signals. The analysis filter of the filter bank is carefully designed to enable maximal decimation of signals without aliasing. Short filters applied within subbands then result in sufficiently long filters in fullband. Using a reasonable number of subbands, the method yields improved speed, stability and performance at an arbitrary sampling frequency.

1 Introduction

Blind separation (BSS) of simultaneously active audio sources is a challenging problem within audio signal processing. The goal is to retrieve d audio sources from their convolutive mixtures recorded by m microphones. The model is described by

$$x_i(n) = \sum_{j=1}^{d} \sum_{\tau=0}^{M_{ij}-1} h_{ij}(\tau) s_j(n-\tau), \quad i = 1, \ldots, m, \tag{1}$$

where $x_1(n), \ldots, x_m(n)$ are the observed signals on microphones and $s_1(n), \ldots, s_d(n)$ are the unknown original (audio) signals. This means that the mixing system is a MIMO (multi-input multi-output) linear filter with source-microphone

* This work was supported by Ministry of Education, Youth and Sports of the Czech Republic through the project 1M0572 and by Grant Agency of the Czech Republic through the project 102/09/1278.

V. Vigneron et al. (Eds.): LVA/ICA 2010, LNCS 6365, pp. 25–32, 2010.

impulse responses h_{ij}'s each of length M_{ij}. Linear separation consists in finding a MIMO filter that inverts the mixing process (1) and yields estimates of the original signals $s_1(n), \ldots, s_d(n)$. It is convenient to assume the independence of $s_1(n), \ldots, s_d(n)$, and the separation can be based on Independent Component Analysis (ICA) [1]. Indeterminacies that are inherent to the ICA cause that the original colorations of $s_1(n), \ldots, s_d(n)$ cannot be retrieved. The goal is therefore to estimate their microphone responses (images), which only have properly defined colorations. The response of the kth source on the ith microphone is

$$s_k^i(n) = \sum_{\tau=0}^{M_{ik}-1} h_{ik}(\tau) s_k(n-\tau). \tag{2}$$

To apply the ICA, the convolutive mixture (1) must be transformed into an instantaneous one. This is done either directly in the time-domain (TD) by decomposing a matrix usually constructed of delayed copies of signals from microphones, or in the frequency-domain (FD) where the signals are transformed by the Short-Time Fourier Transform (STFT) that converts the convolution operation into the ordinary multiplication. Weaknesses of both approaches are well known from literature. The FD approach meets the so-called permutation problem [2] due to inherent indeterminacies in ICA and requires long data to generate sufficient number of samples for each frequency bin. On the other hand, TD methods are computationally more expensive due to simultaneous optimization of all filter coefficients, which restrict their ability to compute long filters.

A reasonable compromise is the subband approach [3] that consists in decomposing the mixed signals into subbands via a filter bank, separating each subband by a TD method, permuting the separated subbands, and synthesizing the final signals. If a moderate number of subbands is chosen, the permutation problem becomes less difficult compared to the FD approach. Since the subband signals are decimated, the length of separating filters is multiplied.

Several subband approaches have already been proposed in literature using various filter banks. The method in [5] uses a uniform DFT filter bank. Araki et al. [3] use a polyphase filter bank with a single side-band modulation. In [6,7], uniform FIR filter banks were used. All the referenced methods do not apply the maximal decimation of signals in order to reduce the aliasing between subbands. This restrains both the computational efficiency and the effective length of separating filters.

We propose a novel subband method designed to be maximally effective in this respect. The signals are decomposed uniformly into 2^M subbands using a two-channel QMF filter bank applied recursively in the full-blown 2-tree structure with M levels [4]. The signals are decimated by 2 in each level of the 2-tree so they are finally decimated by 2^M, which means *maximal decimation*. Through a careful design of a halfband FIR filter, which determines the whole filter bank, the aliasing is avoided. The blind separation within subbands is then carried out independently by the T-ABCD method [10], which is robust and effective in estimating short separating filters. The permutation problem due to the random order of separated signals in each subband is solved by comparing correlations

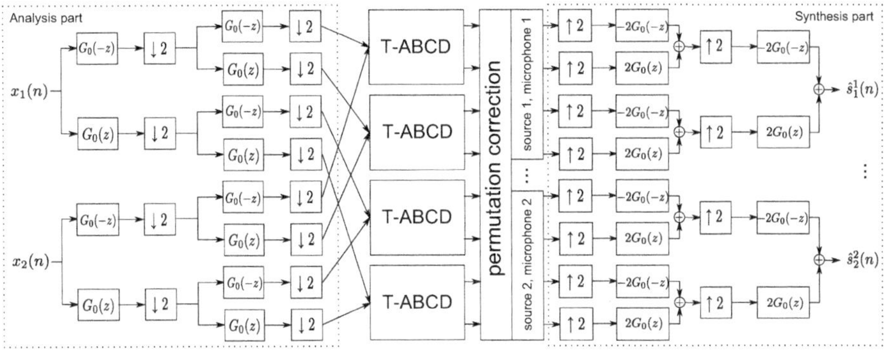

Fig. 1. An illustration of the proposed subband BSS algorithm using the QMF tree-structured filter bank with $M = 2$. Here, two-microphone recording is separated into two responses of each of two original sources.

of absolute values of signals [2]. Finally, the reordered signals are synthesized to yield the estimated responses (2). The flow of the method is illustrated in Fig. 1.

The following section gives more details on the proposed method, and Section 3 demonstrates its performance by experiments done with real-world signals.

2 Proposed Subband BSS Method

2.1 Subband Decomposition

The building block of the tree-structured subband decomposition applied in the proposed method is a two-channel bank that separates the input signals into two bands. In general, a two-channel bank consists of two analysis filters and two synthesis filters whose transfer functions are, respectively, $G_0(z)$, $G_1(z)$, $H_0(z)$ and $H_1(z)$. The input signal is filtered by $G_0(z)$ and $G_1(z)$ in parallel, and the outputs are decimated by 2 giving the subband signals. After the subband processing, the signals are expanded by 2 and passed through the synthesis filters and are added to yield the output signal.

The analysis filters of a Quadrature Mirror Filter (QMF) bank satisfy

$$G_1(z) = G_0(-z). \tag{3}$$

$G_0(z)$ should be a low-pass filter with the pass band $[-\pi/2, \pi/2]$ so that the decimated signals are not aliased. The synthesis filters may be defined as

$$H_0(z) = 2G_1(-z), \qquad H_1(z) = -2G_0(-z). \tag{4}$$

Then the whole two-channel QMF bank is determined by $G_0(z)$.

(4) is a sufficient condition for eliminating the aliasing from synthesized signals provided that no subband processing is done, i.e., when the signals are expanded immediately after the decimation (equation (12.58) in [4]). In such special case,

the transfer function of the two-channel QMF bank is $[G_0(z)]^2 - [G_0(-z)]^2$. It follows that the bank does not possess the perfect reconstruction property in general, which is nevertheless not as important in audio applications. While phase distortions are avoided provided that $G_0(z)$ has a linear phase, amplitude distortions can be made inaudible by a careful design of the filter[1].

To decompose the signal into more than two bands, the analysis part of the two-channel QMF bank can be applied recursively to split each band into two subbands etc. If the depth of the recursion is M, the filter bank splits the spectrum uniformly into 2^M subbands. This approach is utilized in the proposed method as demonstrated by Fig. 1. After the processing of subbands, the synthesis is done backwards then the analysis.

2.2 Separation Algorithm: T-ABCD

T-ABCD is an ICA-based method for blind separation of audio signals working in time-domain. It is based on the estimation of all independent components (ICs) of an *observation space* by an incorporated ICA algorithm. The observation space is spanned by rows of a data matrix $\mathbf{X}$ that may be defined in a general way [10]. For simplicity, we will consider the basic definition that is common to other TD methods [12]: Rows of $\mathbf{X}$ contain L time-shifted copies of each observed signal $x_1(n), \ldots, x_m(n)$. The number of rows of $\mathbf{X}$ is mL, which is the dimension of the observation space. Linear combinations of rows of $\mathbf{X}$ correspond to outputs of FIR MISO filters of the length L (hence also the ICs of $\mathbf{X}$). The steps of T-ABCD are as follows.

1. Find all mL independent components of $\mathbf{X}$ by an ICA algorithm.
2. Group the components into clusters so that each cluster contains components corresponding to the same original source.
3. For each cluster, use components of the cluster to reconstruct microphone responses (images) of a source corresponding to the cluster.

For more details on the method see [9] and [10].

A shortcoming of T-ABCD is that its computational complexity grows rapidly with L. On the other hand, T-ABCD is very powerful when L is reasonably low ($L = 1, \ldots, 40$). This is because all ICs of $\mathbf{X}$ are estimated without applying any constraint to the separating MISO filters (step 1), and all ICs are used to reconstruct the sources' responses (steps 2 and 3). The performance of T-ABCD is robust as it is independent of an initialization provided that the applied ICA algorithm in step 1 is equivariant. Consequently, the use of T-ABCD within the subband separation is desirable, because the separating filters in subbands are shorter than those in fullband [3].

---

[1] We have chosen $G_0(z)$ as an equiripple FIR filter [4] with 159 taps having the minimum attenuation of 60 dB in the stopband. To eliminate the aliasing, the stop-frequency was shifted slightly from $\pi/2$ to the left by $\epsilon \approx 0.01$, which is small enough so that the cut-off band around $\pi/2$ is very narrow and results in inaudible distortions of signals.

2.3 The Permutation Problem

The estimated responses of sources by T-ABCD are randomly permuted due to indeterminacy of ICA or, more specifically, due to the indeterminacy of the order of clusters identified by step 2. Since the permutation might be different in each subband, the estimated signals in subbands must be aligned before synthesizing them.

Let $\hat{s}^i_{k,j}(n)$, $k = 1, \ldots, d$ be the not yet sorted estimates of responses of the sources at the ith microphone in the jth subband. We wish to find permutations $\pi_j(k)$, $j = 1, \ldots, M$ such that $\hat{s}^i_{\pi_j(k),j}(n)$ is the estimated response of the kth source at the microphone in the subband. We shall assume, for convenience, that the order of the components in one, say in the j_1th subband (e.g. $j_1 = 1$), is correct. Therefore we set $\pi_{j_1}(k) = k$, $k = 1, \ldots, d$. Permutations in all other subbands can be found by maximizing the following criterion,

$$d(p, q, r, s) = \sum_{i=1}^{m} \left|\mathrm{cov}\left(|\widehat{s}^i_{p,q}(n)|, |\widehat{s}^i_{r,s}(n)|\right)\right| =$$

$$= \sum_{i=1}^{m} \frac{1}{T} \sum_{n=1}^{T} \left(|\widehat{s}^i_{p,q}(n)| - \frac{1}{T}\sum_{t=1}^{T} |\widehat{s}^i_{p,q}(t)|\right) \left(|\widehat{s}^i_{r,s}(n)| - \frac{1}{T}\sum_{t=1}^{T} |\widehat{s}^i_{r,s}(t)|\right) \quad (5)$$

that compares dynamic profiles (absolute values) of the signals [2], as follows.

1. Put $\mathcal{S} = \{j_1\}$, a set of already permuted subbands.
2. Find $j_2 = \arg\max_{s \notin \mathcal{S}} \{\max_{p,r} d(p, j_1, r, s)\}$.
3. Use the greedy algorithm to find $\pi_{j_2}(\cdot)$ by maximizing $d(\cdot, j_1, \cdot, j_2)$. Namely, define $\mathcal{P} = \emptyset$ and $\mathcal{R} = \emptyset$, and repeat
   (a) $(p, r) = \arg\max_{p \notin \mathcal{P}, r \notin \mathcal{R}} d(p, j_1, r, j_2)$
   (b) put $\pi_{j_2}(p) = r$
   (c) $\mathcal{P} = \mathcal{P} \cup \{p\}$, $\mathcal{R} = \mathcal{R} \cup \{r\}$
   until $\mathcal{P} \subsetneq \{1, \ldots, M\}$
4. $\mathcal{S} = \mathcal{S} \cup \{j_2\}$, $j_1 = j_2$.
5. If $\mathcal{S} \subsetneq \{1, \ldots, M\}$, go to 2.

3 Experiments

To demonstrate the performance of the proposed method, we test it on selected data from the SiSEC 2010 campaign[2]. The data consists of two-microphone real-world recordings of, respectively, two male and two female speakers played over loudspeakers (signal combinations #1 and #2) placed in room #1 in position #1 shown in Fig. 2. Each source was recorded separately to obtain its microphone responses, and the signals were summed to obtain the mixed signals; the original sampling rate was 44.1kHz.

[2] The task "Robust blind linear/non-linear separation of short two-sources-two-microphones recordings" in the "Audio source separation" category; see `http://sisec.wiki.irisa.fr/tiki-index.php`

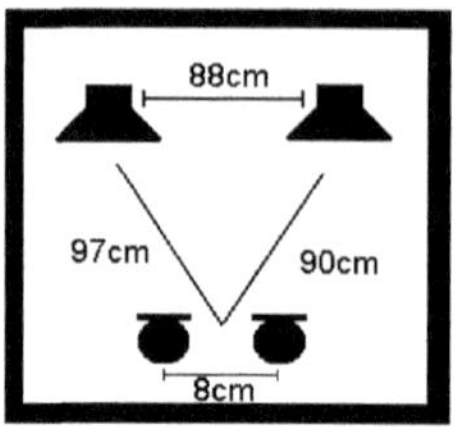

Fig. 2. The position of microphones and loudspeakers in the experiment

For evaluation of the separation, we use the standard Signal-to-Interference Ratio (SIR), as defined in [13]. The evaluation is computed using the full length of recordings, which is about 2 seconds, but only the first second of the data was used for computations of separating filters.

We compare the original T-ABCD from [9] working in fullband with the proposed subband T-ABCD decomposing signals into 2, 4, 8, and 16 subbands, that is, with $M = 1, \ldots, 4$. The fullband T-ABCD is applied with $L = 20$, while in subbands $L = 10$ is taken. The other parameters of T-ABCD are the same both in fullband and subband; namely, the weighting parameter is $\alpha = 1$, and the BGSEP algorithm from [11] is used for finding ICs of $\mathbf{X}$.

Fig. 3 shows the results of experiments done with signals resampled to the sampling rates f_s =8, 16, 32, and 44.1 kHz, respectively. The performance of the fullband T-ABCD decreases with the growing f_s. This is due to the fact that the effective length of separating filters decreases as L is fixed to 20. A comparable length of filters is applied in the 2-subbands method, where $L = 10$ in each subband. The performance of the 2-subbands method is either comparable ($f_s = 8$ and 32 kHz) to the fullband method or even better ($f_s = 16$ and 44.1 kHz) and does not decrease until $f_s \leq 16$. This points to the fact that the fullband method suffers from increased bandwidth of signals when f_s grows.

As can be seen from Fig. 3, the performance of the subband method does not automatically increase with the number of subbands. This is mainly caused by the permutation problem, which becomes more difficult with the growing number of subbands. The results indicate that the optimal bandwidth of subbands is between 2-5 kHz. Namely, (1) the 4-subbands method performs best at $f_s = 16$ and 32 kHz, (2) the 8-subbands method provides the best results when $f_s = 32$ and 44.1 kHz, and (3) the 16-subbands method seems to be effective if $f_s = 44.1$ kHz. On the other hand, the decomposition of signals into 16 subbands seems to be inadequate when $f_s = 8$ or 16 kHz, as the 16-subbands method yields unstable performance here.

3.1 Computational Aspects

The methods were running on a PC with quad-core i7 2.66 GHz processor in Matlab^{TM} with Parallel Computing Toolbox^{TM}. There were four running workers, i.e. one for each core of the processor, which means that up to four T-ABCDs

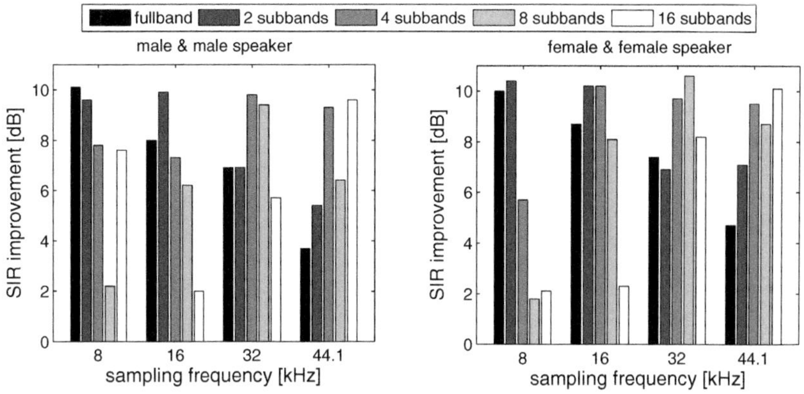

Fig. 3. SIR improvement achieved by the separation. Each value is an average over both separated sources and estimated microphone responses.

may run simultaneously in subbands. The average computational burden summarizes Table 1 in the form A/B, where A and B denote the time needed for separation without and with the aid of parallel computations, respectively. The parallelization was realized through the parallel for-cycle (`parfor`).

The values in Table 1 prove the advantage of the subband method consisting in lower computational complexity. Although the parallelization by means of the Parallel Computing Toolbox^{TM} is not that effective, it points to the potential improvement in terms of speed. For example, the 4-subband method should be almost four-times faster when running in parallel, since about 80% of the computational burden is caused by T-ABCD, while the permutation correction takes about 3% and the rest is due to the filtering operations.

Table 1. Average time needed per separation without and with parallel computations

	computational time [s]			
	8kHz	16kHz	32kHz	44.1kHz
fullband	0.42/ -	0.90/ -	2.03/ -	2.84/ -
2-subband	0.25/0.17	0.46/0.31	0.90/0.69	1.35/0.97
4-subband	0.30/0.19	0.56/0.33	1.06/0.65	1.51/0.97
8-subband	0.40/0.25	0.66/0.42	1.26/0.83	1.83/1.13
16-subband	0.56/0.35	0.87/0.55	1.50/0.99	2.20/1.39

4 Conclusion

The proposed subband T-ABCD was shown to be an improved variant of T-ABCD in terms of speed and separation performance, especially, when working with signals sampled at sampling rates higher than 16 kHz. The method is able to separate one second of data in a lower time, which points to its applicability in a batch-online processing. The optimum number of subbands depends on the sampling frequency, which was shown to correspond to the bandwidth of about 2-5kHz per subband. Experiments not shown here due to lack of space show that the subband T-ABCD might be combined with other filter banks (e.g. [3]) as well, but the analysis filters must be adjusted to avoid the aliasing in maximally decimated signals.

References

1. Comon, P.: Independent component analysis: a new concept? Signal Processing 36, 287–314 (1994)
2. Sawada, H., Mukai, R., Araki, S., Makino, S.: A robust and precise method for solving the permutation problem of frequency-domain blind source separation. IEEE Trans. Speech Audio Processing 12(5), 530–538 (2004)
3. Araki, S., Makino, S., Aichner, R., Nishikawa, T., Saruwatari, H.: Subband-based blind separation for convolutive mixtures of speech. IEICE Trans. Fundamentals E88-A(12), 3593–3603 (2005)
4. Porat, B.: A course in digital signal processing. John Wiley & Sons, New York (1997)
5. Grbić, N., Tao, X.-J., Nordholm, S.E., Claesson, I.: Blind signal separation using overcomplete subband representation. IEEE Transactions on Speech and Audio Processing 9(5), 524–533 (2001)
6. Russell, I., Xi, J., Mertins, A., Chicharo, J.: Blind source separation of nonstationary convolutively mixed signals in the subband domain. In: ICASSP 2004, vol. 5, pp. 481–484 (2004)
7. Kokkinakis, K., Loizou, P.C.: Subband-based blind signal processing for source separation in convolutive mixtures of speech. In: ICASSP 2007, vol. 4, pp. 917–920 (2007)
8. Duplessis-Beaulieu, F., Champagne, B.: Fast convolutive blind speech separation via subband adaptation. In: ICASSP 2003, vol. 5, pp. 513–516 (2003)
9. Koldovský, Z., Tichavský, P.: Time-domain blind audio source separation using advanced component clustering and reconstruction. In: HSCMA 2008, Trento, Italy, pp. 216–219 (2008)
10. Koldovský, Z., Tichavský, P.: Time-domain blind separation of audio sources on the basis of a complete ICA decomposition of an observation space. Accepted for Publication in IEEE Trans. on Audio, Language, and Speech Processing (April 2010)
11. Tichavský, P., Yeredor, A.: Fast approximate joint diagonalization incorporating weight matrices. IEEE Transactions of Signal Processing 57(3), 878–891 (2009)
12. Bousbia-Salah, H., Belouchrani, A., Abed-Meraim, K.: Jacobi-like algorithm for blind signal separation of convolutive mixtures. IEE Elec. Letters 37(16), 1049–1050 (2001)
13. Vincent, E., Févotte, C., Gribonval, R.: Performance measurement in blind audio source separation. IEEE Trans. Audio, Speech and Language Processing 14(4), 1462–1469 (2006)

A General Modular Framework for Audio Source Separation

Alexey Ozerov[1,⋆], Emmanuel Vincent[1], and Frédéric Bimbot[2]

[1] INRIA, Rennes Bretagne Atlantique
[2] IRISA, CNRS - UMR 6074, Campus de Beaulieu, 35042 Rennes cedex, France
{alexey.ozerov,emmanuel.vincent,frederic.bimbot}@irisa.fr

Abstract. Most of audio source separation methods are developed for a particular scenario characterized by the number of sources and channels and the characteristics of the sources and the mixing process. In this paper we introduce a general modular audio source separation framework based on a library of flexible source models that enable the incorporation of prior knowledge about the characteristics of each source. First, this framework generalizes several existing audio source separation methods, while bringing a common formulation for them. Second, it allows to imagine and implement new efficient methods that were not yet reported in the literature. We first introduce the framework by describing the flexible model, explaining its generality, and summarizing our modular implementation using a Generalized Expectation-Maximization algorithm. Finally, we illustrate the above-mentioned capabilities of the framework by applying it in several new and existing configurations to different source separation scenarios.

1 Introduction

Separating audio sources from multichannel mixtures is still challenging in most situations. The main difficulty is that audio source separation problems are usually mathematically ill-posed and to succeed one needs to incorporate additional knowledge about the mixing process and/or the source signals. Thus, efficient source separation methods are usually developed for a particular scenario characterized by *problem dimensionality* ((over)determined case, underdetermined case, and single-channel case), *mixing process characteristics* (synthetic instantaneous, anechoic, and convolutive mixtures, and live recorded mixtures), *source characteristics* (speech, singing voice, drums, bass, and noise (stationary or not, white or colored)). Moreover, there is often no common formulation describing methods applied for different scenarios, and this makes it difficult to reuse a method for a scenario it was not originally conceived for just by modifying some parameters.

The motivation of this work is to design a general audio source separation framework that can be easily applied to several separation scenarios just by

⋆ This work was supported in part by the Quaero Programme, funded by OSEO.

V. Vigneron et al. (Eds.): LVA/ICA 2010, LNCS 6365, pp. 33–40, 2010.

selecting from a library of models a suitable model for each source incorporating *a priori* knowledge about that source. More precisely we wish such a framework to be

- *general*, i.e., generalizing existing methods and making it possible to combine them,
- *flexible*, allowing easy incorporation of the *a priori* information about a particular scenario considered,
- *modular*, allowing an implementation in terms of software blocks addressing the estimation of subsets of parameters.

To achieve the property of generality, we need to find some common formulation for methods we would like to generalize. Several recently proposed methods for source separation and/or characterization [12], [1], [7], [6], [5], [11], [10], [13], [3] (see also [14] and references therein) are based on the same zero-mean Gaussian model describing both the properties of the sources and of the mixing process, and only the global structure of Gaussian covariances differs from one method to another. These methods already cover several possible scenarios, including single-channel [6] or multichannel sources [11], instantaneous [7] or convolutive [11] mixtures of point [7] or diffuse [5] sources, and monophonic (e.g., speech [10]) or polyphonic sources (e.g., polyphonic music [6]). Moreover, a few of these methods have already been combined together, for example hidden Markov model (HMM) (a monophonic source model) and nonnegative matrix factorization (NMF) (a polyphonic source model) were combined in [10], NMF [6] was combined with point and diffuse source models in [11], [3]. We chose this local Gaussian model as the basis of our framework. To achieve flexibility, we leave the global structures of Gaussian covariances be specifiable in every particular case, allowing introduction of knowledge about every particular source and its mixing conditions. Thus, our framework generalizes all the above methods, and, thanks to its flexibility, it becomes applicable in many other scenarios one can imagine. We implement our framework using a Generalized Expectation-Maximization (GEM) algorithm, where the M-step is solved in a modular fashion by alternating between optimization of different parameter subsets.

Our approach is in line with the *library of components* by Cardoso and Martin [4] developed for the separation of components in astrophysical images. However, we consider advanced audio-specific structures (an overview of such structures can be found in [8], [13]) for source spectral power that generalize the structures considered in [4] that assume simply that source power is constant in some predefined region of time and space. In that sense our framework is more flexible than [4].

The rest of this paper is organized as follows. The audio source separation problem considered here is described in section 2. Section 3 is devoted to the presentation of the model at the heart of our framework, and some information about the employed GEM algorithm is given in section 4. The results of several source separation experiments are given in section 5 to illustrate the flexibility of our framework and the resulting performance improvement compared to individual approaches. Conclusions are drawn in section 6.

2 Audio Source Separation

Since we would like to address the separation of both point and diffuse sources, the standard point source based convolutive blind source separation (BSS) problem formulation (see e.g., [11], Eq. (1)) is not suitable in our case. Thus, we rather assume that the observed multichannel time-domain signal, called *mixture*, $\tilde{\mathbf{x}}(t) \in \mathbb{R}^I$ (I being the number of channels, and $t = 1, \ldots, T$) is a sum of J multichannel signals $\tilde{\mathbf{y}}_j(t) \in \mathbb{R}^I$, called *spatial source images* [4], [14]:

$$\tilde{\mathbf{x}}(t) = \sum\nolimits_{j=1}^{J} \tilde{\mathbf{y}}_j(t), \tag{1}$$

and the goal is to estimate the spatial source images $\tilde{\mathbf{y}}_j(t)$, given the mixture $\tilde{\mathbf{x}}(t)$.

Audio signals are usually processed in the time-frequency domain, due to sparsity property of audio signals in such a representation. Thus, we convert all the signals in the short-time Fourier transform (STFT) domain, and equation (1) becomes:

$$\mathbf{x}_{fn} = \sum\nolimits_{j=1}^{J} \mathbf{y}_{j,fn} \tag{2}$$

where $\mathbf{x}_{fn} \in \mathbb{C}^I$ and $\mathbf{y}_{j,fn} \in \mathbb{C}^I$ are I-dimensional complex-valued vectors of STFT coefficients of the corresponding time-domain signals; and $f = 1, \ldots, F$ and $n = 1, \ldots, N$ denote respectively STFT frequency and time indices.

3 Flexible Model

We assume that every vector $\mathbf{y}_{j,fn} \in \mathbb{C}^I$ is a proper complex-valued Gaussian random vector with zero mean and covariance matrix $\boldsymbol{\Sigma}_{\mathbf{y},j,fn} = v_{j,fn}\mathbf{R}_{j,fn}$:

$$\mathbf{y}_{j,fn} \sim \mathcal{N}_c\left(\bar{0}, v_{j,fn}\mathbf{R}_{j,fn}\right) \tag{3}$$

where the matrix $\mathbf{R}_{j,fn} \in \mathbb{C}^{I \times I}$ called *spatial covariance matrix*, represents the spatial characteristics of the source and of the mixing setup, and the non-negative scalar $v_{j,fn} \in \mathbb{R}_+$ called *spectral power* represents the spectral characteristics of the source. Moreover, the random vectors $\mathbf{y}_{j,fn}$ are assumed to be independent given $\boldsymbol{\Sigma}_{\mathbf{y},j,fn}$. The model of the j-th source can be parametrized as $\theta_j = \{v_{j,fn}, \mathbf{R}_{j,fn}\}_{f,n=1}^{F,N}$, and the overall model writes $\theta = \{\theta_j\}_{j=1}^{J}$.

Given the model parameters θ, the sources can be estimated in the minimum mean square error (MMSE) sense via Wiener filtering:

$$\hat{\mathbf{y}}_{j,fn} = v_{j,fn}\mathbf{R}_{j,fn}\boldsymbol{\Sigma}_{\mathbf{x},fn}^{-1}(\theta)\mathbf{x}_{fn}, \tag{4}$$

where $\boldsymbol{\Sigma}_{\mathbf{x},fn}(\theta) \triangleq \sum_{j=1}^{J} v_{j,fn}\mathbf{R}_{j,fn}$.

The model parameters θ are usually not given and should be estimated. It is clear that estimating model parameters in the maximum likelihood (ML) sense would not lead to any consistent estimation, since there are more free parameters in θ than data samples in $[\mathbf{x}_{fn}]_{f,n}$. We hence assume that θ belongs to a subset of admissible parameters Θ (*structural constraints*) and/or we consider that θ follows some *a priori* distribution $p(\theta|\eta)$, where η denotes some hyperparameters. With these assumptions we use the maximum *a posteriori* (MAP) criterion that can be rewritten as [11]:

$$\theta^*, \eta^* = \arg\min_{\theta\in\Theta,\eta} \sum_{f,n} \left[\mathrm{tr}\left(\boldsymbol{\Sigma}_{\mathbf{x},fn}^{-1}(\theta) \mathbf{x}_{fn} \mathbf{x}_{fn}^H \right) + \log |\boldsymbol{\Sigma}_{\mathbf{x},fn}(\theta)| \right] - \log p(\theta|\eta). \quad (5)$$

In this paper we focus on structural constraints (overviewed in the following sections) allowing the incorporation of additional knowledge about the mixing process and the audio source signals, and we leave aside the prior $p(\theta|\eta)$.

3.1 Spatial Covariance Structures

In this work we first assume that the spatial covariances are time invariant, i.e., $\mathbf{R}_{j,fn} = \mathbf{R}_{j,f}$. In the case of audio, it is mostly interesting to consider either rank-1 covariances representing instantaneously mixed (or convolutively mixed with weak reverberation) point sources (see e.g., [11]) or full rank covariances modeling diffuse or reverberated sources [5]. We assume in the rank-1 case that $\mathbf{R}_{j,f} = \mathbf{a}_{j,f}\mathbf{a}_{j,f}^H$, where $\mathbf{a}_{j,f} \in \mathbb{C}^I$ is a column vector, and in the full rank case that $\mathbf{R}_{j,f}$ is a positive definite Hermitian matrix.

Moreover, we assume that for every source j the spatial covariances $\{\mathbf{R}_{j,f}\}_f$ are either linear instantaneous (i.e., constant over frequency: $\mathbf{a}_{j,f} = \mathbf{a}_j$ or $\mathbf{R}_{j,f} = \mathbf{R}_j$) or convolutive (i.e., varying with frequency), and either fixed (i.e., not updated during model estimation) or adaptive.

3.2 Spectral Power Structures

To model spectral power we use non-negative tensor factorization (NTF)-like audio-specific decompositions [8], thus all variables introduced in this section are implicitly assumed to be non-negative. We first model spectral power $v_{j,fn}$ as the product of *excitation spectral power* $v_{j,fn}^{\text{excit}}$ (e.g., representing the excitation of the glottal source for voice or the plucking of the string of a guitar) and *filter spectral power* $v_{j,fn}^{\text{filt}}$ (e.g., representing the vocal tract or the impedance of the guitar body) [8]:

$$v_{j,fn} = v_{j,fn}^{\text{excit}} \times v_{j,fn}^{\text{filt}} \quad (6)$$

The excitation spectral power $[v_{j,fn}^{\text{excit}}]_f$ is modeled as the sum of K_{excit} characteristic spectral patterns $[e_{j,fk}^{\text{excit}}]_f$ modulated in time by $p_{j,kn}^{\text{excit}}$, i.e., $v_{j,fn}^{\text{excit}} = \sum_{k=1}^{K_{\text{excit}}} p_{j,kn}^{\text{excit}} e_{j,fk}^{\text{excit}}$ [6]. In order to further constrain the fine spectral structure of the spectral patterns, they can be represented as linear combinations of L_{excit} *elementary narrowband spectral patterns* $[w_{j,fl}^{\text{excit}}]_f$ [13], i.e., $e_{j,fk}^{\text{excit}} = \sum_{l=1}^{L_{\text{excit}}} u_{j,lk}^{\text{excit}} w_{j,fl}^{\text{excit}}$, where $u_{j,lk}^{\text{excit}}$ are some non-negative weights. These narrowband patterns may be for instance harmonic, inharmonic or noise-like with a smooth spectral envelope.

Following exactly the same idea, we propose to represent the series of time activation coefficients $p^{\text{excit}}_{j,kn}$ as sums of M_{excit} *time localized patterns* to ensure their continuity or some other structure, i.e., $p^{\text{excit}}_{j,kn} = \sum_{m=1}^{M_{\text{excit}}} h^{\text{excit}}_{j,mn} g^{\text{excit}}_{j,km}$. Altogether we have:

$$v^{\text{excit}}_{j,fn} = \sum_{k=1}^{K_{\text{excit}}} \sum_{m=1}^{M_{\text{excit}}} h^{\text{excit}}_{j,mn} g^{\text{excit}}_{j,km} \sum_{l=1}^{L_{\text{excit}}} u^{\text{excit}}_{j,lk} w^{\text{excit}}_{j,fl}, \tag{7}$$

and, introducing matrices $\mathbf{V}^{\text{excit}}_j \triangleq [v^{\text{excit}}_{j,fn}]_{f,n}$, $\mathbf{H}^{\text{excit}}_j \triangleq [h^{\text{excit}}_{j,mn}]_{m,n}$, $\mathbf{G}^{\text{excit}}_j \triangleq [g^{\text{excit}}_{j,km}]_{k,m}$, $\mathbf{U}^{\text{excit}}_j \triangleq [u^{\text{excit}}_{j,lk}]_{l,k}$ and $\mathbf{W}^{\text{excit}}_j \triangleq [w^{\text{excit}}_{j,fl}]_{f,l}$, this equation can be rewritten in matrix form as $\mathbf{V}^{\text{excit}}_j = \mathbf{W}^{\text{excit}}_j \mathbf{U}^{\text{excit}}_j \mathbf{G}^{\text{excit}}_j \mathbf{H}^{\text{excit}}_j$.

Filter spectral power $[v^{\text{filt}}_{j,fn}]_f$ is represented with exactly the same structure as (7), so that to allow modeling time-varying filters as linear combination of some characteristic spectral patterns $[e^{\text{filt}}_{j,fk}]_f$ constrained to be continuous using some smooth narrowband elementary spectral patterns $[w^{\text{filt}}_{j,fl}]_f$.

Altogether spectral power structure can be represented by the following matrix decomposition ($\odot$ denotes element-wise matrix multiplication):

$$\mathbf{V}_j = \mathbf{V}^{\text{excit}}_j \odot \mathbf{V}^{\text{filt}}_j = \left(\mathbf{W}^{\text{excit}}_j \mathbf{U}^{\text{excit}}_j \mathbf{G}^{\text{excit}}_j \mathbf{H}^{\text{excit}}_j\right) \odot \left(\mathbf{W}^{\text{filt}}_j \mathbf{U}^{\text{filt}}_j \mathbf{G}^{\text{filt}} \mathbf{H}^{\text{filt}}_j\right), \tag{8}$$

where each matrix in this decomposition is assumed to be either fixed or adaptive. To cover Gaussian mixture models (GMM), HMM, and scaled versions of these models (SGMM, HSMM) [10], every column $\mathbf{g}^{\text{excit}}_{j,m} = [g^{\text{excit}}_{j,km}]_k$ of matrix $\mathbf{G}^{\text{excit}}_j$ (and similarly for matrix $\mathbf{G}^{\text{filt}}_j$) may further be constrained to have either a single nonzero entry (for SGMM, HSMM) or a single nonzero entry equal to 1 (for GMM, HMM). Mixture component probabilities of GMM and transition probabilities for HMM should be included in hyperparameters η of (5).

3.3 Generality

It can be easily shown that the model structures considered in [12], [1], [7], [6], [5], [11], [10], [13], [3], [14] are particular instances of the proposed general formulation. Let us give some examples.

Pham *et al* [12] assume rank-1 spatial covariances and constant spectral power over time-frequency regions of size (1 frequency bin $\times$ L frames). This structure can be implemented in our framework by choosing rank-1 adaptive spatial covariances and constraining spectral power to $\mathbf{V}_j = \mathbf{W}^{\text{excit}}_j \mathbf{G}^{\text{excit}} \mathbf{H}^{\text{excit}}_j$ [1] with $\mathbf{W}^{\text{excit}}_j$ being the identity ($F \times F$) matrix, $\mathbf{G}^{\text{excit}}$ being ($F \times \lceil N/L \rceil$) adaptive, and $\mathbf{H}^{\text{excit}}_j$ being ($\lceil N/L \rceil \times N$) fixed with entries $h^{\text{excit}}_{j,mn} = 1$ for $n \in \mathcal{L}_m$ and $h^{\text{excit}}_{j,mn} = 0$ for $n \notin \mathcal{L}_m$, where $\mathcal{L}_m$ is the set of time indices of the L-length block.

Multichannel NMF structures with point [11] or diffuse source model [3] can be represented within our framework as $\mathbf{V}_j = \mathbf{W}^{\text{excit}}_j \mathbf{H}^{\text{excit}}_j$ with $\mathbf{W}^{\text{excit}}_j$ of size ($F \times K_{\text{excit}}$) and $\mathbf{H}^{\text{excit}}_j$ of size ($K_{\text{excit}} \times F$), both being adaptive, and rank-1 or full rank adaptive spatial covariances.

[1] The power structure in (8) can be easily reduced to $\mathbf{V}_j = \mathbf{W}^{\text{excit}}_j \mathbf{G}^{\text{excit}} \mathbf{H}^{\text{excit}}_j$ by assuming that all the other matrices $\mathbf{U}^{\text{excit}}_j$, $\mathbf{W}^{\text{filt}}_j$, $\mathbf{U}^{\text{filt}}_j$, $\mathbf{G}^{\text{filt}}$ and $\mathbf{H}^{\text{filt}}_j$ are of sizes ($1 \times K_{\text{excit}}$), ($F \times 1$), ($1 \times 1$), ($1 \times 1$), and ($1 \times N$), fixed, and composed of 1.

4 Modular Implementation

Due to lack of space we here give a very brief overview of the framework implementation via a GEM algorithm that consists in iterating the expectation (E) and maximization (M) steps. The E-step consists in computing conditional expectation $\widehat{\mathbf{T}}$ of a *natural sufficient statistics* $\mathbf{T}$, given the observations $\mathbf{X} = \{\mathbf{x}_{fn}\}_{f,n}$ and current model parameters. The M-step consists in updating model parameters θ so as to increase the conditional expectation of the log-likelihood of the *complete data*. We assume that the J-th source with full rank spatial covariance represents a controllable additive noise needed for *simulated annealing* as in [11]. Let $\mathcal{J}_{\mathrm{r1}}$ and $\mathcal{J}_{\mathrm{rf}}$ be the subsets of the remaining source indices $\{1, \ldots, J-1\}$ corresponding respectively to rank-1 and full rank spatial covariances, and we assume that each source with rank-1 spatial covariance $\mathbf{R}_{j,f} = \mathbf{a}_{j,f}\mathbf{a}_{j,f}^H$ writes $\mathbf{y}_{j,fn} = \mathbf{a}_{j,f}s_{j,fn}$, where $s_{j,fn}$ are the STFT coefficients of a single-channel signal. With these conventions we choose $\mathbf{Z} = \{\mathbf{x}_{fn}, \{\mathbf{y}_{j,fn}\}_{j\in\mathcal{J}_{\mathrm{rf}}}, \{s_{j,fn}\}_{j\in\mathcal{J}_{\mathrm{r1}}}\}_{f,n}$ as the complete data set of the proposed GEM algorithm. The model $\theta = \{\theta_j\}_{j=1}^J$ being a set of source models, each source model is further represented as a set of 9 parameter subsets $\theta_j = \{\theta_j^m\}_{m=1}^9 = \{\mathbf{R}_j, \mathbf{W}_j^{\mathrm{excit}}, \mathbf{U}_j^{\mathrm{excit}}, \mathbf{G}_j^{\mathrm{excit}}, \mathbf{H}_j^{\mathrm{excit}}, \mathbf{W}_j^{\mathrm{filt}}, \mathbf{U}_j^{\mathrm{filt}}, \mathbf{G}_j^{\mathrm{filt}}, \mathbf{H}_j^{\mathrm{filt}}\}$. We implement the M-step via a loop over all $J \times 9$ parameter subsets. Each subset, depending whether it is adaptive or fixed, is updated or not in turn using existing spatial covariance update rules [11], [5] and multiplicative NMF update rules [13] that guarantee that the log-likelihood of the complete data is non-decreasing. Finally, particular constraints (see Sec. 3.1 and 3.2) are applied, if specified.

5 Experimental Illustrations

To illustrate the flexibility, we have evaluated four instances of our framework on the development data of the second community-based Signal Separation Evaluation Campaign (SiSEC 2010) [2] "Underdetermined-speech and music mixtures" task. The former two instances considered are NMF spectral power structures with rank-l [11] and full rank [3] spatial covariances (see Sec. 3.3). The later two instances are similar, except that the spectral power is structured as $\mathbf{V}_j = \mathbf{W}_j^{\mathrm{excit}}\mathbf{U}_j^{\mathrm{excit}}\mathbf{H}_j^{\mathrm{excit}}$ with adaptive $\mathbf{W}_j^{\mathrm{excit}}$ and $\mathbf{H}_j^{\mathrm{excit}}$ of sizes $(F \times L_{\mathrm{excit}})$ and $(K_{\mathrm{excit}} \times F)$, and fixed $\mathbf{U}_j^{\mathrm{excit}}$ of size $(K_{\mathrm{excit}} \times F)$ being composed of harmonic (e.g., to represent voiced speech) and noise-like and smooth (e.g., to represent non-voiced speech) narrowband spectral patterns [13]. Such a spectral structure is simply referred hereafter as *harmonic NMF*. In line with [11], parameter estimation via GEM is very sensitive to initialization for all the configurations we consider. To provide our GEM algorithm with a "good initialization" we used the DEMIX mixing matrix estimation algorithm [2], followed by l_0 norm minimization (see e.g., [14]) to initialize the source spectra, for the instantaneous mixtures. For synthetic convolutive and live recorded mixtures we used Cumulative State Coherence (CSC) transform-based Time Differences Of Arrival (TDOAs)

[2] `http://sisec.wiki.irisa.fr/tiki-index.php`

estimation algorithm [9] to initialize anechoic spatial covariances, followed by binary masking to initialize the source spectra. Source separation results in terms of average Source to Distortion Ratio (SDR) after 200 iterations of the proposed GEM algorithm are summarized in table 1 together with results of the *baseline* used for initialization. As expected, rank-1 spatial covariances perform the best for instantaneous mixtures and full rank spatial covariances perform the best for synthetic convolutive and live recorded mixtures. Moreover, as compared to the NMF spectral power, the harmonic NMF spectral power improves results for speech sources in almost all cases. Thus, we see that each tested configuration is performs the best for some setting. For each setting the configuration preforming the best on the development data was entered to the SiSEC 2010.

Table 1. Average SDRs on subsets of SiSEC 2010 development data

Mixing	instantaneous		synth. convolutif				live recorded			
Sources	speech	music	speech		music		speech		music	
Microphone distance	-	-	5 cm	1 m	5 cm	1 m	5 cm	1 m	5 cm	1 m
baseline (l_0 min. or bin. mask.)	8.6	12.4	0.3	1.4	-0.8	-0.9	1.0	1.4	2.3	0.0
NMF / rank-1 [11]	9.6	**18.4**	1.0	2.3	-0.6	-0.6	2.0	2.4	**3.6**	0.3
NMF / full-rank [3]	8.7	17.9	1.2	2.9	-2.3	-0.5	2.2	2.9	3.3	**0.7**
harmonic NMF / rank-1	**10.6**	15.1	1.0	2.7	**-0.1**	**0.0**	2.2	3.4	2.2	0.6
harmonic NMF / full-rank	10.5	14.3	**1.5**	**3.5**	-1.8	-0.2	**2.5**	**3.9**	1.5	0.4

6 Conclusion

We have introduced a general flexible and modular audio source separation framework that generalizes several existing source separation methods, brings them into a common framework, and allows to imagine and implement new efficient methods. The framework capabilities were illustrated in the experimental part, where we have reproduced two existing methods, namely NMF / rank-1 [11] and NMF / full-rank [3], but also we have tested two new methods, namely harmonic NMF / rank-1 and harmonic NMF / full-rank, that in our best knowledge were not yet reported in the literature. We have observed that adding harmonic and noise-like smooth constraints to NMF allows improving separation results for speech signals. Note also that the proposed framework can also be seen as a statistical implementation of Computational Auditory Scene Analysis (CASA) principles, whereby primitive grouping cues and learned grouping cues are simultaneously used to segregate the sources, thereby avoiding error propagation due to sequential use of grouping cues. Examples primitive grouping cues accounted by our model include harmonicity, spectral smoothness, time continuity, common onset, common amplitude modulation, spectral similarity and spatial similarity.

Acknowledgments

The authors would like to thank S. Arberet and F. Nesta for kindly sharing their implementations of DEMIX [2] and a TDOAs estimation [9] algorithms.

References

1. Abdallah, S.A., Plumbley, M.D.: Polyphonic transcription by nonnegative sparse coding of power spectra. In: Proc. 5th International Symposium Music Information Retrieval (ISMIR 2004), pp. 318–325 (October 2004)
2. Arberet, S., Gribonval, R., Bimbot, F.: A robust method to count and locate audio sources in a multichannel underdetermined mixture. IEEE Transactions on Signal Processing 58(1), 121–133 (2010)
3. Arberet, S., Ozerov, A., Duong, N., Vincent, E., Gribonval, R., Bimbot, F., Vandergheynst, P.: Nonnegative matrix factorization and spatial covariance model for under-determined reverberant audio source separation. In: 10th Int. Conf. on Information Sciences, Signal Proc. and their Applications, ISSPA 2010 (2010)
4. Cardoso, J.F., Martin, M.: A flexible component model for precision ICA. In: Davies, M.E., James, C.J., Abdallah, S.A., Plumbley, M.D. (eds.) ICA 2007. LNCS, vol. 4666, pp. 1–8. Springer, Heidelberg (2007)
5. Duong, N.Q.K., Vincent, E., Gribonval, R.: Under-determined convolutive blind source separation using spatial covariance models. In: IEEE International Conference on Acoustics,Speech, and Signal Processing ICASSP (March 2010)
6. Févotte, C., Bertin, N., Durrieu, J.L.: Nonnegative matrix factorization with the Itakura-Saito divergence. With application to music analysis. Neural Computation 21(3), 793–830 (2009)
7. Févotte, C., Cardoso, J.F.: Maximum likelihood approach for blind audio source separation using time-frequency Gaussian models. In: WASPAA 2005, Mohonk, NY, USA (October 2005)
8. FitzGerald, D., Cranitch, M., Coyle, E.: Extended nonnegative tensor factorisation models for musical sound source separation. In: Computational Intelligence and Neuroscience. Hindawi Publishing Corp. 2008 (2008)
9. Nesta, F., Svaizer, P., Omologo, M.: Cumulative state coherence transform for a robust two-channel multiple source localization. In: Adali, T., Jutten, C., Romano, J.M.T., Barros, A.K. (eds.) ICA 2009. LNCS, vol. 5441, pp. 290–297. Springer, Heidelberg (2009)
10. Ozerov, A., Févotte, C.: Multichannel nonnegative matrix factorization in convolutive mixtures for audio source separation. IEEE Trans. on Audio, Speech and Lang. Proc. 18(3), 550–563 (2010)
11. Ozerov, A., Févotte, C., Charbit, M.: Factorial scaled hidden Markov model for polyphonic audio representation and source separation. In: WASPAA 2009, October 18-21, pp. 121–124 (2009)
12. Pham, D.T., Servière, C., Boumaraf, H.: Blind separation of speech mixtures based on nonstationarity. In: Proceedings of the 7th International Symposium on Signal Processing and its Applications, pp. II–73–76 (2003)
13. Vincent, E., Bertin, N., Badeau, R.: Adaptive harmonic spectral decomposition for multiple pitch estimation. IEEE Trans. on Audio, Speech and Language Processing 18(3), 528–537 (2010)
14. Vincent, E., Jafari, M., Abdallah, S.A., Plumbley, M.D., Davies, M.E.: Probabilistic modeling paradigms for audio source separation. In: Machine Audition: Principles, Algorithms and Systems. IGI Global (2010) (to appear)

Adaptive Segmentation and Separation of Determined Convolutive Mixtures under Dynamic Conditions

Benedikt Loesch and Bin Yang

Chair of System Theory and Signal Processing, University of Stuttgart
{benedikt.loesch,bin.yang}@LSS.uni-stuttgart.de

Abstract. In this paper, we propose a method for blind source separation (BSS) of convolutive audio recordings with short blocks of stationary sources, i.e. dynamically changing source activity but no source movements. It consists of a time-frequency sparseness based localization step to identify segments with stationary sources whose number is equal to the number of microphones. We then use a frequency domain independent component analysis (ICA) algorithm that is robust to short data segments to separate each identified segment. In each segment we solve the permutation problem using the state coherence transform (SCT). Experimental results using real room impulse responses show a good separation performance.

Keywords: Blind source separation, dynamic mixing conditions.

1 Introduction

The task of convolutive blind source separation is to separate M convolutive mixtures into N different source signals. In this paper we consider dynamically changing source activity, i.e. active sources can change at any time during the recording but the sources cannot move.

With stationary mixing conditions we can apply frequency domain ICA with permutation correction to the complete recording (batch processing). However, the performance will be poor if the source positions change during the recording. To overcome this problem we can apply a frame-by-frame or block adaptive processing but performance will be limited by the convergence time and the limited amount of considered data. A better separation can be achieved if we run batch processing on each segment of $N = M$ stationary sources. This is why we propose to first find segments of $N = M$ stationary sources using a TF sparseness based localization step. This is done using source positions and pauses as segmentation cues. Once we have identified the segments, we apply a frequency domain ICA algorithm to each segment that can cope with short data segments. The permutation problem is solved using the state coherence transform (SCT) [1,2] which is also robust to short data lengths.

Some recent works for dynamically changing source activity are [3,4]. [3] models source activity with a hidden Markov model and switches off learning of the

V. Vigneron et al. (Eds.): LVA/ICA 2010, LNCS 6365, pp. 41–48, 2010.

demixing parameters for inactive sources. However, the computation complexity increases exponentially with the number of sources since all possible combinations of source activity need to be modelled. [4] proposes an online Bayesian learning procedure for instantaneous mixtures to incrementally estimate the mixing matrix and source signals in each time frame. This approach greatly reduces the computational complexity. However, it is not the purpose of this paper to compare the different approaches for dynamically changing mixing conditions. Instead we want to propose a simple but effective algorithm to find and separate segments of $N = M$ active sources.

2 Proposed Segmentation Algorithm

After a short-time Fourier transform (STFT), we can approximate the convolutive mixtures in the time-domain as instantaneous mixtures at each time-frequency (TF) point $[k, l]$:

$$\mathbf{X}[k,l] \approx \sum_{n=1}^{\tilde{N}} S_n[k,l]\mathbf{H}_n[k] \tag{1}$$

$k = 1, \cdots, K$ is the frequency bin index, $l = 1, \ldots, L$ is the time frame index. $\mathbf{X} = [X_1, \ldots, X_M]^T$ is called an observation vector, $\mathbf{H}_n = [H_{1n}, \ldots, H_{Mn}]^T$ is the vector of frequency responses from source n to all sensors. $\tilde{N}$ in (1) reflects the total number of sources of which only up to $N = M$ sources are assumed to be active in each time frame l, i.e. the other source signals $S_n[k, l]$ are zero.

We assume that the direct path is stronger than the multipath components. This allows us to exploit the DOA information for segmentation. The proposed algorithm consists of two steps: normalization and segmentation.

2.1 Normalization

From the observation vectors $\mathbf{X}[k, l]$, we derive normalized phase vectors $\bar{\mathbf{X}}[k, l]$ which contain only the phase differences of the elements of $\mathbf{X}[k, l]$ with respect to a reference microphone J:

$$\bar{\mathbf{X}}[k,l] = \left[e^{j\cdot\arg(X_m[k,l]/X_J[k,l])}\right], \quad m = 1, \cdots, M \tag{2}$$

For a single active source, the phase of the ratio of two elements of $\mathbf{X}[k, l]$ is a linear function of the frequency index k (modulo 2π). We use a distance metric that includes mod 2π to estimate the direction-of-arrival (DOA) θ_n of the sources:

$$\|\bar{\mathbf{X}}[k,l] - \mathbf{c}[k,\theta]\|^2 = 2M - 2\cdot\sum\nolimits_{m=1}^{M} \cos\left(\arg\left[\frac{X_m[k,l]}{X_J[k,l]}\right] - 2\pi\Delta f k\tau_m(\theta)\right) \tag{3}$$

Δf is the frequency bin width. $\mathbf{c}[k,\theta] = [c_m]_{1\leq m\leq M} = \left[e^{j2\pi\Delta f k\tau_m(\theta)}\right]_{1\leq m\leq M}$ is a state vector which contains the expected phase differences between the microphones $m = 1, \cdots, M$ and the reference one J for a potential source at DOA θ. Using this distance metric and TF sparseness, we can localize the active sources. For more details, please refer to [5].

2.2 Segmentation Algorithm

After the normalization we calculate the function

$$\mathcal{J}_l(\theta) = \sum_k \rho(\|\bar{\mathbf{X}}[k,l] - \mathbf{c}[k,\theta]\|^2) \tag{4}$$

where $\rho(\cdot)$ is a monotonously decreasing nonlinear function which reduces the influence of outliers and increases DOA resolution. Inspired from [1], we propose to use $\rho(t) = 1 - \tanh(\alpha\sqrt{t})$ in (4). Independently of our research, [6] proposed a similar cost function $\mathcal{J}_l(\theta)$ for only two microphones. In the ideal case, the function $\mathcal{J}_l(\theta)$ in (4) shows maxima at the true source DOAs θ for frame l and a small value for other DOA values.

We want to use this two-dimensional function $\mathcal{J}_l(\theta)$ to detect source position changes and to find segments with $N = M$ stationary sources by looking for the cumulative source activity in the time interval $[l_{\text{start}}, l_{\text{end}}]$. By this we mean how many sources have been active in total during this time interval. For this purpose we define $\mathcal{J}(\theta) = f\left(\mathcal{J}_{l_{\text{start}}}(\theta), \cdots, \mathcal{J}_{l_{\text{end}}}(\theta)\right)$, where the generic function $f(\cdot)$ could be $\underset{l}{\text{mean}}(\cdot)$, $\underset{l}{\text{median}}(\cdot)$, $\underset{l}{\max}(\cdot)$, or $\underset{l}{\max}_q(\cdot)$. The operation $\underset{l}{\max}_q(\cdot)$ selects the q-th largest value from its arguments. The mean and median operation have the disadvantage of a long memory, i.e. they detect a new source too late. The max operation detects a new source very fast, but it is not robust to single spikes of $\mathcal{J}_l(\theta)$. In comparison, the $\max_q$ operation is more robust since $\mathcal{J}_l(\theta)$ should have a large value in at least q frames for a fixed θ before $\mathcal{J}(\theta)$ confirms the source activity with a large value as well for the same θ. However, the $\max_q$ operation detects a new source too late. Hence, we use a combination of the max and $\max_q$ approaches (Algorithm 1):

Algorithm 1. Search for segments with $N = M$ stationary sources

$l_{\text{start}} := 1, l_{\text{end}} := l_{\text{start}} + l_{\min}$, marker:=[ ], $l_{\text{prev}} := 1$, $l_{\text{b}} := 1$
while $l_{\text{end}} < L$ **do**
  Determine $\hat{N}$ using Algorithm 2 with $\mathcal{J}(\theta) := \underset{l}{\max}(\mathcal{J}_{l_{\text{start}}}(\theta), \cdots, \mathcal{J}_{l_{\text{end}}}(\theta))$
  if $\hat{N} \leq M$ **then**
    $l_{\text{end}} := l_{\text{end}} + 1$
  else
    Determine $\hat{N}_2$ using Algorithm 2 with $\tilde{\mathcal{J}}(\theta) = \underset{l}{\max}_q(\mathcal{J}_{l_{\text{prev}}}(\theta), \cdots, \mathcal{J}_{l_{\text{end}}}(\theta))$
    if $\hat{N}_2 > M$ **then**
      Append l_{b} to the list of segment boundaries: marker:=[marker l_{b}], $l_{\text{prev}} := l_{\text{b}}$
    end if
    Start a new segment: $l_{\text{b}} := l_{\text{end}}$, $l_{\text{start}} := l_{\text{b}}, l_{\text{end}} := l_{\text{start}} + l_{\min}$
  end if
end while

The proposed algorithm starts with a short segment of length $l_{\min}$ frames and increases the size of the current segment until $\hat{N} > M$ active sources are detected by $\mathcal{J}(\theta) = \max_l \mathcal{J}_l(\theta)$. We store the current frame as a potential segment boundary in l_b. We then start a new segment and increase this segment until we detect the next potential segment boundary by $\mathcal{J}(\theta) = \max_l \mathcal{J}_l(\theta)$. Now we verify the previously detected segment boundary l_b by checking if $\tilde{\mathcal{J}}(\theta) = \max_{q\,l} \mathcal{J}_l(\theta)$ shows $\hat{N}_2 > M$ maxima for the combined segment $[l_{\text{prev}}, l_{\text{end}}]$ containing the previous and current segment. l_{prev} contains the last but one segment boundary. This process is repeated until the end of the recording. The number of sources $\hat{N}$ for the current segment is determined using Algorithm 2 by looking for the number of significant and distinct maxima of $\mathcal{J}(\theta)$ or $\tilde{\mathcal{J}}(\theta)$.

Algorithm 2. Source number estimation

```
Find all extrema of J(θ)
Find the distance h in height between each maximum and its neighbouring minima
Discard maxima with small h
Sort remaining maxima θ_n in descending order of J(θ_n)
n:=1, max_list:=[ ]
while J(θ_n) > t_1 do
  if (min |max_list − θ_n|) > t_2 then
    max_list:=[max_list θ_n]
  end if
  n:=n+1
end while
N̂ := length(max_list)
```

Fig. 1 illustrates $\mathcal{J}(\theta)$ and $\tilde{\mathcal{J}}(\theta)$ for three segments starting at 0 s and ending at 1.5 s, 3 s and 3.9 s. We want to identify segments with $N = M = 3$ sources. 3 sources at $\theta_{1,2,3} = 30°, 90°, 150°$ are active before 3 s. At the time instant 3 s, a new fourth source at $\theta_4 = 126°$ appears while the third source at $\theta_3 = 150°$ disappears. Clearly $\mathcal{J}(\theta)$ detects the fourth source as soon as it becomes active (at 3 s) since $\mathcal{J}(\theta)$ shows four distinct maxima in Fig. 1(b). $\tilde{\mathcal{J}}(\theta)$ takes additional 0.9 s to verify that it is truely a new source and not a spurious spike since $\tilde{\mathcal{J}}(\theta)$ shows three distinct maxima in Fig. 1(b) and four distinct maxima in Fig. 1(c).

Algorithm 1 works quite well, but sometimes it still detects a segment boundary too late if the newly active source does not start with a frame with high phase coherence, i.e. $\mathcal{J}(\theta)$ is not large enough. This can happen if the newly active source has a smaller power or there is no frame at the beginning of the segment where it is the single dominant source. However, we can use additional information based on pauses in the segmentation process: We first detect pauses of more than T frames by counting the number of consecutive frames where $\max_\theta \mathcal{J}_l(\theta)$ is small. This corresponds to frames with no source activity. We detect a pause end if the maximum of $\mathcal{J}(\theta)$ gets larger than a predefined threshold t_3, i.e. the coherence of the observed phase becomes large. This corresponds to one or multiple active sources. This procedure is summarized in Algorithm 3. Using the detected pauses, we perform a segment verification step: If Algorithm 1

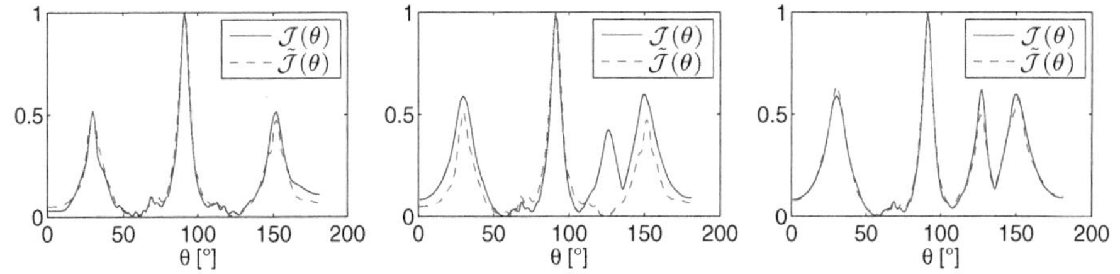

(a) $0 - 1.5\,\mathrm{s}$ (b) $0 - 3\,\mathrm{s}$ (c) $0 - 3.9\,\mathrm{s}$

Fig. 1. Segmentation process using $\mathcal{J}(\theta)$ and $\tilde{\mathcal{J}}(\theta)$, new source becomes active at 3 s

detects a segment boundary shortly after a pause we move the segment boundary to the end of the pause if this yields a segmentation with $N = M$ sources in the previous and current segment.

Algorithm 3. Pause detection

count:=0
for all $l = 1$ to L **do**
  if $\max_\theta \mathcal{J}_l(\theta) < t_3$ **then**
    count:=count+1
  else
    count:=0
  end if
  pause_count[l] := count
end for
pause_end:=$\{l \in [1, \cdots, L] : \text{pause\_count}[l] = 0 \vee \text{pause\_count}[l-1] > T\}$

3 Separation

After we have identified the segments containing $N = M$ active sources, we perform frequency domain ICA to separate the sources in each segment. We have to deal with the following two issues:

- **Choice of the ICA algorithm for short data segments.** It is well known that the performance of most ICA algorithms degrades if only a small amount of data is available. Since we are considering dynamically changing mixing conditions, we have to use an ICA algorithm that can deal with short amounts of data. [7] showed that a recursive initialization of the demixing matrices across frequencies improves the robustness of the scaled Infomax algorithm for short data segments. We use this separation algorithm below.
- **Permutation problem.** Since we are applying ICA to each frequency bin individually, the permutation problem has to be solved. For this task many approaches have been proposed. They can be classified into a family using

properties of the separated signals (e.g. correlation across frequency) and another one based on propagation model parameters or smoothness of the demixing matrices across frequency. Correlation based methods work well if the observable data length is sufficiently long. However, when the data length is short, performance decreases. We have shown in [2] that the multi-dimensional SCT is a robust way to solve the permutation problem even for short data lengths. Hence, we will use it to solve the permutation problem in the given context of short data segments with stationary sources. For more details please refer to [1,2].

4 Experimental Results

4.1 Results Using RWCP Database

We consider two scenarios using impulse responses from the E2A room (T_{60} = 300 ms) of the RWCP database [8]: We use an uniform linear array (ULA) with $M = 2$ or $M = 3$ sensors with a total aperture of $d = 11$ cm and segments from the short stories of the CHAINS database [9]. The source activity and the corresponding source DOAs for the two scenarios are depicted in Fig. 2(a) and (b). Each scenario has 7 segments with different lengths and source DOAs.

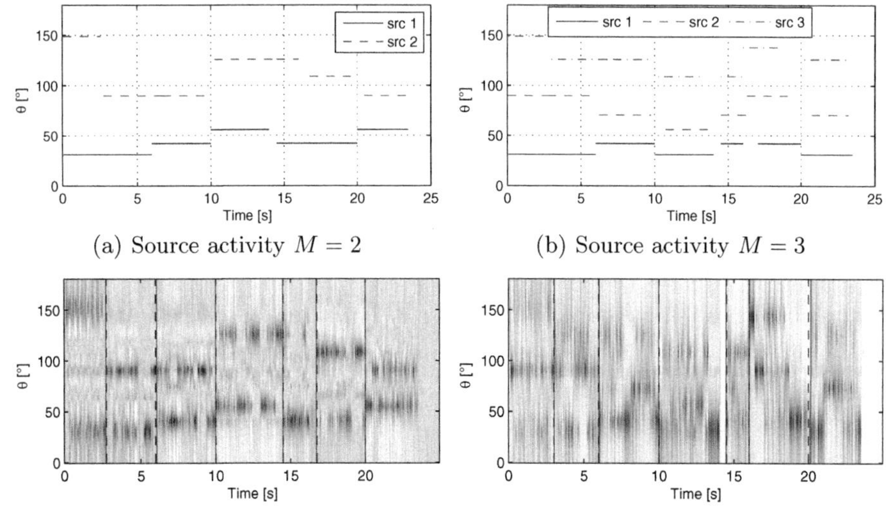

(a) Source activity $M = 2$ (b) Source activity $M = 3$

(c) Segmentation $M = 2$ (d) Segmentation $M = 3$

Fig. 2. Source activity and resulting segmentation using our algorithm

Fig. 2(c) and (d) show $\mathcal{J}_l(\theta)$ from (4) as gray value and the detected segment boundaries (red solid lines) together with the true ones (blue dashed lines).

Clearly our algorithm detects the segments with $N = M$ sources very well since the estimated segment boundaries match the true boundaries.

For each segment found by our proposed segmentation algorithm, we run frequency domain ICA with the SCT for permutation correction. We used an STFT frame size of 4096 with 75% overlap. Evaluation of the separation quality is done using the BSS_EVAL toolbox [10] for each segment where there are $N = M$ active sources. We use the signal-to-interference ratio (SIR), signal-to-distortion ratio (SDR) and signal-to-artifact ratio (SAR) defined in [10] as separation performance measures. As proposed in the SISEC2010 task "Determined Convolutive Mixtures under Dynamic Conditions", we use an A-weighting filter before the evaluation of the performance measures to model the frequency characteristic of the human ear. The separation results for $M = 2$ and $M = 3$ are summarized in Table 1. Clearly, the proposed algorithm is able to separate the sources very well. Separation quality is influenced by the duration of the segments, the amount of activity for each source and the angular spacing between the sources. The more difficult case of $N = M = 3$ shows a slightly lower separation quality than $N = M = 2$.

Table 1. Separation performance for each segment in dB with A-weighting

	segment	1	2	3	4	5	6	7	mean
	SIR	19.4	21.1	21.1	17.5	18.9	16.9	16.9	18.9
$N = M = 2$	SDR	5.9	11.7	8.0	4.3	7.6	10.2	4.3	7.4
	SAR	6.2	12.3	8.3	4.6	8.0	11.4	4.6	7.9
	SIR	18.2	20.0	18.9	13.2	11.8	20.1	17.8	17.2
$N = M = 3$	SDR	6.2	8.8	6.6	4.0	3.4	10.8	7.5	6.8
	SAR	6.6	9.3	6.9	4.9	4.7	11.5	8.0	7.4

4.2 SISEC2010 Data

We have submitted our algorithm for the task "Determined Convolutive Mixtures under Dynamic Conditions" of the SISEC2010 campaign. The task uses impulse responses from a very reverberant room with $T_{60} = 700$ ms and different datasets for a microphone array with $M = 2$ microphones and spacing $d = 2, 6, 10$ cm. Here we show the results for the example dataset for a microphone spacing of $d = 6$ cm. The separation performance for the complete recording using an STFT frame size of 8192 with 75% overlap is summarized in Table 2 where we give the mean values of SIR, SAR and SDR with and without A-weighting and the corresponding standard deviations.

On the test dataset (`http://irisa.fr/metiss/SiSEC10/dynamic/dynamic_task2_all.html`), our algorithm outperforms the other approaches except for the case of $d = 2$ cm. A possible explanation is that localization accuracy for $d = 2$ cm is insuffіcent to yield an accurate segmentation.

Table 2. Mean and standard deviation of separation performance for SISEC2010 example dataset in dB

without A-weighting			with A-weighting		
SIR	SDR	SAR	SIR	SDR	SAR
9.13 ± 2.99	3.21 ± 2.86	6.42 ± 1.54	12.13 ± 3.78	4.47 ± 3.71	6.47 ± 2.36

5 Conclusion

In this paper we have presented a method to separate recordings of short blocks of stationary sources. It is based on a segmentation of the recording into blocks of $N = M$ active sources through a time-frequency sparseness based DOA estimation for each time frame. Through a sliding time window, the change points are detected and the recordings are divided into blocks of $N = M$ active sources. We then use a frequency domain ICA algorithm suited for short data segments [7] together with permutation correction using the state coherence transform [1,2]. Experimental results show that our approach achieves good separation performance even when the source activity changes frequently.

References

1. Nesta, F., Omologo, M., Svaizer, P.: A novel robust solution to the permutation problem based on a joint multiple TDOA estimation. In: Proc. International Workshop for Acoustic Echo and Noise Control, IWAENC (2008)
2. Loesch, B., Nesta, F., Yang, B.: On the robustness of the multidimensional state coherence transform for solving the permutation problem of frequency-domain ICA. In: Proc. ICASSP (2010)
3. Masnadi-Shirazi, A., Zhang, W., Rao, B.D.: Glimpsing indepdendent vector analysis: Separation more sources than sensors using active and inactive states. In: Proc. ICASSP (2010)
4. Hsieh, H.-L., Chien, J.-T.: Online bayesian learning for dynamic source separation. In: Proc. ICASSP (2010)
5. Loesch, B., Yang, B.: Blind source separation based on time-frequency sparseness in the presence of spatial aliasing. Submitted to LVA/ICA (2010)
6. Chami, Z.E., Guerin, A., Pham, A., Serviere, C.: A phase-based dual microphone method to count and locate audio sources in reverberant rooms. In: Proc. IEEE Workshop on Applications of Signal processing to Audio and Acoustics, WASPAA (2009)
7. Nesta, F., Svaizer, P., Omologo, M.: Separating short signals in highly reverberant environment by a recursive frequency domain BSS. In: Proc. Joint Workshop on Hands-free Speech Communication and Microphone Arrays (HSCMA) (May 2008)
8. Real World Computing Partnership, RWCP Sound Scene Database in Real Acoustic Environment (2001), `http://tosa.mri.co.jp/sounddb/indexe.htm`
9. Cummins, F., Grimaldi, M., Leonard, T., Simko, J.: The CHAINS corpus (characterizing individual speakers) (2006), `http://chains.ucd.ie/`
10. Vincent, E., Gribonval, R., Fevotte, C.: Performance measurement in blind audio source separation. IEEE Transactions on Speech and Audio Processing 14(4) (2006)

Blind Speech Extraction Combining Generalized MMSE STSA Estimator and ICA-Based Noise and Speech Probability Density Function Estimations

Hiroshi Saruwatari*, Ryoi Okamoto, Yu Takahashi, and Kiyohiro Shikano

Nara Institute of Science and Technology,
8916-5 Takayama-cho, Ikoma, Nara, 630-0192, Japan

Abstract. In this paper, we propose a new blind speech extraction method combining ICA-based dynamic noise estimation and a generalized minimum mean-square-error short-time spectral amplitude estimator of the target speech. To deal with various types of speech signals with different probability density functions (p.d.f.), we also introduce a spectral-subtraction-based speech p.d.f. estimation and provide a theoretical justification of the proposed approach. We conduct an experiment in an actual railway-station environment, and show the improved noise reduction of the proposed method by objective and subjective evaluations.

Keywords: Blind speech extraction, ICA, Generalized MMSE STSA estimator.

1 Introduction

In recent years, multichannel noise reduction methods have been actively studied. One of the authors has proposed a blind speech extraction method, *blind spatial subtraction array (BSSA)* [1], mainly for hands-free speech recognition systems. BSSA consists of a delay-and-sum (DS)-based primary path and a reference path for independent component analysis (ICA)-based noise estimation. Noise reduction in BSSA is achieved by spectral subtraction (SS) [2], that is, the estimated noise amplitude spectrum is subtracted from the target speech amplitude spectrum partly enhanced by DS. This method is used to extract a target speech source under a nonstationary and non-point-source noise condition. However, BSSA is not appropriate for audio applications because the SS in BSSA generates a large amount of artificial distortion, so-called musical noise.

To improve the sound quality, we have proposed an extension [3] of BSSA, which consists of ICA-based noise estimation and a minimum mean-square error (MMSE) short-time spectral amplitude (STSA) estimator [4] of the target speech. Although the simple MMSE STSA estimator cannot effectively reduce the nonstationary noise because it requires a noise prototype obtained in the nonspeech period, our combined approach of ICA-based dynamic noise estimation and the MMSE STSA estimator can provide better noise reduction.

In [3], we used the nonparametric MMSE STSA estimator under the assumption that the speech probability density function (p.d.f.) has a Gaussian distribution. In the

* This work was partly supported by MIC Strategic Information and Communications R&D Promotion Programme in Japan.

V. Vigneron et al. (Eds.): LVA/ICA 2010, LNCS 6365, pp. 49–56, 2010.

context of ICA, however, the speech signal always obeys more spiky p.d.f., and we have no knowledge of the speech p.d.f. in advance. Therefore, in this paper, we propose a new method combining ICA and a *generalized (parametric)* MMSE STSA estimator [5] that can treat various types of signals with different p.d.f. In addition, we propose a new blind estimation algorithm for estimating the speech p.d.f. prior in order to appropriately control the generalized MMSE STSA estimator.

2 Signal Model and Noise Estimation Based on Independent Component Analysis

We consider an acoustic mixing model where the number of microphones is J and the observed signal contains only one target speech signal, which can be regarded as a point source, and an additive noise signal. This additive noise represents a signal that cannot be regarded as point sources. Hereafter, the observed signal vector in the time-frequency domain, $\boldsymbol{x}(f,\tau) = [x_1(f,\tau),\ldots,x_J(f,\tau)]^{\mathrm{T}}$, is given by

$$\boldsymbol{x}(f,\tau) = \boldsymbol{h}(f)s(f,\tau) + \boldsymbol{n}(f,\tau), \tag{1}$$

where f is the frequency bin number, τ is the time-frame index number, $\boldsymbol{h}(f) = [h_1(f),\ldots,h_J(f)]^{\mathrm{T}}$ is a column vector of transfer functions from the target signal component to each microphone, $s(f,\tau)$ is the target speech signal component, and $\boldsymbol{n}(f,\tau) = [n_1(f,\tau),\ldots,n_J(f,\tau)]^{\mathrm{T}}$ is the column vector of the additive noise signal.

In ICA, we perform signal separation using an unmixing matrix $\boldsymbol{W}(f)$, so that the output signals $\boldsymbol{y}(f,\tau) = [y_s(f,\tau), y_n(f,\tau)]^{\mathrm{T}}$ become mutually independent; in this calculation,

$$\boldsymbol{y}(f,\tau) = [y_s(f,\tau), y_n(f,\tau)]^{\mathrm{T}} = \boldsymbol{W}(f)\boldsymbol{x}(f,\tau), \tag{2}$$

$$\boldsymbol{W}^{[p+1]}(f) = \boldsymbol{W}^{[p]}(f) + \mu\left[\boldsymbol{I} - \left\langle \boldsymbol{\phi}(\boldsymbol{y}(f,\tau))\,\boldsymbol{y}^{\mathrm{H}}(f,\tau)\right\rangle_{\tau}\right]\boldsymbol{W}^{[p]}(f), \tag{3}$$

where μ is the step-size parameter, $[p]$ is used to express the value at the pth step of the iteration, $y_s(f,\tau)$ is the estimated target speech signal, $y_n(f,\tau)$ is the estimated noise signal, and $\boldsymbol{I}$ is the identity matrix. In addition, $\langle\cdot\rangle_{\tau}$ denotes a time-averaging operator, $\boldsymbol{M}^{\mathrm{H}}$ denotes the conjugate transpose of matrix $\boldsymbol{M}$, and $\boldsymbol{\phi}(\cdot)$ is an appropriate nonlinear activation function.

Under the non-point-source noise condition, it has been pointed out that ICA is not proficient in target speech estimation, whereas it is effective in noise estimation [1]. That is, it is strongly recommended that ICA is used for noise estimation under the non-point-source noise condition. ICA is also more proficient in noise estimation under the reverberant condition than the fixed blocking matrix, e.g., the null beamformer (NBF) [6]. On the basis of the above-mentioned fact, one of the authors has proposed BSSA [1], which utilizes ICA as a noise estimator.

3 Proposed Method

3.1 Our Previous Work: BSSA [1]

BSSA consists of a DS-based primary path and a reference path for ICA-based noise estimation. The noise component estimated by ICA is subtracted from the primary path in the amplitude spectral domain, neglecting phase information; this enables BSSA to realize error-robust noise reduction.

First, in the primary path, the observed signal is partly enhanced by DS. This procedure can be given as

$$y_{\mathrm{DS}}(f,\tau) = \boldsymbol{w}_{\mathrm{DS}}^{\mathrm{T}}(f)\boldsymbol{x}(f,\tau), \tag{4}$$

$$\boldsymbol{w}_{\mathrm{DS}}(f) = \frac{1}{J}\left[\exp\left(-i2\pi(f/N)f_s d_1 \sin\theta_s/c\right), \ldots, \exp\left(-i2\pi(f/N)f_s d_J \sin\theta_s/c\right)\right]^{\mathrm{T}}, \tag{5}$$

where $y_{\mathrm{DS}}(f,\tau)$ is a target speech signal slightly enhanced by DS, $\boldsymbol{w}_{\mathrm{DS}}(f)$ is the filter coefficient vector of DS, N is the DFT size, f_s is the sampling frequency, d_j $(j = 1, \ldots, J)$ denote microphone positions, c is the sound velocity, and θ_s represents the estimated direction of arrival of the target speech signal, which is obtained by the ICA part in the reference path. We estimate θ_s in (5) from the unmixing matrix $\boldsymbol{W}(f)$ without *a priori* information of the target speech's location [6].

Next, in the reference path, the estimated target speech signal is discarded as it is not required because we want to estimate only the noise component. Instead, we construct a *noise-only vector* $\boldsymbol{y}^{(\mathrm{noise})}(f,\tau) = [0, y_n(f,\tau)]^{\mathrm{T}}$ from the output signal obtained by ICA using (2). Following this, we apply the projection back operation to remove the ambiguity of amplitude and construct the estimated noise signal $z(f,\tau)$ by applying DS, where

$$z(f,\tau) = \boldsymbol{w}_{\mathrm{DS}}^{\mathrm{T}}(f)\boldsymbol{W}^{-1}(f)\boldsymbol{y}^{(\mathrm{noise})}(f,\tau) = \boldsymbol{w}_{\mathrm{DS}}^{\mathrm{T}}(f)\boldsymbol{W}^{-1}(f)\left[0, y_n(f,\tau)\right]^{\mathrm{T}}. \tag{6}$$

Finally, noise reduction is carried out by SS, in which the estimated noise amplitude spectrum of (6) is subtracted from the partly enhanced target speech amplitude spectrum of (4). This procedure is given as

$$|y_{\mathrm{BSSA}}(f,\tau)| = \begin{cases} |y_{\mathrm{DS}}(f,\tau)| - \beta \cdot |z(f,\tau)| & (\text{if } |y_{\mathrm{DS}}(f,\tau)| - \beta \cdot |z(f,\tau)| \geq 0), \\ g \cdot y_{\mathrm{DS}}(f,\tau) & (\text{otherwise}), \end{cases} \tag{7}$$

where $y_{\mathrm{BSSA}}(f,\tau)$ is the final output of BSSA, β is an oversubtraction parameter, and g is a flooring parameter. The phase of the observed signal is used to construct the resultant speech-enhanced signal.

3.2 Problem of BSSA and Motivation of Study

Although BSSA can reduce a nonstationary and non-point-source noise efficiently, it often causes musical noise. To solve this problem, we have proposed an improved BSSA method [3] that consists of ICA and the original (nonparametric) MMSE STSA estimator [4]. This method still suffers from a mismatch problem owing to the assumption of the speech p.d.f. prior, i.e., the original MMSE STSA estimator assumes that the speech signal obeys a Gaussian distribution. However, it is well known that, in ICA, the speech signal has more spiky p.d.f. similar to a Laplacian distribution. Therefore, we propose a new method combining ICA and the generalized MMSE STSA estimator with a new blind estimation of the speech p.d.f. prior.

3.3 Parametric Modeling of Signals

In this study, we introduce the *generalized gamma distribution* [7] to model the amplitude spectral grid signal in the time-frequency domain. Its p.d.f. is written as

$$p(x) = 2\phi^{\eta}/\Gamma(\eta) \cdot x^{2\eta-1} \cdot \exp\left(-\phi x^2\right), \tag{8}$$

where $\Gamma(\cdot)$ denotes the gamma function, $\phi = \eta/\mathrm{E}\{|x|^2\}$, and η $(0 < \eta \leq 1)$ is a shape parameter; $\eta = 1$ gives a Rayleigh distribution that corresponds to a Gaussian signal and a smaller value of η corresponds to a super-Gaussian signal.

3.4 Proposed Algorithm Combining ICA-Based Noise Estimation and Generalized MMSE STSA Estimator

Figure 1(a) shows a block diagram of the proposed method. The details of signal processing are discussed below.

The proposed method can estimate temporal *a priori* and *a posteriori* SNRs and the spectral gain using the noise signal estimated by ICA. First, the *a posteriori* SNR estimate $\hat{\gamma}(f,\tau)$ is given as

$$\hat{\gamma}(f,\tau) = |y_{\mathrm{DS}}(f,\tau)|^2/\lambda(f,\tau) = \left|\boldsymbol{w}_{\mathrm{DS}}^{\mathrm{T}}(f)\boldsymbol{x}(f,\tau)\right|^2/\mathrm{E}\left\{|\boldsymbol{w}_{\mathrm{DS}}^{\mathrm{T}}(f)\boldsymbol{W}^{-1}(f)\boldsymbol{y}^{(\mathrm{noise})}(f,\tau)|^2\right\}_{(\tau-\tau_{\mathrm{th}})}^{\tau}, \tag{9}$$

where $\lambda(f,\tau)$ is the power spectrum of the estimated noise (6), τ_{th} is a smoothing parameter denoting a certain time frame window, and $\mathrm{E}\{\cdot\}_A^B$ denotes the expectation operator from A to B. Note that we can momentarily estimate the instantaneous *a posteriori* SNR (9) by utilizing the noise signal estimated by ICA (6), unlike the original MMSE STSA estimator. Therefore, it can be considered that our proposed method can suppress nonstationary noise more efficiently than the conventional MMSE STSA estimator.

Next, using (9), the *a priori* SNR estimate $\hat{\xi}(f,\tau)$ is given as [4]

$$\hat{\xi}(f,\tau) = \alpha\hat{\gamma}(f,\tau-1)\tilde{G}^2(f,\tau-1) + (1-\alpha)P\left[\hat{\gamma}(f,\tau)-1\right] \quad (0 \leq \alpha < 1), \tag{10}$$

where α is the weighting factor of the decision-directed estimation, $\tilde{G}(f,\tau)$ is a spectral gain function and the operator $P[\cdot]$ is a flooring function in which the negative input is floored to zero. Also, the spectral gain function is defined by [5]

$$\tilde{G}(f,\tau) = \frac{\sqrt{\tilde{v}(f,\tau)}}{\hat{\gamma}(f,\tau)} \frac{\Gamma(\eta+0.5)}{\Gamma(\eta)} \frac{\Phi(0.5-\eta, 1, -\tilde{v}(f,\tau))}{\Phi(1-\eta, 1, -\tilde{v}(f,\tau))}, \tag{11}$$

where Φ is a confluent hypergeometric function and $\tilde{v} = \hat{\xi}(f,\tau)/(\eta+\hat{\xi}(f,\tau))\hat{\gamma}(f,\tau)$. The gain $\tilde{G}(f,\tau)$ includes a shape parameter η that should represent speech p.d.f. prior, and we discuss how to determine it in Sect. 4.

Finally, noise reduction is carried out as follows:

$$y_{\mathrm{PROP}}(f,\tau) = \tilde{G}(f,\tau)y_{\mathrm{DS}}(f,\tau), \tag{12}$$

where $y_{\mathrm{PROP}}(f,\tau)$ is the final output of this method.

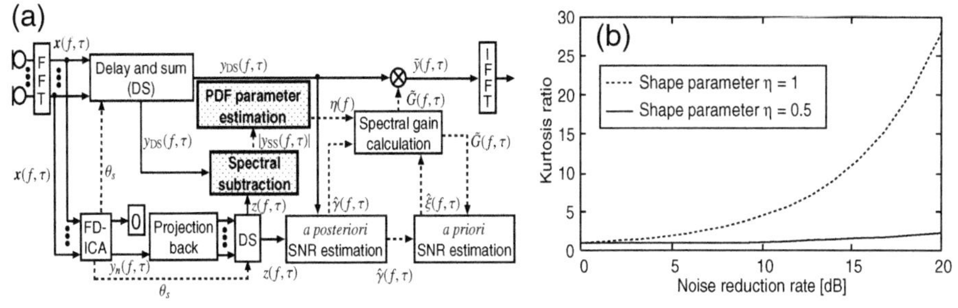

Fig. 1. (a) Block diagram of proposed method. (b) Kurtosis ratio vs. noise reduction rate.

4 Blind Estimation of Shape Parameter of Speech p.d.f.

4.1 Overview

Regarding the gamma distribution $p(x)$ in (8), we have the useful relation $\eta = 1/\{(\mu_4/\mu_2^2) - 1\}$, where μ_4/μ_2^2 is called the *kurtosis* and μ_n is the nth-order moment of the amplitude spectrum. From this relation, the shape parameter of the subjective speech signal can be estimated by obtaining its kurtosis value. In general, however, it is difficult to directly estimate the kurtosis of a speech signal because of the contamination by additive noise. Even using ICA, we cannot sufficiently enhance the target speech signal under diffuse noise environments as described in Sect. 2.

To cope with the above-mentioned kurtosis-estimation problem, we propose to use the SS procedure (i.e., the BSSA procedure) to enhance the target speech's p.d.f. In the following, we theoretically clarify that SS can enhance the target speech signal while maintaining the speech signal's kurtosis. Thus, although BSSA is *not* suitable for human-hearing applications, the kurtosis of its output can be used to obtain a good estimation of the shape parameter of the target speech signal.

4.2 Theoretical Analysis of Noise Reduction and Kurtosis via SS

Owing to the limitation of space in this paper, we hereafter show the results of theoretical analysis under the assumptions that the noise is Gaussian (i.e., $\eta = 1$) and the speech is super-Gaussian with $\eta = 0.5$. These parameter settings greatly simplify the analysis because the equations for kurtosis can be expressed analytically in a form that does not include any integrals. The experimental results for real-world data are given in Sect. 5.

The p.d.f. of the SS-applied amplitude spectral grid signal is formulated by laterally shifting $p(x)$ and stacking the negative probabilities at $x = 0$ (in the case of $g = 0$), given by

$$p_{\mathrm{SS}}(x) = \begin{cases} \frac{2\phi^{\eta}}{\Gamma(\eta)}(x + \beta\mu_1^{(\mathrm{noise})})^{2\eta-1} \exp\left(-\phi(x + \beta\mu_1^{(\mathrm{noise})})^2\right) & (x > 0), \\ \frac{2\phi^{\eta}}{\Gamma(\eta)} \int_0^{\beta\mu_1^{(\mathrm{noise})}} x^{2\eta-1} \exp\left(-\phi x^2\right) & (x = 0), \end{cases} \tag{13}$$

where $\mu_1^{(\mathrm{noise})}$ is the 1st-order moment of $p(x)$ under $\eta = 1$, which corresponds to the average of the noise amplitude spectrum, given by $\mu_1^{(\mathrm{noise})} = \Gamma(3/2)\phi^{-1/2}$.

Based on (13), we first estimate the *noise reduction rate* (NRR) as a measure of the noise reduction performance, which is defined as the output signal-to-noise ratio (SNR) in dB minus the input SNR in dB [6]. This is equivalent to the relative decrease in the noise components, given by

$$10\log_{10}\frac{\int_0^\infty x^2 p(x)dx\mid_{\eta=1}}{\int_0^\infty x^2 p_{\mathrm{SS}}(x)dx\mid_{\eta=1}} = -10\log_{10}\left\{\exp\left(-\frac{\pi}{4}\beta^2\right)-\frac{\pi}{2}\beta\cdot\mathrm{erfc}\left(\frac{\sqrt{\pi}}{2}\beta\right)\right\}, \quad (14)$$

where erfc(·) is the complementary error function, defined as $\mathrm{erfc}(x) = 2/\sqrt{\pi}\int_x^\infty \exp(-u^2)du$.

Next, we formulate the resultant kurtosis of noise after SS, which is given by

$$\frac{\int_0^\infty x^4 p_{\mathrm{SS}}(x)dx\mid_{\eta=1}}{\left(\int_0^\infty x^2 p_{\mathrm{SS}}(x)dx\mid_{\eta=1}\right)^2} = \left\{2\left(1+\frac{\pi}{4}\beta^2\right)\exp\left(-\frac{\pi}{4}\beta^2\right)-\frac{\pi}{2}\beta\left(3+\frac{\pi}{2}\beta^2\right)\mathrm{erfc}\left(\frac{\sqrt{\pi}}{2}\beta\right)\right\}$$
$$\cdot\left\{\exp\left(-\frac{\pi}{4}\beta^2\right)-\frac{\pi}{2}\beta\cdot\mathrm{erfc}\left(\frac{\sqrt{\pi}}{2}\beta\right)\right\}^{-2}. \quad (15)$$

Similarly, the resultant kurtosis of speech after SS is

$$\frac{\int_0^\infty x^4 p_{\mathrm{SS}}(x)dx\mid_{\eta=0.5}}{\left(\int_0^\infty x^2 p_{\mathrm{SS}}(x)dx\mid_{\eta=0.5}\right)^2} = \left\{\left(-\frac{\tilde{\beta}^2}{\pi}-\frac{5}{2}\tilde{\beta}\right)\exp\left(-\frac{\tilde{\beta}^2}{\pi}\right)+\left(\frac{3\pi}{4}+3\tilde{\beta}^2+\frac{\tilde{\beta}^4}{\pi}\right)\mathrm{erfc}\left(\frac{\tilde{\beta}}{\sqrt{\pi}}\right)\right\}$$
$$\cdot\left\{-\frac{\tilde{\beta}}{\sqrt{\pi}}\exp\left(-\frac{\tilde{\beta}^2}{\pi}\right)+\left(\frac{1}{2}+\frac{\tilde{\beta}^2}{\pi}\right)\sqrt{\pi}\mathrm{erfc}\left(\frac{\tilde{\beta}}{\sqrt{\pi}}\right)\right\}^{-2}, \quad (16)$$

where $\tilde{\beta} = \rho\cdot\beta$ and $\rho = \mu_1^{(\mathrm{noise})}/\mu_1^{(\mathrm{speech})}$, which is related to the input SNR.

Overall, since we have obtained analytical equations for the NRR, the noise kurtosis, and the speech kurtosis after SS as parametric functions of η, we can trace the relation between NRR and kurtosis by changing η. Figure 1(b) shows the relation between NRR (14) and the *kurtosis ratio*, i.e., the ratio of kurtosis before/after SS, for noise ($\eta = 1$) and speech ($\eta = 0.5$); kurtosis after SS is calculated by (15) and (16). From this figure, we can confirm that the speech-enhanced signal obtained by SS can maintain the kurtosis of the original speech, even though the kurtosis of the residual noise component drastically changes. This demonstrates that SS is suitable for estimating the parameter of the p.d.f., particularly that for speech.

4.3 Remark on Utilization of Other Noise Reduction Methods

Several other noise reduction methods appear to be possible for the estimation of the speech p.d.f. prior, such as time-frequency masking [8] and the original MMSE STSA estimator with a Gaussian prior (fixed $\eta = 1$) [4]. However, in our previous study [9], it was revealed that such a *gain-control-type* speech enhancer is not appropriate for this task because the above methods often excessively distort the speech signal, leading to a greater change in the speech kurtosis than that resulting from SS. In addition, there has been no theoretical justification that the kurtosis of the processed speech signal is maintained, in contrast to the case of using SS.

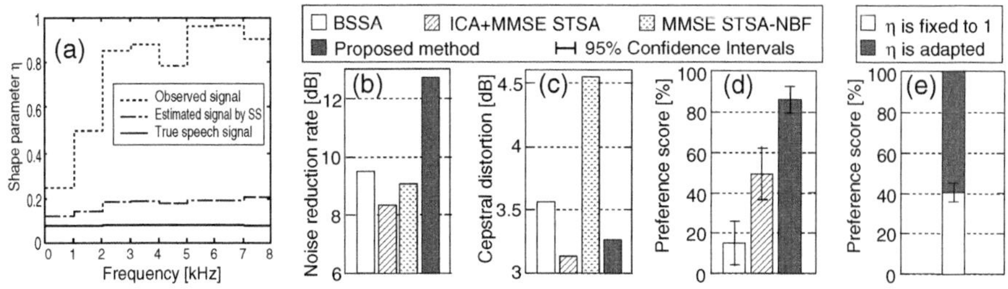

Fig. 2. Experimental results in railway-station environment: (a) example of estimated shape parameter, (b) noise reduction rate, (c) cepstral distortion, (d) preference among competitive methods, and (e) preference with/without speech p.d.f. adaptation in proposed method

5 Experiment in Real Environment

5.1 Experimental Setup

We conducted experiments in an actual railway-station environment to confirm the effectiveness of the proposed method. In this experiment, we compare the following four methods: BSSA, the proposed method, the original MMSE STSA estimator cascaded with ICA (ICA+MMSE STSA), and the MMSE STSA estimator with noise estimation based on NBF (MMSE STSA-NBF). In MMSE STSA-NBF, NBF used for the noise estimation is fixed to steer its spatial null to the assumed target speech direction of 0° (normal to the microphone array).

In the railway-station environment, the reverberation time is about 1000 ms, and the distance between the user and the microphone array is approximately 0.7 m. We used 16 kHz-sampled signals as test data. We used four speakers (two males and two females) as the target speech signals, and five noise signals recorded in distinct time periods in this environment as the noise signals. The noise in this environment is non-stationary and is almost diffused. The input SNR of test data was set to 0 dB. We used a two-element microphone array with an interelement spacing of 2.15 cm. The direction of the target speech was not exactly 0° but fluctuated around 0° depending on the speaker. The oversubtraction parameter β is 2, the flooring parameter g is 0 in BSSA, the smoothing parameter $\tau_{\rm th}$ is 3 time frame windows, corresponding to 96 ms, and the weighting factor α of the decision-directed estimation is 0.98 for ICA+MMSE STSA and 0.97 for the proposed method and MMSE STSA-NBF (these values were optimized in experiments).

5.2 Experimental Results

First, Fig. 2(a) shows an example of the estimated shape parameter, where the estimated values of the subband η are averaged over every 1 kHz range. Although the observed signal becomes Gaussian ($\eta \simeq 1$), particularly in the higher-frequency range with a low input SNR, the SS-applied signal can recover almost the same value of η as that of the original speech.

Next, the experimental results of noise reduction are discussed. Figures 2(b) and (c) respectively show the results for the average NRR and cepstral distortion (CD) (a measure of the degree of spectral envelope distortion) of all the target speakers and noise signals. Figure 2(d) shows the results for a subjective evaluation by the human ear. Six examinees participated in the subjective evaluation, in which a pair of processed speech signals were presented, and participants were asked to select which signal they preferred. From Figs. 2(b) and (c), we can observe that the NRR of the proposed method is significantly superior to those of the conventional methods. Moreover, the CD of the proposed method is slightly inferior to that of ICA+MMSE STSA but significantly superior to those of BSSA and MMSE STSA-NBF. From these results, we can confirm that the proposed method realizes better overall noise reduction in terms of quality. Indeed, from Fig. 2(d), the proposed method had a significantly higher preference score than the conventional methods.

Finally, we evaluate the net efficacy of the adaptation to the speech p.d.f. by comparing two versions of the proposed algorithm: using the original MMSE STSA estimator [3] under the assumption of Gaussian for speech (i.e., always $\eta = 1$) and the generalized MMSE STSA estimator with the blind estimation of the non-Gaussian speech p.d.f. Figure 2(e) shows the result of the preference test, which indicates that the blind adaptation for the speech p.d.f. contributes to the improvement of sound quality.

6 Conclusion

In this paper, we proposed a new blind speech extraction method based on the generalized MMSE STSA estimator and ICA-based noise and speech p.d.f. estimations. We performed an experiment in a real environment, and confirmed the improved noise reduction of the proposed method by objective and subjective evaluations.

References

1. Takahashi, Y., et al.: Blind spatial subtraction array for speech enhancement in noisy environment. IEEE Trans. Audio, Speech and Lang. Process. 17(4), 650–664 (2009)
2. Boll, S.F.: Suppression of acoustic noise in speech using spectral subtraction. IEEE Trans. ASSP-27(2), 113–120 (1979)
3. Okamoto, R., et al.: MMSE STSA estimator with nonstationary noise estimation based on ICA for high-quality speech enhancement. In: Proc. ICASSP, pp. 4778–4781 (2010)
4. Ephraim, Y., et al.: Speech enhancement using a minimum mean-square error short-time spectral amplitude estimator. IEEE Trans. ASSP-32(6), 1109–1121 (1984)
5. Andrianakis, I., et al.: MMSE speech spectral amplitude estimators with chi and gamma speech priors. In: Proc. ICASSP, vol. 1071, pp. III-1068–III-1071 (2006)
6. Saruwatari, H., et al.: Blind source separation combining independent component analysis and beamforming. EURASIP J. Appl. Sig. Process. 2003(11), 1135–1146 (2003)
7. Stacy, E.W.: A generalization of the gamma distribution. Ann. Math. Stat. 33(3), 1187–1192 (1962)
8. Hoffmann, E., et al.: Time frequency masking strategy for blind source separation of acoustic signals based on optimally-modified log-spectral amplitude estimator. In: Adali, T., Jutten, C., Romano, J.M.T., Barros, A.K. (eds.) ICA 2009. LNCS, vol. 5441, pp. 581–588. Springer, Heidelberg (2009)
9. Uemura, Y., et al.: Musical noise generation analysis for noise reduction methods based on spectral subtraction and MMSE STSA estimation. In: Proc. ICASSP, pp. 4433–4436 (2009)

Blind Estimation of Locations and Time Offsets for Distributed Recording Devices

Keisuke Hasegawa, Nobutaka Ono, Shigeki Miyabe, and Shigeki Sagayama

Department of Information Physics and Computing,
Graduate School of Information Science and Technology, The University of Tokyo
7-3-1 Hongo Bunkyo-ku, Tokyo, 113-8656, Japan
{khasegawa,onono,miyabe,sagayama}@hil.t.u-tokyo.ac.jp

Abstract. This paper presents a blind technique to estimate locations and recording time offsets of distributed recording devices from asynchronously recorded signals. In our method, locations of sound sources and recording devices, and the recording time offsets are estimated from observed time differences of arrivals (TDOAs) by decreasing the mean squared errors. The auxiliary-function-based updates guarantee the monotonic decrease of the objective function at each iteration. The TDOAs are estimated by the generalized cross correlation technique. The validity of our approach is shown by experiments in real environment, where locations of seven sound sources and eight microphones and eight time offsets were estimated from signals recorded by four stereo IC recorders in reverberant rooms.

Keywords: Blind alignment, asynchronous recording, generalized cross correlation.

1 Introduction

Microphone array processing is one of the powerful techniques for sound source localization and separation and it has been greatly developed in several decades. The framework has been recently extended from conventional fixed microphone array to *distributed* microphone array [1–4]. It is a new concept to exploit distributed recording devices such as internal microphones of personal computers, voice recorders, mobile phones as channels of array signal processing. Its wireless configuration, flexibility, large number of recording elements will enlarge application. The growth of mobile recording devices and the miniaturization of microphones such as silicon microphones would facilitate this direction.

While, unlike conventional microphone arrays, channels recorded by different devices are not synchronous in most cases due to their different recording time offsets, mismatch of sampling frequencies and so on. The location of their devices would be unknown. Therefore, in order to apply conventional array signal processing technique, it is necessary to estimate them, which is a new kind of blind estimation problem. In our previous work [6], we have proposed a method to estimate positions of sources and microphones and time offsets from observed

V. Vigneron et al. (Eds.): LVA/ICA 2010, LNCS 6365, pp. 57–64, 2010.

time differences of arrivals (TDOAs) for sources in a blind fashion and shown proof of concept by preliminary experiments.

In this paper, we challenge to the blind estimation of locations and time offsets of distributed recording devices in real environments. In order to obtain apparent time differences accurately from asynchronously recorded signals, we present the two-stage estimation method, where rough alignment of entire recorded signals is first performed, and then, the apparent TDOA for each source, (which still includes unknown time offsets,) is estimated by generalized cross correlation method (GCC method [7]). Finally, positions of sources and microphones and time offsets are estimated by minimizing square errors of observation model with auxiliary-function-based updates. We also show the experimental results in real reverberant rooms.

2 Problem Formulation

Suppose K sound sources are observed by L microphones. Let $\boldsymbol{s}_i$ $(i = 1 \ldots, K)$ and $\boldsymbol{r}_k$ $(k = 1, \ldots, L)$ be the locations of the sound sources and the microphones, respectively. In this paper, we assume that the recording devices have the same sampling frequency but different recording time offsets. That is, each device of index k starts its recording at different time t_k. To sum up, The problem here is the estimation of all these parameters $\boldsymbol{s}_i$, $\boldsymbol{r}_k$, and t_k only with observed signals.

Generally, one of the most significant cues for localization of sources and microphones is the TDOA. However, it should be noted that the apparent TDOA include unknown time offsets in this problem. Figure 1 depicts the relationship between the apparent and the true TDOAs. The apparent TDOA between the mth and the nth channels for the ith source can be represented by

$$\tau_{imn} = \frac{1}{c}(||\boldsymbol{s}_i - \boldsymbol{r}_m|| - ||\boldsymbol{s}_i - \boldsymbol{r}_n||) - (t_m - t_n), \tag{1}$$

where c is the speed of sound. Here $||\boldsymbol{x}||$ denotes the Euclidean norm of $\boldsymbol{x}$. In this equation, only τ_{imn} is observable. To estimate unknown variables, the number of the observable variables must exceed the number of the unknown variables. When τ_{imn} are given for all the combination of i,m and n, the necessary condition to solve the problem in 2D case is $(K-3)(L-3) \geq 5$, which is derived in the same manner as in [6].

3 Applying Auxiliary Function Method to Parameter Estimation

Let $\boldsymbol{\Theta}$ be the unknown parameter set: $\boldsymbol{\Theta} := \{\boldsymbol{s}_i, \boldsymbol{r}_m, t_m | i = 1, \ldots, K,\ m = 1, \ldots, L\}$.In order to find $\boldsymbol{\Theta}$, the squared error of Eq. (1) is considered as the objective function:

$$J(\boldsymbol{\Theta}) := \frac{1}{c^2 K L^2} \sum_{i=1}^{K} \sum_{m=1}^{L} \sum_{n=1}^{L} \varepsilon_{imn}^2, \tag{2}$$

$$\varepsilon_{imn} := ||\boldsymbol{s}_i - \boldsymbol{r}_m|| - ||\boldsymbol{s}_i - \boldsymbol{r}_n|| - c(\tau_{imn} + t_m - t_n). \tag{3}$$

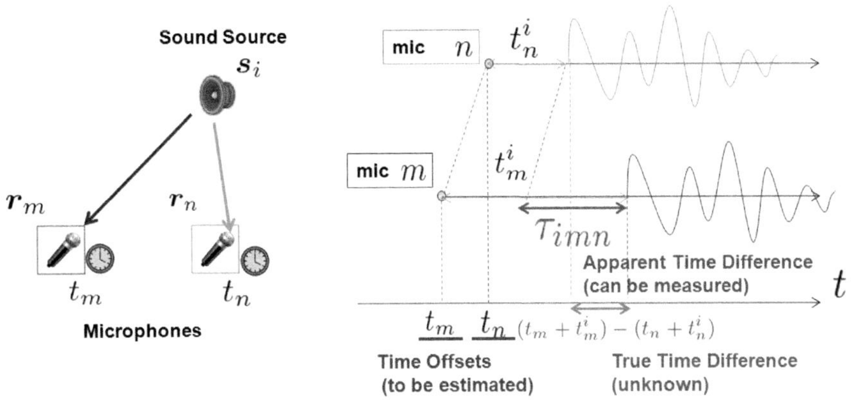

Fig. 1. The relationship between the time offsets t_m, t_n and apparent TDOA τ_{imn}

The objective function derived in Eq. (2) contains some cross terms of the norms of vectors, which make it infeasible to calculate the parameter set of optimization analytically. Here we introduce *auxiliary function method* [5] for solving this problem. Consider the function $J^+(\boldsymbol{\Theta}, \boldsymbol{\Theta}^+)$ which holds

$$J(\boldsymbol{\Theta}) = \min_{\boldsymbol{\Theta}^+} J^+(\boldsymbol{\Theta}, \boldsymbol{\Theta}^+). \tag{4}$$

Here, $J^+(\boldsymbol{\Theta}, \boldsymbol{\Theta}^+)$ is refereed to as an auxiliary function of an objective function $J(\boldsymbol{\Theta})$ and $\boldsymbol{\Theta}^+$ an *auxiliary variable.* $J^+(\boldsymbol{\Theta}, \boldsymbol{\Theta}^+)$ is non-increasing during following update procedure of variables $\boldsymbol{\Theta}$ and $\boldsymbol{\Theta}^+$:

$$\boldsymbol{\Theta}^+_{l+1} = \mathrm{argmin}_{\boldsymbol{\Theta}^+} J^+(\boldsymbol{\Theta}_l, \boldsymbol{\Theta}^+), \tag{5}$$

$$\boldsymbol{\Theta}_{l+1} = \mathrm{argmin}_{\boldsymbol{\Theta}} J^+(\boldsymbol{\Theta}, \boldsymbol{\Theta}^+_{l+1}), \tag{6}$$

where l indicates the index of iterations. A brief proof follows:

1. $J(\boldsymbol{\Theta}_l) = J^+(\boldsymbol{\Theta}_l, \boldsymbol{\Theta}^+_{l+1})$ is held due to Eq. (4) and Eq. (5),
2. $J^+(\boldsymbol{\Theta}_l, \boldsymbol{\Theta}^+_{l+1}) \geq J^+(\boldsymbol{\Theta}_{l+1}, \boldsymbol{\Theta}^+_{l+1})$ from Eq. (6),
3. $J^+(\boldsymbol{\Theta}_{l+1}, \boldsymbol{\Theta}^+_{l+1}) \geq J(\boldsymbol{\Theta}_{l+1})$ from Eq. (4),

then,

$$J(\boldsymbol{\Theta}_l) \geq J(\boldsymbol{\Theta}_{l+1}). \tag{7}$$

The auxiliary function method does not require tuning of parameters such as step size required in gradient descent and many other iterative optimization algorithms. However it is not always guaranteed that an adequate auxiliary function can be easily designed.

The derivative of Eq. (2) cannot be calculated. Here we intend to design an auxiliary function whose derivative is calculable instead. We have the auxiliary function below, which is derived from [6] in detail.

$$J_2(\boldsymbol{\Theta}, \boldsymbol{\mu}, \boldsymbol{e}) = \frac{2}{c^2 K L^2} \sum_{i,m,n} \{(\boldsymbol{s}_i - \boldsymbol{r}_m - \boldsymbol{e}_{im}\mu^m_{imn})^2 + (\boldsymbol{s}_i - \boldsymbol{r}_n - \boldsymbol{e}_{in}\mu^n_{imn})^2\}, \tag{8}$$
$$\text{where } \boldsymbol{\mu} := \{\mu^m_{imn}, \mu^n_{imn} \,|i = 1, \ldots K,\ m, n = 1, \ldots L\}$$
$$\boldsymbol{e} := \{\boldsymbol{e}_{im} | i = 1, \ldots, K,\ m = 1, \ldots, L\}$$

Here we have $\boldsymbol{\mu}, \boldsymbol{e}$ as auxiliary parameters. The whole update rule is written as follows [6]:

$$\varepsilon_{imn} \leftarrow ||\boldsymbol{s}_i - \boldsymbol{r}_m|| - ||\boldsymbol{s}_i - \boldsymbol{r}_n|| - c(\tau_{imn} + t_m - t_n), \tag{9}$$
$$\mu^m_{imn} \leftarrow ||\boldsymbol{s}_i - \boldsymbol{r}_m|| - \frac{1}{2}\varepsilon_{imn}, \tag{10}$$
$$\mu^n_{imn} \leftarrow ||\boldsymbol{s}_i - \boldsymbol{r}_n|| + \frac{1}{2}\varepsilon_{imn}, \tag{11}$$
$$\boldsymbol{e}_{im} \leftarrow (\boldsymbol{s}_i - \boldsymbol{r}_n)/||\boldsymbol{s}_i - \boldsymbol{r}_i||, \tag{12}$$
$$\boldsymbol{s}_i \leftarrow \frac{1}{L^2} \sum_{m=1}^{L} \left(L\boldsymbol{r}_m + \boldsymbol{e}_{im} \sum_{n=1}^{L} \mu^m_{imn} \right), \tag{13}$$
$$\boldsymbol{r}_n \leftarrow \frac{1}{KL} \sum_{i=1}^{K} \left(L\boldsymbol{s}_i - \boldsymbol{e}_{in} \sum_{m=1}^{L} \mu^n_{imn} \right), \tag{14}$$
$$t_n \leftarrow t_n + \frac{1}{cKL} \sum_{i=1}^{K} \left(L||\boldsymbol{s}_i - \boldsymbol{r}_n|| - \sum_{m=1}^{L} \mu^n_{imn} \right). \tag{15}$$

4 Estimation of Time Difference of Arrival Based on Generalized Cross Correlation

In this section we introduce two-stage TDOA analysis to obtain the estimation of the apparent time differences in real room environment, which were assumed to be given in the discussions above. The apparent time differences τ_{imn} are the time differences of the sources in the observation channels, and have to be estimated only with the observed signals. A key to accurate TDOA analysis is the efficient use of the limited number of observation samples, and one approach is frame analysis with many frames. However, we have to use short window to increase the number of the frames, and that disable us to deal with long time differences: Frame-based TDOA analysis can not estimate longer time difference than the frame length. To overcome this problem, the first stage aligns the time differences among the channels roughly in the time domain, and the second stage estimates more detailed TDOA for each source using frame analysis and GCC.

While the detailed TDOA estimation in the second stage is conducted for each of the source-channel pairs, the rough alignment in the first stage is applied for

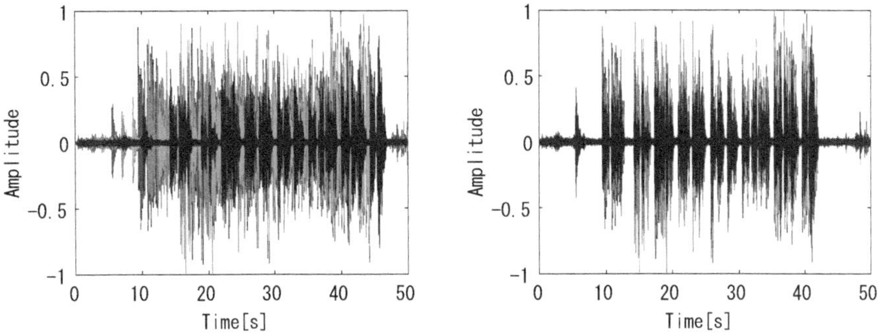

Fig. 2. Asynchronously recorded signals (left) and the same signals after preprocessing, rough alignment (right)

each channel without distinguishing the sources. Here we justify this strategy. As we discussed, each apparent time difference is the sum of the true TDOA and the recording time offset. While the length of the former is limited by the size of the room, the latter can have any length and is often much longer than the former. To cancel such large time differences caused by the different starting times of recording, we obtain average time difference of all the sources for each channel. The rough alignment is obtained by the following procedure. First, we select one channel as a reference channel. Second, we calculate cross correlations between the reference and each rest of the channels and detect its peak. Finally, we can roughly synchronize the channels by shifting the other channels keeping the reference fixed so that the cross correlations have peaks at the zeroth sample. An example of the rough alignment is shown in Fig. 2. It can be seen that the rough alignment successfully cancels the average time difference of all the sources for each channel.

For the TDOA analysis in the second stage, we use GCC method, which is known to be robust against interference. Note that we assume the time durations when only one source is active for each source is given (referred to as double-talk detection), and we analyze the signal in those time durations. GCC gives the maximum likelihood estimate of the TDOA under an observation model, which assumes that only the direct wave reaches the microphones from the desired source, and the desired source and the observation noises are uncorrelated Gaussian. As a result, the estimated TDOA is obtained by detecting a peak of the GCC function, which is a filtered version of the cross correlation function between the analyzed signals. The filter of GCC enhances the peak of the cross correlation by effective whitening and noise suppression.

Although we omit the discussion of probabilistic modeling of the observed signal here, we review the filtering of GCC and its property. Let $X(\omega)$ and $Y(\omega)$ be the Fourier transform of the two observed signals $x(t)$ and $y(t)$ to be analyzed.

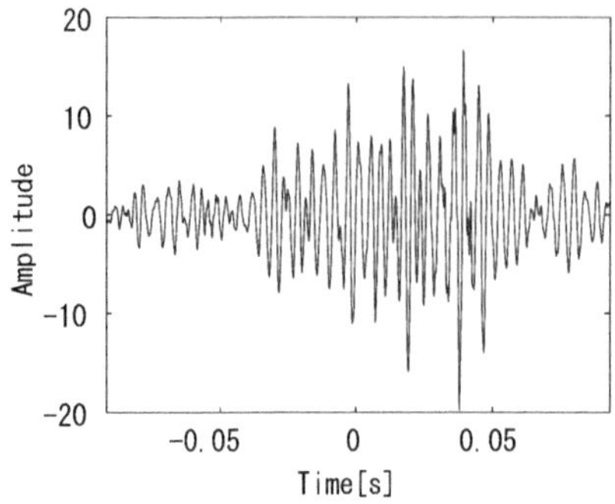

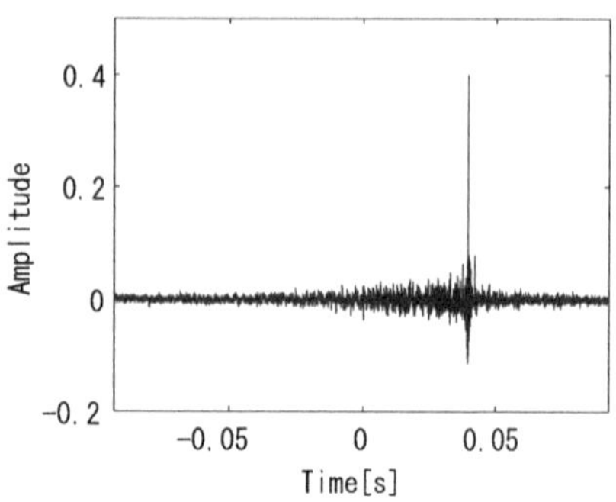

Fig. 3. Conventional cross correlation(left) and generalized cross correlation(right)

According to [7], the filter $U(\omega)$ in the frequency domain is designed as follows:

$$U(\omega) = \frac{1}{|E\left[X(\omega)Y^*(\omega)\right]|} \cdot \frac{|\gamma(\omega)|^2}{1-|\gamma(\omega)|^2}, \tag{16}$$

where

$$\gamma(\omega) = \frac{E\left[X(\omega)Y^*(\omega)\right]}{\sqrt{|E\left[X(\omega)\right]|^2}\sqrt{|E\left[Y(\omega)\right]|^2}} \tag{17}$$

is the coherence of each frequency. According to the coherence, reliable frequency components in the frequencies with low SNR is weighted lightly, and the weighting results as noise suppression. Also, the normalization of the amplitude whitens the cross correlation and enhances its peak.

Figure 3 shows examples of conventional cross correlation and GCC of observed signals after the preprocessing of the rough alignment in the first step. Successful enhancement of the peak by GCC can be seen.

5 Experimental Results

5.1 Experimental Conditions

We executed an experiment in real room environment to evaluate the performance of blind alignment. We placed seven loudspeakers and four stereo IC recorders of the same model number (SANYO ICR-PS603EM) in the identical plane in the room with $6\times7\times2.7\text{m}^3$ (See Fig. 4). As source signals, recorded speech signals whose bit rate is 44100 Hz are used. The reverberation time of the room is approximately 300 ms. There were no temporal overlaps from different sound sources in every observed signal and we manually gave the segmentation in the observed signals corresponding to each sound source. The relative error of sampling rate among devices were up to 2.6×10^{-6}. We set flame length to 8192 samples. The frame analysis with half-overlap was performed to execute the

Fig. 4. IC recorder used in the experiment (left) and The recording room (right)

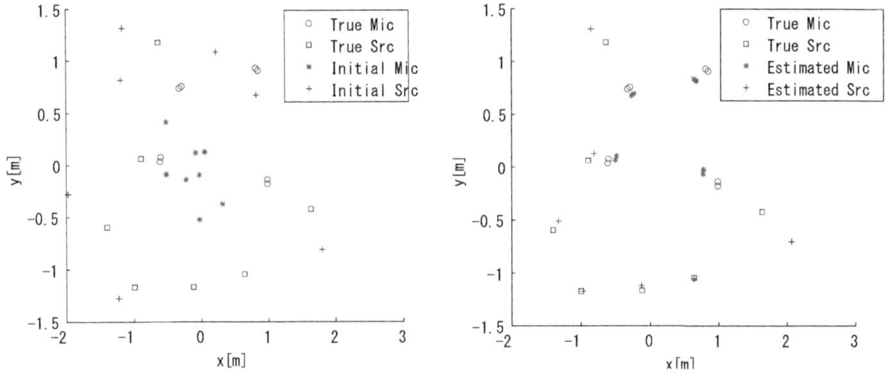

Fig. 5. The initial position (left) and the estimated position (right)

GCC method. Since the objective function in our proposed method has many local minima, initialization of the parameters is an important problem. However, appropriate initialization can be easily designed with some rough assumption on the layout. We gave a true but rough assumption that the microphones are surrounded by the sources. Specifically, the initial $\boldsymbol{r}_m$ were randomly given in a circle whose radius was 1 m. The initial $\boldsymbol{s}_i$ were also randomly given by the rule that the norm of each $\boldsymbol{s}_i$ be from 1 m up to 1.5 m. The estimation of the positions were performed in a two-dimensional plane.

5.2 Experimental Results of Blind Alignment

Figure 5 shows the estimated positions of microphones and sources. After 50000 iterations, the objective function converged toward a satisfactorily small value. Each of microphone pairs with indices $\{1, 2\}$, $\{3, 4\}$, $\{5, 6\}$ and $\{7, 8\}$ is equipped as the stereo microphones in the same recording device. Thus the corresponding time offsets of each pair should be identical, e.g., $t_m = t_{m+1}$ $(m = 1, 3, 5, 7)$. In

Table 1. The estimated time offsets t_m[ms]

m	1	2	3	4	5	6	7	8
t_m	0.097	0.075	-18.22	-18.24	23.32	23.29	20.77	20.74

table 1, it is confirmed that the time offsets of each pair was estimated to be very close. While, the averaged position error of microphones and sources was 15.66 cm. We can see that the proposed method works properly for the blind alignment task in real environment.

6 Conclusion

We presented an auxiliary-function based technique for estimating the locations of microphones and sound sources and time offsets of each microphones by observed signals alone. For the estimation of the apparant time differences used in the auxiliary function method, we used two-stage TDOA analysis; the first stage for rough alignment of the channels with large time differences, and the second stage for the estimation of each apparant time difference using frame-based GCC method. The blind alignment experiment in real reverberant room environment was ascertained to be performed well. To attack the remaining problem that the our current framework reqires the time durations when only a single source is active for each source, our ongoing work is development of a new framework to perform both blind alignment and source separation simultaneously.

References

1. Lienhart, R., Kozintsev, I., Wehr, S., Yeung, M.: On the importance of exact synchronization for distributed audio processing. In: Proc. ICASSP, pp. 840–843 (2003)
2. Bertrand, A., Moonen, M.: Energy-based multi-speaker voice activity detection with an ad hoc microphone array. In: Proc. ICASSP, pp. 85–88 (2010)
3. Brutti, A., Omologo, M., Svaizer, P.: Oriented global coherence field for the estimation of the head orientation in smart rooms equipped with distributed microphone arrays. In: Proc. Interspeech, pp. 2337–2340 (2005)
4. Kobayashi, K., Furuya, K., Kataoka, A.: A blind source localization by using freely positioned microphones. Trans. IEICE J86-A(6), 619–627 (2003) (in Japanese)
5. Kameoka, H., Ono, N., Sagayama, S.: Auxiliary Functional Approach to Parameter Estimation of Constrained Sinusoidal Model for Monaural Speech Separation. In: Proc. ICASSP, pp. 29–32 (March 2008)
6. Ono, N., Kohno, H., Ito, N., Sagayama, S.: Blind Alignment of Asynchronously Recorded Signals for Distributed Microphone Array. In: Proc. WASPAA, pp. 161–164 (October 2009)
7. Knapp, C.H., Cartar, G.C.: The Generalized Correlation Method for Estimation of Time Delay. IEEE Trans. ASSP 24(4), 320–327 (1976)

Speech Separation via Parallel Factor Analysis of Cross-Frequency Covariance Tensor

Xiao-Feng Gong and Qiu-Hua Lin

School of Information and Communication Engineering,
Dalian University of Technology, Dalian 116024, China
xfgong@dlut.edu.cn

Abstract. This paper considers separation of convolutive speech mixtures in frequency-domain within a tensorial framework. By assuming that components associated with neighboring frequency bins of the same source are still correlated, a set of cross-frequency covariance tensors with trilinear structure are established, and an algorithm consisting of consecutive parallel factor (PARAFAC) decompositions is developed. Each PARAFAC decompositon used in the proposed method can simultaneously estimate two neighboring frequency responses, one of which is a common factor with the subsequent cross-frequency covariance tensor, and thus could be used to align the permutations of the estimates in all the PARAFAC decompositions. In addition, the issue of identifiability is addressed, and simulations with synthetic speech signals are provided to verify the efficacy of the proposed method.

Keywords: Blind source separation, Tensor, Parallel factor analysis.

1 Introduction

Independent component analysis (ICA) aims at recovering multiple independent source signals mixed through unknown channels, with only the observations collected at a set of sensors. Since ICA requires little prior information about the source signals and channels, it has become a widely used method for speech separations [1-6]. Earlier works on ICA are mostly focused on instantaneous mixtures. However, the instantaneous mixing model does not always match the practical situations. For example, in reverberation environment which is often encountered in practice, the signals captured by microphones are attenuated and delayed versions of multiple source signals superimposing one another, resulting in a convolutive mixing model. As a result, blind separation of convolutive mixtures has become a key problem in speech processing.

Generally speaking, blind separation of convolutive speech mixtures can be performed in either time-domain or frequency-domain. More exactly, the time-domain methods propose to minimize some independent criterion with respect to the coefficients of mixing channels at each time delay. The problem with time-domain methods is that when the speech signals are recorded in strong reverberation, there would be too many parameters to adjust and this may result in convergence difficulty and heavy

V. Vigneron et al. (Eds.): LVA/ICA 2010, LNCS 6365, pp. 65–72, 2010.

computational burden [1]. The frequency-domain methods, on the other hand, transfer the deconvolution problem into a set of instantaneous mixture separation problems, and hence facilitate the use of well-studied instantaneous mixture separation methods.

However, there are prices to pay for the advantage of frequency-domain methods. Firstly, the fourier transform used in frequency-domain methods tends to generate nearly Gaussian signals. Therefore, many existing methods for instantaneous mixture separation, that require non-Gaussianity, are no longer valid. To solve this problem, the non-stationarity of speech signals is exploited as well as second-order statistics, and joint diagonalization or tensor based methods have been proposed [4-6].

Secondly, the independently implemented separation procedures may yield mismatched permutations and scalings of the estimated sources or mixing matrices associated with different frequency bins, resulting in the so-called permutation and scaling ambiguities. Since the scaling ambiguity can be well solved [2], the frequency ambiguity becomes the main problem with the frequency-domain methods. In existing works, prior information on both the mixing filters (such as the continuity of frequency response) and source signals (such as the covariance across frequency bins) have been used to align the permutations, a detailed survey on permutation correction methods could be found in [1].

In this paper, we propose a new method for convolutive speech separation within the tensorial framework. More precisely, a set of cross-frequency covariance tensors, which incorporate both the non-stationarity and the covariance between neighboring frequency bins, are firstly established. And then an algorithm comprising consecutive parallel factor analysis (PARAFAC) decompositions is proposed. Unlike the existing works that estimate one frequency response associated with a single frequency bin each time, the proposed method could generate paired estimates of two neighboring frequency responses in each run of the PARAFAC decomposition. In addition, we develop a permutation correction scheme by exploiting the common factor shared by neighboring cross-frequency covariance tensors.

2 Problem Statement

We consider M mutually uncorrelated speech signals $\boldsymbol{s}(t) = [s_1(t), s_2(t), ..., s_M(t)]^T$ collected with an array of N microphones, and denote the recorded mixtures by $\boldsymbol{x}(t) = [x_1(t), x_2(t), ..., x_N(t)]^T$. In the noise-free case, $\boldsymbol{x}(t)$ could be modeled as:

$$\boldsymbol{x}(t) = \sum_{\tau=0}^{L-1} \boldsymbol{H}(\tau)\boldsymbol{s}(t-\tau) = \boldsymbol{H}(t) \star \boldsymbol{s}(t) \tag{1}$$

where '$\star$' denotes the operation of linear convolution, the $N \times M$ matrix $\boldsymbol{H}(\tau)$ denotes the impulse response matrix of the mixing filter at time-lag τ. Its elements $h_{nm}(\tau)$ are coefficients of the room impulse response (RIR) between the *mth* source and the *nth* microphone, $n = 1, 2, ..., N$, $m = 1, 2, ..., M$, and L denotes the maximum channel length. To recover the original speech sources, the goal is to find a demixing filter $\boldsymbol{G}(\tau)$ such that:

$$\tilde{s}(t) = \sum_{\tau=0}^{K-1} \boldsymbol{G}(\tau)\boldsymbol{x}(t-\tau) \qquad (2)$$

where K is the length of the demixing filter, and $\tilde{s}(t)$ is the restored source vector.

In frequency-domain methods, the convolutive model in (1) is reduced to a set of instantaneous mixing models by applying the short-time fourier transform (STFT):

$$\boldsymbol{x}(t,f) = \sum_{\tau=1}^{F} \boldsymbol{w}_F(t)\boldsymbol{x}(\tau-t)e^{2\pi i f t} \qquad (3)$$

where F is the frame length, and $\boldsymbol{w}_F(t)$ is an F point Hanning window. Then (1) could be rewritten as:

$$\boldsymbol{x}(t,f) = \boldsymbol{H}_f\boldsymbol{s}(t,f) \qquad (4)$$

where $\boldsymbol{H}_f$ denotes the frequency response of the mixing filter, and $\boldsymbol{s}(t,f)$ is the STFT of $\boldsymbol{s}(t)$. Therefore, the deconvolution problem could be solved by identifying the instantaneous mixture model (4) independently for all the frequency bins. However, the main difficulty with the frequency-domain methods is the necessity to solve the scaling and permutation ambiguities raised in the separation procedure, of which the latter still remains as an open problem. In section 3, we will propose a new permutation correction scheme to solve this problem. Before addressing further the details, we list the assumptions used throughout the whole paper as follows:

A1) The sources are zero-mean and mutually uncorrelated at each frequency bin f. In addition, components associated with adjacent frequency bins of the same source are correlated;

A2) The cross-frequency covariance of different sources for each pair of frequency bins $(f, f+1)$ vary differently with time;

A3) The number of speakers M is known, and not larger than the number of microphones N.

A4) The impulse responses of all mixing filters are constant. In addition, arbitrary M columns of $\boldsymbol{H}_f$ are linearly independent.

3 Proposed Algorithm

3.1 Estimation of Cross Frequency Covariance Tensor

We firstly define the cross frequency covariance matrix as follows:

$$\boldsymbol{R}_{t,f} \triangleq \mathrm{E}\left(\boldsymbol{x}(t,f)\boldsymbol{x}^H(t,f+1)\right) \qquad (5)$$

where 'E' denotes the mathematical expectation. Noting that the speech signals are nonstationary, $\boldsymbol{R}_{t,f}$ actually varies with time, and the cross frequency covariance tensor $\mathcal{R}_f$ could then be obtained by stacking T temporal samples of $\boldsymbol{R}(t,f)$ into a third-order tensor as follows:

$$\mathcal{R}_f(:,:,k) \triangleq \boldsymbol{R}_{t_k,f} \qquad (6)$$

where $\mathcal{R}_f(:,:,k)$ denotes the matrix slice of tensor $\mathcal{R}_f \in \mathbb{C}^{N\times N\times T}$ by fixing its third index to k and varying its first and second indices, $k = 1,2,...,T$. According to (4), assumptions A1) and A3), (6) could be rewritten as:

$$\mathcal{R}_f = \sum_{m=1}^{M} \boldsymbol{h}_{f,m} \circ \boldsymbol{r}_{f,m} \circ \boldsymbol{h}^*_{f+1,m} \tag{7}$$

where $\boldsymbol{h}_{f,m}$ and $\boldsymbol{h}_{f+1,m}$ are the mth column vectors of $\boldsymbol{H}_f$ and $\boldsymbol{H}_{f+1}$, respectively. $\boldsymbol{r}_{f,m} \triangleq \mathrm{E}\left[s_m(t_1,f)s^*_m(t_1,f+1),...,s_m(t_T,f)s^*_m(t_T,f+1)\right]^T$, $s_m(t_k,f)$ is the component of the mth source associated with frequency bin f and time instant t_k , and '$\circ$' denotes tensor outer product[1]. In practice, the cross frequency tensor is unavailable but can be estimated from the collected data sampled at T different time instants. The idea is to average the results obtained from Q successive frames, each overlapping the neighboring one with $3F/4$ samples, where F is the frame length (see Fig. 1). As a result, $\boldsymbol{R}_{t_k,f}$ could be estimated as, $f = 1,2,...,F-1$:

$$\tilde{\boldsymbol{R}}_{t_k,f} = \frac{1}{Q}\sum_{q=1}^{Q} \boldsymbol{x}\left(t_k + q\frac{F}{4}, f\right)\boldsymbol{x}^H\left(t_k + q\frac{F}{4}, f+1\right) \tag{8}$$

Then the estimate of the cross frequency covariance tensor is obtained by replacing $\boldsymbol{R}_{t_k,f}$ in (6) by $\tilde{\boldsymbol{R}}_{t_k,f}$.

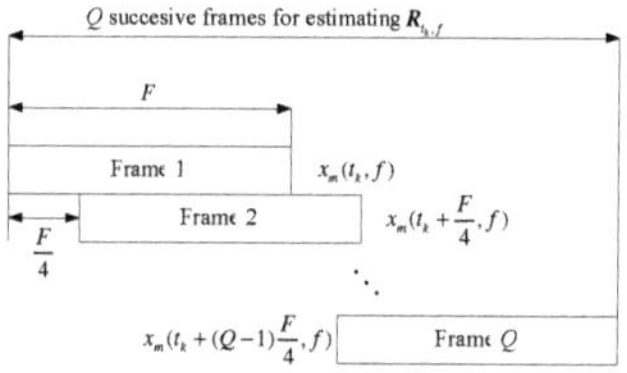

Fig. 1. Q succesive frames for estimating $\boldsymbol{R}_{t_k,f}$

3.2 The Algorithm

When the identifiability condition for the unique (up to scaling and permutation ambiguities) PARAFAC decomposition is met (this issue is to be addressed in subsection 3.3), we could obtain the estimate of $\{\boldsymbol{h}_{f,m}, \boldsymbol{h}_{f+1,m}\}$ by fitting the trilinear structure of $\mathcal{R}_f$. The PARAFAC decomposition is usually carried out under an alternating least squares (ALS) principle [7]. Some other methods such as the simultaneous matrix diagonalization or optimization based approaches may also be used, one can refer to [8] for a detailed comparison of PARAFAC fitting methods. In addition, methods for accelerating ALS based algorithm such as compression based PARAFAC (COMFAC) have also been proposed in literature [9]. We herein adopt the scheme introduced in [10] along with COMFAC to speed up the PARAFAC decomposition.

---

[1] The outer product of three vectors $\boldsymbol{a} \in \mathbb{C}^I$, $\boldsymbol{b} \in \mathbb{C}^K$ and $\boldsymbol{c} \in \mathbb{C}^L$ is a tensor $\mathcal{T} \in \mathbb{C}^{I\times K\times L}$ given by $t_{i,k,l} \triangleq a_i b_k c_l$.

Denote the estimate of $\boldsymbol{h}_{f+1,m}$ by $\tilde{\boldsymbol{h}}_{f+1,m}$, and $\tilde{\boldsymbol{H}}_{f+1} \triangleq [\tilde{\boldsymbol{h}}_{f+1,1},...,\tilde{\boldsymbol{h}}_{f+1,M}]$, then $\boldsymbol{H}_{f+1}\boldsymbol{P}_f\boldsymbol{\Lambda}_f = \tilde{\boldsymbol{H}}_{f+1}$, where $\boldsymbol{P}_f \in \mathbb{R}^{M\times M}$ and $\boldsymbol{\Lambda}_f \in \mathbb{R}^{M\times M}$ are the permutation matrix and scaling diagonal matrix associated with frequency bin f, respectively. Furthermore, by denoting $\boldsymbol{B}_f \triangleq [\boldsymbol{r}_{f,1},\boldsymbol{r}_{f,2},...,\boldsymbol{r}_{f,M}]$ and letting $\boldsymbol{D}_k(\boldsymbol{B}_f)$ be a diagonal matrix containing the *kth* row of $\boldsymbol{B}_f$, $k=1,2,...,T$, we could further define a new tensor $\mathcal{C}_{f+1} \in \mathbb{C}^{N\times N\times T}$ as follows:

$$\begin{aligned}\mathcal{C}_{f+1}(:,:,k) &\triangleq \left(\tilde{\boldsymbol{H}}_{f+1}^H\tilde{\boldsymbol{H}}_{f+1}\right)^{-1}\tilde{\boldsymbol{H}}_{f+1}^H\mathcal{R}_{f+1}(:,:,k)\\ &= \left(\tilde{\boldsymbol{H}}_{f+1}^H\tilde{\boldsymbol{H}}_{f+1}\right)^{-1}\tilde{\boldsymbol{H}}_{f+1}^H\boldsymbol{H}_{f+1}\boldsymbol{D}_k(\boldsymbol{B}_{f+1})\boldsymbol{H}_{f+2}^H = \boldsymbol{D}_k(\boldsymbol{B}_{f+1})(\boldsymbol{H}_{f+2}\boldsymbol{P}_f\boldsymbol{\Lambda}_f)^H\end{aligned} \tag{9}$$

The equation above could be rewritten into the standard PARAFAC formulism:

$$\mathcal{C}_{f+1} = \sum_{m=1}^{M}\boldsymbol{e}_m \circ \boldsymbol{r}_{f+1,m} \circ \left(\boldsymbol{h}'_{f+2,m}\right)^* \tag{10}$$

where $\boldsymbol{e}_m \in \mathbb{C}^M$ is the *mth* column vector of the $M\times M$ identity matrix $\boldsymbol{I}_M$, and $\boldsymbol{h}'_{f+2,m}$ is the *mth* column vector of $\boldsymbol{H}_{f+2}\boldsymbol{P}_f\boldsymbol{\Lambda}_f$. By implementing PARAFAC decomposition to $\mathcal{C}_{f+1}$ we could obtain $\boldsymbol{E}_{f+1} \triangleq \left[\tilde{\boldsymbol{e}}_{f+1,1},...,\tilde{\boldsymbol{e}}_{f+1,M}\right]$ and $\tilde{\boldsymbol{H}}_{f+2} \triangleq \left[\tilde{\boldsymbol{h}}_{f+2,1},...,\tilde{\boldsymbol{h}}_{f+2,M}\right]$, where $\tilde{\boldsymbol{e}}_{f+1,m}$ and $\tilde{\boldsymbol{h}}_{f+2,m}$ are estimates of $\boldsymbol{e}_m$ and $\boldsymbol{h}'_{f+2,m}$, respectively. In addition, since the uniqueness of PARAFAC decomposition does not take the scaling and permutation ambiguities into consideration, we have:

$$\begin{cases}\boldsymbol{E}_{f+1} = \boldsymbol{P}_{f+1}\boldsymbol{\Lambda}_{f+1}\\ \tilde{\boldsymbol{H}}_{f+2} = \boldsymbol{H}_{f+2}\boldsymbol{P}_f\boldsymbol{P}_{f+1}\boldsymbol{\Lambda}_f\boldsymbol{\Lambda}_{f+1}\end{cases} \tag{11}$$

where $\boldsymbol{\Lambda}_{f+1}$ and $\boldsymbol{P}_{f+1}$ are the scaling and permutation matrices associated with frequency bin $f+1$, respectively. If we constrain $\boldsymbol{E}_{f+1}$ to be an identity matrix, then from (11) and noting that $\boldsymbol{P}_{f+1}$ is either non-diagonal or equal to identity matrix, we have $\boldsymbol{P}_{f+1} = \boldsymbol{\Lambda}_{f+1} = \boldsymbol{I}_M$ and $\tilde{\boldsymbol{H}}_{f+2} = \boldsymbol{H}_{f+2}\boldsymbol{P}_f\boldsymbol{\Lambda}_f$. We note that $\boldsymbol{P}_f$ and $\boldsymbol{\Lambda}_f$ are associated with the previous frequency bin, indicating $\{\tilde{\boldsymbol{h}}_{f,m}\}$, $\{\tilde{\boldsymbol{h}}_{f+1,m}\}$, and $\{\tilde{\boldsymbol{h}}_{f+2,m}\}$ are permuted from $\{\boldsymbol{h}_{f,m}\}$, $\{\boldsymbol{h}_{f+1,m}\}$, and $\{\boldsymbol{h}_{f+2,m}\}$ in the same manner, $m=1,2,...,M$, and thus the permutations for $\{\tilde{\boldsymbol{h}}_{f,m}\},\{\tilde{\boldsymbol{h}}_{f+1,m}\},\{\tilde{\boldsymbol{h}}_{f+2,m}\}$ are aligned. By implementing the aforementioned scheme successively for all the frequency bins, we could finally solve the permutation ambiguity problem. As for the scaling ambiguity, we could refer to the principle introduced in [2] for its solution.

The potential advantage of the proposed method is that the use of cross-frequency covariance tensors could result in an aligned pair of frequency responses in each run of PARAFAC decomposition, and this property could be used to enable a more reliable permutation correction scheme (not limited to the one proposed herein). However, we note that the proposed permutation aligning strategy is sequential, and thus consecutive errors may occur if the permutation correction scheme fails at a certain frequency index. Moreover, by noting that equation (9) implies more microphones than speakers, the proposed algorithm could only be used in the over-determined case although it has been addressed in literature that the powerful uniqueness properties of PARAFAC could be used to tackle underdetermined problems [10]. These problems will be the focus of our future works.

3.3 Identifiability

We firstly introduce the following theorem.

Theorem 1 (Kruskal theorem [9]): Given a PARAFAC model $\mathcal{T} = \sum_{r=1}^{R} \boldsymbol{a}_r \circ \boldsymbol{b}_r \circ \boldsymbol{c}_r$, denote $\boldsymbol{A} = [\boldsymbol{a}_1,...,\boldsymbol{a}_R]$, $\boldsymbol{B} = [\boldsymbol{b}_1,...,\boldsymbol{b}_R]$, and $\boldsymbol{C} = [\boldsymbol{c}_1,...,\boldsymbol{c}_R]$. Then the decomposition of this PARAFAC model is unique up to permutation and scaling when:

$$k_A + k_B + k_C = 2k_A + k_B \geq 2M + 2 \tag{12}$$

where k_A, k_B, and k_C denote the Kruskal rank of $\boldsymbol{A}$, $\boldsymbol{B}$, and $\boldsymbol{C}$, respectively[2].

We then use Theorem 1 to analyze the identifiability issue of the proposed algorithm. It is required that the PARAFAC decomposition for each frequency bin should be unique for the proposed method to be valid, and therefore the models given in (7) and (10) should both satisfy the identifiability condition. That is:

$$\begin{aligned} k_{\boldsymbol{H}_f} + k_{\boldsymbol{B}_f} + k_{\boldsymbol{H}_{f+1}} &\geq 2M + 2 \\ k_{\boldsymbol{I}_M} + k_{\boldsymbol{B}_f} + k_{\boldsymbol{H}_{f+1}} &\geq 2M + 2 \end{aligned} \tag{13}$$

Obviously, $k_{\boldsymbol{I}_M} = M$. Moreover, by assumptions A2) and A3), we have $k_{\boldsymbol{H}_f} = k_{\boldsymbol{H}_{f+1}} = M$, amd (13) can be reduced to $k_{\boldsymbol{B}_f} \geq 2$, which is satisfied if and only if $\boldsymbol{B}_f$ does not contain collinear columns. Recall that the mth column of $\boldsymbol{B}_f$ represent the cross-frequency covariance of the mth source for the frequency pair $(f, f+1)$ at different time instants, then by assumption A2), we know that arbitrary two columns of $\boldsymbol{B}_f$ are not collinear. Then $k_{\boldsymbol{B}_f} \geq 2$ is also satisfied. As a result, the identifiability conditions for both (7) and (10) are met, and the proposed algorithm is valid under assumptions A1) – A4).

4 Simulation Results

In this section, we use simulations with synthetic speech signals to demonstrate the performance of the proposed method. The overall signal-to-interference ratio (SIR) is used as the measure of performance:

$$\chi = 10\sum_{m=1}^{M} \log \frac{\sum_t \tilde{s}_{mm}^2(t)}{\sum_t \sum_{k \neq m} \tilde{s}_{mk}^2(t)} \tag{14}$$

where $\tilde{s}_{mm}(t)$ is the component coming from the mth source in its estimate $\tilde{s}_m(t)$, and $\tilde{s}_{mk}(t)$ is the cross-talk from the kth source. Since the observations are synthetic, we have access to the microphone signals $x_{nm}(t)$, $n = 1, 2, ..., N$, recorded when only the mth source is present. Therefore, (14) could be rewritten as:

$$\chi = 10\sum_{m=1}^{M} \log \frac{\sum_t \left(\sum_{n=1}^{N} g_{mn}(t) \star x_{nm}(t)\right)^2}{\sum_t \sum_{k \neq m} \left(\sum_{n=1}^{N} g_{mn}(t) \star x_{nk}(t)\right)} \tag{15}$$

---

[2] The Kruscal rank of matrix $\boldsymbol{A}$, denoted by k_A, is the maximal number r such that any set of r columns of $\boldsymbol{A}$ is linearly independent.

where $g_{mn}(t)$ is the $(m,n)th$ entry of the demixing filter $G(t)$.

We compare the proposed method with PARAFAC method based on stacking the time varying covariance matrices into a covariance tensor for each frequency bin [5]. Moreover, to enable a clearer comparison, we modify the permutation correction method used in [5] by exploiting continuity of frequency responses with the scheme introduced in [6]. As a result, the two compared methods are similar in form except that different tensors are used for PARAFAC decompositions. We consider the scenario that two speech signals, sampled at 16kHz with duration of 10 seconds, are mixed with filters with order 128 and 256. The overall SIR curves of the proposed cross frequency tensor based PARAFAC (CFT-PARAFAC) and existing covariance tensor based PARAFAC (CT-PARAFAC) are plotted in Fig. 2.

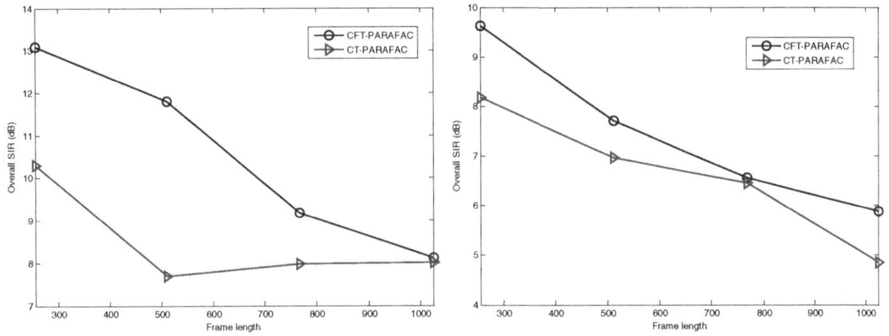

(a) Mixing filter with order 128 (b) Mixing filter with order 256

Fig. 2. Overall SIR versus the frame length with mixing filters of order 128 and 256

From Fig. 2, we can see that the proposed CFT-PARAFAC method offers larger overall SIR than CT-PARAFAC. In addition, we note that the overall SIR curves of both methods decrease as the frame length increases. This is because a larger frame length could result in an increased number of mismatched frequency responses, which would add to the difficulty of permutation correction. Recall that the frame length should not be too small for a desired frequency resolution, hence, we need to select a proper frame length for the proposed method.

5 Conclusion

In this paper, we considered the problem of blind separation for convolutive speech mixtures. The covariance tensors across frequency bins are used instead of the covariance tensors within a single frequency bin, and a tensorial scheme consisting of successive PARAFAC decompositions is developed. The problem of permutation correction is solved in the proposed method by simultaneously extracting and pairing two adjacent frequency responses, and exploiting the common factors shared by neighboring cross frequency covariance tensors, under the assumptions that components of the same source associated with adjacent frequency bins are correlated.

Simulation results have shown that, for mixing filters with order 128 and 256, the overall signal-to-interference ratio of the proposed method is larger than the existing covariance tensor based PARAFAC method. This observation further implies that exploiting covariance across frequency bins in-process (contrary to post-processing) may be advantageous than merely using continuity of frequency responses in-process, and the cross frequency covariance tensors may just provide such a way.

Acknowledgments. This work was supported by the National Natural Science Foundation of China under Grant No. 60971097.

References

1. Pedersen, M.S., Larsen, J., Kjems, U., Parra, L.C.: A survey of convolutive blind source separation methods. Springer Handbook on Speech Processing and Speech Communication, 1–34 (2007)
2. Murata, N., Ikeda, S., Ziehe, A.: An approach to blind source separation based on temporal structure of speech signal. Neurocomputing 41, 1–24 (2001)
3. Wang, L.D., Lin, Q.H.: Frequency-domain blind separation of convolutive speech mixturees with energy correlation-based permutation correction. In: Zhang, L., Lu, B.-L., Kwok, J. (eds.) ISNN 2010. LNCS, vol. 6063, Springer, Heidelberg (2010)
4. Parra, L.C., Spence, C.: Convolutive blind separation of non-stationary sources. IEEE Transactions on Speech and Audio Processing 8, 320–327 (2000)
5. Nion, D., Mokios, K. N., Sidiropoulos, N. D., Potamianos, A.C.: Batch and adaptive PARAFAC-based blind separation of convolutive speech mixtures. IEEE Transactions on Audio, Speech and Language Processing (to appear)
6. Serviere, C., Pham, D.T.: Permutation correction in the frequency domain in blind separation of speech mixtures. EURASIP Journal on Applied Signal Processing Article ID 75206, 1–16 (2006)
7. Sidiropoulos, N.D., Bro, R., Giannakis, G.B.: Parallel factor analysis in sensor array processing. IEEE Transactions on Signal Processing 48, 2377–2388 (2000)
8. Tomasi, G., Bro, R.: A comparison of algorithms for fitting the PARAFAC model. Computational Statistics and Data Analysis 50, 1700–1734 (2006)
9. Sidiropoulos, N.D., Giannakis, G.B., Bro, R.: Blind PARAFAC receivers for DS-CDMA systems. IEEE Transactions on Signal Processing 48, 810–823 (2000)
10. De Lathauwer, L., Castaing, J.: Blind identification of underdetermined mixtures by simultaneous matrix diagonalization. IEEE Transactions on Signal Processing 56, 1096–1105 (2008)

Under-Determined Reverberant Audio Source Separation Using Local Observed Covariance and Auditory-Motivated Time-Frequency Representation

Ngoc Q.K. Duong, Emmanuel Vincent, and Rémi Gribonval

INRIA, Centre Inria Rennes - Bretagne Atlantique, France
qduong@irisa.fr, emmanuel.vincent@inria.fr,
remi.gribonval@inria.fr

Abstract. We consider the local Gaussian modeling framework for under-determined convolutive audio source separation, where the spatial image of each source is modeled as a zero-mean Gaussian variable with full-rank time- and frequency-dependent covariance. We investigate two methods to improve the accuracy of parameter estimation, based on the use of local observed covariance and auditory-motivated time-frequency representation. We derive an iterative expectation-maximization (EM) algorithm with a suitable initialization scheme. Experimental results over stereo synthetic reverberant mixtures of speech show the effectiveness of the proposed methods.

Keywords: Under-determined convolutive source separation, nonuniform time-frequency representation, ERB transform, full-rank spatial covariance model.

1 Introduction

Multichannel under-determined source separation is the problem of recovering from a vector of I observed mixture channels $\mathbf{x}(t)$ either the J source signals $s_j(t)$ or their *spatial images* $\mathbf{c}_j(t)$, *i.e.* the vector of their contributions to all mixture channels, with $1 < I < J$. The mixture of several sound sources, *e.g.* musical instruments, speakers and background noise, recorded by the microphone array is then expressed as

$$\mathbf{x}(t) = \sum_{j=1}^{J} \mathbf{c}_j(t). \tag{1}$$

We consider a reverberant environment where $\mathbf{c}_j(t)$ can be modeled via the convolutive mixing process

$$\mathbf{c}_j(t) = \sum_{\tau} \mathbf{h}_j(\tau) s_j(t-\tau) \tag{2}$$

where $\mathbf{h}_j(\tau)$ is the vector of filter coefficients modeling the acoustic path from source j to all microphones.

Most existing approaches, *e.g.* [1], first transform the signals into the time-frequency (T-F) domain via the short time Fourier transform (STFT) and approximate the convolutive mixing process (2) under a narrowband assumption by complex-valued multiplication in each frequency bin f and time frame n

$$\mathbf{c}_j(n,f) \approx \mathbf{h}_j(f) s_j(n,f) \tag{3}$$

V. Vigneron et al. (Eds.): LVA/ICA 2010, LNCS 6365, pp. 73–80, 2010.

where $s_j(n,f)$ and $\mathbf{c}_j(n,f)$ are the STFT coefficients of $s_j(t)$ and $\mathbf{c}_j(t)$, respectively, and the mixing vector $\mathbf{h}_j(f)$ is the Fourier transform of $\mathbf{h}_j(\tau)$. In [2,3] we considered a different framework where $\mathbf{c}_j(n,f)$ is modeled as a zero-mean Gaussian vector random variable with covariance matrix

$$\mathbf{R}_{\mathbf{c}_j}(n,f) = v_j(n,f)\,\mathbf{R}_j(f) \tag{4}$$

where $v_j(n,f)$ are scalar time-varying *variances* encoding the spectro-temporal power of the sources and $\mathbf{R}_j(f)$ are time-invariant *spatial covariance matrices* encoding their spatial position and spatial spread. We proposed to represent $\mathbf{R}_j(f)$ as a full-rank matrix and showed that this better models reverberation and improves separation compared to the rank-1 spatial covariance model resulting from the narrowband assumption, *i.e.* $\mathbf{R}_j(f) = \mathbf{h}_j(f)\mathbf{h}_j^H(f)$ with ^{H} denoting matrix conjugate transposition [3]. Nevertheless, the *model parameters* $v_j(n,f)$ and $\mathbf{R}_j(f)$ are often inaccurately estimated due to T-F overlap between sources. In this paper, we present two complementary methods to further enhance separation performance by respectively improving the robustness of parameter estimation to source overlap and reducing this overlap.

Firstly, instead of estimating the model parameters from the mixture STFT coefficients $\mathbf{x}(n,f)$ as in [3], we infer them from the observed mixture covariance $\widehat{\mathbf{R}}_{\mathbf{x}}(n,f)$ in the neighborhood of each T-F bin in a maximum likelihood (ML) sense. This method was introduced in [4] in the context of instantaneous audio source separation as a sliding window variant of the T-F patch-based model in [5]. Besides the interchannel phase and intensity differences encoded by $\mathbf{x}(n,f)$, $\widehat{\mathbf{R}}_{\mathbf{x}}(n,f)$ also encodes correlation between the mixture channels which decreases with the larger number of active sources and the larger angle between these sources. This additional information results in improved separation of instantaneous mixtures: for instance, time-frequency bins involving a single active source are perfectly estimated since they are characterized by rank-1 local observed covariance *i.e.* maximum interchannel correlation [6,4]. In the following, we show that this method also improves separation performance on reverberant mixtures, despite the fact that interchannel correlation is intrinsically lower in this context.

Secondly, we investigate the use of a nonuniform T-F representation on the auditory-motivated equivalent rectangular bandwidth (ERB) scale. ERB-scale representations have been widely used in computational auditory scene analysis (CASA) [7] and in a few other studies [8,9] since they provide finer spectral resolution than the STFT at low frequencies, hence decrease the overlap between sources in this crucial frequency range where most sound energy lies. These representations were shown to improve separation performance based on ℓ_1-norm minimization for instantaneous mixtures [9] or single-channel Wiener filtering for convolutive mixtures [8]. Yet, they failed so far to improve separation performance based on more powerful multichannel filtering techniques for reverberant mixtures, due to the fact that the narrowband approximation does not hold because of their coarse spectral resolution at high frequencies [8]. In the following, we show that ERB-scale representations are also beneficial for multichannel convolutive source separation provided that the full-rank covariance model is used.

The structure of the rest of the paper is as follows. We define a ML criterion exploiting local observed covariance and applicable to auditory-motivated T-F representations in Section 2. We then derive an associated source separation algorithm in Section 3 and report some experimental results in Section 4. Finally we conclude in Section 5.

2 ML Criterion Exploiting Local Observed Covariance and Auditory-Motivated T-F Representation

We first derive the likelihood using the local observed covariance in the neighborhood of each STFT bin. We then extend this criterion to an auditory-motivated T-F representation resulting from the ERB scale.

2.1 ML Inference Employing Local Observed Covariance

Let us assume that the source image STFT coefficients $\mathbf{c}_j(n', f')$ follow a zero-mean Gaussian distribution over some neighborhood of (n, f) with covariance matrix $\mathbf{R}_{\mathbf{c}_j}(n, f)$ defined in (4). If the sources are uncorrelated, the mixture STFT coefficients $\mathbf{x}(n, f)$ also follow a zero-mean Gaussian distribution over this neighborhood with covariance

$$\mathbf{R}_{\mathbf{x}}(n, f) = \sum_{j=1}^{J} v_j(n, f)\, \mathbf{R}_j(f). \tag{5}$$

The log-likelihood of the data can then be defined as [5, 4]

$$\log \mathcal{L}_w = \sum_{n,f} \sum_{n',f'} w_{nf}^2(n', f') \log \mathcal{N}(\mathbf{x}(n', f')|0, \mathbf{R}_{\mathbf{x}}(n, f)) \tag{6}$$

where w_{nf} is a bi-dimensional window specifying the shape of the neighborhood such that $\sum_{n',f'} w_{nf}^2(n', f') = 1$. This model aims to improve the robustness of parameter estimation by locally exploiting the observed data in several T-F bins instead of a single one. Simple computation leads to

$$\log \mathcal{L}_w = -\sum_{n,f} \mathrm{tr}(\mathbf{R}_{\mathbf{x}}^{-1}(n, f) \widehat{\mathbf{R}}_{\mathbf{x}}(n, f)) + \log \det(\pi\, \mathbf{R}_{\mathbf{x}}(n, f)) \tag{7}$$

where $\mathrm{tr}(.)$ denotes the trace of a square matrix and $\widehat{\mathbf{R}}_{\mathbf{x}}(n, f)$ is the *local observed mixture covariance* in the neighborhood of (n, f) given by [4]

$$\widehat{\mathbf{R}}_{\mathbf{x}}(n, f) = \sum_{n',f'} w_{nf}^2(n', f') \mathbf{x}(n', f') \mathbf{x}^H(n', f'). \tag{8}$$

ML inference of the model parameters can now be achieved by maximizing (7), which does not require knowledge of $\mathbf{x}(n, f)$ anymore but of $\widehat{\mathbf{R}}_{\mathbf{x}}(n, f)$ only.

2.2 Auditory-Motivated T-F Representation

The ERB scale is an auditory motivated frequency scale defined by [7]

$$f_{ERB} = 9.26 \log(0.00437 f_{Hz} + 1). \tag{9}$$

For a given number of frequency bands, ERB-scale T-F representations exhibit higher frequency resolution than the STFT at low frequencies and lower resolution at high frequencies. We use the representation in [8] which differs from those used in CASA [7] in two ways. Firstly, the bandwidth of each subband is not set to that of the cochlea but to a narrower value resulting in better separation performance. Secondly, the representation itself does not consist of a set of subband signals but of a set of local observed covariance matrices. Therefore, although it has never been employed in this context, it leads to straightforward computation of the ML criterion in (7).

The ERB-scale representation is computed as follows[1]. Let $H_f(t)$ be a bank of centered complex-valued bandpass filters associated with individual frequency bands f. These filters are defined as modulated Hanning windows. Their center frequencies are linearly spaced on the ERB scale between zero and the Nyquist frequency and the width of their main lobe is set to four times the difference between the central frequencies of adjacent filters. The mixture signal $\mathbf{x}(t)$ is split into subband signals $\mathbf{x}_f(t)$ defined by

$$\mathbf{x}_f(t) = \sum_{\tau} H_f(\tau)\mathbf{x}(t-\tau). \tag{10}$$

The local observed covariance is then computed from windowed subband signals by

$$\widehat{\mathbf{R}}_{\mathbf{x}}(n,f) = \sum_{t,f'} w_f^2(f')w_{\text{ana}}^2(t-nL)\mathbf{x}_{f'}(t)\mathbf{x}_{f'}^H(t) \tag{11}$$

where $w_{\text{ana}}(t)$ is an analysis window, $w_f(f')$ is a window specifying the shape of the frequency neighborhood and L the step between successive time frames.

3 Blind Source Separation Algorithm

We now propose a source separation algorithm using the above ML criterion and applicable to ERB-scale time-frequency representations. This algorithm consists of three successive steps: initialization of the model parameters $v_j(n,f)$ and $\mathbf{R}_j(f)$ by hierarchical clustering and permutation alignment, ML parameter estimation by iterative expectation-maximization (EM), and derivation of the source images by Wiener filtering. We present the details of this algorithm for full-rank spatial covariances $\mathbf{R}_j(f)$ and briefly explain how to add the rank-1 constraint $\mathbf{R}_j(f) = \mathbf{h}_j(f)\mathbf{h}_j^H(f)$.

3.1 Parameter Initialization and Permutation Alignment in the STFT Domain

It has been shown that parameter initialization affects the source separation performance resulting from the EM algorithm [3]. In the case of a STFT representation, we follow the hierarchical clustering-based initialization [3] in which we have awared that the initialization robustness decreases when the number of active sources increases. We first normalize the vectors of mixture STFT coefficients so that they have unit Euclidean norm and the first entry has zero phase. In a given frequency bin f, each normalized vector of mixture STFT coefficients $\bar{\mathbf{x}}(n,f)$ is first considered as a cluster containing a

[1] Matlab software implementing this representation as well as ERB-scale multichannel filtering is available at `http://www.irisa.fr/metiss/members/evincent/software`

single item. The distance between each pair of clusters is computed as the average distance between the associated normalized mixture STFT coefficients and the two clusters with the smallest distance are merged [3]. This linking process is repeated until the number of clusters is smaller than a fixed threshold K, which is much larger than the number of sources J [10], so as to eliminate outliers. We finally choose the J clusters C_j with the largest number of items denoted by $|C_j|$ and compute the associated spatial covariance matrices by

$$\mathbf{R}_j^{\text{init}}(f) = \frac{1}{|C_j|} \sum_{\mathbf{x}(n,f) \in C_j} \mathbf{x}(n,f)\mathbf{x}^H(n,f). \tag{12}$$

The mixing vectors $\mathbf{h}_j^{\text{init}}(f)$ are initialized in a similar fashion [3]. The order of $\mathbf{R}_j^{\text{init}}(f)$ and $\mathbf{h}_j^{\text{init}}(f)$ is then aligned across all frequency bins based on the DOA information carried by $\mathbf{h}_j^{\text{init}}(f)$ [11] or by the first principal component of $\mathbf{R}_j^{\text{init}}(f)$ [3] so as to correspond to the same source signal j. Finally, the source variances are simply initialized to $v_j^{\text{init}}(n,f) = 1$.This basic initialization scheme performed as well as the more advanced but slower schemes in our experiment.

3.2 Parameter Initialization and Permutation Alignment in the ERB Domain

The above initialization procedure does not readily extend to ERB-scale representations for which T-F coefficients $\mathbf{x}(n,f)$ are not available. While this procedure can be applied to cluster the first principal components of the local observed covariances $\widehat{\mathbf{R}}_{\mathbf{x}}(n,f)$, we found that this did not result in good performance in the high frequency range. Indeed, due to broadness of high frequency subbands, the reduction of local observed covariances to their first principal components results in some information loss.

In order to ensure a fair comparison of both representations independently of the parameter initialization procedure, we derive the initial parameters in the ERB domain from those in the STFT domain as follows. Given the estimated clusters C_j, initial estimates of the source images in the STFT domain are obtained by binary masking as

$$\mathbf{c}_j^{\text{init}}(n,f) = \begin{cases} \mathbf{x}(n,f) & \text{if } \bar{\mathbf{x}}(n,f) \in C_j \\ \mathbf{0} & \text{otherwise} \end{cases} \tag{13}$$

The order of $\mathbf{c}_j^{\text{init}}(n,f)$ is aligned across frequency based on $\mathbf{h}_j^{\text{init}}(f)$ as above. ERB-domain initial local observed covariances $\widehat{\mathbf{R}}_{\mathbf{c}_j}^{\text{init}}(n,f)$ are then derived by STFT inversion of $\mathbf{c}_j^{\text{init}}(n,f)$ followed by ERB-scale representation. Finally the spatial covariance matrices $\mathbf{R}_j^{\text{init}}(f)$ are initialized by averaging $\widehat{\mathbf{R}}_{\mathbf{c}_j}^{\text{init}}(n,f)$ over all time frames n, while the initial mixing vectors $\mathbf{h}_j^{\text{init}}(f)$ are taken as the first principal component of $\mathbf{R}_j^{\text{init}}(f)$. Again, the source variances are initialized to $v_j^{\text{init}}(n,f) = 1$.

3.3 ML Parameter Estimation by EM

We now wish to optimize the ML criterion in (7). Since this criterion does not assume a particular representation, a single parameter estimation algorithm can be designed both the the STFT and the ERB-scale representation. The EM algorithm is well known as an appropriate choice in this case. We reuse the M-step of the EM algorithm derived for rank-1 and full-rank covariance models in [3] but adapt the E-step to account for the use of the local observed covariance. Note that, strictly speaking, this algorithm and that in [3] are generalized forms of EM since the M-step increases but does not maximize the likelihood of the complete data.

For the full-rank model, we consider the set of T-F coefficients of all source images on all channels and all time frames $\{\mathbf{c}_j(n,f)\ \forall j,n\}$ as the EM *complete data*. The details of one EM iteration for each source j in frequency bin f are as follows. In the E-step, the Wiener filter $\mathbf{W}_j(n,f)$ and the covariance matrix $\widehat{\mathbf{R}}_{\mathbf{c}_j}(n,f)$ of the estimated source image are computed as

$$\mathbf{W}_j(n,f) = \mathbf{R}_{\mathbf{c}_j}(n,f)\mathbf{R}_{\mathbf{x}}^{-1}(n,f) \tag{14}$$

$$\widehat{\mathbf{R}}_{\mathbf{c}_j}(n,f) = \mathbf{W}_j(n,f)\widehat{\mathbf{R}}_{\mathbf{x}}(n,f)\mathbf{W}_j^H(n,f) + (\mathbf{I} - \mathbf{W}_j(n,f))\mathbf{R}_{\mathbf{c}_j}(n,f) \tag{15}$$

where $\mathbf{I}$ is the $I \times I$ identity matrix, $\mathbf{R}_{\mathbf{c}_j}(n,f)$ is defined in (4), $\mathbf{R}_{\mathbf{x}}(n,f)$ in (5) and $\widehat{\mathbf{R}}_{\mathbf{x}}(n,f)$ in (8). In the M-step, $\mathbf{R}_j(f)$ and $v_j(n,f)$ are updated as [3]

$$v_j(n,f) = \frac{1}{I}\operatorname{tr}(\mathbf{R}_j^{-1}(f)\widehat{\mathbf{R}}_{\mathbf{c}_j}(n,f)) \tag{16}$$

$$\mathbf{R}_j(f) = \frac{1}{N}\sum_{n=1}^{N}\frac{1}{v_j(n,f)}\widehat{\mathbf{R}}_{\mathbf{c}_j}(n,f) \tag{17}$$

Note that, compared to the EM algorithm in [3], $\widehat{\mathbf{R}}_{\mathbf{c}_j}(n,f)$ is computed directly from $\widehat{\mathbf{R}}_{\mathbf{x}}(n,f)$ in the E-step instead of the estimated source images.

This modification of the E-step is also applied to the rank-1 model. Due to the lack of space we do not describe the details of the EM algorithm for the rank-1 model here. EM improves the separation performance compared to the initialization alone [3].

3.4 Source Separation by Wiener Filtering

The spatial images of all sources can be estimated given $v_j(n,f)$ and $\mathbf{R}_j(f)$ by Wiener filtering. In the STFT domain, this filter is applied to the mixture STFT coefficients as

$$\widehat{\mathbf{c}}_j(n,f) = v_j(n,f)\mathbf{R}_j(f)\mathbf{R}_{\mathbf{x}}^{-1}(n,f)\mathbf{x}(n,f) \tag{18}$$

and time-domain signals are obtained by STFT inversion. In the ERB domain, this filter is applied to the windowed subband signals $w_{\text{ana}}(t-nL)\mathbf{x}_f(t)$ instead. Subband source image signals are obtained by overlap-and-add via

$$\widehat{\mathbf{c}}_{jf}(t) = \sum_n w_{\text{syn}}(t-nL)w_{\text{ana}}(t-nL)v_j(n,f)\mathbf{R}_j(f)\mathbf{R}_{\mathbf{x}}^{-1}(n,f)\mathbf{x}_f(t) \tag{19}$$

where $w_{\text{syn}}(t)$ is a synthesis window satisfying $\sum_t w_{\text{syn}}(t-nL)w_{\text{ana}}(t-nL) = 1$. Fullband signals are finally recovered by least-squares filterbank inversion [8].

Table 1. Average source separation performance (dB) of stereo mixtures of 3 speech sources

Approach	SDR	SIR	SAR	ISR
Full-rank model + LOC + ERB	6.6	10.1	9.3	10.9
Full-rank model + LOC	6.1	8.9	9.0	10.2
Full-rank model	5.8	8.7	8.6	9.9
Rank-1 model + LOC + ERB	4.1	7.7	8.4	8.9
Rank-1 model + LOC	4.6	8.1	8.5	9.2
Rank-1 model	4.3	7.9	7.9	9.2
Binary masking	4.8	10.2	5.6	10.3
ℓ_0-norm minimization	3.8	7.9	6.9	9.2

4 Experimental Results

We compared the performance of the blind source separation algorithms presented in this paper and those in [3], based either on a rank-1 or a full-rank model, on the use of the STFT or the ERB-scale representation as input, and on the choice of the ML criterion based on the mixture STFT coefficients or on the local observed covariance (LOC). We generated three stereo synthetic mixtures of three speech sources by convolving different sets of 10 s speech signals sampled at 16 kHz with room impulse responses simulated via the source image method for a room with reverberation time $T_{60} = 250$ ms. The distance between the microphones was 5 cm and that from the sources to microphones was 50 cm. The STFT was computed with half-overlapping sine windows of length 1024 and the bi-directional window w defining the shape of neighborhood was the outer product of two Hanning windows of length 3. The ERB-scale representation was computed with a bank of 250 filters and half-overlapping sine analysis and synthesis windows, *i.e.* w_{ana} and w_{syn}, of length 1024. The number of EM iterations for all algorithms was 10. Note that the optimal number of frequency bin for the STFT and the ERB-scale representation are different. Binary masking [1] and ℓ_0-norm minimization [2] were also evaluated as baseline approaches with the same mixing vectors used for the rank-1 model. Separation performance was evaluated using the signal-to-distortion ratio (SDR) criterion measuring overall distortion, as well as the signal-to-interference ratio (SIR), signal-to-artifact ratio (SAR) and source image-to-spatial distortion ratio (ISR) criteria in [12], averaged over all sources and all mixtures.

The results in Table 1 indicate that the full-rank model always provides higher SDR and SAR than both the rank-1 model and baseline approaches which comforts our claim that it better approximates the reverberant mixing process [3], even with other representations than the STFT. The exploitation of local observed covariance is shown to be beneficial for both rank-1 and full-rank models since it provides 0.3 dB SDR improvement in both cases and also increases all other performance criteria, especially the SAR. The ERB-scale T-F representation provides an additional 0.5 dB SDR enhancement compared to sole use of the local observed covariance for the full-rank model. However, this representation results instead in a decrease of the SDR when the rank-1 model is used. This suggests that full-rank models are needed from now to achieve further improvements in ERB scale-based source separation, since they partially overcome the narrowband assumption and do not suffer from the large bandwidth at high frequencies.

In the end, the combination of local observed covariance and ERB-scale representation improves the SDR, SIR, SAR, and ISR by 0.8 dB, 1.4 dB, 0.7 dB, and 1 dB, respectively compared to the full-rank model based algorithm proposed in [3] and outperforms all other approaches. This improvement satisfies our theoretical expectations and even appears more positively when considering the fact that the new algorithm is faster due to the processing of 250 ERB-scale frequency bins instead of 513 STFT bins.

5 Conclusion

In this paper, we presented an under-determined convolutive source separation algorithm offered by the combination of local observed covariance, auditory-motivated T-F representation and the full-rank covariance model. We addressed the estimation of model parameters by EM algorithm with a proper parameter initialization scheme. Experimental results over speech mixtures confirm that both local covariance model and auditory-motivated T-F representation are beneficial for full-rank model based source separation. Consequently, the proposed approach outperforms both the full-rank model and rank-1 model in [3] and baseline approaches in a reverberant environment.

References

1. Yılmaz, O., Rickard, S.T.: Blind separation of speech mixtures via time-frequency masking. IEEE Trans. on Signal Processing 52(7), 1830–1847 (2004)
2. Vincent, E., Jafari, M.G., Abdallah, S.A., Plumbley, M.D., Davies, M.E.: Probabilistic modeling paradigms for audio source separation. In: Wang, W. (ed.) Machine Audition: Principles, Algorithms and Systems. GI Global (to appear)
3. Duong, N.Q.K., Vincent, E., Gribonval, R.: Under-determined reverberant audio source separation using a full-rank spatial covariance model. IEEE Trans. on Audio, Speech and Language Processing (2010) (to appear)
4. Vincent, E., Arberet, S., Gribonval, R.: Underdetermined instantaneous audio source separation via local Gaussian modeling. In: Proc. ICA, pp. 775–782 (2009)
5. Févotte, C., Cardoso, J.F.: Maximum likelihood approach for blind audio source separation using time-frequency Gaussian models. In: Proc. WASPAA, pp. 78–81 (2005)
6. Deville, Y.: Temporal and time-frequency correlation-based blind source separation methods. In: Proc. ICA, pp. 1059–1064 (2003)
7. Roman, N., Wang, D., Brown, G.: Speech segregation based on sound localization. Journal of the ASA 114(4), 2236–2252 (2003)
8. Vincent, E.: Musical source separation using time-frequency source priors. IEEE Trans. on Audio, Speech and Language Processing 14 (1), 91–98 (2006)
9. Burred, J., Sikora, T.: Comparison of frequency-warped representations for source separation of stereo mixtures. In: Proc. 121st AES Convention (October 2006)
10. Winter, S., Kellermann, W., Sawada, H., Makino, S.: MAP-based underdetermined blind source separation of convolutive mixtures by hierarchical clustering and ℓ_1-norm minimization. EURASIP Journal on Advances in Signal Processing,, 2007, article ID 24717 (2007)
11. Sawada, H., Araki, S., Mukai, R., Makino, S.: Grouping separated frequency components by estimating propagation model parameters in frequency-domain blind source separation. IEEE Trans. on Audio, Speech, and Language Processing 15(5), 1592–1604 (2007)
12. Vincent, E., Sawada, H., Bofill, P., Makino, S., Rosca, J.: First stereo audio source separation evaluation campaign: data, algorithms and results. In: Davies, M.E., James, C.J., Abdallah, S.A., Plumbley, M.D. (eds.) ICA 2007. LNCS, vol. 4666, pp. 552–559. Springer, Heidelberg (2007)

Crystal-MUSIC: Accurate Localization of Multiple Sources in Diffuse Noise Environments Using Crystal-Shaped Microphone Arrays

Nobutaka Ito[1,2], Emmanuel Vincent[1], Nobutaka Ono[2], Rémi Gribonval[1], and Shigeki Sagayama[2]

[1] INRIA, Centre de Rennes - Bretagne Atlantique
Campus de Beaulieu, 35042 Rennes Cedex, France
{nobutaka.ito,emmanuel.vincent,remi.gribonval}@inria.fr
[2] The University of Tokyo
7-3-1 Hongo, Bunkyo-ku, Tokyo, Japan
{ito,onono,sagayama}@hil.t.u-tokyo.ac.jp

Abstract. This paper presents crystal-MUSIC, a method for DOA estimation of multiple sources in the presence of diffuse noise. MUSIC is well known as a method for the estimation of the DOAs of multiple sources but *is not very robust to diffuse noise from many directions, because the covariance structure of such noise is not spherical.* Our method makes it possible for MUSIC to accurately estimate the DOAs by removing the contribution of diffuse noise from the spatial covariance matrix. This denoising is performed in two steps: 1) denoising of the off-diagonal entries via a blind noise decorrelation using crystal-shaped arrays, and 2) denoising of the diagonal entries through a low-rank matrix completion technique. *The denoising process does not require the spatial covariance matrix of diffuse noise to be known, but relies only on an isotropy feature of diffuse noise.* Experimental results with real-world noise show that the DOA estimation accuracy is substantially improved compared to the conventional MUSIC.

Keywords: Diffuse noise, DOA estimation, microphone arrays, MUSIC, source localization.

1 Introduction

DOA estimation of sound sources is an important issue with many applications such as beamforming and speaker tracking. Real-world sound environments typically contain multiple directional sounds as well as diffuse noise, which comes from many directions like in vehicles or cafeterias. In this paper, we present crystal-MUSIC, a method for accurate estimation of the DOAs of multiple sources in the presence of diffuse noise.

One of the most fundamental approaches to DOA estimation is to maximize the output power of the delay-and-sum or other fixed beamformers with respect to the steering direction. However, since a sharp beam cannot be achieved with a

V. Vigneron et al. (Eds.): LVA/ICA 2010, LNCS 6365, pp. 81–88, 2010.

practical small-sized array, DOA estimates may be inaccurate in the presence of multiple sources. Methods based on Time Delay Of Arrival (TDOA) estimates [1] widely in use today assume a single target source and again the performance can be degraded when more than one source is present. In comparison, MUSIC [2–4] estimates the DOAs of multiple sources as directions in which the corresponding steering vector becomes most nearly orthogonal to the noise subspace.

It is important in MUSIC to accurately identify the noise subspace. When there is no noise, it is easily obtained as the null space of the observed covariance matrix. It can be obtained also in the presence of spatially white noise, since such noise only adds its power to all eigenvalues uniformly without changing the eigenvectors because of its spherical covariance structure. Therefore, the basis vectors of the noise subspace coincide with the eigenvectors of the observed covariance matrix belonging to the smallest eigenvalue. Directional noise can be dealt with as well, for it can be regarded as one of the directional signals. In contrast, diffuse noise can significantly degrade the identification of the noise subspace, because the noise spans the whole observation space, and unlike spatially white noise, its covariance structure is not spherical.

Aiming to make MUSIC robust to diffuse noise, this paper proposes a method for removing the contribution of diffuse noise from the spatial covariance matrix. The denoising is performed in two steps. In the first step, the contribution of diffuse noise is removed from the off-diagonal entries through the diagonalization of the covariance matrix of diffuse noise. This is performed through a technique of Blind Noise Decorrelation (BND) [5, 6], in which any covariance matrix of isotropic noise is diagonalized by a single unitary matrix based on the use of symmetrical arrays called crystal arrays. We do not assume the coherence matrix of diffuse noise like in Ref. [7], but only assume an isotropy defined later, aiming to adapt more to various environments in the real world. In the second step of the denoising, thus obtained off-diagonal entries are completed to be the full matrix with the diagonal entries filled in via a low-rank matrix completion technique [8–10]. We present a modified version of the method in Ref. [8] with a positive semi-definite constraint on the estimated covariance matrix.

Throughout, the superscript ^{H} denotes Hermitian transposition. Signals are represented in the time-frequency domain with τ and ω denoting the frame index and the angular frequency. The covariance matrix of a zero-mean random vector $\gamma(\tau,\omega)$ is denoted by $\boldsymbol{\Phi}_{\gamma\gamma}(\tau,\omega) \triangleq \mathcal{E}[\gamma(\tau,\omega)\gamma^{\mathsf{H}}(\tau,\omega)]$, where $\mathcal{E}[\cdot]$ is expectation.

2 Review of MUSIC

2.1 Observation Model

We assume that an array of M microphones receives $L(< M)$ directional signals (some of them can be unwanted directional interferences) from unknown directions in the presence of diffuse noise. We assume the number of directional sources, L, to be known in this paper. Let $\boldsymbol{s}(\tau,\omega) \in \mathbb{C}^L$ be the vector comprising the directional signals observed at a reference point (*e.g.* the array centroid), and $\boldsymbol{x}(\tau,\omega) \in \mathbb{C}^M$ and $\boldsymbol{v}(\tau,\omega) \in \mathbb{C}^M$ be the vectors comprising the observed signals and the diffuse noise at the microphones, respectively. Assuming planewave propagation and static sources of the directional signals, we can model the transfer function from $s_l(\tau,\omega)$ to $x_m(\tau,\omega)$ as $D_{ml}(\omega) \triangleq \mathrm{e}^{-\mathrm{j}\omega\delta_{ml}}$, where δ_{ml} denotes the delay in arrival of the directional signal $s_l(\tau,\omega)$ from the reference point to the m-th microphone. Consequently, our observation model is given by

$$\boldsymbol{x}(\tau,\omega) = \boldsymbol{D}(\omega)\boldsymbol{s}(\tau,\omega) + \boldsymbol{v}(\tau,\omega) \tag{1}$$

$$= \sum_{l=1}^{L} \boldsymbol{d}(\omega;\theta_l)s_l(\tau,\omega) + \boldsymbol{v}(\tau,\omega), \tag{2}$$

where θ_l denotes the DOA of the l-th directional sound,

$$\boldsymbol{d}(\omega;\theta) \triangleq \left[\mathrm{e}^{-\mathrm{j}\omega\delta_1(\theta)}\ \mathrm{e}^{-\mathrm{j}\omega\delta_2(\theta)}\ \ldots\ \mathrm{e}^{-\mathrm{j}\omega\delta_M(\theta)}\right]^{\mathsf{T}} \tag{3}$$

denotes the steering vector corresponding to DOA θ, and $\delta_m(\theta)$ denotes the time delay of arrival for DOA θ from the reference point to the m-th microphone. We assume $\boldsymbol{s}(\tau,\omega)$ and $\boldsymbol{v}(\tau,\omega)$ to be uncorrelated zero-mean random vectors. As a result, $\boldsymbol{x}(\tau,\omega)$ is a zero-mean random vector with covariance matrix

$$\boldsymbol{\Phi}_{\boldsymbol{xx}}(\tau,\omega) = \boldsymbol{D}(\omega)\boldsymbol{\Phi}_{\boldsymbol{ss}}(\tau,\omega)\boldsymbol{D}^{\mathsf{H}}(\omega) + \boldsymbol{\Phi}_{\boldsymbol{vv}}(\tau,\omega). \tag{4}$$

2.2 DOA Estimation

The orthogonal projection of $\boldsymbol{d}(\omega;\theta)$ onto the noise subspace, *i.e.* the orthogonal complement of $\mathrm{span}\{\boldsymbol{d}(\omega;\theta_l)\}_{l=1}^{L}$, becomes zero when θ coincides with θ_l. Therefore, the MUSIC spectrum

$$f_{\mathrm{MUSIC}}(\omega;\theta) \triangleq \|\boldsymbol{V}^{\mathsf{H}}(\omega)\boldsymbol{d}(\omega;\theta)\|_2^{-2} \tag{5}$$

attains peaks at θ_l, where $\boldsymbol{V}$ is a matrix whose columns are orthonormal basis vectors of the noise subspace. Since the MUSIC spectrum (5) is defined for each ω, it is needed to integrate the information from all frequency bins in order to obtain a single estimate of the DOAs. A common approach is to average Eq. (5) over frequencies [3, 4]. For example, the geometric mean [3] gives

$$\overline{f}_{\mathrm{MUSIC}}(\theta) \triangleq \left[\prod_{\omega} f_{\mathrm{MUSIC}}(\omega;\theta)\right]^{\frac{1}{K}}, \tag{6}$$

with K denoting the number of averaged frequency bins. The DOAs are estimated as peaks in $\overline{f}_{\mathrm{MUSIC}}(\theta)$.

3 Denoising of the Spatial Covariance Matrix

To calculate (5), it is important to accurately estimate $\boldsymbol{V}(\omega)$, namely basis vectors of the noise subspace. However, diffuse noise can significantly degrade the estimation, because it spans the whole observation space, and unlike spatially white noise, its covariance structure is not spherical. Our idea therefore consists in restoring the covariance matrix of the directional signals, $\boldsymbol{D}(\omega)\boldsymbol{\Phi}_{\boldsymbol{ss}}(\tau,\omega)\boldsymbol{D}^{\mathsf{H}}(\omega)$, from the observed covariance matrix $\boldsymbol{\Phi}_{\boldsymbol{xx}}(\tau,\omega)$ contaminated by diffuse noise, so that we can obtain $\boldsymbol{V}(\omega)$ as eigenvectors of the restored matrix belonging to the eigenvalue 0. The matrix denoising is performed in two steps. Firstly, the contribution of diffuse noise to the off-diagonal entries is removed using BND [5, 6] as explained in Section 3.1. Secondly, the diagonal entries are denoised via a low-rank matrix completion technique [8–10] as explained in Section 3.2.

3.1 Diffuse Noise Removal from the Off-Diagonal Entries

Coming from many directions, diffuse noise can be regarded as more isotropic than directional. Therefore, we make the following assumptions:

1) Diffuse noise has the same power spectrogram at all microphones:

$$[\boldsymbol{\Phi}_{\boldsymbol{vv}}]_{11}(\tau,\omega) = [\boldsymbol{\Phi}_{\boldsymbol{vv}}]_{22}(\tau,\omega) = \cdots = [\boldsymbol{\Phi}_{\boldsymbol{vv}}]_{MM}(\tau,\omega). \tag{7}$$

2) The inter-channel cross-spectrogram of diffuse noise is identical for all microphone pairs with an equal distance:

$$r_{mn} = r_{pq} \Rightarrow [\boldsymbol{\Phi}_{\boldsymbol{vv}}]_{mn}(\tau,\omega) = [\boldsymbol{\Phi}_{\boldsymbol{vv}}]_{pq}(\tau,\omega), \tag{8}$$

where r_{mn} is the distance between the m-th and n-th microphones. It was shown that there exist such array geometries that any $\boldsymbol{\Phi}_{\boldsymbol{vv}}(\tau,\omega)$ satisfying these assumptions is diagonalized by a single unitary matrix [5, 6]. So far, we have found five classes of geometries enabling such diagonalization, namely, regular polygonal, (twisted) rectangular, (twisted) regular polygonal prism, rectangular solid, and regular polyhedral arrays. They are called crystal arrays from their shapes.

Making use of a crystal array, we can remove the contribution of the diffuse noise to the off-diagonal entries as follows:

$$\boldsymbol{P}^{\mathsf{H}}\boldsymbol{\Phi}_{\boldsymbol{xx}}(\tau,\omega)\boldsymbol{P} = \boldsymbol{P}^{\mathsf{H}}\boldsymbol{D}(\omega)\boldsymbol{\Phi}_{\boldsymbol{ss}}(\tau,\omega)\boldsymbol{D}^{\mathsf{H}}(\omega)\boldsymbol{P} + \boldsymbol{P}^{\mathsf{H}}\boldsymbol{\Phi}_{\boldsymbol{vv}}(\tau,\omega)\boldsymbol{P} \tag{9}$$

where $\boldsymbol{P}$ is a unitary diagonalization matrix of $\boldsymbol{\Phi}_{\boldsymbol{vv}}(\tau,\omega)$.

3.2 Denoising of the Diagonal Entries

Now that the off-diagonal entries of $\boldsymbol{P}^{\mathsf{H}}\boldsymbol{D}(\omega)\boldsymbol{\Phi}_{\boldsymbol{ss}}(\tau,\omega)\boldsymbol{D}^{\mathsf{H}}(\omega)\boldsymbol{P}$ has been obtained, the problem has reduced to that of completing its missing diagonal elements. Once this is done, the desired matrix $\boldsymbol{D}(\omega)\boldsymbol{\Phi}_{\boldsymbol{ss}}(\tau,\omega)\boldsymbol{D}^{\mathsf{H}}(\omega)$ will be computed by the transformation $\boldsymbol{P}(\cdot)\boldsymbol{P}^{\mathsf{H}}$. Since $\boldsymbol{P}^{\mathsf{H}}\boldsymbol{D}(\omega)\boldsymbol{\Phi}_{\boldsymbol{ss}}(\tau,\omega)\boldsymbol{D}^{\mathsf{H}}(\omega)\boldsymbol{P}$ is of rank at most L, the technique of low-rank matrix completion [8–10] can be applied. We present here a variant of an EM-based method by Srebro *et al.* [8] with a positive semi-definite constraint on the matrix to be completed. This is because MUSIC identifies the noise subspace based on the property that the eigenvectors of $\boldsymbol{D}(\omega)\boldsymbol{\Phi}_{\boldsymbol{ss}}(\tau,\omega)\boldsymbol{D}^{\mathsf{H}}(\omega)$ belonging to the positive and zero eigenvalues form bases of the signal and noise subspaces, respectively. Therefore, if the estimated matrix has some negative eigenvalues, there is no way of assigning the corresponding eigenvectors to one of these subspaces in a reasonable way.

We consider that we obtain via BND an imcomplete observation $\boldsymbol{Y}$ of $\boldsymbol{\Theta} \triangleq \boldsymbol{P}^{\mathsf{H}}\boldsymbol{D}(\omega)\boldsymbol{\Phi}_{\boldsymbol{ss}}(\tau,\omega)\boldsymbol{D}^{\mathsf{H}}(\omega)\boldsymbol{P}$, where the obtained off-diagonal elements $\{x_{mn}\}$ are regarded as observables, and the missing diagonal elements $\{z_{mm}\}$ as latent variables. (Therefore, the diagonal entries of the basis-transformed observed covariance matrix $\boldsymbol{P}^{\mathsf{H}}\boldsymbol{\Phi}_{\boldsymbol{xx}}(\tau,\omega)\boldsymbol{P}$ is just abandoned here.) The observation $\boldsymbol{Y}$ is considered to contain some errors because BND is generally not perfect. Therefore, $\boldsymbol{Y}$ is modeled as follows:

$$\underbrace{\begin{bmatrix} z_{11} & y_{12} & \cdots & y_{1M} \\ y_{21} & z_{22} & \cdots & y_{2M} \\ \vdots & \vdots & \ddots & \vdots \\ y_{M1} & y_{M2} & \cdots & z_{MM} \end{bmatrix}}_{\boldsymbol{Y}} = \underbrace{\begin{bmatrix} \theta_{11} & \theta_{12} & \cdots & \theta_{1M} \\ \theta_{21} & \theta_{22} & \cdots & \theta_{2M} \\ \vdots & \vdots & \ddots & \vdots \\ \theta_{M1} & \theta_{M2} & \cdots & \theta_{MM} \end{bmatrix}}_{\boldsymbol{\Theta}} + \underbrace{\begin{bmatrix} \epsilon_{11} & \epsilon_{12} & \cdots & \epsilon_{1M} \\ \epsilon_{21} & \epsilon_{22} & \cdots & \epsilon_{2M} \\ \vdots & \vdots & \ddots & \vdots \\ \epsilon_{M1} & \epsilon_{M2} & \cdots & \epsilon_{MM} \end{bmatrix}}_{\boldsymbol{E}}, \tag{10}$$

where $\boldsymbol{E}$ is the error term and its entries ϵ_{mn} are assumed to be i.i.d. complex-valued Gaussian random variables. The criterion is the maximization of the log-likelihood of the observed data subject to the constraint that $\boldsymbol{\Theta}$ is positive semi-definite and of rank at most L:

$$\hat{\boldsymbol{\Theta}} = \arg\max_{\boldsymbol{\Theta}\in\Omega} \ln P(\{y_{mn}\}_{m\neq n}|\boldsymbol{\Theta}), \tag{11}$$

where Ω denotes the set of the $M \times M$ positive semi-definite matrices of rank at most L.

The E-step amounts to the calculating the new estimate $\hat{\boldsymbol{Y}}^{(i+1)}$ of $\boldsymbol{Y}$ by completing the diagonal entries of $\boldsymbol{Y}$ by those of the current estimate $\hat{\boldsymbol{\Theta}}^{(i)}$ of $\boldsymbol{\Theta}$:

$$\hat{\boldsymbol{Y}}^{(i+1)} = \begin{bmatrix} \hat{\theta}_{11}^{(i)} & y_{12} & \cdots & y_{1M} \\ y_{21} & \hat{\theta}_{22}^{(i)} & \cdots & y_{2M} \\ \vdots & \vdots & \ddots & \vdots \\ y_{M1} & y_{M2} & \cdots & \hat{\theta}_{MM}^{(i)} \end{bmatrix}. \tag{12}$$

The M-step amounts to calculating the new estimate $\hat{\boldsymbol{\Theta}}^{(i+1)}$ of $\boldsymbol{\Theta}$ as the best approximation of $\hat{\boldsymbol{Y}}^{(i+1)}$ in the Frobenius sense subject to $\boldsymbol{\Theta} \in \Omega$:

$$\hat{\boldsymbol{\Theta}}^{(i+1)} = \arg\min_{\boldsymbol{\Theta}\in\Omega} \|\hat{\boldsymbol{Y}}^{(i+1)} - \boldsymbol{\Theta}\|_{\mathrm{F}}. \tag{13}$$

The solution is written explicitly using the eigenvalue decomposition of $\hat{\boldsymbol{Y}}^{(i+1)}$:

$$\hat{\boldsymbol{Y}}^{(i+1)} = \boldsymbol{U}^{(i+1)}\boldsymbol{\Lambda}^{(i+1)}\boldsymbol{U}^{\mathsf{H}(i+1)}, \tag{14}$$

where $\boldsymbol{U}^{(i+1)}$ is unitary and the eigenvalues in the diagonal of $\boldsymbol{\Lambda}^{(i+1)}$ are ordered from largest to smallest (possibly negative). Then, the solution to (13) is

$$\hat{\boldsymbol{\Theta}}^{(i+1)} = \boldsymbol{U}^{(i+1)}\boldsymbol{\Lambda}_{\mathrm{T}}^{(i+1)}\boldsymbol{U}^{\mathsf{H}(i+1)}. \tag{15}$$

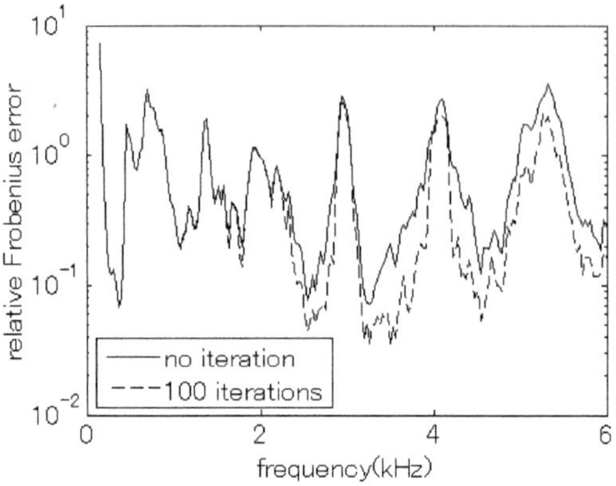

Fig. 1. The relative Frobenius error as a function of the frequency before and after the the covariance matrix denoising

Here, $\boldsymbol{\Lambda}_{\mathrm{T}}^{(i+1)}$ is the truncated version of $\boldsymbol{\Lambda}^{(i+1)}$ whose diagonal entry (eigenvalue of $\hat{\boldsymbol{Y}}^{(i+1)}$) is kept if and only if it is positive and among the L largest and replaced by zero otherwise. The parameters are initialized by $\hat{\boldsymbol{Y}}^{(0)} = \hat{\boldsymbol{\Theta}}^{(0)} = \boldsymbol{P}^{\mathsf{H}}\boldsymbol{\Phi}_{\boldsymbol{xx}}\boldsymbol{P}$. Using the resulting estimate $\hat{\boldsymbol{\Theta}}$, the estimate of $\boldsymbol{D}\boldsymbol{\Phi}_{\boldsymbol{ss}}\boldsymbol{D}^{\mathsf{H}}$ is given by $\boldsymbol{P}\hat{\boldsymbol{\Theta}}\boldsymbol{P}^{\mathsf{H}}$, from which the vector $\boldsymbol{V}$ in Eq. (5) is calculated.

4 Experimental Results

We present experimental results to show the effectiveness of crystal-MUSIC. We recorded noise in a station building in Tokyo with a square array of diameter 5 cm [11]. Two target speech signals were added to this noise recording under the assumption of plane wave propagation. The speech data were taken from the ATR Japanese speech database [12]. The duration of thus generated observed signals was 7 s, and the sampling frequency was 16 kHz. We used STFT for subband decomposition, where the frame length and the frame shift were 512 and 64, respectively, and the Hamming window was used. $\boldsymbol{\Phi}_{\boldsymbol{xx}}$ for both methods was calculated by averaging $\boldsymbol{x}(\tau,\omega)\boldsymbol{x}^{\mathsf{H}}(\tau,\omega)$ temporally over all frames.

To see how well the covariance matrix denoising works, we plot in Fig. 1 a relative Frobenius error defined by $\| \cdot - \boldsymbol{D}\boldsymbol{\Phi}_{\boldsymbol{ss}}\boldsymbol{D}^{\mathsf{H}}\|_{\mathrm{F}}/\|\boldsymbol{D}\boldsymbol{\Phi}_{\boldsymbol{ss}}\boldsymbol{D}^{\mathsf{H}}\|_{\mathrm{F}}$ as a function of the frequency. The solid and dashed lines are the results for the covariance matrices before and after the denoising (*i.e.* $\boldsymbol{\Phi}_{\boldsymbol{xx}}$ and $\boldsymbol{P}\hat{\boldsymbol{\Theta}}\boldsymbol{P}^{\mathsf{H}}$). The SNR at the first microphone was adjusted to -5 dB and the number of iterations of the EM algorithm was 100. The true DOAs of the target signals were 200° and 260°. We see that the error was effectively reduced through the denoising.

Figure 2 is an example of (a) the conventional MUSIC spectra and (b) the crystal-MUSIC spectra, for each frequency. The SNR at the first microphone was adjusted to -5 dB and the number of iterations of the EM was 100. The lines show the true DOAs, namely 200° and 260°. We see that the crystal-MUSIC gave more accurate peak positions and much less spurious peaks.

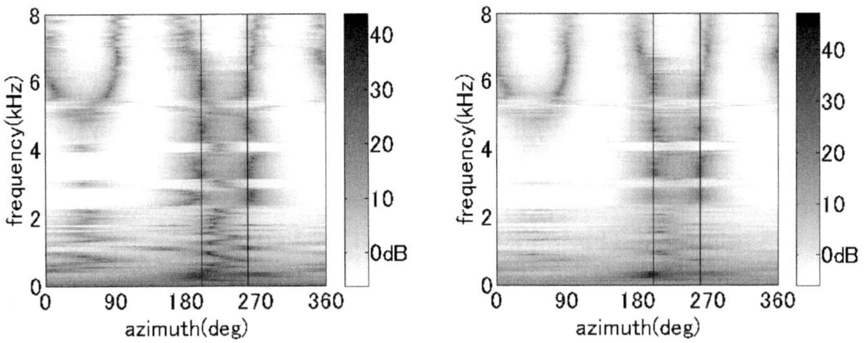

(a) subband MUSIC spectra (b) subband crystal-MUSIC spectra

Fig. 2. An example of the subband MUSIC spectra for (a) conventional MUSIC and (b) crystal-MUSIC

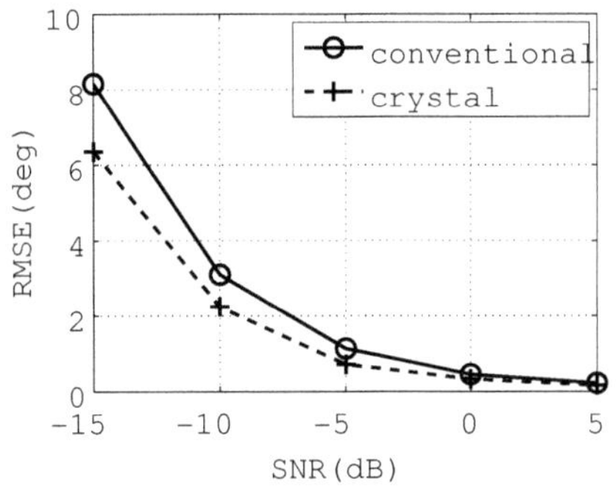

Fig. 3. The RMSE of DOA estimation as a function of SNR

Finally, we compare the accuracy of DOA estimation by MUSIC and crystal-MUSIC statistically. Fig. 3 shows the Root Mean Square Error (RMSE) of the DOA estimation by the methods as a function of the SNR at the first microphone. The estimates were obtained from the geometric mean of narrowband MUSIC spectra as in Eq. (6). The range of averaging was 80th to 150th frequency bins (approximately 2.5 to 4.7 kHz). The range was determined so as to avoid using low frequencies with a very low SNR and high frequencies with spatial aliasing. The RMSE was calculated from an experiment with various source DOAs, where all the 15 DOA combination from the set $\{0°, 60°, 120°, \cdots, 300°\}$ were tested. The figure shows a substantial improvement in RMSE by crystal-MUSIC.

5 Conclusion

We described crystal-MUSIC, an accurate method for estimating DOAs of multiple sounds in a diffuse noise field. It is based on removal of the contribution of diffuse noise from the observation covariance matrix via BND using crystal arrays and a low-rank matrix completion technique. We presented a new matrix completion method with a positive semi-definite constraint, which is more suitable to MUSIC. The experiment using real-world noise showed the effectiveness of the covariance matrix denoising and the substantial improvement in the DOA estimation accuracy by crystal-MUSIC.

Acknowledgments. This work is supported by INRIA under the Associate Team Program VERSAMUS and by Grant-in-Aid for Young Scientists (B) 21760309 from MEXT, Japan.

References

1. Knapp, C.H., Carter, G.C.: The generalized correlation method for estimation of time delay. IEEE Trans. Acoust., Speech, Signal Process. (4), 320–327 (August 1976)
2. Schmidt, R.O.: Multiple emitter location and signal parameter estimation. IEEE Trans. Antennas Propag., 276–280 (March 1986)
3. Wax, M., Shan, T.-J., Kailath, T.: Spatio-temporal spectral analysis by eignestructure methods. IEEE Trans. Acoust., Speech, Signal Process., 817–827 (August 1984)
4. Pham, T., Sadler, B.M.: Adaptive wideband aeroacoustic array processing. In: Proceedings of the 8th IEEE Signal Processing Workshop on Stastical Signal and Array Processing, Corfu, Greece, pp. 295–298 (June 1996)
5. Shimizu, H., Ono, N., Matsumoto, K., Sagayama, S.: Isotropic noise suppression in the power spectrum domain by symmetric microphone arrays. In: Proc. WASPAA, New Paltz, NY, pp. 54–57 (October 2007)
6. Ono, N., Ito, N., Sagayama, S.: Five classes of crystal arrays for blind decorrelation of diffuse noise. In: Proc. SAM, Darmstadt, Germany, pp. 151–154 (July 2008)
7. Bitzer, J., Simmer, K.U.: Superdirective microphone arrays. In: Brandstein, M., Ward, D. (eds.) Microphone Arrays: Signal Processing Techniques and Applications, ch. 2, pp. 19–38. Springer, Berlin (2001)
8. Srebro, N., Jaakkola, T.: Weighted low-rank approximations. In: 20th International Conference on Machine Learning, pp. 720–727. AAAI Press, Menlo Park (2003)
9. Candès, E.J., Recht, B.: Exact matrix completion via convex optimization. The Journal of the Society for the Foundations of Computational Mathematics (9), 717–772 (April 2009)
10. Ji, S., Ye, J.: An accelerated gradient method for trace norm minimization. In: Proceedings of the 26th International Conference on Machine Learning, Montreal, Canada, pp. 457–464 (2009)
11. Ito, N., Ono, N., Vincent, E., Sagayama, S.: Designing the Wiener post-filter for diffuse noise suppression using imaginary parts of inter-channel cross-spectra. In: Proc. ICASSP 2010, Dallas, USA (March 2010)
12. Kurematsu, A., Takeda, K., Sagisaka, Y., Katagiri, S., Kuwabara, H., Shikano, K.: ATR Japanese speech database as a tool of speech recognition and synthesis. Speech Communication 9(4), 357–363 (1990)

Consistent Wiener Filtering: Generalized Time-Frequency Masking Respecting Spectrogram Consistency

Jonathan Le Roux[1], Emmanuel Vincent[2], Yuu Mizuno[3], Hirokazu Kameoka[1], Nobutaka Ono[3], and Shigeki Sagayama[3]

[1] NTT Communication Science Laboratories, NTT Corporation, 3-1 Morinosato Wakamiya, Atsugi, Kanagawa 243-0198, Japan
[2] INRIA, Centre Inria Rennes - Bretagne Atlantique, Campus de Beaulieu, 35042 Rennes Cedex, France
[3] Graduate School of Information Science and Technology, The University of Tokyo, 7-3-1 Hongo, Bunkyo-ku, Tokyo 113-8656, Japan
{leroux,kameoka}@cs.brl.ntt.co.jp, emmanuel.vincent@inria.fr, {mizuno,onono,sagayama}@hil.t.u-tokyo.ac.jp

Abstract. Wiener filtering is one of the most widely used methods in audio source separation. It is often applied on time-frequency representations of signals, such as the short-time Fourier transform (STFT), to exploit their short-term stationarity, but so far the design of the Wiener time-frequency mask did not take into account the necessity for the output spectrograms to be consistent, i.e., to correspond to the STFT of a time-domain signal. In this paper, we generalize the concept of Wiener filtering to time-frequency masks which can involve manipulation of the phase as well by formulating the problem as a consistency-constrained Maximum-Likelihood one. We present two methods to solve the problem, one looking for the optimal time-domain signal, the other promoting consistency through a penalty function directly in the time-frequency domain. We show through experimental evaluation that, both in oracle conditions and combined with spectral subtraction, our method outperforms classical Wiener filtering.

Keywords: Wiener filtering, Short-time Fourier transform, Spectrogram consistency, Source separation, Spectral subtraction.

1 Introduction

Wiener filtering has been one of the most widely used methods for source separation for several decades, in particular in audio signal processing. To exploit the short-term stationarity of audio signals, it is very often applied on time-frequency representations [1], especially the short-time Fourier transform (STFT). However, classical Wiener filtering does not take into account the intrinsically redundant structure of STFT spectrograms, and its output is actually in general not the optimal solution. We show here that by ensuring that the output spectrograms are "consistent", i.e., that they correspond to actual time-domain signals, we can obtain a more efficient filtering. Many of the most promising methods for

V. Vigneron et al. (Eds.): LVA/ICA 2010, LNCS 6365, pp. 89–96, 2010.

source separation exploit spectral models of the sources (non-negative matrix factorization, Gaussian mixture models, autoregressive modeling, etc.) and, as these models are often based on Gaussian assumptions, they commonly involve Wiener filtering as a post-processing [2]. It is thus of tremendous importance to ensure that the information gathered by these algorithms is best exploited.

Wiener filtering can be formulated as the solution of a Maximum-Likelihood (ML) problem in the time-frequency domain without constraint on the space of admissible solutions. The classical solution then only involves a manipulation on the magnitude part of the spectrograms, leading in general to arrays of complex numbers which do not correspond to any time-domain signal. We generalize here the concept of Wiener filtering to time-frequency masks which can involve a manipulation of the phase as well in order to find the ML solution among consistent spectrograms. Formulating the problem as the minimization of an objective function derived from the Wiener likelihood and explicitely taking into account consistency, we present two methods to solve it: one consists in computing the exact optimum by solving the problem in the time domain; the other relies on a relaxation of the consistency constraints through the introduction of a penalty function promoting consistency of the output spectrogram. We already exploited the idea of consistency-promoting penalty functions for fast signal reconstruction from modified magnitude spectrograms [3] and to improve the modeling accuracy of the complex non-negative matrix factorization framework [4]. It enables us here to develop an efficient algorithm which computes an approximate solution close to the true optimum obtained with the time-domain method.

We evaluate the performance of the proposed methods compared to classical Wiener filtering on two tasks: separation of concurrent speech by two speakers under oracle conditions, and denoising of speech mixed with synthetic and real-world background noises where only the noise mean power spectra are known and the speech spectrum is estimated through spectral subtraction [5].

2 Wiener Filtering and Consistency

2.1 Maximum-Likelihood Formulation

We assume that the observed signal x is the mixture of two signals, a target s_1 and an interference signal s_2. We further assume that the STFT coefficients S_1 and S_2 of the signals s_1 and s_2 at each time frame t and frequency bin ω are modeled as statistically independent Gaussian random variables with variance σ_1^2 and σ_2^2 respectively. For convenience of notation, we shall write $\nu^{(i)} = 1/\sigma_i^2$. Note that the case of several interference signals can be reduced, without loss of generality, to that of two sources only, as we assume in particular that the sources are not correlated.

Denoting by X the spectrogram of x, classical Wiener filtering consists in maximizing the log-likelihood of the STFT coefficients S_1 and S_2, which can be written, under the constraint that $X = S_1 + S_2$, as a function of $S = S_1$ only:

$$\mathcal{L}(S) = -\frac{1}{2}\Big(\sum_{\omega,t} \nu_{\omega,t}^{(1)}|S_{\omega,t}|^2 + \sum_{\omega,t} \nu_{\omega,t}^{(2)}|X_{\omega,t} - S_{\omega,t}|^2\Big) + C(\nu^{(1)}, \nu^{(2)}) \ , \qquad (1)$$

where C is a constant depending only on $\nu^{(1)}$, $\nu^{(2)}$. Introducing the classical Wiener filtering estimate for S_1,

$$\hat{S}_{\omega,t} = \frac{\nu^{(2)}_{\omega,t}}{\nu^{(1)}_{\omega,t} + \nu^{(2)}_{\omega,t}} X_{\omega,t} \ , \tag{2}$$

the ML problem can be reformulated as the minimization of the objective function

$$\psi(S) = \sum_{\omega,t} \alpha_{\omega,t} |S_{\omega,t} - \hat{S}_{\omega,t}|^2, \text{ where } \alpha_{\omega,t} = \nu^{(1)}_{\omega,t} + \nu^{(2)}_{\omega,t} \ . \tag{3}$$

2.2 Wiener Filtering with Consistency Constraint

If no further constraint is assumed on S, the objective function is obviously minimized for $S = \hat{S}$. However, we need to keep in mind that the STFT is a redundant representation with a particular structure. Denoting by N the number of frequency bins and T the number of frames, STFT spectrograms of time-domain signals are elements of $\mathbb{C}^{NT}$, which we shall call "consistent spectrograms", but one of the fundamental points of this paper is that not all elements of $\mathbb{C}^{NT}$ can be obtained as such [6,3]. If we assume that inverse STFT is performed in such a way that a signal can be exactly reconstructed from its spectrogram through inverse STFT, then we showed in [3] that a necessary and sufficient condition for an array W to be a consistent spectrogram is for it to be equal to the STFT of its inverse STFT. The set of consistent spectrograms can thus be described as the null space $\mathrm{Ker}(\mathcal{F})$ of the $\mathbb{R}$-linear operator $\mathcal{F}$ from $\mathbb{C}^{NT}$ to itself defined by

$$\mathcal{F}(W) = \mathcal{G}(W) - W, \text{ where } \mathcal{G}(W) = \mathrm{STFT}(\mathrm{iSTFT}(W)) \ . \tag{4}$$

Going back to the Wiener filtering problem, if we now impose that the solution be consistent, the problem amounts to finding a consistent spectrogram S minimizing ψ, or in other words to minimize ψ under the constraint that $\mathcal{F}(S) = 0$. Imposing consistency is not a mere elegance or theory-oriented concern, but a truly fundamental problem. Indeed, the spectrogram of the signal resynthesized from the classical Wiener filter spectrogram $\hat{S}$ is actually different in general from $\hat{S}$, and no longer maximizing the Wiener log-likelihood (or minimizing ψ), so that the final result of the processing that we are listening to is in fact not the optimal solution. What we really want to do is to find a signal such that its spectrogram minimizes the Wiener criterion ψ, or, formulating this in the time-frequency domain, to minimize the following "true" objective function

$$\tilde{\psi}(S) = \sum_{\omega,t} \alpha_{\omega,t} \left| \mathcal{G}(S)_{\omega,t} - \hat{S}_{\omega,t} \right|^2 \ , \tag{5}$$

where $\mathcal{G}(S)$ is again the spectrogram of the signal resynthesized from S by inverse STFT. We can try to solve the problem directly in the time domain by minimizing $\psi(\mathrm{STFT}(s))$ w.r.t. the time-domain signal s. Another possibility is to relax the consistency constraint by introducing it as a penalty function: if the weight of the penalty is chosen sufficiently large, or is increased during the course of the optimization, the estimated spectrogram should finally be both consistent and minimizing ψ among the consistent spectrograms.

3 Optimization Algorithms

3.1 Time-Domain Formulation

The consistent Wiener filtering optimization problem amounts to minimizing $\sum_{\omega,t} \alpha_{\omega,t}|S_{\omega,t} - \hat{S}_{\omega,t}|^2$ on the subspace of consistent spectrograms, while that of estimating the signal whose STFT spectrogram is closest to the modified STFT spectrogram $\hat{S}$ amounts to minimizing $\sum_{\omega,t} |S_{\omega,t} - \hat{S}_{\omega,t}|^2$ on the same subspace [6]. The latter problem can be transformed through Parseval's theorem into the minimization of a simple quadratic form on the time signal parameters, but the weights α make here the computation of the optimal signal cumbersome as they hinder us from simplifying the product of the Fourier matrix and its transpose. Let A_t be the $N \times N$ diagonal matrix with diagonal coefficients $\alpha_{\omega,t}$, F the $N \times N$ Fourier transform matrix, w_t the $N \times L$ matrix which computes the t-th windowed frame of the signal x (of length L), and $\hat{s}_t$ the inverse transform of the t-th STFT frame of $\hat{S}$. We can show that the optimal signal x is given by

$$\hat{x} = \Big(\sum_t w_t^H F^H A_t F w_t\Big)^{-1} \sum_t w_t^H F^H A_t F \hat{s}_t \ . \tag{6}$$

If A_t were not present, as in the latter problem, then $F^H F$ would simplify to $N\mathrm{Id}$ and we would get the simple weighted overlap-add estimation $x = \sum_t w_t^H \hat{s}_t / \sum_t w_t^H w_t$. However, the simplification cannot be done here, leading to a very large $(L \times L)$ matrix inversion problem. Still, this matrix is band-diagonal (and Hermitian), and solving the system is possible in a reasonable amount of time and using a reasonable amount of memory space. To reduce in particular the memory requirements, we can split in practice the estimation of the signal on overlapping blocks of a few frames, and reconstruct an approximate solution on the whole interval by overlap-add from the locally optimal signals.

3.2 Consistency as a Penalty Function

For an array of complex numbers $W \in \mathbb{C}^{NT}$, $\mathcal{F}(W)$ represents the relation between W and the STFT of its inverse STFT. Instead of enforcing consistency through the "hard" constraint $\mathcal{F}(W) = 0$, which may be difficult to handle, we can relax that constraint by using any vector norm of $\mathcal{F}(W)$ to derive a numerical criterion quantifying how far W is from being consistent. We consider here the L^2 norm of $\mathcal{F}(W)$, which leads, as shown in [3], to a criterion related to that used by Griffin and Lim to derive their iterative STFT algorithm [6]. Introducing the consistency penalty in (3), the new objective function to minimize reads

$$\psi_\gamma(S) = \psi(S) + \gamma \sum_{\omega,t} \left|\mathcal{G}(S)_{\omega,t} - S_{\omega,t}\right|^2 \ . \tag{7}$$

An efficient optimization algorithm for ψ_γ can be derived through the auxiliary function method [7]. A function $\psi_\gamma^+(S, \bar{S})$ is called an auxiliary function for $\psi_\gamma(S)$ and $\bar{S}$ an auxiliary variable if $\psi_\gamma(S) = \min_{\bar{S}} \psi_\gamma^+(S, \bar{S})$, $\forall S$. The minimization of ψ_γ can be performed indirectly by alternately minimizing ψ_γ^+ w.r.t. S and $\bar{S}$.

If we assume, as we shall do, that the inverse STFT is performed using the windowed overlap-add procedure with the synthesis window before normalization equal to the analysis window, it results from [6] that $\mathcal{G}(S)$ is the closest consistent spectrogram to S in a least-squares sense:

$$\sum_{\omega,t} \left|\mathcal{G}(S)_{\omega,t} - S_{\omega,t}\right|^2 = \min_{\bar{S}\in\mathrm{Ker}(\mathcal{F})} \sum_{\omega,t} \left|\bar{S}_{\omega,t} - S_{\omega,t}\right|^2, \ \forall S \ . \tag{8}$$

If we now define the function $\psi_\gamma^+ : \mathbb{C}^{NT} \times \mathrm{Ker}(\mathcal{F}) \to \mathbb{R}$ such that

$$\forall S \in \mathbb{C}^{NT}, \forall \bar{S} \in \mathrm{Ker}(\mathcal{F}),\ \psi_\gamma^+(S,\bar{S}) = \psi(S) + \gamma \sum_{\omega,t} \left|S_{\omega,t} - \bar{S}_{\omega,t}\right|^2 \ , \tag{9}$$

we easily see from (8) that ψ_γ^+ is an auxiliary function for ψ_γ. This leads to an iterative optimization scheme in which, starting at step p from a spectrogram $S^{(p)}$, $\bar{S}$ is first updated to $\mathcal{G}(S^{(p)})$, and the new estimate $S^{(p+1)}$ is simply estimated as the minimum of a second-order form with diagonal coefficients, altogether resulting in the following update equation:

$$S_{\omega,t}^{(p+1)} \leftarrow \frac{\alpha_{\omega,t}\hat{S}_{\omega,t} + \gamma\mathcal{G}(S_{\omega,t}^{(p)})}{\alpha_{\omega,t} + \gamma} \ . \tag{10}$$

4 Experimental Evaluation

4.1 Settings and Implementations

The sampling rate was 16 kHz. All spectrograms were built with a frame length $N = 1024$, a frame shift $R = 512$ and a sine window for analysis and synthesis.

The time-domain method was implemented as follows: the analytical solution is computed separately on blocks of 64 STFT frames; the blocks have a 50 % overlap, and the resulting short-time signals are cross-faded on a small region (here 16 frames) around the center of the overlap regions in order to discard portions of signal near the block boundaries, expected to suffer from boundary effects. The above values for the block size and the amount of overlap and cross-fade were determined experimentally so as to minimize computation and memory costs while still obtaining solutions with a true Wiener criterion very close to that of the analytical solution computed on the whole interval.

For the penalty-based algorithm, heuristically, the larger γ, the slower the convergence, but the better the solution. We noticed experimentally that $\tilde{\psi}$ monotonically decreased through the update (10) with γ fixed when starting from a point obtained through updates with a smaller γ. We thus designed an update scheme for γ: starting from a small value γ_0 (typically 10^{-5}) for γ, we update S through (10) while slightly increasing γ by δ (initially set to γ_0 as well) until the decrease of $\tilde{\psi}$ becomes slower than 1 %, in which case we update δ to 2δ and restart the process. We stop after two increases of δ without significant improvement of $\tilde{\psi}$, which typically occurred after around 200 iterations.

4.2 Speech Separation under Oracle Conditions

We evaluate here the performance of the proposed methods for the separation of 10 mixtures of two speakers under oracle conditions, i.e., assuming that the true

Table 1. Performance comparison for speech separation under oracle conditions

Method	SDR	ISR	SIR	SAR	$\psi(S)$	Time
Wiener	15.0 dB	25.0 dB	24.6 dB	15.6 dB	2.0×10^9	0.05 s
Griffin-Lim	11.4 dB	21.9 dB	27.6 dB	11.4 dB	6.8×10^{12}	18.1 s
Time domain	17.1 dB	28.2 dB	27.5 dB	17.7 dB	6.2×10^4	1423.0 s
Penalty	16.5 dB	27.1 dB	26.7 dB	17.1 dB	2.2×10^5	8.1 s

power spectrograms of both sources are known. The speech signals were taken from the BSS Oracle Toolbox data [8], downsampled to 16 kHz and downmixed to mono before being mixed together to obtain 12 s long 0 dB Signal to Distortion Ratio (SDR) mixtures. For comparison, we also give the results for the classical Wiener filter output $\hat{S}$ ("Wiener") and for the spectrogram whose magnitude is closest to the magnitude of the classical Wiener filter, computed through Griffin and Lim's iterative STFT algorithm [6] run for 400 iterations ("Griffin-Lim"). This way of obtaining a consistent spectrogram through post-processing of the classical Wiener filter magnitude seems indeed a natural method to attempt.

The results are summarized in Table 1. For each method are reported four commonly used objective source separation performance criteria [9], namely the Signal to Distortion ratio, the source Image to Spatial distortion Ratio (ISR), the Signal to Interference Ratio (SIR) and the Signal to Artifacts Ratio (SAR), as well as the computation time and the final value of the "true" Wiener criterion $\tilde{\psi}$. Although the performance of the classical Wiener filter is already very good, with 15.0 dB output SDR, we can see that the proposed methods all lead to significant improvements in both the true Wiener criterion $\tilde{\psi}$ and the objective performance criteria, with in particular the output SDR raised to 17.1 dB for the time-domain method and 16.5 dB for the penalty-based one, while simply reconstructing the phase as a post-processing does not solve the problem (higher $\tilde{\psi}$, lower SDR). The increase in SDR may not seem straightforward, but it can be understood as a result of the fact that with our methods the spectrogram of the resynthesized signal is closer to the intended ML solution. Computation of the analytical time-domain solution is very costly, but enables us to see that the solution obtained in much less time with the penalty-based algorithm is close to optimal. We will use this last algorithm for the noise reduction experiments below.

We also studied the influence of the frame shift on performance, and noticed that the output SDR increases with the amount of overlap between frames in the STFT, especially for the analytical solution. This could be expected as consistency constraints become stronger when overlap increases. Computation time of course increases as well, roughly linearly with the total number of spectrogram frames for all the methods. Detailed results are skipped due to space constraints.

4.3 Real-World Background Noise Reduction

In order to test our method in more realistic conditions, we performed noise reduction experiments on speech mixed with various types of noise, assuming that only the average power spectrum of the noise is known and that the power

Table 2. Performance comparison for noise reduction. Values are in dB.

Input SDR		-10 dB				0 dB				+10 dB			
		SDR	ISR	SIR	SAR	SDR	ISR	SIR	SAR	SDR	ISR	SIR	SAR
Ga	Wiener	-3.1	12.9	-3.2	6.1	6.2	20.3	7.3	12.0	14.8	29.0	16.5	20.0
	Penalty	3.4	7.4	11.6	2.5	9.4	15.6	18.8	10.2	15.9	24.7	23.9	17.1
Su	Wiener	-5.9	18.2	-5.4	6.5	3.9	26.9	4.9	11.5	13.6	34.9	14.8	20.3
	Penalty	-2.8	13.6	-0.7	2.3	6.7	23.5	9.8	10.0	15.6	31.9	19.0	19.0
Sq	Wiener	-4.6	14.6	-4.4	5.7	5.1	23.6	6.3	11.5	14.6	32.5	16.0	20.4
	Penalty	-1.7	9.5	0.1	1.0	7.1	17.8	11.4	9.2	15.6	27.1	20.3	18.1
Ca	Wiener	-4.8	9.7	-5.9	5.6	4.7	18.5	5.4	11.0	13.8	29.1	15.1	19.7
	Penalty	-1.0	4.5	-0.6	-0.9	6.1	12.2	11.2	7.2	14.0	24.0	19.2	16.0

spectrogram of speech is estimated by spectral subtraction [5]. We considered four noise signals: a synthetic Gaussian white noise, and three real-world background noises from SiSEC 2010's "Source separation in the presence of real-world background noise" task [10] recorded near a subway car ("Su"), on a square ("Sq") and in a cafeteria ("Ca"). The stereo signals were downmixed to mono and cut to 10 s length to match the speech signals, which were the same as above. We considered 30 mixtures for each noise, with 10 different speech signals and at three input SDRs: -10 dB, 0 dB and 10 dB. The results for the penalty-based algorithm and the classical Wiener filter, averaged for each noise and input SDR on the 10 corresponding mixtures, are summarized in Table 2.

We can see that the proposed method leads to a significant improvement over Wiener filtering in terms of output SDR, with, averaged on all noises, further gains of 4.1 dB, 2.4 dB and 1.1 dB respectively for -10 dB, 0 dB and 10 dB input SDRs. This can be further analyzed as a strong improvement of the SIR, offset, to a lesser extent, by a deterioration of the SAR and ISR. Note that this trade-off between improvement of SIR and deterioration of SAR and ISR can be tuned through the penalty weight γ depending on the application, as classical Wiener filtering indeed corresponds to $\gamma = 0$. Perceptually, although there remains some musical noise, the residual noise present in the Wiener filter estimates is much weaker with the proposed method.

We believe that the tendency of our algorithm to further suppress the interference compared with the classical Wiener filter is related to the distribution of the time-frequency bins whose power has not been canceled through spectral subtraction. This can be simply understood in the particular case where speech is replaced by silence. If most bins are set to zero, our algorithm will tend to cancel the remaining ones as well, as a consistent solution with most bins equal to zero in a given neighborhood is likely to be zero on the whole neighborhood, an effect similar to block-thresholding [11], shown to be one of the most effective denoising methods to date. This is first confirmed by the fact that our algorithm seems to perform quite well on Gaussian noise, whose power is exponentially distributed and for which 63 % of the bins are thus set to zero when subtracting the mean power. We tested this hypothesis informally by looking at synthetic noises with various power distribution: the improvements of our algorithm over classical Wiener filtering decreased as the proportion of bins above the mean power increased, although we shall skip the details here for the sake of brevity.

Finally, we note that by comparing the spectral subtraction results, obtained with a rather crude estimate for the noise power spectrum, with the oracle ones, we can expect our method's performance to depend on the reliability of the power spectrum estimates.

5 Conclusion

We presented a new framework for Wiener filtering and more generally time-frequency masking which takes into account the consistency of spectrograms to compute the true optimal solution to the Wiener filtering problem. We presented two methods to find optimal or near optimal solutions, investigated their performance in comparison with previous works, and showed in particular that our method combined with spectral subtraction outperforms classical Wiener filtering. Future works include combining our method with more sophisticated algorithms for the estimation of the noise power spectrum, and extension of the framework to the multichannel case.

References

1. Diethorn, E.J.: Subband noise reduction methods for speech enhancement. In: Huang, Y., Benesty, J. (eds.) Audio Signal Processing for Next-Generation Multimedia Communication Systems, pp. 91–115. Kluwer, Dordrecht (2004)
2. Vincent, E., Jafari, M.G., Abdallah, S.A., Plumbley, M.D., Davies, M.E.: Probabilistic modeling paradigms for audio source separation. In: Machine Audition: Principles, Algorithms and Systems. IGI Global (to appear)
3. Le Roux, J., Ono, N., Sagayama, S.: Explicit consistency constraints for STFT spectrograms and their application to phase reconstruction. In: Proc. SAPA, pp. 23–28 (September 2008)
4. Le Roux, J., Kameoka, H., Vincent, E., Ono, N., Kashino, K., Sagayama, S.: Complex NMF under spectrogram consistency constraints. In: Proc. ASJ Autumn Meeting, (2-4-5) (September 2009)
5. Boll, S.: Suppression of acoustic noise in speech using spectral subtraction. IEEE Trans. ASSP 27, 113–120 (1979)
6. Griffin, D.W., Lim, J.S.: Signal estimation from modified short-time Fourier transform. IEEE Trans. ASSP 32(2), 236–243 (1984)
7. Lee, D.D., Seung, H.S.: Algorithms for non-negative matrix factorization. In: Proc. NIPS*2000, pp. 556–562. The MIT Press, Cambridge (2001)
8. Vincent, E., Gribonval, R., Plumbley, M.D.: BSS Oracle Toolbox Version 2.1, `http://bass-db.gforge.inria.fr/bssoracle/`
9. Vincent, E., Sawada, H., Bofill, P., Makino, S., Rosca, J.P.: First stereo audio source separation evaluation campaign: Data, algorithms and results. In: Proc. ICA, pp. 552–559 (September 2007)
10. Araki, S., Ozerov, A., Gowreesunker, V., Sawada, H., Theis, F., Nolte, G., Lutter, D., Duong, N.Q.: The 2010 signal separation evaluation campaign (SiSEC 2010) –Part II–: Audio source separation challenges. In: Proc. LVA/ICA (2010)
11. Yu, G., Mallat, S., Bacry, E.: Audio denoising by time-frequency block thresholding. IEEE Trans. Signal Process. 56(5), 1830–1839 (2008)

Blind Separation of Convolutive Mixtures of Non-stationary Sources Using Joint Block Diagonalization in the Frequency Domain

Hicham Saylani[1], Shahram Hosseini[2], and Yannick Deville[2]

[1] Laboratoire des Systèmes de Télécommunications et Ingénierie de la Décision, Faculté des Sciences, Université Ibn Tofaïl, BP. 133, 14000 Kénitra, Maroc
[2] Laboratoire d'Astrophysique de Toulouse-Tarbes
Université de Toulouse, CNRS, 14 Av. Edouard Belin, 31400 Toulouse, France
hsaylani@gmail.com, {shosseini,ydeville}@ast.obs-mip.fr

Abstract. We recently proposed a new method based on spectral decorrelation for blindly separating linear instantaneous mixtures of non-stationary sources. In this paper, we propose a generalization of this method to FIR convolutive mixtures using a second-order approach based on block-diagonalization of covariance matrices in the frequency domain. Contrary to similar time or time-frequency domain methods, our approach requires neither the piecewise stationarity of the sources nor their sparseness. The simulation results show the better performance of our approach compared to these methods.

1 Introduction

This paper deals with the blind separation of Finite Impulse Response (FIR) convolutive mixtures. Consider M mixtures $x_i(n)$ of N discrete-time sources $s_j(n)$. Denoting by $A_{ij}(z) = \sum_{k=0}^{K} a_{ij}(k)z^{-k}$ the transfer function of each mixing filter where K is the order of the longest filter, we can write

$$x_i(n) = \sum_{j=1}^{N}\sum_{k=0}^{K} a_{ij}(k)s_j(n-k), \quad i = 1, ..., M. \tag{1}$$

One of the possible solutions for blindly separating such a convolutive mixture consists in reformulating it as an instantaneous mixture [1–3] : considering delayed versions of the mixtures, i.e. $x_i(n-l)$ $(l = 0, 1, ..., L-1)$, Eq. (1) reads

$$x_i(n-l) = \sum_{j=1}^{N}\sum_{k=0}^{K} a_{ij}(k)s_j(n-(k+l)), \ (i, l) \in [1, M] \times [0, L-1]. \tag{2}$$

These ML "generalized observations" $x_{il}(n) = x_i(n-l), (i, l) \in [1, M] \times [0, L-1]$ can be then considered as instantaneous mixtures of $N(K+L)$ "generalized sources" $s_{jr}(n) = s_j(n-r) = s_j(n-(k+l)), (j, r) \in [1, N] \times [0, K+L-1]$.

V. Vigneron et al. (Eds.): LVA/ICA 2010, LNCS 6365, pp. 97–105, 2010.

This mixture is (over-)determined if $ML \geqslant N(K+L)$. It is clear that this condition may be satisfied only if $M > N$ i.e. if the original convolutive mixture is strictly over-determined. In this case, by choosing the integer number L so that $L \geqslant \frac{NK}{M-N}$, the reformulated instantaneous mixture (2) is (over-)determined. To represent the reformulated mixture in vector form, we define $\tilde{\mathbf{x}}(n) = [\mathbf{x}_1^T(n), ..., \mathbf{x}_M^T(n)]^T$ and $\tilde{\mathbf{s}}(n) = [\mathbf{s}_1^T(n), ..., \mathbf{s}_N^T(n)]^T$ where

$$\mathbf{x}_i(n) = [x_i(n), x_i(n-1), ..., x_i(n-(L-1))]^T, \ \forall i \in [1, M], \tag{3}$$

$$\mathbf{s}_j(n) = [s_j(n), s_j(n-1), ..., s_j(n-(K+L-1))]^T, \ \forall j \in [1, N], \tag{4}$$

which yield using (2):

$$\tilde{\mathbf{x}}(n) = \tilde{\mathbf{A}}\tilde{\mathbf{s}}(n), \tag{5}$$

where

$$\tilde{\mathbf{A}} = \begin{pmatrix} \mathbf{A}_{11} & \dots & \mathbf{A}_{1N} \\ \vdots & \ddots & \vdots \\ \mathbf{A}_{M1} & \dots & \mathbf{A}_{MN} \end{pmatrix} \tag{6}$$

and $\mathbf{A}_{ij}$ is a matrix of dimension $L \times (K+L)$ defined by

$$\mathbf{A}_{ij} = \begin{pmatrix} a_{ij}(0) \dots a_{ij}(K) & 0 & \dots & 0 \\ \ddots & & \ddots & \\ & \ddots & & \ddots \\ 0 \quad \dots \quad 0 & a_{ij}(0) & \dots & a_{ij}(K) \end{pmatrix}. \tag{7}$$

Then, Eq. (5) models an (over-)determined instantaneous mixture with $M' = ML$ observations $x_{il}(n)$ and $N' = N(K+L)$ sources $s_{jr}(n)$. The $M' \times N'$ mixing matrix $\tilde{\mathbf{A}}$ is supposed to admit a *pseudo-inverse* $\tilde{\mathbf{A}}^+$ that we want to estimate for retrieving the "generalized source vector" $\tilde{\mathbf{s}}(n)$.

In Section 2, a review of the existing approaches based on the above mentioned reformulation is presented. In Section 3, we propose a new approach developed for FIR convolutive mixtures of non-stationary sources which consists in mapping the reformulated mixture (2) in the frequency domain before source separation. Simulation results are presented in Section 4 and we conclude in Section 5.

2 Existing Approaches

Bousbia-Salah et al. were among the first researchers who reformulated the problem as explained in Section 1 and used a block-diagonalization algorithm to solve it [1, 2]. In [1], they generalized the ***SOBI*** algorithm [4] supposing (as in the initial *SOBI*) that the original sources $s_j(n)$ are **stationary**, **autocorrelated** and **mutually uncorrelated**, i.e. $\underline{\forall\, j \neq k,\ \ \forall\, \tau,}\ \ E\left[s_j(n)s_k(n-\tau)\right] = 0$, so that the generalized sources $s_{jr}(n)$ verify the following relation

$$\underline{\forall j \neq k}, \ \forall r, d, \ \forall \tau \ \ E\left[s_{jr}(n)s_{kd}(n-\tau)\right] = E\left[s_j(n-r)s_k(n-\tau-d)\right] = 0. \tag{8}$$

However, since the generalized sources $s_{jr}(n) = s_j(n-r)$ and $s_{jd}(n) = s_j(n-d)$, generated from the same original source $s_j(n)$, are not mutually uncorrelated, the matrix $\mathcal{R}_{\mathbf{s}_j}(\tau) = E\left[\mathbf{s}_j(n)\mathbf{s}_j^H(n-\tau)\right]$ (the vector $\mathbf{s}_j(n)$ being defined by (4)) is not diagonal and the covariance matrix of the generalized source vector $\tilde{\mathbf{s}}(n)$ is **block-diagonal** :

$$\mathcal{R}_{\tilde{\mathbf{s}}}(\tau) = E\left[\tilde{\mathbf{s}}(n)\tilde{\mathbf{s}}^H(n-\tau)\right] = \begin{pmatrix} \mathcal{R}_{\mathbf{s}_1}(\tau) & \dots & 0 \\ \vdots & \ddots & \vdots \\ 0 & \dots & \mathcal{R}_{\mathbf{s}_N}(\tau) \end{pmatrix}. \tag{9}$$

As a result, the generalization of the *SOBI* algorithm to FIR convolutive mixtures proposed by Bousbia-Salah et al. in [1], called ***SOBI-C***, consists in estimating the separating matrix $\tilde{\mathbf{A}}^+$ thanks to **J**oint **B**lock-**D**iagonalization (**JBD**) of several covariance matrices corresponding to different time lags and estimated over whole whitened mixtures [1]. In fact, if $\tilde{\mathbf{W}}$ is the $N' \times M'$ matrix which permits one to obtain the vector of whitened observations $\tilde{\mathbf{z}}(n) = \tilde{\mathbf{W}}\tilde{\mathbf{x}}(n)$, and which is estimated by diagonalizing the matrix $\widehat{\mathcal{R}}_{\tilde{\mathbf{x}}}(0)$ and if $\tilde{\mathbf{U}}$ is the ***unitary*** matrix which permits one to ***jointly block diagonalize*** several covariance matrices $\widehat{\mathcal{R}}_{\tilde{\mathbf{z}}}(\tau)$ $(\tau = 1, 2, ...)$, then an estimate of the separating matrix $\tilde{\mathbf{A}}^+$, denoted $\tilde{\mathbf{A}}^+_{est}$, is given by

$$\tilde{\mathbf{A}}^+_{est} = \tilde{\mathbf{U}}^H\tilde{\mathbf{W}} \simeq \tilde{\mathbf{P}}\tilde{\boldsymbol{\Omega}}\tilde{\mathbf{A}}^+, \tag{10}$$

where $\tilde{\boldsymbol{\Omega}}$ is a *block-diagonal* matrix, composed of N non-zero blocks of dimension $(K+L) \times (K+L)$, which will be denoted $\tilde{\boldsymbol{\Omega}}_1, ..., \tilde{\boldsymbol{\Omega}}_N$, and $\tilde{\mathbf{P}}$ is a *permutation* matrix of these blocks [1]. Hence, using Equations (5) and (10), an estimate of the generalized source vector $\tilde{\mathbf{s}}(n)$, denoted $\tilde{\mathbf{s}}_{est}(n)$, is given by

$$\tilde{\mathbf{s}}_{est}(n) = \tilde{\mathbf{A}}^+_{est}\tilde{\mathbf{x}}(n) \simeq \tilde{\mathbf{P}}\tilde{\boldsymbol{\Omega}}\tilde{\mathbf{s}}(n). \tag{11}$$

Denoting $\tilde{\mathbf{s}}_{est}(n) = \left[\hat{\mathbf{s}}_1^T(n), ..., \hat{\mathbf{s}}_N^T(n)\right]^T$, where $\hat{\mathbf{s}}_j(n)$ are the estimates of the vectors $\mathbf{s}_j(n)$, defined by (4), and taking into account the permutation in (11), we obtain

$$\forall j \in [1, N], \ \exists k \in [1, N] \ / \ \hat{\mathbf{s}}_j(n) \simeq \tilde{\boldsymbol{\Omega}}_k\mathbf{s}_j(n). \tag{12}$$

Thus, following (12), each component of the vector $\hat{\mathbf{s}}_j(n)$ results from filtering of the source $s_j(n)$ by an *FIR* filter whose coefficients are contained in the corresponding raw of the *block* $\tilde{\boldsymbol{\Omega}}_k$. As a result, the *SOBI-C* algorithm [1] finally provides $K+L$ filtered versions of each initial source $s_j(n)$.

Another approach proposed by the same authors [2] consists in generalizing the ***BGML*** algorithm[1] of *Pham and Cardoso* [5] to convolutive mixtures, and supposes that the sources are **non-stationary** and **mutually uncorrelated**. Once more, since the covariance matrix of the generalized source vector $\mathcal{R}_{\tilde{\mathbf{s}}}(n, \tau)$ is block-diagonal, this generalization is based on JBD of covariance matrices as in [1]. However, contrary to the approach proposed in [1], the covariance matrices $\mathcal{R}_{\tilde{\mathbf{x}}}(n, \tau)$ are here estimated only for $\tau = 0$ (i.e. zero lag) but over several adjacent

[1] For Block-Gaussian Maximum Likelihood.

frames of the non-whitened mixtures, supposing the signals are stationary over each of them.

A few time-frequency approaches exploiting the sparseness of signals and developed for instantaneous mixtures of non-stationary sources, like *TFBSS* method proposed in [3], have also been generalized to FIR convolutive mixtures using JBD of matrices. For example, the generalization of the *TFBSS* method (called *TFBSS-C* in the following) is based on the JBD of the matrices obtained from spatial Wigner-Ville spectra of the mixtures [3].

All the above-mentioned approaches [1–3] being based on the JBD of the matrices, they finally estimate $K+L$ filtered versions of each of the initial sources $s_j(n)$. Then, as proposed by Févotte in [3], it is possible to estimate each of the generalized sources $s_{jr}(n)$, and in particular each of the initial sources $s_j(n)$, up to a scale factor by using a blind deconvolution algorithm. This is a great advantage compared to the classical convolutive Blind Source Separation (BSS) methods which estimate each source $s_j(n)$ up to a filter.

The approach that we propose in the following is a frequency-domain approach for non-stationary signals, also based on block diagonalization, which requires neither the piecewise stationarity of the sources nor their sparseness, contrary to the temporal or time-frequency approaches mentioned above. It is a generalization to convolutive mixtures of a new method based on spectral decorrelation for instantaneous mixtures that we recently developed [6].

3 Proposed Approach

Our original method in [6] was first proposed for separating instantaneous mixtures of non-stationary, temporally uncorrelated and mutually uncorrelated sources. It is based on the following theorem.

Theorem 1. *If $u(n)$ is a temporally uncorrelated, real, zero-mean random signal with a (possibly non-stationary) variance $\gamma(n)$, i.e. $E\left[u(n_1)u(n_2)\right] = \gamma(n_1)\delta(n_1-n_2)$, then its Fourier transform*[2] *$U(\omega)$ is a wide-sense stationary process with autocorrelation $\Gamma(\nu)$ which is the Fourier transform of $\gamma(n)$, i.e.*[3]

$$E\left[U(\omega)U^*(\omega-\nu)\right] = \Gamma(\nu) = \sum_{n=-\infty}^{+\infty} \gamma(n)\, e^{-j\nu n}. \tag{13}$$

Moreover, if $u(n)$ is non-stationary with respect to its variance, i.e. if $\gamma(n)$ is not constant, then the process $U(\omega)$ is autocorrelated.

The key idea in this method is that *uncorrelatedness* and *non-stationarity* in time domain are transformed into *wide-sense stationarity* and *autocorrelation* in the frequency domain. Thus, by processing non-stationary, temporally uncorrelated

[2] The Fourier transform of a discrete-time stochastic process $u(n)$ is a stochastic process $U(\omega)$ defined by $U(\omega) = \sum_{n=-\infty}^{\infty} u(n)e^{-j\omega n}$.

[3] In the following * stands for conjugate.

signals in the frequency domain, we can separate them using the classical BSS algorithms initially developed for separating mixtures of time-domain wide-sense stationary, time-correlated signals like *SOBI*. The main advantage of this method is that the piecewise stationarity of signals and their sparseness are not required for applying it unlike in the approaches described in Section 2. Since this method assumes that the source signals are temporally uncorrelated, in the following, we first present an extension of this method for separating FIR convolutive mixtures of non-stationary temporally uncorrelated signals, then propose a solution for applying it to non-stationary autocorrelated signals.

3.1 Non-stationary and Temporally Uncorrelated Sources

If the original sources $s_j(n)$ are non-stationary and temporally uncorrelated, then the generalized sources $s_{jr}(n)$ are too. Denote $X_{il}(\omega)$ and $S_{jr}(\omega)$ the Fourier transforms of the generalized observations $x_{il}(n)$ and the generalized sources $s_{jr}(n)$. By computing the Fourier transform of (5), we obtain a new instantaneous mixture with respect to the frequency-domain generalized sources $S_{jr}(\omega)$ with the same mixing matrix $\tilde{\mathbf{A}}$, defined by

$$\tilde{\mathbf{X}}(\omega) = \tilde{\mathbf{A}}\tilde{\mathbf{S}}(\omega). \tag{14}$$

where $\tilde{\mathbf{X}}(\omega) = \left[\mathbf{X}_1^T(\omega), ..., \mathbf{X}_M^T(\omega)\right]^T$ and $\tilde{\mathbf{S}}(\omega) = \left[\mathbf{S}_1^T(\omega), ..., \mathbf{S}_N^T(\omega)\right]^T$, with $\mathbf{X}_i(\omega) = \left[X_{i0}(\omega), ..., X_{i(L-1)}(\omega)\right]^T$ and $\mathbf{S}_j(\omega) = \left[S_{j0}(\omega), ..., S_{j(K+L-1)}(\omega)\right]^T$. Moreover,

1. from Theorem 1, since the generalized sources $s_{jr}(n)$ are temporally uncorrelated and non-stationary, the **frequency-domain sources** $S_{jr}(\omega)$ are **wide-sense stationary** and **autocorrelated**,
2. since the mutual uncorrelatedness is preserved by Fourier transform [6], Equation (8) yields

$$\underline{\forall j \neq k}, \quad \forall r, d, \quad \forall \nu \quad E\left[S_{jr}(\omega) S_{kd}^*(\omega - \nu)\right] = 0, \tag{15}$$

so that the covariance matrix of $\tilde{\mathbf{S}}(\omega)$ is **block-diagonal**:

$$\mathcal{R}_{\tilde{\mathbf{S}}}(\nu) = E\left[\tilde{\mathbf{S}}(\omega)\tilde{\mathbf{S}}^H(\omega - \nu)\right] = \begin{pmatrix} \mathcal{R}_{\mathbf{S}_1}(\nu) & \dots & 0 \\ \vdots & \ddots & \vdots \\ 0 & \dots & \mathcal{R}_{\mathbf{S}_N}(\nu) \end{pmatrix}. \tag{16}$$

Hence, we find the same type of covariance matrix (see Eq. (9)) and the same source properties as in [1]. The only difference concerns the working domain: in [1] the signals are processed in the *time domain*, the components of the generalized source vector $\tilde{\mathbf{s}}(n)$ (i.e. $s_{jr}(n)$) are *stationary* and *autocorrelated*, and its covariance matrix $\mathcal{R}_{\tilde{\mathbf{s}}}(\tau)$ is *block-diagonal*, while here the signals are processed in the *frequency domain*, the components of the vector $\tilde{\mathbf{S}}(\omega)$ (i.e. $S_{jr}(\omega)$) are *wide-sense stationary* and *autocorrelated*, and its covariance matrix $\mathcal{R}_{\tilde{\mathbf{S}}}(\nu)$, is *block-diagonal*. Thus, for estimating the separating matrix $\tilde{\mathbf{A}}^+$ we can use the *SOBI-C* algorithm [1] in the **frequency domain** which will be called ***SOBI-F-C*** in the following. Denote by $\tilde{\mathbf{W}}$ the $N' \times M'$ whitening matrix producing the whitened observations $\tilde{\mathbf{Z}}(\omega) = \tilde{\mathbf{W}}\tilde{\mathbf{X}}(\omega)$ which is obtained by diagonalizing the matrix $\widehat{\mathcal{R}}_{\tilde{\mathbf{X}}}(0)$. If $\tilde{\mathbf{U}}$ is the **unitary** matrix permitting us to **jointly block-diagonalize** several covariance matrices $\widehat{\mathcal{R}}_{\tilde{\mathbf{Z}}}(\nu)$ corresponding to different values of ν, then an estimate of the separating matrix $\tilde{\mathbf{A}}^+$, denoted $\tilde{\mathbf{A}}^+_{est}$, is given by

$$\tilde{\mathbf{A}}^+_{est} = \Re\{\tilde{\mathbf{U}}^H \tilde{\mathbf{W}}\} \simeq \tilde{\mathbf{P}}\tilde{\mathbf{\Omega}}\tilde{\mathbf{A}}^+, \tag{17}$$

where $\tilde{\mathbf{\Omega}}$ is a $(K+L) \times (K+L)$ *block-diagonal* real matrix, composed of N non-zero blocks, and $\tilde{\mathbf{P}}$ is a *permutation* matrix of these blocks. Because of working in the frequency domain, the matrix $\tilde{\mathbf{U}}^H\tilde{\mathbf{W}}$ may be complex, that is why $\tilde{\mathbf{A}}^+_{est}$ is equal to the real part of this matrix (see [6] for more details). After estimating the separating matrix $\tilde{\mathbf{A}}^+_{est}$ in the frequency domain, we can retrieve the generalized source vector $\tilde{\mathbf{s}}(n)$ in the time domain using Eq. (5) in the same manner as explained in Section 2 for *SOBI-C* [1]. Thus, the estimates of the vectors $\tilde{\mathbf{s}}(n)$ and $\mathbf{s}_j(n)$ are given, as in (11) and (12), respectively by

$$\tilde{\mathbf{s}}_{est}(n) = \tilde{\mathbf{A}}^+_{est}\tilde{\mathbf{x}}(n) \simeq \tilde{\mathbf{P}}\tilde{\mathbf{\Omega}}\tilde{\mathbf{s}}(n). \tag{18}$$

$$\forall j \in [1,N],\ \exists k \in [1,N]\ /\ \hat{\mathbf{s}}_j(n) \simeq \tilde{\mathbf{\Omega}}_k \mathbf{s}_j(n). \tag{19}$$

where $\tilde{\mathbf{\Omega}}_k$ are the non-zero blocks of the block-diagonal matrix $\tilde{\mathbf{\Omega}}$, of dimension $(K+L) \times (K+L)$, so that our *SOBI-F-C* algorithm finally provides $K+L$ filtered versions of each original source $s_j(n)$, exactly like other BSS approaches based on JBD of matrices (for example [1–3]).

3.2 Non-stationary and Autocorrelated Sources

If the original sources $s_j(n)$ are non-stationary and autocorrelated, then the related generalized sources $s_{jr}(n)$ are too. Our solution to cope with this problem is based on the following theorem proved in [6].

Theorem 2. *Let $u_p(t)$ $(p = 1, ..., \mathcal{N})$ be $\mathcal{N}$ autocorrelated, real, zero-mean, mutually uncorrelated random signals. Suppose $g(n)$ is a temporally uncorrelated, stationary random signal, independent from $u_p(n)$, $\forall p$. Then, the signals $u'_p(n) = g(n)u_p(n)$ $(p = 1, ..., \mathcal{N})$ are temporally and mutually uncorrelated. Moreover, each new source $u'_p(n)$ has the same normalized variance profile as the original source $u_p(n)$.*

Multiplying the generalized observation vector $\tilde{\mathbf{x}}(n)$ in (5) by a random signal $g(n)$ satisfying the conditions of Theorem 2, we obtain

$$\tilde{\mathbf{x}}'(n) = \tilde{\mathbf{A}}\tilde{\mathbf{s}}'(n), \tag{20}$$

where $\tilde{\mathbf{x}}'(n) = g(n)\tilde{\mathbf{x}}(n)$ and $\tilde{\mathbf{s}}'(n) = g(n)\tilde{\mathbf{s}}(n)$ are two vectors with respective components $x'_{il}(n) = g(n)x_{il}(n)$ and $s'_{jr}(n) = g(n)s_{jr}(n)$. According to Theorem 2, the new sources $s'_{jr}(n)$ are temporally uncorrelated and have the same normalized variance profiles as the old sources $s_{jr}(n)$. Moreover, the new sources are also mutually uncorrelated, i.e. the relation (8) is verified by the new sources $s'_{jr}(n)$ too. Thus, we obtain a new instantaneous mixture with the same mixing matrix $\tilde{\mathbf{A}}$, modeled by (20), which verifies the working hypotheses used in Section 3.1. By mapping this new mixture in the frequency domain, we can then estimate the separating matrix $\tilde{\mathbf{A}}^+$ exactly as in Section 3.1.

4 Simulation Results

In this section, we present our simulation results using three artificial FIR convolutive mixtures of two speech signals containing $N_e = 32768$ samples. The mixtures are generated using FIR filters of order $K \in \{1, 3, 7\}$. The coefficients $a_{ij}(k)$ of each transfer function $A_{ij}(z) = \sum_{k=0}^{K} a_{ij}(k)z^{-k}$ are generated randomly. For each value of K we choose in the model (2) the integer L equal to $2K$. This choice provides $M' = 6K$ generalized observations and $N' = 6K$ generalized sources so that the matrix $\tilde{\mathbf{A}}$ is square[4]. Since the speech signals are non-stationary and autocorrelated, we use the approach proposed in Section 3.2 which consists in transforming them into non-stationary and temporally uncorrelated signals. To this end, we multiply all the generalized observations $x_{il}(n)$ by an i.i.d., real, zero-mean and uniformly distributed signal, independent from the generalized sources. We compare our method with *SOBI-C* [1] and *BGML-C* [2] mentioned in Section 2. For our *SOBI-F-C* method as well as the *SOBI-C* method we consider 4 covariance matrices, while for the *BGML-C* method we consider 64 covariance matrices computed over 64 adjacent frames of 512 samples. To block-diagonalize these matrices, we use the orthogonal algorithm proposed by Févotte et al. in [3]. After the JBD stage, we obtain $K + L$ filtered versions of each initial source $s_j(n)$. Then, we use a blind deconvolution method proposed in which allows us to estimate each of the generalized sources $s_{jr}(n)$, and in particular each initial source $s_j(n)$, up to a scaling factor. Performance is measured using the Signal to Interference Ratio (SIR) defined as $SIR = \frac{1}{2}\sum_{j=1}^{2} SIR_j$ where

$$SIR_j = 10\log_{10}\left[\frac{E\{s_j(n)^2\}}{E\{(\hat{s}_j(n) - s_j(n))^2\}}\right], \quad j = 1, 2, \tag{21}$$

after normalizing the estimated sources $\hat{s}_j(n)$ so that they have the same variances and signs as the original sources $s_j(n)$. Table 1 presents the obtained results. It can be seen that our *SOBI-F-C* method is more efficient than *SOBI-C*.

[4] Having originally 3 FIR mixtures of 2 sources, i.e. $M = 3$ et $N = 2$, we obtain $M' = ML = 6K$ and $N' = N(K + L) = 6K$ after reformulating the problem as in (2).

Table 1. SIR in dB as a function of the parameter K for each tested method

Method	$SIR(dB)$, $K=1$	$SIR(dB)$, $K=3$	$SIR(dB)$, $K=7$
SOBI-F-C	42.8	33.4	24.6
BGML-C	35.7	30.9	23.9
SOBI-C	30.9	29.2	20.1

This result is not surprising because, in the time domain, the stationarity assumption required for *SOBI-C* is not verified by speech signals while in the frequency domain the frequency-domain transformed sources are wide-sense stationary. Our method also outperforms *BGML-C* even if the performances are comparable for $K=3$ and $K=7$. This may be explained by the fact that speech signals verify approximately the piecewise stationarity assumption required by *BGML-C*. We did not test the *TFBSS-C* method [3] because following the results reported in [3], *SOBI-C* outperforms this method for audio signals.

5 Conclusion and Perspectives

We proposed a new approach for separating convolutive mixtures of non-stationary sources. The approach starts by reformulating the problem in terms of an over-determined instantaneous model, then mapping it in the frequency domain to obtain new stationary frequency-domain signals. Afterwards, by jointly block diagonalizing several covariance matrices followed by a blind deconvolution, it provides an estimate of each original source up to a scaling factor. The main advantage of our approach compared to other methods based on the JBD of the matrices is that the piecewise stationarity or sparseness of the sources are not required for using it. The simulation results confirmed the better performance of our approach with respect to these methods. It is however interesting to evaluate the performance with more statistical tests[5], or with noisy mixtures.

References

1. Bousbia-Salah, H., Belouchrani, A., Abed-Meraim, K.: Blind separation of convolutive mixtures using joint block diagonalization. In: International Symposium on Signal Processing and its Applications (ISSPA), Malaysia, vol. 1, pp. 13–16 (August 2001)
2. Bousbia-Salah, H., Belouchrani, A., Abed-Meraim, K.: Blind separation of non stationary sources using joint block diagonalization. In: Proceedings of the 11th IEEE Signal Processing Workshop on Statistical Signal Processing, pp. 448–451 (August 2001)

[5] That is by changing all the related parameters : source signals, their number N, the number of mixtures M, order K of the FIR filters, the value of the integer L used for reformulating the model, etc.

3. Févotte, C.: Approche temps-fréquence pour la séparation aveugle de sources non-stationnaires. PHD Thesis, École Centrale de Nantes et Université de Nantes (October 2003), http://www.tsi.enst.fr/~cfevotte/
4. Belouchrani, A., Abed-Meraim, K., Cardoso, J.-F., Moulines, E.: A blind source separation technique based on second order statistics. IEEE Trans. on Signal Processing 45, 434–444 (1997)
5. Pham, D.-T., Cardoso, J.-F.: Blind separation of instantaneous mixtures of non stationary source. IEEE Trans. on Signal Processing 49(9), 1837–1848 (2001)
6. Hosseini, S., Deville, Y., Saylani, H.: Blind separation of linear instantaneous mixtures of non-stationary signals in the frequency domain. Signal Processing 89(5), 819–830 (2009)

Single Microphone Blind Audio Source Separation Using EM-Kalman Filter and Short+Long Term AR Modeling

Siouar Bensaid, Antony Schutz, and Dirk T.M. Slock*

EURECOM
2229 route des Crêtes, B.P. 193, 06904 Sophia Antipolis Cedex, France
{siouar.bensaid,antony.schutz,dirk.slock}@eurecom.fr
http://www.eurecom.fr

Abstract. Blind Source Separation (BSS) arises in a variety of fields in speech processing such as speech enhancement, speakers diarization and identification. Generally, methods for BSS consider several observations of the same recording. Single microphone analysis is the worst underdetermined case, but, it is also the more realistic one. In this article, the autoregressive structure (short term prediction) and the periodic signature (long term prediction) of voiced speech signal are modeled and a linear state space model with unknown parameters is derived. The Expectation Maximization (EM) algorithm is used to estimate these unknown parameters and therefore help source separation.

Keywords: Blind audio source separation, EM, Kalman, speech processing, autoregressive.

1 Introduction

Blind Source Separation is an important issue in audio processing. It helps solving "'the cocktail party problem"' where each speaker needs to be retrieved independently. Several works exploit the temporal structure of speech signal to help separation. In literature, three categories can be listed : The first exploits only the short term correlation in speech signal and models it with a short term Auto-Regressive (AR) process [2]. A second category models the quasi-periodicity of speech by introducing the fundamental frequency (or pitch) in the analysis [3,4]. Finally, few works combine the two aspects [5]. This article is classified in the last category. In [5], The problem is presented like an over-determined instantaneous model where the aim is to estimate jointly the long term (LT)and short term (ST) AR coefficients, as well as the demixing matrix in order to retrieve the speakers in a deflation scheme. An ascendant gradient algorithm is used to minimize the mean square of the total estimation error (short term and long term), and thus learn the parameters recursively. Our case

* EURECOM's research is partially supported by its industrial partners: BMW, Cisco Systems, France Télécom, Hitachi Europe, SFR, Sharp, ST Microelectronics, Swisscom, Thales. This research has also been partially supported by Project SELIA.

V. Vigneron et al. (Eds.): LVA/ICA 2010, LNCS 6365, pp. 106–113, 2010.

is more difficult, since only a single sensor is used. Therefore, the proposed model of speech propagation is rather simplified (the observation is the instantaneous sum of sources). Nevertheless, this model is still relevant in several scenarios. Using some mathematical manipulation, a state space model with unknown parameters is derived. Since the involved signals are Gaussians, Kalman filtering can be used in the EM algorithm (Expectation step) to estimate the state. This paper is organized as follows: The state space model is introduced in section 2. A small recapitulation of the EM-Kalman algorithm is presented in section 3 and the estimators' expressions are then computed. Numerical results are provided in section 4, and conclusions are drawn in section 5.

2 State Space Model Formulation

We consider the problem of estimating N_s mixed Gaussian sources. We use a voice production model [6], that can be described by filtering an excitation signal with long term prediction filter followed by a short term filter and which is mathematically formulated

$$\begin{aligned} y_t &= \sum_{k=1}^{N_s} s_{k,t} + n_t, \\ s_{k,t} &= \sum_{n=1}^{p_k} a_{k,n}\, s_{k,t-n} + \tilde{s}_{k,t} \\ \tilde{s}_{k,t} &= b_k\, \tilde{s}_{k,t-T_k} + e_{k,t} \end{aligned} \tag{1}$$

where

- y_t is the scalar observation.
- $s_{k,t}$ is the kth source at time t, an AR process of order p_k
- $a_{k,n}$ is the nth short term coefficient of the kth source
- $\tilde{s}_{k,t}$ is the short term prediction error of the kth source
- b_k is the long term prediction coefficient of the kth source
- T_k is the period of the kth source, not necessary an integer
- $\{e_{k,t}\}_{k=1..N_s}$ are the independent Gaussian distributed innovation sequences with variance ρ_k
- $\{n_t\}$ is a white Gaussian process with variance σ_n^2, independent of the innovations $\{e_{k,t}\}_{k=1..N_s}$

This model seems to describe more faithfully the speech signal, especially the voiced part (the most energetic part of speech). It is because it uses the short term auto-regressive model (AR) to describe the correlation between the signal samples jointly with the long term AR model to depict the harmonic structure of speech, rather than being restricted to just one of both [2,3]. Let $\mathbf{x}_{k,t}$ be the vector of length $(N+p_k+2)$, defined like $\mathbf{x}_{k,t} = [s_k(t)\ s_k(t-1)\cdots s_k(t-p_k-1) \mid \tilde{s}_k(t)\ \tilde{s}_k(t-1)\cdots\tilde{s}_k(t-\lfloor T_k\rfloor)\cdots\tilde{s}_k(t-N+1)]^T$. This vector can be written in terms of $\mathbf{x}_{k,t-1}$ as the following

$$\mathbf{x}_{k,t} = \mathbf{F}_k\, \mathbf{x}_{k,t-1} + \mathbf{g}_k\, e_{k,t} \tag{2}$$

where $\mathbf{g}_k$ is the $(N+p_k+2)$ length vector defined as $\mathbf{g}_k = [\,1\,0 \cdots 0 \mid 1\,0 \,\cdots\cdots\, 0]^T$. The second non null component is at the position (p_k+3). The $(N+p_k+2) \times (N+p_k+2)$ matrix $\mathbf{F}_k$ has got the following structure

$$\mathbf{F}_k = \begin{bmatrix} \mathbf{F}_{11,k} & \mathbf{F}_{12,k} \\ \mathbf{O} & \mathbf{F}_{22,k} \end{bmatrix}$$

where the $(p_k+2) \times (p_k+2)$ matrix $\mathbf{F}_{11,k}$, the $(p_k+2) \times N$ matrix $\mathbf{F}_{12,k}$ and the $N \times N$ matrix $\mathbf{F}_{22,k}$ are given by

$$\mathbf{F}_{11,k} = \begin{bmatrix} a_{k,1} \; a_{k,2} \; \cdots \; a_{k,p_k} \; 0 & 0 \\ & \vdots \\ I_{(p_k+1)} & \vdots \\ & \vdots \\ & 0 \end{bmatrix}$$

$$\mathbf{F}_{12,k} = \begin{bmatrix} 0 \cdots & (1-\alpha_k)\, b_k & \alpha_k\, b_k & 0 \cdots 0 \\ 0 \cdots & 0 & 0 & 0 \cdots 0 \\ \vdots \;\ddots & \vdots & \vdots & \vdots \;\; \vdots \\ 0 \cdots & 0 & 0 & 0 \cdots 0 \end{bmatrix}$$

$$\mathbf{F}_{22,k} = \begin{bmatrix} 0 \cdots \; (1-\alpha_k)\, b_k \;\; \alpha_k\, b_k \;\; 0 \cdots & 0 \\ & \vdots \\ I_{(N-1)} & \vdots \\ & \vdots \\ & 0 \end{bmatrix}$$

In matrices $\mathbf{F}_{12,k}$ and $\mathbf{F}_{22,k}$, the variable α_k, given by $\alpha_k = (1 - \frac{\lfloor (T_k) \rfloor}{T_k})$, is present to consider the case where pitches are not integer. It is noteworthy that the choice of the $\mathbf{F}_{22,k}$ matrix size N should be done carefully. In fact, the value of N should be superior to the maximum value of pitches T_k in order to detect the long-term aspect. It can be noticed that the coefficients $(1-\alpha_k)\, b_k$ and $\alpha_k\, b_k$ are situated respectively in the $\lfloor T_k \rfloor^{th}$ and $\lceil T_k \rceil^{th}$ columns of $\mathbf{F}_{22,k}$ and $\mathbf{F}_{12,k}$. Since N_s sources are present, we introduce the vector $\mathbf{x}_t$ that consists of the concatenation of the $\{\mathbf{x}_{k,t}\}_{k=1:N_s}$ vectors ($\mathbf{x}_t = [\mathbf{x}_{1,t}^T \; \mathbf{x}_{2,t}^T \; \cdots \; \mathbf{x}_{N_s,t}^T]^T$) which results in the time update equation 3. Moreover, by reformulating the expression of $\{y_t\}$, we introduce the observation equation 4. We obtain the following state space model

$$\mathbf{x}_t = \mathbf{F}\, \mathbf{x}_{t-1} + \mathbf{G}\, \mathbf{e}_t \tag{3}$$

$$y_t = \mathbf{h}^T \mathbf{x}_t + n_t \tag{4}$$

where

- $\mathbf{e}_t = [e_{1,t}\; e_{2,t} \; \cdots \; e_{N_s,t}]^T$ is the $N_s \times 1$ column vector resulting of the concatenation of the N_s innovations. Its covariance matrix is the $N_s \times N_s$ diagonal matrix $\mathbf{Q} = \mathrm{diag}(\rho_1, \cdots, \rho_{N_s})$.

- $\mathbf{F}$ is the $\sum_{k=1}^{N_s}(p_k+N+2) \times \sum_{k=1}^{N_s}(p_k+N+2)$ block diagonal matrix given by $\mathbf{F} = \text{blockdiag}(\mathbf{F}_1, \cdots, \mathbf{F}_{N_s})$.
- $\mathbf{G}$ is the $\sum_{k=1}^{N_s}(p_k+N+2) \times N_s$ matrix given by $\mathbf{G} = \text{block diag}(\mathbf{g}_1, \cdots, \mathbf{g}_{N_s})$
- $\mathbf{h}$ is the $\sum_{k=1}^{N_s}(p_k+N+2) \times 1$ column vector given by $\mathbf{h} = [\mathbf{h}_1^T \cdots \mathbf{h}_{N_s}^T]^T$ where $\mathbf{h}_i = [1\ 0 \cdots 0]^T$ of length $(N+p_k+2)$.

It is obvious that the linear dynamic system derived before depends on unknown parameters recapitulated in the variable $\theta = \left\{ \{a_{k,n}\}_{\substack{k\in\{1,\ldots,N_s\} \\ n\in\{1,\ldots,p_k\}}}, \{b_k\}_{k\in\{1,\ldots,N_s\}}, \{\rho_k\}_{k\in\{1,\ldots,N_s\}}, \sigma_n^2 \right\}$. Hence, a joint estimation of sources (the state) and θ is required. In literature([11,10,7]), the EM-Kalman algorithm presents an efficient approach for estimating iteratively parameters and its convergence to the Maximum Likelihood solution is proved [9]. In the next section, the application of this algorithm to our case is developed.

3 EM-Kalman Filter

The EM-Kalman algorithm permits to estimate iteratively parameters and sources by alternating two steps : E-step and M-step [9]. In the M-step, an estimate of the parameters $\hat{\theta}$ is computed. In our problem, there are two types of parameters: the parameters of the time update equation 3 which consist on the short term and long term coefficients and the innovation power of all the N_s sources, and one parameter of the observation equation 4, the observation noise power. From the state space model presented in the first part, and for each source k, the relation between the innovation process at time $t-1$ and the LT+ST coefficients could be written as

$$e_{k,t-1} = \mathbf{v}_k^T \breve{\mathbf{x}}_{k,t-1} \tag{5}$$

where $\mathbf{v}_k = [1\ -a_{k,1} \cdots -a_{k,p_k}\ -(1-\alpha_k)\ b_k\ -\alpha_k\ b_k]^T$ is a $(p_k+3)\times 1$ column vector and $\breve{\mathbf{x}}_{k,t-1} = [s_k(t-1,\theta) \cdots s_k(t-p_k-1,\theta)\ \tilde{s}_k(t-\lfloor T_k \rfloor -1,\theta)\ \tilde{s}_k(t-\lfloor T_k \rfloor -2,\theta)]^T$ is called the partial state deduced from the full state $\mathbf{x}_t$ with the help of a selection matrix $\mathbf{S}_k$. This lag of one time sample between the full and partial state is justified later. After multiplying (5) by $\breve{\mathbf{x}}_{k,t-1}^T$ in the two sides, applying the operator $E\{\ |y_{1:t}\}$ and doing a matrix inversion, the following relation between the vector of coefficients and the innovation power is deduced

$$\mathbf{v}_k = \rho_k \mathbf{R}_{k,t-1}^{-1}[1,0\cdots 0]^T \tag{6}$$

where the covariance matrix $\mathbf{R}_{k,t-1}$ is defined as $E\left\{\breve{\mathbf{x}}_{k,t-1}\breve{\mathbf{x}}_{k,t-1}^T|y_{1:t}\right\}$. It is important to notice that the estimation of $\mathbf{R}_{k,t-1}$ is done using observations till time t, which consists on a fixed-lag smoothing treatment with $lag = 1$. As mentioned previously, the relation between the partial state at time $t-1$ and the full state at time t is $\breve{\mathbf{x}}_{k,t-1} = \mathbf{S}_k\mathbf{x}_t$. This key relation is used in the partial state covariance matrix computation

$$\mathbf{R}_{k,t-1}^{-1} = \mathbf{S}_k E\left\{\mathbf{x}_t\mathbf{x}_t^T|y_{1:t}\right\}\mathbf{S}_k^T \tag{7}$$

Notice here the transition from the fixed lag smoothing with the partial state to the simple filtering with the full state. This fact justifies the selection of the partial state at time $t-1$ from the full state at time t. This selection is possible due to the augmented form matrix F_k or more precisely $\mathbf{F}_{11,k}$. The innovation power is simply deduced as the first component of the matrix $\mathbf{R}_{k,t-1}^{-1}$. The estimation of the observation noise power σ_n^2 is achieved by maximizing the loglikewood function $\log P\left(y_t|\mathbf{x}_t, \sigma_n^2\right)$ relative to σ_n^2. The optimal value can be easily proved equal to

$$\hat{\sigma_n^2}^{(t)} = y_t^2 - 2y_t\mathbf{h}^T\hat{\mathbf{x}}_{t|t} + \mathbf{h}^T\left(\hat{\mathbf{x}}_{t|t}\hat{\mathbf{x}}_{t|t}^T + \mathbf{P}_{t|t}\right)\mathbf{h} \tag{8}$$

The time index in (t) in $\hat{\sigma_n^2}^{(t)}$ is to denote the iteration number. The computation of the partial covariance matrix $\mathbf{R}_{k,t-1}$ is achieved in the $E-step$. This matrix depends on the quantity $E\left\{\mathbf{x}_{k,t}\mathbf{x}_{k,t}^T|y_{1:t}\right\}$ the definition of which is

$$E\left\{\mathbf{x}_t\mathbf{x}_t^T|y_{1:t}\right\} = \hat{\mathbf{x}}_{t|t}\hat{\mathbf{x}}_{t|t}^T + \mathbf{P}_{t|t} \tag{9}$$

where the quantities $\hat{\mathbf{x}}_{t|t}$ and $\hat{\mathbf{P}}_{t|t}$ are respectively the full estimated state and the full estimation error covariance computed using Kalman filtering equations.

Adaptive EM Kalman Algorithm

– E-Step. Estimation of the sources covariance

$$\begin{aligned}
\mathbf{K}_t &= \mathbf{P}_{t|t-1}\mathbf{h}(\mathbf{h}^T\mathbf{P}_{t|t-1}\mathbf{h} + \hat{\sigma_n^2})^{-1}\\
\hat{\mathbf{x}}_{t|t} &= \hat{\mathbf{x}}_{t|t-1} + \mathbf{K}_t(\mathbf{y}_t - \mathbf{h}^T\hat{\mathbf{x}}_{t|t-1})\\
\mathbf{P}_{t|t} &= \mathbf{P}_{t|t-1} - \mathbf{K}_t\mathbf{h}^T\mathbf{P}_{t|t-1}\\
\hat{\mathbf{x}}_{t+1|t} &= \hat{\mathbf{F}}\hat{\mathbf{x}}_{t|t}\\
\mathbf{P}_{t+1|t} &= \hat{\mathbf{F}}\mathbf{P}_{t|t}\hat{\mathbf{F}}^T + \mathbf{G}\hat{\mathbf{Q}}\mathbf{G}^T
\end{aligned}$$

– M-Step. Estimation of the AR parameters using linear prediction. $k = 1, ..., N_s$

$$\begin{aligned}
\hat{\mathbf{s}}_{k,t} &= (\hat{\mathbf{x}}_{k,t|t})_{[1,1]}\\
\mathbf{R}_{k,t-1} &= \lambda\mathbf{R}_{k,t-2} + (1-\lambda)\mathbf{S}_k(\mathbf{x}_{t|t}\mathbf{x}_{t|t}^T + \mathbf{P}_{t|t})\mathbf{S}_k^T\\
\rho_k^{(t)} &= (\mathbf{R}_{k,t-1}^{-1})_{(1,1)}^{-1}\\
\mathbf{v}_k^{(t)} &= \rho_k\mathbf{R}_{k,t-1}^{-1}[1, 0\cdots 0]^T\\
\hat{\sigma_n^2}^{(t)} &= y_t^2 - 2y_t\mathbf{h}^T\hat{\mathbf{x}}_{t|t} + \mathbf{h}^T\left(\hat{\mathbf{x}}_{t|t}\hat{\mathbf{x}}_{t|t}^T + \mathbf{P}_{t|t}\right)\mathbf{h}
\end{aligned}$$

The adaptive algorithm is presented as Algorithm 1. The algorithm needs an accurate initialization, which will be discussed afterward. In the algorithm $\hat{\mathbf{s}}_{k,t}$ is the estimation of the source k at time t.

The estimation of the pitches $\{T_k\}_{k=1:N_s}$ is done along with this algorithm using a multipitch estimation algorithm [12].

4 Simulations

In this section, we present some results of our source separation algorithm. We assume that the maximum number of sources is known. We limit our analysis to the case of a mixture of two simultaneous sources corrupted by white noise. In the most energetic parts of the mixture, the inconstant SNR is about 20 dB as shown in Fig. 1. We work with real speech data to which we add artificially the observation noise. The mixture consists of two voiced speech signals and is of 10 s duration. The parameters are initialized randomly, except the periods where we use a multipitch algorithm [12] running in parallel to our main algorithm. The estimated periods from the multipitch algorithm are updated in the main algorithm every 64 ms. We do two experiments. The first one is the filtering case in which all the parameters are initialized with values close to the true ones. The results are close to perfect and are shown in Fig 2. In the second experiment, the parameters are initialized randomly and estimated adaptively in the M-Step. We can see the results in Fig 3. The separation looks not very good but, when listening to the estimated sources, we find that they are under-estimated, leading to a mixture of the original sources in which the interferences are reduced. This is, in part, due to the fact that at a given moment we don't know the number of sources, so even when only one source is present, the algorithm seeks to estimate two sources. During the separation process, the estimated correlations are still polluted by the other source but the desired source is enhanced. The results can be listened on the first author personal page [1].

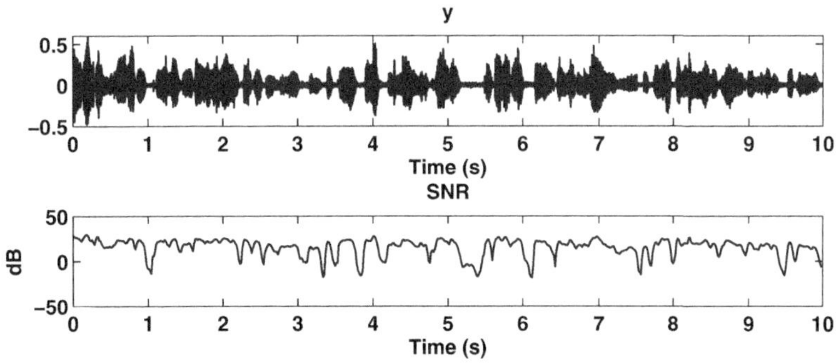

Fig. 1. Mixture and SNR evolution

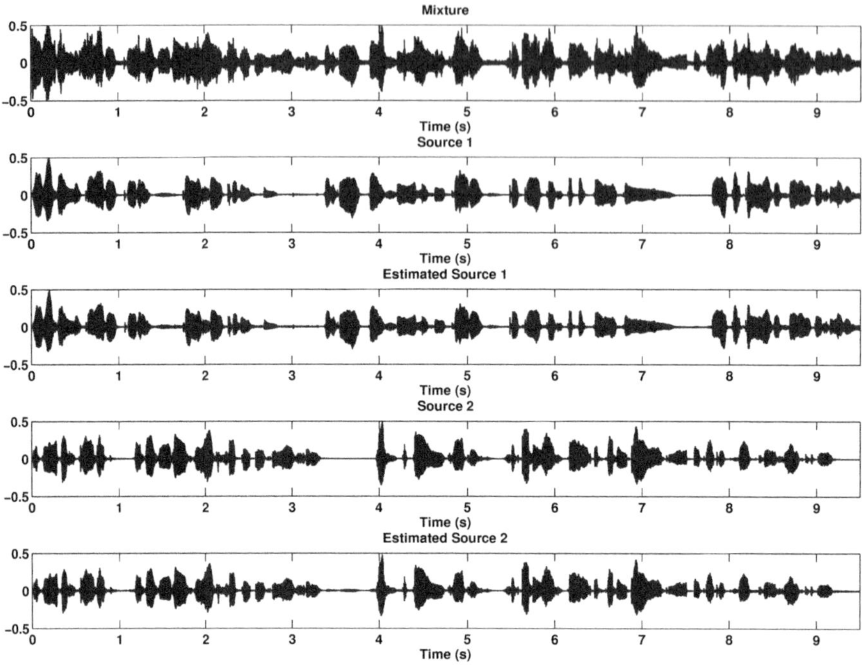

Fig. 2. Source separation with fixed and known parameters

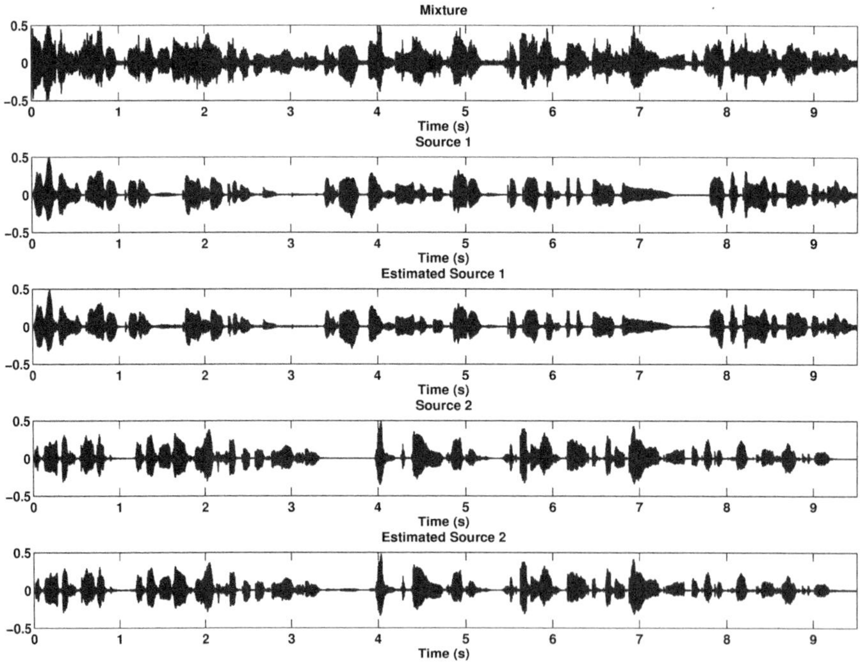

Fig. 3. Source separation adaptive estimation of the parameters

5 Conclusion

In this paper we use the adaptive EM-Kalman algorithm for the blind audio source separation problem. The model takes into account the different aspects of speech signals production and sources are jointly estimated. The traditional smoothing step is included into the algorithm and is not an additional step. Simulations show the potential of the algorithm for real data. Yet, this performance depends a lot on the multipitch estimation quality. An error on tracking the pitches may induce the performance decreasing drastically. This work would be more complete if an other process aiming to estimate the number of active sources is working in parallel.

References

1. http://www.eurecom.fr/~bensaid/ICA10
2. Cichocki, A., Thawonmas, R.: On-line algorithm for blind signal extraction of arbitrarily distributed, but temporally correlated sources using second order statistics Neural Process. Neural Process. Lett. 12(1), 91–98 (2000)
3. Barros, A.K., Cichocki, A.: Extraction of specific signals with temporal structure. Neural Comput. 13(9), 1995–2003 (2001)
4. Tordini, F., Piazza, F.: A semi-blind approach to the separation of real world speech mixtures. In: Proceedings of the 2002 International Joint Conference on Neural Networks. IJCNN 2002, vol. 2, pp. 1293–1298 (2002)
5. Smith, D., Lukasiak, J., Burnett, I.: Blind speech separation using a joint model of speech production. IEEE Signal Processing Letters 12(11), 784–787 (2005)
6. Chu, W.C.: Speech coding algorithms-foundation and evolution of standardized coders. John Wiley and Sons, NewYork (2003)
7. Feder, M., Weinstein, E.: Parameter estimation of superimposed signals using the EM algorithm. IEEE Trans. Acoust., Speech, Signal Processing 36, 477–489 (1988)
8. Gannot, S., Burshtein, D., Weinstein, E.: Iterative-batch and sequential algorithms for single microphone speech enhancement. In: ICASSP 1998, pp. 1215–1218. IEEE, Los Alamitos (1998)
9. Dempster, A.P., Laird, N.M., Rubin, D.B.: Maximum Likelihood from Incomplete Data via the EM Algorithm. Journal of the Royal Society B 39, 1–38 (1977)
10. Gao, W., Tsai, S., Lehnert, J.: Diversity combining for ds/ss systems with time-varying, correlated fading branches. IEEE Transactions on Communications 51(2), 284–295 (2003)
11. Couvreur, C., Bresler, Y.: Decomposition of a mixture of Gaussian AR processes, Acoustics, Speech, and Signal Processing. In: 1995 International Conference on ICASSP 1995, vol. 3, pp. 1605–1608 (1995)
12. Christensen, M., Jakobsson, A., Juang, B.H.: Multi-pitch estimation, Morgan & Claypool (2009)

The 2010 Signal Separation Evaluation Campaign (SiSEC2010): Audio Source Separation

Shoko Araki[1], Alexey Ozerov[2], Vikrham Gowreesunker[3], Hiroshi Sawada[1], Fabian Theis[4], Guido Nolte[5], Dominik Lutter[4], and Ngoc Q.K. Duong[2]

[1] NTT Communication Science Labs., NTT Corporation, Japan
[2] INRIA, Centre Inria Rennes - Bretagne Atlantique, France
[3] DSPS R&D Center, Texas Instruments Inc., USA
[4] IBIS, Helmholtz Zentrum München, Germany
[5] Fraunhofer Institute FIRST IDA, Germany

Abstract. This paper introduces the audio part of the 2010 community-based Signal Separation Evaluation Campaign (SiSEC2010). Seven speech and music datasets were contributed, which include datasets recorded in noisy or dynamic environments, in addition to the SiSEC2008 datasets. The source separation problems were split into five tasks, and the results for each task were evaluated using different objective performance criteria. We provide an overview of the audio datasets, tasks and criteria. We also report the results achieved with the submitted systems, and discuss organization strategies for future campaigns.

1 Introduction

SiSEC2010 aims to be a large-scale regular campaign that builds on the experience of previous evaluation campaigns (e.g., the MLSP'05 Data Analysis Competition[1], the PASCAL Speech Separation Challenge [1], and the Stereo Audio Source Separation Evaluation Campaign (SASSEC) [2]), and the first community-based Signal Separation Evaluation Campaign (SiSEC2008) [3]. The unique aspect of this campaign is that SiSEC is not a competition but a scientific evaluation from which we can draw rigorous scientific conclusions.

This article introduces the audio part of SiSEC2010. In response to the feedback received at SiSEC2008, SiSEC2010 was designed to contain more realistic, and consequently, more challenging datasets, which had not previously been proposed for a large scale evaluation. Such datasets include recordings made in more reverberant rooms, under diffused noise conditions, or under dynamic conditions. We also repeated some of the typical tasks employed in SiSEC2008 (e.g, underdetermined and determined mixtures) with some fresh datasets.

Datasets and tasks are specified in Section 2 and the obtained outcomes in Section 3. Due to the variety of the submissions, we focus on the general outcomes of the campaign and ask readers to refer to `http://sisec.wiki.irisa.fr/` for further detail.

[1] `http://mlsp2005.conwiz.dk/index.php@id=30.html`

V. Vigneron et al. (Eds.): LVA/ICA 2010, LNCS 6365, pp. 114–122, 2010.

2 Specifications

This section describes the datasets, tasks and evaluation criteria, which were specified in a collaborative fashion. A few initial specifications were first suggested by the organizers. Potential participants were then invited to provide feedback and contribute additional specifications. All materials are available at `http://sisec.wiki.irisa.fr/`.

2.1 Datasets

The data consisted of audio signals spanning a range of mixing conditions. The channels $x_i(t)$ $(1 \leq i \leq I)$ of each mixture signal were generally obtained as [3]: $x_i(t) = \sum_{j=1}^{J} s_{ij}^{\text{img}}(t)$, where $s_{ij}^{\text{img}}(t) = \sum_{j=1}^{J} \sum_{\tau} a_{ij}(\tau) s_j(t-\tau)$ is the *spatial image* of source j $(1 \leq j \leq J)$ on channel i, namely the contribution of source i to the mixture in channel j, $s_j(t)$ are source signals, and a_{ij} are mixing gains. Seven distinct datasets were provided for SiSEC2010:

D1 Under-determined speech and music mixtures
This dataset includes dataset D1 from SiSEC2008 [3], and a fresh dataset consisting of 30 stereo mixtures of three to four audio sources of 10 s duration, sampled at 16 kHz. The room reverberation time (RT) for the fresh dataset was 130 ms or 380 ms.

D2 Determined and over-determined speech and music mixtures
For this category, SiSEC2010 had the following four datasets.

D2-1 Determined and over-determined speech and music mixtures
This dataset contains two sets of 2×2 (#sources $\times$ #microphones), 3×3 and 4×4 mixtures. The RT of the recording room was about 500 ms, and microphones were arranged as a linear array with a spacing of approximately 10 cm. This dataset also includes dataset D2 from SiSEC2008 [3], which consists of 21 four-channel recordings of two to four speech or music sources acquired in four different rooms.

D2-2 Robust blind linear/non-linear separation of short two-source-two-microphone recordings
This dataset consists of 36 mixtures of different sources, located at three different positions in two environments. Each recording is 1 second long. The mixtures are six combinations consisting of a target speech signal and a jammer source (male/female speech, sneeze, laugh, glass break or TV sport noise). The data were recorded with directional microphones that were 8 cm apart, in an ordinary living room or study room.

D2-3 Overdetermined speech and music mixtures for human-robot interaction
These data include 27 recordings of three sources, which were recorded with five microphones attached to a dummy head. We consider three different microphone configurations, in three rooms: an anechoic laboratory room, a fully equipped office and a cafeteria. The source signals include male and female speech and music.

D2-4 Determined convolutive mixtures under dynamic conditions
This dataset consists of two kinds of scenarios. One comprises the short (1-2 seconds) mixtures of two sources obtained with a stereo microphone (**D2-4i**). The other is a sequence of audio mixtures obtained by the random combination of source locations and utterances (**D2-4ii**). Here, up to two sources are active at the same time. The components are generated by convolving random utterances with measured impulse responses, which were measured in a real room (RT≈700-800 ms). The microphone spacings were 2, 6, and 10 cm.

D3 Professionally produced music recordings
This dataset contains five stereo music signals sampled at 44.1 kHz, which include the same data as these in D4 used in SiSEC2008 [3], and new recordings for SiSEC2010. In addition to 20-second snips to be separated, full-length recordings are provided as well. The mixtures were created by sound engineers, and the ways of mixing and the mixing effects applied are unknown.

D4 Source separation in the presence of real-world background noise
These data consist of 80 multichannel speech mixtures in the presence of several kinds of real-world diffused noise. Noise signals were recorded in three different real-world noise environments (in a subway car, cafeterias, and squares), and each noise signal was recorded at two different microphone positions, the center or corner of the environment. There were one or three sources. Two types of microphone arrays, a stereo or a 4-element uniform linear array, were employed.

Datasets D1, D3 and D4 include both test and development data. The true source signals and source positions underlying the test data were hidden to the participants, while they were provided for the development data. The true number of speech/music sources was always available.

2.2 Tasks

We specified the following five tasks:

- T1 Source counting
- T2 Mixing system estimation
- T3 Source signal estimation
- T4 Source spatial image estimation
- T5 Source DOA estimation

These tasks consist of finding, respectively: (T1) the number of sources J, (T2) the mixing gains a_{ij} or the discrete Fourier transform $a_{ij}(\nu)$ of the mixing filters, (T3) the source signals $s_j(t)$, (T4) the spatial images $s_{ij}^{\mathrm{img}}(t)$ of the sources for all channels i, and (T5) the direction of arrival (DOA) of each source. Participants were asked to submit the results of their systems for T3 and/or T4, and optionally for T1 and/or T2 and/or T5.

Two oracle systems were also considered for benchmarking task T4: ideal binary masking over a short-time Fourier transform (STFT) [4] (O1) and over a cochleagram [5] (O2). These systems require true source spatial images and provide upper performance bounds for binary masking-based systems.

Table 1. Average performance for tasks T3 or T4 for instantaneous dataset D1. Figures relate to T4 when the ISR is reported and to T3 otherwise.

dataset	Test								Test2					
System	[9]		[10][2]		O1		O2		[9]		O1		O2	
SDR OPS	13.5	60.7	7.7	36.2	10.5	55.0	8.1	39.3	12.8	53.2	9.3	40.5	8.4	33.9
ISR TPS	24.0	75.2		58.3	20.0	68.1	14.4	49.4	22.8	70.6	17.0	55.5	14.4	46.4
SIR IPS	20.5	77.1	18.2	67.8	21.6	79.3	17.4	71.4	19.8	70.7	18.7	65.6	17.0	61.8
SAR APS	14.8	73.2	8.4	36.9	11.5	60.3	9.1	45.5	14.1	64.8	10.2	49.6	9.6	35.7

Table 2. Average performance for task T4 for convolutive dataset D1. Correlation between the classical (SDRs etc.) and the auditory-motivated measures (OPS etc.) was 0.6-0.68.

System	Test:RT=130ms				Test:RT=250ms				Test2:RT=130ms				Test2:RT=380ms			
	SDR	ISR	SIR	SAR	SDR	ISR	SIR	SAR	SDR	ISR	SIR	SAR	SDR	ISR	SIR	SAR
	OPS	TPS	IPS	APS	OPS	TPS	IPS	APS	OPS	TPS	IPS	APS	OPS	TPS	IPS	APS
[9][2]	2.6	7.3	4.8	7.5	1.4	5.6	2.5	7.4	2.9	8.6	6.3	9.2	0.6	5.0	0.5	6.8
	32.2	55.2	50.4	59.7	25.7	46.2	41.0	52.5	34.8	58.4	54.2	60.6	15.8	38.3	27.2	41.1
[11]	5.3	10.2	9.9	7.1	3.9	8.5	7.3	6.9	3.6	8.4	6.5	7.1	2.1	5.9	3.3	6.0
	25.7	53.8	49.8	40.4	21.5	50.1	43.7	35.6	20.5	45.2	37.2	40.0	11.6	36.2	22.9	26.2
[12][2,3]	3.1	6.4	3.8	7.9	1.9	5.4	1.9	8.3	2.5	6.2	3.4	10.1	0.2	3.4	-0.8	8.2
	30.3	55.2	44.4	64.6	21.2	45.3	29.8	59.8	27.9	52.1	36.3	76.5	14.3	32.3	14.5	68.5
[13][3]	2.7	6.2	4.1	8.3	1.7	5.5	2.7	8.8	3.2	6.7	4.3	9.1	0.9	4.0	-0.6	7.4
	33.5	55.0	45.1	78.0	27.5	48.3	36.2	77.0	32.9	56.9	49.0	61.7	11.7	31.3	13.8	50.0
O1	9.7	18.3	19.9	10.2	8.7	16.2	19.4	10.4	10.2	18.1	19.6	11.0	9.2	16.8	18.5	9.9
	52.4	68.9	79.5	57.2	50.6	63.8	78.3	56.4	52.0	59.8	77.9	59.2	41.6	50.0	72.5	48.8
O2	6.9	12.1	16.5	7.6	6.6	11.5	16.0	7.9	7.3	13.6	16.0	7.9	6.2	11.8	14.4	6.5
	34.7	48.2	69.3	39.6	31.8	41.9	65.1	37.9	27.5	41.0	61.3	32.2	18.2	28.2	49.6	20.7

2.3 Evaluation Criteria

Task T2 was designed to be evaluated by using the mixing error ratio (MER), which was proposed in SiSEC2008 [3]. However, because we had no entrant for task T2, we skipped the T2 evaluation.

Tasks T3 and T4 were evaluated via the criteria in the BSS_EVAL [6,2], termed the signal to distortion ratio (SDR), source image to spatial distortion ratio (ISR), signal to interference ratio (SIR) and signal to artifacts ratio (SAR).

In addition, new auditory-motivated objective measures [7] were used to assess the quality of the estimated signals for T3 and T4 in the stereo cases. Four performance measures akin to SDR, ISR, SIR and SAR are given: overall perceptual score (OPS), target-related perceptual score (TPS), interference-related perceptual score (IPS) and artifact-related perceptual score (APS). Here, a new method for estimating the distortion components is employed based on a gammatone filterbank, and the salience of the target, interference and artifact distortions are calculated by using the PEMO-Q measure [8]. It has been confirmed

[2] The system is an extended version of the reference.
[3] Figure computed by averaging over an incomplete set of mixtures or sources.

that these auditory-motivated measures improve the correlation to subjective measures compared to classical SDR, ISR, SIR and SAR [7]. The new measures are expressed in terms of a figure between 0 and 100 (not in dB).

Task T5 was evaluated from the absolute difference between the true and estimated DOAs.

3 Results

A total of 38 submissions received for the audio tasks. Their average performance values are given in Tables 1 to 8. The details and all the results are available at `http://sisec.wiki.irisa.fr/`. It should be noted that the presented values are the absolute values, not the improvements from the values for mixtures. Although a close analysis of each table is beyond the scope of this paper, what we observed was following. We can obtain good results for instantaneous/anechoic mixtures (e.g., Tables 1 and 5), however, the separation of reverberant mixtures still remains challenging, especially in underdetermined scenarios (e.g., Table 2). The realistic datasets (D2-2, D2-4 and D4) attracted a relatively large number of participants as shown in Tables 4, 6 and 8.

Some tasks were evaluated with the new auditory-motivated objective measures (e.g., Tables 1, 2, 7 and 8). Sometimes, the performance tendency was the opposite of that with BSS_EVAL. More detailed investigations are required, however, it seems that binary mask based methods tend to achieve a poorer grade than non-binary mask approaches.

Table 3. Average SIR for task T3 over dataset D2-1

System	Cushioned rooms			Office/lab rooms			Conference room			New office room		
	$J=2$	$J=3$	$J=4$	$J=2$	$J=3$	$J=4$	$J=2$	$J=3$	$J=4$	$J=2$	$J=3$	$J=4$
[14]	8.0[3]	6.8[3]	4.5[3]							11.6	7.8	9.0
[15]	3.2	0.9	0.1	9.0	2.0	-2.0	4.3	-0.3	-3.1	5.2	2.6	1.5
[15][2]	4.2	4.0	4.4	6.4	4.4	-1.2	4.9	0.9	-2.2	9.3	6.0	4.7
[16]										13.7	10.4	10.0

Table 4. Average performance for dataset D2-2. SIR* and SDR* in [17] for linear systems were also evaluated according to the dataset providers' proposal.

System	speech+speech					speech+(sneeze or laugh)					speech+(glass or TV noise)				
	SIR*	SDR*	SDR	SIR	SAR	SIR*	SDR*	SDR	SIR	SAR	SIR*	SDR*	SDR	SIR	SAR
[18]	8.5	5.0	3.7	8.7	18.6	6.1	3.3	1.9	6.8	16.7	9.4	5.3	4.7	9.8	19.1
[19]	8.1	0.8	-0.9	8.4	23.2	6.3	0.0	-1.8	7.3	27.2	11.3	0.6	-1.2	11.3	18.4
[20]	7.9	6.4	5.9	9.4	9.7	7.4	5.0	3.5	8.6	7.8	8.1	5.6	5.6	9.5	9.3
[21]	11.5	8.2	7.7	12.2	18.8	10.6	6.2	4.0	11.3	17.0	9.9	6.4	6.1	10.7	15.5
[21][2]			3.1	16.3	10.1			2.7	14.0	10.3			2.6	12.8	8.1
[16]			2.0	5.5	10.5			4.5	9.4	12.3			0.4	2.5	8.0
[14]			0.6	6.2	5.0			0.5	7.0	3.7			-0.6	5.5	4.3

Table 5. Average SIR for dataset D2-3 for three microphone configurations C1-C3

	Anechoic			Cafeteria			Office		
System	C1	C2	C3	C1	C2	C3	C1	C2	C3
[16]	24.6	23.4	22.8	9.2	15.0	12.9	11.9	12.7	13.3
[14]	5.5	13.6	12.5	3.7	4.8	5.7	2.5	3.6	4.3

Table 6. Average performance for dataset D2-4. For D2-4ii, the evaluation was performed only considering the segments where two sources overlap.

dataset D2-4i								dataset D2-4ii			
System	SDR	SIR	SAR	System	SDR	SIR	SAR	System	SDR	SIR	SAR
[22]	3.2	9.9	5.2	[23]	3.2	8.1	6.1	[23]	2.5	8.6	5.0
[22][2]	4.2	11.7	5.8	[21]	5.6	11.3	7.6	[21]	4.6	10.7	6.8
[22][2]	3.7	10.6	5.5	[21][2]	6.2	13.8	7.4	[24]	4.5	9.9	7.2
								[25][2]	2.1	9.2	3.9

Table 7. Average performance for dataset D3

System	Development data								Test data							
	SDR	ISR	SIR	SAR	OPS	TPS	IPS	APS	SDR	ISR	SIR	SAR	OPS	TPS	IPS	APS
[26]	2.1	6.5	7.9	2.9	26.6	49.9	51.5	43.7	1.3	6.7	4.7	0.9	31.7	49.0	49.9	46.3
[27][2,3]	0.9	11.2	∞	-0.8	24.9	56.4	89.6	6.0	-1.1	7.5	∞	-2.1	21.3	39.0	89.6	3.7
[28]	3.3	8.0	7.2	5.3	17.3	45.0	27.4	48.1	2.2	6.2	6.2	3.9	18.7	40.9	30.4	45.4
O1	6.5	18.3	20.0	6.7	31.8	64.3	70.0	30.7	5.8	20.8	21.2	5.6	24.9	53.6	62.0	23.2
O2	3.5	8.4	17.0	3.3	29.2	47.5	63.8	30.2	2.9	8.9	14.9	1.7	20.9	34.2	53.0	21.8

Table 8. Average performance for source separation of dataset D4. Figures relate to T4 when the ISR is reported and to T3 otherwise. [29], [14], [30] and [31] also addressed task T5. Results can be found at `http://sisec.wiki.irisa.fr/`.

System	2-ch×1-src				2-ch×3-srcs				4-ch×1-src				4-ch×3-srcs			
	SDR	ISR	SIR	SAR	SDR	ISR	SIR	SAR	SDR	ISR	SIR	SAR	SDR	ISR	SIR	SAR
	OPS	TPS	IPS	APS	OPS	TPS	IPS	APS	OPS	TPS	IPS	APS	OPS	TPS	IPS	APS
[13]	2.7	16.1	4.4	11.9	1.3	7.0	1.9	7.7	3.0	15.7	4.0	12.8	2.6	7.7	4.3	9.1
	36.3	69.9	45.2	88.3	18.6	42.3	23.7	63.4								
[29]	-9.3	-7.2	14.1	8.0					-8.1	-4.3	12.2	8.1				
	33.5	49.9	55.5	53.8												
[14]	3.2		∞	3.2	0.5		5.2	5.0	7.4		∞	7.4	4.8		13.7	6.3
[30][2]	-12.2	-7.7	6.2	8.8												
	19.3	35.2	34.8	40.6												
[31][2]	-8.6	-3.2	7.5	5.4												
	16.3	39.4	35.2	28.3												
[16]	2.5	16.0	4.0	12.8					7.5	17.1	10.4	14.3	3.5	13.8	5.3	11.2
	38.4	73.4	51.2	78.4												

4 Conclusion

This paper presented the specifications and results of SiSEC2010. We hope that SiSEC2010 will provide a common platform for the source separation research field. This time, we welcomed all proposed tasks and datasets and finally we had seven datasets. This increase in the number of active participants is the product of a series of evaluation campaigns. On the other hand, we think that seven datasets may have been too large to collect sufficient number of participants for each task, and to evaluate all the submissions in detail. We may need a framework for pre-selecting the task/dataset proposals. In addition, perhaps it is time to reorganize the specifications and datasets in a series of SiSECs for prospective evaluation campaigns and future source separation research. We invite all willing participants to join a continuous collaborative discussion on the future of source separation evaluation.

Acknowledgments. We thank all the participants, as well as B. Lösch, Z. Koldovsky, P. Tichavsky, M. Durkovic, M. Kleinsteuber, M. Rothbucher, H. Shen, F. Nesta, O. Le Blouch, N. Ito, L.C. Parra, M. Dyrholm, K.E. Hild II, M. Vinyes Raso and J. Woodruff, for sharing their datasets and evaluation codes. Our special thanks go to V. Emiya, who helped us to evaluate all the stereo submissions with the new auditory-motivated objective measures.

References

1. Cooke, M.P., Hershey, J., Rennie, S.: Monaural speech separation and recognition challenge. Computer Speech and Language 24, 1–15 (2010)
2. Vincent, E., Sawada, H., Bofill, P., Makino, S., Rosca, J.: First stereo audio source separation evaluation campaign: Data, algorithms and results. In: Davies, M.E., James, C.J., Abdallah, S.A., Plumbley, M.D. (eds.) ICA 2007. LNCS, vol. 4666, pp. 552–559. Springer, Heidelberg (2007)
3. Vincent, E., Araki, S., Bofill, P.: The signal separation evaluation campaign: A community-based approach to large-scale evaluation. In: Adali, T., Jutten, C., Romano, J.M.T., Barros, A.K. (eds.) ICA 2009. LNCS, vol. 5441, pp. 734–741. Springer, Heidelberg (2009)
4. Vincent, E., Gribonval, R., Plumbley, M.D.: Oracle estimators for the benchmarking of source separation algorithms. Signal Processing 87(8), 1933–1950 (2007)
5. Wang, D.L.: On ideal binary mask as the computational goal of auditory scene analysis. In: Speech Separation by Humans and Machines. Springer, Heidelberg (2005)
6. Vincent, E., Gribonval, R., Févotte, C.: Performance measurement in blind audio source separation. IEEE Trans. on Audio, Speech and Language Processing 14(4), 1462–1469 (2006)
7. Emiya, V., Vincent, E., Harlander, N., Hohmann, V.: Subjective and objective quality assessment of audio source separation. IEEE Trans. on Audio, Speech and Language Processing (submitted)
8. Huber, R., Kollmeier, B.: PEMO-Q – a new method for objective audio quality assessment using a model of auditory perception. IEEE Trans. on Audio, Speech, and Language Processing 14(6), 1902–1911 (2006)

9. Ozerov, A., Vincent, E., Bimbot, F.: A general modular framework for audio source separation. In: Vigneron, V. (ed.) LVA/ICA 2010. LNCS, vol. 6365, pp. 33–40. Springer, Heidelberg (2010)
10. Rickard, S.: The DUET blind source separation algorithm. In: Blind Speech Separation, Springer, Heidelberg (2007)
11. Sawada, H., Araki, S., Makino, S.: Underdetermined convolutive blind source separation via frequency bin-wise clustering and permutation alignment. IEEE Trans. on Audio, Speech and Language Processing (in press)
12. Arberet, S., Ozerov, A., Duong, N.Q.K., Vincent, E., Gribonval, R., Bimbot, F., Vandergheynst, P.: Nonnegative matrix factorization and spatial covariance model for under-determined reverberant audio source separation. In: Proc. ISSPA (2010)
13. Duong, N.Q.K., Vincent, E., Gribonval, R.: Under-determined reverberant audio source separation using local observed covariance and auditory-motivated time-frequency representation. In: Vigneron, V. (ed.) LVA/ICA 2010. LNCS, vol. 6365, pp. 73–80. Springer, Heidelberg (2010)
14. Vu, D.H.T., Haeb-Umbach, R.: Blind speech separation employing directional statistics in an expectation maximization framework. In: Proc. ICASSP, pp. 241–244 (2010)
15. Ono, N., Miyabe, S.: Auxiliary-function-based independent component analysis for super-gaussian sources. In: Vigneron, V. (ed.) LVA/ICA 2010. LNCS, vol. 6365, pp. 165–172. Springer, Heidelberg (2010)
16. Sawada, H., Araki, S., Makino, S.: Measuring dependence of bin-wise separated signals for permutation alignment in frequency-domain BSS. In: Proc. ISCAS, pp. 3247–3250 (2007)
17. Schobben, D., Torkkola, K., Smaragdis, P.: Evaluation of blind signal separation methods. In: Proc. ICA, pp. 261–266 (1999)
18. Koldovsky, Z., Tichavsky, P., Malek, J.: Time-domain blind audio source separation method producing separating filters of generalized feedforward structure. In: Vigneron, V. (ed.) LVA/ICA 2010. LNCS, vol. 6365, pp. 17–24. Springer, Heidelberg (2010)
19. Koldovsky, Z., Tichavsky, P., Malek, J.: Subband blind audio source separation using a time-domain algorithm and tree-structured QMF filter bank. In: Vigneron, V. (ed.) LVA/ICA 2010. LNCS, vol. 6365, pp. 25–32. Springer, Heidelberg (2010)
20. Loesch, B., Yang, B.: Blind source separation based on time-frequency sparseness in the presence of spatial aliasing. In: Vigneron, V. (ed.) LVA/ICA 2010. LNCS, vol. 6365, pp. 1–9. Springer, Heidelberg (2010)
21. Nesta, F., Svaizer, P., Omologo, M.: Convolutive BSS of short mixtures by ICA recursively regularized across frequencies. IEEE Trans. on Audio, Speech, and Language Processing 14 (2006)
22. Yoshioka, T., Nakatani, T., Miyoshi, M., Okuno, H.G.: Blind separation and dereverberation of speech mixtures by joint optimization. IEEE Trans. on Audio, Speech, and Language Processing (2010) (in press)
23. Málek, J., Koldovský, Z., Tichavský, P.: Adaptive time-domain blind separation of speech signals. In: Vigneron, V. (ed.) LVA/ICA 2010. LNCS, vol. 6365, pp. 9–16. Springer, Heidelberg (2010)
24. Loesch, B., Yang, B.: Adaptive segmentation and separation of determined convolutive mixtures under dynamic conditions. In: Vigneron, V. (ed.) LVA/ICA 2010. LNCS, vol. 6365, pp. 41–48. Springer, Heidelberg (2010)
25. Araki, S., Sawada, H., Makino, S.: Blind speech separation in a meeting situation. In: Proc. ICASSP, pp. 41–45 (2007)

26. Ozerov, A., Fevotte, C.: Multichannel nonnegative matrix factorization in convolutive mixtures for audio source separation. IEEE Trans. on Audio, Speech and Language Processing 18(3), 550–563 (2010)
27. Bonada, J., Loscos, A., Vinyes Raso, M.: Demixing commercial music productions via human-assisted time-frequency masking. In: Proc. AES (2006)
28. Spiertz, M., Gnann, V.: Unsupervised note clustering for multichannel blind source separation. In: Proc. LVA/ICA (2010) (submitted)
29. Even, J., Saruwatari, H., Shikano, K., Takatani, T.: Speech enhancement in presence of diffuse background noise: Why using blind signal extraction? In: Proc. ICASSP, pp. 4770–4773 (2010)
30. Okamoto, R., Takahashi, Y., Saruwatari, H., Shikano, K.: MMSE STSA estimator with nonstationary noise estimation based on ICA for high-quality speech enhancement. In: Proc. ICASSP, pp. 4778–4781 (2010)
31. Takahashi, Y., Takatani, T., Osako, K., Saruwatari, H., Shikano, K.: Blind spatial subtraction array for speech enhancement in noisy environment. IEEE Trans. on Audio, Speech and Language Processing 17(4), 650–664 (2009)

The 2010 Signal Separation Evaluation Campaign (SiSEC2010): Biomedical Source Separation

Shoko Araki[1], Fabian Theis[2], Guido Nolte[3], Dominik Lutter[2], Alexey Ozerov[4], Vikrham Gowreesunker[5], Hiroshi Sawada[1], and Ngoc Q.K. Duong[4]

[1] NTT Communication Science Labs., NTT Corporation, Japan
[2] IBIS, Helmholtz Zentrum München, Germany
[3] Fraunhofer Institute FIRST IDA, Germany
[4] INRIA, Centre Inria Rennes - Bretagne Atlantique, France
[5] DSPS R&D Center, Texas Instruments Inc., USA

Abstract. We present an overview of the biomedical part of the 2010 community-based Signal Separation Evaluation Campaign (SiSEC2010), coordinated by the authors. In addition to the audio tasks which have been evaluated in the previous SiSEC, SiSEC2010 considered several biomedical tasks. Here, three biomedical datasets from molecular biology (gene expression profiles) and neuroscience (EEG) were contributed. This paper describes the biomedical datasets, tasks and evaluation criteria. This paper also reports the results of the biomedical part of SiSEC2010 achieved by participants.

1 Introduction

Large-scale evaluations are a key ingredient to scientific and technological maturation by revealing the effects of different system designs, promoting common test specifications and attracting the interest of industries and funding bodies. Recent evaluations of source separation systems include a series of the BCI (Brain Computer Interface) competitions [1,2,3].

The former community-based Signal Separation Evaluation Campaign (SiSEC 2008) [4] was designed based on the panel discussion at the 7th International Conference on Independent Component Analysis and Signal Separation (ICA 2007) which featured the Stereo Audio Source Separation Evaluation Campaign [5]. The general principles of the SiSEC aim to facilitate the entrance of researchers addressing different tasks and to enable detailed diagnosis of the submitted systems. The unique aspect of the SiSEC is that, SiSEC is not a competition but a scientific evaluation from which we can draw rigorous scientific conclusions.

The former SiSEC2008 attracted around 30 entrants, however, it consisted solely of audio datasets and tasks. Obviously, in addition to audio, there are many important areas where signal separation techniques are contributing to successful analyses. According to such a feedback to SiSEC2008, it was decided

V. Vigneron et al. (Eds.): LVA/ICA 2010, LNCS 6365, pp. 123–130, 2010.

that SiSEC2010 included not only the audio datasets, but also the biomedical datasets and tasks. For the first biomedical evaluation campaign of in SiSEC, we had three datasets: a set of gene expression profiles and two EEG datasets.

With the advent of high-throughput technologies in molecular biology, genome-wide measurements of gene transcript levels have become available. It has become clear that efficient computational tools are needed to successfully interpret the information buried in those large scale gene expression patterns. One commonly taken approach is to apply exploratory machine learning to detect similarities in multiple expression profiles in order to group genes into functional categories — for example, genes that are expressed to a greater or lesser extent in response to a drug or an existing disease. Although classically hierarchical clustering techniques are mostly used for these analyses, it has recently been shown [6, 7] that factorization techniques such as ICA can successfully separate meaningful categories by considering gene expression patterns as a superposition of independent expression modes, which are considered putative independent biological processes. Here we want to evaluate blind source separation problems applied to a set of microarrays in which we can quantify the expected cluster properties. We will see that inclusion of additional biological information as prior will be key to successful separations.

Most commonly, biomedical campaigns for electrophysiological data are formulated for BCI tasks, where an algorithm has to estimate e.g. the side of actual movement or imagined movement of left or right fingers of a subject from EEG data [1, 2, 3]. Here, the objective is clear and can be formulated using real EEG data only. The situation is more difficult in the case of decomposition of real data because the ground truth is not known and any simulation is prone to miss important aspects of real data. Here, two approaches are possible: a) one puts much effort to simulate *all* aspects of real EEG data [8], as imperfect as that may be, or b) one modifies real data as little as possible to construct some ground truth which can be tested for. We here take the second approach, specifically addressing the question whether independent sources can be distinguished from dependent ones.

This article describes the biomedical tasks, which are newly considered in the SiSEC2010. The traditional "audio source separation" task in SiSEC2010 is described in [9]. Section 2 describes the biomedical datasets, tasks, criteria. We summarize the results for each task in Section 3. In this paper, we focus on the general specifications and outcomes of the campaign and let readers refer to `http://sisec.wiki.irisa.fr/` for more detail.

2 Biomedical Separation Task Specifications

SiSEC2010 includes three biomedical tasks. The remaining part of this section provides the explanations about the datasets, tasks and evaluation criteria for each task. All materials, including data and reference codes, are available on the website at `http://sisec.wiki.irisa.fr/`.

2.1 Task 1: Source Identification in Microarray Data

Dataset. Since microarray technology has become one of the most popular approaches in the field of gene expression analysis, numerous statistical methods have been used to provide insights into the biological mechanisms of gene expression regulation [10, 11]. A common microarray dataset consists of multiple gene expression profiles. Each expression profile $\mathbf{x}_i$ mirrors the expression of N genes via measuring the level of the corresponding messenger RNA (mRNA) under a specific condition. In our case, mRNA was extracted from $i = 189$ invasive breast carcinomes [12] and measured using Affymetrix U133A Gene-chips. The Affymetrix raw data was normalized using the RMA algorithm [13] from the R Bioconductor package *simpleaffy*. Non-expressed genes were filtered out and Affymetrix probe sets were mapped to Gene Symbols. This resulted in a total of $N = 11815$ expressed genes.

The task. Cell signaling pathways play a major role in the formation of cancer. Understanding the biology of cell signaling helps to understand cancer and to develop new therapies. The regulation of these signaling pathways takes place on multiple layers, one of those is the regulation of gene expression or transcription. Single genes can take part in more than one pathway and the expression profiles can be regarded as linear superpositions of different signaling pathways or more generally biological processes. Using blind source separation (BSS) techniques, a linear mixture model can be decomposed to reconstruct source signals, which can be interpreted as these signaling pathways. A more detailed discussion of the linear factor model can be found in [6, 7].

The task is now to reconstruct these signaling pathways or parts of it from the microarray expression profiles using BSS techniques. Here, we approximate signaling pathways as simple gene lists. These pathway gene lists were taken from NETPATH (www.netpath.org).

Evaluation. To evaluate the quality of the reconstructed pathways, we tested for the significance of enriched genes that can be mapped to the pathways. For each source signal or estimated pathway we identify the number of genes that map to the distinct pathways and calculate p-values using Fisher's exact test. To correct for multiple testing we use the Benjamini-Hochberg procedure to estimate false positive rates (FDR). Now, after Benjamini-Hochberg correction a reconstructed pathway was declared as enriched if the p-value was below 0.05. We then count the number of all different significantly reconstructed pathways.

2.2 Task 2: Dependent Component Extraction

Dataset. In this set, data are constructed which are as close as possible to real EEG data with minor changes to ensure some ground truth which is, in principle, detectable. While independent sources are often a useful assumption recent research has focused on the analysis of brain connectivity. Therefore, the objective of the research is the detection of dependent sources.

The data were constructed as a superposition of $N = 19$ sources measured in as many sensors. Out of the 19 sources two were dependent and all others were mutually independent. The construction was done in the following way. We decomposed EEG data $\mathbf{x}(t) = (x_1(t), ..., x_N(t))^T$ from one subject using the TDSEP algorithm [14] as a Blind Source Separation in the standard way as

$$\mathbf{x}(t) = A\mathbf{s}(t) \tag{1}$$

with A being the mixing matrix and $\mathbf{s}(t)$ being the estimated independent source activities. Two of the 19 sources were selected as being dependent. For notational simplicity we denote the respective source indices as 1 and 2. The criterion for dependence was the fact that the imaginary part of the coherency showed a clear signal at around 10Hz. Generally, coherency between sources i and j at frequency f is defined as

$$C_{ij}(f) = \frac{< \hat{s}_i(f)\hat{s}_j^*(f) >}{\sqrt{< |\hat{s}_i(f)|^2 >< |\hat{s}_j(f)|^2 >}} \tag{2}$$

where $\hat{s}_i(f)$ is the short-time Fourier-transform of $s_i(t)$ for each trial (or segment) and $<>$ denotes averaging over all trials/segments. For these data we chose segments of 1 second duration giving a total of around 600 segments.

It can easily be shown that a significant non-vanishing imaginary part of $C_{ij}(f)$ cannot be explained by a mixture of independent sources [15]. The reason is that for mixtures of independent sources any coupling is caused by contributions of the same source which has a phase delay of 0 or π if the mixing coefficients have equal or opposite sign, respectively. In either case, the imaginary part vanishes.

For the present data the spatial patterns of the two ICA-components, i.e. the first two columns of the mixing matrix A, are shown in the upper panels of Fig.1. The patterns indicate clear signals from frontal and occipital parts of the brain, respectively. In the lower panels we show the power spectrum of the two ICA-components and the imaginary part of coherency between the components. Interestingly, the power at 10Hz is almost not visible in the frontal source, but the imaginary part of coherency still indicates a clear interaction between the sources at 10Hz.

The data $\mathbf{y}(t)$ were then constructed as

$$\mathbf{y}(t) = \sum_{i=1}^{2} \mathbf{a}_i s_i(t) + \sum_{i=3}^{N} \mathbf{a}_i \tilde{s}_i(t) \tag{3}$$

where $\mathbf{a}_i$ is the $ith.$ column of the mixing matrix A. The time series $\tilde{s}_i(t)$ were all taken from real data from different subjects. For each subject the data were decomposed using ICA and the $i.th$ original source $s_i(t)$ (for $i > 2$) was replaced by the $i.th$ source of the $i.th$ subject with ordering according to magnitude of the ICA-components.

The task. The task was to recover from $\mathbf{y}(t)$ the space spanned by the two columns $\mathbf{a}_1$ and $\mathbf{a}_2$. It was not the task to recover the two columns separately because for an interacting system the information given was not sufficient. It would have been necessary to make additional, e.g. spatial assumptions, on the nature of the sources to uniquely decompose the subspace into separate sources. Although in this special case, having distinct topographies at opposite sides of the scalp, such assumptions would have been both reasonable and easy to implement we preferred to avoid additional complications.

Evaluation. The geometrical relations between two subspaces can be defined in terms of the respective projectors on the respective subspace. If $\hat{A} = (\mathbf{a}_1, \mathbf{a}_2)$ then

$$P_A = \hat{A}\left(\hat{A}^T\hat{A}\right)^{-1}\hat{A}^T \tag{4}$$

is a projector onto the space spanned by the columns $\mathbf{a}_1$ and $\mathbf{a}_2$, i.e. P_A is a projector onto the true subspace. Similarly, let P_B be the projector on the estimated 2-dimensional subspace. We then calculate the eigenvalues of

$$D = P_A P_B P_A \tag{5}$$

Writing the eigenvalues in descending order for two-dimensional subspaces only the first two eigenvalues can be non-vanishing and all eigenvalues are in the interval $[0, 1]$. The subspaces are identical if and only if the second eigenvalue is equal to 1. For the data evaluation the value of this second eigenvalue was used to asses how accurately the true subspace was recovered.

2.3 Task 3: Artifact Removal in EEG Data

Dataset. This task contains two data sets: (1) 8-ch newborn EEG data, that are effectively only six channels due to the connection of electrodes, not containing any obvious artifact, and (2) an artificial data that represent artifacts of different kind: unleaded electrode, eye blinking. Some example figures can be found in `http://sisec.wiki.irisa.fr/`.

The data from active sleep of a newborn individual were sampled at 128 Hz and have total length of two minutes. There is a mutual dependence within the first four channels and also within the last four channels due to montage of electrodes, so that only six channels of eight are really independent.

The task. The data contains separately artifact and clean data, denoted a and x. The task is to blindly separate a and x from $y = a + x$. Participants have to estimate $\hat{x}$ that minimizes the norm $||\hat{x} - x||$. Of course, just y is available for the task.

Evaluation. We measure the Euclidean distance between the original and the reconstructed data.

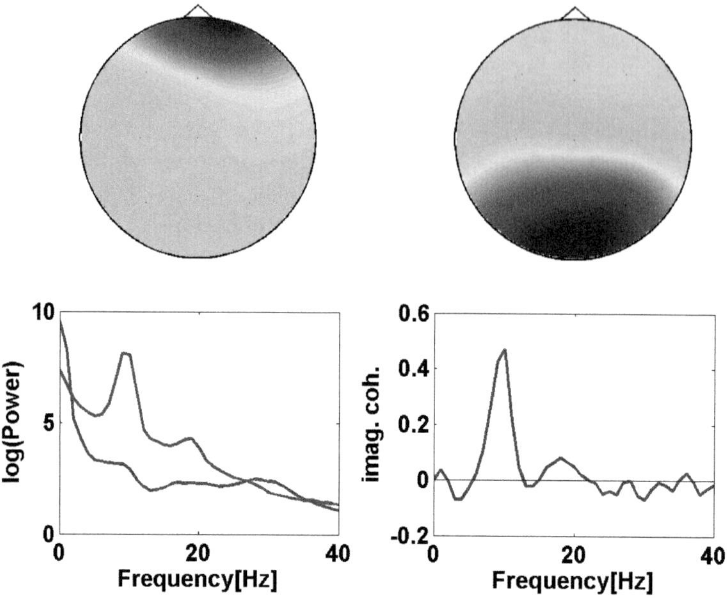

Fig. 1. Upper panels: topographies of selected ICA components chosen as interacting sources. Lower left panel: power spectrum of interacting sources. The blue line corresponds to the frontal source (upper left panel) and the red line, having a pronounced peak at around 10Hz, to the occipital source (upper right panel). Lower right panel: Imaginary part of coherency between these two sources.

3 Results

3.1 Task 1: Source Identification in Microarray Data

The task was processed by two independent groups. Chen et al. included prior information by using network component analysis (NCA) [16] to reconstruct the pathways. Blöchl et al. applied a matrix factorization method using prior knowledge encoded in a graph model (GraDe) [17], which was also presented at the LVA/ICA workshop 2010. Chen et al. obtained 58 cancer relevant pathways, out of which 8 have highly correlated expression profiles and were removed from the analysis to decrease redundancy. Blöchl et al. reconstructed 194 source profiles or pathways (one for each microarray measurement).

Both groups separated all 10 pathways with at least one source profile with a p-value below 0.05 (Fisher's exact test). Reduced to source profiles that specifically match to one pathway only, Blöchl et al. separated 7 of 10 pathways and Chen et al. 5 of 10. After FDR correction the number of enriched pathways reconstructed using GraDe reduced to 5 whereas whereas the NCA approach could not deliver any significantly enriched pathway. According to this task one can assess that the GraDe approach outperformed the NCA approach. We hypothesize that the better performance of the GraDe methods arises from the use of a knowledge-based transcription factor network including pathway information (TRANSPATH [18]) in contrast to a transcription factor network as prior knowledge (TRANSFAC [19]) used in the NCA approach.

3.2 Task 2: Dependent Component Extraction

About task 2, we had one submission by Petr Tichavsky and Zbynel Koldovsky. The contributers applied ICA to the data and analyzed the dependence between the ICA components. Although they reported that the dependence was very weak they were able to pick the right components. With a second eigenvalue of 0.9998 the space was reconstructed almost perfectly.

3.3 Task 3: Artifact Removal in EEG Data

Unfortunately, task 3 did not receive any attempt to solve it. Perhaps, task 3 was too difficult, because a few channels were available and the dataset included a large number of artifacts.

4 Conclusion

This paper provided the specifications and results of the biomedical part of the SiSEC2010. Our dataset included not only the EEG data that were commonly addressed in previous biomedical campaigns, but also a microarray dataset. This time, we had three entrants. In order to obtain more participants, we have to reconsider how we can encourage to collaborate each other more efficiently. We invite all the willing participants to the continuous collaborative discussion on future of, both biomedical source separation and other separation tasks than biomedical and audio.

Acknowledgments. We would like to thank all the entrants, as well as P. Tichavsky for sharing his dataset and evaluation code.

References

1. Sajda, P., Gerson, A., Müller, K.R., Blankertz, B., Parra, L.: A data analysis competition to evaluate machine learning algorithms for use in brain-computer interfaces. IEEE Trans. Neural Sys. Rehab. Eng. 11(2), 184–185 (2003)
2. Blankertz, B., Müller, K.R., Curio, G., Vaughan, T.M., Schalk, G., Wolpaw, J.R., Schlögl, A., Neuper, C., Pfurtscheller, G., Hinterberger, T., Schröder, M., Birbaumer, N.: The BCI competition 2003: Progress and perspectives in detection and discrimination of EEG single trials. IEEE Trans. Biomed. Eng. 51(6), 1044–1051 (2004)
3. Blankertz, B., Müller, K.R., Krusienski, D., Schalk, G., Wolpaw, J.R., Schlögl, A., Pfurtscheller, G., Millán, J.R., Schröder, M., Birbaumer, N.: The BCI competition III: Validating alternative approachs to actual BCI problems. IEEE Trans. Neural Sys. Rehab. 14(2), 153–159 (2006)
4. Vincent, E., Araki, S., Bofill, P.: The 2008 signal separation evaluation campaign (SiSEC 2008. In: Adali, T., Jutten, C., Romano, J.M.T., Barros, A.K. (eds.) ICA 2009. LNCS, vol. 5441, pp. 734–741. Springer, Heidelberg (2009)

5. Vincent, E., Sawada, H., Bofill, P., Makino, S., Rosca, J.: First stereo audio source separation evaluation campaign: Data, algorithms and results. In: Davies, M.E., James, C.J., Abdallah, S.A., Plumbley, M.D. (eds.) ICA 2007. LNCS, vol. 4666, pp. 552–559. Springer, Heidelberg (2007)
6. Lutter, D., Ugocsai, P., Grandl, M., Orso, E., Theis, F., Lang, E., Schmitz, G.: Analyzing m-csf dependent monocyte/macrophage diffrentiation: expression modes and meta-modes derived from an independent component analysis. BMC Bioinformatics 9(100) (2008)
7. Teschendorff, A.E., Journ'ee, M., Absil, P.A., Sepulchre, R., Caldas, C.: Elucidating the altered transcriptional programs in breast cancer using independent component analysis. PLoS Computational Biology 3(8) (2007)
8. Vidaurre, C., Nolte, G.: Generating synthetic EEG. In: Blankertz, B. (ed.) The BCI competition IV. LNCS, Springer, Heidelberg (to appear)
9. Araki, S., Ozerov, A., Gowreesunker, V., Sawada, H., Theis, F., Nolte, G., Lutter, D., Duong, N.Q.K.: The 2010 signal separation evaluation campaign (SiSEC2010): - audio source separation -. In: Vigneron, V. (ed.) LVA/ICA 2010. LNCS, vol. 6365, pp. 114–122. Springer, Heidelberg (2010)
10. Quackenbush, J.: Computational analysis of microarray data. Nature 2, 418–427 (2001)
11. Schachtner, R., Lutter, D., Knollmller, P., Tom, A.M., Theis, F.J., Schmitz, G., Stetter, M., Vilda, P.G., Lang, E.W.: Knowledge-based gene expression classification via matrix factorization. Bioinformatics 24, 1688–1697 (2008)
12. Sotiriou, C., Wirapati, P., Loi, S., Harris, A., Fox, S., Smeds, J., Nordgren, H., Farmer, P., Praz, V., Haibe-Kains, B., Desmedt, C., Larsimont, D., Cardoso, F., Peterse, H., Nuyten, D., Buyse, M., de Vijver, M.V., Bergh, J., Piccart, M., Delorenzi, M.: Gene expression profiling in breast cancer: understanding the molecular basis of histologic grade to improve prognosis. Journal of Natl. Cancer Inst. 98(4), 262–272 (2006)
13. Irizarry, R., Bolstad, B., Collin, F., Cope, L., Hobbs, B., Speed, T.: Summaries of affymetrix genechip probe level data. Nucleic Acid Research Journal 31(4) (2003)
14. Ziehe, A., Mueller, K.: TDSEP - an effient algorithm for blind separation using time structure. In: Proc. of ICANN 1998, pp. 675–680 (1998)
15. Nolte, G., Bai, U., Weathon, L., Mari, Z., Vorbach, S., Hallet, M.: Identifying true brain interaction from EEG data using the imaginary part of coherency. Clinical Neurophysiology (115) (2004)
16. Chen, W., Chang, C., Hung, Y.: Transcription factor activity estimation based on particle swarm optimization and fast network component analysis. IEEE EMBC 2010 (2010) (submitted)
17. Blöchl, F., Kowarsch, A., Theis, F.J.: Second-order source separation based on prior knowledge realized in a graph model. In: Vigneron, V. (ed.) LVA/ICA 2010. LNCS, vol. 6365, pp. 434–441. Springer, Heidelberg (2010)
18. Krull, M., Pistor, S., Voss, N., Kel, A., Reuter, I., Kronenberg, D., Michael, H., Schwarzer, K., Potapov, A., Choi, C., Kel-Margoulis, O., Wingender, E.: Transpath: an information resource for storing and visualizing signaling pathways and their pathological aberrations. Nucleic Acids Res. 34(Database issue), D546–D551 (2006)
19. Matys, V., Fricke, E., Geffers, R., Gssling, E., Haubrock, M., Hehl, R., Hornischer, K., Karas, D., Kel, A.E., Kel-Margoulis, O.V., Kloos, D.U., Land, S., Lewicki-Potapov, B., Michael, H., Mnch, R., Reuter, I., Rotert, S., Saxel, H., Scheer, M., Thiele, S., Wingender, E.: Transfac: transcriptional regulation, from patterns to profiles. Nucleic Acids Res. 31(1), 374–378 (2003)

Use of Bimodal Coherence to Resolve Spectral Indeterminacy in Convolutive BSS*

Qingju Liu, Wenwu Wang, and Philip Jackson

Centre for Vision, Speech and Signal Processing,
Faculty of Engineering and Physical Sciences,
University of Surrey, Guildford, GU2 7XH, United Kingdom
{Q.Liu,W.Wang,P.Jackson}@surrey.ac.uk

Abstract. Recent studies show that visual information contained in visual speech can be helpful for the performance enhancement of audio-only blind source separation (BSS) algorithms. Such information is exploited through the statistical characterisation of the coherence between the audio and visual speech using, e.g. a Gaussian mixture model (GMM). In this paper, we present two new contributions. An adapted expectation maximization (AEM) algorithm is proposed in the training process to model the audio-visual coherence upon the extracted features. The coherence is exploited to solve the permutation problem in the frequency domain using a new sorting scheme. We test our algorithm on the XM2VTS multimodal database. The experimental results show that our proposed algorithm outperforms traditional audio-only BSS.

Keywords: Convolutive blind source separation (BSS), audio-visual coherence, Gaussian mixture model (GMM), feature extraction and fusion, adapted expectation maximization (AEM), indeterminacy.

1 Introduction

Human speech perception is essentially bimodal as speech is perceived by auditory and visual senses. In traditional blind source separation (BSS) for auditory mixtures, only audio signals are considered. With the independence assumption, many algorithms have been proposed, e.g. [1]-[4]. The use of visual stimuli in BSS represents a recent development in multi-modal signal processing. Sodoyer et al. [5] addressed the separation problem for an instantaneous stationary mixture of decorrelated sources, with no further assumptions on independence or non-Gaussianity. Wang et al. [6] implemented a similar idea by applying the Bayesian framework to the fused feature observations for both instantaneous and convolutive mixtures. Rivet et al. [7] proposed a new statistical tool utilizing the log-Rayleigh distribution for modeling the audio-visual coherence, and then used the coherence to address the permutation and scale ambiguities in the spectral

* This work was supported by the Engineering and Physical Sciences Research Council (EPSRC) (Grant number EP/H012842/1) and the MOD University Defence Research Centre on Signal Processing (UDRC).

V. Vigneron et al. (Eds.): LVA/ICA 2010, LNCS 6365, pp. 131–139, 2010.

domain. However, the algorithm proposed in [5] used simple visual stimuli with only plosive consonants and vowels and worked for only instantaneous mixtures; the method in [6] considered a convolutive model with a relatively small number of taps for the mixing filters; the approach in [7] trained the audio-visual coherence with high dimensional audio feature vectors, thus the coherence model was sensitive to outliers.

In this paper, we consider the convolutive model [6]-[11] with the assumption of non-Gaussianity and independence constraints of the sources. We synchronize and merge the modified Mel-frequency cepstrum coefficients (MFCCs) as audio features and some geometric-type features from the video stream to obtain the audio-visual features for the estimation of the parameters of the bimodal coherence. A GMM model is trained on the audio-visual features using the adapted expectation maximization (AEM) algorithm that considers the different influences of the audio features on the model. The audio-visual coherence is then applied to address the permutation indeterminacy in the frequency domain based on an iterative sorting scheme. The remainder of the paper is organised as follows. An overview of convolutive BSS is presented in Section 2. Section 3 introduces our bimodal feature extraction and fusion method. Detailed indeterminacies cancellation algorithm exploiting the audio-visual coherence is presented in Section 4. The simulation results are analyzed and discussed in Section 5. Finally Section 6 concludes the paper.

2 BSS for Convolutive Mixtures

BSS aims to recover sources from their mixtures without any or with little prior knowledge about the sources or the mixing process. Consider the convolutive model:

$$x_p(n) = \sum_{k=1}^{K} \sum_{m=0}^{+\infty} h_{pk}(m) s_k(n-m) + \xi_p(n), \qquad \text{for } p = 1, ..., P \qquad (1)$$

or in matrix form: $\mathbf{x}(n) = \mathbf{H}(n) * \mathbf{s}(n) + \boldsymbol{\xi}(n)$, where $\mathbf{x}(n) = [x_1(n), ..., x_P(n)]^T$ are P observations obtained from K sources $\mathbf{s}(n) = [s_1(n), ..., s_K(n)]^T$ and $*$ denotes a convolution; $\mathbf{H}(n)$ is the mixing matrix whose entry $h_{pk}(n)$ represents the impulse response from source k to sensor p; $\boldsymbol{\xi}(n)$ is the additive noise vector; n is the discrete time index. The objective of convolutive BSS is to find a set of separation filters $\{w_{kp}(n)\}$ that satisfy:

$$\hat{s}_k(n) = y_k(n) = \sum_{p=1}^{P} \sum_{m=0}^{+\infty} w_{kp}(m) x_p(n-m). \qquad \text{for } k = 1, ..., K \qquad (2)$$

The matrix form of the separation process is $\hat{\mathbf{s}}(n) = \mathbf{y}(n) = \mathbf{W}(n) * \mathbf{x}(n)$ where $\mathbf{W}(n)$ is the separation matrix whose entries are the impulse responses $w_{kp}(n)$.

Convolutive BSS is often performed in the frequency domain as depicted by the upper dashed box in Fig.1. After applying the short-time Fourier transform (STFT) to the observations, the convolutive mixture in the time domain is transformed to a set of instantaneous mixtures in the frequency domain. Then ICA is applied to the spectral components $\mathbf{X}(f,t) = [X_1(f,t), ..., X_P(f,t)]^T$ in each frequency bin f to obtain the independent outputs $\mathbf{Y}(f,t) = [Y_1(f,t), ..., Y_K(f,t)]^T$, and t is the time-frame index. In matrix form $\mathbf{Y}(f,t) = \mathbf{W}(f)\mathbf{X}(f,t) = \hat{\mathbf{S}}(f,t)$, where $\mathbf{W}(f)$ is the separation filter, assumed to be linear time-invariant (LTI). It would be ideal if we could exactly recover the original sources that $\hat{\mathbf{S}}(f,t) = \mathbf{Y}(f,t) = \mathbf{S}(f,t)$. However, the ICA algorithms can estimate the sources only up to a permutation matrix $\mathbf{P}(f)$ and a diagonal matrix of gains $\mathbf{D}(f)$:

$$\hat{\mathbf{S}}(f,t) = \mathbf{Y}(f,t) = \mathbf{P}(f)\mathbf{D}(f)\mathbf{S}(f,t). \tag{3}$$

The permutation and scale ambiguities at each frequency bin present severe problems when reconstructing the separated sources in the time domain:

1. Recovered signal $Y_k(f,t)$ may not correspond to the same source $s_k(n)$ at some frequency bins, caused by $\mathbf{P}(f_i) \neq \mathbf{P}(f_j), i \neq j$.........**permutation indeterminacy**
2. Spectral components of $Y_k(f,t)$ coming from $s_k(n)$ are amplified at different frequency bins, caused by $\mathbf{D}(f_i) \neq \mathbf{D}(f_j), i \neq j$......................**scale indeterminacy**

To solve the permutation ambiguity, there are traditionally two methods. The first method is based on the continuity of adjacent bins, also known as the correlation approach [8]. The other method uses beam-forming theory [3] such as directional pattern estimation for permutation alignment [9]. As for the scale ambiguity, the minimum distortion principle can be applied to reduce the influence of a scale factor [10].

Previous algorithms using only audio streams have some drawbacks. The correlation approach may lead to continuous alignment errors; the beam-forming approach requires prior knowledge about the microphone array arrangement and some constraints on the spacing of microphones; the scale ambiguity is not perfectly solved. To solve the indeterminacies, as motivated by the work in [6] and

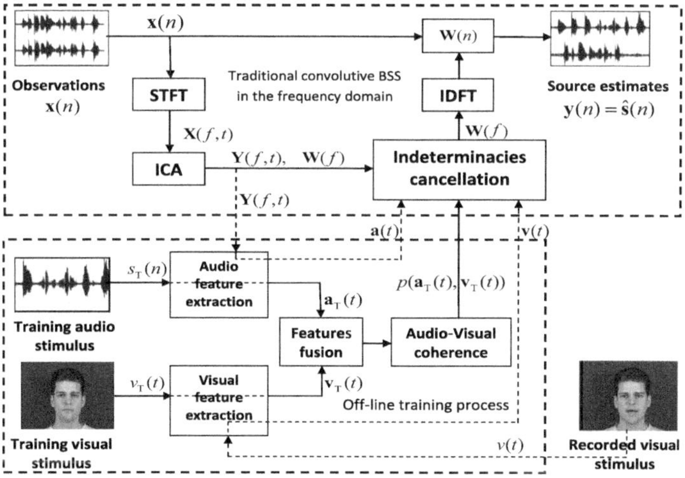

Fig. 1. Flow of a typical audio-visual BSS system

[7], we use the bimodal coherence of audio-visual features described in detail in the next section (shown in Fig.1).

3 Feature Extraction and Fusion

3.1 Audio and Visual Feature Extraction

We take the Mel-frequency cepstrum coefficients (MFCCs) as audio features as in [6] with some modifications. The MFCCs exploit the non-linear resolution of the human auditory system across an audio spectrum, which are the Discrete Cosine Transform (DCT) results of the logarithm of the short term power spectrum on a Mel-scale frequency. To avoid inverse DFT to $\mathbf{Y}(f,t)$ in the separation process described in Section 4, we replace the first component of MFCCs with the logarithmic power of spectral data. We obtain the modified L-dimensional MFCCs $\mathbf{a}_{\mathrm{T}}(t) = [\log\mathrm{E}(t), c_1(t)..., c_{L-1}(t)]^T$ (Fig.2). For simplicity, we denote the audio feature vector as $\mathbf{a}_{\mathrm{T}}(t) = [a_{\mathrm{T}1}(t), ..., a_{\mathrm{T}L}(t)]^T$.

Unlike the appearance-based visual features used in [6], we use the same front geometric visual features as in [5][7]: the lip width (LW) and height (LH) from the internal labial contour. Fig.3 shows the method for obtaining the 2-dimensional visual feature vector $\mathbf{v}_{\mathrm{T}}(t) = [\mathrm{LW}(t), \mathrm{LH}(t)]^T$.

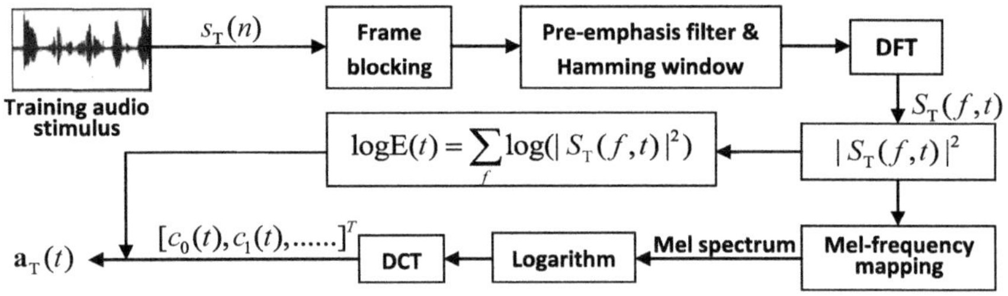

Fig. 2. Audio feature extraction in the training process

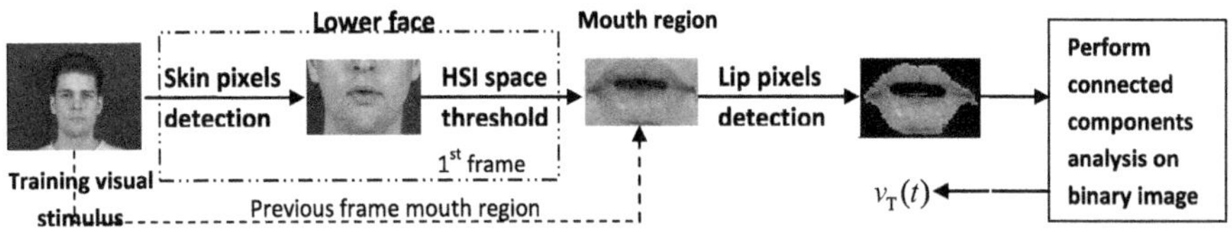

Fig. 3. Visual feature extraction in the training process

3.2 Feature-Level Fusion

We concatenate the audio and visual features after synchronization to get the $(L+2)$-dimensional audio-visual vector $\mathbf{u}_{\mathrm{T}}(t) = [\mathbf{a}_{\mathrm{T}}(t); \mathbf{v}_{\mathrm{T}}(t)]$, which will be used for training. The audio-visual coherence can be statistically characterized as a GMM with I kernels:

$$p_{AV}(\mathbf{u_T}(t)) = \sum_{i=1}^{I} \gamma_i p_G(\mathbf{u_T}(t) \mid \boldsymbol{\mu}_i, \boldsymbol{\Sigma}_i), \tag{4}$$

where γ_i is the weighting parameter, $\boldsymbol{\mu}_i$ is the mean vector and $\boldsymbol{\Sigma}_i$ is the covariance matrix of the i-th kernel. Every kernel of this mixture represents one cluster of the audio-visual data modeled by a joint Gaussian distribution:

$$p_G(\mathbf{u_T}(t) \mid \boldsymbol{\mu}_i, \boldsymbol{\Sigma}_i) = \frac{\exp\{-\frac{1}{2}(\mathbf{u_T}(t) - \boldsymbol{\mu}_i)^T \boldsymbol{\Sigma}_i^{-1}(\mathbf{u_T}(t) - \boldsymbol{\mu}_i)\}}{\sqrt{(2\pi)^{L+2} \mid \boldsymbol{\Sigma}_i \mid}}. \tag{5}$$

We denote $\lambda_i = \{\gamma_i, \boldsymbol{\mu}_i, \boldsymbol{\Sigma}_i\}$ as the parameter set, and it can be estimated by the expectation maximization (EM) algorithm. In the traditional EM training process, all the components of the training data are treated equally whatever their magnitudes. Therefore if we train the data $\mathbf{u}_\dagger(t) = [\mathbf{a_T}(t); \mathbf{v_T}(t)]$ and $\mathbf{u}_\ddagger(t) = [2a_{\mathbf{T}1}(t), a_{\mathbf{T}2}(t), ...a_{\mathbf{T}L}(t), \mathbf{v_T}(t)^T]^T$ respectively, we get two joint distributions $p_{AV\dagger}(\cdot)$ and $p_{AV\ddagger}(\cdot)$ with two sets of parameters $\{\lambda_{i\dagger}\}$ and $\{\lambda_{i\ddagger}\}$. However, these joint distributions are identical:

$$p_{AV\dagger}(\mathbf{u}_\dagger(t)) = p_{AV\ddagger}(\mathbf{u}_\ddagger(t)). \tag{6}$$

Thus the influence of $a_{\mathbf{T}1}(t)$ on the final probability does not change even its magnitude is doubled. Nevertheless, some components of the audio vector with large magnitudes are actually more informative about the audio-visual coherence than the remaining components (consider, for instance, the case of lossy compression of audio and images using DCT where small components can be discarded). For example, the first component of the audio vector ($a_{\mathbf{T}1}(t)$) should play a more dominant role in affecting the probability $p_{AV}(\mathbf{u_T}(t))$ than the last one. Also, the components of the audio vector having very small magnitudes are likely to be affected by noise. Therefore, considering these factors, we propose an adapted expectation maximization (AEM) algorithm.

I. Initialize the parameter set $\{\lambda_i\}$ with the K-means algorithm.
II. Run the following iterative process:

i. Compute the influence parameters $\beta_i(\cdot)$ of $\mathbf{u_T}(t)$ for $i = 1, ...I$.

$$\beta_i(\mathbf{u_T}(t)) = 1 - \frac{\parallel \mathbf{u_T}(t) - \boldsymbol{\mu}_i \parallel}{\sum_{j=1}^{I} \parallel \mathbf{u_T}(t) - \boldsymbol{\mu}_j \parallel}, \tag{7}$$

where $\parallel \cdot \parallel$ denotes the squared Euclidean distance.

ii. Calculate the probability of each cluster given $\mathbf{u_T}(t)$.

$$p_i(\mathbf{u_T}(t)) = \frac{\gamma_i p_G(\mathbf{u_T}(t) \mid \boldsymbol{\mu}_i, \Sigma_i)\beta_i(\mathbf{u_T}(t))}{\sum_{j=1}^{I} \gamma_j p_G(\mathbf{u_T}(t) \mid \boldsymbol{\mu}_j, \Sigma_j)\beta_j(\mathbf{u_T}(t))}. \tag{8}$$

iii. Update the parameter set $\{\lambda_i\}$:

$$\boldsymbol{\mu}_i = \frac{\sum_t \mathbf{u_T}(t) p_i(\mathbf{u_T}(t))}{\sum_t p_i(\mathbf{u_T}(t))}, \gamma_i = \frac{\sum_t p_i(\mathbf{u_T}(t))}{\sum_t}, \Sigma_i = \frac{\sum_t (\mathbf{u_T}(t) - \boldsymbol{\mu}_i)^2 p_i(\mathbf{u_T}(t))}{\sum_t p_i(\mathbf{u_T}(t))}. \tag{9}$$

---

4 Resolution of Spectral Indeterminacy

As $y_k(n)$ is the estimate of $s_k(n)$, $y_k(n)$ will have maximum coherence with the corresponding video signal $v_k(t)$. Therefore we can maximize the following criterion in the frequency domain to address the indeterminacies as mentioned in Section 2:

$$[\hat{\mathbf{P}}(f), \hat{\mathbf{D}}(f)] = \arg \max_{\mathbf{P}(f), \mathbf{D}(f)} \sum_t \sum_{k=1}^{K} p_{AV}(\mathbf{u}_k(t)), \tag{10}$$

where $\mathbf{u}_k(t) = [\mathbf{a}_k(t); \mathbf{v}_k(t)]$ is the audio-visual feature extracted from the profile $\hat{S}_k(\cdot, t) = Y_k(\cdot, t)$ of the k-th source estimate and the recorded video associated with the k-th speaker at time-frame t. If we are just interested in an estimate of $s_1(n)$ from the observations, we can get the first separation vector $\mathbf{p}(f)$ (note it is not the separation matrix) and the scale parameter $\alpha(f)$ by maximizing:

$$[\hat{\mathbf{p}}(f), \hat{\alpha}(f)] = \arg \max_{\mathbf{p}(f), \alpha(f)} \sum_t p_{AV}(\mathbf{u}_1(t)). \tag{11}$$

Since the permutation problem is the main factor in the degradation of the recovered sources, we focus on permutation indeterminacy cancellation for a two-source and two-mixture case detailed as follows. Suppose there are $2M$ frequency bins in total. Based on the symmetry, we will only consider the positive M bins. Denote $\mathbf{v}_1(t)$ as the visual feature that we have extracted from the recorded video signal associated with the target speaker. Generate an intermediate variable $Y_1^\dagger(f, t)$ spanning the same frequency and time-frame space as $Y_1(f, t)$ (or $Y_2(f, t)$). Initialize $\mathbf{P}(f)$ with identity matrices for $f = f_1, ..., f_M$.

I. Test which profile, $Y_1(\cdot, t)$ or $Y_2(\cdot, t)$, is more coherent with $\mathbf{v}_1(t)$.

1. For $f = f_1, ..., f_M$, let $Y_1^\dagger(f, \cdot) = Y_2(f, \cdot)$.
2. Extract the audio feature $\mathbf{a}_1(t)$ and $\mathbf{a}_1^\dagger(t)$ from $Y_1(\cdot, t)$ and $Y_1^\dagger(\cdot, t)$. Let $\mathbf{u}_1(t) = [\mathbf{a}_1(t); \mathbf{v}_1(t)]$, $\mathbf{u}_1^\dagger(t) = [\mathbf{a}_1^\dagger(t); \mathbf{v}_1(t)]$, and then calculate the audio-visual probability $p_{AV}(\mathbf{u}_1(t))$ and $p_{AV}(\mathbf{u}_1^\dagger(t))$ respectively based on the GMM model in equation (4) and the parameter set λ_i that has been estimated with the AEM algorithm.
3. If $\sum_t p_{AV}(\mathbf{u}_1(t)) > \sum_t p_{AV}(\mathbf{u}_1^\dagger(t))$, do nothing; otherwise, swap the rows of $\mathbf{P}(f)$ (i.e. $\mathbf{P}(f) \longleftarrow \left[\begin{smallmatrix} 0 & 1 \\ 1 & 0 \end{smallmatrix}\right] \mathbf{P}(f)$), $\mathbf{W}(f)$ and $\mathbf{Y}(f, \cdot)$ for $f = f_1, ..., f_M$.

II. Equally divided M bins into 2 parts.

II.i. 1. For $f = f_1, ... f_{M/2}$, $Y_1^\dagger(f, \cdot) = Y_2(f, \cdot)$; for the remaining bins, $Y_1^\dagger(f, \cdot) = Y_1(f, \cdot)$.
2. Extract $\mathbf{u}_1(t)$ and $\mathbf{u}_1^\dagger(t)$, and then calculate $p_{AV}(\mathbf{u}_1(t))$ and $p_{AV}(\mathbf{u}_1^\dagger(t))$.
3. If $\sum_t p_{AV}(\mathbf{u}_1(t)) > \sum_t p_{AV}(\mathbf{u}_1^\dagger(t))$ do nothing; otherwise, update $\mathbf{P}(f)$, $\mathbf{W}(f)$ and $\mathbf{Y}(f, \cdot)$ as in step **I** for $f = f_1, ... f_{M/2}$.

II.ii. Repeat the replacement, calculation, comparison and update as in step **II.i** for $f = f_{M/2+1}, ... f_M$.

III. Divide M bins into 4 (8, 16,...) parts, and continue the progressive scheme.

This scheme can reach a high resolution, which is determined by the number of partitions at the final division, and the larger the number, the higher the resolution. However, most permutations occur continuously in practical situations, therefore even if we stop running the algorithm at a very 'coarse' resolution, the permutation ambiguity can still be well reduced.

The scale indeterminacy can be addressed by some gradient algorithms [4]. However, estimating the gradient of $\sum_t p_{AV}(\mathbf{u}_1(t))$ is computationally demanding, and it remains an issue in our future work.

5 Experimental Results

The proposed method was tested on the XM2VTS [12] multi-modal database, in which the speech data were recorded 4 times at approximately one month intervals, with continuous sentences of words and digits in mono, 16 bit, 32 kHz, PCM wave files, and the frontal face videos were captured at 25 fps.

We trained the audio-visual coherence model of the target speaker with concatenated audio and visual speech signals lasting for approximately 40 seconds. The audio was downsampled to 16 kHz, and the 32 ms (512 points) Hamming window with 12 ms overlap was applied in STFT. 5-dimensional ($L = 5$) MFCCs as audio features were extracted from 24 mel-scaled filter banks. The visual features were upsampled to 50 Hz to be synchronized with the audio features. Thus the audio-visual data were 7-dimensional. For simplicity, we only used 5 ($I = 5$) kernels to approximate the audio-visual coherence.

The algorithm was tested on convolutive mixtures synthesized on computer. The filters $\{h_{pk}(n)\}$ were generated by the system utilizing the impulse response measurements of a conference room [13] with various positions of the microphones and the speakers. We resampled those filters and used the beginning 256 measurements (the reverberation time was 16 ms) as the final mixing filters. Two audio signals with each lasting 4 s were convolved with the filters to generate the mixtures, and Gaussian white noise (GWN) was added to both mixtures.

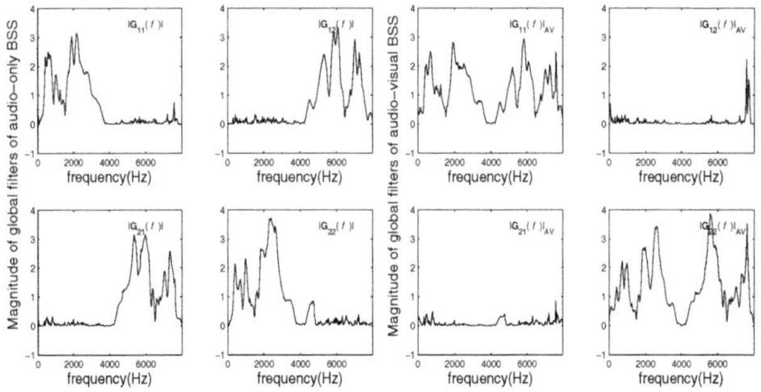

Fig. 4. Global filters comparison

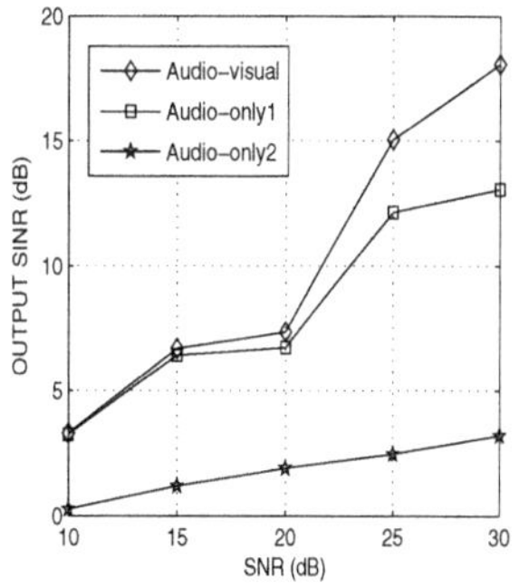

Fig. 5. SINR comparison

We use the global filters $\mathbf{G}(f) = \begin{bmatrix} \mathbf{G}_{11}(f) & \mathbf{G}_{12}(f) \\ \mathbf{G}_{21}(f) & \mathbf{G}_{22}(f) \end{bmatrix} = \mathbf{W}(f)\mathbf{H}(f)$ in the frequency domain and signal to interference and noise ratio (SINR) at different signal to noise ratios (SNRs) as criteria to evaluate the performance of our bimodal BSS algorithm. Suppose $s_1(n)$ is the target source, then

$$\mathrm{SINR} = 10\log\frac{P_{s_1}}{P_{\hat{s}_1 - s_1}} = 10\log\frac{\sum_n \parallel \sum_{p=1}^{P} w_{1p}(n) * h_{p1}(n) * s_1(n) \parallel}{\sum_n \parallel \hat{s}_1(n) - \sum_{p=1}^{P} w_{1p}(n) * h_{p1}(n) * s_1(n) \parallel}. \tag{12}$$

Fig.4 is the comparison of the global filters between the frequency-domain audio-only BSS using the correlation method [8] (left half) and audio-visual BSS (right half). It shows that our algorithm has corrected the permutation ambiguity at most frequency bins, while the permutation ambiguities in a large number of bins has not been resolved with the correlation method [8]. Fig.5 shows the SINR over different input SNRs. The SINR is calculated over a total of 100 independent runs with different convolutive filters. In the figure, Audio-only1 and Audio-only2 are two algorithms using only audio signals. Audio-only1 is a frequency-domain BSS algorithm, exploiting the correlation method [8]. Audio-only2 is a time-domain fast fixed-point BSS algorithm based on a convolutive sphering process [11], and when the order of the mixing filters is high (e.g., 256 in our simulation), it may not converge. However, the ICA technique degrades in adverse conditions, and as a result the improvement of bimodal BSS disappears at low SNRs. When it is noise-free, the SINR of bimodal BSS is 21.9dB.

6 Conclusions

We have presented a new audio-visual convolutive BSS system. In this system, we have used the modified MFCCs as audio features, which were combined with geometric visual features to form an audio-visual feature space. An adapted EM algorithm is then proposed to exploit the different influences of the audio features on the statistically modeling of the audio-visual coherence. A new sorting scheme exploiting the audio-visual coherence to solve the spectral indeterminacy problem has also been presented. Our algorithm has been tested on the XM2VTS database and has shown improved performance over audio-only BSS systems. In the future, we will consider using some dynamic features in video as well, instead of just static features. In addition, we will increase the number of kernels to improve the accuracy of the audio-visual model.

References

1. Jutten, C., Herault, J.: Blind Separation of Sources, Part I: An Adaptive Algorithm Based on Neuromimetic Architecture. Signal Process. 24(1), 1–10 (1991)
2. Comon, P.: Independent Component Analysis, a New Concept? Signal Process. 36(3), 287–314 (1994)
3. Cardoso, J.F., Souloumiac, A.: Blind Beamforming for Non-Gaussian Signals. IEEE Proc.-F 140(6), 362–370 (1993)
4. Hyvärinen, A., Karhunen, J., Oja, E.: Independent Component Analysis. John Wiley & Sons, New York (2001)

5. Sodoyer, D., Schwartz, J.L., Girin, L., Klinkisch, J., Jutten, C.: Separation of Audio-Visual Speech Sources: a New Approach Exploiting the Audio-Visual Coherence of Speech Stimuli. EURASIP J. Appl. Signal Process. 11, 1165–1173 (2002)
6. Wang, W., Cosker, D., Hicks, Y., Sanei, S., Chambers, J.: Video Assisted Speech Source Separation. In: Proc. IEEE ICASSP, pp. 425–428 (2005)
7. Rivet, B., Girin, L., Jutten, C.: Mixing Audiovisual Apeech Processing and Blind Source Separation for the Extraction of Speech Signals from Convolutive Mixtures. IEEE Trans. Audio Speech Lang. Process. 15(1), 96–108 (2009)
8. Anemüller, J., Kollmeier, B.: Amplitude Modulation Decorrelation for Convolutive Blind Source Separation. In: Proc. ICA, pp. 215–220 (2000)
9. Ikram, M.Z., Morgan, D.R.: A Beamforming Approach to Permutation Alignment for Multichannel Frequency-Domain Blind Speech Separation. In: Proc. IEEE ICASSP, pp. 881–884 (2002)
10. Matsuoka, K., Nakashima, S.: Minimal Distortion Principle for Blind Source Separation. In: Proc. ICA, pp. 722–727 (2001)
11. Thomas, J., Deville, Y., Hosseini, S.: Time-Domain Fast Fixed-Point Algorithms for Convolutive ICA. IEEE Signal Process. Lett. 13(4), 228–231 (2006)
12. Messer, K., Matas, J., Kittler, J., Luettin, J., Maitre, G.: XM2VTSDB: The Extended M2VTS Database. In: AVBPA (1999), `http://www.ee.surrey.ac.uk/CVSSP/xm2vtsdb/`
13. Westner, A.: Room Impulse Responses (1998), `http://alumni.media.mit.edu/~westner/papers/ica99/node2.html`

Non-negative Hidden Markov Modeling of Audio with Application to Source Separation

Gautham J. Mysore[1,⋆], Paris Smaragdis[2], and Bhiksha Raj[3]

[1] Center for Computer Research in Music and Acoustics, Stanford University
[2] Advanced Technology Labs, Adobe Systems Inc.
[3] School of Computer Science, Carnegie Mellon University

Abstract. In recent years, there has been a great deal of work in modeling audio using non-negative matrix factorization and its probabilistic counterparts as they yield rich models that are very useful for source separation and automatic music transcription. Given a sound source, these algorithms learn a dictionary of spectral vectors to best explain it. This dictionary is however learned in a manner that disregards a very important aspect of sound, its temporal structure. We propose a novel algorithm, the non-negative hidden Markov model (N-HMM), that extends the aforementioned models by jointly learning several small spectral dictionaries as well as a Markov chain that describes the structure of changes between these dictionaries. We also extend this algorithm to the non-negative factorial hidden Markov model (N-FHMM) to model sound mixtures, and demonstrate that it yields superior performance in single channel source separation tasks.

1 Introduction

A common theme in most good strategies to modeling audio is the ability to make use of structure. Non-negative factorizations such as non-negative matrix factorization (NMF) and probabilistic latent component analysis (PLCA) have been shown to be powerful in representing spectra as a linear combination of vectors from a dictionary [1]. Such models take advantage of the inherent low-rank nature of magnitude spectrograms to provide compact and informative descriptions. Hidden Markov models (HMMs) have instead made use of the inherent temporal structure of audio and have shown to be particularly powerful in modeling sounds in which temporal structure is important, such as speech [2]. In this work, we propose a new model that combines the rich spectral representative power of non-negative factorizations and the temporal structure modeling of HMMs.

In [3], ideas from non-negative factorizations and HMMs have been used by representing sound mixtures as a linear combination of spectral vectors and also modeling the temporal structure of each source. However, at a given time frame, each source is represented by a single spectral vector rather than a linear combination of multiple spectral vectors. As pointed out, this has some virtue in speech

⋆ This work was performed while interning at Adobe Systems Inc.

V. Vigneron et al. (Eds.): LVA/ICA 2010, LNCS 6365, pp. 140–148, 2010.

as it is monophonic but it can break down when representing rich polyphonic sources such as music, for which one would resort to using standard NMF. In our proposed method, at a given time frame, a given source is represented as a linear combination of multiple spectral vectors from one (of the many) dictionaries of the source. This allows us to model finer details in an input such as variations in a phoneme or a note. As shown in the results section, the performance improves even for speech.

2 Models of Single Sources

In this section, we first briefly describe probabilistic spectrogram factorization for modeling a single source. We then describe the non-negative hidden Markov model (N-HMM) and parameter estimation for the model.

2.1 Probabilistic Spectrogram Factorization

The magnitude spectrogram of a sound source can be viewed as a histogram of "sound quanta" across time and frequency. With this view, probabilistic factorization [4], which is a type of non-negative factorization, has been used to model a magnitude spectrogram as a linear combination of spectral vectors from a dictionary. The model is defined by two sets of parameters:

1. $P(f|z)$ is a multinomial distribution of frequencies for latent component z. It can be viewed as a spectral vector from a dictionary.
2. $P(z_t)$ is a multinomial distribution of weights for the aforementioned dictionary elements at time t.

Given a magnitude spectrogram, these parameters can be jointly estimated using the Expectation–Maximization (EM) algorithm. As can be seen in the graphical model representation in Fig. 1a, the weights of each time frame are estimated independently of the other time frames, therefore failing to capture the temporal structure of the sound source.

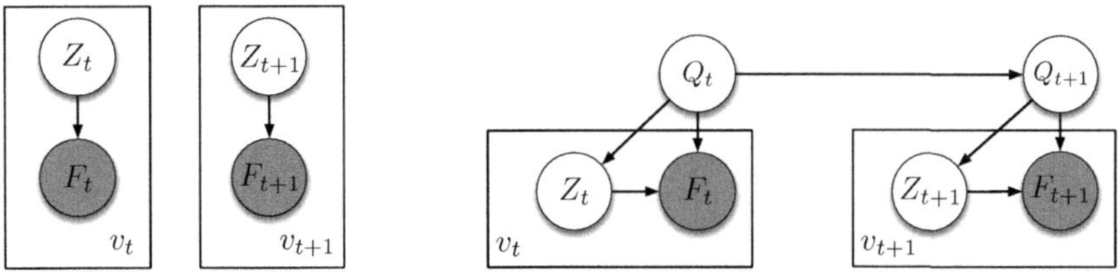

(a) Factorization (b) N-HMM

Fig. 1. Probabilistic factorization models each time frame independently, whereas the N-HMM models the transitions between successive time frames

2.2 Non-negative Hidden Markov Model

There is a temporal aspect to the proposed model, the N-HMM, shown in Fig. 1b. The model has a number of states, q, which can be interpreted as individual dictionaries. Each dictionary has a number of latent components, z, which can be interpreted as spectral vectors from the given dictionary. The spectral vector z of state q is defined by the multinomial distribution, $P(f|z,q)$.

As in traditional HMMs, in a given time frame, only one of the states is active. The given magnitude spectrogram at that time frame is modeled as a linear combination of the spectral vectors of the corresponding dictionary (state), q. At time t, the weights are defined by the multinomial distribution $P(z_t|q_t)$.

This notion of modeling a given time frame with one (of many) dictionaries rather than using a single large dictionary globally caters well to the non-stationarity of audio signals. The idea is that as an audio signal dynamically changes towards a new state, a new and appropriate dictionary should be used. We capture the temporal structure of these changes with a transition matrix, defined by, $P(q_{t+1}|q_t)$. The initial state probabilities (priors) are defined by $P(q_1)$. We also define a distribution $P(v|q)$ which is a distribution of the energy of a given state. It is modeled as a Gaussian distribution. It has been left out of the graphical model for clarity. The overall generative process is as follows:

1. Set $t = 1$ and choose a state according to the initial state distribution $P(q_1)$.
2. Choose the number of draws (energy) for the given time frame according to $P(v_t|q_t)$.
3. Repeat the following steps v_t times:
   (a) Choose a latent component according to $P(z_t|q_t)$.
   (b) Choose a frequency according to $P(f_t|z_t, q_t)$.
4. Transit to a new state q_{t+1} according to $P(q_{t+1}|q_t)$
5. Set $t = t + 1$ and go to step 2 if $t < T$.

2.3 Parameter Estimation for the N-HMM

Given the scaled magnitude spectrogram, V_{ft}, of a sound source[1], we use the EM algorithm to estimate the model parameters of the N-HMM. The E-step is computed as follows:

$$P(z_t, q_t|f_t, \overline{\mathbf{f}}, \overline{\mathbf{v}}) = \frac{\alpha(q_t)\beta(q_t)}{\sum_{q_t} \alpha(q_t)\beta(q_t)} P(z_t|f_t, q_t) , \tag{1}$$

where

$$P(z_t|f_t, q_t) = \frac{P(z_t|q_t)P(f_t|z_t, q_t)}{\sum_{z_t} P(z_t|q_t)P(f_t|z_t, q_t)} . \tag{2}$$

$P(q_t, z_t|f_t, \overline{\mathbf{f}}, \overline{\mathbf{v}})$ is the posterior distribution that is used to estimate the dictionary elements and the weights vectors. $\overline{\mathbf{f}}$ denotes the observations across all time frames[2], which is the entire spectrogram. $\overline{\mathbf{v}}$ denotes the number of draws over all

[1] Since the magnitude spectrogram is modeled as a histogram, the entries should be integers. To account for this, we weight it by an appropriate scaling factor.

[2] It should be noted that f_t is part of $\overline{\mathbf{f}}$. It is however mentioned separately to indicate that the posterior over z_t and q_t is computed separately for each f_t.

time frames. The forward/backward variables $\alpha(q_t)$ and $\beta(q_t)$ are computed using the likelihoods of the data, $P(\mathbf{f_t}, v_t|q_t)$, for each state (as in standard HMMs [2]). The likelihoods are computed as follows:

$$P(\mathbf{f_t}, v_t|q_t) = P(v_t|q_t) \prod_{f_t} \left(\sum_{z_t} P(f_t|z_t, q_t) P(z_t|q_t) \right)^{V_{ft}}, \tag{3}$$

where $\mathbf{f_t}$ represents the observations at at time t, which is the magnitude spectrum at that time frame. The dictionary elements and their weights are estimated in the M-step as follows:

$$P(f|z,q) = \frac{\sum_t V_{ft} P(z_t, q_t|f_t, \bar{\mathbf{f}}, \bar{\mathbf{v}})}{\sum_{f_t} \sum_t V_{ft} P(z_t, q_t|f_t, \bar{\mathbf{f}}, \bar{\mathbf{v}}))}, \tag{4}$$

$$P(z_t|q_t) = \frac{\sum_{f_t} V_{ft} P(z_t, q_t|f_t, \bar{\mathbf{f}}, \bar{\mathbf{v}}))}{\sum_{z_t} \sum_{f_t} V_{ft} P(z_t, q_t|f_t, \bar{\mathbf{f}}, \bar{\mathbf{v}}))}. \tag{5}$$

The transition matrix, $P(q_{t+1}|q_t)$, and priors, $P(q_1)$, are computed exactly as in standard HMMs [2]. The mean and variance of $P(v|q)$ are also estimated from the data. The learned dictionaries and transition matrix for an instance of speech data can be seen in Fig. 2. This model can be interpreted as an HMM in which the observation model $P(f_t|q_t)$ is a multinomial mixture model:

$$P(f_t|q_t) = \sum_{z_t} P(f_t|z_t, q_t) P(z_t|q_t). \tag{6}$$

However, this implies that for a given state, q, there is a single set of spectral vectors $P(f|z,q)$ and a single set of weights $P(z|q)$. If the weights did not change

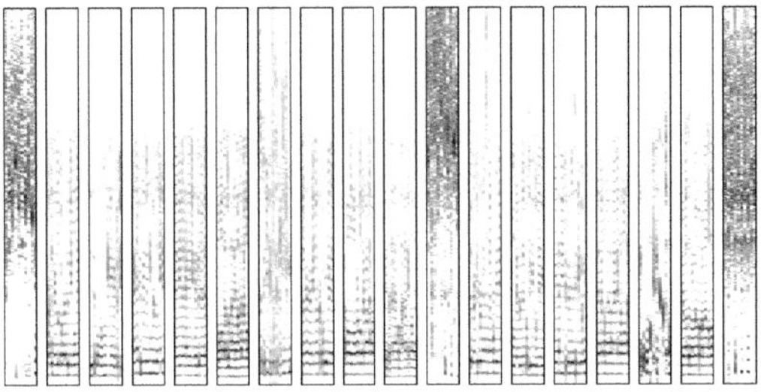

(a) Learned dictionaries

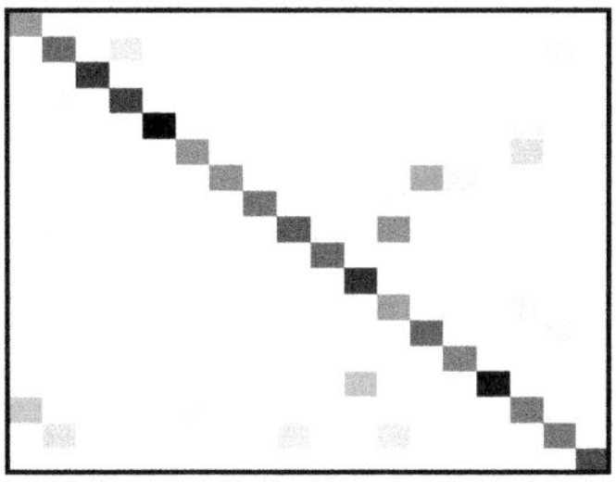

(b) Transition matrix

Fig. 2. (a) Dictionaries were learned from speech data of a single speaker. Shown are the dictionaries for 18 states, each state dictionary comprised of 10 elements. Each of these dictionaries roughly corresponds to a subunit of speech, either a voiced or unvoiced phoneme. (b) The learned transition matrix describes the transitions between the learned dictionaries in Fig. 2a. As can be seen by the strong diagonal, the algorithm correctly learns a model with state persistence.

across time, the observation model would collapse to a single spectral vector per state. In the proposed model however, the weights $P(z_t|q_t)$ change with time. This flexible observation model allows us to model variations in the occurrences of a given state. This idea has previously been explored for Gaussian mixture models [5]. It should be noted that the proposed model collapses to a regular non-negative factorization if we use only a single state, therefore only one dictionary.

3 Model for Sound Mixtures

In this section, we describe the non-negative factorial hidden Markov model (N-FHMM) for modeling sound mixtures. We then describe how to perform source separation using the model.

As shown in the two-source graphical model in Fig. 3, the N-FHMM combines multiple N-HMMs of single sources. The interaction model introduces a new variable s_t that indicates the source. In the generative process, for each draw of each time frame, we first choose the source and then choose the latent component as before. In order to perform separation, we use trained models of individual sources. We train an N-HMM and learn the dictionaries and the transition matrix for each class of sound we expect to encounter in a mixture. We then use the a priori source information in these trained models to resolve mixtures that involve such sources. The dictionaries and transition matrices of the N-FHMM will therefore already be defined, and one will only need to estimate the appropriate weights from the mixture.

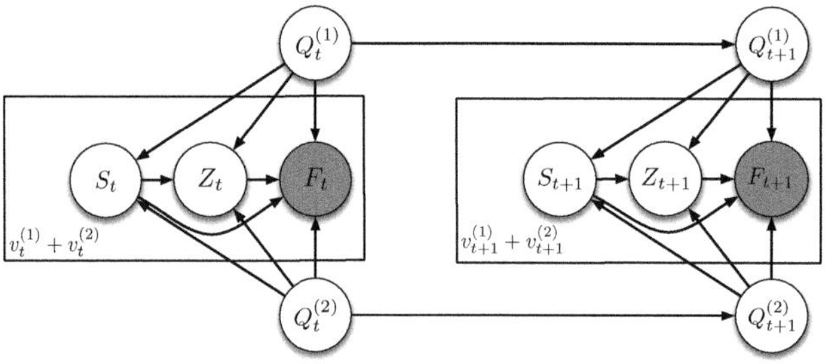

Fig. 3. N-FHMM. The structure of two individual N-HMMs (one in the upper half and one in the lower half) can be seen in this model.

In a given time frame t, each source is explained by one of its dictionaries. Therefore, a given mixture is modeled by a pair of dictionaries, $\{q_t^{(1)}, q_t^{(2)}\}$, one for each source (superscripts indicate the source). For a given pair of dictionaries, the mixture spectrum is defined by the following interaction model:

$$P(f_t|q_t^{(1)}, q_t^{(2)}) = \sum_{s_t} \sum_{z_t} P(f_t|z_t, s_t, q_t^{(s_t)}) P(z_t, s_t|q_t^{(1)}, q_t^{(2)}) . \tag{7}$$

As can be seen, the mixture spectrum is modeled as a linear combination of the individual sources which are in turn modeled as a linear combination of spectral vectors from the given dictionaries. This allows us to model the mixture as a linear combination of the spectral vectors from the given pair of dictionaries [3].

3.1 Source Separation

In order to perform separation, we need to first estimate the mixture weights, $P(z_t, s_t|q_t^{(1)}, q_t^{(2)})$ for each pair of states. That can be done using the EM algorithm. The E-step is computed as follows:

$$P(z_t, s_t, q_t^{(1)}, q_t^{(2)}|f_t, \bar{\mathbf{f}}, \bar{\mathbf{v}}) = \frac{\alpha(q_t^{(1)}, q_t^{(2)})\beta(q_t^{(1)}, q_t^{(2)})}{\sum_{q_t^{(1)}}\sum_{q_t^{(2)}} \alpha(q_t^{(1)}, q_t^{(2)})\beta(q_t^{(1)}, q_t^{(2)})} P(z_t, s_t|f_t, q_t^{(1)}, q_t^{(2)}), \tag{8}$$

where

$$P(z_t, s_t|f_t, q_t^{(1)}, q_t^{(2)}) = \frac{P(f|z_t, s_t, q_t^{(s_t)})P(z_t, s_t|q_t^{(1)}, q_t^{(2)})}{\sum_{s_t}\sum_{z_t} P(f|z_t, s_t, q_t^{(s_t)})P(z_t, s_t|q_t^{(1)}, q_t^{(2)})}. \tag{9}$$

$\alpha(q_t^{(1)}, q_t^{(2)})$ and $\beta(q_t^{(1)}, q_t^{(2)})$ are computed with a two dimensional forward–backward algorithm [6] using the likelihoods of the data, $P(\mathbf{f_t}, v_t|q_t^{(1)}, q_t^{(2)})$, for each pair of states. The likelihoods are computed as follows:

$$P(\mathbf{f_t}, v_t|q_t^{(1)}, q_t^{(2)}) = P(v_t|q_t^{(1)}, q_t^{(2)}) \prod_{f_t}\left(\sum_{s_t}\sum_{z_t} P(f_t|z_t, s_t, q_t^{(s_t)})P(z_t, s_t|q_t^{(1)}, q_t^{(2)})\right)^{V_{ft}}. \tag{10}$$

The weights are computed in the M-step as follows:

$$P(z_t, s_t|q_t^{(1)}, q_t^{(2)}) = \frac{\sum_{f_t} V_{ft}P(z_t, s_t, q_t^{(1)}, q_t^{(2)}|f_t, \bar{\mathbf{f}}, \bar{\mathbf{v}})}{\sum_{s_t}\sum_{z_t}\sum_{f_t} V_{ft}P(z_t, s_t, q_t^{(1)}, q_t^{(2)}|f_t, \bar{\mathbf{f}}, \bar{\mathbf{v}})}. \tag{11}$$

Once we estimate the weights using the EM algorithm, we compute the proportion of the contribution of each source at each time–frequency bin as follows:

$$P(s_t|f_t, \bar{\mathbf{f}}, \bar{\mathbf{v}}) = \frac{\sum_{q_t^{(1)}}\sum_{q_t^{(2)}} P(q_t^{(1)}, q_t^{(2)}|\bar{\mathbf{f}}, \bar{\mathbf{v}})\sum_{z_t} P(f|z_t, s, q_t^{(s)})P(z_t, s|q_t^{(1)}, q_t^{(2)})}{\sum_{s_t}\sum_{q_t^{(1)}}\sum_{q_t^{(2)}} P(q_t^{(1)}, q_t^{(2)}|\bar{\mathbf{f}}, \bar{\mathbf{v}})\sum_{z_t} P(f|z_t, s, q_t^{(s)})P(z_t, s|q_t^{(1)}, q_t^{(2)})}, \tag{12}$$

where

$$P(q_t^{(1)}, q_t^{(2)}|\bar{\mathbf{f}}, \bar{\mathbf{v}}) = \frac{\alpha(q_t^{(1)}, q_t^{(2)})\beta(q_t^{(1)}, q_t^{(2)})}{\sum_{q_t^{(1)}}\sum_{q_t^{(2)}} \alpha(q_t^{(1)}, q_t^{(2)})\beta(q_t^{(1)}, q_t^{(2)})}. \tag{13}$$

---

[3] We deal with $P(z_t, s_t|q_t^{(1)}, q_t^{(2)})$ rather than dealing with $P(z_t|s_t, q_t^{(1)}, q_t^{(2)})$ and $P(s_t|q_t^{(1)}, q_t^{(2)})$ individually (as shown in the graphical model) so that we will have a single set of mixture weights over both sources.

This effectively gives us a soft mask with which we modulate the mixture spectrogram to obtain the separated spectrograms of the individual sources. In Eq. 12, we sum the contribution of every pair of states. This implies that the reconstruction of each source has contributions from each of its dictionaries. However, in practice, $P(q_t^{(1)}, q_t^{(2)}|\bar{\mathbf{f}}, \bar{\mathbf{v}})$ tends to zero for all but one $\{q_t^{(1)}, q_t^{(2)}\}$ pair, effectively using only one dictionary per time frame per source. This happens because the dictionaries of individual source models are learned in such a way that each time frame is explained almost exclusively by one dictionary. The provision of having a small contribution from more than one dictionary is sometimes helpful in modeling the decay of the active dictionary in the previous time frame.

4 Experimental Results

We performed speech separation experiments on data from the TIMIT database. Specifically, we performed separation on eight pairs of speakers. Each speaker pair consists of one male and one female speaker. We first used nine sentences of each speaker as training data and learned individual N-HMM model parameters as described in Sec. 2.3. Specifically, for each speaker, we first obtained a spectrogram with a window size of 1024 and a hop size of 256 (at F_s=16,000). We then learned a model of the spectrogram with 40 dictionaries with 10 latent components each (K=10). We then repeated the experiment with 1 latent component per dictionary (K=1). After training, we combined the models into a joint model as described in Sec. 3. We constructed test data by artificially mixing one unseen sentence from each speaker at 0dB and performed separation[4]. The separation yields estimated magnitude spectrograms for each source. We used the phase of the mixture and resynthesized each source.

As a comparison, we then performed the same experiments using a traditional non-negative factorization approach. The experimental procedure as well as the training and test data are the same as above. After thorough testing, we found that the optimal results were obtained in the non-negative factorization approach by using 30 components per speaker and we therefore used this for the comparison to the proposed model. The separation performance increases up to using 30 components. When more components are used, the dictionary of one source starts to explain the other source and the separation performance goes down. It should be noted that this is equivalent to using the proposed models with 1 dictionary of 30 components per speaker.

We evaluated the separation performance in terms of the metrics defined in [7]. The averaged results over the eight pairs of speakers are as follows:

	SDR (dB)	SIR (dB)	SAR (dB)
N-FHMM (K=10)	6.49	14.07	7.74
N-FHMM (K=1)	5.58	12.07	7.26
Factorization	4.82	8.65	7.95

[4] Examples at https://ccrma.stanford.edu/~gautham/Site/lva_ica_2010.html

As shown in the table, the performance of the N-FHMM (by all metrics) is better when we use 10 components rather than 1 component. This shows the need to use a dictionary to model each state rather than a single component. We found no appreciable improvement in performance by using more than 10 components per dictionary. We see an improvement over factorizations in the overall performance of the N-FHMM (SDR). Specifically, we see a large improvement in the actual suppression of the unwanted source (SIR). We however see a small increase in the introduced artifacts (SAR). The results intuitively make sense. The N-FHMM performs better in suppressing the competing source by enforcing a reasonable temporal arrangement of each speaker's dictionary elements, therefore not simultaneously using dictionary elements that can describe both speakers. On the other hand, this exclusive usage of smaller dictionaries doesn't allow us to model the source as well as we would otherwise (with 1 component per dictionary being the extreme case). There is therefore an inherent trade-off in the suppression of the unwanted source and the reduction of artifacts.

Traditional factorial HMMs that use a Gaussian for the observation model have also been used for source separation [8,9]. As in [3], these methods model each time frame of each source with a single spectral vector. The proposed model, on the other hand, extends non-negative factorizations by modeling each time frame of each source as a linear combination of spectral vectors. As shown above, this type of modeling can be advantageous for source separation.

5 Conclusions

We have presented new models and associated estimation algorithms that model the non-stationarity and temporal structure of audio. We presented a model for single sources and a model for sound mixtures. The performance of the proposed model was demonstrated on single channel source separation and was shown to have a much higher suppression capability than similar approaches that do not incorporate temporal information. The computational complexity is exponential in the number of sources (as with traditional factorial HMMs). Therefore, approximate inference algorithms [6] such as variational inference is an area for future work. Although the model was only demonstrated on source separation in this paper, it can be useful for various applications that deal with sound mixtures such as concurrent speech recognition and automatic music transcription.

References

1. Smaragdis, P., Brown, J.C.: Non-negative matrix factorization for polyphonic music transcription. In: WASPAA (2003)
2. Rabiner, L.R.: A tutorial on hidden Markov models and selected applications in speech recognition. Proceedings of the IEEE 77(2) (1989)
3. Ozerov, A., Fevotte, C., Charbit, M.: Factorial scaled hidden Markov model for polyphonic audio representation and source separation. In: WASPAA (2009)
4. Smaragdis, P., Raj, B., Shashanka, M.: A probabilistic latent variable model for acoustic modeling. In: Advances in models for acoustic processing, NIPS (2006)

5. Benaroya, L., Bimbot, F., Gribonval, R.: Audio source separation with a single sensor. IEEE TASLP 14(1) (2006)
6. Ghahramani, Z., Jordan, M.: Factorial hidden Markov models. Machine Learning 29 (1997)
7. Vincent, E., Gribonval, R., Févotte, C.: Performance measurement in blind audio source separation. IEEE TASLP 14(4) (2006)
8. Hershey, J.R., Kristjansson, T., Rennie, S., Olsen, P.A.: Single channel speech separation using factorial dynamics. In: NIPS (2007)
9. Virtanen, T.: Speech recognition using factorial hidden Markov models for separation in the feature space. In: Proceedings of Interspeech (2006)

Nonnegative Matrix Factorization with Markov-Chained Bases for Modeling Time-Varying Patterns in Music Spectrograms

Masahiro Nakano[1], Jonathan Le Roux[2], Hirokazu Kameoka[2], Yu Kitano[1], Nobutaka Ono[1], and Shigeki Sagayama[1]

[1] Graduate School of Information Science and Technology, The University of Tokyo
[2] NTT Communication Science Laboratories, NTT Corporation

Abstract. This paper presents a new sparse representation for polyphonic music signals. The goal is to learn the time-varying spectral patterns of musical instruments, such as attack of the piano or vibrato of the violin in polyphonic music signals without any prior information. We model the spectrogram of music signals under the assumption that they are composed of a limited number of components which are composed of Markov-chained spectral patterns. The proposed model is an extension of nonnegative matrix factorization (NMF). An efficient algorithm is derived based on the auxiliary function method.

Keywords: Nonnegative matrix factorization, Sparse signal representation, Source separation, Markov chain, Auxiliary function.

1 Introduction

The use of sparse representation in acoustic signal processing has been a very active area of research in recent years, with very effective algorithms based on nonnegative matrix factorization (NMF) [1] and sparse coding. These are typically based on a simple linear model.

NMF, in particular, has been applied extensively with considerable success to various problems including automatic music transcription, monaural sound source separation [2]. NMF is able to project all signals that have the same spectral shape on a single basis, allowing one to represent a variety of phenomena efficiently using a very compact set of spectrum bases. However, because NMF is also fundamentally a dimension reduction technique, a lot of information on the original signal is lost. This is in particular what happens when assuming that the spectrum of the note of a musical instrument can be represented through a single spectral basis whose amplitude is modulated in time, while its variations in time are actually much richer. For example, a piano note would be more accurately characterized by a succession of several spectral patterns such as "attack", "decay", "sustain" and "release". As another example, singing voices and stringed instruments feature a particular musical effect, vibrato, which can

V. Vigneron et al. (Eds.): LVA/ICA 2010, LNCS 6365, pp. 149–156, 2010.

be characterized by its "depth", (the amount of pitch variation), and its "speed", (the speed at which the pitch varies). Learning such time-varying spectra with standard NMF would require to use a large number of bases, and some post-processing to group the bases into single events.

In this paper, we propose a new sparse representation, "NMF with Markov-chained bases" for modeling time-varying patterns in music spectrograms. Our model represents a single event as a succession of spectral patterns. The proposed model is presented in Section 2, together with the derivation of an efficient algorithm to optimize its parameters. We present basic experimental results in Section 3.

2 NMF with Markov-Chained Bases

2.1 Presentation of the Model

Most algorithms for unsupervised sound source separation are based on a signal model where the magnitude or power spectrogram $\mathbf{Y} = (Y_{\omega,t})_{\Omega \times T} \in \mathbb{R}^{\geq 0, \Omega \times T}$, where $\omega = 1, \cdots, \Omega$ is a frequency bin index, and $t = 1, \cdots, T$ is a time frame index, is factorized into nonnegative parameters, $\mathbf{H} = (H_{\omega,d})_{\Omega \times D} \in \mathbb{R}^{\geq 0, \Omega \times D}$ and $\mathbf{U} = (U_{d,t})_{D \times T} \in \mathbb{R}^{\geq 0, D \times T}$. This can be written as

$$Y_{\omega,t} = \sum_{d=1}^{D} H_{\omega,d} U_{d,t} \ , \tag{1}$$

where D is the number of bases $\mathbf{h}_d = [H_{1,d}, \cdots, H_{\Omega,d}]$. The term *component* is used to refer to one basis $\mathbf{h}_d$ and its time-varying gain $U_{d,t}$. The bases can be considered as spectral patterns which are frequently observed.

Hopefully, one component should represent a single event. However, the spectrum of instrument sounds is actually in general nonstationary. Each source will thus tend to be modeled as a sum several components, leading to the difficult problem of determining which source each component belongs to. Recently, an extension of NMF where temporal activations become time/frequency activations based on a source/filter model [3] have been proposed to overcome this problem. However, it has not been clarified whether the source/filter model is fit for auditory stream composed of various origin, such as the piano. For example, "attack" caused by a keystroke has energies on wide-band spectrum, while "sustain" has a harmonic structure.

In contrast with the above approaches, we focus on the hierarchical structure of the sounds produced by musical instruments, and model the spectrogram of music signals under the assumption that they are composed of spectral patterns which have themselves a limited number of Markov-chained states. Concretely, we assume that each basis $\mathbf{h}_d$ has Φ states, the transitions between those states being constrained and only one state being activated at each time t.

We attempt to model the spectrogram again based on $\mathbf{H} = (H_{\omega,\phi,d})_{\Omega \times \Phi \times D} \in \mathbb{R}^{\geq 0, \Omega \times \Phi \times D}$ Here, $\mathbf{P} = (P_{\phi,t,d})_{\Phi \times T \times D} \in \mathbb{R}^{\geq 0, \Phi \times T \times D}$ is binary to show which basis is activated at time t, i.e., $P_{\phi,t,d} = 1$ if $\mathbf{h}_d^{(\phi)}$ is activated at time t, and $P_{\phi,t,d} = 0$ otherwise. Then, the proposed model can be written as

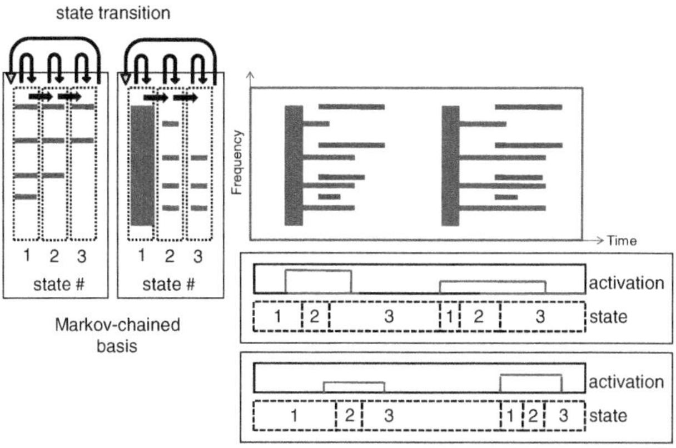

Fig. 1. Diagram of NMF with Markov-chained bases

$$Y_{\omega,t} = \sum_{d,\phi} H_{\omega,\phi,d} P_{\phi,t,d} U_{d,t} \quad . \tag{2}$$

Note that $\mathbf{P}$ does not show the probabilities that state ϕ is activated at time t and the class $\mathfrak{S}$ of all possible $\mathbf{P}$ is defined by the topology of the Markov chain, for example left-to-right model or ergodic model.

2.2 Problem Setting

Given an observed spectrogram, we would like to find the optimal estimates of $\mathbf{H}$, $\mathbf{U}$ and $\mathbf{P}$. The standard NMF algorithm proposed by Lee and Seung [1] performs the decomposition by minimizing the reconstruction error between the observation and the model while constraining the matrices to be entry-wise non-negative. Our estimation can also be written as optimization problem, similarly to standard NMF. Various measures for standard NMF have been proposed. We choose here the following β-divergence, which has been widely used,

$$\mathcal{D}_\beta(y|x) = \frac{\left(y^\beta + (\beta-1)x^\beta - \beta y x^{\beta-1}\right)}{\beta(\beta-1)} \quad . \tag{3}$$

Note that the above definition can be properly extended by continuity to $\beta = 0$ and $\beta = 1$ [5]. Let θ denote the set of all parameters $\{(H_{\omega,\phi,d})_{\Omega\times\Phi\times D}, (U_{d,t})_{D\times T}, (P_{\phi,t,d})_{\Phi\times T\times D}\}$. We then need to solve the following optimization problem:

$$\begin{array}{l} \text{minimize} \;\; \mathcal{J}(\theta) = \sum_{\omega,t} \mathcal{D}_\beta\left(Y_{\omega,t} \middle| \sum_{d,\phi} H_{\omega,\phi,d} P_{\phi,t,d} U_{d,t}\right) \\ \text{subject to} \; \forall\omega,\phi,d,\; H_{\omega,\phi,d} \geq 0,\; \forall d,t,\; U_{d,t} \geq 0,\; \mathbf{P} \in \mathfrak{S} \; . \end{array} \tag{4}$$

In this paper, we assume that the transition probabilities of the Markov-chained bases are uniform. Thus, the cost of each of the paths is regarded as constant and ignored in Eq. (4). In the following, we will refer to the algorithm to solve NMF with Markov-chained bases as "MNMF".

Note that our model can also be expressed as Factorial Hidden Markov Model for specific βs. The case of $\beta = 0$ reduces to Factorial scaled Hidden Markov Model which reduces to NMF with Itakura-Saito (IS) divergence and Gaussian Scaled Mixture Model [4]. These probabilistic model may achieve the extension of NMF with the Euclidean distance ($\beta = 2$), the generalized Kullback-Leibler divergence ($\beta = 1$) and IS divergence ($\beta = 0$) based on a statistical approach to NMF [5,6]. However, it is not clarified whether statistical approach can apply to NMF with various measures, such as β-divergence.

2.3 Iterative Algorithm

Our derivation is based on a principle called the auxiliary function method, similar to [1]. Let $G(\theta)$ denote an objective function to be minimized w.r.t. a parameter θ. A function $G^+(\theta,\hat{\theta})$ which satisfies $G(\theta) = \min_{\hat{\theta}} G^+(\theta,\hat{\theta})$ is then called an auxiliary function for $G(\theta)$, and $\hat{\theta}$ an auxiliary variable. The function $G(\theta)$ is easily shown to be non-increasing through the following iterative update rules: $\hat{\theta}^{(s+1)} \leftarrow \operatorname{argmin}_{\hat{\theta}} G^+(\theta^{(s)},\hat{\theta})$ and $\theta^{(s+1)} \leftarrow \operatorname{argmin}_{\theta} G^+(\theta,\hat{\theta}^{(s+1)})$, where $\hat{\theta}^{(s+1)}$ and $\theta^{(s+1)}$ denote the updated values of $\hat{\theta}$ and θ after the s-th step.

An auxiliary function for standard NMF with β-divergence has been proposed [8]. This strategy can apply to our problem. $\hat{\theta}$ denotes auxiliary variables $\{(\lambda_{\omega,t,k})_{\Omega\times T\times K},(Z_{\omega,t})_{\Omega\times T}\}$ ($\forall k, \lambda_{\omega,t,k} \geq 0, \sum_k \lambda_{\omega,t,k} = 1, Z_{\omega,t} \in \mathbb{R}$) for convenience. We obtain the following auxiliary function:

$$\mathcal{J}^+(\theta,\hat{\theta}) = \sum_{\omega,t} \frac{Y_{\omega,t}}{\beta(\beta-1)} + \sum_{\omega,t} \begin{cases} \mathcal{R}^{(\beta)}_{\omega,t} - Y_{\omega,t}\mathcal{Q}^{(\beta-1)}_{\omega,t} & (\beta < 1) \\ \mathcal{Q}^{(\beta)}_{\omega,t} - Y_{\omega,t}\mathcal{Q}^{(\beta-1)}_{\omega,t} & (1 \leq \beta \leq 2) \\ \mathcal{Q}^{(\beta)}_{\omega,t} - Y_{\omega,t}\mathcal{R}^{(\beta-1)}_{\omega,t} & (\beta > 2) \end{cases} , \tag{5}$$

where $\mathcal{Q}^{(\beta)}_{\omega,t} = (1/\beta)\sum_{d,\phi} \lambda_{\omega,t,d} P_{\phi,t,d}(H_{\omega,\phi,d}U_{d,t}/\lambda_{\omega,t,d})^{\beta}$ and $\mathcal{R}^{(\beta)}_{\omega,t} = (1/\beta)Z^{\beta}_{\omega,t}$ $+Z^{\beta-1}_{\omega,t}(\sum_{d,\phi} H_{\omega,\phi,d}P_{\phi,t,d}U_{d,t} - Z_{\omega,t})$. $\mathcal{J}^+(\theta,\hat{\theta})$ is minimized w.r.t. $\hat{\theta}$ when

$$\lambda_{\omega,t,d} = \frac{\sum_{\phi} H_{\omega,\phi,d}P_{\phi,t,d}U_{d,t}}{\sum_{\phi,d} H_{\omega,\phi,d}P_{\phi,t,d}U_{d,t}} , \quad Z_{\omega,t} = \sum_{\phi,d} H_{\omega,\phi,d}P_{\phi,t,d}U_{d,t} . \tag{6}$$

Minimizing $\mathcal{J}^+(\theta,\hat{\theta})$ w.r.t. $\mathbf{P} \in \mathfrak{S}$ is a search problem

$$\mathbf{P} \leftarrow \operatorname*{argmin}_{\mathbf{P}\in\mathfrak{S}} \left\{\mathcal{J}^+(\theta,\hat{\theta})\right\} , \tag{7}$$

which can be straightforwardly solved using the Viterbi algorithm. Differentiating $\mathcal{J}^+(\theta,\hat{\theta})$ partially w.r.t. $H_{\omega,\phi,t}$ and $U_{d,t}$, and setting to zero, we obtain update rules for $H_{\omega,\phi,d}$ and $U_{d,t}$:

$$H_{\omega,\phi,d} \leftarrow \left(\frac{\sum_t \alpha^{\beta-2}_{\omega,t} Y_{\omega,t} \gamma^{\mathrm{Max}\{2-\beta,\ 0\}}_{\omega,t,d} P_{\phi,t,d}U_{d,t}}{\sum_t \alpha^{\beta-1}_{\omega,t} \gamma^{\mathrm{Min}\{1-\beta,\ 0\}}_{\omega,t,d} P_{\phi,t,d}U_{d,t}} \right)^{\varphi(\beta)} , \tag{8}$$

$$U_{d,t} \leftarrow U_{d,t} \left(\frac{\sum_{\omega} \alpha^{\beta-2}_{\omega,t} Y_{\omega,t}\gamma_{\omega,t,d}}{\sum_{\omega} \alpha^{\beta-1}_{\omega,t} \gamma_{\omega,t,d}} \right)^{\varphi(\beta)} , \tag{9}$$

where $\alpha_{\omega,t} = \sum_{d,\phi} H_{\omega,\phi,d}P_{\phi,t,d}U_{d,t}$, $\gamma_{\omega,t,d} = \sum_{\phi} H_{\omega,\phi,d}P_{\phi,t,d}$ and $\varphi(\beta) = 1/(2-\beta)$ $(\beta < 1)$, 1 $(1 \leq \beta \leq 2)$, $1/(\beta-1)$ $(\beta > 2)$.

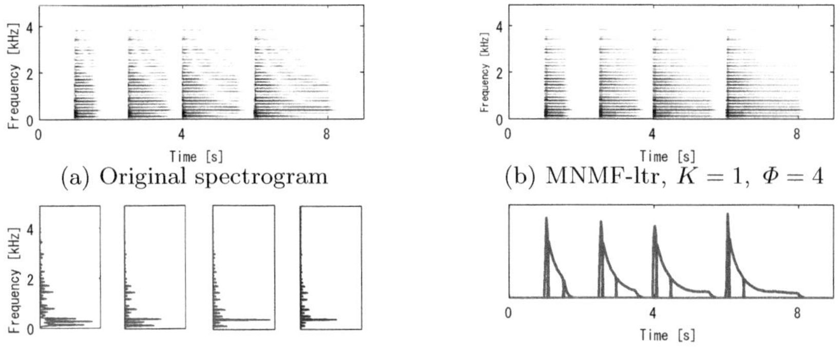

(a) Original spectrogram

(b) MNMF-ltr, $K = 1,\ \Phi = 4$

(c) Markov-chained bases, $K = 1,\ \Phi = 4$

(d) Activation, $K = 1,\ \Phi = 4$

Fig. 2. Original spectrogram of the extract of the piano (MIDI) (a), reconstructed spectrograms (b), the bases (c) and the activation (d) learned by MNMF. MNMF was able to decompose the evolution of the spectrum as a succession of several part, "attack", "sustain", "decay" and "release".

2.4 Update Scheduling

The algorithm described above converges quickly. However, it often falls into unexpected stationary points. **H** and **U** at the early stage of iterations are not useful for estimating **P**. **P** fixed using unrealistic **H** and **U** induces in turn **H** and **U** in unexpected directions. We thus improve the update rule for **P** by introducing an updating schedule. Here, let the scheduling parameter, $k_\phi^{(s)}$, at s-th iteration satisfy $\forall\phi,\ k_\phi \geq 0,\ \sum_\phi k_\phi = 1,\ k_1^{(s+1)} \geq k_1^{(s)},\ k_1^{(S)} = 1,\ k_\phi^{(s+1)} \leq k_\phi^{(s)},\ k_\phi^{(S)} = 0\ (\phi = 2, \cdots, \Phi)$. We replace the update rule, Eq. (7) by

$$P_{\phi,t,d}^{(s+1)} \leftarrow \sum_{n=1}^{\Phi} k_n^{(s+1)} \hat{P}_{\phi-(n-1),t,d}\ , \tag{10}$$

where $\forall n,\ \hat{P}_{-(n-1),t,d} = \hat{P}_{\Phi-(n-1),t,d}$ and $(\hat{P}_{\phi,t,d})_{\Phi\times T\times D} = \mathrm{argmin}_{\mathbf{P}\in\mathfrak{S}}\, \mathcal{J}^+(\theta, \hat{\theta})$. As a result, at each iteration the auxiliary function is not minimized anymore. However, convergence is guaranteed.

3 Simulation Results

In this section, some results on the application of our algorithm to audio signals. All data were downmixed to mono and downsampled to 16kHz. The magnitude spectrogram was computed using the short time Fourier transform with 32 ms long Hanning window and with 16 ms overlap. The state transitions of the bases in MNMF were modeled using left-to-right (-ltr) and ergodic (-erg) models.

At first, we tested whether the algorithm was able to learn in an unsupervised way the time-varying spectral patterns from notes with a unique pitch. The proposed method was applied to a piano note (C3) synthesized from MIDI, a piano note (C3) recorded from RWC-MDB-I-2001 No.1 [7] and a violin note

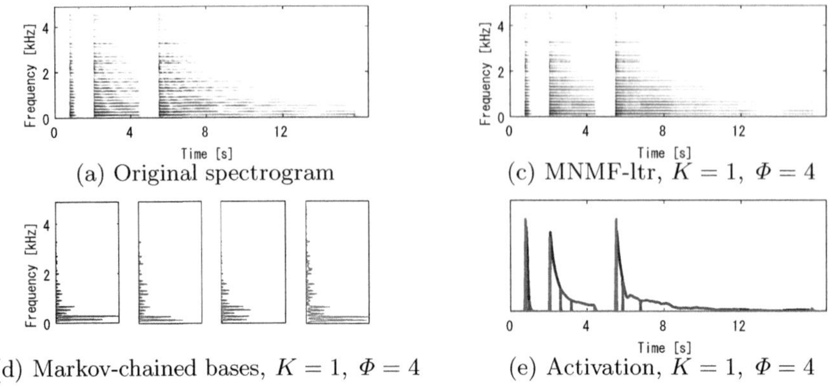

Fig. 3. Original spectrogram of the extract of the piano (RWC database) (a), reconstructed spectrograms (b) and (d), the bases (c) and the activation (d) learned by MNMF. Time-varying spectral patterns are also learned.

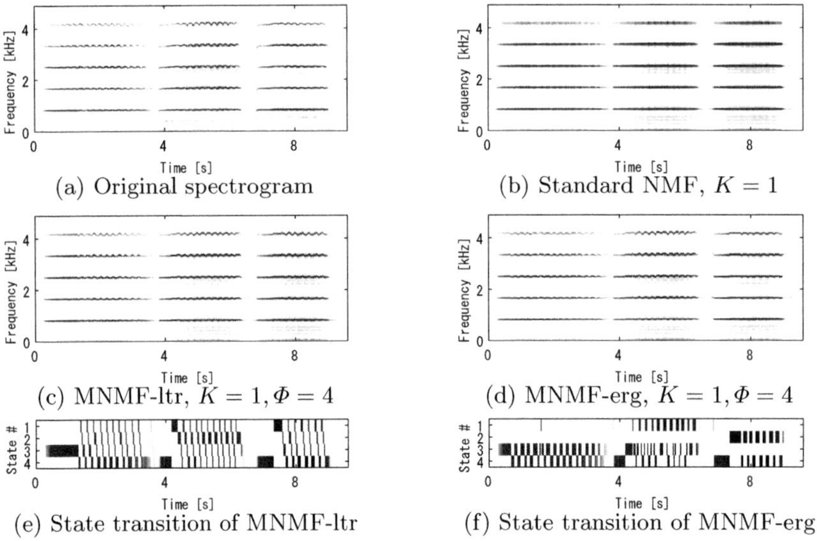

Fig. 4. Original spectrogram of the extract of the violin (RWC database) (a), reconstructed spectrograms (b), (c) and (d), and the activation (e) and (f) learned by MNMF. The topology of the Markov chain affects the state transitions.

(A♭) recorded from RWC-MDB-I-2001 No.15. We used MNMF with $\beta = 1$ (the generalized Kullback-Leibler divergence). As shown in Fig. 2, 3 and 4, time-varying spectral patterns are learned in an unsupervised way.

Next, we applied our model to a mixture of vocal signals taken from RWC-MDB-I-2001 No.45. The sequence is composed of 3 notes (D♭CFCA♭): first, each note is played alone in turn, then all combination of two notes are played and finally all notes are played simultaneously. The result is shown in Fig. 5.

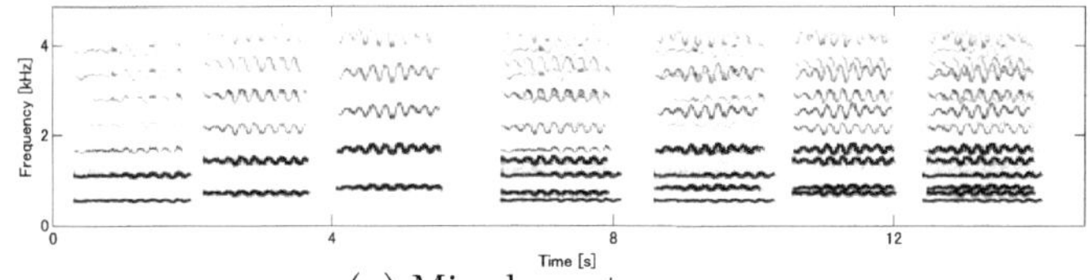

(a) Mixed spectrogram

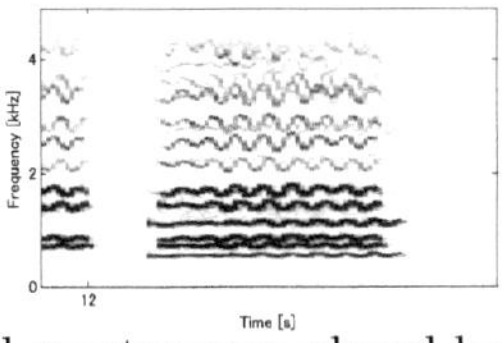

(b) Partial spectrogram played by all notes.

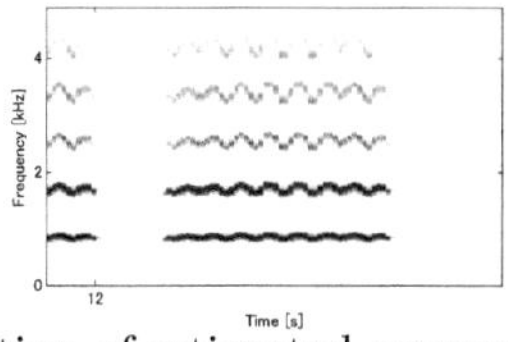

(c) Portion of estimated component: A♭

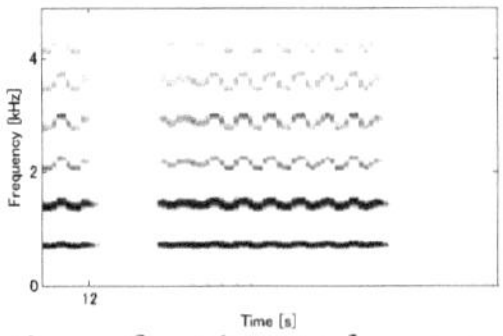

(d) Portion of estimated component: F

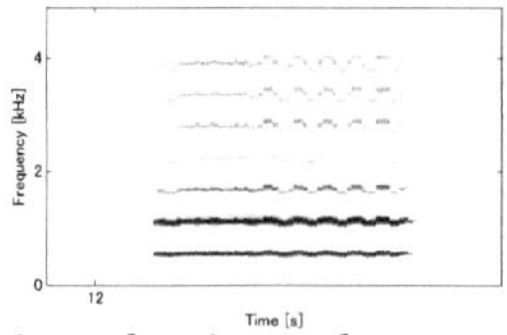

(e) Portion of estimated component: D♭

Fig. 5. Mixed spectrogram (a) and (b), and one portion of estimated spectrogram of each component (c), (d) and (e)

Table 1. Source Separation Performance

Algorithms	RWC signal		Real-world audio signal	
	Number of **H** and **U**	SNR(dB)	Number of **H** and **U**	SNR(dB)
NMF($D=3$)	2919	3.55	2871	6.49
NMF($D=6$)	5838	7.11	5742	9.61
NMF($D=9$)	**8757**	**10.63**	**8613**	**10.50**
NMF($D=12$)	11676	13.51	11484	11.70
MNMF-erg($\Phi=4$)	7536	9.62	7488	9.57
MNMF-erg($\Phi=5$)	**9075**	**11.02**	**9027**	**9.84**

Finally, our model was applied to sound source separation. The tested signals are RWC signal (as shown in Fig. 5) and audio data recorded in real-world conditions from male vocal in the room size of 5.5m × 3.5m × 3m by IC recorder, whose sequence is composed similar to RWC signal (A, D♭ and E instead of D♭CF and A♭). We use the signal-to-noise ratio (SNR) between each component and source as the measure for assigning components to sources. The measure was calculated between the magnitude spectrograms $I_{\omega,t}^{(m)}$ and $\hat{I}_{\omega,t}^{(n)}$ of the mth reference and the nth separated component, respectively, $\mathrm{SNR} = 10\log_{10}(\sum_{\omega,t}(I_{\omega,t}^{(m)})^2/\sum_{\omega,t}(I_{\omega,t}^{(m)}-\hat{I}_{\omega,t}^{(n)})^2)$. The SNR was averaged over all the sources to get the separation performance and each algorithm was run 10 times with random initializations. We set the number of bases to $D=3$

for MNMF. This means that one component with Φ spectral patterns may be expected to represent one source having the perceveid pitch of a note with vibrato. Standard NMF was used as the baseline. NMF with $D = 3$ was expected to separate each source into a single component, while for NMF with $D > 3$ each source was expected to be split into the sum of several components. A component n is assigned to the source m which leads to the highest SNR. As reported by [2], using NMF with a large number of components and clustering them using the original signals as references may produce unrealistically good results. One should thus keep in mind when comparing the results of our method with those of NMF that the experimental conditions were strongly biased in favor of NMF. As shown in Table 1, we can see that MNMF performs as well as standard NMF when the total number of parameters $\mathbf{H}$ and $\mathbf{H}$ is similar ($\mathbf{P}$ is excluded), although again for NMF the original signals need to be used to cluster the extracted components.

4 Concluding Remarks

We developed a new framework for the sparse representation of audio signals. The proposed model is extension of NMF, in which the bases consist of state transition. We derived an efficient algorithm for the optimization of the model based on the auxiliary function method. Future work will includes the extension of our model to automatic estimation of the number of bases and states.

Acknowledgements

This research was partially supported by CrestMuse Project under JST from Mext, Japan, and Grant-in-Aid for Scientific Research (KAKENHI) (A) 20240017.

References

1. Lee, D.D., Seung, H.S.: Algorithms for non-negative matrix factorization. In Proc. NIPS, vol. 13, pp. 556–562 (December 2001)
2. Virtanen, T.: Monaural sound source separation by nonnegative matrix factorization with temporal continuity and sparseness criteria. IEEE Transactions on Audio, Speech, and Language Processing 15, 1066–1074 (2007)
3. Hennequin, R., Badeau, R., David, B.: NMF with time-frequency activations to model non stationary audio events. In: Proc. ICASSP, pp. 445–448 (March 2010)
4. Ozerov, A., Fevotte, C., Charbit, M.: Factorial scaled hidden Markov model for polyphonic audio representation and source separation. In Proc. WASPAA (2009)
5. Févotte, C., Bertin, N., Durrieu, J.-L.: Nonnegative matrix factorization with the itakura-saito divergence with application to music analysis. Neural Computation 21, 793–830 (2009)
6. Févotte, C., Cemgil, A.T.: Nonnegative matrix factorizations as probabilistic inference in composite models. In: Proc. EUSIPCO, vol. 47, pp. 1913–1917 (2009)
7. Goto, M., Hashiguchi, H., Nishimura, T., Oka, R.: RWC music database: Popular, classical, and jazz music database. In: Proc. ISMIR, pp. 287–288 (2002)
8. Nakano, M., Kameoka, H., Le Roux, J., Kitano, Y., Ono, N., Sagayama, S.: Convergence-guaranteed multiplicative algorithms for nonnegative matrix factorization with β-divergence. In: Proc. MLSP (2010)

An Experimental Evaluation of Wiener Filter Smoothing Techniques Applied to Under-Determined Audio Source Separation

Emmanuel Vincent

INRIA, Centre de Rennes - Bretagne Atlantique
Campus de Beaulieu, 35042 Rennes Cedex, France
emmanuel.vincent@inria.fr

Abstract. Multichannel under-determined source separation is often carried out in the time-frequency domain by estimating the source coefficients in each time-frequency bin based on some sparsity assumption. Due to the limited amount of data, this estimation is often inaccurate and results in musical noise artifacts. A number of single- and multichannel smoothing techniques have been introduced to reduce such artifacts in the context of speech denoising but have not yet been systematically applied to under-determined source separation. We present some of these techniques, extend them to multichannel input when needed, and compare them on a set of speech and music mixtures. Many techniques initially designed for diffuse and/or stationary interference appear to fail with directional nonstationary interference. Temporal covariance smoothing provides the best tradeoff between artifacts and interference and increases the overall signal-to-distortion ratio by up to 3 dB.

1 Introduction

Source separation is the task of recovering the contribution or *spatial image* $\mathbf{c}_j(t)$ of each source indexed by j, $1 \leq j \leq J$, within a multichannel mixture signal $\mathbf{x}(t)$ with

$$\mathbf{x}(t) = \sum_{j=1}^{J} \mathbf{c}_j(t). \tag{1}$$

In this paper, we focus on audio source separation [14] in the under-determined setting when the number of mixture channels I is such that $1 < I < J$. This task is typically addressed in the time-frequency domain via the Short-Term Fourier Transform (STFT). A popular approach consists of exploiting the sparsity of audio sources in this domain so as to estimate the STFT coefficients of the most prominent sources in each time-frequency bin and set the STFT coefficients of the other sources to zero. Depending on the assumed number of active sources from 1 to I and on the chosen estimation criterion, this leads to algorithms such as *binary masking* [11], soft masking [3] or ℓ_1-*norm minimization* [17].

V. Vigneron et al. (Eds.): LVA/ICA 2010, LNCS 6365, pp. 157–164, 2010.

Although these separation algorithms succeed at reducing *interference* from unwanted sources, they generate a significant amount of time- and frequency-localized *artifacts* also known as musical noise [15]. These artifacts are particularly annoying in scenarios such as hearing-aid speech processing or high-fidelity music processing where fewer artifacts are preferred at the expense of increased interference. Adaptive time-frequency representations with maximal sparsity or STFTs with increased frame overlap only moderately reduce artifacts [10,2]. Indeed, the localized nature of artifacts is due to the limited amount of data available for estimation in each time-frequency bin, such that similar mixture STFT coefficients in neighboring time-frequency bins may result in very dissimilar estimated source STFT coefficients. This causes strong discontinuities independently of the chosen representation. Joint processing of several time-frequency bins is needed to further reduce artifacts.

One approach consists of modeling the dependencies between the STFT coefficients of the spatial image of each source via some joint probabilistic prior. For instance, these coefficients may be locally modeled as zero-mean Gaussian vector random variables whose covariance matrices are either constant over neighboring time-frequency bins [8,13] or subject to more advanced spectral models including constraints such as harmonicity [14]. These constraints increase the *smoothness* of the estimated source covariance matrices hence of the estimated source STFT coefficients derived by Wiener filtering. Although they typically reduce both interference and artifacts compared to sparsity-based algorithms, these algorithms still result in a significant level of artifacts [15,12].

In this paper, we explore a complementary approach whereby initial estimates of the source covariance matrices obtained via any source separation algorithm are post-processed by some smoothing technique so as to reduce artifacts. Several such techniques have been introduced in the context of speech denoising or beamforming [7,1,6,4,16] and employed for the post-filtering of linear source estimates in the context of determined audio source separation [9]. However, they have not yet been systematically studied in the context of under-determined source separation involving directional interference instead of a somewhat diffuse background. Also, most of these techniques are specifically designed for single-channel input. In the following, we propose multichannel extensions of three single-channel smoothing techniques [7,1,16] and compare them with two existing multichannel techniques [6,4] on a set of speech and music mixtures.

The structure of the paper is as follows. We explain how to initially estimate the source covariance matrices and present five multichannel smoothing techniques in Section 2. We assess the performance of each technique for various source separation algorithms in Section 3 and conclude in Section 4.

2 Source Covariance Estimation and Smoothing

Let us denote by $\mathbf{x}(n, f)$ and $\mathbf{c}_j(n, f)$ the $I \times 1$ vectors of STFT coefficients of the mixture and the spatial image of source j respectively. We presume that estimates of the source spatial images or their parameters have been obtained via any source separation algorithm and apply the following three-step post-processing. Firstly, assuming that $\mathbf{c}_j(n,f)$ follows a zero-mean Gaussian distribution with *local covariance matrix* $\mathbf{R}_{\mathbf{c}_j}(n,f)$ [14], we derive initial estimates $\widehat{\mathbf{R}}_{\mathbf{c}_j}(n,f)$ of these covariance matrices. Secondly, we replace the classical *multichannel Wiener filter* [4,14]

$$\widehat{\mathbf{W}}_j(n,f) = \widehat{\mathbf{R}}_{\mathbf{c}_j}(n,f)\widehat{\mathbf{R}}_{\mathbf{x}}^{-1}(n,f), \tag{2}$$

where $\widehat{\mathbf{R}}_{\mathbf{x}}(n,f) = \sum_{j=1}^{J} \widehat{\mathbf{R}}_{\mathbf{c}_j}(n,f)$ is the estimated mixture covariance matrix, by a smooth filter $\widetilde{\mathbf{W}}_j(n,f)$. Finally, the source spatial images are recovered by

$$\widetilde{\mathbf{c}}_j(n,f) = \widetilde{\mathbf{W}}_j(n,f)\mathbf{x}(n,f). \tag{3}$$

In the following, we discuss the first two steps in more detail.

2.1 Initial Source Covariance Estimation

Source separation algorithms can be broadly divided into two categories: linear *vs.* variance model-based algorithms [14]. Linear model-based algorithms such as binary masking [11] or ℓ_1-norm minimization [17] directly operate on the mixture STFT coefficients and provide estimates $\widehat{\mathbf{c}}_j(n,f)$ of the source spatial images. The source covariance matrices can then be naturally initialized as $\widehat{\mathbf{R}}_{\mathbf{c}_j}(n,f) = \widehat{\mathbf{c}}_j(n,f)\widehat{\mathbf{c}}_j^H(n,f)$ where ^{H} denotes Hermitian transposition. By contrast, variance-model based algorithms [14] represent the mixture by some parametric distribution and operate on the parameters of this distribution. Initial estimates of the sources covariances may then be derived from the estimated parameters. In the particular case when a Gaussian distribution is chosen [8,13], the source covariances are readily estimated as the output of the algorithm.

In both cases, we add a small regularization term $\epsilon\,\mathbf{I}$ to the initial covariance matrices, where $\mathbf{I}$ is the $I \times I$ identity matrix. This term ensures that the matrix inversions in (2), (5) and (8) can always be computed even when a single source is active. The regularization factor is set to $\epsilon = 10^{-6} \times \operatorname{tr}\widehat{\mathbf{R}}_{\mathbf{c}_j}(n,f)$.

2.2 Spatial Smoothing

A few multichannel smoothing techniques have been proposed in the beamforming literature in order to widen the spatial response of the Wiener filter, so as to reduce artifacts supposedly located close to the target source direction. While these techniques were originally formulated for a single source in the presence of background noise, their application to multiple sources is straightforward. One technique proposed in [4, eq. 55] amounts to interpolating the Wiener filter as

$$\widetilde{\mathbf{W}}_j^{\mathrm{SFS}}(n,f) = (1-\mu)\widehat{\mathbf{W}}_j(n,f) + \mu\,\mathbf{I}. \tag{4}$$

This is equivalent to time-domain interpolation of the estimated source spatial image signals with the mixture signal as suggested in [11]. Another technique stemming from a weighted likelihood model results in a distinct interpolation [6]

$$\widetilde{\mathbf{W}}_j^{\mathrm{SCS}}(n,f) = \widehat{\mathbf{R}}_{\mathbf{c}_j}(n,f)[(1-\mu)\widehat{\mathbf{R}}_{\mathbf{x}}(n,f) + \mu\widehat{\mathbf{R}}_{\mathbf{c}_j}(n,f)]^{-1}. \tag{5}$$

We refer to the techniques in (4) and (5) as *spatial filter smoothing* (SFS) and *spatial covariance smoothing* (SCS) respectively. In both cases, the smoothness of the resulting filter increases with μ, so that it is equal to the conventional Wiener filter for $\mu = 0$ and to the identity filter for $\mu = 1$.

2.3 Temporal Smoothing

Many techniques based on temporal smoothing of the source variances have also been proposed for single-channel speech denoising [16]. However, their extension to multichannel source separation is not straightforward. Two approaches may be taken: either split the source covariance matrices into a spectral power $v_j(n,f) = \operatorname{tr} \mathbf{R}_{\mathbf{c}_j}(n,f)$ and a spatial covariance matrix $\mathbf{R}_j(n,f) = v_j^{-1}(n,f)\mathbf{R}_{\mathbf{c}_j}(n,f)$ [13], process the spectral power alone via a single-channel technique and recombine it with the spatial covariance matrix, or design new smoothing equations that process spectral power and spatial covariance at the same time. Our preliminary experiments showed that the latter approach always performed better. Hence, we only present the new proposed smoothing equations below.

The most popular technique consists of smoothing the initial Signal-to-Noise Ratio (SNR) in a causal [7] or noncausal [5] fashion, with the latter resulting in better onset preservation. Numerous variants of this so-called decision-directed technique have been proposed [9]. By replacing variances by covariance matrices and ratios by matrix inversion, we extend it to source separation as

$$\begin{aligned}
\widehat{\mathbf{G}}_j(n,f) &= \widehat{\mathbf{R}}_{\mathbf{c}_j}(n,f)[\widehat{\mathbf{R}}_{\mathbf{x}}(n,f) - \widehat{\mathbf{R}}_{\mathbf{c}_j}(n,f)]^{-1} \\
\widetilde{\mathbf{G}}_j(n,f) &= \frac{1}{L+1} \sum_{l=-L/2}^{L/2} \widehat{\mathbf{G}}_j(n+l,f) \\
\widetilde{\mathbf{W}}_j^{\mathrm{TRS}}(n,f) &= \mathbf{I} - [\widetilde{\mathbf{G}}_j(n,f) + \mathbf{I}]^{-1}
\end{aligned} \tag{6}$$

where $\mathbf{G}_j(n,f)$ is a multichannel generalization of the SNR and we assume a noncausal rectangular smoothing window of length $L+1$. Note that this technique does not apply to binary masking, since $\widehat{\mathbf{G}}_j(n,f)$ is infinite in that case.

A simpler technique consists of smoothing the conventional single-channel Wiener filter [1], which easily extends to a multichannel setting as

$$\widetilde{\mathbf{W}}_j^{\mathrm{TFS}}(n,f) = \frac{1}{L+1} \sum_{l=-L/2}^{L/2} \widehat{\mathbf{W}}_j(n+l,f). \tag{7}$$

Finally, one may also compute the Wiener filter from smoothed source variances [16]. By smoothing the source covariances instead, we obtain

$$\begin{aligned}
\widetilde{\mathbf{R}}_{\mathbf{c}_j}(n,f) &= \frac{1}{L+1} \sum_{l=-L/2}^{L/2} \widehat{\mathbf{R}}_{\mathbf{c}_j}(n+l,f) \\
\widetilde{\mathbf{W}}_j^{\mathrm{TCS}}(n,f) &= \widetilde{\mathbf{R}}_{\mathbf{c}_j}(n,f)\widetilde{\mathbf{R}}_{\mathbf{x}}^{-1}(n,f)
\end{aligned} \tag{8}$$

with $\widetilde{\mathbf{R}}_{\mathbf{x}}(n,f) = \sum_{j=1}^{J} \widetilde{\mathbf{R}}_{\mathbf{c}_j}(n,f)$. We here consider sliding smoothing windows instead of disjoint windows as in [16].

We call the techniques in (6), (7) and (8) *temporal SNR smoothing* (TRS), *temporal filter smoothing* (TFS) and *temporal covariance smoothing* (TCS) respectively. Smoothness increases with L and the conventional Wiener filter is obtained for $L = 0$. Note that, contrary to spatial smoothing, the filter does not tend to identity when $L \to \infty$ but to a stationary Wiener filter instead.

3 Experimental Evaluation

We applied the five above smoothing techniques for the post-processing of three separation algorithms, namely binary masking [11], ℓ_1-norm minimization with two active sources per time-frequency bin [17] and local Gaussian modeling [13], over four instantaneous stereo ($I = 2$) mixtures of $J = 3$ sources. These mixtures were taken from the 2008 Signal Separation Evaluation Campaign (SiSEC) [12]

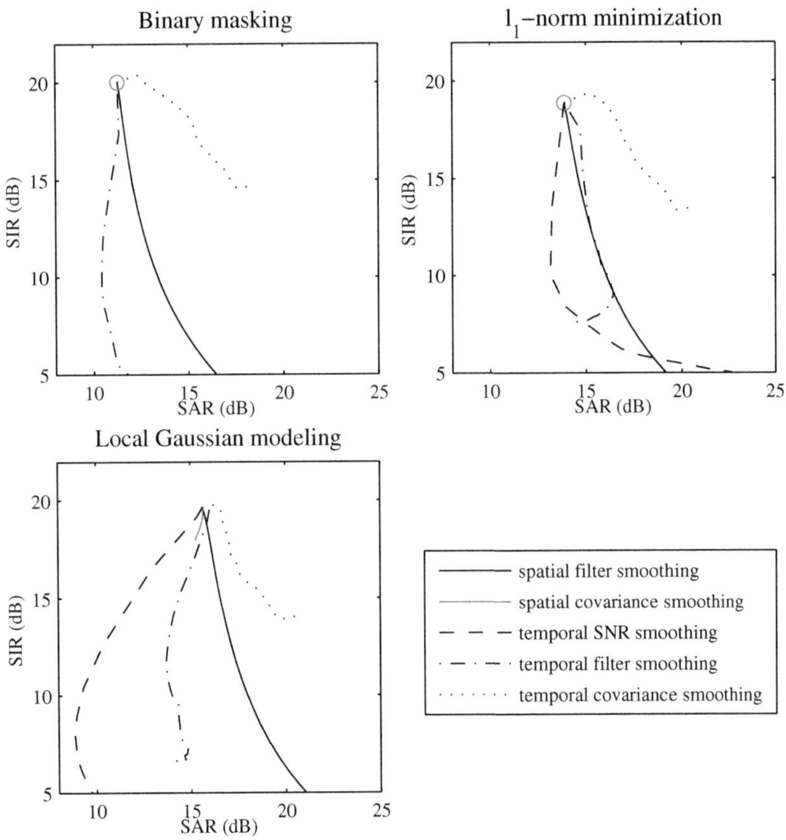

Fig. 1. Average tradeoff between SAR and SIR achieved by each separation algorithm and each smoothing technique

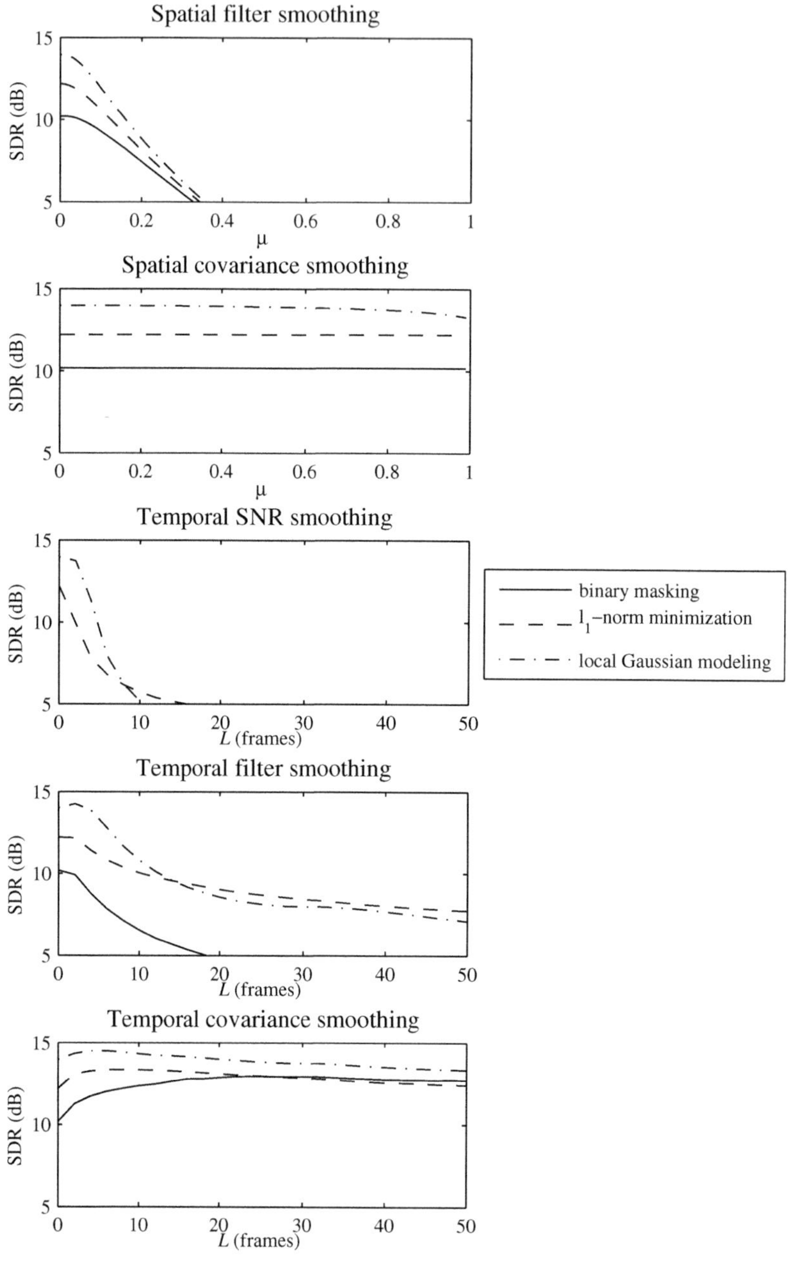

Fig. 2. Average SDR achieved by each separation algorithm and each smoothing technique as a function of the smoothing parameter

and cover both male and female speech, percussive and non-percussive music. *The mixing matrices were known.* Performance was evaluated using the overall Signal-to-Distortion Ratio (SDR) as well as the Signal-to-Interference Ratio (SIR) and the Signal-to-Artifacts Ratio (SAR) in [15], averaged over all sources and all mixtures. The choice of instantaneous mixing was dictated by the limited accuracy of these criteria in a convolutive setting. Indeed, while they are accurate for instantaneous mixtures, they do not yet provide sufficiently precise distinction of interference and artifacts on convolutive mixtures for this study.

The tradeoff between SAR and SIR as a function of μ and L is shown in Figure 1. Temporal covariance smoothing provides the best tradeoff independently of the considered separation algorithm. The resulting SIR decreases in similar proportion to the increase of the SAR and a small increase of the SIR is even observed for small μ or L. Spatial filter smoothing also improves the SAR but results in a much steeper decrease of the SIR. All other methods fail to reduce artifacts in a predictable manner and result either in non-monotonous increase or decrease of the SAR. This indicates that many state-of-the-art smoothing techniques initially designed for diffuse and/or stationary noise can fail in the presence of directional nonstationary sources. In particular, temporal SNR smoothing appears extremely sensitive to the initial estimation of the variances, while spatial covariance smoothing results in conventional Wiener filtering for all $0 \leq \mu < 1$ both for binary masking and ℓ_1-norm minimization[1] and in small deviation from conventional Wiener filtering for local Gaussian modeling.

These conclusions are further supported by the SDR curves in Figure 2, which decrease quickly for all techniques except for temporal covariance smoothing due to its good tradeoff between interference and artifacts and for spatial covariance smoothing as explained above. A SDR increase is even observed with temporal covariance smoothing, which is equal to 3 dB for binary masking and less for the two other algorithms.

4 Conclusion and Perspectives

We reformulated state-of-the-art Wiener filter smoothing techniques in the context of under-determined audio source separation. Experimental results indicate that many techniques thought for spatially diffuse and/or stationary noise fail with directional nonstationary sources. Temporal covariance smoothing provides the best tradeoff between SAR and SIR and also potentially increases the overall SDR. Future work will concentrate on assessing robustness to mixing matrix estimation errors and adaptively estimating the optimal size L of the smoothing window in each time-frequency bin for that technique.

References

1. Adami, A., Burget, L., Dupont, S., Garudadri, H., Grezl, F., et al.: Qualcomm-ICSI-OGI features for ASR. In: Proc. 7th Int. Conf. on Spoken Language Processing, pp. 21–24 (2002)

[1] It can be shown that SCS amounts to conventional Wiener filtering for all $0 \leq \mu < 1$ when $\epsilon \to 0$ as soon as at most two sources are active in each time-frequency bin.

2. Araki, S., Makino, S., Sawada, H., Mukai, R.: Reducing musical noise by a fine-shift overlap-add method applied to source separation using a time-frequency mask. In: Proc. 2005 IEEE Int. Conf. on Acoustics, Speech and Signal Processing, pp. III-81–III-84 (2005)
3. Araki, S., Sawada, H., Mukai, R., Makino, S.: Blind sparse source separation with spatially smoothed time-frequency masking. In: Proc. 2006 Int. Workshop on Acoustic Echo and Noise Control (2006)
4. Chen, J., Benesty, J., Huang, Y., Doclo, S.: New insights into the noise reduction Wiener filter. IEEE Transactions on Audio, Speech and Language Processing 14(4), 1218–1234 (2006)
5. Cohen, I.: Speech enhancement using a noncausal a priori SNR estimator. IEEE Signal Processing Letters 11(9), 725–728 (2004)
6. Doclo, S., Moonen, M.: On the output SNR of the speech-distortion weighted multichannel Wiener filter. IEEE Signal Processing Letters 12(12), 809–811 (2005)
7. Ephraim, Y., Malah, D.: Speech enhancement using a minimum mean square error short-time spectral amplitude estimator. IEEE Transactions on Acoustics, Speech and Signal Processing 32(6), 1109–1121 (1984)
8. Févotte, C., Cardoso, J.F.: Maximum likelihood approach for blind audio source separation using time-frequency Gaussian models. In: Proc. 2005 IEEE Workshop on Applications of Signal Processing to Audio and Acoustics. pp. 78–81 (2005)
9. Hoffmann, E., Kolossa, D., Orglmeister, R.: Time frequency masking strategy for blind source separation of acoustic signals based on optimally-modified log-spectral amplitude estimator. In: Proc. 8th Int. Conf. on Independent Component Analysis and Signal Separation. pp. 581–588 (2009)
10. Nesbit, A., Jafari, M.G., Vincent, E., Plumbley, M.D.: Audio source separation using sparse representations. In: Machine Audition: Priciples, Algorithms and Systems. IGI Global (2010)
11. Rickard, S.T.: The DUET blind source separation algorithm. In: Blind Speech Separation, pp. 217–237. Springer, Heidelberg (2007)
12. Vincent, E., Araki, S., Bofill, P.: The 2008 Signal Separation Evaluation Campaign: A community-based approach to large-scale evaluation. In: Proc. 8th Int. Conf. on Independent Component Analysis and Signal Separation (ICA), pp. 734–741 (2009)
13. Vincent, E., Arberet, S., Gribonval, R.: Underdetermined instantaneous audio source separation via local Gaussian modeling. In: Proc. 8th Int. Conf. on Independent Component Analysis and Signal Separation. pp. 775–782 (2009)
14. Vincent, E., Jafari, M.G., Abdallah, S.A., Plumbley, M.D., Davies, M.E.: Probabilistic modeling paradigms for audio source separation. In: Machine Audition: Priciples, Algorithms and Systems. IGI Global (2010)
15. Vincent, E., Sawada, H., Bofill, P., Makino, S., Rosca, J.P.: First stereo audio source separation evaluation campaign: Data, algorithms and results. In: Proc. 7th Int. Conf. on Independent Component Analysis and Signal Separation. pp. 552–559 (2007)
16. Yu, G., Mallat, S., Bacry, E.: Audio denoising by time-frequency block thresholding. IEEE Transactions on Signal Processing 56(5), 1830–1839 (2008)
17. Zibulevsky, M., Pearlmutter, B.A., Bofill, P., Kisilev, P.: Blind source separation by sparse decomposition in a signal dictionary. In: Independent Component Analysis: Principles and Practice, pp. 181–208. Cambridge Press, New York (2001)

Auxiliary-Function-Based Independent Component Analysis for Super-Gaussian Sources

Nobutaka Ono and Shigeki Miyabe

Graduate School of Information Science and Technology,
The University of Tokyo
7-3-1 Hongo, Bunkyo-ku, Tokyo, 113-8656, Japan
{onono,miyabe}@hil.t.u-tokyo.ac.jp

Abstract. This paper presents new algorithms of independent component analysis (ICA) for super-Gaussian sources based on auxiliary function technique. The algorithms consist of two alternative updates: 1) update of demixing matrix and 2) update of weighted covariance matrix, which include no tuning parameters such as step size. The monotonic decrease of the objective function at each update is guaranteed. The experimental results show that the derived algorithms are robust to nonstationary data and outliers, and the convergence is faster than natural-gradient-based algorithm.

Keywords: Independent component analysis, Infomax, blind source separation, auxiliary function, super-Gaussian.

1 Introduction

Independent component analysis (ICA) is a powerful technique to find independent components from mixtures without mixing information, which has been widely used for blind source separation [1]. One of the popular algorithms of ICA is Infomax [2], where the demixing matrix is estimated iteratively by applying natural-gradient-based update [3]. However, like other kinds of gradient-based optimization, there is a tradeoff between the convergence of speed and stability. The larger step size would make the convergence faster, but it may leads diverge. For improving robustness, modification of the natural gradient algorithm has been investigated in [4].

In this paper, we derive another kind of iterative solution for optimization of Infomax-type objective function based on auxiliary function approach, which is a framework to find efficient iterative solution for nonlinear optimization problem. In signal processing field, it has been used in well-known EM algorithm, and recently applied to various kinds of optimization problems such as nonnegative matrix factorization (NMF) [5], multi pitch analysis [6], sinusoidal parameter estimation [7], music signal separation [8], source localization [9], etc. In this paper, for a class of contrast functions related with super-Gaussianity, efficient iterative update rules of ICA are derived, which have no tuning parameters such as step size, and the monotonic decrease of the objective function is guaranteed. The experimental comparisons with existing other methods are also shown.

V. Vigneron et al. (Eds.): LVA/ICA 2010, LNCS 6365, pp. 165–172, 2010.

2 Objective Function of Infomax

In this paper, we generally assume that all variables can take complex values. A real-valued case is easily obtained by just replacing Hermitian conjugate ^{h} by transpose ^{t}. In Infomax algorithm, the demixing matrix is estimated by minimizing the following objective function:

$$J(W) = \sum_{k=1}^{K} E[G(\boldsymbol{w}_k^h \boldsymbol{x})] - \log|\det W|, \tag{1}$$

where

$$W = (\boldsymbol{w}_1 \ \cdots \ \boldsymbol{w}_K)^h \tag{2}$$

is a demixing matrix, $\boldsymbol{x} = (x_1 \ \cdots \ x_K)^t$ is an observed random vector, $E[\cdot]$ denotes expectation, and $G(y)$ is called contrast function [2]. The goal of this paper is to derive efficient iterative algorithms to find W to minimize eq. (1).

3 Auxiliary Function of Contrast Function

3.1 Auxiliary Function Approach

In order to introduce auxiliary function approach, let's consider a general optimization problem to find a parameter vector $\boldsymbol{\theta} = \boldsymbol{\theta}^\dagger$ satisfying

$$\boldsymbol{\theta}^\dagger = \mathrm{argmin}_{\boldsymbol{\theta}} \ J(\boldsymbol{\theta}), \tag{3}$$

where $J(\boldsymbol{\theta})$ is an objective function.

In the auxiliary function approach, a function $Q(\boldsymbol{\theta}, \tilde{\boldsymbol{\theta}})$ is designed such that it satisfies

$$J(\boldsymbol{\theta}) = \min_{\tilde{\boldsymbol{\theta}}} \ Q(\boldsymbol{\theta}, \tilde{\boldsymbol{\theta}}). \tag{4}$$

$Q(\boldsymbol{\theta}, \tilde{\boldsymbol{\theta}})$ is called an auxiliary function for $J(\boldsymbol{\theta})$, and $\tilde{\boldsymbol{\theta}}$ are called auxiliary variables. Then, instead of directly minimizing the objective function $J(\boldsymbol{\theta})$, the auxiliary function $Q(\boldsymbol{\theta}, \tilde{\boldsymbol{\theta}})$ is minimized in terms of $\boldsymbol{\theta}$ and $\tilde{\boldsymbol{\theta}}$, alternatively, the variables being iteratively updated as

$$\tilde{\boldsymbol{\theta}}^{(i+1)} = \mathrm{argmin}_{\tilde{\boldsymbol{\theta}}} \ Q(\boldsymbol{\theta}^{(i)}, \tilde{\boldsymbol{\theta}}), \tag{5}$$

$$\boldsymbol{\theta}^{(i+1)} = \mathrm{argmin}_{\boldsymbol{\theta}} \ Q(\boldsymbol{\theta}, \tilde{\boldsymbol{\theta}}^{(i+1)}), \tag{6}$$

where i denotes the iteration index. The monotonic decrease of $J(\boldsymbol{\theta})$ under the above updates is guaranteed [5,6,7,8,9].

Note that even if eq. (3) has no closed-form solutions, there could exist $Q(\boldsymbol{\theta}, \tilde{\boldsymbol{\theta}})$ satisfying eq. (4) such that both eq. (5) and eq. (6) have closed-form solutions. In such situations, the auxiliary function approach gives us efficient iterative update rules. However, how to find appropriate $Q(\boldsymbol{\theta}, \tilde{\boldsymbol{\theta}})$ is problem-dependent.

3.2 Auxiliary Function of Contrast Function

A good candidate of auxiliary functions is given by a quadratic function because it can be easily minimized. For exploiting a quadratic function as an auxiliary function, eq. (4) requires that an objective function has to grow up more slowly than a quadratic function when $|\boldsymbol{\theta}| \to \infty$. From this point, we found a class of contrast functions which a quadratic function is applicable for, which is tightly related with super-Gaussianity. We first begin with two definitions.

Definition 1. *A set of real-valued functions of a real- or complex- variable z, S_G, is defined as*

$$S_G = \{G(z)|\ G(z) = G_R(|z|)\} \tag{7}$$

where $G_R(r)$ is a continuous and differentiable function of a real variable r satisfying that $G'_R(r)/r$ is continuous everywhere and it is monotonically decreasing in $r \geq 0$.

Definition 2. *([10] pp. 60-61) If a random variable r has a probability density of the form $p(r) = e^{-G_R(r)}$ where $G_R(r)$ is an even function which is differentiable, except possibly at the origin, $G_R(r)$ is strictly increasing in R_+, and $G'_R(r)/r$ is strictly decreasing in R_+, then, we shall say that it is super-Gaussian.*

From definition 2, it is clear that $G(z) = G_R(|z|)$ is a member of S_G if the distribution $p(r) = e^{-G_R(r)}$ is super-Gaussian and $G'_R(r)/r$ is continuous at the origin. Note that several well-used contrast functions such as $G_1(r) = \frac{1}{a_1} \log\cosh(a_1 r)$ and $G_2(r) = -\frac{1}{a_2} e^{-a_2 r^2/2}$, where $1 \leq a_1 \leq 2$, $a_2 \simeq 1$ [1], and their polar-type contrast functions [11] are included in S_G. Based on this definition of S_G, an explicit auxiliary function for the objective function of Infomax is obtained by the following two theorems.

Theorem 1. *For any $G(z) = G_R(|z|) \in S_G$,*

$$G(z) \leq \frac{G'_R(|z_0|)}{2|z_0|}|z|^2 + \left(G_R(|z_0|) - \frac{|z_0| G'_R(|z_0|)}{2}\right) \tag{8}$$

holds for any z. The equality sign is satisfied if and only if $|z| = |z_0|$.

Proof. First, let $r = |z|$ and $r_0 = |z_0|$, and consider the following function:

$$F(r) = \frac{G'_R(r_0)}{2r_0} r^2 + \left(G_R(r_0) - \frac{r_0 G'_R(r_0)}{2}\right) - G_R(r). \tag{9}$$

Differentiating $F(r)$, we have

$$F'(r) = \frac{G'_R(r_0)}{r_0} r - G'_R(r) = r\left(\frac{G'_R(r_0)}{r_0} - \frac{G'_R(r)}{r}\right). \tag{10}$$

Note that $G'_R(r)/r$ monotonically decreases in $r > 0$ and $F'(r_0) = 0$. Then, $F(r)$ has the unique minimum value at $r = r_0$ since $F(r)$ is continuous everywhere, and $F(r_0) = 0$. Consequently, it is clear that eq. (8) holds for any z. ∎

Theorem 2. *For any* $G(z) = G_R(|z|) \in S_G$, *let*

$$Q(W, \tilde{W}) = \frac{1}{2}\sum_{k=1}^{K} \boldsymbol{w}_k^h V_k \boldsymbol{w}_k - \log|\det W| + R, \tag{11}$$

where $\tilde{W} = (\tilde{\boldsymbol{w}}_1 \cdots \tilde{\boldsymbol{w}}_K)^h$, $r_k = |\tilde{\boldsymbol{w}}_k^h \boldsymbol{x}|$,

$$V_k = E\left[\frac{G'(r_k)}{r_k}\boldsymbol{x}\boldsymbol{x}^h\right], \tag{12}$$

and R *is a constant independent of* W. *Then,*

$$J(W) \leq Q(W, \tilde{W}) \tag{13}$$

holds for any W. *The equality sign holds if and only if* $\tilde{\boldsymbol{w}}_k = e^{j\phi_k}\boldsymbol{w}_k$ *for* $^\forall k$ *where* ϕ_k *denotes an arbitrary phase.*

Proof. Eq. (13) is directly obtained by substituting $z = \boldsymbol{w}_k^h \boldsymbol{x}$ and $z_0 = \tilde{\boldsymbol{w}}_k^h \boldsymbol{x}$ into eq. (8), summing them up from $k = 1$ to $k = K$, and taking expectation. ■

4 Derivation of Update Rules

4.1 Derivative of Auxiliary Function

Based on the principle of the auxiliary function approach, update rules should be obtained by minimizing $Q(W, \tilde{W})$ in terms of W and $\tilde{W}$, alternatively. From Theorem 2, the minimization of Q in terms of $\tilde{W}$ is easily obtained by just setting $\tilde{W} = W$. Then, let's focus on minimizing Q in terms of W. It can be done by solving $\partial Q(W, \tilde{W})/\partial \boldsymbol{w}_k^* = 0$ $(1 \leq k \leq K)$, where * denotes complex conjugate. They yield

$$\frac{1}{2}V_k \boldsymbol{w}_k - \frac{\partial}{\partial \boldsymbol{w}_k^*}\log|\det W| = 0 \quad (1 \leq k \leq K). \tag{14}$$

Rearranging eq. (14) with using $(\partial/\partial W)\det W = W^{-t}\det W$, the following simultaneous vector equations are derived.

$$\boldsymbol{w}_l^h V_k \boldsymbol{w}_k = \delta_{lk} \quad (1 \leq k \leq K,\ 1 \leq l \leq K). \tag{15}$$

For updating all of $\boldsymbol{w}_k$ simultaneously, it is necessary to solve eq. (15), and especially, a closed form solution is desirable for efficient updates. If all of V_k are commutable, their matrices would share eigenvectors and they would give the solutions of eq. (15). However, V_ks are generally not commutable. In that case, to the authors' knowledge, there are no closed-form solutions for eq. (15) except when $K = 2$. Instead of simultaneously updating all of $\boldsymbol{w}_k$, we here propose two kinds of different update rules: 1) sequentially updating each of $\boldsymbol{w}_k$, and 2) sequentially updating each pair of $\boldsymbol{w}_k$s, which can be both performed in closed-form manners.

4.2 AuxICA1: Sequential Update Rules

The first method updates each of $\boldsymbol{w}_k$ sequentially. Consider an update of $\boldsymbol{w}_k$ with keeping other $\boldsymbol{w}_l$s $(l \neq k)$ fixed. In this case, $\partial Q(W, \tilde{W})/\partial \boldsymbol{w}_k^* = 0$ yields

$$\boldsymbol{w}_k^h V_k \boldsymbol{w}_k = 1, \tag{16}$$

$$\boldsymbol{w}_l^h V_k \boldsymbol{w}_k = 0 \quad (l \neq k). \tag{17}$$

From eq. (17), $\boldsymbol{w}_k$ has to be orthogonal to all of $V_k \boldsymbol{w}_l$ $(l \neq k)$ at least. Such a vector can be obtained by projecting the estimate of $\boldsymbol{w}_k$ at the previous iteration into the complementary space of the space spanned by all of $V_k \boldsymbol{w}_l$ $(l \neq k)$. Then, the normalization should be performed to satisfy eq. (16). While, the auxiliary variables $\tilde{\boldsymbol{w}}_k$ are only included in V_k. Hence, updates of the auxiliary variables $\tilde{\boldsymbol{w}}_k$ are equivalent to calculating V_k. Consequently, The algorithm is summarized as follows.

AuxICA1: The following alternative updates are applied for all k in order until convergence.

Auxiliary variable updates: Calculate the following matrices:

$$V_k = E\left[\frac{G'(r_k)}{r_k}\boldsymbol{x}\boldsymbol{x}^h\right], \tag{18}$$

$$P = V_k(\boldsymbol{w}_1 \ \cdots \ \boldsymbol{w}_{k-1} \ \boldsymbol{w}_{k+1} \ \cdots \ \boldsymbol{w}_K), \tag{19}$$

where $r_k = |\boldsymbol{w}_k^h \boldsymbol{x}|$.

Demixing matrix updates: Apply the following updates in order.

$$\boldsymbol{w}_k \leftarrow \boldsymbol{w}_k - P(P^h P)^{-1} P^h \boldsymbol{w}_k, \tag{20}$$

$$\boldsymbol{w}_k \leftarrow \boldsymbol{w}_k / \sqrt{\boldsymbol{w}_k^h V_k \boldsymbol{w}_k}. \tag{21}$$

4.3 AuxICA2: Pairwise Update Rules

The second update rule is based on the closed form solution of eq. (15) in $K = 2$. When $K = 2$, eq. (15) indicates that both of $V_1 \boldsymbol{w}_1$ and $V_2 \boldsymbol{w}_1$ are orthogonal to $\boldsymbol{w}_2$. Because the direction orthogonal to $\boldsymbol{w}_2$ is uniquely determined in the two dimensional space, $V_1 \boldsymbol{w}_1$ and $V_2 \boldsymbol{w}_1$ have to be parallel such as

$$V_1 \boldsymbol{w}_1 = \gamma V_2 \boldsymbol{w}_1, \tag{22}$$

where γ is a constant. In the same way, $V_1 \boldsymbol{w}_2$ and $V_2 \boldsymbol{w}_2$ are also parallel. Such vectors are obtained as solutions of eq. (22), which is a generalized eigenvalue problem. If V_1 and V_2 are not singular, they are simply obtained as eigen vectors of $V_2^{-1} V_1$. Even if $K > 2$, it is possible to apply this pairwise updates for all possible pairs. Then, the algorithm is summarized as follows.

AuxICA2: The following alternative updates are applied for all pairs of m and n such that $m < n$ in order until convergence.

Auxiliary variable updates: Calculate the following covariance matrices:

$$U_m = E\left[\frac{G'_R(r_m)}{r_m}\boldsymbol{u}\boldsymbol{u}^h\right], \quad U_n = E\left[\frac{G'_R(r_n)}{r_n}\boldsymbol{u}\boldsymbol{u}^h\right], \tag{23}$$

where $r_m = |\boldsymbol{w}_m^h \boldsymbol{x}|$, $r_n = |\boldsymbol{w}_n^h \boldsymbol{x}|$, and $\boldsymbol{u} = (\boldsymbol{w}_m^h \boldsymbol{x}\ \boldsymbol{w}_n^h \boldsymbol{x})^t$.

Demixing matrix updates: Find two solutions $\boldsymbol{h}_m$ and $\boldsymbol{h}_n$ of the generalized eigenvalue problem of 2×2 matrix: $U_m \boldsymbol{h} = \gamma U_n \boldsymbol{h}$. Then, apply the following updates in order.

$$\boldsymbol{h}_m \leftarrow \boldsymbol{h}_m / \sqrt{\boldsymbol{h}_m^h U_m \boldsymbol{h}_m}, \quad \boldsymbol{h}_n \leftarrow \boldsymbol{h}_n / \sqrt{\boldsymbol{h}_n^h U_n \boldsymbol{h}_n}, \tag{24}$$

$$(\boldsymbol{w}_m\ \boldsymbol{w}_n) \leftarrow (\boldsymbol{w}_m\ \boldsymbol{w}_n)(\boldsymbol{h}_m\ \boldsymbol{h}_n). \tag{25}$$

5 Experimental Evaluations

In order to evaluate the performance of estimating demixing matrix and the convergence speed of AuxICA1 and AuxICA2, three artificial sources were prepared here. All of them were complex-valued random process with independent amplitude and phase. The phase followed the uniform distribution from 0 to 2π, and the amplitude $a \geq 0$ followed $p_1(a) = e^{-a}$, $p_2(a) = (3/4)\delta(a) + (1/4)e^{-a}$, and $p_3(a) = \arctan a_0/\pi(1 + a^2)$ $(0 \leq a \leq a_0)$ where $a_0 = 1000$, by which we intended to simulate 1) stationary source, 2) nonstationary source (each source is silent with probability 3/4), and 3) spiky source including outliers, respectively. In each case, the numbers of sources are $K = 2$ or $K = 6$ and the data length is $N = 1000$. The observed mixtures were made with instantaneous mixing matrices, where each element was independently generated by complex-valued Gaussian random process with zero mean and unity variance.

For these mixtures, we compared the proposed algorithms with popular existing methods such as polar-type Infomax [11], scaled Infomax [4] and FastICA [12]. In all of the algorithms, $G(z) = \log\cosh|z|$ was used as a contrast function. The initial value of the demixing matrix was given by data whitening.

The performance was evaluated by averaged signal to noise ratio:

$$\mathrm{SNR} = \frac{1}{K}\sum_k 10\log_{10}\frac{\sum_t |s_k(t)|^2}{\sum_t |s_k(t) - y_k(t)|^2}, \tag{26}$$

which was calculated by using demixing matrix iteratively-estimated at each iteration, where the correct permutation was given and the appropriate scaling was estimated by projection back [13]. The experiments were performed in Matlab ver. 7.9 on a laptop PC with 2.66GHz CPU. Fig. 1 shows the relationship between actual computational time and resultant SNR obtained by averaging 100 trials. AuxICA1 or AuxICA2 shows the best convergence speed in most cases, and the obtained solutions are robust to nonstationary signal and outliers.

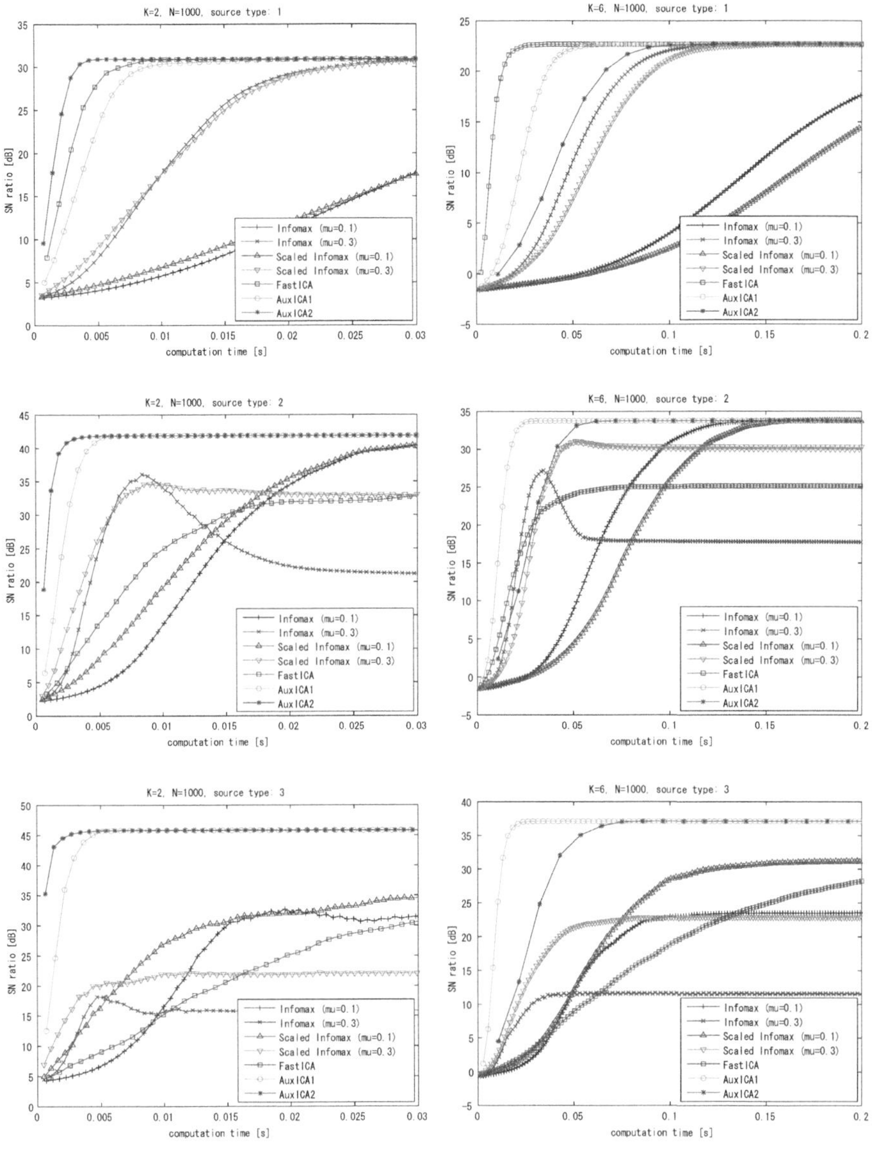

Fig. 1. The relationship between actual computational time and resultant SNR obtained by averaging 100 trials. The left and the right columns represent the results for $K = 2$ and $K = 6$, respectively, where K denotes the number of sources. The amplitude of sources follow $p_1(a)$, $p_2(a)$, and $p_3(a)$ from top to bottom, respectively. In the legend, "mu" denotes a step size. In conventional Infomax, there is a tradeoff between the convergence speed and stability. The scaled Infomax improves them, but the tradeoff still remains. FastICA gives the fast convergence for stationary signal, but it becomes rather slow for nonstationay signal or outliers. While, the proposed methods (AuxICA1 or AuxICA2) show the best convergence speed in most cases and robust to nonstationary signal and outliers.

6 Conclusion

In this paper, we presented new algorithms of ICA for super-Gaussian sources based on auxiliary function technique. The experimental results showed that the derived algorithms give the faster convergence than natural-gradient-based Infomax, and are robust to nonstationary data and outliers. Applying the proposed algorithms to blind source separation for convolutive-mixture is ongoing.

References

1. Hyvärinen, A., Karhunen, J., Oja, E.: Independent Component Analysis. John Wiley & Sons, New York (2001)
2. Bell, A.J., Sejnowski, T.J.: An Information-Maximization Approach to Blind Separation and Blind Deconvolution. Neural Computation 7(6), 1129–1159 (1995)
3. Amari, S., Cichocki, A., Yang, H.H.: A New Learning Algorithm for Blind Signal Separation. In: Touretzky, D., Mozer, M., Hasselmo, M. (eds.) Advances in Neural Information Processing Systems, pp. 757–763. MIT Press, Cambridge (1996)
4. Douglas, S.C., Gupta, M.: Scaled Natural Gradient Algorithms for Instantaneous and Convolutive Blind Source Separation. In: Proc. ICASSP, pp. 637–640 (2007)
5. Lee, D.D., Seung, H.S.: Algorithms for Non-Negative Matrix Factorization. In: Proc. NIPS, pp. 556–562 (2000)
6. Kameoka, H., Nishimoto, T., Sagayama, S.: A Multipitch Analyzer Based on Harmonic Temporal Structured Clustering. IEEE Trans. ASLP 15(3), 982–994 (2007)
7. Kameoka, H., Ono, N., Sagayama, S.: Auxiliary Function Approach to Parameter Estimation of Constrained Sinusoidal Model. In: Proc. ICASSP, pp. 29–32 (April 2008)
8. Ono, N., Miyamoto, K., Kameoka, H., Sagayama, S.: A Real-time Equalizer of Harmonic and Percussive Components in Music Signals. In: Proc. ISMIR, pp. 139–144 (September 2008)
9. Ono, N., Sagayama, S.: R-Means Localization: A Simple Iterative Algorithm for Source Localization Based on Time Difference of Arrival. In: Proc. ICASSP, pp. 2718–2721 (March 2010)
10. Benveniste, A., M'etivier, M., Priouret, P.: Adaptive algorithms and stochastic approximations. Springer, Heidelberg (1990)
11. Sawada, H., Mukai, R., Araki, S., Makino, S.: Polar Coordinate Based Nonlinear Function for Frequency-domain Blind Source Separation. IEICE Trans. Fundamentals E86-A(3), 590–596 (2003)
12. Bingham, E., Hyvärinen, A.: A Fast Fixed-Point Algorithm for Independent Component Analysis of Complex Valued Signals. International Journal of Neural Systems 10(1), 1–8 (2000)
13. Murata, N., Ikeda, S., Ziehe, A.: An Approach to Blind Source Separation Based on Temporal Structure of Speech Signals. Neurocomputing 41(1-4), 1–24 (2001)

ICA Separability of Nonlinear Models with References: General Properties and Application to Heisenberg-Coupled Quantum States (Qubits)

Yannick Deville

Laboratoire d'Astrophysique de Toulouse-Tarbes, Université de Toulouse, CNRS, 14 Av. Edouard Belin, 31400 Toulouse, France
ydeville@ast.obs-mip.fr

Abstract. Relatively few results were reported about the separability of given classes of nonlinear mixtures by means of the ICA criterion. We here prove the separability of a wide class of nonlinear global (i.e. mixing + separating) models involving "reference signals", i.e. unmixed signals. This work therefore concerns a nonlinear extension of linear adaptive noise cancellation (ANC). We then illustrate the usefulness of our general results by applying them to a model of Heisenberg-coupled quantum states. This paper opens the way to practical ICA methods for nonlinear mixtures encountered in various applications.

1 Introduction

A generic signal processing problem consists in extracting one or several unknown source signals of interest from several observations, which are mixtures of these signals of interest and possibly of additional, undesired, source signals. A first generation of such problems was especially studied by Widrow et al. and e.g. reported in 1975 in [6]. It is known as adaptive noise cancellation (ANC). It typically corresponds to configurations where almost all observations are "reference signals", i.e. unmixed signals. An extended version of this problem, known as blind source separation (BSS), has then been widely studied since the 1980's [1], [4]. It concerns cases when *all* observations are mixtures of all source signals.

The "mixing model" involved in these problems is most often the functional form which defines the expression of the vector of observed signals with respect to the vector of source signals and to the parameters of that functional form. The values of these parameters are unknown in the *blind* version of the source separation problem. The development of a complete BSS method for a given mixing model typically consists of the following steps. Step 1: analyze the invertibility of this mixing model if possible, and define a separating model, which essentially aims at implementing the inverse of the considered mixing model. Step 2: select a separation criterion for estimating the values of the parameters of the separating model. Step 3 (closely linked to Step 2): determine if this criterion ensures the separability of the considered models (at least for some classes of source signals). This consists in determining if this criterion is met only when

V. Vigneron et al. (Eds.): LVA/ICA 2010, LNCS 6365, pp. 173–180, 2010.

the outputs of the separating system are equal to the sources up to acceptable indeterminacies (such as permutation, scale factors, additive constants ...). Step 4: develop practical estimation algorithms associated to the considered criterion.

The above procedure has been widely applied to simple mixing models, i.e. linear (and especially instantaneous) ones. It has been much less explored and is much tougher for nonlinear models (see e.g. the survey in [5]). Its Step 1, i.e. the definition of separating *structures*, has e.g. been addressed for a wide class of nonlinear models in [3]. A natural way to tackle its Step 2 consists in considering Independent Component Analysis (ICA) methods, which have been widely used for linear mixtures. The relevance of these methods should then be proved for the considered nonlinear models, by investigating *separability*, which corresponds to Step 3 of the above procedure. Although some general ICA (non-)separability properties have been reported, very few results are available for the specific models which have been considered in the literature. This separability issue is the topic that we address in this paper, by first deriving general properties for the class of models defined in the next section. This work therefore concerns a nonlinear extension of linear ANC. We then apply these results to a specific model encountered in an application dealing with quantum signals.

2 Considered Global (Mixing + Separating) Model

We only consider memoryless mixing and separating models, i.e. models whose outputs at a given time t only depend on their inputs at the same time. Moreover, we do not require the signals to have any temporal structure (nor do we use it if it exists). Therefore, we omit the argument "(t)" in all signal notations hereafter. All configurations studied below involve N source signals s_i which form a vector $s = [s_1, \ldots, s_N]^T$, where ^{T} stands for transpose. These signals are transferred through a mixing operator M, which belongs to a given class and depends on a set of parameters which form a vector θ, whose value is unknown. This yields N observed signals x_i which form a vector $x = [x_1, \ldots, x_N]^T$, defined as

$$x = M(s, \theta). \tag{1}$$

These signals are then processed by a separating or unmixing system, which corresponds to an operator U. This operator belongs to a fixed class and depends on a set of parameters which form a vector ϕ. The N output signals y_i of the unmixing system form a vector $y = [y_1, \ldots, y_N]^T$, defined as

$$y = U(x, \phi). \tag{2}$$

The class of operator U is "matched" to the class of operator M in the sense that there exists at least one value ϕ_{opt} of ϕ which depends on the considered value of θ and which is such that, when $\phi = \phi_{opt}$, the output signals y_i of the unmixing system are respectively equal to the source signals s_j, up to a set of acceptable indeterminacies. Combining (1) and (2), the global model from the source signals s_j to the unmixing system outputs y_i reads

$$y = G(s, \theta, \phi) \tag{3}$$

where $G = U \circ M$ is the global operator. In the separability analysis presented hereafter, we only have to consider the global model G, i.e. we need not define the mixing and unmixing models M and U from which it is derived. We set the following conditions on G. Whatever θ and ϕ, only one of the output signals of the unmixing system may be a mixture of source signals, whereas each of all other output signals only depends on a single source. The possibly mixed output is always the same, and we assign it index no. 1, i.e. the corresponding signal is y_1. Moreover, we request that the components of the global model G may first be expressed as follows, with adequate ordering of the other outputs with respect to the sources:

$$y_1 = T(s_1; \theta, \phi) + I(s_2, \dots, s_N; \theta, \phi) \tag{4}$$

$$y_i = H_i(s_i; \theta, \phi) \qquad \forall\, i = 2, \dots, N. \tag{5}$$

The output of interest, y_1, is thus the sum of (i) a term $T(s_1; \theta, \phi)$, which only depends on the *target* source, i.e. on the source s_1 that we aim at extracting from y_1, and (ii) an *interference* term $I(s_2, \dots, s_N; \theta, \phi)$, which may depend on all sources except the target source, and that we aim at cancelling, by properly adjusting ϕ. Each signal $H_i(s_i; \theta, \phi)$ in (5) is thus a transformed version of source signal s_i. We denote it as s'_i, for $i = 2, \dots, N$. For $i = 1$, we define

$$s'_1 = s_1. \tag{6}$$

Moreover, we assume that the considered model G is such that the term $I(s_2, \dots, s_N; \theta, \phi)$ may also be expressed with respect to the transformed sources s'_2 to s'_N, instead of the original source signals s_2 to s_N (this is especially true when all operators H_i are invertible). The global model (4)-(5) may then be reformulated as

$$y_1 = T(s_1; \theta, \phi) + I'(s'_2, \dots, s'_N; \theta, \phi) \tag{7}$$

$$y_i = s'_i \qquad \forall\, i = 2, \dots, N. \tag{8}$$

Considering this data model, BSS/ANC aims at adjusting ϕ so as to extract the additive term $T(s_1; \theta, \phi)$ associated to the target source s_1 from the output signal y_1 defined in (7). The source signal s_1 may then be derived from $T(s_1; \theta, \phi)$ if T is invertible.

3 Source Properties and Goal of Investigation

We aim at determining if the above global model is separable in the sense of ICA, for given source statistics and operators T and I'. Using the standard ICA formulation for an N-input to N-output global model, this ICA-separability problem may be defined as follows. We consider the situation when the random variables defined at time t by all original source signals $s_1(t)$ to $s_N(t)$ have given marginal statistics and are mutually statistically independent. The random variables defined at time t by all transformed source signals $s'_1(t)$ to $s'_N(t)$ are then also mutually statistically independent. We consider the random variables defined at time t by the separating system outputs $y_1(t)$ to $y_N(t)$. We denote them as Y_1 to Y_N, and we aim at determining all cases when they are mutually statistically independent. If this only includes cases when the output signals are equal to the source signals up to acceptable indeterminacies, then the considered global model is said to be ICA-separable, for the considered type of sources.

In all this paper, the source and output random variables are assumed to be *continuous*. Their statistics may therefore be defined by their probability density *functions* (pdf), i.e. without having to resort to representations based on distributions. Each of these pdf is non-zero on at least one interval. For such random variables, statistical independence and the associated ICA-separability criterion may be analyzed by considering the joint and marginal pdf of Y_1 to Y_N. Therefore, we first derive the expressions of these pdf hereafter.

4 Analysis of ICA Separability

Joint pdf of output random variables. When expressing the outputs y_i with respect to the *transformed* sources s'_j, the global model (3) becomes

$$y = G'(s', \theta, \phi) \tag{9}$$

and is explicitly defined by (7)-(8). We consider the case when the operator G' is bijective, from s' to y (in their considered domains) and for fixed θ and ϕ. The pdf of the random vector Y composed of the output random variables Y_i may then be expressed vs the pdf of the random vector S' composed of the random variables S'_i associated to the transformed sources, as

$$f_Y(y) = \frac{f_{S'}(s')}{|J_{G'}(s')|} \tag{10}$$

where y and s' are linked by (9) and $J_{G'}(s')$ is the Jacobian of G' (see e.g. [4]). The expression of $J_{G'}(s')$ is derived from the explicit form (7)-(8) of the considered operator G'. It may be shown that this yields

$$J_{G'}(s') = \frac{\partial T}{\partial s_1}(s_1; \theta, \phi). \tag{11}$$

Moreover, since all transformed sources S'_i are independent, we have

$$f_{S'}(s') = \prod_{i=1}^{N} f_{S'_i}(s'_i). \tag{12}$$

Taking into account (6), (12) and denoting

$$O_1(s_1; \theta, \phi) = \frac{f_{S'_1}(s_1)}{|\frac{\partial T}{\partial s_1}(s_1; \theta, \phi)|}, \tag{13}$$

Eq. (10) becomes

$$f_Y(y) = O_1(s_1; \theta, \phi) \prod_{i=2}^{N} f_{S'_i}(s'_i). \tag{14}$$

The right-hand term of (14) should be expressed with respect to the output signals y_i. To this end, we first use (8), which yields

$$f_{S'_i}(s'_i) = f_{S'_i}(y_i) = f_{Y_i}(y_i) \qquad \forall\, i = 2, \ldots, N. \tag{15}$$

Moreover, we consider the case when operator T, which is a function of s_1 with parameters θ and ϕ, is invertible. Denoting T^{-1} the inverse of this operator, (7) and (8) then yield

$$s_1 = T^{-1}(y_1 - I'(y_2, \ldots, y_N; \theta, \phi); \theta, \phi). \tag{16}$$

Eq. (14) thus becomes

$$f_Y(y) = O_2(y_1 - I'(y_2, \ldots, y_N; \theta, \phi); \theta, \phi) \prod_{i=2}^{N} f_{S'_i}(y_i) \tag{17}$$

where we define operator O_2 by

$$O_2(v; \theta, \phi) = O_1(T^{-1}(v; \theta, \phi); \theta, \phi). \tag{18}$$

It may be shown that O_2 is equal to the pdf of the random variable $T(S'_1; \theta, \phi)$.

Marginal pdf of output random variables. Due to (8), the pdf of Y_2 to Y_N are equal to those of the corresponding transformed sources. The pdf of Y_1 is then obtained by integrating the joint pdf of all output random variables, i.e.

$$f_{Y_1}(y_1) = \int_{\mathbb{R}^{N-1}} f_Y(y) \prod_{i=2}^{N} dy_i. \tag{19}$$

Inserting (17) in (19), we obtain

$$f_{Y_1}(y_1) = \int_{\mathbb{R}^{N-1}} O_2(y_1 - I'(v_2, \ldots, v_N; \theta, \phi); \theta, \phi) \prod_{i=2}^{N} f_{S'_i}(v_i) \prod_{i=2}^{N} dv_i. \tag{20}$$

Condition for independent outputs. The output random variables Y_i are statistically independent if and only if

$$f_Y(y) = \prod_{i=1}^{N} f_{Y_i}(y_i) \qquad \forall\, y \in \mathbb{R}^N. \tag{21}$$

Taking into account (15) and (17), Eq. (21) becomes

$$O_2(y_1 - I'(y_2, \ldots, y_N; \theta, \phi); \theta, \phi) \prod_{i=2}^{N} f_{Y_i}(y_i) = \prod_{i=1}^{N} f_{Y_i}(y_i) \quad \forall\, y \in \mathbb{R}^N. \tag{22}$$

Let us denote as $\mathcal{D}$ the subset of $\mathbb{R}^{N-1}$ composed of all vectors $[y_2, \ldots, y_N]^T$ which are such that

$$f_{Y_i}(y_i) \neq 0 \qquad \forall\, i = 2, \ldots, N. \tag{23}$$

Eq. (22) is always met for any y_1 and any vector $[y_2, \ldots, y_N]^T$ which does not belong to $\mathcal{D}$, because its left-hand and right-hand terms are then both equal to zero. The constraint actually set by (22) therefore reduces to

$$O_2(y_1 - I'(y_2, \ldots, y_N; \theta, \phi); \theta, \phi) = f_{Y_1}(y_1) \quad \forall\, y_1 \in \mathbb{R}, \forall\, [y_2, \ldots, y_N]^T \in \mathcal{D}. \tag{24}$$

Consequence for ICA separability. Let us consider the case when the output random variables Y_i are independent. Then, as shown above, condition (24) is met. This condition concerns the domain $\mathcal{D}$, which is based on the constraint $f_{Y_i}(y_i) \neq 0$, as shown by (23). We explained in Section 3 that this constraint $f_{Y_i}(y_i) \neq 0$ is met at least over one interval, separately for each y_i with $i = 2, \ldots, N$, in the configuration studied here. Let us consider the situation when these variables y_2 to y_N are varied within such intervals where $f_{Y_i}(y_i) \neq 0$, while y_1 takes an arbitrary, fixed, value. The major phenomenon in Eq. (24) is then that (i) the arguments y_2 to y_N of $I'(y_2, \ldots, y_N; \theta, \phi)$ in the left-hand term of (24) vary, and (ii) meanwhile, the complete right-hand term of (24) remains constant: that term does not depend on y_2 to y_N, as shown not only by the notation $f_{Y_1}(y_1)$ used for that term, but also by its explicit expression (20).

We can then derive consequences of this phenomenon by also taking into account the fact that operator O_2 is a pdf: therefore, it sums to one over $\mathbb{R}$, so that it cannot be constant over $\mathbb{R}$, i.e. $O_2(v; \theta, \phi)$ varies at least over one interval of values of v. We then consider values of y_1 to y_N such that, when y_2 to y_N are varied (in intervals such that $f_{Y_i}(y_i) \neq 0$), the resulting value $v = y_1 - I'(y_2, \ldots, y_N; \theta, \phi)$ is situated inside an interval where $O_2(v; \theta, \phi)$ is not constant with respect to v. As explained above, the right-hand term of (24) thus remains constant, and therefore its left-hand term $O_2(v; \theta, \phi)$ also remains constant. This implies that, although y_2 to y_N are varied, v remains constant, and therefore $I'(y_2, \ldots, y_N; \theta, \phi)$ also remains constant. Eq. (7) then shows that the output signal y_1 is equal to the target term $T(s_1; \theta, \phi)$ that we aim at extracting, up to a constant term[1], which is equal to $I'(s'_2, \ldots, s'_N; \theta, \phi)$. This proves that, whatever the source pdf, the class of global models analyzed in this paper is ICA-separable, i.e. output independence implies that the output signals are equal to the transformed source signals s'_i (up to an additive constant and an invertible function for the first output).

5 Application to Coupled Quantum Bits

We introduced the Blind Quantum Source Separation (BQSS) field e.g. in [2], by considering the situation when two quantum bits (qubits) of a physical system

[1] The case when any of the random variables Y_2 to Y_N has a non-zero pdf on several disjoint intervals would deserve some comments but yields no major issue.

are separately initialized with pure states and then get "mixed" (in the BSS sense) due to the undesired coupling effect which exists in the considered system (Heisenberg coupling in our case). We proposed an approach which consists in repeatedly initializing the two qubits and later measuring spin components associated to the system composed of these two coupled qubits. We showed that the resulting data obey the following nonlinear "mixing model" (in BSS terms)[2]:

$$p_1 = r_1^2 r_2^2 \tag{25}$$

$$p_2 = r_1^2(1-r_2^2)(1-v^2) + (1-r_1^2)r_2^2v^2 - 2r_1r_2\sqrt{1-r_1^2}\sqrt{1-r_2^2}\sqrt{1-v^2}v\sin\Delta_I \tag{26}$$

$$p_4 = (1-r_1^2)(1-r_2^2) \tag{27}$$

where (i) the observation vector is $x = [x_1, x_2, x_3]^T$ with $x_1 = p_1, x_2 = p_2$ and $x_3 = p_4$, (ii) the source vector is $s = [s_1, s_2, s_3]^T$ with $s_1 = r_1, s_2 = r_2$ and $s_3 = \Delta_I$ and (iii) the only unknown mixing parameter is v. The physical meaning and properties of all these quantities are provided in [2] and need not be detailed here, except the conditions met in the considered configurations, i.e.

$$0 < r_1 < \tfrac{1}{2} < r_2 < 1 \tag{28}$$

$$-\tfrac{\pi}{2} \le \Delta_I \le \tfrac{\pi}{2} \tag{29}$$

$$0 < v^2 < 1. \tag{30}$$

In [2], we showed that the above mixing model is invertible in these conditions. The separating structure proposed in [2] for retrieving the sources then yields an output vector $y = [y_1, y_2, y_3]^T$ which reads

$$y_1 = \sqrt{\frac{1}{2}\left[(1+p_1-p_4) - \sqrt{(1+p_1-p_4)^2 - 4p_1}\right]} \tag{31}$$

$$y_2 = \sqrt{\frac{1}{2}\left[(1+p_1-p_4) + \sqrt{(1+p_1-p_4)^2 - 4p_1}\right]} \tag{32}$$

$$y_3 = \arcsin\left[\frac{y_1^2(1-y_2^2)(1-\hat{v}^2) + (1-y_1^2)y_2^2\hat{v}^2 - p_2}{2y_1y_2\sqrt{1-y_1^2}\sqrt{1-y_2^2}\sqrt{1-\hat{v}^2}\hat{v}}\right] \tag{33}$$

where $\hat{v}$ is the estimate of v used in the separating system.

The resulting global model is here derived by combining the mixing model (25)-(27) and the separating model (31)-(33). It may be shown that this yields

$$y_1 = s_1 \tag{34}$$

$$y_2 = s_2 \tag{35}$$

$$y_3 = \arcsin\left[\frac{(\hat{v}^2 - v^2)(s_2^2 - s_1^2)}{2s_1s_2\sqrt{1-s_1^2}\sqrt{1-s_2^2}\sqrt{1-\hat{v}^2}\hat{v}} + \frac{\sqrt{1-v^2}v}{\sqrt{1-\hat{v}^2}\hat{v}}\sin s_3\right]. \tag{36}$$

[2] We here use different signal numbering as compared to [2].

The class of global models (7)-(8) does not include the specific quantum model defined by (34), (35) and (36) in its original form, but contains a slightly reformulated version of this model, obtained as follows. We split the above-defined separating system as the cascade of two sub-systems. The first, and main, subsystem derives (i) the signals y_1 and y_2 respectively defined in (31) and (32), and (ii) the argument of the arcsin() function in the right-hand term of (33). The second sub-system is a fixed "post-distortion" stage which only consists in deriving y_3 by computing the arcsin() of the third output of the first sub-system, while leaving its first two outputs, i.e. y_1 and y_2, unchanged. The mutual independence of the outputs of the overall separating system is equivalent to that of the outputs of its first stage, and the same equivalence therefore holds for the ICA separability of these two systems. Hence, we only consider the first stage of the separating system hereafter. The resulting global model is defined by (34), (35) and by (36) without arcsin() in its right-hand term. It clearly belongs to the class of models defined by (7)-(8), after adequate signal reordering, and it meets the above-defined hypotheses. Therefore, the general results that we derived in Section 4 directly show that this quantum global model is ICA-separable.

6 Conclusion

In this paper, we proved the ICA separability of a wide class of nonlinear global models involving reference signals. This directly opens the way to practical ICA methods for nonlinear mixtures encountered in various applications. We especially plan to develop such methods for the quantum model whose ICA-separability was shown above as an example.

Acknowledgment. The author would like to thank A. Deville for his participation in the development of the quantum model defined in [2], reused here.

References

1. Comon, P., Jutten, C. (eds.): Handbook of blind source separation, independent component analysis and applications. Academic Press, London (2010)
2. Deville, Y., Deville, A.: Maximum likelihood blind separation of two quantum states (qubits) with cylindrical-symmetry Heisenberg spin coupling. In: Proceedings of ICASSP 2008, Las Vegas, Nevada, USA, March 30 - April 4, pp. 3497–3500 (2008)
3. Deville, Y., Hosseini, S.: Recurrent networks for separating extractable-target nonlinear mixtures. Part I: non-blind configurations. Signal Processing 89(4), 378–393 (2009), `http://dx.doi.org/10.1016/j.sigpro.2008.09.016`
4. Hyvarinen, A., Karhunen, J., Oja, E.: Independent Component Analysis. Wiley, New York (2001)
5. Jutten, C., Karhunen, J.: Advances in Blind Source Separation (BSS) and Independent Component Analysis (ICA) for Nonlinear Mixtures. International Journal of Neural Systems 14(5), 267–292 (2004)
6. Widrow, B., Glover, J.R., McCool, J.M., Kaunitz, J., Williams, C.S., Hearn, R.H., Zeidler, J.R., Dong, E., Goodlin, R.C.: Adaptive noise cancelling: principles and applications. Proceedings of the IEEE 63(12), 1692–1716 (1975)

Adaptive Underdetermined ICA for Handling an Unknown Number of Sources

Andreas Sandmair, Alam Zaib, and Fernando Puente León

Karlsruhe Institute of Technology
Institute of Industrial Information Technology
Hertzstr. 16, 76187 Karlsruhe, Germany
{sandmair,puente}@kit.edu
http://www.iiit.kit.edu

Abstract. Independent Component Analysis is the best known method for solving blind source separation problems. In general, the number of sources must be known in advance. In many cases, previous assumption is not justified. To overcome difficulties caused by an unknown number of sources, an adaptive algorithm based on a simple geometric approach for Independent Component Analysis is presented. By adding a learning rule for the number of sources, the complete method is a two-step algorithm, adapting alternately the number of sources and the mixing matrix. The independent components are estimated in a separate source inference step as required for underdetermined mixtures.

Keywords: Underdetermined blind source separation, independent component analysis.

1 Introduction

Since its inception, Independent Component Analysis (ICA) has become a fundamental tool for solving Blind Signal Separation (BSS) problems [2]. BSS involves extracting the source signals from multiple sensor observations which are (linear) mixtures of unobserved source signals. Based on the principle of statistical independence, ICA renders output signals as independent as possible by evaluating e.g. higher order statistics (HOS). Originally, ICA was designed to solve determined linear systems of equations (number of sources is equal to number of sensors). If there are fewer sensors than sources, the problem is referred to as underdetermined or overcomplete and more difficult to solve. Therefore, several methods based on classical [5] and geometric approaches [9], [10] have been proposed. In order to solve the BSS problems, whether in the determined or in the underdetermined case, the number of sources should be known in advance. The problem of unknown number of sources in BSS has received little attention in the past, and current approaches are still less developed. Therefore, the problem of unknown source numbers will be addressed in this paper. In the following, a new approach based on a geometric algorithm is presented.

V. Vigneron et al. (Eds.): LVA/ICA 2010, LNCS 6365, pp. 181–188, 2010.

2 Independent Component Analysis

2.1 Basic Principles

In ICA, it is assumed that the observed m-dimensional (sensor) data $\mathbf{x}(t) = [x_1(t), \ldots, x_m(t)]^{\mathrm{T}}$ has been generated from the model

$$\mathbf{x}(t) = \mathbf{A}\,\mathbf{s}(t), \tag{1}$$

where $\mathbf{A}$ is some unknown mixing matrix of dimensions $m \times n$ and $\mathbf{s}(t) = [s_1(t), \ldots, s_n(t)]^{\mathrm{T}}$ is the n-dimensional source data. In terms of column vectors $\mathbf{a}_i$ of the matrix $\mathbf{A}$, eq. (1) can be rewritten as

$$\mathbf{x}(t) = \sum_{i=1}^{n} \mathbf{a_i}\, s_i(t). \tag{2}$$

The goal of ICA is to estimate both the mixing matrix $\mathbf{A}$ and the independent components $s_i(t)$ given only the observed data $\mathbf{x}(t)$. In the determined case, where the number of sources is equal to the number of sensors ($n = m$), this problem can be rephrased as finding an inverse transformation $\mathbf{W}$ such that the original signals $s_i(t)$ can be reconstructed as $\mathbf{s}(t) = \mathbf{W}\,\mathbf{x}(t)$. To apply Independent Component Analysis, the signals must be statistically independent and non-Gaussian distributed. Based on these assumptions, the mixing matrix can be estimated by means of measures describing the independence of the components [3].

To sum up, ICA essentially represents a linear transformation of multivariate data, that captures the underlying structure in the data. This is exploited in many applications including BSS and feature extraction.

2.2 Underdetermined ICA

For underdetermined systems, the estimation of the source signals is more complex, because the source recovery problem is ill-posed ($n > m$). However, after estimating the mixing matrix, the original signals can be recontructed by exploiting the underlying statistical structure. Unfortunately, finding the 'best' representation in terms of an overcomplete basis is a challenging problem because the signal representation is not a unique combination of the basis functions (vectors). The problem of estimating original sources from sensor observations now involves two separate problems. One is to estimate the mixing matrix, referred to as matrix recovery step, and the other is to estimate the original sources also called source inference step. This is in sheer contrast with the determined case, where source inference is trivially done by inverting the mixing matrix. It is also worth mentioning that even if the mixing matrix is perfectly estimated, original sources cannot be recovered perfectly, because some information is permanently lost in representation.

3 Adaptive Geometrical Approach

A geometric approach to ICA was first proposed by Puntonet [8] and since that time successfully used for separating real-world data in determined and underdetermined cases. Because the geometrical approaches do not require estimation of HOS, both the matrix recovery and source inference steps are decoupled and independent of each other. Source inference can be obtained by maximum likelihood approaches or linear programming [10], but will not be described here in detail. In the following, the matrix recovery step for a geometric algorithm [10] is presented, before a extension of the algorithm to an unknown number of sources is illustrated. Following considerations are restricted to two sensors ($m = 2$.)

3.1 Geometrical Approach

The basic idea of geometric approaches is to use the concept of independence from a geometrical point of view. As the mixing process can be regarded as a geometrical transformation of a rectangular area into a parallelogram, the angle of rotation can be identified either in the mixture or in the whitened space, in order to recover the original sources using ordinary geometric algorithms [7]. The theoretical background for geometric ICA has been studied in detail and a convergence criterion has been derived, which resulted in a faster geometric algorithm [9]. Ideas of geometric algorithms have been successfully generalized to overcomplete and higher-dimensional systems [10].

The first step of the geometrical algorithm is to project the mixture or the observed data onto the unit sphere. The task is then to locate the axis of the maxima of the distributions on the unit sphere, which correspond to the original basis vectors, thus solving the separation problem. The idea of identifying the axis of maximum distributions is implemented as an unsupervised neural net with competitive learning which contains $2\,n$ elements (neurons).

As follows, the method will be introduced according to the steps in the flow diagram shown in figure 1. The key elements of the algorithm are: initialization (random) of $2\,n$ elements, calculating the proximity of the input data sample from each element w.r.t. the Euclidean metric and then applying following update rule to the closest or winning neuron:

$$\begin{aligned}\mathbf{w}_i(t+1) &= \mathrm{Pr}[\mathbf{w}_i(t) + \eta(t)\mathrm{sgn}(\mathbf{y}(t) - \mathbf{w}_i(t))] \\ \mathbf{w}_i^{'}(t+1) &= -\mathbf{w}_i(t+1)\end{aligned} \tag{3}$$

where 'Pr' denotes the projection onto the unit sphere. All other neurons are not moved in this iteration. A frequency f_i is assigned to each element (neuron), which counts the number of times each neuron ($\mathbf{w}_i$) has won. The step size is then modified according to:

$$\eta(t+1) = \eta_0\, \mathrm{c}^{f_i(t)/\tau}. \tag{4}$$

To prevent the network from becoming stuck in a meta-stable state, the learning rate is maintained at a certain low level η_f.

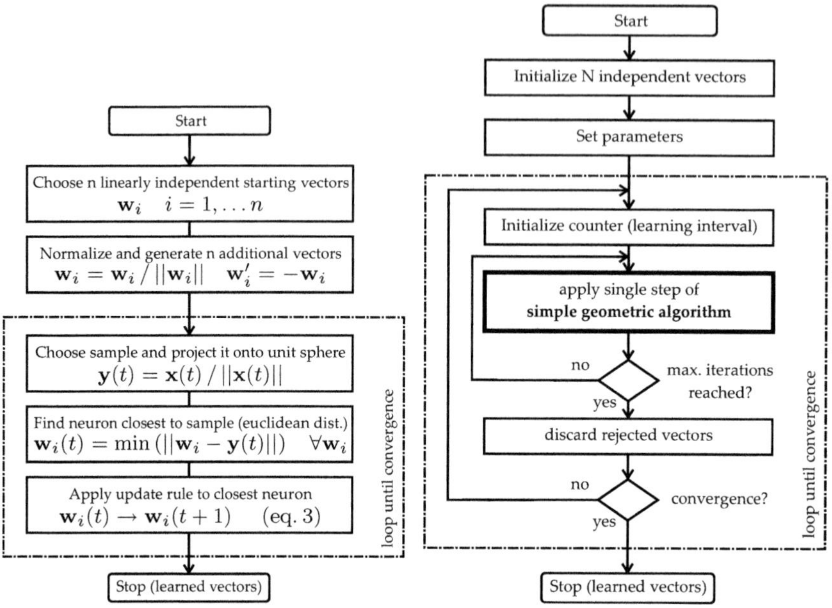

(a) geometric algorithm (b) modified geometric algorithm

Fig. 1. Flowchart for two different geometric algorithms

3.2 Handling an Unknown Number of Sources

A limitation of most ICA algorithms is that the number of sources n must be known in advance. To become independent of this constraint, an extension of the geometric algorithm presented in previous section is proposed. It combines the source number estimation and the geometric learning procedure to recover the mixing matrix. Therefor, the independence of the matrix recovery step and the source inference step is needed.

Recall that in the geometric algorithm the observed data is first projected onto the unit sphere, which results in an asymmetric distribution. But unlike the previous case, not only the locations of the axis of maximum distribution that correspond to the true basis vectors has to be estimated, but also the number of maxima, which correspond to the number of sources.

The idea is to start with a large number of independent basis vectors, or neurons N that span the whole data space. A basic assumption is that N must be greater than the actual number of unknown sources n, which can be easily satisfied because for m sensors. The maximum number of signals that can be separated using m sensors can be assessed knowing the nature of the distributions of the sources and the degree of their sparseness. After N is fixed, the geometric algorithm is applied as usual and the network is gradually pruned iteratively by comparing the neuron frequencies with a predefined threshold. The key point is that frequencies greater than the threshold suggests that the maximum of a data distribution is in its direction, which is also an indication of the presence of a source signal. If two parameters – interval length and threshold – are appropriately chosen, the algorithm would stabilize with no further pruning of the network. At that point the algorithm would not only have converged to the true number of sources but also would have learned the true directions of the distributions. The mixing matrix can be recovered in a single step by unifying the source number estimation and the learning of the basis vectors.

3.3 Description of Algorithm

For applicability of our algorithm, the following assumption is made: $N > n\,(unknown)$. The algorithm will be presented by explaining the separate steps of the blockdiagram in figure 1.

- Initialize N independent vectors symmetrically according to $\mathbf{w}_i = e^{j\frac{\pi i}{N}}$, $i = 1, 2, \ldots, N$
- Set the parameters: Interval length Δ_{win}, threshold p_{thr}, learning rate parameters (η_0, η_f, τ)
- Execute iteratively:
  - Start outer loop and initialize the counter ($n_C = 0$)
  - Apply the geometric algorithm to each data sample and increment the counter
  - Exit inner loop if $n_C = \Delta_{\text{win}}$
  - Discard vectors $\mathbf{w}_i$ if $p_i = \frac{f_i}{\sum f_i} \leq p_{\text{thr}}$
  - Abort outer loop if convergence is reached (and no more vectors discarded)

The interval length Δ_{win} and the threshold p_{thr} are key parameters of the algorithm, which are responsible for controlling the accuracy and the stability. Appropriate values can be found empirically. The algorithm delivers good results with a low initial value. It is slightly linearly increased after every iteration step as given by following relation:

$$p_{\text{thr}}(i) = p_{\text{thr}}(i-1) + (i-1)\Delta p_{\text{thr}}, \qquad i = 2, 3, \ldots \tag{5}$$

where $p_{\text{thr}}(i-1)$ is the threshold value of the previous iteration.

4 Simulation and Results

4.1 Performance Measure

To evaluate the proposed algorithm and to test the quality of the reconstruction, two measures are defined. To compare the quality of the matrix recovery, the *generalized crosstalking error* $E(\mathbf{A}, \mathbf{B})$ is used [10]. To analyze the source recovery step, the crosstalking error $E_1(\mathbf{C})$ of the correlation matrix ($\mathbf{C} = \text{Cor}(\mathbf{s}(t), \hat{\mathbf{s}}(t))$) of the original signals $\mathbf{s}(t)$ and the recovered signals $\hat{\mathbf{s}}(t)$, defined in [1] is calculated.

4.2 Test Setup

To demonstrate the algorithm, the simulation example of [4] with three speech sources and two mixtures is presented, assuming that the number of sources is not known a priori. The algorithm is initialized with following parameters: $N = 20$; $\Delta_{\text{win}} = 3833$; $\eta_0 = 0.1$, $\eta_f = 0.0002$, $\tau = 1000$ (learning rate parameters); $p_{\text{thr}} = 0.05$, $\Delta p_{\text{thr}} = 0.013$ (threshold parameters).

The original mixing matrix was:

$$\mathbf{A} = \begin{pmatrix} 0 & 0.7071 & 0.7071 \\ 1 & 0.7071 & -0.7071 \end{pmatrix}$$

The simulation result is shown in figure 2 after 46000 samples were presented to the algorithm. As it besomes evident from figure 2(b) the number of vectors are gradually reduced and the algorithm stabilized to the actual number of sources after some iterations. The learned basis vectors are shown in figure 2(a), matching almost exactly the underlying structure.

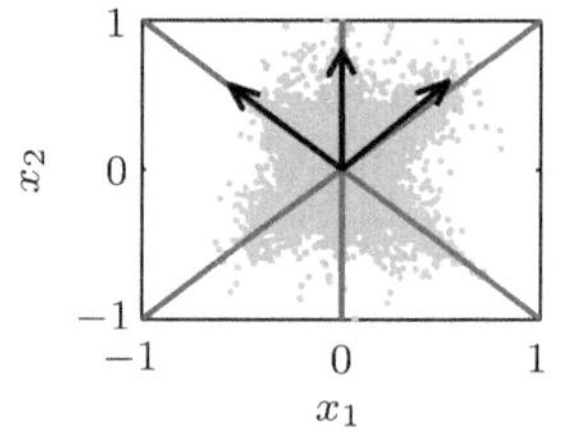

(a) Estimated vectors

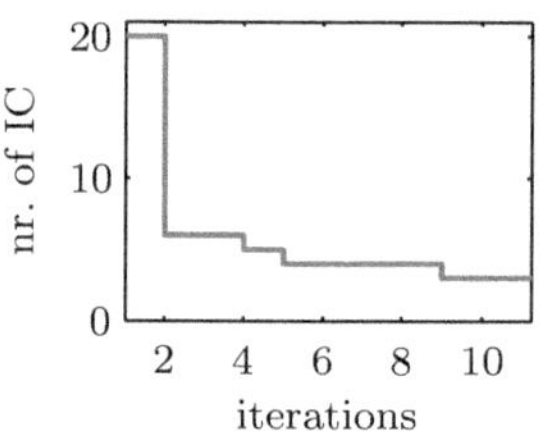

(b) Learning of vectors

Fig. 2. Evaluation of extended geometric algorithm

The recovered mixing matrix was:

$$\mathbf{B} = \begin{pmatrix} -0.0034 & 0.6970 & -0.7116 \\ 1.0000 & 0.7171 & 0.7026 \end{pmatrix}$$

which is very close to the original matrix. Also the cross-talking error $E(\mathbf{A}, \mathbf{B})$ is very close to zero, showing good estimation quality. Additionally, the sources were reconstructed using linear programming. The obtained correlation matrix was:

$$\mathbf{C} = \text{Cor}(\mathbf{s}, \hat{\mathbf{s}}) = \begin{pmatrix} 0.8572 & 0.1574 & 0.1513 \\ 0.2216 & 0.9364 & -0.1456 \\ -0.2084 & 0.1210 & -0.9458 \end{pmatrix}$$

with $E_1(\mathbf{C}) = 2.2135$, which shows high correlation values between the original and the estimated sources and low cross-correlation values indicating a high signal independence. The complexity of the algorithm still remains the same. The approach requires little more samples, because of the additional learning process.

4.3 Results

In order to verify the effectiveness of the proposed method, it is necessary to assess the performance of the algorithm with several simulations under different conditions. In this section, the performance of our algorithm is evaluated with: (a) real speech signals, so that a comparison can be drawn with previous approaches using known source number; (b) sparse sources having nearly delta-like distributions, so that our algorithm can be compared with the results of [6].

For real speech signals, the values of the error index $E(\mathbf{A}, \mathbf{B})$ for 100 independent trials are shown as box plots in figure 3(a) for a varying number of sources. As we can recover up to four speech sources from two mixtures, the maximum number of sources or independent components (ICs) is taken as 4. The error index values were calculated after the algorithm stabilizes to the true number of sources. The median values of $E(\mathbf{A}, \mathbf{B})$ are also tabulated in table 1. Furthermore, it is possible that the algorithm does not stabilize exactly to the actual number of sources. This accounts for errors in the source number estimation. The error values (absolute) are shown in table 1 over 100 trials in each case. Even though the algorithm might have recovered some of the sources successfully, it is treated strictly as an error.

Additionally, the validity of our algorithm for sparse delta-like distributed sources treated in [6], is shown. For simulation purpose, artificial source signals with high sparse distributions were generated. Following the same simulation setup as above, the results illustrated in figure 3(b) were achieved, which represent the accuracy of the algorithm over 100 independent trials with a varying number of sources respectively. The median values of the generalized crosstalking error are also tabulated in table 1, which can be compared with other algorithms.

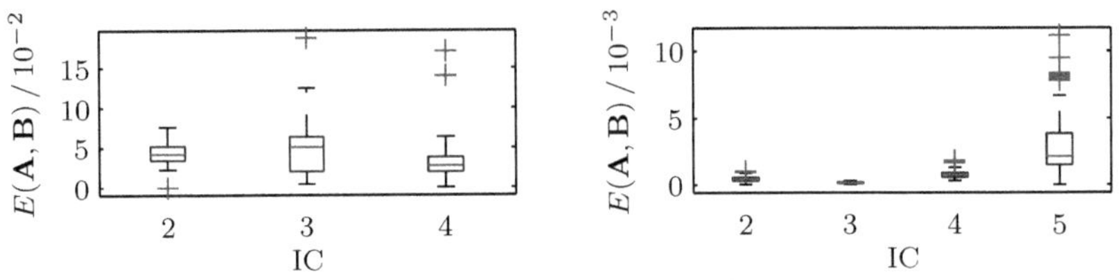

(a) Quality of estimation (b) Quality of estimation

Fig. 3. Evaluation of signals, 100 trials

Table 1. Generalized crosstalking error of speech signals (left table) and sparse signals (right table)

sources	$E(\mathbf{A}, \mathbf{B})$	errors
2	$4.246 \cdot 10^{-2}$	2
3	$5.129 \cdot 10^{-2}$	0
4	$2.792 \cdot 10^{-2}$	8
–	–	–

sources	$E(\mathbf{A}, \mathbf{B})$	errors
2	$4.213 \cdot 10^{-4}$	0
3	$1.507 \cdot 10^{-4}$	0
4	$7.310 \cdot 10^{-4}$	0
5	$2.153 \cdot 10^{-3}$	8

5 Conclusion

In this paper, a modification of a geometrical ICA algorithm was presented in order to address the problem of unknown number of in underdetermined BSS. The simulation demonstrates the efficiency of our algorithm to handle different types of sources. We can see that the variance of the generalized cross-talking error is quite small with no outliers in the real sense, which shows that once the algorithm stabilizes to actual source numbers, it always converges to the original mixing matrix. The number of errors in the estimation of the number of sources can be reduced further by judicious choice of parameters, thus increasing the reliability of our method. For sparse sources, the accuracy of our algorithm is extremely high with error index values almost zero, which means that the mixing matrix is perfectly recovered. We conclude that for sparsely distributed sources, our algorithm stabilizes quickly to actual number of sources in few iterations which extends the learning time for the vectors and improves accuracy and convergence. The algorithm is essentially a top-down approach where a large network is initialized and gradually pruned based on some criterion. The final structure of the network after convergence is what truly defines the actual model.

References

1. Amari, S., Cichocki, A., Yang, H.: A new learning algorithm for blind signal separation. In: Advances in neural information processing systems (1996)
2. Comon, P., et al.: Independent component analysis, a new concept? Signal processing 36(3), 287–314 (1994)
3. Hyvarinen, A., Karhunen, J., Oja, E. (eds.): Independent Component Analysis. Wiley Interscience, Hoboken (2001)
4. Lee, T.-W., Lewicki, M., Girolami, M., Sejnowski, T.: Blind source separation of more sources than mixtures using overcomplete representations. IEEE Signal Processing Letters 6(4), 87–90 (1999)
5. Lewicki, M., Sejnowski, T.: Learning overcomplete representations. Neural computation 12(2), 337–365 (2000)
6. Li, R., Tan, B.: Estimation of Source Signals Number and Underdetermined Blind Separation Based on Sparse Representation. In: Wang, Y., Cheung, Y.-m., Liu, H. (eds.) CIS 2006. LNCS (LNAI), vol. 4456, p. 943. Springer, Heidelberg (2007)
7. Mansour, A., Puntonet, C., Ohnishi, N.: A simple ICA algorithm based on geometrical approach. In: Sixth International Symposium on Signal Processing and its Applications (ISSPA 2001), pp. 9–12 (2001) (Citeseer)
8. Puntonet, C., Prieto, A.: An adaptive geometrical procedure for blind separation of sources. Neural Processing Letters 2(5), 23–27 (1995)
9. Theis, F., Jung, A., Puntonet, C., Lang, E.: Linear geometric ICA: Fundamentals and algorithms. Neural Computation 15(2), 419–439 (2003)
10. Theis, F., Lang, E., Puntonet, C.: A geometric algorithm for overcomplete linear ICA. Neurocomputing 56, 381–398 (2004)

Independent Phase Analysis: Separating Phase-Locked Subspaces

Miguel Almeida[1,2,⋆], José Bioucas-Dias[1], and Ricardo Vigário[2]

[1] Institute of Telecommunications
Instituto Superior Técnico, Lisbon, Portugal
http://www.ist.utl.pt/en/
[2] Adaptive Informatics Research Centre
Aalto University School of Science and Technology, Finland
miguel.almeida@lx.it.pt, jose.bioucas@lx.it.pt, ricardo.vigario@hut.fi
http://www.aalto.fi/en/

Abstract. We present a two-stage algorithm to perform blind source separation of sources organized in subspaces, where sources in different subspaces have zero phase synchrony and sources in the same subspace have full phase synchrony. Typical separation techniques such as ICA are not adequate for such signals, because phase-locked signals are not independent. We demonstrate the usefulness of this algorithm on a simulated dataset. The results show that the algorithm works very well in low-noise situations. We also discuss the necessary improvements to be made before the algorithm is able to deal with real-world signals.

Keywords: Phase-locking factor (PLF), synchrony, blind source separation (BSS), independent component analysis (ICA), subspaces, temporal decorrelation separation (TDSEP).

1 Introduction

Recently, synchrony phenomena have been studied with increasing frequency by the scientific community. Such phenomena have been observed in many different physical systems, including electric circuits, laser beams and human neurons [1].

Synchrony is believed to play a relevant role in the way different parts of human brain interact. For example, it is known that when humans engage in a motor task, several brain regions oscillate coherently [2,3]. Also, several pathologies such as autism, Alzheimer and Parkinson are thought to be associated with a disruption in the synchronization profile of the brain (see [4] for a review).

To perform inference on the networks present in the brain or in other real-world systems, it is important to have access to the dynamics of the individual oscillators (which we will call "sources"). Unfortunately, in many real applications including brain electrophysiological signals (EEG and MEG), the signals from individual oscillators are not directly measurable, and one only has access

⋆ Corresponding author.

V. Vigneron et al. (Eds.): LVA/ICA 2010, LNCS 6365, pp. 189–196, 2010.

to a superposition of the sources. For example, in EEG and MEG the signals measured in one sensor contain components coming from several brain regions [5]. In this case, spurious synchrony occurs, as we will show in this paper.

Undoing this superposition is typically called a blind source separation (BSS) problem. One usually assumes that the mixing is linear and instantaneous, which is a valid approximation in brain signals [6]. However, independence of the sources is not a valid assumption, because phase-locked sources are highly dependent. In this paper we address the problem of how to separate phase-locked sources. We have previously addressed this problem with relative success [7], and in this paper we propose an improved version of that approach which yields better results, is faster and has fewer limitations. The separation algorithm we propose uses TDSEP [8] as an initialization and then uses only the phase information of the signals. The amplitude information is discarded, because signals may exhibit synchrony even when their amplitudes are uncorrelated [9].

The algorithm presented here assumes nothing specific of brain signals, and should work in any situation where phase-locked sources are mixed approximately linearly and noise levels are low.

2 Background, Notation and Algorithm

Given two oscillators with phases $\phi_j(t)$ and $\phi_k(t)$ for $t = 1, \ldots, T$, the Phase Locking Factor (PLF) between those two oscillators is defined as

$$\varrho_{jk} = \left| \frac{1}{T} \sum_{t=1}^{T} e^{i[\phi_j(t) - \phi_k(t)]} \right| = \left| \left\langle e^{i(\phi_j - \phi_k)} \right\rangle \right|, \tag{1}$$

where $\langle \cdot \rangle$ is the time average operator. The PLF obeys $0 \leq \varrho_{jk} \leq 1$. The value $\varrho_{jk} = 1$ corresponds to two oscillators that are perfectly synchronized, i.e., that have a constant phase lag. The value $\varrho_{jk} = 0$ is attained if the two oscillators' phases are not correlated, as long as the observation period T is sufficiently long. The oscillators can be represented by real signals, since techniques such as the Hilbert Transform [10] can be used to extract the phase of real signals.[1] Typically, the PLF values are stored in a PLF matrix $\mathbf{Q}$ such that $\mathbf{Q}(j,k) = \varrho_{jk}$.

We assume that we have a number of signals (called "sources") that are organized in subspaces, such that the PLF between sources belonging to the same subspace is high, and the PLF between sources in different subspaces is low. Let $\mathbf{s}(t)$, for $t = 1, \ldots, T$, denote the vector of true sources and $\mathbf{y}(t) = \mathbf{A}\mathbf{s}(t)$ denote the mixed signals, where $\mathbf{A}$ is the mixing matrix, which is assumed to be square and non-singular. Our goal is to find an unmixing matrix $\mathbf{W}$ such that the estimated sources $\mathbf{x}(t) = \mathbf{W}^{\mathsf{T}}\mathbf{A}\mathbf{s}(t)$ are as close to the true sources as possible, up to permutation, scaling, and sign.

We now illustrate the effect of a linear mixture on PLF values. If the sources have very low or very high synchrony (PLF ≈ 0 or 1), their mixtures will have intermediate values of synchrony. This is illustrated in the top two rows of Fig. 1. Note that significant partial synchrony is present in all pairs of mixture signals.

[1] We assume that the real signals are bandpass.

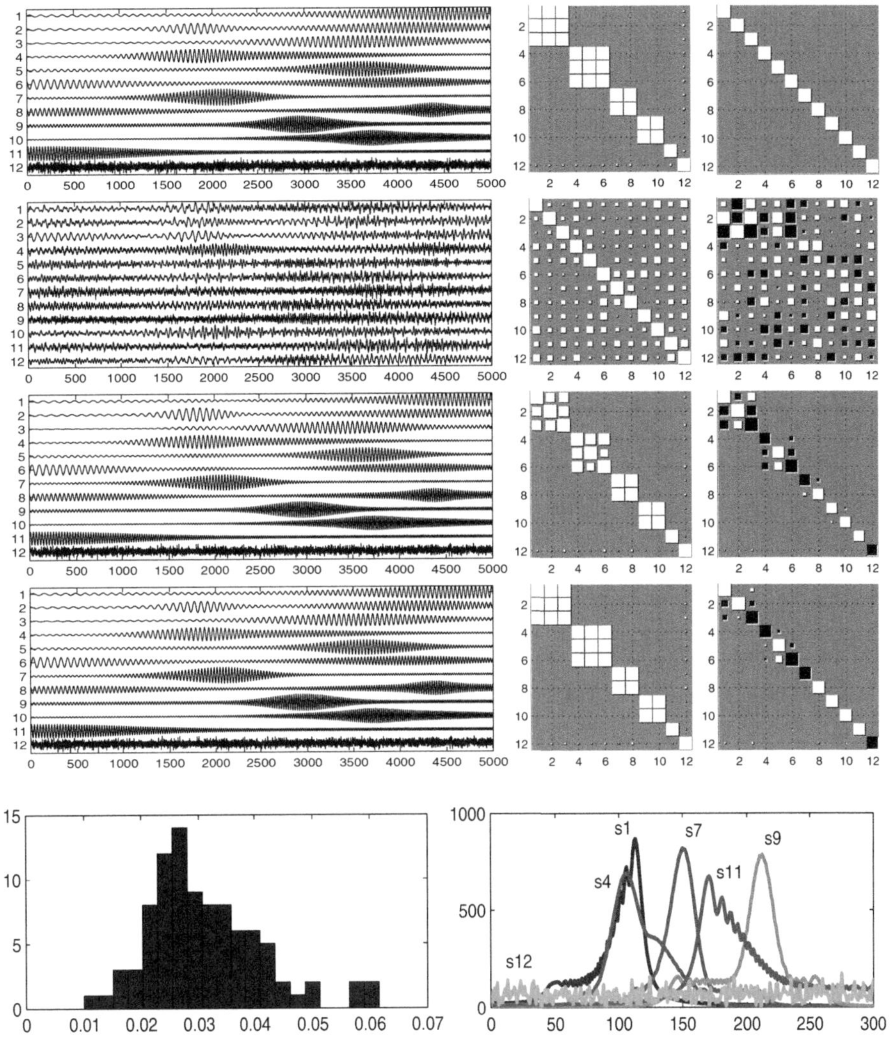

Fig. 1. The dataset used throughout this paper. *(First row)* Original sources *(left)*, PLFs between them *(middle)* and the identity matrix *(right)*, symbolizing that these are the true sources. *(Second row)* Mixed signals, PLFs between them and the mixing matrix **A**. *(Third row)* Sources resulting from TDSEP *(left)*. Note that the inter-subspace PLFs *(middle)* are very close to zero, but the intra-subspace PLFs are not all close to 1. Further, the intra-space separation is poor, as can be seen from inspection of the product $\mathbf{W}_{tdsep}^{\mathsf{T}}\mathbf{A}$ *(right)*. *(Fourth row)* Results found after the second stage of the algorithm. The estimated sources *(left)* are very similar to the original ones. This is corroborated by the PLFs between the estimated sources *(middle)* and the final unmixing matrix *(right)*. In the third and fourth rows, the permutation was corrected manually. *(Bottom row)* Histogram of the Amari Performance Index for 100 runs, corresponding to 100 random mixing matrices for these sources *(left)*, and zoom-in of the Discrete Fourier Transform of the first, fourth, seventh, ninth, eleventh and twelfth sources, one from each subspace *(right)*.

Since linearly mixing sources which have PLFs close to 1 or 0 yields signals which have partial synchrony, one can reason that finding an unmixing matrix such that the estimated sources have PLFs of 1 or 0 can separate such sources. We have previously used such an approach, motivated by ICA, to separate this kind of sources [7]. That approach, which we called Independent Phase Analysis (IPA), showed decent results, but was limited to near-orthogonal mixing matrices. A non-near-orthogonal mixing matrix yielded, however, poor results. This poor performance is closely related to the fact that, given two phase-locked signals, it is always possible to construct a linear combination of them which has a PLF of zero with those signals.[2] This implies that our problem is ill-posed, when the goal is to find an unmixing matrix such that the PLF matrix has only zeroes and ones. Some zero PLFs may, in fact, correspond to phase-locked signals.

To avoid the above referred ill-posedness, we introduce in this paper a different unmixing criterion which consists in dividing the problem into two subproblems: first, separate the subspaces from one another, even if within the subspaces some mixing remains. Second, unmix the sources within each subspace. We now discuss each of these subproblems in detail and explain why this new problem is no longer ill-posed. Since this method is an improvement of the previous approach [7], we will continue to name it Independent Phase Analysis.

2.1 Inter-subspace Separation and Subspace Detection

The objective of the first stage is to find an unmixing matrix $\mathbf{W}$ such that the estimated subspaces are correct, even if sources within each subspace are still mixed. We assume that signals in different subspaces have little interaction with each other, which should usually correspond to a distinct time structure. Therefore, techniques that use temporal structure to perform separation should be adequate to this first stage. We chose to use Ziehe *et. al.*'s implementation of TDSEP [8] for this first subproblem, but SOBI [11] can be used instead. Although we don't know any theoretical results that support TDSEP's adequacy to this task, we have repeatedly observed that it separates subspaces quite well.

A non-trivial step is the detection of subspaces from the result of TDSEP. From our experience, TDSEP can perform the inter-subspace separation very well but cannot adequately do the intra-subspace separation. This means that PLF values within each subspace will be underestimated. Naively, one can arbitrarely define a hard PLF threshold, below which signals are considered not synchronized and above which signals are considered in full phase synchrony.

The matrix resulting from this hard thresholding should be block-diagonal, with each block having all elements equal to 1. If this is the case, we can group the found signals into subspaces, and we further unmix each of these subspaces at a time (see section 2.2). By inverting and transposing the mixing matrix estimated by TDSEP, we have a first estimate of the unmixing matrix $\mathbf{W}_{tdsep}$.

[2] Unfortunately, the proof of this claim is too lengthy to show here.

If the matrix resulting from the thresholding is not block-diagonal with blocks with all elements equal to 1, our algorithm considers that the subspaces were wrongly detected and fails. See section 4 for possible improvements on this.

2.2 Intra-subspace Separation

In the second stage of IPA, we begin by selecting the subset of columns of the unmixing matrix $\mathbf{W}_{tdsep}$ found by TDSEP that form the l-th subspace, which we denote by S_l. We construct a rectangular matrix $\mathbf{W}_{tdsep,l}$ from those columns. Denote the number of signals in the l-th subspace by N_l, and let $\mathbf{z}_l(t) = \mathbf{W}_{tdsep,l}^{\mathsf{T}}\mathbf{y}(t)$ be the vector of the sources of subspace S_l estimated by TDSEP.

In this second stage, our goal is to separate each of these subsets of sources. As explained above, the true sources should have a PLF of 1 with the sources in the same subspace. We should therefore unmix the N_i sources found by TDSEP such that their PLFs are as high as possible. Mathematically, this corresponds to finding a N_l by N_l matrix $\mathbf{W}_l$ such that the estimated sources in the l-th subspace, $\mathbf{x}_l(t) = \mathbf{W}_l^{\mathsf{T}}\mathbf{W}_{tdsep,l}^{\mathsf{T}}\mathbf{y}(t) = \mathbf{W}_l^{\mathsf{T}}\mathbf{z}_l(t)$, have the highest possible PLFs.

In this second stage, for each subspace l, the objective function to be maximized is

$$J_l = (1-\lambda)\sum_{j,k\in S_l}\varrho_{jk}^2 + \lambda\log|\det\mathbf{W}_l|, \tag{2}$$

where $\sum_{j,k\in S_l}\varrho_{jk}^2$ is a sum over all pairs of sources in subspace S_l of the PLF between those sources squared. We use the Hilbert Transform [10] to obtain the phase values of the estimated sources. The second term, similarly to ICA [12], penalizes unmixing matrices that are close to singular, and λ is a parameter controlling the relative weight of the two terms. The second term, already present in the previous version of IPA [7], serves the purpose of preventing the algorithm from finding solutions which trivially have $\varrho_{jk} = 1$. Each column of $\mathbf{W}$ must be constrained to have unit norm to prevent trivial decreases of the penalty term.

With this formulation, the problem is no longer ill-posed as described above. Furthermore, we now need only optimize a subset of parameters at a time. This can drastically reduce the time needed to separate a set of sources.

The gradient of J_l relative to an entry w_{ij} of the weight matrix $\mathbf{W}_l$ is given by (we omit many dependences on l for clarity)

$$\frac{\partial J_l}{\partial w_{ij}} = (1-\lambda)4\pi\sum_{k=1}^{N}[2\varrho_{jk}]\left\langle\sin(\Psi_{jk}-\Delta\phi_{jk})\frac{Z_i}{X_j}\sin(\psi_i-\phi_j)\right\rangle - \lambda\left[\mathbf{W}_l^{-\mathsf{T}}\right]_{ij} \tag{3}$$

where ϱ_{jk} is the PLF between estimated sources j and k, $Z_i = |\tilde{z}_i|$ where $\tilde{z}_i$ is the analytic signal of the i-th source estimated by TDSEP, $X_j = |\tilde{x}_j|$ where $\tilde{x}_j$ is the analytic signal of the j-th estimated source, $\psi_i = \text{angle}(\tilde{z}_i)$ is the phase of the i-th source found by TDSEP, $\phi_j = \text{angle}(\tilde{x}_j)$ is the phase of the j-th estimated source, $\Delta\phi_{jk} = \phi_j - \phi_k$ is the phase difference of estimated sources j and k, $\Psi_{jk} = \text{angle}\left(\langle e^{i\Delta\phi_{jk}}\rangle\right)$ is the average phase difference between estimated sources j and k, and $\left[\mathbf{W}_l^{-\mathsf{T}}\right]_{ij}$ is the (i,j) element of the inverse of $\mathbf{W}_l^{\mathsf{T}}$.

3 Results

We present results showing that this new approach provides drastic improvements on the separation quality. The optimization was done using a gradient algorithm with adaptive step sizes, running up to 600 gradient iterations until the average PLF within the subspace is greater than 0.9999. λ was hand tuned for optimal performance. However, the algorithm yields similar results for λ within a factor of 2 of the optimal one, which is $\lambda = 0.1$ in this case.

We simulate the noiseless instantaneous linear mixture of 12 sources depicted in the first row of Fig. 1. These sources belong to 6 clusters of sizes 3, 2 and 1. We generate 100 datasets from these sources, by generating 100 mixing matrices, each of which with elements i.i.d. from the Uniform(-1,1) distribution. We then run the algorithm once for each mixing matrix, for a total of 100 runs. Each run took about 1 minute on a modern laptop computer.

The second row of Fig. 1 shows the mixed signals which are the input to our algorithm. The third row shows the sources estimated by TDSEP. Inspection of the PLFs between these sources (shown on the subfigure in the second column, third row) shows that some of the estimated sources do not have high PLFs, and an inspection of the product $\mathbf{W}_{tdsep}^{\mathsf{T}}\mathbf{A}$ reveals that the inter-subspace separation was very good, but the intra-subspace separation was poor.

The fourth row of Fig. 1 shows that by maximizing the intra-subspace PLFs we can significantly improve the separation within each subspace. This is clearly visible in the product $\mathbf{W}^{\mathsf{T}}\mathbf{A}$, depicted on the third column.

We measure the performance of IPA using the Amari Performance Index (API) [13], which measures the average relative contamination in each estimated source from all other sources. The API is non-negative and decreases to zero as the separation quality increases. A histogram of the API for the 100 runs of IPA is shown in Fig. 1. The mode of this histogram corresponds to an API of 0.0267. This is very similar to the example in Fig. 1, which has an API of 0.0265.

We used a threshold of 0.1 on the PLF matrix for the detection of subspaces. In 7% of the runs, the resulting matrix is not block-diagonal with each block full of ones, and therefore the algorithm stops.

4 Discussion

The above results demonstrate that IPA can successfully extract mixed sources based on their phase synchrony values and that it performs considerably better than the previous version. The previous version had 71% of separations with API below 0.3 [7], while this new version has all runs below 0.06, which is a remarkable improvement. Furthermore, the new version of IPA works well with any mixing matrix, even if it is close to singular. The previous version worked well only if the mixing matrix was close to orthogonal [7].

It should be noted that a PLF value of zero does not imply that the signals have distinct frequency spectra. In fact, the subspaces used here have overlapping frequency spectra (shown in the bottom right corner of Fig. 1), and an attempt at separating the subspaces based on Fourier transforms alone would fail.

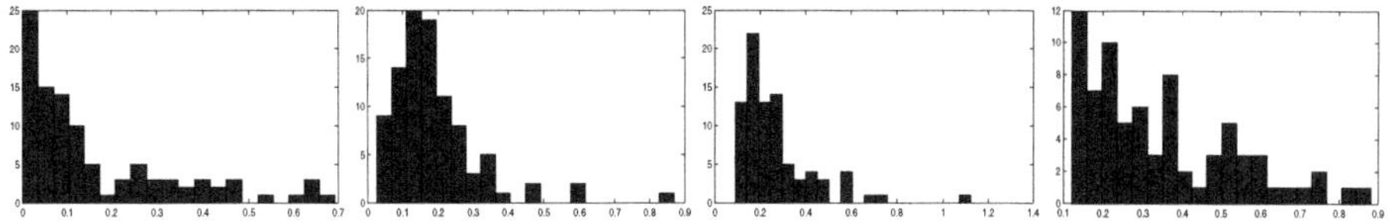

Fig. 2. Histograms of API for 100 runs of separate datasets containing a single cluster of 2 *(far left)*, 3 *(middle left)*, 4 *(middle right)* and 5 *(far right)* signals

Our selection of the value 0.1 for the PLF threshold was empirical. Although all the successful separations were very good, 7% of the runs yielded cases where our (admittedly crude) subspace detection procedure failed. This suggests that an improved subspace detection procedure is the most important improvement to be made, and it is one we are actively working on. Also, although the applicability of IPA is rather general, further research is warranted to search for TDSEP's "blind spots": signals that have similar temporal structure but low PLF values, and signals that have different temporal structure and high PLF values. The possible existence of such signals might make TDSEP identify the wrong subspaces, and the optimization stage will produce erroneous results.

Although this algorithm works quite well for this simulated data, several improvements must be made for it to be usable in real data. First of all, we have noted that the performance of IPA degrades when it is used with large subspaces. Subspaces of 1, 2 or 3 signals can be separated quite well, but on subspaces of 4 signals the performance begins to decrease. For 5 or more signals in a subspace, performance becomes considerably worse. To illustrate this limitation, we present in Fig. 2 histograms of 100 runs of the algorithm on datasets with 2, 3, 4 and 5 signals, all belonging to a single subspace.

Another unclear aspect is how this algorithm will perform if the true sources do not have extreme values of PLF but more intermediate values. It is possible that the second stage of the algorithm will overfit in such situations.

The major hindrance to be overcome before IPA can be applied to real signals is to make it work under noisy conditions. IPA performs quite well for noiseless mixtures, but real signals will always have some amount of noise. Preliminary results show that the algorithm can tolerate small amounts of noise (up to 1% of noise amplitude relative to the signal amplitude), but this still needs to be improved prior to its application to real world signals.

5 Conclusion

We have presented a two-stage algorithm, called Independent Phase Analysis (IPA), to separate phase-locked subspaces from linear mixtures. We have shown that this approach yields much better results than the previous version of IPA, and that it is no longer limited to near-orthogonal mixing matrices. Our results show that although TDSEP alone is not enough to adequately separate phase-locked sources, its conjunction with a subsequent intra-subspace separation gives very good separation quality in low-noise situations. Nevertheless, improvements are necessary before this algorithm can be applied to real-world signals.

Acknowledgments. MA is funded by scholarship SFRH/BD/28834/2006 of the Portuguese Foundation for Science and Technology. This study was partially funded by the Academy of Finland through its Centres of Excellence Program 2006-2011.

References

1. Pikovsky, A., Rosenblum, M., Kurths, J.: Synchronization: A Universal Concept in Nonlinear Sciences. Cambridge University Press, Cambridge (2001)
2. Palva, J.M., Palva, S., Kaila, K.: Phase Synchrony Among Neuronal Oscillations in the Human Cortex. Journal of Neuroscience 25, 3962–3972 (2005)
3. Schoffelen, J.M., Oostenveld, R., Fries, P.: Imaging the Human Motor System's Beta-Band Synchronization During Isometric Contraction. NeuroImage 41, 437–447 (2008)
4. Uhlhaas, P.J., Singer, W.: Neural Synchrony in Brain Disorders: Relevance for Cognitive Dysfunctions and Pathophysiology. Neuron 52, 155–168 (2006)
5. Nunez, P.L., Srinivasan, R., Westdorp, A.F., Wijesinghe, R.S., Tucker, D.M., Silberstein, R.B., Cadusch, P.J.: EEG Coherency I: Statistics, Reference Electrode, Volume Conduction, Laplacians, Cortical Imaging, and Interpretation at Multiple Scales. Electroencephalography and clinical Neurophysiology 103, 499–515 (1997)
6. Vigário, R., Särelä, J., Jousmäki, V., Hämäläinen, M., Oja, E.: Independent Component Approach to the Analysis of EEG and MEG Recordings. IEEE Transactions On Biomedical Engineering 47, 589–593 (2000)
7. Almeida, M., Vigário, R.: Source-Separation of Phase-Locked Subspaces. In: Proceedings of the Independent Component Analysis Conference (2009)
8. Ziehe, A., Müller, K.-R.: TDSEP - an Efficient Algorithm for Blind Separation Using Time Structure. In: Proceedings of the International Conference on Artificial Neural Networks (1998)
9. Rosenblum, M.G., Pikovsky, A.S., Kurths, J.: Phase Synchronization of Chaotic Oscillators. Physical Review Letters 76, 1804–1807 (1996)
10. Oppenheim, A.V., Schafer, R.W., Buck, J.R.: Discrete-Time Signal Processing. Prentice-Hall International Editions (1999)
11. Belouchrani, A., Abed-Meraim, K., Cardoso, J.-F., Moulines, E.: A Blind Source Separation Technique Using Second Order Statistics. IEEE Transactions on Signal Processing 45, 434–444 (1997)
12. Hyvärinen, A., Karhunen, J., Oja, E.: Independent Component Analysis. John Wiley & Sons, Chichester (2001)
13. Amari, S., Cichocki, A., Yang, H.H.: A New Learning Algorithm for Blind Signal Separation. Advances in Neural Information Processing Systems 8, 757–763 (1996)

Second and Higher-Order Correlation Analysis of Multiple Multidimensional Variables by Joint Diagonalization

Xi-Lin Li, Matthew Anderson, and Tülay Adalı*

Machine Learning for Signal Processing Laboratory
University of Maryland Baltimore County, Baltimore, MD 21250
lixilin@umbc.edu, matt.anderson@ieee.org, adali@umbc.edu

Abstract. In this paper, we introduce two efficient methods for second and higher-order correlation, or linear and nonlinear dependence, analysis of several multidimensional variables. We show that both the second and higher-order correlation analysis can be cast into a specific joint diagonalization problem. Compared with existing multiset canonical correlation analysis (MCCA) and independent vector analysis (IVA) algorithms, desired features of the new methods are that they can exploit the nonwhiteness of observations, they do not assume a specific density model, and they use simultaneous separation and thus are free of error accumulation arising in deflationary separation. Simulation results are presented to show the performance gain of the new methods over MCCA and IVA approaches.

Keywords: Canonical correlation analysis (CCA); independent vector analysis (IVA); second-order statistics (SOS); higher-order statistics (HOS); joint diagonalization.

1 Introduction

Correlation or dependence analysis among several sets of multidimensional variables is widely used in practice [1,2,4,5,3]. Canonical correlation analysis (CCA) [1] is used to identify linear dependence between two sets of multidimensional variables. In [2], CCA is extended to linear dependence analysis among several multidimensional variables, and in [3], this extended multiset CCA (MCCA) is applied to joint blind source separation. Both CCA and MCCA only consider second-order correlation, or linear dependence. Although MCCA can find linear dependence among several multidimensional variables, it uses deflationary separation, and thus suffers from error accumulation. On the other hand, independent vector analysis (IVA) introduced in [4,5] can be used to study higher-order correlation, or nonlinear dependence, among multiple multidimensional observations. In the original IVA algorithm proposed in [4, 5], the dependent latent

* This work is supported by the NSF grants NSF-CCF 0635129 and NSF-IIS 0612076.

V. Vigneron et al. (Eds.): LVA/ICA 2010, LNCS 6365, pp. 197–204, 2010.

components are assumed to be uncorrelated, and jointly super-Gaussian. Such assumptions may be undesired in certain applications.

In this paper, we introduce two methods for the second and higher-order correlation analysis of multiple multidimensional variables. The joint diagonalization structure of (time-delayed) correlation and cumulant matrices are used to identify the mixing matrices and latent components up to scaling and order ambiguities. Unlike the MCCA algorithm proposed in [2, 3], we use simultaneous separation, and thus there is no error accumulation arising in deflationary separation. Furthermore, unlike the IVA algorithm proposed in [4, 5], we do not assume a specific density model for the latent components. Thus the proposed methods are expected to be applicable to a wider range of problems. Simulation results are presented to demonstrate the superior performance of the new methods.

2 Signal Model and Problem Statement

We consider K, $K \geq 2$, real-valued or complex-valued multidimensional random variables $\mathbf{x}^{[k]}$, $1 \leq k \leq K$. We assume that they are of zero mean, and have the same dimensionality and the following linear decomposition form:

$$\mathbf{x}^{[k]} = \mathbf{A}^{[k]}\mathbf{s}^{[k]}, \quad 1 \leq k \leq K \tag{1}$$

where the mixing matrix $\mathbf{A}^{[k]}$ is an invertible $N \times N$ matrix, $\mathbf{s}^{[k]} = \left[s_1^{[k]}, \ldots, s_N^{[k]}\right]^T$ is the $N \times 1$ latent component vector or source vector, and superscript T denotes the transpose. In this signal model, we assume that $s_m^{[k]}$ and $s_n^{[k]}$ are independent for $1 \leq m \neq n \leq N$ and $1 \leq k \leq K$, while $s_n^{[k_1]}$ and $s_n^{[k_2]}$ are dependent for any $1 \leq k_1 \neq k_2 \leq K$ and $1 \leq n \leq N$. The problem is to solve for the mixing matrix $\mathbf{A}^{[k]}$ or its inverse, and the source vector $\mathbf{s}^{[k]}$ when T snapshots of $\mathbf{x}^{[k]}$, say $\mathbf{x}^{[k]}(t)$, $1 \leq t \leq T$, are observed.

This signal model includes several existing ones as special cases. In CCA [1], $K = 2$, and $s_n^{[1]}$ and $s_n^{[2]}$ are assumed to be linearly dependent for any $1 \leq n \leq N$. In MCCA [2,3], linear dependence among several multidimensional variables are considered, and a deflationary separation is used such that at each stage, one of the several ad hoc source extraction criteria proposed in [2] is optimized. It is clear that the deflationary separation suffers from error accumulation, especially when N is large. In the specific IVA algorithm proposed in [4,5], $s_n^{[k_1]}$ and $s_n^{[k_2]}$ are assumed to be uncorrelated for $k_1 \neq k_2$, and $\left[s_n^{[1]}, \ldots, s_n^{[K]}\right]^T$ is assumed to have a multivariate Laplace distribution [6] to capture the higher-order correlation, or nonlinear dependence, among the sources in the same group.

In this paper, we show that for the signal model given in (1), the (time-delayed) correlation and cumulant matrices of $\mathbf{x}^{[k]}$ have a joint diagonalization structure, which is similar to, but different from, the one used in the blind source separation (BSS) algorithms for identifying the mixing matrix and sources [8, 7]. We show that the special joint diagonalization structure for (1) can be exploited to identify $\mathbf{A}^{[k]}$ and $\mathbf{s}^{[k]}$ by using a joint diagonalization algorithm. In the next section, we first consider the second-order correlation analysis by joint diagonalization.

3 Second-Order Correlation Analysis

We consider the following correlation matrix:

$$\mathbf{R}^{[k_1,k_2]} = E\left[\mathbf{x}^{[k_1]}\left(\mathbf{x}^{[k_2]}\right)^H\right] = \mathbf{A}^{[k_1]}E\left[\mathbf{s}^{[k_1]}\left(\mathbf{s}^{[k_2]}\right)^H\right]\left(\mathbf{A}^{[k_2]}\right)^H \tag{2}$$

where superscript H denotes the conjugate transpose. According to the signal model given in (1), $E\left[\mathbf{s}^{[k_1]}\left(\mathbf{s}^{[k_2]}\right)^H\right]$ is a diagonal matrix. If we let $\mathbf{W}^{[k]} = \left(\mathbf{A}^{[k]}\right)^{-1}$, (2) suggests that $\mathbf{W}^{[k_1]}\mathbf{R}^{[k_1,k_2]}\left(\mathbf{W}^{[k_2]}\right)^H$ is a diagonal matrix. By varying k_1 and k_2, we can obtain K^2 correlation matrices with this joint diagonalization structure, and $K(K+1)/2$ of them are distinct.[1]

In many applications, the order of samples is important since $s_n^{[k_1]}(t)$ and $s_n^{[k_2]}(t-\tau)$ can be highly dependent for a small time-delay τ. To exploit the temporal structure of sources, we consider the following time-delayed correlation matrix:

$$\begin{aligned}\mathbf{R}^{[k_1,k_2]}(t,\tau) =& E\left[\mathbf{x}^{[k_1]}(t)\left(\mathbf{x}^{[k_2]}(t-\tau)\right)^H\right]\\ =& \mathbf{A}^{[k_1]}E\left[\mathbf{s}^{[k_1]}(t)\left(\mathbf{s}^{[k_2]}(t-\tau)\right)^H\right]\left(\mathbf{A}^{[k_2]}\right)^H\end{aligned} \tag{3}$$

where $E\left[\mathbf{s}^{[k_1]}(t)\left(\mathbf{s}^{[k_2]}(t-\tau)\right)^H\right]$ is a diagonal matrix according to the signal model given in (1). From (3), we know that $\mathbf{W}^{[k_1]}\mathbf{R}^{[k_1,k_2]}(t,\tau)\left(\mathbf{W}^{[k_2]}\right)^H$ is a diagonal matrix. For stationary sources, $\mathbf{R}^{[k_1,k_2]}(t,\tau)$ is independent of t, we can write it simply as $\mathbf{R}^{[k_1,k_2]}(\tau)$. In this paper, we assume that the sources are stationary, although it is straightforward to extend the discussion to the separation of nonstationary sources.

By fixing τ and varying k_1 and k_2, we can obtain K^2 correlation matrices that can be jointly diagonalized. We say that all these K^2 matrices form a slice at time lag τ. By selecting L time-delays τ_ℓ, $\ell = 1,\ldots,L$, we can obtain L slices of matrices having the same joint diagonalization structure. The proposed second-order correlation analysis method identifies $\mathbf{A}^{[k]}$ and $\mathbf{s}^{[k]}$ by jointly diagonalizing matrices of these L slices.

4 Higher-Order Correlation Analysis

The (time-delayed) cumulant matrices of $\mathbf{x}^{[k]}$ have the same joint diagonalization structure as (time-delayed) correlation matrices. To simplify the discussion, we

[1] Note that $\mathbf{R}^{[k_1,k_2]} = \left(\mathbf{R}^{[k_2,k_1]}\right)^H$, and they contain the same statistics.

only consider the fourth-order cumulant matrices of $\mathbf{x}^{[k]}$, and do not consider the cumulant matrices at nonzero time-delay. The fourth order cumulant of four zero mean random variables, say v_i, $i = 1, \ldots, 4$, is defined as

$$\begin{aligned}&\mathrm{cum}\,(v_1, v_2, v_3, v_4)\\&\quad\triangleq E[v_1v_2v_3v_4] - E[v_1v_2]E[v_3v_4] - E[v_1v_3]E[v_2v_4] - E[v_1v_4]E[v_2v_3].\end{aligned}$$

An important property of $\mathrm{cum}\,(v_1, v_2, v_3, v_4)$ is that if v_i and v_j are independent for any $1 \le i \ne j \le 4$, $\mathrm{cum}\,(v_1, v_2, v_3, v_4) = 0$. By using this property and our assumptions on the signal model given in (1), we have

$$\begin{aligned}&\mathrm{cum}\left[x_m^{[k_1]}, \left(x_n^{[k_2]}\right)^*, x_i^{[k_3]}, \left(x_j^{[k_4]}\right)^*\right]\\
&= \mathrm{cum}\left[\sum_{m_1=1}^{N} a_{mm_1}^{[k_1]} s_{m_1}^{[k_1]}, \left(\sum_{n_1=1}^{N} a_{nn_1}^{[k_2]} s_{n_1}^{[k_2]}\right)^*, \sum_{i_1=1}^{N} a_{ii_1}^{[k_3]} s_{i_1}^{[k_3]}, \left(\sum_{j_1=1}^{N} a_{jj_1}^{[k_4]} s_{j_1}^{[k_4]}\right)^*\right]\\
&= \sum_{m_1=1}^{N}\sum_{n_1=1}^{N}\sum_{i_1=1}^{N}\sum_{j_1=1}^{N} a_{mm_1}^{[k_1]} \left(a_{nn_1}^{[k_2]}\right)^* a_{ii_1}^{[k_3]} \left(a_{jj_1}^{[k_4]}\right)^*\\
&\qquad\qquad \times \mathrm{cum}\left[s_{m_1}^{[k_1]}, \left(s_{n_1}^{[k_2]}\right)^*, s_{i_1}^{[k_3]}, \left(s_{j_1}^{[k_4]}\right)^*\right]\\
&= \sum_{h=1}^{N} a_{mh}^{[k_1]} \left(a_{nh}^{[k_2]}\right)^* a_{ih}^{[k_3]} \left(a_{jh}^{[k_4]}\right)^* \mathrm{cum}\left[s_h^{[k_1]}, \left(s_h^{[k_2]}\right)^*, s_h^{[k_3]}, \left(s_h^{[k_4]}\right)^*\right] \qquad (4)\end{aligned}$$

where superscript $*$ denotes complex conjugate, $x_n^{[k]}$ denotes the nth element of $\mathbf{x}^{[k]}$, and $a_{mn}^{[k]}$ denotes the (m, n)th element of $\mathbf{A}^{[k]}$. Now, we can introduce a cumulant matrix $\mathbf{C}^{[k_1,\, k_2,\, (i,\, j,\, k_3,\, k_4)]}$, whose (m, n)th element $c_{mn}^{[k_1,\, k_2,\, (i,\, j,\, k_3,\, k_4)]}$ is defined as

$$c_{mn}^{[k_1,\, k_2,\, (i,\, j,\, k_3,\, k_4)]} = \mathrm{cum}\left[x_m^{[k_1]}, \left(x_n^{[k_2]}\right)^*, x_i^{[k_3]}, \left(x_j^{[k_4]}\right)^*\right].$$

Then from (4), we can write $\mathbf{C}^{[k_1,\, k_2,\, (i,\, j,\, k_3,\, k_4)]}$ as

$$\mathbf{C}^{[k_1,\, k_2,\, (i,\, j,\, k_3,\, k_4)]} = \mathbf{A}^{[k_1]}\mathbf{D}^{[k_1,\, k_2,\, (i,\, j,\, k_3,\, k_4)]}\left(\mathbf{A}^{[k_2]}\right)^H \qquad (5)$$

where $\mathbf{D}^{[k_1,\, k_2,\, (i,\, j,\, k_3,\, k_4)]}$ is a diagonal matrix, and its hth diagonal element is

$$d_h^{[k_1,\, k_2,\, (i,\, j,\, k_3,\, k_4)]} = a_{ih}^{[k_3]}\left(a_{jh}^{[k_4]}\right)^* \mathrm{cum}\left[s_h^{[k_1]}, \left(s_h^{[k_2]}\right)^*, s_h^{[k_3]}, \left(s_h^{[k_4]}\right)^*\right].$$

From (5), we know that the cumulant matrices $\mathbf{C}^{[k_1,\, k_2,\, (i,\, j,\, k_3,\, k_4)]}$ have a joint diagonalization structure, and can be used to identify $\mathbf{A}^{[k]}$ and $\mathbf{s}^{[k]}$.

By varying i, j, k_3 and k_4, we can obtain K^2N^2 slices of cumulant matrices. However, since the cumulant operator is not sensitive to the order of its arguments, we constrain $1 \le i \le j \le N$ and $1 \le k_3 \le k_4 \le K$ to prevent repeat counting of matrices with the same statistics. Then we obtain $KN(K+1)(N+1)/4$ distinct slices of cumulant matrices for joint diagonalization. Even for moderate N and K, $KN(K+1)(N+1)/4$ can be a very large number. To reduce the number of slices, we can constrain $1 \leq i = j \leq N$ and $1 \leq k_3 = k_4 \leq K$, and then the number of slices is reduced to NK.

5 A Gradient Search Joint Diagonalization Algorithm

From the discussions in sections 3 and 4 we know that both the second and higher-order correlation analysis can be cast into a joint diagonalization problem, which is similar to, but still different from, the joint diagonalization problem in BSS [8, 7]. Note that in practice, we usually speak of approximate joint diagonalization, since the target matrices cannot be exactly jointly diagonalized due to estimation errors. We summarize this approximate joint diagonalization problem as below. Given a set of target matrices $\mathbf{T}^{[p,q,\ell]}$, the problem is to find a set of matrices $\mathbf{W}^{[k]}$, $1 \leq k \leq K$, such that all $\mathbf{W}^{[p]}\mathbf{T}^{[p,q,\ell]}\left(\mathbf{W}^{[q]}\right)^H$ are approximately diagonal matrices, where ℓ, $1 \leq \ell \leq L$, is the slice index, $1 \leq p,\ q,\ k \leq K$, and both $\mathbf{T}^{[p,q,\ell]}$ and $\mathbf{W}^{[k]}$ are $N \times N$ matrices.

As in BSS, a pre-whitening step can greatly simplify the design of separation algorithm. Thus in this section, we assume that $\mathbf{x}^{[k]}$ has been pre-whitened, and $\mathbf{W}^{[k]}$ is constrained to be a unitary matrix. We consider the following cost function of $\mathbf{W}^{[k]}$:

$$\begin{aligned} \min \quad & J\left(\mathbf{W}^{[1]}, \ldots, \mathbf{W}^{[K]}\right) = \sum_{\ell=1}^{L}\sum_{p=1}^{K}\sum_{q=1}^{K}\left\|\text{off}\left[\mathbf{W}^{[p]}\mathbf{T}^{[p,q,\ell]}\left(\mathbf{W}^{[q]}\right)^H\right]\right\|_{\text{F}}^2 \\ \text{s.t.} \quad & \mathbf{W}^{[k]}\left(\mathbf{W}^{[k]}\right)^H = \mathbf{I},\ k = 1, \ldots, K \end{aligned} \tag{6}$$

where $\text{off}(\mathbf{X}) = \mathbf{X} - \text{diag}(\mathbf{X})$ sets the diagonal elements of $\mathbf{X}$ to zeros, $\|\cdot\|_{\text{F}}$ denotes the Frobenius norm, and $\mathbf{I}$ denotes the identity matrix. The gradient of J with respect to $\mathbf{W}^{[p]}$ is given by

$$\begin{aligned} \frac{\partial J}{\partial\left(\mathbf{W}^{[p]}\right)^*} = & \sum_{\ell=1}^{L}\sum_{q=1}^{K}\text{off}\left[\mathbf{W}^{[p]}\mathbf{T}^{[p,q,\ell]}\left(\mathbf{W}^{[q]}\right)^H\right]\mathbf{W}^{[q]}\left(\mathbf{T}^{[p,q,\ell]}\right)^H \\ & + \sum_{\ell=1}^{L}\sum_{q=1}^{K}\text{off}\left[\mathbf{W}^{[p]}\left(\mathbf{T}^{[q,p,\ell]}\right)^H\left(\mathbf{W}^{[q]}\right)^H\right]\mathbf{W}^{[q]}\mathbf{T}^{[q,p,\ell]}. \end{aligned}$$

Thus the gradient descent learning rule for $\mathbf{W}^{[p]}$ is:

$$\begin{aligned} \mathbf{W}_+^{[p]} & = \mathbf{W}^{[p]} - \mu_p\frac{\partial J}{\partial\left(\mathbf{W}^{[p]}\right)^*} \\ \left(\mathbf{W}^{[p]}\right)^{[\text{new}]} & = \left[\mathbf{W}_+^{[p]}\left(\mathbf{W}_+^{[p]}\right)^H\right]^{-1/2}\mathbf{W}_+^{[p]} \end{aligned} \tag{7}$$

where $\mu_p > 0$ is a positive step size. The proposed joint diagonalization algorithm optimizes $\mathbf{W}^{[k]}$ alternatively by using the learning rule given in (7).

6 Simulation Results

In this section, we present a simple simulation example for the separation of artificial mixtures of natural color images to demonstrate the performance of the proposed algorithms. Twelve natural color images from the ICALAB benchmark are used as the sources [9]. These images are shown in Fig. 1. The red, green and blue components of these images naturally form $K = 3$ sets of sources. At each run, we randomly select $N < 12$ images, and mix their red, green and blue components by three random $N \times N$ matrices after stacking all the rows of the image components to form the long row vectors to obtain three sets of N-dimensionality mixtures.

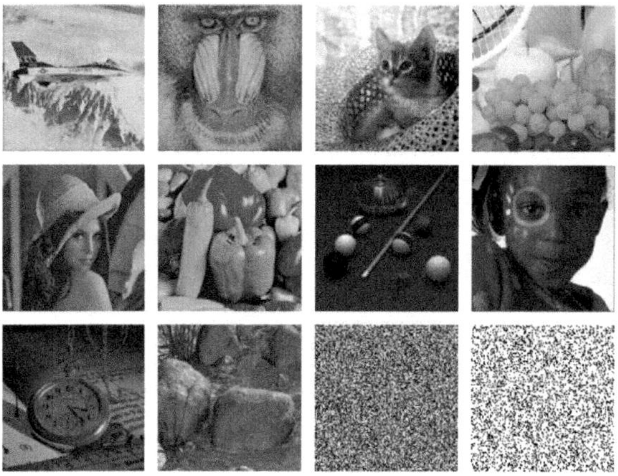

Fig. 1. Twelve natural color images from the ICALAB benchmark

Three types of algorithms are compared: MCCA, the IVA algorithm proposed in [4, 5], and the proposed joint diagonalization based algorithms. For MCCA, the two source extraction criteria, MAXVAR and GENVAR, are considered [2, 3], and we refer to these two algorithms as MCCA-MAXVAR and MCCA-GENVAR, respectively. The original IVA algorithm proposed in [4, 5] assumes a multivariate Laplace distribution for source vector $\left[s_n^{[1]}, \ldots, s_n^{[K]}\right]^T$, and thus we refer to this algorithm as IVA-ML (IVA-Multivariate Laplace). Note that the multivariate Laplace distribution assumption may be not true for certain images. We refer to the proposed second-order correlation analysis algorithm as DIAG-SOS. For DIAG-SOS, two sets of time-delays are considered, *i.e.*, $\{0\}$ and $\{0, 1, \ldots, 4\}$. We refer to the proposed higher-order correlation analysis algorithm as DIAG-CUM4. Two versions of DIAG-CUM4 are considered, *i.e.*, the simplified version with NK slices of cumulant matrices for joint diagonalization and the full version with all $NK(N + 1)(K + 1)/4$ slices of cumulant matrices for joint diagonalization.

The normalized joint inter-symbol-interference (ISI), which is similar to the index used in [3], is used as the performance index, and it is defined as

$$\frac{1}{2N(N-1)}\left[\left(\sum_{i=1}^{N}\sum_{j=1}^{N}\frac{|g_{ij}|}{\max_k |g_{ik}|}-1\right)+\left(\sum_{j=1}^{N}\sum_{i=1}^{N}\frac{|g_{ij}|}{\max_k |g_{kj}|}-1\right)\right]$$

where g_{ij} is the (i,j)th element of matrix $\mathbf{G}=\sum_{k=1}^{K}\left|\mathbf{W}^{[k]}\mathbf{A}^{[k]}\right|$, $|\cdot|$ denotes the absolute value of each element of a matrix, and $\mathbf{W}^{[k]}\mathbf{A}^{[k]}$ is the kth combined demixing-mixing matrix. Note that the joint ISI performance index is sensitive to the scaling of $\mathbf{W}^{[k]}$ and $\mathbf{A}^{[k]}$. To remove this scaling ambiguity, we assume that both $\mathbf{W}^{[k]}$ and $\mathbf{A}^{[k]}$ are normalized such that the sources and their estimations have the same variance.

Fig. 2 summarizes the joint ISI indices averaged over 100 trials. From Fig. 2, we observe that IVA-ML performs the poorest, since certain images in this benchmark are sub-Gaussian, and thus the multivariate Laplace distribution assumption in IVA-ML does not always hold. Although both the MCCA and the DIAG-SOS $L=1$ approaches use the same second-order statistics, the two MCCA algorithms show poorer performance than the proposed one, primarily due to the error accumulation of deflation-based separation algorithms. By exploiting the nonwhiteness of images, DIAG-SOS can jointly diagonalize multiple slices of correlation matrices and provides better performance. The proposed

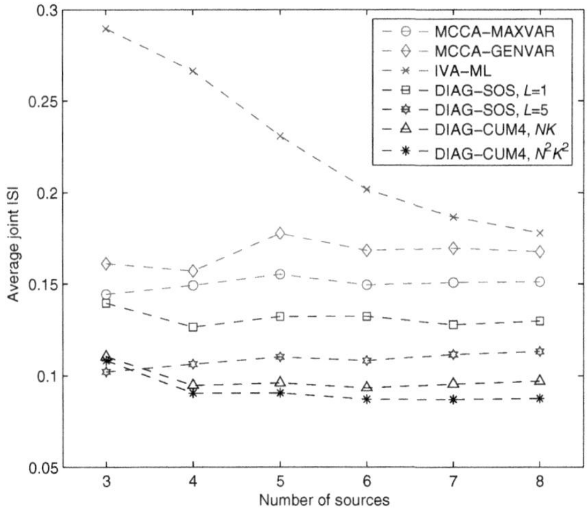

Fig. 2. Each simulation point is averaged over 100 trials. DIAG-SOS $L=1$ is the DIAG-SOS algorithm using one slice of correlation matrices at time-delay zero; DIAG-SOS $L=5$ is the DIAG-SOS algorithm using five slice of correlation matrices at time-delays $\{0,1,\ldots,4\}$; DIAG-CUM4 NK is the simplified DIAG-CUM4 algorithm; DIAG-CUM4 N^2K^2 is the full DIAG-CUM4 algorithm.

higher-order correlation analysis algorithms that jointly diagonalizes cumulant matrices perform the best for this benchmark at the expense of higher computational load.

7 Conclusions

In this paper, several second-order and higher-order correlation analysis methods are proposed for joint blind source separation. They all can be solved by using an approximate joint diagonalization algorithm. Compared with the independent vector analysis (IVA) method proposed in [5, 4], the new methods minimize assumptions on the source distribution, and thus promise to be more widely applicable in practice. Simulation results on the separation of artificial mixtures of color images are presented to show the desirable performance of new methods.

The future work will include the extensions to nonorthogonal joint diagonalization, the exploitation of full statistics, such as by using pseudo-covariance matrices, in the separation of noncircular complex-valued signals, and applications of the new methods to real-world data, such as analysis of multi-subject functional magnetic resonance imaging (fMRI) data, as in [3].

References

1. Hotelling, H.: Relation Between Two Sets of Variables. Biometrika 28(3/4), 321–377 (1936)
2. Kettenring, J.R.: Canonical Analysis of Several Sets of Variables. Biometrika 58(3), 433–451 (1971)
3. Li, Y.O., Adalı, T., Wang, W., Calhoun, V.D.: Joint Blind Source Separation by Multiset Canonical Correlation Analysis. IEEE Trans. Signal Process. 57(10), 3918–3929 (2009)
4. Kim, T., Lee, I., Lee, T.W.: Independent Vector Analysis: Definition and Algorithms. In: Fortieth Asilomar Conference on Signals, Systems and Computers 2006. pp. 1393–1396 (2006)
5. Lee, J.H., Lee, T.W., Jolesz, F.A., Yoo, S.S.: Independent Vector Analysis (IVA): Multivariate Approach for fMRI Group Study. NeuroImage 40(1), 86–109 (2008)
6. Eltoft, T., Kim, T., Lee, T.W.: On the Multivariate Laplace Distribution. IEEE Signal Process. Letters 13(5), 300–303 (2006)
7. Belouchrani, A., Abed Meraim, K., Cardoso, J.F., Moulines, E.: A Blind Source Separation Technique Based on Second Order Statistics. IEEE Trans. Signal Process. 45(2), 434–444 (1997)
8. Cardoso, J.F., Souloumiac, A.: Blind Beamforming for non-Gaussian Signals. IEE Proceedings F 140(6), 362–370 (1993)
9. Cichocki, A., Amari, S., Siwek, K., et al.: ICALAB Toolboxes., `http://www.bsp.brain.riken.jp/ICALAB`

Independent Component Analysis of Time/Position Varying Mixtures

Michael Shamis and Yehoshua Y. Zeevi

Department of Electrical Engineering, Technion - Israel Institute of Technology, Haifa, Israel
smisha1@tx.technion.ac.il, zeevi@ee.technion.ac.il

Abstract. Blind Source Separation (BSS) is a well known problem that has been addressed in numerous studies in the last few decades. Most of the studies in this field address the problem of time/position invariant mixtures of multiple sources. Real problems are however usually not time and/or position invariant, and much more complicated. We present an extension of the Maximum Likelihood (ML) Independent Component Analysis (ICA) approach to time variant instantaneous mixtures.

Keywords: Blind Source Separation, Independent Component Analysis, Time/Position Varying Mixtures, Instantaneous Mixtures, Maximum Likelihood.

1 Introduction

Blind Source Separation (BSS) is a well known problem that has been addressed in numerous studies in the last few decades. Most of the studies in this field address the problem of time/position invariant mixtures of multiple sources. Many real problems are not, however, time invariant. Time/position varying BSS problems with *a priori* known parametric family of mixtures were recently addressed in [1] and [2], using Geometric Source Separation (GSS) and with unknown parametric families in [3], using block decorrelation.

Instantaneous time and/or position varying mixtures can arise in various applications, such as separation of images taken through a non-flat medium, under non constant lightning condition, or in the case of moving objects. In some applications the nature of the mixtures is unknown, but in a wide range of applications the parametric family of the mixtures is known, and the problem is to estimate the parameters of the signal reconstruction (time invariant BSS is a simple example of such a family) . We propose a method for reconstruction of the signals in cases where the parametric family is known. To this end we propose and algorithm that performs well on numerous problems, but in general the optimization problem is non-convex and might have multiple extrema, which prevents in some problems the algorithm from converging to the true solutions.

We briefly review first, the time invariant BSS problem and its solution (section 1.2). Later, in section 2, we extend the time invariant problem to time varying. Then, in section 3, we present reconstruction results of a few parametric families.

V. Vigneron et al. (Eds.): LVA/ICA 2010, LNCS 6365, pp. 205–212, 2010.

1.1 Notations

- A - capital letters represent matrices
- $\mathbf{a}$ - bold letters represent column vectors
- A_i - capital letters with subscript i represent the ith row of the matrix (as a column vector)
- $\mathbf{a}_i$ - bold letters with subscript i represent the ith element of the vector.
- $\mathbf{a}(t)$ - represents a vector of signals samples $\mathbf{a}$ at time t.
- $\mathbf{a}_i(t)$ - represents the scalar sample of signal $\mathbf{a}_i$ at time t.
- $A_{i,j}$ - represents the scalar located at the index i,j of the matrix
- $diag(\mathbf{a})$ - represents a diagonal matrix whose elements are given by vector $\mathbf{a}$.

1.2 ICA of Time Invariant Mixtures

Problem Definition. Time invariant BSS is a well studied problem. We assume that there exist l samples of n independent signals, $\mathbf{s}(t) = (\mathbf{s}_1(t), \mathbf{s}_2(t), ..., \mathbf{s}_n(t))^T$ where $t \in 1..l$. The samples of the signals are unknown, but instead we are given a sample of their linear mixtures $\mathbf{x}(t) = (\mathbf{x}_1(t), \mathbf{x}_2(t), ..., \mathbf{x}_n(t))^T$ mixed by unknown mixing matrix A :

$$\mathbf{x}(t) = A\mathbf{s}(t). \tag{1}$$

The problem is to estimate the original signals based on these given samples of the mixtures. Maximum Likelihood(ML) approach attempts to reconstruct these signals by finding a reconstruction matrix B such that:

$$\mathbf{y}(t) = B\mathbf{x}(t) \simeq \mathbf{s}(t). \tag{2}$$

The original signals can only be reconstructed up to their magnitude (matrix A can amplify the magnitude by any value that can't be found) and permutation (there is no meaning to signals order). For this reason, and for simplicity of the solution of the above problem, it is usually solved under the constraint that the variance of the reconstructed values is 1 and their mean is 0. The condition of mean 0 can be easily resolved for signals which do not satisfy this assumption, using whitening [6].

ML Reconstruction. Let's assume that the probability density function of the original signals $p_s(\mathbf{s}_i(t))$ is known. In addition we assume that the signals are independent. It can be easily shown that in the case under consideration (using discrete probability function approximation):

$$p(\mathbf{x}_1(t), \mathbf{x}_2(t), ..., \mathbf{x}_n(t)) = |det(A^{-1})| \prod_{i=1}^{n} p_s(B_i^T \mathbf{x}_i(t)). \tag{3}$$

Since B is an estimate of A^{-1}, it can be substituted in the expression above. The probability that the set of mixture samples $\{\mathbf{x}_i(t)\}$ represents the signals mixed by the given matrix B^{-1} is:

$$P(B) = \prod_{t=1}^{l} |det(B)| \prod_{i=1}^{n} p_s(B_i^T \mathbf{x}_i(t)), \tag{4}$$

and its log is given by

$$L(B) \triangleq \log P(B) = l \log |det(B)| + \sum_{t=1}^{l} \sum_{i=1}^{n} \log p_x(B_i^T \mathbf{x}_i(t)). \qquad (5)$$

$L(B)$ is the log likelihood of the given samples. This expression is usually optimized for matrix B to achieve reconstruction. The right part of $L(B)$ can also be viewed as an estimation of entropy of the signals. The same problem can then be reformulated as entropy maximization of the reconstructed signals under the constraint of their independence.

In the above derivations we assumed that the probability density function is known *a priori*, but in practice a simple function can be used for sub gaussian signals (images can usually be considered as such):

$$f(v) \equiv \log p_s(v) \equiv \log cosh(v) - \frac{v^2}{2} + c. \qquad (6)$$

The constant c does not affect the optimization and, thus, can be omitted. It is important to note that in this case the function is non-positive. This fact is later used in our derivations. For further motivation using such probability estimation function see [6].

2 ICA of Time/Position Varying Mixtures

2.1 Problem Definition

The problem of BSS for time/position varying mixtures is formulated in almost the same way as for time/position invariant, with the difference that now we assume that the mixing matrix is dependent on time or position, and on some vector θ of k parameters which defines the parametric family[1] of the mixtures:

$$\mathbf{x}(t) = A(t, \theta_{\mathbf{mix}})\mathbf{s}(t) \qquad (7)$$

Given the vector θ, the matrix $B(t, \theta) = A(t, \theta)^{-1}$ is also known.

For given θ we define the reconstructed approximated signals $\mathbf{z}_1(t), \mathbf{z}_2(t), ..., \mathbf{z}_n(t)$ as

$$\mathbf{z}(t, \theta) = B(t, \theta)\mathbf{x}(t). \qquad (8)$$

As in the time invariant case, there are given only a set of samples of the mixed signals $\mathbf{x}_1(t), \mathbf{x}_2(t), ..., \mathbf{x}_n(t)$, where $t \in 1..l$. The goal is to find the vector of parameters θ^* for which the signals $\mathbf{z}_1(t), \mathbf{z}_2(t), ..., \mathbf{z}_n(t)$ will best approximate $\mathbf{s}_1(t), \mathbf{s}_2(t), ..., \mathbf{s}_n(t)$, up to scaling and permutations (if only one such vector exists then, of course, $\theta^* = \theta_{\mathbf{mix}}$).

[1] The number of parameters k depends on the parametric family and varies from one parametric family to another.

Note that in the time/position varying problem, whitening is not an option since after whitening the reconstruction matrix may no longer belong to our parametric family.

It can be easily observed that the problem formulated here is indeed a generalization of the Time Invariant BSS (TIBSS). The latter can be formulated in terms of time varying definitions as

$$A(t,\theta) \equiv \begin{pmatrix} \theta_1 & \dots & \theta_{\mathbf{n}} \\ \dots & \dots & \\ \theta_{n^2-n+1} & \dots & \theta_{n^2} \end{pmatrix}. \tag{9}$$

If the parametric family contains some redundancies (reconstruction is the same for multiple values of θ), it is usually advisable to assume another parametric family that is included in the original one, but is of lower redundancy.

2.2 ML Reconstruction

As mentioned above, we constrain ourselves to reconstruction of zero mean and unit variance signals in order to avoid an ill-posed problem. Since whitening is impossible for time/position varying mixtures, we overcome this problem by normalizing the signals. For this purpose we define the mean vector

$$\mathbf{m}(\theta) \equiv \frac{1}{l}\sum_{t=1}^{l} \mathbf{z}(t,\theta), \tag{10}$$

and variance normalization matrix

$$N(\theta) = diag\begin{pmatrix} \sqrt{\frac{l-1}{\Sigma_{t=1}^{l}\mathbf{x}(t)^T B_1(t,\theta)B_1(t,\theta)^T\mathbf{x}(t) - l\mathbf{m}_1(\theta)^2}} \\ \sqrt{\frac{l-1}{\Sigma_{t=1}^{l}\mathbf{x}(t)^T B_2(t,\theta)B_2(t,\theta)^T\mathbf{x}(t) - l\mathbf{m}_2(\theta)^2}} \\ \cdot \\ \cdot \\ \sqrt{\frac{l-1}{\Sigma_{t=1}^{l}\mathbf{x}(t)^T B_n(t,\theta)B_n(t,\theta)^T\mathbf{x}(t) - l\mathbf{m}_n(\theta)^2}} \end{pmatrix}. \tag{11}$$

We then define a normalized reconstructed signals vector $\mathbf{y}(t,\theta)$ as

$$\mathbf{y}(t,\theta) \equiv N(\theta)\left(\mathbf{z}(t,\theta) - \mathbf{m}(\theta)\right). \tag{12}$$

It can be easily shown, based on, the definition of the signals $\mathbf{y}(t,\theta)$ that their mean over time is zero and variance of each signal is equal to one.

Assuming again that the probability density function of original signals p_s $(\mathbf{s}_i(t))$ is known, estimating the probability of the reconstructed signal yields[2]:

$$p(\mathbf{y}_1(t,\theta), \mathbf{y}_2(t,\theta), ..., \mathbf{y}_n(t,\theta)) = \prod_{i=1}^{n} p_s(\mathbf{y}_i(t,\theta)). \tag{13}$$

[2] Note the difference - in time invariant case we have maximized the probability of the samples vector, whereas here we maximize the probability of the reconstructed signal samples. It's not exactly the same problem due to the fact that we use a discrete probability function.

The probability that the set of reconstructed samples $\{\mathbf{y}_i(t)\}$ represents the samples of the original signals is now given by

$$P_\theta(\theta) = \prod_{t=1}^{l}\prod_{i=1}^{n} p_s(\mathbf{y}_i(t,\theta)). \tag{14}$$

For which the log likelihood is now given by:

$$L_\theta(\theta) = \sum_{t=1}^{l}\sum_{i=1}^{n} \log p_s(\mathbf{y}_i(t,\theta)) \tag{15}$$

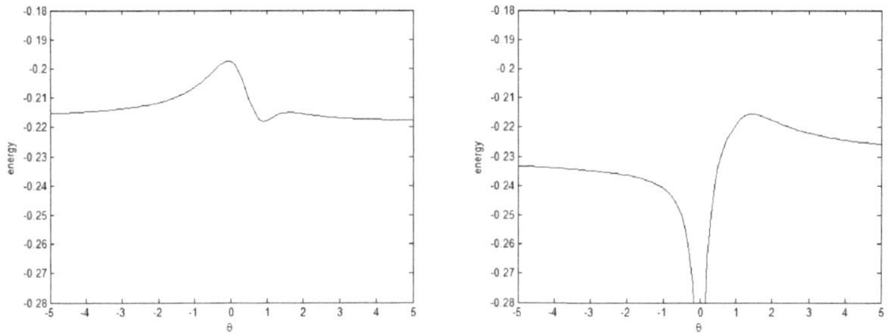

Fig. 1. Graphs of energy with and without regularization. Energies are given for parametric family with single parameter θ. The mixing parameter is equal to 1. There is a deep spike due to the fact that this is a singularity point of the covariance matrix. When regularization is applied (right) the maximal value is achieved close to 1. Without regularization (left) the maximum is at 0.

Maximizing $L_\theta(\theta)$ without constraints can be problematic for numerous parametric families. The problem arises from the fact that multiple signals of the reconstructed vector can be same signals ($\mathbf{y}_i(t,\theta) = \mathbf{y}_j(t,\theta)$ where $i \neq j$) - there is no constraint on the independence of the reconstructed signals. In fact, for parametric families for which such reconstruction is possible, all of the reconstructed signals will be practically identical. An illustration of the worst case of the problem for the parametric family

$$B\,(t,(\theta) = \begin{pmatrix} \theta t & (1-\theta)t \\ 0 & t \end{pmatrix} \tag{16}$$

is depicted at Fig.1. Although in the given example the correct reconstruction is obtained for $\theta = 1$, when no correlation constraints applied, the maximal probability value is achieved when both original signals are reconstructed to the same signal. This happens for value $\theta = 0$. To overcome this problem, we introduce the covariance matrix of the reconstructed signals

$$C(\theta) = \frac{1}{l-1}\sum_{t=1}^{l} \mathbf{y}(t,\theta)\mathbf{y}(t,\theta)^T. \tag{17}$$

Clearly the covariance matrix is singular in the above cases. By adding a factor that penalizes the singularity of this matrix we can avoid solutions in which the same signal is reconstructed more than once. For this purpose we use the modified(penalized) function

$$E(\theta) = L_\theta(\theta) - \lambda|L_\theta(\theta)\log\det C(\theta)| = L_\theta(\theta)(1+\lambda|\log\det C(\theta)|). \quad (18)$$

The second equality follows from the fact that $L_\theta(\theta)$ is a non-positive expression. It's clear that $\lambda|\log\det C(\theta)|$ equals to zero if the reconstructed signals are uncorrelated, since the variance of each signal is equal to 1. The λ factor is the weight of the penalty factor, in our experiments we mostly used values around 0.5. Note that the second factor is not added directly but multiplied by the value of $L_\theta(\theta)$. This is done in order to normalize the variables. Without this normalization, the second factor becomes usually, much larger and the optimization just becomes decorrelation. This is, of course, an unwanted result.

Note that the regularization term used in Eq.18 is probably not optimal and can add additional extrema points. Instead of this regularization other penalties on signals correlation might be used. Another way to avoid the signals correlation problem is just to solve constrained optimization problem, under the constraint that the signals must be uncorrelated. We don't use the constrained optimization approach since it demands much more computational and time complexity.

3 Results

For reconstruction of the signals we used gradient ascent algorithm on the energy $E(\theta)$. Since for some parametric families this problem has more than one local maxima, we used techniques such as multiple random restarts and simulated annealing to avoid local maxima. For additional information about these methods see [7] and [8]. From our experiments with many parametric families, $\theta_{\mathbf{mix}}$ is a global maximum, and, thus, techniques that avoid local maxima provide good reconstruction.

3.1 Time/Position Invariant Mixtures

It is of course a desirable property of the algorithm that separates time/position varying mixtures, that it separates also time invariant mixtures. As demonstrated by the example shown in Fig. 3, the separation of mixtures obtained by mixing with a constant matrix, works as well as other existing techniques. As mentioned above we often wish to reduce the parametric family space. Although a general random matrix was used for mixing of the signals, for reconstruction we used the simplified parametric family(with less redundancy) where

$$A(\theta) = \begin{pmatrix} 1 & \theta_1 \\ \theta_2 & 1 \end{pmatrix}. \quad (19)$$

It is easily observed that with this mixing matrix, it is possible to reconstruct the original signals up to permutation.

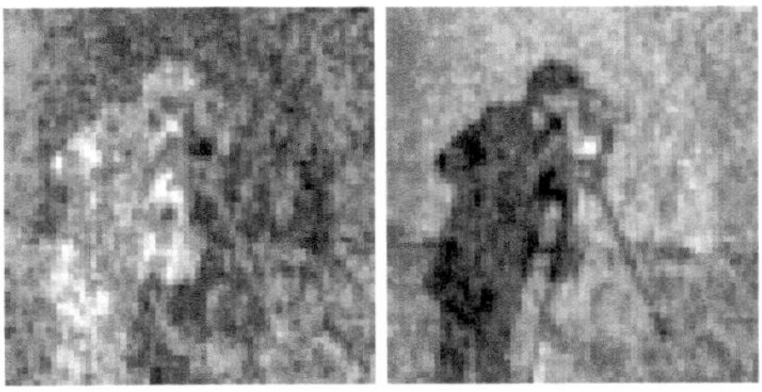

Fig. 2. Two mixtures of two images obtained by mixing with a constant mixture matrix

Fig. 3. "Source" images separated from the two mixtures shown in Fig. 2

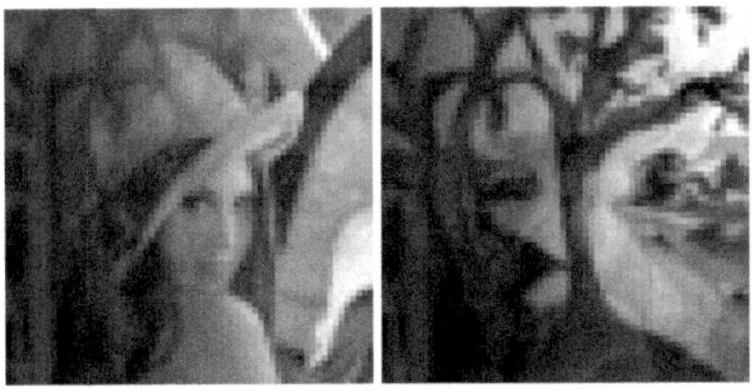

Fig. 4. Two mixtures of two images obtained by mixing with a mixture that varies linearly along spatial coordinate

Fig. 5. "Source" images separated from the two mixtures shown in Fig. 4

3.2 Separation of Time/Position Varying Mixtures

We have applied the algorithm to various time/position varying parametric families. In Fig. 4 we show an example of images that are mixed by matrices that vary linearly along the spatial coordinate, and in Fig. 5 we show the images reconstructed by using our algorithm. Reconstruction from such mixtures is, of course, impossible with existing techniques of time/position invariant mixtures separation. Although linear mixing may be considered as simple case, the energy function defined above still has a lot of local maxima, mostly near points where reconstruction matrix is singular. Thus, we need to use multiple restart points and simulated annealing to find global maxima.

In general it seems that there are parametric families on which the algorithm failed to converge due to the fact that such families allow too flexible changes of the image histograms (probability distributions), which introduces additional points of extrema. In some cases, the global maxima may even be located at points which correspond to very bad reconstructions. But, for a large number of parametric families and for most parametric families that can be obtained from real world simulated examples, the results are good and the original signals can be reconstructed with small distortions. We thus conclude that the proposed approach is applicable to a wide range of time varying mixtures, and provides a promising technique for separation of such mixtures.

Acknowledgments. This research has been supported in part by the Ollendorff Minerva Center for Vision and Image Sciences.

References

1. Kaftory, R., Zeevi, Y.Y.: Blind separation of images obtained by spatially-varying mixing system. In: ICIP 2008, pp. 2604–2607 (2008)
2. Kaftory, R., Zeevi, Y.Y.: Probabilistic Geometric Approach to Blind Separation of Time-Varying Mixtures. In: Davies, M.E., James, C.J., Abdallah, S.A., Plumbley, M.D. (eds.) ICA 2007. LNCS, vol. 4666, pp. 373–380. Springer, Heidelberg (2007)
3. Sarel, B., Irani, M.: Separating Transparent Layers through Layer Information Exchange. In: Pajdla, T., Matas, J(G.) (eds.) ECCV 2004. LNCS, vol. 3024, pp. 328–341. Springer, Heidelberg (2004)
4. Zibulevsky, M., Pearlmutter, B.A.: Blind Source Separation by Sparse Decomposition in a Signal Dictionary. Neural Computation 13(4), 863–882 (2001); An information-maximization approach to blind separation and blind deconvolution. Neural Comput. 7, 1129–1159 (1995)
5. Cardoso, J.F.: Informax and maximum likelihood for blind source separation. IEEE Signal Process. Lett. 4(4), 112–114 (1997)
6. Hyvarienen, A., Karhunen, J., Oja, E.: Independ Component Analysis. In: Haykin, S. (Series ed.) Communications, and Control. A Volume in the Wiley series on adaptive and Learning Systems for Signal Processing
7. Kirkpatrick, S., Gelatt, C.D., Vecchi, M.P.: Optimization by Simulated Annealing, Science. New Series 220(4598), 671–680 (1983)
8. Hickernell, F.J., Yuan, Y.-x.: A Simple Multistart Algorithm for Global Optimization. OR Transactions 1(2) (December 1997)

Random Pruning of Blockwise Stationary Mixtures for Online BSS

Alessandro Adamo and Giuliano Grossi

Dipartimento di Scienze dell'Informazione
Università degli Studi di Milano
Via Comelico 39, 20135 Milano, Italy
alessandro.adamo@unimi.it,
grossi@dsi.unimi.it

Abstract. We explore information redundancy of linearly mixed sources in order to accomplish the demixing task (BSS) by ICA techniques in real-time. Assuming piecewise stationarity of the sources, the idea is to prune uniformly and independently most of sample data while preserving the ability of Kurtosis-based algorithms to reconstruct the original sources using pruned mixtures instead of original ones. The mainstay of this method is to control the sub-mixtures size so that the Kurtosis is sharply concentrated about that of the entire mixtures with exponentially small error probabilities. Referring to the FastICA algorithm, it is shown that the dimensionality reduction proposed while assuring high quality of the source estimate yields to a significant reduction of the demixing time. In particular, it is experimentally shown that, in case of online applications, the pruning of blockwise stationary data is not only essential for guarantying the time-constraints keeping, but it is also effective.

1 Introduction

The goal of independent component analysis (ICA) is to describe very large sets of data in terms of latent variables better capturing the essential structure of the problem. One of the main application is instantaneous blind source separation (BSS) which arises in many areas such as speech recognition, sensor signal processing, feature extraction and medical science.

In many cases, due to the huge amount of sample data or real-time constraints, it is crucial to make ICA analysis as fast as possible. However, among the several algorithms proposed for solving this problem, because of elevated computation time, most of them can only work off-line. In this regard, one of the most popular algorithm is FastICA [5], which is based on the optimization of some nonlinear contrast functions [4] characterizing the non-Gaussianity of the components. Even though this algorithm is one of the fastest to converge, it has cubic time-complexity [5].

The purpose of this paper is to look at the performance of the FastICA algorithm when a controlled random pruning of the input mixtures is done, both

V. Vigneron et al. (Eds.): LVA/ICA 2010, LNCS 6365, pp. 213–220, 2010.

on the entire mixture available and when they are segmented into fixed-size blocks. Naturally, it is likely to speed up FastICA preserving, at the same time, the demixture quality. The method proposed consists in randomly select a sufficiently small subset of sample data in such a way that its sample kurtosis is not too far from those of the entire observations. In other words, we perform an analysis of the kurtosis estimator on the subsample with the purpose to find the maximum reduction which guarantees a narrow confidence interval with high confidence level.

The aim of this study is also in line with other works recently developed to effectively exploit blockwise ICA mainly to face the nonstationarity problem, as done in [7,8]. Even if the purpose of these two works was to develop separation algorithm asymptotically efficient in the Cramér Rao lower bound sense, they consider nonstationary signals (time varying distribution) that have however constant variance in time-intervals (piecewise stationarity).

As shown in the last section, related to tests based on real signals, this attitude to prune the most of data sample provide good results even in case of blockwise analysis, especially because the dimensionality reduction of the FastICA input data is, in some case, up to two orders of magnitude.

2 Basic ICA Model

In the ICA model we are dealing with, the latent random vector $\mathbf{s} = [s_1, \dots, s_n]^T$ of source signals is assumed to be statistically independent, zero-mean and at most one Gaussian. A random vector of their instantaneous linear mixtures $\mathbf{x} = [x_1, \dots, x_n]^T$ is a standard data model used in ICA and BSS [2,6] given by:

$$\mathbf{x} = A\,\mathbf{s}, \tag{1}$$

where A is an unknown, nonsingular $n \times n$ scalar matrix, as well as unknown are the source distributions p_{s_i}, with $i = 1, \dots, n$.

Given D i.i.d. realizations of $\mathbf{x}$ arranged as columns of the data matrix $X_D = [\mathbf{x}(1), \dots, \mathbf{x}(D)]$, the goal of ICA is to estimate the matrix A which makes the sources as independent as possible. Thus, denoting with $W = A^{-1}$ such a matrix, ICA algorithms yield a separating matrix $\hat{W} \approx A^{-1}$ providing estimates of the latent components

$$\hat{\mathbf{s}} = \hat{W}\,\mathbf{x}. \tag{2}$$

A more suitable model for blockwise stationary components must consider probability density functions $p_{s_i}^{(k)}$ different for each block k. As a consequence, given L-length data blocks $X_L^{(k)}$, the estimate of latent components take the form

$$\hat{\mathbf{s}}^{(k)} = \hat{W}^{(k)}\,\mathbf{x}^{(k)}, \qquad \text{for } k = 1, 2, \dots\,. \tag{3}$$

3 Probability Bounds for Random Pruned Mixtures

The method proposed in this section consists on discarding or pruning the most of the D samples $\mathbf{x}(1), \dots, \mathbf{x}(D)$ by means of random subsampling, but guarantying a sharp concentration of the subsample Kurtosis about that of the entire mixture.

Let us formalize the idea in statistical sense for a given observation $\mathbf{x}_i(t)$ taken at times $t = \tau, 2\tau, \ldots, D\tau$, where τ is the sampling period.

Let y be a random variable assuming values over the set of observed data $\mathbf{x}_i = \{x_i(1), \ldots, x_i(D)\}$ with uniform distribution. Assuming zero mean data $\mathbf{x}_i$, the moment of order p of y and its kurtosis are defined respectively as

$$\mu_p = \mathsf{E}[y^p] = \frac{1}{D}\sum_{t=1}^{D}(x_i(t))^p \quad \text{and} \quad \mathsf{K}[y] = \mu_4 - 3\mu_2^2.$$

Fixed a dimension $d \leq D$, let $Y = \{y_1, \ldots, y_d\}$ be a set of iid random variables distributed as y. Let us now consider the following non linear function (sample kurtosis) of the variables in Y:

$$\phi(\mathbf{y}) = \frac{1}{d}\sum_{i=1}^{d} y_i^4 - 3\left(\frac{\sum_{i=1}^{d} y_i^2}{d}\right)^2. \tag{4}$$

It can be stated the following concentration result:

Theorem 1. *Given the finite zero mean sample data* $\Omega = \{\omega_1, \ldots, \omega_n\}$ *such that* $|\omega_i| \leq 1$ *and the set* $Y = \{y_1, \ldots, y_d\}$ *of iid uniform random variables assuming values on* Ω. *If* $\phi(\mathbf{y})$ *is defined as in (4), then it holds:*

$$\Pr\{|\phi(\mathbf{y}) - \mathsf{K}[y]| \geq \rho\mathsf{K}[y]\} \leq 4\exp\left\{-\frac{(\rho\mathsf{K}[y])^2 d}{2(6\sigma^2+1)^2}\right\}$$

where σ^2 *is the variance of* y_i.

Proof. Let $m_p = \frac{1}{d}\sum_{i=1}^{d} y_i^p$ denote the sample moment of order p of the variable set Y, $\omega_m = \min\Omega$ and $\omega_M = \max\Omega$ such that $|\omega_M - \omega_m| \leq 2$ because of the bound $|\omega_i| \leq 1$. Since it holds that $\Pr\{\omega_m \leq y_i \leq \omega_M\} = 1$, by applying Hoeffding inequality [3] we obtain:

$$\Pr\{|m_p - \mu_p| \geq \varepsilon\} \leq 2\exp\left\{\frac{-2\varepsilon^2 d^2}{\sum_{i=1}^{d}(\omega_M - \omega_m)^2}\right\} \leq 2\exp\left\{\frac{-\varepsilon^2 d}{2}\right\}. \tag{5}$$

Let's now consider the functional

$$\kappa(m_4, m_2)(\mathbf{y}) = \phi(\mathbf{y}) = m_4 - 3m_2^2. \tag{6}$$

In order to establish how $\kappa(m_4, m_2)$ concentrates around the kurtosis of the entire sample space $\mathsf{K}[y]$, i.e., the probability

$$\Pr\{|\kappa(m_4, m_2) - \mathsf{K}[y]| \geq \rho\mathsf{K}[y]\},$$

we use a linear approximation around the point (μ_4, μ_2), obtaining

$$\begin{aligned}\kappa(m_4, m_2) &\approx \kappa(\mu_4, \mu_2) + \nabla\kappa|_{(\mu_4,\mu_2)} \cdot [(m_4, m_2) - (\mu_4, \mu_2)] \\ &= \mu_4 - 3\mu_2^2 + m_4 - \mu_4 - 6\mu_2(m_2 - \mu_2).\end{aligned}$$

Therefore, substituting the event $|m_4 - \mu_4| + 6\mu_2|(m_2 - \mu_2)|$ to the event $|m_4 - \mu_4 - 6\mu_2(m_2 - \mu_2)|$, we derive the inequality:

$$\Pr\{|\kappa(m_4, m_2) - \mathsf{K}[\mathbf{y}]| \geq \rho\mathsf{K}[y]\} \leq \Pr\{|m_4 - \mu_4| + 6\mu_2|(m_2 - \mu_2)| \geq \rho\mathsf{K}[y]\}.$$

To balance the error between the two events we introduce two parameters α and β such that $\alpha + \beta = 1$ and

$$\Pr\{|m_4 - \mu_4| + 6\mu_2|(m_2 - \mu_2)| \geq \rho\mathsf{K}[y]\} = \\ \Pr\{|m_4 - \mu_4| \geq \alpha\rho\mathsf{K}[y]\} + \Pr\{6\mu_2|(m_2 - \mu_2)| \geq \beta\rho\mathsf{K}[y]\}.$$

By using (5), after some algebras it results:

$$\alpha = \frac{1}{6\mu_2 + 1} \qquad \text{and} \qquad \beta = \frac{6\mu_2}{6\mu_2 + 1}.$$

Finally, since sample data in Ω are zero mean, i.e., $\sigma^2 = \mu_2$, by inequality (5) it holds

$$\Pr\{|m_4 - \mu_4| + 6\mu_2|(m_2 - \mu_2)| \geq \rho\mathsf{K}[y]\} \leq 4\exp\left\{-\frac{(\rho\mathsf{K}[y])^2 d}{2(6\sigma^2 + 1)^2}\right\}. \qquad \square$$

The main reason of Theorem 1 is to explore uniform pruning operation on the given samples $\mathbf{x}(1), \ldots, \mathbf{x}(D)$ without weakening too much the algorithm separation ability. This can be done by fixing an error $\varepsilon = \rho\mathsf{K}[y]$ and a precision δ such that

$$\Pr\{|\phi(\mathbf{y}) - \mathsf{K}[y]| \geq \varepsilon\} \qquad \text{and} \qquad \delta = 4\exp\left\{-\frac{\varepsilon^2 d}{2(6\sigma^2 + 1)^2}\right\},$$

in order to derive the minimum dimension d of the subsample respecting the error and confidence requested, i.e,

$$d = \frac{2(6\sigma^2 + 1)^2}{\varepsilon^2} \log\frac{4}{\delta}. \tag{7}$$

Recalling the Kurtosis property of linear combination of a set of independent variables, in previous section it has been shown that

$$\phi(\mathbf{y}) \approx \mathsf{K}[x_i] = \sum_{j=1}^{n} a_{ij}^4 \mathsf{K}[s_j],$$

where $(a_{ij})_{1\times n}$ is the i-th column of the mixture matrix A appearing in eq. (1) and $\mathsf{K}[s_j]$ the Kurtosis of the j-th source. Therefore, having chosen of normalizing the sample data, i.e., rescaling each mixtures in the unit interval, it results that $\phi(\mathbf{y})$ is sharply concentrated around the Kurtosis of all mixtures.

4 Simulation Results

We have carried out a vast bulk of simulation experiments using data obtained from a variety of source signals, mainly derived by audio an speech real signals. The sources that we used included essentially unimodal sub and supergaussian distributions having zero mean, as well as some examples of distributions that are nearly Gaussian (interpreted as noise). We varied the number of mixed components, from 2 to 30 and studied the average behavior of the algorithm Pruning + FastICA (P-FastICA) to vary the confidence parameters which control the dimension d of the pruned mixtures.

To measure the accuracy of the demixing matrix we use a metric called *performance index* [1] defined as:

$$\mathcal{E} = \sum_{i=1}^{n} \left(\sum_{j=1}^{n} \frac{|\rho_{ij}|}{\max_k |\rho_{ik}|} - 1 \right) + \sum_{j=1}^{n} \left(\sum_{i=1}^{n} \frac{|\rho_{ij}|}{\max_k |\rho_{kj}|} - 1 \right),$$

where $(\rho_{ij})_{n\times n} = A\hat{W}$. Such a matrix should then be close to the identity (up to a permutation), and the variances of its nondiagonal elements $\mathrm{Var}[\rho_{ij}]$, for $i \neq j$, reflect the mean value of residual interference between the separated signals $\hat{W}X$.

To experimentally establish the worth of the pruning preprocessing for FastICA (but it could be extended to different kurtosis-based algorithms for ICA) we conduct two kinds of experiments, devoted respectively to offline and online applications whose results are discussed in the next two subsections. All experiments are done on an Intel Core 2 processor of 1.6 GHz, equipped with 2 GB of RAM and running MATLAB implementation of the algorithms.

4.1 Offline Experiments

In offline tests we directly compare the performances of FastICA and P-FastICA in terms of the performance index and execution times using mixtures of 10^6 samples. The huge amount of sample data considered is mainly due to two reasons: on one hand there is experimental evidence that FastICA suffers of instability and lack of convergence against small data, on the other hand the method proposed has a sense only when high frequency rates are involved in the mixing process.

The experiments carried out with the two algorithms have shown that the performance indexes are comparable in all executions while the times are smaller up to two orders of magnitude (Fig. 1 reports the obtained results). The parameters chosen for confidence interval ε and δ assume values on $[0.05, 0.1]$ and $[0.1, 0.2]$ respectively.

4.2 Online Experiments

In order to apply the pruning technique in real-time applications, we experimentally study how the separation operation is performed by P-FastICA on blocks

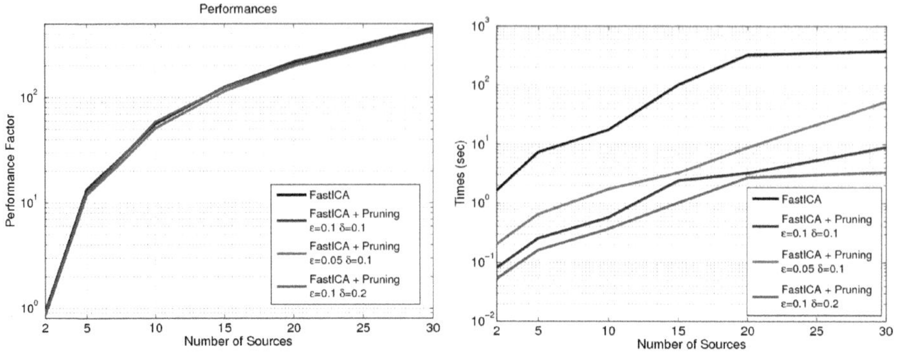

Fig. 1. Log-scale average performance index and computation times over 100 executions of FastICA and P-FastICA algorithms. The pruning parameters vary in the range $[0.05, 0.1]$ for ε and $[0.1, 0.2]$ for δ.

of data. The long (theoretically infinite) input mixtures must be segmented to fixed-size blocks of length imposed by the sampling ratio, here denoted by L. If there are n sensors, at every time unit we must arrange everything for processing n mixtures of L samples each one. There are two main operations to be performed: the first is the demixing task while the second is the block fitting to form an estimate of the original sources.

To spare execution time, the first task is preceded by pruning that allows to demix the blocks in a time unit fraction. The size of the subsample is indicated by parameter d given in (7), achieved by setting the confidence parameters explained in Section 2. The second task is critical because it is affected by the two well known indeterminacy problems of ICA: scaling factor and permutation ambiguity of the separated sources. For the type of signals we use, i.e., speech and audio signals, the scaling factor seems not to be a serious matter since the fitting of the output data blocks together do not let perceive trouble at all. As far as the permutation ambiguity is concerned, in Fig. 2 it is shown how the block-demixing and the reassembling process are tackled. All mixtures are synchronously break up every L samples and buffered on a block-matrix $X^{(k)}$ (depicted in light gray in the figure) of size $n \times L$, where index $k = 1, 2, \ldots$ denotes the block sequence. P-FastICA is then applied on a bigger block-matrix $Y^{(k)}$ (depicted in dark grey in the figure) of dimension $D > L$, obtained partially overlapping previous block of $D - L$ points. Overlapping is mainly motivated by the nonstationary of mixed sources across blocks showing that to correlate the samples of two adjacent blocks and initializing the optimization algorithm on the same point helps to locally rearrange in the right order the separated blocks. So, for every mixture-block k we find the estimate sources

$$\hat{S}^{(k)} = P^{(k)} \hat{W}^{(k)} X^{(k)},$$

where $\hat{W}^{(k)}$ is the demixing matrix carried out by P-FastICA with input matrix $Y^{(k)}$ and $P^{(k)}$ is a permutation matrix. The permutation allows to locally solve

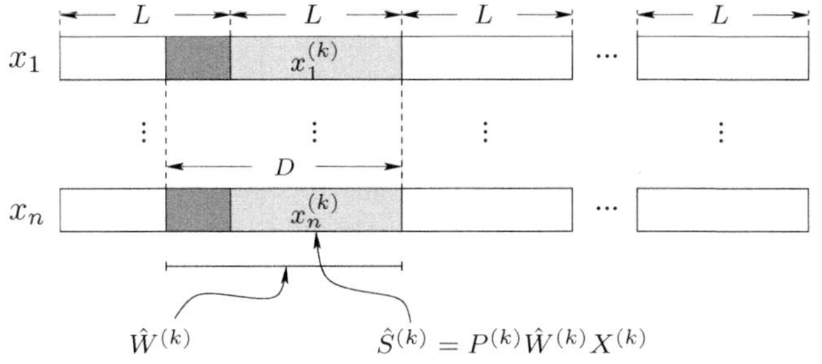

Fig. 2. Mixture segmentation in fixed-size blocks of length L and overlapping block of length $D > L$

the reassembling operation between blocks already separated and to be joined. It is defined by element-wise thresholding the matrix product $(\rho_{ij}(k))_{n\times n} = A^{(k-1)}W^{(k)}$, where $A^{(k-1)}$ is the inverse of $\hat{W}^{(k-1)}$, with threshold 0.5:

$$p_{ij}^{(k)} = \begin{cases} 1, & \text{if } \rho_{ij}(k) \geq 0.5 \\ 0, & \text{otherwise} \end{cases} \qquad \text{for } k > 0,$$

where $W^{(0)} = I_n$.

Fig. 3 shows the average performance index and the computation times over 100 executions of FastICA and P-FastICA algorithms. The pruning parameters used here also vary in the range $[0.05, 0.1]$ for ε and $[0.1, 0.2]$ for δ. Observe that, even if our hw/sw architecture is not a real-time oriented apparatus, the performance index of the two algorithms are comparable while the computation time of the more pruned mixtures are under the unitary computation time.

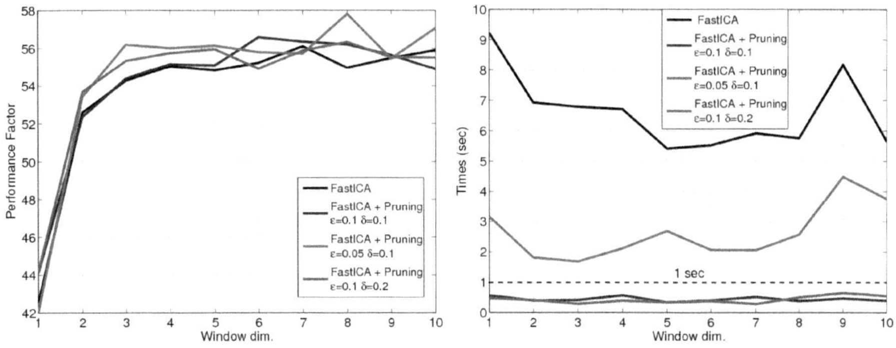

Fig. 3. Average performance index and computation times over 100 executions of blockwise FastICA and P-FastICA algorithms applied on 10 sources. The pruning parameters vary in the range $[0.05, 0.1]$ for ε and $[0.1, 0.2]$ for δ.

5 Conclusion

In this paper we explore the ability of a kurtosis-based algorithm like FastICA to carry out demixing task with a reduced number of observations with respect to those available. We exhibit a probability concentration result which allows to control a synchronized random pruning of the mixtures in order to hardly reduce the data dimension and consequently speed up the demixing process without too many losses in estimation quality.

In simulations, the algorithm is shown to be applicable in blind separation of a linear mixture of audio signals both offline and online.

References

1. Amari, S., Cichocki, A.: Recurrent neural networks for blind separation of sources. In: Proceedings of International Symposium on Nonlinear Theory and Applications, vol. I, pp. 37–42 (1995)
2. Comon, P.: Independent component analysis - a new concept? Signal Processing 36, 287–314 (1994)
3. Hoeffding, W.: Probability inequalities for sums of bounded random variables. Journal of the American Statistical Association 58(301), 13–30 (1963)
4. Hyvärinen, A.: Fast and robust fixed-point algorithms for independent component analysis. IEEE Transactions on Neural Networks 10(3), 626–634 (1999)
5. Hyvärinen, A., Oja, E.: A fast fixed-point algorithm for independent component analysis. Neural Computation 9, 1483–1492 (1997)
6. Jutten, C., Herault, J.: Blind separation of sources, part i: An adaptive algorithm based on neuromimetic architecture. Signal Processing 24, 1–10 (1991)
7. Koldovský, Z., Málek, J., Tichavský, P., Deville, Y., Hosseini, S.: Blind separation of piecewise stationary non-gaussian sources. Signal Processing 89(12), 2570–2584 (2009)
8. Tichavsky, P., Yeredor, A., Koldovsky, Z.: A fast asymptotically efficient algorithm for blind separation of a linear mixture of block-wise stationary autoregressive processes. In: ICASSP 2009: Proceedings of the 2009 IEEE International Conference on Acoustics, Speech and Signal Processing, pp. 3133–3136. IEEE Computer Society, Los Alamitos (2009)

Use of Prior Knowledge in a Non-Gaussian Method for Learning Linear Structural Equation Models

Takanori Inazumi, Shohei Shimizu, and Takashi Washio

The Institute of Scientific and Industrial Research, Osaka University, Japan

Abstract. We discuss causal structure learning based on linear structural equation models. Conventional learning methods most often assume Gaussianity and create many indistinguishable models. Therefore, in many cases it is difficult to obtain much information on the structure. Recently, a non-Gaussian learning method called LiNGAM has been proposed to identify the model structure without using prior knowledge on the structure. However, more efficient learning can be achieved if some prior knowledge on a part of the structure is available. In this paper, we propose to use prior knowledge to improve the performance of a state-of-art non-Gaussian method. Experiments on artificial data show that the accuracy and computational time are significantly improved even if the amount of prior knowledge is not so large.

Keywords: Structural equation models, Bayesian networks, independent component analysis, non-Gaussianity.

1 Introduction

Structural equation models [1] and Bayesian networks [2,3] can be used to analyze causal relationships and have been widely applied in many fields. Many methods [2,3] have been proposed to learn such a causal model when no prior knowledge on the causal structure is available. Most of conventional learning methods implicitly assume Gaussianity and often provide many indistinguishable models.

Recently, a non-Gaussian learning method called LiNGAM [4] was proposed. The new method is based on independent component analysis (ICA) and estimates a causal ordering of variables using passive observational data alone. The estimated ordering is correct if the causal relations form a linear structural equation model with non-Gaussian external influence variables and the sample size is sufficiently large.

However, the LiNGAM method has several potential problems [5]. Most ICA algorithms including FastICA [6] and gradient-based algorithms [7] may not converge to a correct solution in a finite number of steps if the initially guessed state is not well chosen or if the step size is not suitably selected for those gradient-based methods. It is not so easy to appropriately select such algorithmic

V. Vigneron et al. (Eds.): LVA/ICA 2010, LNCS 6365, pp. 221–228, 2010.

parameters. Therefore, as an alternative, a direct method called DirectLiNGAM was proposed in [5] that is guaranteed to converge in a fixed number of steps if the data strictly follows the model. Simulations showed that DirectLiNGAM provided more accurate variable orderings for larger numbers of variables [5].

Although DirectLiNGAM requires no prior knowledge on the structure, more efficient learning methods can be developed if some prior knowledge is utilized since the number of variable orderings and connection strengths to be estimated gets smaller. In this paper, we propose such a method that uses prior knowledge to improve the performance of the state-of-art method DirectLiNGAM.

2 Background

2.1 A Linear Non-Gaussian Acyclic Model: LiNGAM Model

In [4], a non-Gaussian variant of structural equation models and Bayesian networks, which is called LiNGAM, was proposed. Assume that observed data are generated from a process represented graphically by a directed acyclic graph, *i.e.*, DAG. Let us represent this DAG by a $p \times p$ adjacency matrix $\mathbf{B}=\{b_{ij}\}$ where every b_{ij} represents the direct causal effect (connection strength) from a variable x_j to another x_i. Moreover, let us denote by $k(i)$ a causal order of variables x_i so that no later variable has a directed path[1] to any earlier variable in the DAG. In other words, $k(i)<k(j)$ means that x_i could have a directed path to x_j, but not vice versa. Further, assume that the relations between variables are linear. Without loss of generality, each variable x_i is assumed to have zero mean. Then we have

$$x_i = \sum_{k(j)<k(i)} b_{ij} x_j + e_i, \tag{1}$$

where e_i is an external influence. All external influences e_i are continuous random variables having *non-Gaussian* distributions with zero means and non-zero variances, and e_i are independent of each other so that there is no unobserved confounding variables [3]. We rewrite the model (1) in a matrix form as follows:

$$\mathbf{x} = \mathbf{B}\mathbf{x} + \mathbf{e}, \tag{2}$$

where $\mathbf{x}$ is a p-dimensional variable vector, and $\mathbf{B}$ could be permuted by simultaneous equal row and column permutations to be lower triangular with all zeros on the diagonal due to the acyclicity assumption [1]. The model (1) is also written using $\mathbf{A}=(\mathbf{I}-\mathbf{B})^{-1}$ as follows:

$$\mathbf{x} = \mathbf{A}\mathbf{e} = (\mathbf{I}-\mathbf{B})^{-1}\mathbf{e}, \tag{3}$$

which defines the linear ICA model [8] since e_i are non-Gaussian and mutually independent. Every element a_{ij} represents the total causal effect from x_j to x_i. The non-Gaussian model (1) is identifiable similarity to the ICA model [4].

[1] A directed path from x_i to x_j is a sequence of directed edges such that x_j is reachable from x_i.

We emphasize that x_i is equal to e_i if any other observed variable x_j ($j \neq i$) inside the model does not have a directed path to x_i. That is, an external influence e_i is observed as x_i. Then the x_i is called an exogenous observed variable. Otherwise, e_i is called an error.

2.2 A Non-Gaussian Direct Method: DirectLiNGAM

In [5], a direct method called DirectLiNGAM was proposed. We first quote the definition of a concept called correlation-faithfulness [5]:

Definition 1 (correlation-faithfulness). *The distribution of* $\mathbf{x}$ *is said to be correlation-faithful to the generating graph if correlation and conditional correlation of* x_i *are entailed by the graph structure, i.e., the zero/non-zero status of* b_{ij}*, but not by special parameter values of* b_{ij}*.*

In the below, we quote two lemmas and a corollary proved in [5]. The basic idea of DirectLiNGAM is as follows. One first finds an exogenous variable as the top variable in the causal order by applying the following Lemma 1:

Lemma 1. *Assume that the input data* $\mathbf{x}$ *follows the LiNGAM model (1). Further, assume that the distribution of* $\mathbf{x}$ *is correlation-faithful to the generating graph. Denote by* $r_i^{(j)}$ *the residuals when* x_i *are regressed on* x_j*:* $r_i^{(j)} = x_i - \frac{\text{cov}(x_i, x_j)}{\text{var}(x_j)} x_j$ $(i \neq j)$*. Then a variable* x_j *is exogenous if and only if* x_j *is independent of its residuals* $r_i^{(j)}$ *for all* $i \neq j$*.* □

Next, the component of the exogenous variable from the other variables is removed using least squares regression. Then, it can be shown in the following Lemma 2 that a LiNGAM model (1) also holds for the residuals. Further, it can be proved in the following Corollary 1 that the ordering of the residuals is *equivalent* to that of the corresponding original observed variables:

Lemma 2. *Assume the assumptions of Lemma 1. Further, assume that a variable* x_j *is exogenous. Denote by* $\mathbf{r}^{(j)}$ *a (p-1)-dimensional vector that collects the residuals* $r_i^{(j)}$ *when all* x_i *of* $\mathbf{x}$ *are regressed on* x_j $(i \neq j)$*. Then a LiNGAM model holds for the residual vector* $\mathbf{r}^{(j)}$*:* $\mathbf{r}^{(j)} = \mathbf{B}^{(j)}\mathbf{r}^{(j)} + \mathbf{e}^{(j)}$*, where* $\mathbf{B}^{(j)}$ *is a matrix that can be permuted to be lower-triangular with all zeros on the diagonal by a simultaneous row and column permutation, and elements of* $\mathbf{e}^{(j)}$ *are non-Gaussian and mutually independent.* □

Corollary 1. *Assume the assumptions in Lemma 2. Denote by* $k_{r^{(j)}}(i)$ *the order of* $r_i^{(j)}$*. Recall that* $k(i)$ *denotes the order of* x_i*. Then, the ordering of the residuals is equivalent to that of corresponding original observed variables:* $k_{r^{(j)}}(l) < k_{r^{(j)}}(m) \Leftrightarrow k(l) < k(m)$*.* □

Thus, one can find the second top variable in the causal ordering of the original observed variables by analyzing the residuals and their LiNGAM model, *i.e.*, by applying Lemma 1 to the residuals and finding an "exogenous" residual. The repeat of these component removal and causal ordering derives the causal order of the original variables.

3 A Non-Gaussian Direct Method That Uses Prior Knowledge

3.1 Use of Prior Knowledge in the DirectLiNGAM Algorithm

We present three lemmas that ensure the validity of a variant of DirectLiNGAM proposed in Section 3.3. Let us first define a matrix $\mathbf{A}^{pk}{=}[a_{ji}^{pk}]$ that collects prior knowledge *under the assumptions of Lemma 1* as follows:

$$a_{ji}^{pk} = \begin{cases} 0 & \text{if } x_i \text{ does } \textit{not} \text{ have a directed path to } x_j, \textit{i.e.}, \; a_{ji} = 0 \\ 1 & \text{if } x_i \text{ has a directed path to } x_j, \textit{i.e.}, a_{ji} \neq 0 \\ -1 & \text{if no prior knowledge is available to know if either of} \\ & \text{the two cases above (0 or 1) is true.} \end{cases} \quad (4)$$

Due to the definition of exogenous variables and that of prior knowledge matrix $\mathbf{A}^{pk}$, we readily obtain the following three lemmas.

Lemma 3. *Assume the assumptions of Lemma 1. An observed variable x_j is exogenous if a_{ji}^{pk} is zero for all $i{\neq}j$.*

Lemma 4. *Assume the assumptions of Lemma 1. An observed variable x_j is endogenous, i.e., not exogenous, if there exist such $i{\neq}j$ that a_{ji}^{pk} is unity.*

Lemma 5. *Assume the assumptions of Lemma 1. An observed variable x_j does not have the component of x_i if a_{ji}^{pk} is zero.*

The idea of our algorithm in Section 3.3 is as follows. We first find an exogenous variable as the top variable in the causal order by applying Lemma 3 instead of Lemma 1 if an exogenous variable is identified based on prior knowledge. Then, we do not have to compute residuals in linear regression or evaluate independence between any observed variable and its residuals. If no exogenous variable is identified based on prior knowledge, we next find endogenous (non-exogenous) variables by applying Lemma 4. Since endogenous variables are never exogenous we can narrow down the search space to find an exogenous variable based on Lemma 1. We can further skip to compute the residual of an observed variable and take the variable itself as the residual if its regressor does not have the component of the variable due to Lemma 5. Thus, we can decrease the number of variable orders and connection strengths to be estimated, and it improves the accuracy and computational time as empirically demonstrated in Section 4.

3.2 Evaluation of Independence

It is necessary to define an independence (not merely uncorrelatedness) measure between an observed variable and its residuals in DirectLiNGAM and our variant. We use the following independence measure used in [5]. Let us denote by U the set of the subscripts of variables x_i, *i.e.*, $U{=}\{1, \cdots, p\}$. The following statistic evaluates nonlinear correlation between a variable x_j and its residuals $r_i^{(j)} = x_i - \frac{\text{cov}(x_i,x_j)}{\text{var}(x_j)}x_j$ when x_i is regressed on x_j:

$$T(x_j; U) = \sum_{i \in U, i \neq j} \left[|\text{corr}\{g(r_i^{(j)}), x_j\}| + |\text{corr}\{r_i^{(j)}, g(x_j)\}| \right], \tag{5}$$

where g is a nonlinear and non-quadratic function, here $g(\cdot)=tanh(\cdot)$. The statistic T in Eq. (5) is zero if x_j and $r_i^{(j)}$ are independent. Strictly speaking, independence is a much stronger condition than requiring the statistic T to be zero. However, in many cases evaluating such a nonlinear correlation as Eq. (5) would work well enough as implied in the ICA literature [8]. We can use more sophisticated nonparametric independence measures [9,10] instead of the statistic T in Eq. (5) to evaluate independence when needed.

3.3 pk-DirectLiNGAM Algorithm

We now propose a new variant of the DirectLiNGAM algorithm that utilizes prior knowledge under the correlation-faithfulness, which we call pk-DirectLiNGAM.

---

pk-DirectLiNGAM algorithm

1. Given a p-dimensional variable vector $\mathbf{x}$, a set of its variable subscripts U, a $p \times n$ data matrix of the variable vector as $\mathbf{X}$ and a $p \times p$ prior knowledge matrix $\mathbf{A}^{pk}$, an ordered list of variables $K = \emptyset$ and $m := 1$.
2. Repeat until $p-1$ subscripts are appended to K:
   (a) Find such a variable(s) x_j ($j \in U - K$) that the j-th row of $\mathbf{A}^{pk}$ has zero in the i-th column for all $i \in U - K$ ($i \neq j$) and denote the set of such variables by U_{exo}. If U_{exo} is not empty, set $U_c = U_{exo}$. If U_{exo} is empty, find such a variable(s) x_j ($j \in U - K$) that the j-th row of $\mathbf{A}^{pk}$ has unity in the i-th column for at least one of $i \in U - K$ ($i \neq j$), denote the set of such variables by U_{end} and set $U_c = U - K - U_{end}$.
   (b) Denote by $V^{(j)}$ a set of such a variable subscript $i \in U - K$ ($i \neq j$) that $a_{ij}^{pk} = 0$ for all $j \in U_c$. First set $\mathbf{r}_i^{(j)} = \mathbf{x}_i$ for all $i \in V^{(j)}$, next perform least squares regressions of x_i on x_j for all $i \in U - K - V^{(j)}$ ($i \neq j$) and derive the residual vectors $\mathbf{r}^{(j)}$ and its residual data matrix $\mathbf{R}^{(j)}$ from the data matrix $\mathbf{X}$ for all $j \in U_c$. If U_c has a single variable, set the variable to be x_m. Otherwise, find a variable x_m in U_c that is most independent of its residuals:
   $$x_m = \arg \min_{j \in U_c} T(x_j; U - K), \tag{6}$$
   where T is the independence measure defined in Eq. (5).
   (c) Append m to the end of K.
   (d) Let $\mathbf{x} := \mathbf{r}^{(j)}$, $\mathbf{X} := \mathbf{R}^{(j)}$ and $m := m + 1$.
3. Append the remaining variable to the end of K.

4. Construct a strictly lower triangular matrix $\mathbf{B}$ by following the order in K, and estimate the connection strengths b_{ij} by using some conventional covariance-based regression such as least squares and maximum likelihood approaches on the original variable vector $\mathbf{x}$ and the original data matrix $\mathbf{X}$.

Note that pk-DirectLiNGAM is equivalent to the original DirectLiNGAM if no prior knowledge is given. MATLAB codes implementing this algorithm is available online: `http://www.ar.sanken.osaka-u.ac.jp/~inazumi/`.

4 Simulations

We randomly generated 51 datasets under each combination of number of variables p and sample size n (p=5, 10, 30, 50; n=100, 300, 500, 1000) as follows:

1. We randomly constructed a $p \times p$ strictly lower-triangular matrix $\mathbf{B}$ so that standard deviations of variables x_i owing to parent variables ranged in the interval $[0.5, 1.5]$. Either of a fully connected network or a sparse network

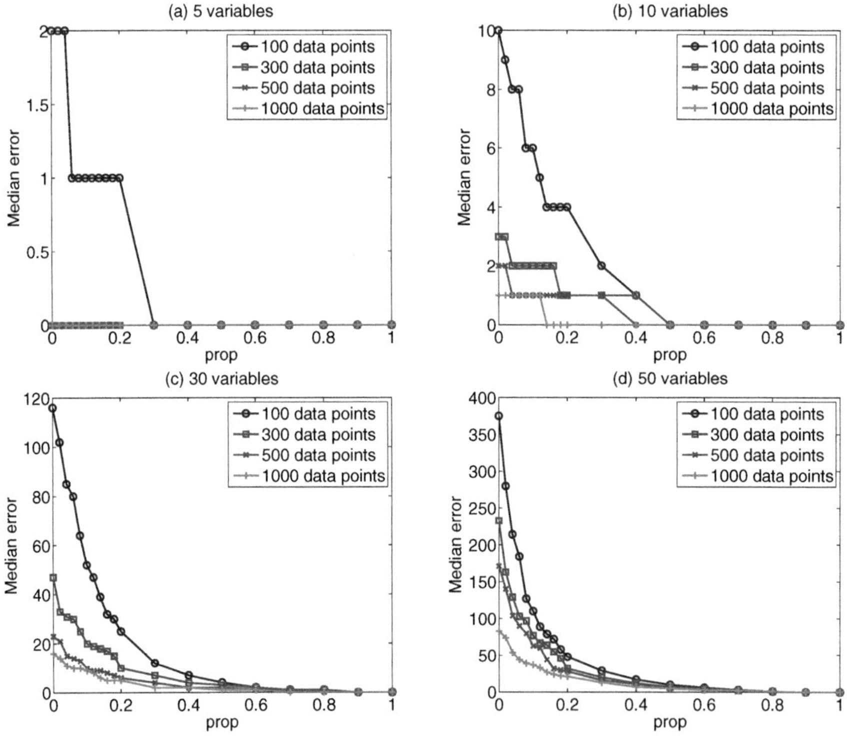

Fig. 1. Median numbers of errors under (a) 5 variables; (b) 10 variables; (c) 30 variables; (d) 50 variables. Horizontal axis: Amount of prior knowledge (*prop*). Vertical axis: Median number of errors.

was randomly created. We also randomly selected standard deviations of the external influences e_i from $[0.5, 1.5]$.
2. We generated data with sample size n by independently drawing the external influence variables e_i from various non-Gaussian distributions with zero mean and unit variance. We first generated Gaussian variables z_i with zero means and unit variances and subsequently transformed it to non-Gaussian variables by $e_i = \text{sign}(z_i)|z_i|^{q_i}$. The nonlinear exponents q_i were randomly selected from the interval $[0.5, 0.8] \cup [1.2, 2.0]$. Nonlinear exponents q_i selected from $[0.5, 0.8]$ gave sub-Gaussian variables, and exponents selected from $[1.2, 2.0]$ provided super-Gaussian variables. Finally, the transformed variables were standardized to have zero means and unit variances.
3. The values of the observed variables x_i were generated according to the LiNGAM model (1). Then we randomly permuted the order of x_i.
4. We created several prior knowledge matrices $\mathbf{A}^{pk}$ by first replacing every non-zero element by unity and every diagonal element by zero in $\mathbf{A}=(\mathbf{I} - \mathbf{B})^{-1}$ and subsequently hiding each of the off-diagonal elements, *i.e.*, replacing it by -1, with probability $1-prop$, where $prop$=0.0, 0.02, $\cdots$, 0.18, 0.2, 0.3, $\cdots$, 0.9, 1.0. The *prop* represents the proportion of the elements *not* hided.

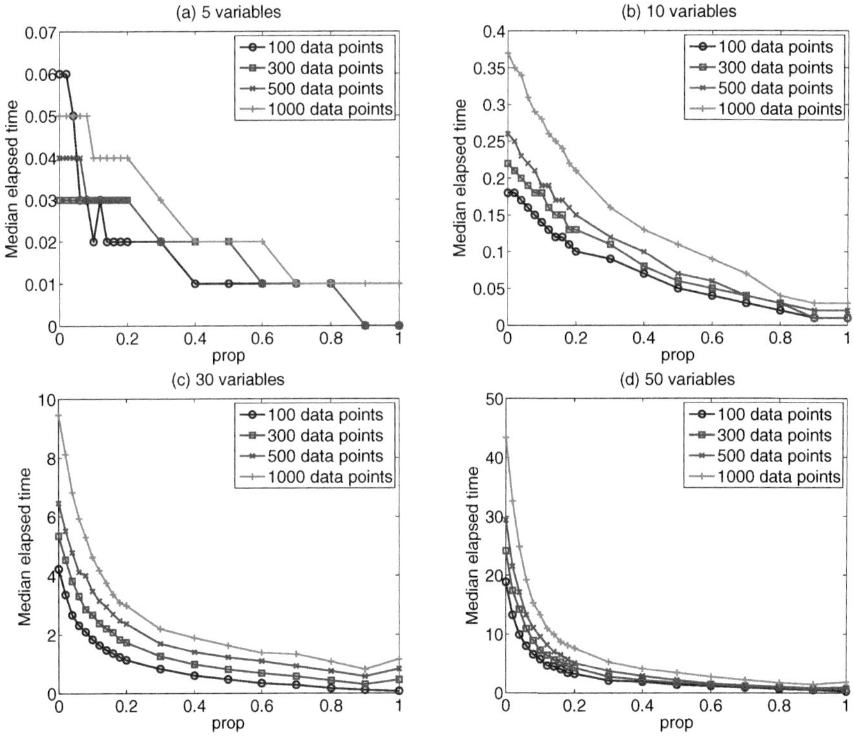

Fig. 2. Median computational times (sec.) under (a) 5 variables; (b) 10 variables; (c) 30 variables; (d) 50 variables. Horizontal axis: Amount of prior knowledge (*prop*). Vertical axis: Median computational time.

This way of generating x_i is the same as the one used in [4], and nothing is done to make parameter values satisfy the correlation-faithfulness assumption.

Then we tested our pk-DirectLiNGAM on the datasets. For each trial, we first permuted the *true* connection strength matrix $\mathbf{B}$ according to the estimated ordering K. We then counted the number of errors, *i.e.*, how many elements in its strictly upper triangular part are not zeros. If the ordering is correctly estimated, the elements in the strictly upper triangular part are *all zeros*.

The medians of the numbers of non-zero strictly upper triangular elements were plotted in Fig. 1. The median errors were significantly smaller when more prior knowledge was given. For example, in most of the cases, the median errors were approximately more than 50% smaller even when only 10% of the structure was given to the algorithm. Furthermore, median computational times were shown in Fig. 2. In most of the conditions, the algorithm worked considerably faster even when the amount of prior knowledge was not so large.

5 Conclusions

We proposed to use prior knowledge in a state-of-art estimation method for linear non-Gaussian causal models. In simulations, the estimation accuracy and computational time were greatly improved even if the amount of prior knowledge was not so large.

Acknowledgments. This work was partially supported by MEXT Grant-in-Aid for Young Scientists #21700302 and Grant-in-Aid for Scientific Research (B) #22300054. We thank Akihiro Inokuchi for helpful comments.

References

1. Bollen, K.A.: Structural Equations with Latent Variables. Wiley, Chichester (1989)
2. Pearl, J.: Causality: Models, Reasoning, and Inference. Camb. Univ. Pres., New York (2000)
3. Spirtes, P., Glymour, C., Scheines, R.: Causation, Prediction, and Search. Springer, Heidelberg (1993); 2nd edn. MIT Press (2000)
4. Shimizu, S., Hoyer, P.O., Hyvärinen, A., Kerminen, A.: A linear non-gaussian acyclic model for causal discovery. J. Mach. Lear. Res. 7, 2003–2030 (2006)
5. Shimizu, S., Hyvärinen, A., Kawahara, Y., Washio, T.: A direct method for estimating a causal ordering in a linear non-gaussian acyclic model. In: Proc. the 25th Conf. on Uncertainty in Artificial Intelligence, UAI 2009 (2009)
6. Hyvärinen, A.: Fast and robust fixed-point algorithms for independent component analysis. IEEE Trans. on Neural Networks 10, 626–634 (1999)
7. Amari, S.: Natural gradient learning works efficiently in learning. Neu. Comp. 10, 251–276 (1998)
8. Hyvärinen, A., Karhunen, J., Oja, E.: Independent component analysis. Wiley, New York (2001)
9. Kraskov, A., Stögbauer, H., Grassberger, P.: Estimating mutual information. Physical Review E 69(6), 066138 (2004)
10. Bach, F.R., Jordan, M.I.: Kernel independent component analysis. J. Machine Learning Research 3, 1–48 (2002)

A New Performance Index for ICA: Properties, Computation and Asymptotic Analysis

Pauliina Ilmonen[1], Klaus Nordhausen[1], Hannu Oja[1], and Esa Ollila[2,3]

[1] Tampere School of Public Health, University of Tampere
[2] Department of Mathematical Sciences, University of Oulu
[3] Department of Signal Processing and Acoustics,
Aalto University School of Science and Technology

Abstract. In the independent component (IC) model it is assumed that the components of the observed p-variate random vector $\boldsymbol{x}$ are linear combinations of the components of a latent p-vector $\boldsymbol{z}$ such that the p components of $\boldsymbol{z}$ are independent. Then $\boldsymbol{x} = \boldsymbol{\Omega}\boldsymbol{z}$ where $\boldsymbol{\Omega}$ is a full-rank $p \times p$ mixing matrix. In the independent component analysis (ICA) the aim is to estimate an unmixing matrix $\boldsymbol{\Gamma}$ such that $\boldsymbol{\Gamma}\boldsymbol{x}$ has independent components. The comparison of the performances of different unmixing matrix estimates $\hat{\boldsymbol{\Gamma}}$ in the simulations is then difficult as the estimates are for different population quantities $\boldsymbol{\Gamma}$. In this paper we suggest a new natural performance index which finds the shortest distance (using Frobenius norm) between the identity matrix and the set of matrices equivalent to the gain matrix $\hat{\boldsymbol{\Gamma}}\boldsymbol{\Omega}$. The index is shown to possess several nice properties, and it is easy and fast to compute. Also, the limiting behavior of the index as the sample size approaches infinity can be easily derived if the limiting behavior of the estimate $\hat{\boldsymbol{\Gamma}}$ is known.

1 Introduction

In the independent component (IC) model we assume that the components of the observed p-variate random vector $\boldsymbol{x}$ are linear combinations of the components of a latent p-vector $\boldsymbol{z}$ such that the p components of $\boldsymbol{z}$ are independent. Then

$$\boldsymbol{x} = \boldsymbol{\Omega}\boldsymbol{z} \tag{1}$$

where $\boldsymbol{\Omega}$ is a full-rank $p \times p$ mixing matrix. The IC model can be formulated in several ways: If the independent components are permuted or multiplied by nonzero scalars they still remain independent. In (1) we choose any of the formulations to fix $\boldsymbol{\Omega}$. In the independent component analysis (ICA) the aim is to find an estimate for an unmixing matrix $\boldsymbol{\Gamma}$ such that $\boldsymbol{\Gamma}\boldsymbol{x}$ has independent components. Naturally $\boldsymbol{\Gamma} = \boldsymbol{\Omega}^{-1}$ is just one possible unmixing matrix and the ICA problem reduces to estimating an unmixing matrix $\boldsymbol{\Omega}^{-1}$ only up to the order, signs and scales of the row vectors. In order to be able to identify a mixing matrix one has to assume that at most one of the components of $\boldsymbol{z}$ is normally distributed [6]. In the literature there is a large number of algorithms for the

V. Vigneron et al. (Eds.): LVA/ICA 2010, LNCS 6365, pp. 229–236, 2010.

ICA problem. Well-known algorithms are for example FOBI, SOBI, JADE and fastICA just to mention a few. Good overviews are given in [4, 6].

Several performance indices have been proposed to compare different ICA algorithms in different IC models used in simulation studies. First, one can compare the true sources $\boldsymbol{z}$ (known in simulations) and the estimated sources $\hat{\boldsymbol{z}} = \hat{\boldsymbol{\Gamma}}\boldsymbol{x}$. Second, one can measure the closeness of the true unmixing matrix $\boldsymbol{\Omega}^{-1}$ and the estimated unmixing matrix $\hat{\boldsymbol{\Gamma}}$. In both cases the problem is that order, signs and scales of the estimated quantities may not match as $\hat{\boldsymbol{\Gamma}}$ is not necessarily an estimate for $\boldsymbol{\Omega}^{-1}$. Key quantities in the second approach are the gain matrix $\hat{\boldsymbol{G}} = \hat{\boldsymbol{\Gamma}}\boldsymbol{\Omega}$ and its normalized versions. Moreau and Macchi [8] proposed the index called ISI (intersymbol interference) which is based on squares of the elements of the normalized gain matrix. One of the most popular indices, Amari index [1], uses the absolute values of the elements of the normalized gain matrix. ISR and Amari index are based on row-wise and column-wise comparisons, while indices such as ISR [11] or ICI [5] use only row-wise comparisons of the normalized gain matrix. There are several other slight modifications of the indices based on normalized gain matrix. Theis et al. [14] proposed the index called the generalized crosstalking error which is the closest distance (measured by a matrix norm) between the mixing matrix $\boldsymbol{\Omega}$ and the set of estimates equivalent to $\hat{\boldsymbol{\Gamma}}^{-1}$.

In this paper we introduce a new index based on the second approach. The index finds the shortest distance (using Frobenius norm) between the identity matrix and the set of matrices equivalent to the gain matrix $\hat{\boldsymbol{\Gamma}}\boldsymbol{\Omega}$. We organize the paper as follows. First, in Section 2, we give a formal (mathematical) definition of the IC functional which is independent of the model formulation. Then in Section 3 we consider a class of IC functionals which are based on two scatter matrices. This class of functionals is important as the limiting behavior of the corresponding estimates is known in this class. The new index is proposed and its properties are discussed in Section 4. In Section 5 the computation of the new index is discussed and it is shown to be straightforward and easy. Section 6 considers the limiting behavior of the index as the sample size approaches infinity; the asymptotic properties of the index are naturally determined by the asymptotic properties of the estimate $\hat{\boldsymbol{\Gamma}}$. The last section illustrates the finite sample vs. asymptotic behavior of the index for the FOBI estimate with known asymptotics in a small simulation study.

2 IC Functionals

Let $\mathcal{G}$ be the set of full-rank $p \times p$ matrices. Then naturally all unmixing matrices $\boldsymbol{\Gamma} \in \mathcal{G}$. Let $\boldsymbol{P}$ denote a permutation matrix (obtained from $\boldsymbol{I}_p$ by permuting its rows or columns), $\boldsymbol{J}$ a sign-change matrix (a diagonal matrix with diagonal elements ± 1), and $\boldsymbol{D}$ a rescaling matrix (a diagonal matrix with positive diagonal elements). For the definition of the IC functional we need the subset

$$\mathcal{C} = \{\boldsymbol{C} \in \mathcal{G} : \boldsymbol{C} = \boldsymbol{PJD} \text{ for some } \boldsymbol{P}, \boldsymbol{J}, \text{ and } \boldsymbol{D}\}.$$

We say that two matrices $\boldsymbol{\Gamma}_1$ and $\boldsymbol{\Gamma}_2$ are equivalent if $\boldsymbol{\Gamma}_1 = \boldsymbol{C}\boldsymbol{\Gamma}_2$ for some $\boldsymbol{C} \in \mathcal{C}$. We then write $\boldsymbol{\Gamma}_1 \sim \boldsymbol{\Gamma}_2$.

We then give the following.

Definition 1. *Let $F_{\boldsymbol{x}}$ denote the cdf $\boldsymbol{x}$. The functional $\boldsymbol{\Gamma}(F_{\boldsymbol{x}}) \in \mathcal{G}$ is an IC functional in the IC model if (i) $\boldsymbol{\Gamma}(F_{\boldsymbol{x}})\boldsymbol{\Omega} \sim \boldsymbol{I}_p$ and if (ii) it is affine equivariant in the sense that $\boldsymbol{\Gamma}(F_{\boldsymbol{A}\boldsymbol{x}}) = \boldsymbol{\Gamma}(F_{\boldsymbol{x}})\boldsymbol{A}^{-1}$ for all nonsingular $p \times p$ matrices $\boldsymbol{A}$.*

If $\boldsymbol{z}$ has independent components, then so has $\boldsymbol{C}\boldsymbol{z}$ for all $\boldsymbol{C} \in \mathcal{C}$. Then, for any $\boldsymbol{C} \in \mathcal{C}$, the IC model can be reformulated as

$$\boldsymbol{x} = (\boldsymbol{\Omega}\boldsymbol{C}^{-1})(\boldsymbol{C}\boldsymbol{z}) = \boldsymbol{\Omega}^*\boldsymbol{z}^*$$

where $\boldsymbol{\Omega}^*$ is a new mixing matrix and $\boldsymbol{z}^*$ is a new (transformed) vector of independent components. (Matrix $\boldsymbol{C}$ is used in the transformation.) Note that $\boldsymbol{\Gamma}(F_{\boldsymbol{x}})\boldsymbol{\Omega} \sim \boldsymbol{\Gamma}(F_{\boldsymbol{x}})\boldsymbol{\Omega}^*$ so that the definition of the IC functional does not depend on the formulation of the model.

The functional $\boldsymbol{C}(F_{\boldsymbol{x}}) = \boldsymbol{\Gamma}(F_{\boldsymbol{x}})\boldsymbol{\Omega}$ with values in $\mathcal{C}$ depends on the distribution of $\boldsymbol{z}$ but not on the value of $\boldsymbol{\Omega}$. If we fix the model $\boldsymbol{x} = \boldsymbol{\Omega}^*\boldsymbol{z}^*$ by choosing $\boldsymbol{z}^* = \boldsymbol{C}(F_{\boldsymbol{x}})\boldsymbol{z}$ then $\boldsymbol{\Omega}^* = \boldsymbol{\Gamma}(F_{\boldsymbol{x}})^{-1}$. This formulation of the model is then most natural (canonical) for functional $\boldsymbol{\Gamma}(F_{\boldsymbol{x}})$.

3 IC Functionals Based on Two Scatter Matrices

A *scatter functional* $\boldsymbol{S}(F_{\boldsymbol{x}})$ is a $p \times p$ -matrix-valued functional which is positive definite and affine equivariant in the sense $\boldsymbol{S}(F_{\boldsymbol{A}\boldsymbol{x}+\boldsymbol{b}}) = \boldsymbol{A}\boldsymbol{S}(F_{\boldsymbol{x}})\boldsymbol{A}^T$ for all nonsingular $p \times p$ matrices $\boldsymbol{A}$ and for all p-vectors $\boldsymbol{b}$. A scatter functional $\boldsymbol{S}$ is said to possess the *independence property* if $\boldsymbol{S}(F_{\boldsymbol{x}})$ is a diagonal matrix for all $\boldsymbol{x}$ with independent components. Naturally, the regular covariance matrix $\boldsymbol{S}_1(F_{\boldsymbol{x}}) = E\left((\boldsymbol{x} - E(\boldsymbol{x}))(\boldsymbol{x} - E(\boldsymbol{x}))^T\right)$ is a scatter matrix with the independence property. Another scatter matrix with the independence property is the matrix based on fourth moments, namely,

$$\boldsymbol{S}_2(F_{\boldsymbol{x}}) = \frac{1}{p+2} E\left((\boldsymbol{x} - E(\boldsymbol{x}))(\boldsymbol{x} - E(\boldsymbol{x}))^T \boldsymbol{S}_1(F_{\boldsymbol{x}})^{-1}(\boldsymbol{x} - E(\boldsymbol{x}))(\boldsymbol{x} - E(\boldsymbol{x}))^T\right).$$

For any scatter matrix $\boldsymbol{S}(F_{\boldsymbol{x}})$, its symmetrized version $\boldsymbol{S}_{sym}(F_{\boldsymbol{x}}) = \boldsymbol{S}(F_{\boldsymbol{x}_1 - \boldsymbol{x}_2})$, where $\boldsymbol{x}_1$ and $\boldsymbol{x}_2$ are independent copies of $\boldsymbol{x}$, has the independence property [10, 15].

The IC functional $\boldsymbol{\Gamma}(F_{\boldsymbol{x}})$ based on the scatter matrix functionals $\boldsymbol{S}_1(F_{\boldsymbol{x}})$ and $\boldsymbol{S}_2(F_{\boldsymbol{x}})$ is defined as a solution of the equations

$$\boldsymbol{\Gamma}\boldsymbol{S}_1\boldsymbol{\Gamma}^T = \boldsymbol{I}_p \quad \text{and} \quad \boldsymbol{\Gamma}\boldsymbol{S}_2\boldsymbol{\Gamma}^T = \boldsymbol{\Lambda}$$

where $\boldsymbol{\Lambda} = \boldsymbol{\Lambda}(F_{\boldsymbol{x}})$ is a diagonal matrix with diagonal elements $\lambda_1 \geq ... \geq \lambda_p > 0$. The FOBI functional [3] is obtained if the scatter functionals $\boldsymbol{S}_1$ and $\boldsymbol{S}_2$ are the scatter matrices based on the second and fourth moments, respectively. The use of two scatter matrices in ICA has been studied in [9, 10] (real data) and in [12] (complex data).

Assume now (wlog) that $\boldsymbol{\Omega} = \boldsymbol{I}_p$ and that $\boldsymbol{S}_1(F_{\boldsymbol{x}}) = \boldsymbol{I}_p$ and $\boldsymbol{S}_2(F_{\boldsymbol{x}}) = \boldsymbol{\Lambda}$ where now $\lambda_1 > ... > \lambda_p > 0$. Assume also that both $\boldsymbol{S}_1$ and $\boldsymbol{S}_2$ have the independence property. If the two scatter matrix estimates $\hat{\boldsymbol{S}}_1 = \boldsymbol{S}_1(F_n)$ and $\hat{\boldsymbol{S}}_2 = \boldsymbol{S}_2(F_n)$ (values of the functionals at the empirical cdf F_n) are root-n consistent, that is,

$$\sqrt{n}(\hat{\boldsymbol{S}}_1 - \boldsymbol{I}_p) = O_p(1) \text{ and } \sqrt{n}(\hat{\boldsymbol{S}}_2 - \boldsymbol{\Lambda}) = O_p(1).$$

then we have the following result [7]. The estimates $\hat{\boldsymbol{\Gamma}}$ and $\hat{\boldsymbol{\Lambda}}$ are given by

$$\hat{\boldsymbol{\Gamma}}\hat{\boldsymbol{S}}_1\hat{\boldsymbol{\Gamma}}^T = \boldsymbol{I}_p \text{ and } \hat{\boldsymbol{\Gamma}}\hat{\boldsymbol{S}}_2\hat{\boldsymbol{\Gamma}}^T = \hat{\boldsymbol{\Lambda}}.$$

Then

$$\sqrt{n}\ \mathrm{off}(\hat{\boldsymbol{\Gamma}}) = \sqrt{n}\ \boldsymbol{H} \odot \Big((\hat{\boldsymbol{S}}_2 - \boldsymbol{\Lambda}) - (\hat{\boldsymbol{S}}_1 - \boldsymbol{I}_p)\boldsymbol{\Lambda}\Big) + o_p(1),$$

where $\boldsymbol{H}$ is a $p \times p$ matrix with elements

$$H_{ij} = 0, \text{ if } i = j, \text{ and } H_{ij} = (\lambda_i - \lambda_j)^{-1}, \text{ if } i \neq j.$$

Here $\mathrm{off}(\boldsymbol{\Gamma}) = \boldsymbol{\Gamma} - \mathrm{diag}(\boldsymbol{\Gamma})$ where $\mathrm{diag}(\boldsymbol{\Gamma})$ is a diagonal matrix with the same diagonal elements as $\boldsymbol{\Gamma}$, and $\odot$ denotes the Hadamard (entrywise) product. See [7] for the limiting distribution of the FOBI estimate.

Related approaches like JADE [2] or the matrix-pencil approach [16] (approximately) diagonalize jointly two or more data matrices (not necessarily scatter matrices). The estimates are typically not affine equivariant and their asymptotic behavior is unknown.

4 New Performance Index for ICA

Let $\boldsymbol{A}$ be a $p \times p$ matrix. The shortest squared distance between the set $\{\boldsymbol{C}\boldsymbol{A} : \boldsymbol{C} \in \mathcal{C}\}$ of equivalent matrices (to $\boldsymbol{A}$) and $\boldsymbol{I}_p$ is given by

$$D^2(\boldsymbol{A}) = \frac{1}{p-1} \inf_{\boldsymbol{C} \in \mathcal{C}} \|\boldsymbol{C}\boldsymbol{A} - \boldsymbol{I}_p\|^2$$

where $\|\cdot\|$ is the matrix (Frobenius) norm. It is straightforward to see that

1. $1 \geq D^2(\boldsymbol{A}) \geq 0$ for all $\boldsymbol{A}$,
2. $D^2(\boldsymbol{A}) = 0$ if and only if $\boldsymbol{A} \sim \boldsymbol{I}_p$,
3. $D^2(\boldsymbol{A}) = 1$ if and only if $\boldsymbol{A} \sim \boldsymbol{1}_p\boldsymbol{a}^T$ for some p-vector $\boldsymbol{a}$, and
4. the function $c \to D^2(\boldsymbol{I}_p + c\ \mathrm{off}(\boldsymbol{A}))$ is increasing in $c \in [0, 1]$ for all $p \times p$ matrices $\boldsymbol{A}$ such that $A_{ij}^2 \leq 1$, $i \neq j$.

Let $\boldsymbol{X} = (\boldsymbol{x}_1, ..., \boldsymbol{x}_n)^T$ be a random sample from a distribution $F_{\boldsymbol{x}}$ where $\boldsymbol{x}$ obeys the IC model (1) with unknown mixing matrix $\boldsymbol{\Omega}$. Let $\boldsymbol{\Gamma}(F_{\boldsymbol{x}})$ be an unmixing matrix functional. Then clearly $D^2(\boldsymbol{\Gamma}(F_{\boldsymbol{x}})\boldsymbol{\Omega}) = 0$. If F_n is the empirical cumulative distribution function based on $\boldsymbol{X}$ then

$$\hat{\boldsymbol{\Gamma}} = \hat{\boldsymbol{\Gamma}}(\boldsymbol{X}) = \boldsymbol{\Gamma}(F_n)$$

is the unmixing matrix estimate based on the functional $\boldsymbol{\Gamma}(F_{\boldsymbol{x}})$.

The shortest distance between the identity matrix and the set of matrices $\{\boldsymbol{C}\hat{\boldsymbol{\Gamma}}\boldsymbol{\Omega} : \boldsymbol{C} \in \mathcal{C}\}$ equivalent to the gain matrix $\hat{\boldsymbol{G}} = \hat{\boldsymbol{\Gamma}}\boldsymbol{\Omega}$ is as given in the following definition.

Definition 2. *The minimum distance index for* $\hat{\boldsymbol{\Gamma}}$ *is*

$$\hat{D} = D(\hat{\boldsymbol{\Gamma}}\boldsymbol{\Omega}) = \frac{1}{\sqrt{p-1}} \inf_{\boldsymbol{C}\in\mathcal{C}} \|\boldsymbol{C}\hat{\boldsymbol{\Gamma}}\boldsymbol{\Omega} - \boldsymbol{I}_p\|.$$

Clearly, $1 \geq \hat{D} \geq 0$, and $\hat{D} = 0$ if and only if $\hat{\boldsymbol{\Gamma}} \sim \boldsymbol{\Omega}^{-1}$. The worst case with $\hat{D} = 1$ is obtained if all the row vectors of $\hat{\boldsymbol{\Gamma}}\boldsymbol{\Omega}$ point in the same direction.

Recall that the generalized crosstalking error in [14] is defined as

$$E(\boldsymbol{\Omega}, \hat{\boldsymbol{\Gamma}}) = \inf_{\boldsymbol{C}\in\mathcal{C}} \|\boldsymbol{\Omega} - \hat{\boldsymbol{\Gamma}}^{-1}\boldsymbol{C}\|$$

where $\|\cdot\|$ denotes a matrix norm. If one uses the Frobenius norm here, our criterion may be seen as a standardized version of the generalized crosstalking error as

$$\hat{D} = \inf_{\boldsymbol{C}\in\mathcal{C}} \|\boldsymbol{C}^{-1}\hat{\boldsymbol{\Gamma}}\left(\boldsymbol{\Omega} - \hat{\boldsymbol{\Gamma}}^{-1}\boldsymbol{C}\right)\|.$$

The value of $\hat{D}$ does not depend on the model formulation at all, and it is rescaled so that $1 \geq \hat{D} \geq 0$. This is an advantage when compared to the Amari index, since the Amari index is not affine invariant and therefore the values of it depend on the model formulation [9]. Therefore there might be pitfalls when different algorithms are compared using the Amari index.

5 Computation of the Value of the Index

At first glance the index $\hat{D} = D(\hat{\boldsymbol{\Gamma}}\boldsymbol{\Omega})$ seems difficult to compute in practice since one has to minimize over all choices $\boldsymbol{C} \in \mathcal{C}$. The minimization can be done, however, with two easy steps. Write first $\hat{\boldsymbol{G}} = \hat{\boldsymbol{\Gamma}}\boldsymbol{\Omega}$ for the gain matrix, and define then the matrix $\tilde{\boldsymbol{G}}$ by $\tilde{G}_{ij} = \hat{G}_{ij}^2 / \sum_{k=1}^{p} \hat{G}_{ik}^2$, $i, j = 1, ..., p$. Then the minimum distance index can be written as

$$\hat{D} = D(\hat{\boldsymbol{G}}) = \frac{1}{\sqrt{p-1}} \left(p - \max_{\boldsymbol{P}} \left(\mathrm{tr}(\boldsymbol{P}\tilde{\boldsymbol{G}})\right)\right)^{1/2}.$$

The maximization problem

$$\max_{\boldsymbol{P}} \left(\mathrm{tr}(\boldsymbol{P}\tilde{\boldsymbol{G}})\right)$$

over all permutation matrices $\boldsymbol{P}$ can be expressed as a linear programming problem. In a personal communication Ravi Varadhan pointed out that it can be seen also as a linear sum assignment problem (LASP). The cost matrix $\boldsymbol{\Delta}$ of the LASP in this case is given by $\Delta_{ij} = \sum_{k=1}^{p}(\boldsymbol{I}_{jk} - \tilde{G}_{ik})^2$, $i, j, = 1, ..., p$, and for example the Hungarian method (see e.g. [13]) can then be used to find the maximizer $\hat{\boldsymbol{P}}$, and $\hat{D}$ itself can be computed. The ease of computations is demonstrated in Table 1 where we give the computation time of thousand indices for randomly generated $p \times p$ matrices in different dimensions. The computations were performed on Intel Core 2 Duo T9600, 2.80 GHz, 4GB Ram using R 2.11.1 on Windows 7.

Table 1. Computation time in seconds for 1000 indices for different dimensions p

p	3	5	10	25	50	100
Time	0.53	0.53	1.46	7.15	36.63	181.15

6 Asymptotical Behavior of the Index

As $\hat{\boldsymbol{\Gamma}}$ is equivariant in the sense that $\hat{\boldsymbol{\Gamma}}(\boldsymbol{X}\boldsymbol{A}^T) = \hat{\boldsymbol{\Gamma}}(\boldsymbol{X})\boldsymbol{A}^{-1}$, it is clear that the value of $\hat{D}^2$ does not depend on the model formulation and mixing matrix $\boldsymbol{\Omega}$ at all. In this Section we consider the limiting behavior of $\hat{D}^2$.

The model is written as $\boldsymbol{x} = \boldsymbol{\Omega}\boldsymbol{z}$, where now $\boldsymbol{z}$ is standardized so that $\boldsymbol{C}(F_{\boldsymbol{z}}) = \boldsymbol{I}_p$. Then $\boldsymbol{\Gamma}(F_{\boldsymbol{x}}) = \boldsymbol{\Omega}^{-1}$, and without any loss of generality we can assume that $\boldsymbol{\Gamma}(F_{\boldsymbol{x}}) = \boldsymbol{\Omega} = \boldsymbol{C}(F_{\boldsymbol{x}}) = \boldsymbol{I}_p$. We then have the following.

Theorem 1. *Assume that the model is fixed so that* $\boldsymbol{\Gamma}(F_{\boldsymbol{x}}) = \boldsymbol{\Omega} = \boldsymbol{C}(F_{\boldsymbol{x}}) = \boldsymbol{I}_p$ *and that* $\sqrt{n}\ \mathrm{vec}(\hat{\boldsymbol{\Gamma}} - \boldsymbol{I}_p) \to_d N_{p^2}(\boldsymbol{0}, \boldsymbol{\Sigma})$. *Then*

$$n\hat{D}^2 = \frac{n}{p-1}\|\mathrm{off}(\hat{\boldsymbol{\Gamma}})\|^2 + o_P(1)$$

and the limiting distribution of $n\hat{D}^2$ *is that of* $(p-1)^{-1}\sum_{i=1}^{k}\delta_i\chi_i^2$ *where* $\chi_1^2, ..., \chi_k^2$ *are independent chi squared variables with one degree of freedom, and* $\delta_1, ..., \delta_k$ *are the* k *nonzero eigenvalues (including all algebraic multiplicities) of*

$$ASCOV(\sqrt{n}\ vec(\mathrm{off}(\hat{\Gamma}))) = (\boldsymbol{I}_{p^2} - \boldsymbol{D}_{p,p})\boldsymbol{\Sigma}(\boldsymbol{I}_{p^2} - \boldsymbol{D}_{p,p})$$

with $\boldsymbol{D}_{p,p} = \sum_{i=1}^{p}(\boldsymbol{e}_i\boldsymbol{e}_i^T) \otimes (\boldsymbol{e}_i\boldsymbol{e}_i^T)$.

The proof of Theorem 1 is lengthy and not trivial. Due to space limitations the proof will be given in an extended version of this paper.

Note that the asymptotic distribution of $n\hat{D}^2$ is naturally independent on the "true" model with choices of $\boldsymbol{\Omega}$ and $\boldsymbol{z}$. For the theorem we fix the model in a specific way (canonical formulation) only to find the limiting distributions. Note that

$$n(p-1)E(\hat{D}^2) \to tr\left(ASCOV(\sqrt{n}\ \mathrm{vec}(\mathrm{off}(\hat{\boldsymbol{\Gamma}})))\right)$$

which is just a regular global measure of the asymptotic accuracy of the estimate $\hat{\boldsymbol{\Gamma}}$ in a model where it is estimating the identity matrix.

We also would like to point out that similar asymptotics for the Amari index is not possible, since the Amari index lacks affine invariance and it is based on the use of l_1 norms.

7 A Simulation Study

To consider the finite samples vs. asymptotic behavior of the index $\hat{D}$ we performed a small simulation study. We use the FOBI estimate in our simulations as, to our knowledge, it is the only estimate so far with proven asymptotic normality. See [7]. For the FastICA estimate, for example, the asymptotic covariance matrix is known [11], but there is no proof for the asymptotic normality.

In our simulation we had three independent components with uniform, normal and t_{10} distributions. The sample sizes (N) were 500, 5000, 10000, 25000, 50000, 75000, 100000 with 10000 repetitions, and for each repetition the value of $\hat{D}$ was computed. The true six nonzero eigenvalues of $(\boldsymbol{I}_{p^2} - \boldsymbol{D}_{p,p})\boldsymbol{\Sigma}(\boldsymbol{I}_{p^2} - \boldsymbol{D}_{p,p})$ as defined in Theorem 1 are $\delta_1 = 66.7782$, $\delta_2 = 28.2656$, $\delta_3 = 9.8899$, $\delta_4 = 0.5018$, $\delta_5 = 0.1721$ and $\delta_6 = 0.0817$. Figure 1 shows, for each sample size, the boxplots

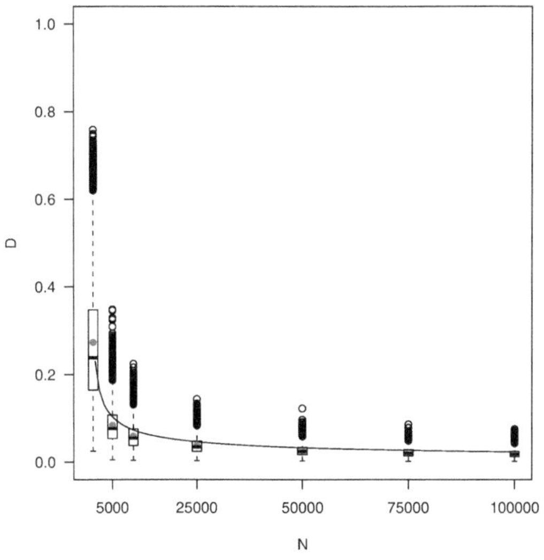

Fig. 1. The boxplots for $\hat{D}$ based on the FOBI estimate for different sample sizes and 10000 repetitions in the case of three independent components. The grey dots give the sample means over repetitions and the black line their approximations based on the asymptotics.

of the observed values of $\hat{D}$ over all repetitions. The observed sample means are indicated by grey dots, and the black line gives their asymptotic approximations $[\sum \delta_i/(N(p-1))]^{1/2}$ as explained in Section 6. The FOBI estimate converges in distribution to a multivariate normal distribution, but the convergence is very slow. It would be interesting to compare the behavior of the FOBI estimate to other IC estimates in a similar way but this will be possible only after their asymptotic behaviors have been found out.

References

[1] Amari, S.I., Cichocki, A., Yang, H.H.: A new learning algorithm for blind source separation. Advances in Neural Information Processing Systems 8, 757–763 (1996)

[2] Cardoso, J.F., Souloumiac, A.: Blind beamforming for non Gaussian signals. IEEE Proceedings-F 140, 362–370 (1993)

[3] Cardoso, J.F.: Source separation using higher moments. In: Proceedings of IEEE international conference on acustics, speech and signal processing, pp. 2109–2112 (1989)

[4] Cichocki, A., Amari, S.I.: Adaptive Blind Signal and Image Processing. John Wiley & Sons, Chichester (2006)

[5] Douglas, S.C.: Fixed-point algorithms for the blind separation of arbitrary complex-valued non-gaussian signal mixtures. EURASIP Journal on Advances in Signal Processing 1, 83–83 (2007)

[6] Hyvärinen, A., Karhunen, J., Oja, E.: Independent Component Analysis. John Wiley & Sons, New York (2001)

[7] Ilmonen, P., Nevalainen, J., Oja, H.: Characteristics of multivariate distributions and the invariant coordinate system (2010) (submitted)

[8] Moreau, E., Macchi, O.: A one stage self-adaptive algorithm for source separation. In: IEEE International Conference on Acoustics, Speech and Signal Processing, Adelaide, Australia, pp. 49–52 (1994)

[9] Nordhausen, K., Oja, H., Ollila, E.: Robust independent component analysis based on two scatter matrices. Austrian Journal of Statistics 37, 91–100 (2008)

[10] Oja, H., Sirkiä, S., Eriksson, J.: Scatter matrices and independent component analysis. Austrian Journal of Statistics 35, 175–189 (2006)

[11] Ollila, E.: The deflation-based FastICA estimator: statistical analysis revisited. IEEE Transactions in Signal Processing 58, 1527–1541 (2010)

[12] Ollila, E., Oja, H., Koivunen, V.: Complex-valued ICA based on a pair of generalized covariance matrices. CSDA 52, 3789–3805 (2008)

[13] Papadimitriou, C., Steiglitz, K.: Combinatorial Optimization: Algorithms and Complexity. Prentice Hall, Englewood Cliffs (1982)

[14] Theis, F.J., Lang, E.W., Puntonet, C.G.: A geometric algorithm for overcomplete linear ICA. Neurocomputing 56, 381–398 (2004)

[15] Tyler, D.E., Critchley, F., Dümbgen, L., Oja, H.: Invariant coordinate selection. Journal of Royal Statistical Society, Series B 71, 549–592 (2009)

[16] Yeredor, A.: On Optimal Selection of Correlation Matrices for Matrix-Pencil-Based Separation. In: Adali, T., Jutten, C., Romano, J.M.T., Barros, A.K. (eds.) ICA 2009. LNCS, vol. 5441, pp. 187–194. Springer, Heidelberg (2009)

Blind Operation of a Recurrent Neural Network for Linear-Quadratic Source Separation: Fixed Points, Stabilization and Adaptation Scheme

Yannick Deville and Shahram Hosseini

Laboratoire d'Astrophysique de Toulouse-Tarbes, Université de Toulouse, CNRS, 14 Av. Edouard Belin, 31400 Toulouse, France
ydeville@ast.obs-mip.fr, shahram.hosseini@ast.obs-mip.fr

Abstract. Retrieving unknown sources from *nonlinear* mixtures of them requires one to define a separating structure, before proceeding to methods for estimating mixing or separating parameters in blind configurations. Recurrent neural networks are attractive separating structures for a wide range of nonlinear mixing models. In a previous paper, we proposed such a network for the *non-blind* version of linear-quadratic separation. We here extend this approach to the more difficult *blind* case. We optimize the fixed points and stability of this structure thanks to its free weights. We define the general architecture of future adaptation algorithms that will be able to take advantage of these free weights. Numerical results illustrate the theoretical properties derived in this paper.

1 Introduction

Blind source separation (BSS) methods aim at restoring a set of unknown source signals from a set of observed signals which are mixtures of these source signals. Whereas most reported methods are intended for linear mixtures, nonlinear mixtures have also been studied recently (see e.g. the survey in [4]). More specifically, several papers addressed linear-quadratic mixtures (see e.g. [1], [2], [3], [5]).

For nonlinear mixtures, even when considering a given class of mixing models including some parameters, and when restricting oneself to the non-blind configuration (i.e. with known mixing parameter values), defining a corresponding separating *structure* is not trivial, unlike in the linear case. This results from the fact that, for most nonlinear mixing models, the analytical form of the *inverse* of such a model cannot be derived. In our previous papers (see especially [1]), we proposed a general solution to this problem, based on recurrent neural networks. We especially analyzed the properties of these networks (e.g. their stability) for linear-quadratic mixtures. That analysis was only performed for the *non-blind* configuration, i.e. when the mixing parameter values are known and the parameter values of the separating system are matched to these mixing parameters, so that the separating system implements the exact inverse of the mixing model (up to scale factors and additive constants).

Beyond the above non-blind case, we aim at deriving a complete solution for the *blind* configuration. The ultimate stage of that investigation will therefore

V. Vigneron et al. (Eds.): LVA/ICA 2010, LNCS 6365, pp. 237–244, 2010.

consist in defining a separation criterion and associated estimation algorithm, for adapting the estimated mixing or unmixing parameters in such a way that the mapping achieved by the separating system converges to the inverse of the mixing mapping. This means that the separating neural network will thus be used with "arbitrary" parameter values (i.e. network weights) while adapting this network, as opposed to their specific values "matched" to the mixing model which were mentioned above. Considering such arbitrary parameter values raises new questions about the equilibrium points and stability of the considered neural network, since these questions were only solved for the non-blind configuration, i.e. for the "matched" parameter values, in our previous investigation.

Separating network equilibrium and stability in the nonlinear blind case is therefore a research topic in itself, before proceeding to parameter estimation. This is the main topic that we address in this paper, for linear-quadratic mixtures. Moreover, we anticipate on the future development of associated estimation algorithms. On-line algorithms only consider a single observed vector at a time and therefore only require network stability for that vector. On the contrary, batch algorithms use a complete set of observed vectors. They therefore require us to analyze the equilibrium points and stability of the proposed network (for a single set of parameters) over a complete set of observed vectors. We therefore consider that more general case of batch algorithms hereafter and we describe the general structure of the operation of the considered network (initialization and overall adaptation scheme) which results from the use of its free parameters, without yet focusing on a specific adaptation algorithm.

2 Mixing and Separating Models

We here consider the linear-quadratic instantaneous mixing model that we e.g. defined in [1], [3]. Its rescaled version reads

$$x_1(n) = s_1(n) - L_{12}s_2(n) - Q_1 s_1(n)s_2(n) \tag{1}$$
$$x_2(n) = -L_{21}s_1(n) + s_2(n) - Q_2 s_1(n)s_2(n) \tag{2}$$

where $x_1(n)$ and $x_2(n)$ are the observed signals, whereas $s_1(n)$ and $s_2(n)$ are the two source signals. L_{ij} and Q_i are the mixing parameters. Their values are assumed to be fixed and unknown in this paper, unlike in [1].

These observations are processed by the *extended* feedback network that we e.g. introduced in [1], which operates as follows. For each time n, it performs a recurrence to compute the values of its outputs y_i. We denote as m the index associated to this recurrence and $y_i(m)$ the successive values of each output in this recurrence at time n. This recurrence reads

$$y_1(m+1) = x_1(n) + l_{11}y_1(m) + l_{12}y_2(m) + q_1 y_1(m)y_2(m) \tag{3}$$
$$y_2(m+1) = x_2(n) + l_{21}y_1(m) + l_{22}y_2(m) + q_2 y_1(m)y_2(m) \tag{4}$$

where l_{ij} and q_i are the adaptive weights of the proposed neural network. In [1], we focused on specific values of these weights, corresponding to separating points in the non-blind configuration: we selected l_{11} and l_{22} as explained in [1] and we then set the other network weights to

$$l_{12} = L_{12}l'_{22} \qquad q_1 = Q_1 l'_{11} l'_{22} \qquad l_{21} = L_{21} l'_{11} \qquad q_2 = Q_2 l'_{11} l'_{22} \quad (5)$$

$$\text{where} \qquad l'_{11} = 1 - l_{11} \quad \text{and} \quad l'_{22} = 1 - l_{22}. \quad (6)$$

On the contrary, we here aim at extending this investigation to the blind configuration and we therefore consider all possible values of l_{ij} and q_i. These weights are of two different natures. On the one hand, l_{12}, l_{21}, q_1 and q_2 must be used to allow this network to achieve the inverse of the mixing model (up to usual indeterminacies). On the other hand, l_{11} and l_{22} might be removed, i.e. set to zero, thus yielding the basic network that we initially introduced e.g. in [3]. However, in [1] we showed that these weights l_{11} and l_{22} are attractive because they are free parameters, which may be used to improve the equilibrium and stability of the resulting extended network in the non-blind configuration (they also have an influence on output signal scales, but one may then compensate for it). We here aim at again taking advantage of these free weights, now in the blind case.

3 Fixed Points

Existence and locations of fixed points. The stability of the recurrence (3)-(4) is analyzed for fixed points (i.e. equilibrium points) of this recurrence. The first step of this investigation therefore consists in determining these fixed points, i.e. all the points (y_1^E, y_2^E) which are such that

$$y_1(m+1) = y_1(m) = y_1^E \quad \text{and} \quad y_2(m+1) = y_2(m) = y_2^E . \quad (7)$$

This analysis is performed for given, arbitrary, values of the following quantities: (i) the network inputs $x_i(n)$, which depend on the source values and mixing parameters through (1)-(2) and (ii) the network weights l_{ij} and q_i. These fixed points are derived by combining (3)-(4) with (7), which yields

$$x_1(n) - l'_{11} y_1^E + l_{12} y_2^E + q_1 y_1^E y_2^E = 0 \quad (8)$$

$$x_2(n) + l_{21} y_1^E - l'_{22} y_2^E + q_2 y_1^E y_2^E = 0. \quad (9)$$

These equations may e.g. be solved by first deriving a linear combination of them which cancels their cross-terms $y_1^E y_2^E$, such as $q_2 \times (8) - q_1 \times (9)$, where we assume that both q_i are non-zero hereafter. This yields a linear relationship between y_1^E and y_2^E, from which one then derives the expression of y_2^E vs y_1^E, i.e.

$$y_2^E = \frac{(q_2 l'_{11} + q_1 l_{21}) y_1^E - [q_2 x_1(n) - q_1 x_2(n)]}{q_2 l_{12} + q_1 l'_{22}}, \quad (10)$$

assuming that the denominator of this expression is non-zero. Inserting (10) in (8) and only considering the case when $(q_2 l'_{11} + q_1 l_{21})$ is non-zero, one gets a second-order polynomial equation for y_1^E, which yields

$$y_1^E = \frac{-[q_2 x_1(n) - q_1 x_2(n) + l'_{11} l'_{22} - l_{12} l_{21}] + \epsilon_{y_1} \Delta_{y_1}^{1/2}}{-2(q_2 l'_{11} + q_1 l_{21})}, \quad (11)$$

$$\text{where} \qquad \epsilon_{y_1} = \pm 1 \tag{12}$$

defines which solution of the second-order polynomial equation in y_1^E is considered, and Δ_{y_1} is the discriminant of this equation, which reads

$$\begin{aligned} \Delta_{y_1} = & \left[q_2 x_1(n) - q_1 x_2(n) + l'_{11} l'_{22} - l_{12} l_{21}\right]^2 \\ & -4(q_2 l'_{11} + q_1 l_{21})[l'_{22} x_1(n) + l_{12} x_2(n)]. \end{aligned} \tag{13}$$

Inserting (11) in (10) yields the corresponding expressions of y_2^E, i.e.

$$y_2^E = \frac{q_2 x_1(n) - q_1 x_2(n) - (l'_{11} l'_{22} - l_{12} l_{21}) + \epsilon_{y_1} \Delta_{y_1}^{1/2}}{-2(q_2 l_{12} + q_1 l'_{22})}. \tag{14}$$

This analysis shows that the considered recurrence has at most two real-valued fixed points (y_1^E, y_2^E), depending on the sign of Δ_{y_1}. Moreover, based on the expression in (13), we cannot currently exclude that Δ_{y_1} might indeed be either positive or negative, depending on the values of the observations and network weights. This general property should be contrasted with the results that we derived in [1] by only considering the non-blind configuration: for those specific values of the network weights, Δ_{y_1} was shown to always be non-negative, so this network has real-valued fixed point(s). For the general (blind) case considered in this paper, we hereafter define how to operate the network in practice in order to optimize the existence of real-valued fixed points.

Practical operation of network, considering fixed points. The above results may be used in several aspects of the operation of the considered network. This is based on the fact that, for a given observed vector $[x_1(n), x_2(n)]^T$ (where ^{T} stands for transpose) and for given network weights, we are able to determine if this network has real-valued fixed points: we just have to test if Δ_{y_1}, defined by (13), is non-negative. It should be noted that this expression (13) can be used because it only involves quantities which are known in practice in the blind configuration, i.e. observations and network weights, as opposed to sources and mixing coefficients. This approach for testing if the network has fixed points then directly extends to a set of observed vectors.

The above test may be applied as follows in the blind case, using a batch algorithm for adapting the estimates $\hat{L}_{12}$, $\hat{L}_{21}$, $\hat{Q}_1$ and $\hat{Q}_2$ of the mixing parameters. All required parameters should first be initialized. To this end, a simple procedure consists in setting $\hat{L}_{12}$, $\hat{L}_{21}$, $\hat{Q}_1$, $\hat{Q}_2$, l_{11} and l_{22} to random values and assigning the other network weights according to Eq. (5'), which is derived from (5) by replacing the mixing parameters by their above estimates. One then tests if these values are such that $\Delta_{y_1} \geq 0$ for all observed vectors (or at least for a number of them: one may start network adaptation only with a reduced set of vectors, while keeping the others for subsequent adaptation steps if they become such that $\Delta_{y_1} \geq 0$ with the network weights obtained after adaptation). If this condition is not met, a new set of random values is created and tested.

One then performs adaptation steps. Each step consists of two-substeps. One first updates $\hat{L}_{12}$, $\hat{L}_{21}$, $\hat{Q}_1$ and $\hat{Q}_2$, with the considered BSS algorithm (not fixed in this paper), operated over the (sub)set of observations such that the network has real fixed points for the weight values before that update. One then selects l_{11} and l_{22} so that the network has real-valued fixed points for as many observed vectors of the available set as possible, for the values of $\hat{L}_{12}$, $\hat{L}_{21}$, $\hat{Q}_1$ and $\hat{Q}_2$ obtained after the above update and for the corresponding values of l_{12}, l_{21}, q_1 and q_2 derived from (5'). A simple procedure for setting l_{11} and l_{22} consists of systematic search: l_{11} and l_{22} are varied with small stepsizes and, for each value of this couple of parameters, one sets l_{12}, l_{21}, q_1 and q_2 and counts the number of observed vectors which are such that $\Delta_{y_1} \geq 0$.

4 Stability Analysis

Stability condition. The considered recurrent network is a two-dimensional nonlinear dynamic system, since the evolution of its state vector $[y_1(m), y_2(m)]^T$ is defined by the nonlinear equations (3)-(4). The local linear stability of any such system at a given fixed point may be analyzed by considering a first-order approximation of its evolution equations at that point. The new value of a small disturbance added to the state vector is thus expressed as the product of a matrix $\mathbf{H}$, which defines the first-order approximation of the system, by the former value of that disturbance (see e.g. details in [6]). The (asymptotic) stability of the system at the considered point is guaranteed by constraining the moduli of both eigenvalues of the corresponding matrix $\mathbf{H}$ to be lower than 1.

Whereas general information about this approach may e.g. be found in [6], its specific application to the network addressed here was studied in [1]. Although most of that investigation was only applicable to the non-blind case, i.e. to specific network weights, its very first steps were developed for arbitrary values of l_{ij} and q_i. The very first results of [1] may therefore be reused here and read as follows. We showed that the fixed point corresponding to $\epsilon_{y_1} = -1$ is always unstable. We therefore only investigate the fixed point corresponding to $\epsilon_{y_1} = 1$ hereafter. Assuming Δ_{y_1} is strictly positive, we showed that the above-defined stability condition reads as follows at that fixed point:

$$2T + \sqrt{\Delta_{y_1}} > 0 \tag{15}$$

$$T + \sqrt{\Delta_{y_1}} < 2 \tag{16}$$

where

$$T = q_1 y_2^E + q_2 y_1^E + l_{11} + l_{22}. \tag{17}$$

Beyond the above results, we here derive the explicit expression of T for the general case considered in this paper, by combining (6), (11), (14), and (17). It may be shown that, for $\epsilon_{y_1} = 1$, this yields

$$T = E\sqrt{\Delta_{y_1}} + F + 2 - l'_{11} - l'_{22} \tag{18}$$

with

$$E = -\frac{q_1(q_2 l'_{11} + q_1 l_{21}) + q_2(q_2 l_{12} + q_1 l'_{22})}{2(q_2 l'_{11} + q_1 l_{21})(q_2 l_{12} + q_1 l'_{22})} \tag{19}$$

$$F = -\frac{1}{2(q_2 l'_{11} + q_1 l_{21})(q_2 l_{12} + q_1 l'_{22})} \{ [q_2 x_1(n) - q_1 x_2(n)][q_1(q_2 l'_{11} + q_1 l_{21}) - q_2(q_2 l_{12} + q_1 l'_{22})] -(l'_{11} l'_{22} - l_{12} l_{21})[q_1(q_2 l'_{11} + q_1 l_{21}) + q_2(q_2 l_{12} + q_1 l'_{22})]\} . \quad (20)$$

The stability condition may then be rewritten by inserting (18)-(20) into (15)-(16) with Δ_{y_1} defined by (13). It may then be transformed into polynomial inequalities with respect to l'_{11} and l'_{22} as follows: (i) only keep the terms which depend on Δ_{y_1} on one side of the inequalities, (ii) take the square of these inequalities while taking into account the signs of the considered quantities, (iii) multiply inequalities by the square of the denominator of E. One may then even reduce these expressions to polynomials of l'_{11} only, with a parameter λ, by considering l'_{11} as the primary variable and l'_{22} as the secondary variable, and by expressing it as $l'_{22} = \lambda l'_{11}$, as in [1]. For the specific case considered in [1], using these variables made it possible to analytically solve the considered inequalities and to derive values of l'_{11} and l'_{22} which always guarantee stability in the non-blind configuration. On the contrary, for the blind case studied here, the polynomials have high orders, unless simplifications will be derived. So, it is currently not guaranteed that a completely analytical solution may be derived in this case for selecting l'_{11} and l'_{22} so as to ensure stability. Instead, we hereafter propose two numerical approaches for addressing stability issues in practice.

Practical operation of network, considering stability. A first possible use of the above results consists of a direct extension of the procedure described in Section 3. This extended procedure makes it possible to test not only the existence of the fixed point with $\epsilon_{y_1} = 1$ but also its stability. To this end, one uses the same procedure as in Section 3, except that one not only tests if $\Delta_{y_1} > 0$ but also if conditions (15) and (16) are met for the considered weight values and observed vector(s), when initializing and adapting the network.

The basic version of the above approach uses systematic search for l_{11} and l_{22}. This may be avoided by using another approach, based on the above description about how to express the stability condition as a set of polynomial inequalities with respect to l'_{11}, involving the parameter λ. In this second approach, the sweep on l_{11} and l_{22} is replaced by a sweep on λ, again with a small stepsize. For each value of λ, each inequality involved in the stability solution is first solved separately and for one observed vector. This yields interval(s) of values of l'_{11} for which the inequality is met. One then derives the intersections of the intervals corresponding to these inequalities. All observed vectors are then gathered, again to determine the intersection of all intervals or, if it is empty, to derive a subset (preferably the largest) of the observed vectors which yields a non-empty interval intersection. This overall process is repeated for several values of λ, to derive a non-empty intersection for the largest subset of observed vectors. One thus obtains (i) a value of λ, (ii) one (or several) interval of suitable values of l'_{11}, from which one select a particular value of l'_{11}, i.e. typically the middle of a suitable interval, (iii) the value $l'_{22} = \lambda l'_{11}$. This completely addresses the stability condition. In order to check that the considered fixed points are

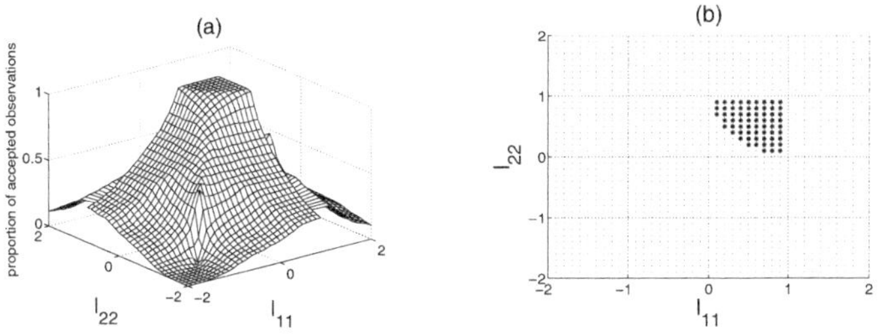

Fig. 1. (a) Proportion of accepted samples in the first experiment. (b) Region where all the samples are accepted.

real-valued, one also tests if $\Delta_{y_1} > 0$. This may be achieved for the solution (i.e. parameter values) eventually selected above, based on stability. If the test $\Delta_{y_1} > 0$ thus fails, one removes that potential solution and applies that test to other values of λ and l'_{11} which are suitable (possibly for fewer observations).

5 Simulation Results

In a first experiment, we generated two independent zero-mean sources s_1 and s_2, uniformly distributed over [-2, 2] and containing 1000 samples, then mixed them using the model (1)-(2) with $L_{12} = L_{21} = 0$ and $Q_1 = Q_2 = 0.5$ (as in Sec. 4.1 of [1]). Afterwards, we initialized the estimated parameters $\hat{L}_{12}$, $\hat{L}_{21}$, $\hat{Q}_1$ and $\hat{Q}_2$ with their above actual values. We varied the two parameters l_{11} and l_{22} between -2 and 2 and verified, for each of their values, the non-negativity of Δ_{y_1} and the stability conditions over all the samples of the mixed signals. Fig. 1.a shows the proportion of the samples verifying all three conditions and Fig. 1.b shows the region where all the samples verify these conditions. In another experiment, we generated the mixtures as above but using sources uniformly distributed over [-0.5, 0.5] and the coefficients $L_{12} = -0.2$, $L_{21} = 0.2$, $Q_1 = -0.8$, $Q_2 = 0.8$. This time, we initialized the estimated parameters to "their actual values multiplied by a varying coefficient k". For each initial value (corresponding to a particular value of k), we varied the parameters l_{11} and l_{22} between -2 and 2 and measured the maximum proportion of accepted observations in this interval. Fig. 2 shows this maximum proportion as a function of the variable k. As can be seen, in a relatively large interval, corresponding to $k \in [-0.5, 1.06]$, it is possible to tune the parameters l_{11} and l_{22} so that all observations satisfy the conditions. Even outside this interval, it is always possible to use at least 75% of data samples.

We also repeated this experiment using 100 random initializations of the parameters in a circle of radius 4, centered at the actual parameters. For each initial value, we found the parameters l_{11} and l_{22} (in the interval $[-5, 5]$) providing as many accepted observations as possible. For 35% of the initial values, all the samples were accepted. On the average, 83% of the observations per initial value are accepted, the worst case being 46% for a single initial value.

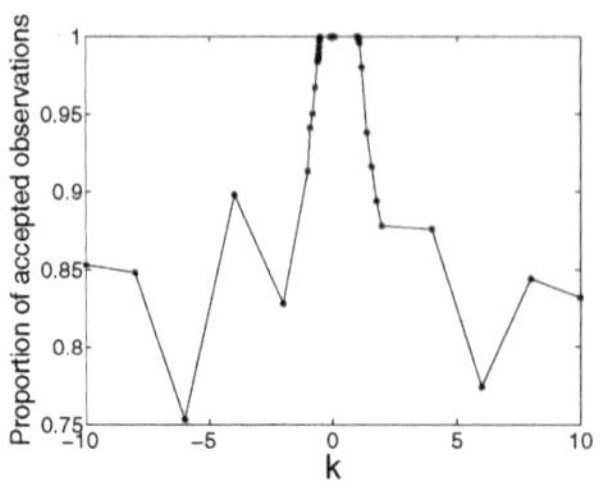

Fig. 2. Maximum proportion of accepted observations as a function of the variable k used in the second experiment

6 Conclusion and Future Work

The definition of a separating structure is an issue for *nonlinear* mixtures. The extended neural network considered in this paper for linear-quadratic mixtures is attractive, because its free weights may be used to optimize its stability. We here proposed a procedure for exploiting this flexibility in *blind* configurations. This opens the way to the development of various practical model estimation algorithms for linear-quadratic mixtures (e.g. using the maximum likelihood framework). They will be "plugged" into the general adaptation architecture that we defined above (taking into account that the evolution of l_{11} and l_{22} has an influence on signal scales: see [1]). We will also study analytical solutions of the stability condition.

References

1. Deville, Y., Hosseini, S.: Recurrent networks for separating extractable-target nonlinear mixtures. Part I: non-blind configurations. Signal Processing 89(4), 378–393 (2009), `http://dx.doi.org/10.1016/j.sigpro.2008.09.016`
2. Duarte, L.T., Jutten, C., Moussaoui, S.: Bayesian source separation of linear-quadratic and linear mixtures through a MCMC method. In: Proceedings of IEEE MLSP, Grenoble, France, September 2-4 (2009)
3. Hosseini, S., Deville, Y.: Blind maximum likelihood separation of a linear-quadratic mixture. In: Puntonet, C.G., Prieto, A.G. (eds.) ICA 2004. LNCS, vol. 3195, pp. 694–701. Springer, Heidelberg (2004), `http://arxiv.org/abs/1001.0863`
4. Jutten, C., Karhunen, J.: Advances in Blind Source Separation (BSS) and Independent Component Analysis (ICA) for Nonlinear Mixtures. International Journal of Neural Systems 14(5), 267–292 (2004)
5. Mokhtari, F., Babaie-Zadeh, M., Jutten, C.: Blind separation of bilinear mixtures using mutual information minimization. In: Proceedings of IEEE MLSP, Grenoble, France, September 2-4 (2009)
6. Thompson, J.M.T., Stewart, H.B.: Nonlinear dynamics and chaos. Wiley, Chichester (2002)

Statistical Model of Speech Signals Based on Composite Autoregressive System with Application to Blind Source Separation

Hirokazu Kameoka, Takuya Yoshioka, Mariko Hamamura, Jonathan Le Roux, and Kunio Kashino

NTT Communication Science Laboratories, NTT Corporation,
3-1 Morinosato Wakamiya, Atsugi, Kanagawa 243-0198, Japan
{kameoka,leroux,kunio}@cs.brl.ntt.co.jp, takuya@cslab.kecl.ntt.co.jp

Abstract. This paper presents a new statistical model for speech signals, which consists of a time-invariant dictionary incorporating a set of the power spectral densities of excitation signals and a set of all-pole filters where the gain of each pair of excitation and filter elements is allowed to vary over time. We use this model to develop a combined blind separation and dereverberation method for speech. Reasonably good separations were obtained under a highly reverberant condition.

Keywords: Blind source separation, composite autoregressive system.

1 Introduction

When a set of observed data is considered to be a random sample drawn from a population that follows a particular family of probability distributions, we often refer to the distribution family as a statistical model. In this paper, we present a new statistical model of speech signals, suitable for applying to blind signal processing problems.

The aim of blind source separation (BSS) or blind dereverberation is to detect each source component from observed signals captured by one or several microphones without using any information about the transfer characteristic of the path from each source to each microphone. We thus usually make some assumption about the sources and then formulate an optimization problem based on a criterion that measures the consistency with the assumption. One approach to formulating the problem would be to define a likelihood function of a source signal by employing a statistical model assumption. When choosing which statistical model to invoke, it is important to take account of whether it agrees well with the actual behaviors of the source of interest and whether it leads to a mathematically tractable form of the optimization problem.

Let us now briefly review the BSS problem. BSS algorithms derived from a convolutive mixture model in the time domain such as [1] are fine for short mixing filters, but when it comes to realistically long filters, they can be unrealizable because of computational requirements. In the STFT domain where the frame

V. Vigneron et al. (Eds.): LVA/ICA 2010, LNCS 6365, pp. 245–253, 2010.

size is assumed to be sufficiently larger than the filter length, a convolutive mixture signal can be approximated by an instantaneous mixture in the frequency domain, thus allowing for an efficient implementation of BSS algorithms [2]. However, when reverberation comes into play, the filter length can often be larger than the frame size and so the above approximation becomes relatively less accurate. According to several studies such as [3], reverberation can be modeled fairly well as a convolution for each frequency-band in the STFT domain. Therefore, under highly reverberant conditions, we can expect a convolutive mixture model in the time-frequency domain [4] to be a better approximation.

We consider the situation where we observe signals emanating from M sources and captured by M microphones. Let $Y_{k,n}^{l}$ be the STFT of the signal observed at the l-th microphone and $\boldsymbol{Y}_{k,n} = (Y_{k,n}^{1}, \cdots, Y_{k,n}^{M})^{\mathrm{T}}$ be a set of observed data, where k and n are the frequency and time indices, respectively. We use $\boldsymbol{S}_{k,n} = (S_{k,n}^{1}, \cdots, S_{k,n}^{M})^{\mathrm{T}}$ to denote a set of M source components. In this paper, we use a separation system that has a multichannel finite-impulse-response form in the time-frequency domain such that:

$$\boldsymbol{S}_{k,n} = \sum_{\tau=0}^{n_k} \boldsymbol{W}_{k,\tau} \boldsymbol{Y}_{k,n-\tau}, \tag{1}$$

where $\boldsymbol{W}_{k,\tau}$, $0 \le \tau \le n_k$ are matrix coefficients of size $M \times M$. Let us assume that $S_{k,n}^{m}$ is a random variable and that $S_{k,n}^{m}$ and $S_{k',n'}^{m'}$ are statistically independent when $(k,n,m) \neq (k',n',m')$. The definition of the probability density function (pdf) for $S_{k,n}^{m}$, denoted by $f_{S_{k,n}^{m}|\theta^m}(s_{k,n}^{m}|\theta^m)$, is the main subject of the following section. This pdf corresponds to a statistical model of the m-th source signal and θ^m is the set of parameters characterizing its distribution. Once we define its specific form, the joint pdf of $\boldsymbol{Y} := \{\boldsymbol{Y}_{k,n}\}_{k,n}$ can be written in terms of $f_{S_{k,n}^{m}|\theta^m}(\,\cdot\,|\theta^m)$ explicitly as

$$f_{\boldsymbol{Y}|\Theta}(\boldsymbol{Y}|\Theta) = \prod_{k} |\det \boldsymbol{W}_{k,0}| \prod_{t,m} f_{S_{k,n}^{m}|\theta^m}\left(\bar{S}_{k,n}^{m}|\theta^m\right), \tag{2}$$

where $\bar{S}_{k,n}^{m}$ is the m-th element of the separated signal vector, given by $\sum_{\tau} \boldsymbol{W}_{k,\tau} \boldsymbol{Y}_{k,n-\tau}$. The joint pdf $f_{\boldsymbol{Y}|\Theta}(\boldsymbol{Y}|\Theta)$ is the likelihood of the unknown variables $\Theta := \{\{\theta^m\}_m, \{\boldsymbol{W}_{k,\tau}\}_{k,\tau}\}$ given observation $\boldsymbol{Y}$, which is an objective function for achieving separation and dereverberation in a joint manner, as with [4].

It has been shown that the speech production system can be well modeled on a frame-by-frame basis by a linear system comprising a glottal excitation input and a vocal-tract resonance filter that respectively determine the degree of periodicity and the phoneme of the voice. One of the most frequently used models for short-term speech signals is the autoregressive (AR) model, which models the signal as the output of an all-pole system. [5] was among the first to propose a BSS system in which a Gaussian AR source model is incorporated in $f_{S_{k,n}^{m}|\theta}(s_{k,n}^{m}|\theta^m)$, where a complex spectrum at each frame is assumed to be a set of random samples drawn from a different AR system with a Gaussian white noise input. This type of statistical model for STFT spectrograms of speech has later been shown to work successfully for BSS in highly reverberant environments [4]. The objective of this paper is to investigate the possibility to improve the performance of this state-of-the-art BSS system [4] by replacing the Gaussian AR source model with the speech model we proposed previously [6], which will be reviewed in the next section.

2 Proposed Statistical Model of Speech

The white noise assumption as regards the excitation inputs underlying the standard AR model is known to not hold especially for voices with high fundamental frequencies (F_0s). This is because when F_0 increases the spacing between the harmonics of the excitation spectrum increases correspondingly, thus departing from a white (flat) spectrum. Rather than fixing the characteristics of the excitation input, we would therefore, if possible, like to estimate them in the same way as the vocal-tract characteristics. However, if we simply treat both of the characteristics as separate variables for each frame, it would no longer be possible to determine these characteristics uniquely since there are an infinite number of combinations giving the same filter output. Some additional constraint is needed to avoid this indeterminacy.

There are a wide variety of regularities in speech that can be exploited to constrain speech models. For example, it may be reasonable to assume that every short-term signal of a speech utterance can be represented by a combination of elements drawn from two dictionaries, one consisting of a small number of vocal-tract characteristics, and the other of a small number of excitation characteristics. This is because the phoneme number and periodicity range of speech during an entire utterance are both usually limited. We thus assume here that speech signals have been generated by a compound linear system composed of the direct product of a limited set of time-invariant excitation characteristics and a limited set of time-invariant vocal-tract characteristics where each output associated with an excitation and filter element pair is activated by a time-varying gain. A signal at a particular frame is thus assumed to be characterized by the volume levels of all the filter outputs. Note that since the dictionary elements are in general unknown, they need to be estimated in a data-driven manner.

In this section, we focus on a particular source and so the index m will be omitted for simplicity of notation. We start by modeling an output signal characterized by a particular excitation and filter element pair. Let us assume that a signal in the n-th frame, $\{x_n[t]\}_{t=1}^{K}$, is a sampled sequence drawn from the P-order AR process with a set of AR parameters common over n such that

$$x_n[t] = \sum_{p=1}^{P} a_p x_n[t-p] + \epsilon_n[t], \tag{3}$$

where $\epsilon_n[t]$ is an excitation input signal that is assumed to be a zero-mean stationary Gaussian noise. Its autocorrelation function, $\nu_n[t]$, is constant up to the gain over all the frames such that $\nu_n[t] = U_n h[t]$. U_n is assumed to be the energy of $\epsilon_n[t]$ within the frame. We note here that $\epsilon_n[t]$ is not restricted to being white noise. Let $\boldsymbol{x}_n = (x_n[1], \cdots, x_n[K])^{\mathrm{T}} \in \mathbb{R}^K$ and its discrete Fourier transform (DFT) be $\boldsymbol{X}_n = (X_{1,n}, \cdots, X_{K,n})^{\mathrm{T}} \in \mathbb{C}^K$. Then, according to Eq. (3), $\boldsymbol{X}_n$ follows a multivariate complex Gaussian distribution $\boldsymbol{X}_n \sim \mathcal{N}_{\mathbb{C}}(\boldsymbol{0}, \boldsymbol{\Lambda}_n)$, with a diagonal covariance matrix $\boldsymbol{\Lambda}_n = \mathrm{diag}\big(\lambda_{1,n}, \cdots, \lambda_{K,n}\big)$ whose elements are

$$\lambda_{k,n} = \frac{H_k U_n}{|A(e^{\mathrm{j}2\pi k/K})|^2}, \tag{4}$$

$$A(z) = 1 - a_1 z^{-1} - \cdots - a_P z^{-P}, \tag{5}$$

where k is the frequency index and j is the imaginary unit. $\{H_k\}_{k=1}^K$ is the DFT of $\{h[k]\}_{k=1}^K$, which represents the power spectral density (PSD) of the excitation source signal $\epsilon_n[t]$ (namely, the spectral fine structure), which can have any shape and is not necessarily flat. On the other hand, $1/|A(e^{\mathrm{j}2\pi k/K})|^2$ corresponds to a spectral envelope expressed as the PSD of the all-pole transfer function.

We can now construct a "composite" autoregressive model by extending the above model. The composite autoregressive system is assumed to consist of a dictionary of I excitation PSDs and a dictionary of J all-pole filters. Subsequently, we use superscripts i and j to denote the indices of the excitation PSDs and the all-pole filters, respectively, and we denote the ith excitation PSD and the jth all-pole transfer function by H_k^i and $1/A^j(e^{\mathrm{j}2\pi k/K})$. The system is able to generate $I \times J$ different voice components each of which is characterized by combining elements drawn from the respective dictionaries. If we assume that only one of the $I \times J$ voice components is active at each frame, the source pdf can be defined using a Gaussian scaled mixture model [7]. By contrast, we assume that all the voice components are always active with different volume levels. Let $U_n^{i,j}$ denote the volume level of the $\{i,j\}$-th voice component at the n-th frame. By following the discussion above, the DFT of the $\{i,j\}$-th voice component at the nth frame, $\boldsymbol{X}_n^{i,j} = (X_{1,n}^{i,j}, \cdots, X_{K,n}^{i,j})^{\mathrm{T}}$, follows a multivariate complex Gaussian distribution $\mathcal{N}_{\mathbb{C}}(\boldsymbol{0}, \boldsymbol{\Lambda}_n^{i,j})$ with a diagonal covariance matrix $\boldsymbol{\Lambda}_n^{i,j} = \mathrm{diag}\big(\lambda_{1,n}^{i,j}, \cdots, \lambda_{K,n}^{i,j}\big)$ whose elements are

$$\lambda_{k,n}^{i,j} = \frac{H_k^i U_n^{i,j}}{|A^j(e^{\mathrm{j}2\pi k/K})|^2}, \tag{6}$$

$$A^j(z) = 1 - a_1^j z^{-1} - \cdots - a_P^j z^{-P}. \tag{7}$$

If we now assume that $\boldsymbol{X}_n^{1,1}, \cdots, \boldsymbol{X}_n^{I,J}$ are mutually independent, it follows that

$$\boldsymbol{S}_n = \sum_i \sum_j \boldsymbol{X}_n^{i,j} \sim \mathcal{N}_{\mathbb{C}}\big(\boldsymbol{0}, \boldsymbol{\Phi}_n\big), \quad \boldsymbol{\Phi}_n = \sum_i \sum_j \boldsymbol{\Lambda}_n^{i,j}, \tag{8}$$

where $\boldsymbol{S}_n \in \mathbb{C}^K$ denotes the DFT of the speech signal at the n-th frame. The statistical model, $f_{S_{k,n}|\theta}(S_{k,n}|\theta)$, is thus given concisely as

$$f_{S_{k,n}|\theta}(S_{k,n}|\theta) = \frac{1}{\pi \phi_{k,n}} \exp\left(-\frac{|S_{k,n}|^2}{\phi_{k,n}}\right), \tag{9}$$

where θ contains all the unknown parameters of the present system: $\theta := \{H_k^i, a_p^j, U_n^{i,j}\}_{i,j,p,k,n}$. The diagonal element of $\boldsymbol{\Phi}_n$, i.e. $\phi_{k,n}$, corresponds to the PSD of the output signal produced by the present system such that

$$\phi_{k,n} = \sum_i \sum_j \frac{H_k^i U_n^{i,j}}{|A^j(e^{j2\pi k/K})|^2}. \tag{10}$$

It is important to note that in the special case where $J = 1$ and $P = 0$, the PSD $\phi_{k,n}$ reduces to the form of matrix multiplication $\phi_{k,n} = \sum_i H_k^i U_n^{i,1}$ and Eq. (9) thus reduces to the likelihood function under the statistical interpretation of non-negative matrix factorization (NMF) given in [8]. This fact suggests that the present entire BSS system is structurally related to multichannel NMF [9].

3 Optimization Process

Given a set of observed STFT components $\boldsymbol{Y}$, we want to find the estimate of $\Theta = \{\{\theta^m\}_m, \{\boldsymbol{W}_{k,\tau}\}_{k,\tau}\}$ that maximizes the posterior density $f_{\Theta|\boldsymbol{Y}}(\Theta; \boldsymbol{Y}) \propto f_{\boldsymbol{Y}|\Theta}(\boldsymbol{Y}; \Theta)\, f_\Theta(\Theta)$, or equivalently, the log posterior density

$$L(\theta) := \log f_{\boldsymbol{Y}|\Theta}(\boldsymbol{Y}|\Theta) + \log f_\Theta(\Theta). \tag{11}$$

Eq. (11) can be iteratively increased by using a coordinate descent method in which each iteration comprises the following three maximization steps: (S1) $\theta^m \leftarrow \text{argmax}_{\theta^m} L(\theta)$ for all m, (S2) $\boldsymbol{W}_{k,0} \leftarrow \text{argmax}_{\boldsymbol{W}_{k,0}} L(\theta)$ for all k, and (S3) $\{\boldsymbol{W}_{k,\tau}\}_{\tau=1}^{n_k} \leftarrow \text{argmax}_{\{\boldsymbol{W}_{k,\tau}\}_{\tau=1}^{n_k}} L(\theta)$ for all k. If we were able to obtain an estimate of the PSD of each source, namely $\phi_{k,n}^m$, we could invoke [4] to perform (S2) and (S3). Therefore, obtaining the update formula of (S1) will suffice to complete the derivation of the entire optimization process. It should be noted that (S1) amounts to maximizing

$$\sum_{k,n} \log f_{S_{k,n}^m|\theta^m}(\bar{S}_{k,n}^m|\theta^m) + \log f_{\theta^m}(\theta^m) \tag{12}$$

with respect to θ_m where $\bar{S}_{k,n}^m$ is the m-th vector element of $\sum_\tau \boldsymbol{W}_{k,\tau} \boldsymbol{Y}_{k,n-\tau}$. As this maximization is carried out for each m separately, the index m is omitted again in the following.

$\log f_{S_{k,n}|\theta}(\bar{S}_{k,n}|\theta)$ is equal up to constant terms to the goodness of fit between $|\bar{S}_{k,n}|^2$ and $\phi_{k,n}$ defined by the Itakura-Saito divergence. We are thus led to obtain a PSD model with as small a reconstruction error as possible. On the other hand, as with the sparse coding concept [10], we would like to keep the voice components as sparse as possible. The prior term $\log f_\theta(\theta)$ can be used to promote the sparseness of $U_n^{i,j}$. In the subsequent analysis, for convenience we use an exponential prior (a folded Laplacian prior) defined over $U_n^{i,j} \geq 0$

$$f_\theta(\theta) = \prod_{i,j,n} \alpha \exp(-\alpha U_n^{i,j}), \tag{13}$$

which promotes sparsity when α is large. Maximizing Eq. (12) thus combines the goals of a small reconstruction error and sparseness. As a consequence, the more frequently a certain spectral fine/envelope structure emerges in $|\bar{S}_{k,n}|^2$, the more likely it is to be captured in the excitation/filter dictionary.

Although it is difficult to obtain a closed-form solution for maximizing Eq. (12), we can develop a computationally efficient scheme for its estimation based on the Generalized Expectation-Maximization (GEM) algorithm. When applying the GEM algorithm to the current "partial" MAP estimation problem (that is, (S1)), the first step is to define the "complete data". As the "observed" source component $\bar{S}_{k,n}$ is assumed to contain $I \times J$ concurrent voice components, a natural choice for the complete data is the corresponding hidden components, that is, $\boldsymbol{X}_{k,n} = (X^{1,1}_{k,n}, \cdots, X^{I,J}_{k,n})^{\mathrm{T}}$ with $X^{i,j}_{k,n} \sim \mathcal{N}_{\mathbb{C}}(0, \lambda^{i,j}_{k,n})$. From Eq. (8), the many-to-one relationship between $\boldsymbol{X}_{k,n}$ and $\bar{S}_{k,n}$ is described as $\bar{S}_{k,n} = \mathbf{1}^{\mathrm{T}}\boldsymbol{X}_{k,n}$ with $\mathbf{1} = (1, \cdots, 1)^{\mathrm{T}}$. In Sec. 2, we have already assumed that $X^{i,j}_{k,n}$ is independent of all other $X^{i',j'}_{k',n'}$, so the log-likelihood of the complete data $\boldsymbol{X} := \{\boldsymbol{X}_{k,n}\}_{k,n}$ is

$$\log f_{\boldsymbol{X}|\theta}(\boldsymbol{X}|\theta) = -\sum_{k,n}\Big[\log\det \pi \boldsymbol{\Lambda}_{k,n} + \mathrm{tr}\big(\boldsymbol{\Lambda}^{-1}_{k,n}\boldsymbol{X}_{k,n}\boldsymbol{X}^{\mathrm{H}}_{k,n}\big)\Big], \tag{14}$$

where $\boldsymbol{\Lambda}_{k,n} = \mathrm{diag}\big(\lambda^{1,1}_{k,n}, \cdots, \lambda^{I,J}_{k,n}\big)$. Taking the conditional expectation of Eq. (14) given $\bar{S}_{k,n}$ and $\theta = \theta'$ and then adding $\log f_\theta(\theta)$ to both sides, we obtain

$$\begin{aligned} Q(\theta, \theta') = \log f_\theta(\theta) - \sum_{k,n}\Big[&\log\det \pi \boldsymbol{\Lambda}_{k,n} \\ &+ \mathrm{tr}\big(\boldsymbol{\Lambda}^{-1}_{k,n}\mathbb{E}\big[\boldsymbol{X}_{k,n}\boldsymbol{X}^{\mathrm{H}}_{k,n}|S_{k,n} = \bar{S}_{k,n}, \theta = \theta'\big]\big)\Big], \end{aligned} \tag{15}$$

where $\mathbb{E}\big[\boldsymbol{X}_{k,n}\boldsymbol{X}^{\mathrm{H}}_{k,n}|S_{k,n} = \bar{S}_{k,n}, \theta = \theta'\big] = \boldsymbol{\Lambda}'_{k,n} - \boldsymbol{\Lambda}'_{k,n}\mathbf{1}\big(\mathbf{1}^{\mathrm{T}}\boldsymbol{\Lambda}'_{k,n}\mathbf{1}\big)^{-1}\mathbf{1}^{\mathrm{T}}\boldsymbol{\Lambda}'_{k,n} + |\bar{S}_{k,n}|^2\boldsymbol{\Lambda}'_{k,n}\mathbf{1}\big(\mathbf{1}^{\mathrm{T}}\boldsymbol{\Lambda}'_{k,n}\mathbf{1}\big)^{-1}\big(\mathbf{1}^{\mathrm{T}}\boldsymbol{\Lambda}'_{k,n}\mathbf{1}\big)^{-1}\mathbf{1}^{\mathrm{T}}\boldsymbol{\Lambda}'_{k,n}$. Writing it in an element-wise expression, we obtain

$$Q(\theta, \theta') \stackrel{c}{=} -\sum_{k,n}\sum_{i,j}\left[\log H^i_k U^{i,j}_n + \frac{\Psi^{i,j}_{k,n}|A^j(e^{\mathrm{j}2\pi k/K})|^2}{H^i_k U^{i,j}_n}\right] - \alpha\sum_n\sum_{i,j}U^{i,j}_n, \tag{16}$$

where $\Psi^{i,j}_{k,n}$ represents the PSD estimate of the $\{i,j\}$th voice components,

$$\Psi^{i,j}_{k,n} = \frac{{\lambda'}^{i,j}_{k,n}}{{\phi'}_{k,n}}\Big({\phi'}_{k,n} - {\lambda'}^{i,j}_{k,n} + \frac{{\lambda'}^{i,j}_{k,n}}{{\phi'}_{k,n}}|\bar{S}_{k,n}|^2\Big). \tag{17}$$

The notation $\stackrel{c}{=}$ denotes equality up to constant terms. By setting the partial derivatives of $Q(\theta, \theta')$ with respect to H^i_k and $U^{i,j}_n$ at zero, we obtain the following update formulae for H^i_k and $U^{i,j}_n$

$$H_k^i = \frac{1}{NJ}\sum_n\sum_j \Psi_{k,n}^{i,j}|A^j(e^{\mathrm{j}2\pi k/K})|^2/U_n^{i,j}, \tag{18}$$

$$\alpha U_n^{i,j2} + KU_n^{i,j} - \sum_k \Psi_{k,n}^{i,j}|A^j(e^{\mathrm{j}2\pi k/K})|^2/H_k^i = 0, \quad U_n^{i,j} \geq 0 \tag{19}$$

By setting the partial derivatives of $Q(\theta, \theta')$ with respect to $a_1^j, \cdots, a_P^j$ at zero, we obtain the Yule-Walker equations

$$r_p^j = \sum_{q=1}^{P} a_q^j r_{p-q}^j \ (p = 1, \cdots, P), \tag{20}$$

where r_p^j is defined by the inverse DFT of the average spectral envelope over all the voice components with index j such that

$$r_p^j = \sum_k \sum_{n,i} \frac{\Psi_{k,n}^{i,j}}{H'^i_k U'^{i,j}_n} e^{p\mathrm{j}2\pi k/K}. \tag{21}$$

The update formula for the autoregressive parameters of the jth all-pole filter can therefore be calculated using the well-known Levinson-Durbin algorithm.

Here it is important to note that when a sparse constraint comes into play, there is a need for some constraint on the scales of the factorized elements in order to avoid an indeterminacy in the scaling. We thus adopt a simple procedure that consists of calculating Eq. (18) and then projecting it onto the unit norm space: $H_k^i \leftarrow H_k^i / \sum_{k'} H_{k'}^i$.

4 Experimental Results

We present here the separation results we obtained with the present BSS algorithm. All of the examples use the same two-input four-output impulse response, which was measured in a varechoic chamber where the reverberation time was 0.5 sec. With this impulse response, we mixed two speech signals into four mixtures. The two speech signals of female speakers, taken from the ATR speech database, were sampled at 16 kHz and band limited to the 50 Hz to 7 kHz frequency range. The input Signal-to-Interference ratios (SIRs) are shown in Tab. 1. Time-frequency representations were obtained using the polyphase filterbank analysis with a frame length of 32 ms and a hop size of 8 ms. The filter length n_k was set as follows: $n_k = 25$ for $F_k < 0.8$; $n_k = 20$ for $0.8 \leq F_k < 1.5$; $n_k = 15$ for $1.5 \leq F_k < 3$; $n_k = 10$ for $F_k \geq 3$, where F_k is the frequency in kHz of the k-th frequency bin. The AR order P was set at 12, and α at $10K$. The iterative algorithm comprising (S1)–(S3) was run for 3 iterations. For each step of (S1), the GEM algorithm was run for 100 iterations.

We tested the performance of the present method with different I and J settings. Tab. 2 lists the SIRs obtained with the proposed method for various settings of I and J, along with those obtained with the baseline methods, where

Table 1. Input SIRs (dB)

	Channel			
Source	#1	#2	#3	#4
#1	-0.6	-0.3	-0.1	0.6
#2	0.6	0.3	0.1	-0.6

Table 2. Output SIRs (dB) for various settings

	$I=2$		$I=5$		$I=10$		Base-line1	Base-line2
Source	$J=12$	$J=15$	$J=8$	$J=10$	$J=8$	$J=12$		
#1	18.0	16.8	19.9	18.5	17.7	17.6	11.6	17.2
#2	11.7	10.9	14.1	13.9	13.6	14.5	11.0	13.9

Baseline1 and Baseline2 refer to Sawada's method [11] and Yoshioka's method [4], respectively. For Baseline1, we performed time-frequency analysis with a frame length of 256 ms and a hop size of 64 ms. For Baseline2, the frame length and the hop size were set at 16 ms and 8 ms. The best SIR result by the present method was obtained when I and J were set at 5 and 8, which significantly outperforms both of the baseline methods. The results are very preliminary and they need to be confirmed by a more thorough analysis in the future.

5 Conclusion

This paper described a statistical model called the "composite autoregressive system", which consists of a time-invariant dictionary incorporating a set of PSDs of excitation signals and a set of all-pole filters. Under this model, speech signals are assumed to be characterized by the volume levels of the excitation and filter element pairs, that vary over time. We proposed to use this model to develop a combined blind separation and dereverberation method for speech and reasonably good separations were obtained for four mixtures of two speech signals under a reverberant condition.

References

1. Douglas, S., Sawada, H., Makino, S.: Natural gradient multichannel blind deconvolution and speech separation using causal FIR filters. IEEE Trans. Speech, Audio Process. 13(1), 92–104 (2005)
2. Smaragdis, P.: Blind separation of convolved mixtures in the frequency domain. Neur. Comp. 22, 21–34 (1998)
3. Nakatani, T., Yoshioka, T., Kinoshita, K., Miyoshi, M., Juang, B.-H.: Blind speech dereverberation with multi-channel linear prediction based on short time Fourier transform representation. In: Proc. Int'l. Conf. Acoust., Speech, Signal Process., pp. 85–88 (2008)
4. Yoshioka, T., Nakatani, T., Miyoshi, M., Okuno, H.G.: Blind separation and dereverberation of speech mixtures by joint optimization. IEEE Trans. Audio, Speech, Language Process (2010) (accepted for publication)
5. Dégerine, S., Zaïdi, A.: Separation of an instantaneous mixture of Gaussian autoregressive sources by the exact maximum likelihood approach. IEEE Trans. Signal Processing 52(6), 1499–1512 (2004)
6. Kameoka, H., Kashino, K.: Composite Autoregressive System for Sparse Source-Filter Representation of Speech. In: Proc. 2009 IEEE International Symposium on Circuits and Systems (ISCAS 2009), pp. 2477–2480 (2009)

7. Benaroya, L., Bimbot, F., Gribonval, R.: Audio source separation with a single sensor. IEEE Trans. Audio Speech Language Processing 14(1), 191–199 (2006)
8. Févotte, C., Bertin, N., Durrieu, J.-L.: Nonnegative matrix factorization,with the Itakura-Saito divergence. With application to music analysis. Neural Comput. 21(3), 793–830 (2009)
9. Ozerov, A., Févotte, C.: Multichannel nonnegative matrix factorization in convolutive mixtures for audio source separation. IEEE Trans. Audio, Speech, Language Process. 18(3), 550–563 (2010)
10. Olshausen, B.A., Field, D.J.: Emergence of simple-cell receptive field properties by learning a sparse code for natural images. Nature 381, 607–609 (1996)
11. Sawada, H., Araki, S., Makino, S.: Measuring dependence of binwise separated signals for permutation alignment in frequency-domain BSS. In: Proc. Int'l. Symp. Circ., Syst., pp. 3247–3250 (2007)

Information-Theoretic Model Selection for Independent Components

Claudia Plant[1], Fabian J. Theis[2], Anke Meyer-Baese[1], and Christian Böhm[3]

[1] Florida State University
[2] Helmholtz Zentrum München
[3] University of Munich

Abstract. Independent Component Analysis (ICA) is an essential building block for data analysis in many applications. Selecting the truly meaningful components from the result of an ICA algorithm, or comparing the results of different algorithms, however, are non-trivial problems. We introduce a very general technique for evaluating ICA results rooted in information-theoretic model selection. The basic idea is to exploit the natural link between non-Gaussianity and data compression: The better the data transformation represented by one or several ICs improves the effectiveness of data compression, the higher is the relevance of the ICs. In an extensive experimental evaluation we demonstrate that our novel information-theoretic measure robustly selects the most interesting components from data without requiring any assumptions or thresholds.

The evaluation and interpretation of the result of ICA is often difficult for two major reasons: First, most ICA algorithms yield always a result, even if the underlying assumption (e.g. Gaussianity for FastICA) is not fulfilled. Thus, many ICA algorithms extract as many independent sources as there are mixed signals in the data set, no matter how many of them really fulfill the underlying assumption. Second, ICA has no unique and natural evaluation criterion to assess the relevance or strength of the detected result (like for instance the variance criterion for PCA). Different ICA algorithms use different objective functions, and to select one of them as an *objective* criterion would give unjustified preference to the result of that specific algorithm. Moreover, if the user is interested in comparing an ICA result to completely different modeling techniques like PCA, regression, mixture models etc, these ICA-internal criteria are obviously unsuitable. In this paper, we investigate the *compressability* of the data as a more objective criterion for the quality of single components or the overall ICA result.

1 ICA and Data Compression

One of the fundamental assumptions of many important ICA algorithms is that independent sources can be found by searching for maximal non-Gaussian directions in the data space. Non-Gaussianity leads to a decrease in entropy, and therefore, to a potential improvement of the efficiency of data compression. In

V. Vigneron et al. (Eds.): LVA/ICA 2010, LNCS 6365, pp. 254–262, 2010.

principle, the achievable compression rate of a data set, after ICA, is higher compared to the original data set. The principle of Minimum Description Length (MDL) uses the probability $P(x)$ of a data object x to represent it according to Huffman coding. Huffman coding gives a lower bound for the compression rate of a data set D achievable by any concrete coding scheme, as follows: $\sum_{x \in D} -\log_2(P(x))$. If x is taken from a continuous domain (e.g. the vector space $\mathcal{R}^d$ for the blind source separation of a number d of signals), the relative probability given by a probability density function $p(x)$ is applied instead of the absolute probability. The relative and absolute log-likelihood (which could be obtained by discretizing x) are identical up to a constant value which can be safely ignored. Note that ignoring this constant value may result in a code-length with negative sign. This is a bit counter-intuitive from a data compression perspective, but absolutely no problem for model selection: when comparing two models, always the one having smaller code length is the better. For a complete description of the data set (allowing decompression) the parameters of the probability density function (PDF) such as mean and variance for Gaussian PDFs need to be coded and their code lengths added to the negative log-likelihood of the data. We call this term the code book. For each parameter, a number of bits equal to $\frac{1}{2}\log_2(n)$ where n is the number of objects in D is required, as fundamental results from information theory have proven [1]. Intuitively, the term $\frac{1}{2}\log_2(n)$ reflects the fact that the parameters need to be coded more precisely when a higher number n of data objects is modeled by the PDF. The MDL principle is often applied for model selection of parametric models like Gaussians, or Gaussian Mixture Models (GMM). Gaussian Mixture Models vary in the model complexity, i.e. the number of parameters needed for modeling. MDL-based techniques are well able to compare models of different complexity. The main purpose of the code book is to punish complexity in order to avoid overly complex, over-fitted models (like a GMM having one component exactly at the position of each data object: such a model would yield a minimal Huffman coding, but also a maximal code-book length). By the two concepts, Huffman coding using the negative log-likelihood of a PDF and the code-book for the parameters of the PDF, the principle of MDL provides a very general framework which allows the comparison of very different modeling techniques like Principal Component Analysis (based on a Gaussian PDF model), Clustering [2], regression [3] for continuous domains, but, in principle, also for discrete or mixed domains. At the same time, model complexity is punished and, therefore, overfitting avoided. Related criteria for general model selection include e.g. the Bayesian Information Criterion and the Aikake Information Criterion. However, these criteria are not adapted to the ICA model. In the following section, we discuss how to apply the MDL principle in the context of ICA.

1.1 Nonparametric Data Modeling for ICA

Techniques like OCI [4] or [5] successfully use the exponential power distribution (EPD), a generalized distribution function including Gaussian, Laplacian, uniform and many other distribution functions for assessing the ICA result using MDL. The reduced entropy of non-Gaussian projections in the data allows a higher compression rate and thus favors a good ICA result. However, the selection of EPD is a rather random choice and overly restrictive. For instance, multi-modal and asymmetric distributions cannot be well represented by EPD but are highly relevant to ICA. In the following we describe an alternative representation of the PDF which is efficient if (and only if) the data is considerably different from Gaussian. Besides the non-Gaussianity we have no additional assumption (like for instance EPD) on the data. To achieve this, we tentatively assume Gaussianity in each of our d obtained signals of length n, and transform the assumed Gaussian distribution into a uniform distribution in the interval $(0, 1)$ by applying the Gaussian cumulative distribution function $\Phi(\frac{x_i - \mu_i}{\sigma_i})$ to each dimension i (where $1 \leq i \leq d$) of the data representation to be tested (for instance, after projection on the independent components). Then, the resulting distribution is represented by a histogram $(H_1, ..., H_b)$ with a number b of equidistant bins where b is optimized as we will show later. $H_j (1 \leq j \leq b)$ is the number of objects falling in the corresponding half open interval $[\frac{j-1}{b}, \frac{j}{b})$. Using this coding scheme the overall code length ($CLRG_i(D, b)$, Code Length Relative to Gaussianity) of the data's projection on dimension i is provided by:

$$CLRG_i(D,b) = \overbrace{\sum_{1 \leq j \leq b} H_j \cdot \log_2 \frac{n}{H_j}}^{\text{log-likelihood}} + \overbrace{\frac{b-1}{2} \log n}^{\text{code book}} - \overbrace{n \cdot \log_2 b}^{\text{offset cost}} + \overbrace{PRE_i}^{\text{preproc.}} .$$

As introduced in Section 1, the first term represents the negative log-likelihood of the data given the histogram and the second term is the cost required for the code book: To completely describe b histogram bins it is sufficient to use $b - 1$ codewords since the remaining probability is implicitly specified. The third term refers to the offset cost: when coding the same data with a varying number of histogram bins, the resulting log-likelihoods are based on different basic resolutions of the data space (a grid with a number b of partitions). Although the choice of a particular basic resolution is irrelevant for the end-result, for comparability all alternative solutions must be based on a common resolution. We choose $b = 1$ as basic resolution, and subtract for each object the number of bits by which we know the position of the object more precisely than in the basic resolution. The trivial histogram having $b = 1$ represents the case where the data is assumed to be Gaussian: Since the Gaussian cumulative distribution function $\Phi(\frac{x_i - \mu_i}{\sigma_i})$ has been applied to the data, Gaussian data are transformed into uniform data, and our histograms have an implicit assumption of uniformity *inside* each bin. If some choice of $b \neq 1$ leads to smaller $CLRG_i(D, b)$, we have evidence that the dimension i of the data is different from Gaussian. It is easy to see that in the case $b = 1$ the code length is $CLRG_i(D, b) = 0$, which is a consequence of our definition of the offset cost. Therefore, we call our cost function Code Length Relative to Gaussianity, ($CLRG$). If no choice of $b \neq 1$ leads to $CLRG_i(D, b) < 0$, then either the data is truly Gaussian or the number of data objects is not high enough to give evidence for non-Gaussianity. In the

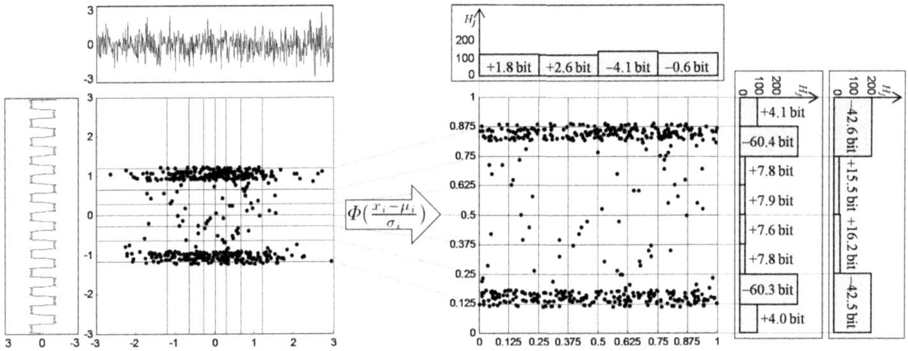

Fig. 1. Overview of the computation of $CLRG_i(D, b)$

latter case we use Gaussianity as the safe default-assumption. We assume that our cost function is usually applied after some transformations by e.g. centering, whitening, some ICA algorithm etc. Some of these transformations need also be represented in the code book, when the model is used in comparison with completely different models. The details of this are beyond the scope of this paper, but the last term in $CLRG_i(D, b)$ is a placeholder for the code length of the description of these preprocessing operations. For comparing different independent components, and for comparing different runs of the same or different ICA algorithms, these cost are constant, and we can set $PRE_i = 0$. Figure 1 gives an overview and example of our method: On the left side, the result of an ICA run is depicted which has successfully separated a number $d = 2$ of signals (each having $n = 500$ points). The corresponding scatter-plot shows a Gaussian signal on the x-axis, a rectangular signal on the y-axis (note that the corresponding signal plots on the axes are actually transposed for better visibility). On the right side, we see the result after applying the Gaussian CDF. Some histograms with different resolutions are also shown: On the x-axis, the histogram with $b = 4$ bins is approximately uniformly filled (like also most other histograms with a different selection of b). Consequently, only a very small number of bits is saved compared to Gaussianity (e.g. only 4.1 bits for the complete signal part falling in the third bin H_3) by applying this histogram as PDF in Huffman coding (here, the cost per bin are reported including log-likelihood and offset-cost). The overall saving of 0.29 bit are contrasted by a code-book length of $\frac{3}{2}\log_2 n = 13.4$, so the histogram representation does not pay off. In contrast, the two histograms on the y-axis do pay off, since for $b = 8$, we have overall savings over Gaussianity of 81.3 bit by Huffman coding, but only $\frac{7}{2}\log_2 n = 31.4$ bits of codebook.

1.2 An Optimization Heuristic for the Histogram Resolution

We need to optimize b individually for each dimension i such that the overall coding cost $CLRG_i(D, b)$ is minimized. As an efficient and effective heuristic we propose to only consider histogram resolutions where b is a power of 2. This is time efficient since the number of alternative results is logarithmic in n (as we will show), and the next coarser histogram can be intelligently gained from the previous. In addition, the strategy is effective since a sufficient number of alternative results is examined.

We start with a histogram resolution based on the worst-case assumption that (almost) all objects fall into the same histogram bin of a histogram of very high resolution b_m. That means that the log-likelihood approaches 0. The offset cost corresponds to $-n \log_2 b_m$ but the parameter cost are very high: $\frac{b_m-1}{2} \log_2 n$. The other extreme case is the model with the lowest possible resolution $b = 1$ having no log-likelihood, no offset-cost and no parameter cost. The histogram with resolution b_m can pay off only if the following condition holds:

$$n \log_2 b_m \geq \frac{b_m - 1}{2} \log_2 n,$$

which is certainly true if $b_m \leq n/2$. We use $b_m = 2^{\lfloor \log_2 n \rfloor}$, the first power of two less or equal n as starting resolution. Then, in each step, the algorithm generates a new histogram $H' = (H'_1, ..., H'_{b/2})$ from the previous histogram $H = (H_1, ..., H_b)$ by merging each pair of adjacent bins using $H'_j = H_{2j-1} + H_{2j}$ for all j having $1 \leq j \leq b/2$. The overall number of adding operations for histogram bins starting from the histogram $H^{\text{start}} = (H_1^{\text{start}}, ..., H_{b_m}^{\text{start}})$ to the final histogram $H^{\text{end}} = (H_1^{\text{end}})$ corresponds to:

$$\sum_{1 \leq i \leq \frac{b_m}{2}} i \;=\; b_m - 1 \;=\; 2^{\lfloor \log_2 n \rfloor} - 1 \;\in\; O(n).$$

The coding cost of the data with respect to each alternative histogram is evaluated as described in Section 1.1 and the histogram with resolution b_{opt} providing the best compression is reported as result for dimension i. In the case of $b_{\text{opt}} = 1$, no compression was achieved by assuming non-Gaussianity. After having optimized b_{opt} for each dimension i separately, $CLRG_i(D, b)$, the coding costs of the data for dimension i are provided as in Section 1.1 applying b_{opt}. To measure the overall improvement in compression achieved by ICA $CLRG_i(D, b)$ is summed up across all dimensions i:

$$CLRG(D) = \sum_{1 \leq i \leq d} \left(\min_{0 \leq \log_2 b \leq \lfloor \log_2 n \rfloor} CLRG_i(D, b) \right).$$

2 Experiments

Selection of the Relevant Dimensions. Which ICs truly represent meaningful signals? Measures like kurtosis, skewness and other approximations of neg-entropy are often used for selecting the relevant ICs but need to be suitably thresholded, which is a non-trivial task. Figures 2(a)-(b) display the recall of signal identification for a data set consisting of highly non-Gaussian sawtooth signals and a varying number of noise dimensions for various thresholds of kurtosis. Kurtosis is measured as the absolute deviation from Gaussianity. The

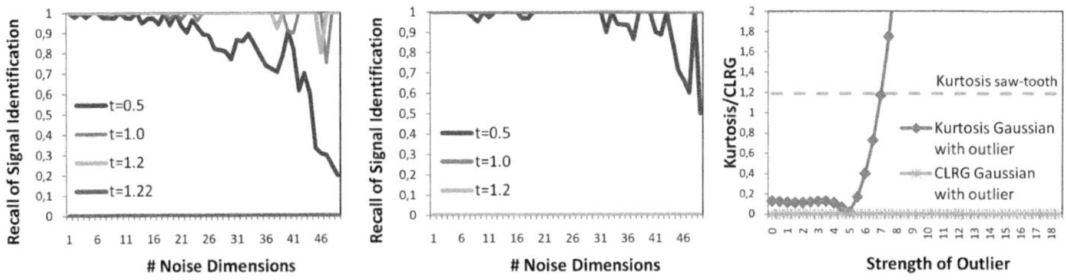

(a) Dimensionality: 200 samples (b) Dimensionality: 500 samples (c) Outlier Robustness

Fig. 2. Comparison of CLRG to kurtosis for selection of relevant ICs from high-dimensional data (a)-(b) and for outlier-robust estimation of IC quality (c)

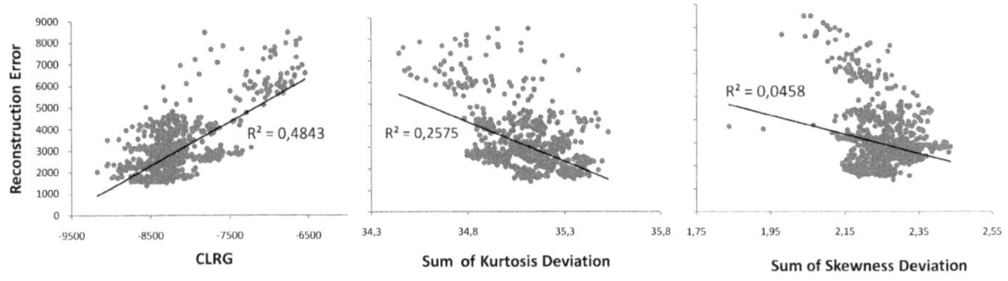

Fig. 3. CLRG in comparison to kurtosis and skewness for assessing the quality of ICA-results. For 1,000 results obtained with FastIca on de-mixing speech signals, CLRG best correlates with the reconstruction error of the ICs.

recall of signal identification is defined as the number of signals which have been correctly identified by the selection criterion divided by the overall number of signals. Figure 2 (a) displays the results for various thresholds on a data set with 200 samples. For this signal length, a threshold of $t = 1.2$ offers the best recall in signal identification for various numbers of noise dimensions. A slightly higher threshold of 1.22 leads to a complete break down in recall to 0, which implies that all noise signals are rated as interesting by kurtosis. For the data set of 500 samples, however, $t = 1.0$ is a suitable threshold and for $t = 1.2$ we can observe a complete break down in recall. Even on these synthetic examples with a very clear distinction into interesting highly non-Gaussian signals and Gaussian noise, the range for suitable thresholding is very narrow. Moreover, the threshold depends on the signal length and of course strongly on the type of the particular signal. Supported by information theory, CLRG automatically identifies the relevant dimensions without requiring any parameters or thresholds. For all examples, CLRG identifies the relevant dimensions as those dimensions allowing data compression with a precision and a recall of 100%.

Stable Estimation of the IC Quality. Commonly used approximations of negentropy are sensitive to single outliers. Outliers may cause an over-estimation of the quality of the IC. CLRG is an outlier-robust measure for the interestingness of a signal. Figure 2(c) displays the influence of one single outlier on the kurtosis (displayed in terms of deviation from Gaussian) and CLRG of a Gaussian noise signal with 500 samples w.r.t. various outlier strengths (displayed in units of standard deviation). For reference, also the kurtosis of a highly non-Gaussian saw-tooth signal is displayed with a dotted line. Already for moderate outlier strength, the estimation of kurtosis becomes unstable. In case of a strong single outlier, kurtosis severely overestimates the interestingness of the signal. CLRG is not sensitive w.r.t. single outliers: Even for strongest outliers, the noise signal is scored as not interesting with a CLRG of zero. For comparison, the saw-tooth curve allows an effective data compression with a CLRG of -553.

Comparing ICA Results. CLRG is a very general criterion for assessing the quality of ICA results which does not rely on any assumptions specific to certain algorithms. In this experiment, we compare CLRG to kurtosis and skewness on the benchmark data set acspeec16 form ICALAB (`http://www.bsp.brain.riken.go.jp/ICALAB/ICALABSignalProc/benchmarks/`). This data set consists of 16 speech signals which we mixed with a uniform random mixing matrix. Figure 3 displays 1,000 results of FastIca [6] generated with the non-linearity tanh and different random starting conditions. For each result, we computed the reconstruction error as the sum of squared deviations of the ICs found by FastIca to the original source signals. For each IC we used the best matching source signal (corrected for sign ambiguity) and summed up the squared deviations.

Figure 3 (left) shows that CLRG correlates best with the reconstruction error. Especially ICA results with a low reconstruction error also allow effective data compression.

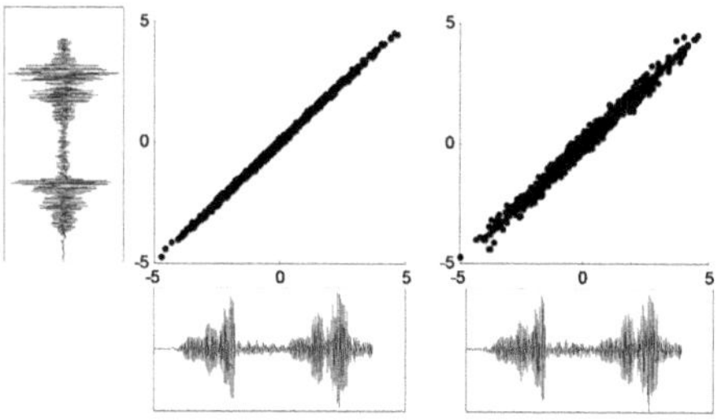

Fig. 4. Reconstruction error of first IC from the result best scored by CLRG and matching IC from the result best scored by kurtosis

For comparison, we computed the sum of kurtosis deviations and the sum of skewness deviations from Gaussianity. Kurtosis and even more skewness show only a slight correlation with the reconstruction error. As an example, Figure 4 (left) shows the first extracted IC from the result best scored by CLRG and the corresponding IC (right) from the result best scored by kurtosis. For each of the two ICs the scatter plots with the original signal are displayed. Obviously, the left IC better matches the true signal than the right IC resulting in a lower reconstruction error.

Selecting Interesting Components from fMRI Data. Functional magnetic resonance imaging (fMRI) yields time series of 3-dimensional volume images allowing to study the brain activity, usually while the subject is performing some task. In this experiment, a subject has been visually stimulated in a block-design

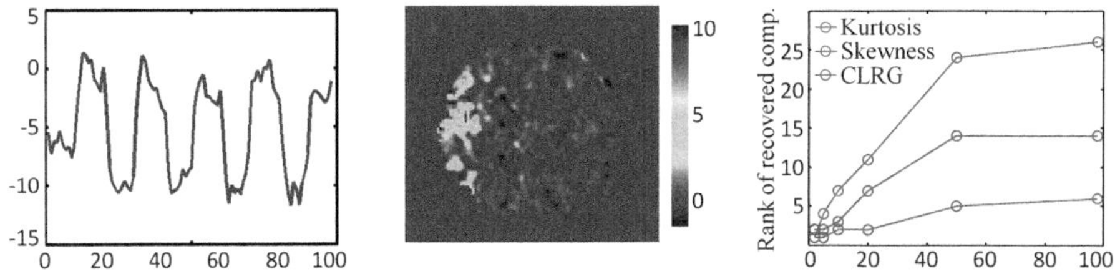

(a) Task-related component (b) Component localization (c) Varying dim. reduction

Fig. 5. fMRI experiment

by alternately displaying a checkerboard stimulus and a central fixation point on a dark background as control condition [7]. fMRI data with 98 images (TR/TE = 3000/60 msec) were acquired with five stimulation and rest periods having each a duration of 30s. After standard pre-processing the dimensionality has been reduced with PCA. FastIca has been applied to extract the task-related component. Figure 5(a) displays an example component with strong correlation to the experimental paradigm. This component is localized in the visual cortex which is responsible for processing photic stimuli, see Figure 5(b). We compared CLRG to kurtosis and skewness w.r.t. their scoring of the task-related component. In particular, we performed PCA reductions with varying dimensionality and identified the component with the strongest correlation to the stimulus protocol. Figure 5(c) shows that CLRG scores the task-related component much better than skewness and kurtosis. Regardless of the dimensionality, the task-related component is always among the top-ranked components by CLRG, in most cases among the top 3 to 5. By kurtosis and skewness, the interestingness of task-related component often rated close to the average.

3 Conclusions and Outlook

In this paper, we introduced CLRG (Code Length Relative to Gaussianity) as an information-theoretic measure to evaluate the quality of single independent components as well as complete ICA results. The basic idea that a good model provides efficient data compression is very general. Therefore, not only different ICs and ICA results obtained by different algorithms can be unbiasedly compared. Given a data set, we can also compare the quality completely different models, e.g. obtained by ICA, PCA and Projection Pursuit. Moreover, it might lead to the best data compression to apply different models to different subsets of the dimensions as well as different subsets of the data objects. In our ongoing and future work we will extend CLRG to support various models and will explore algorithms for finding subsets of objects and dimensions which can be effectively compressed together.

References

1. Barron, A.R., Rissanen, J., Yu, B.: The minimum description length principle in coding and modeling. IEEE Transactions on Information Theory 44(6), 2743 (1998)
2. Pelleg, D., Moore, A.: X-means: Extending K-means with efficient estimation of the number of clusters. In: ICML Conference, pp. 727–734 (2000)
3. Lee, T.C.M.: Regression spline smoothing using the minimum description length principle. Statistics and Probability Letters 48, 71–82 (2000)
4. Böhm, C., Faloutsos, C., Plant, C.: Outlier-robust clustering using independent components. In: SIGMOD Conference, pp. 185–198 (2008)
5. Lee, T.W., Lewicki, M.S.: The generalized gaussian mixture model using ICA. In: ICA Conference (2000)
6. Hyvärinen, A.: Fast and robust fixed-point algorithms for independent component analysis. IEEE Transactions on Neural Networks 10(3), 626–634 (1999)
7. Wismüller, A., Lange, O., Dersch, D.R., Leinsinger, G.L., Hahn, K., Pütz, B., Auer, D.: Cluster analysis of biomedical image time-series. Int. J. Comput. Vision 46(2), 103–128 (2002)

Blind Source Separation of Overdetermined Linear-Quadratic Mixtures

Leonardo T. Duarte[1*], Ricardo Suyama[2], Romis Attux[1], Yannick Deville[3], João M.T. Romano[1], and Christian Jutten[4]

[1] DSPCom Lab - Universtity of Campinas (Unicamp), Campinas, Brazil
{ltduarte,romano}@dmo.fee.unicamp.br, attux@dca.fee.unicamp.br
[2] Enginnering Modeling and Applied Social Sciences, UFABC, Santo André, Brazil
ricardo.suyama@ufabc.edu.br
[3] LATT, Université de Toulouse, CNRS, Toulouse, France
ydeville@ast.obs-mip.fr
[4] GIPSA-Lab, CNRS UMR-5216, Grenoble, and Institut Universitaire de France
christian.jutten@gipsa-lab.grenoble-inp.fr

Abstract. This work deals with the problem of source separation in overdetermined linear-quadratic (LQ) models. Although the mixing model in this situation can be inverted by linear structures, we show that some simple independent component analysis (ICA) strategies that are often employed in the linear case cannot be used with the studied model. Motivated by this fact, we consider the more complex yet more robust ICA framework based on the minimization of the mutual information. Special attention is given to the development of a solution that be as robust as possible to suboptimal convergences. This is achieved by defining a method composed of a global optimization step followed by a local search procedure. Simulations confirm the effectiveness of the proposal.

1 Introduction

An interesting extension of the classical Blind Source Separation (BSS) framework concerns the case in which the mixing model is nonlinear [1]. One of the motivations for studying nonlinear BSS comes from the observation that, in some applications, the mixing process is clearly nonlinear. This is common, for instance, in chemical sensor arrays [2,3].

Nonlinear BSS, in its most general formulation, cannot be dealt with using independent component analysis (ICA) methods [1,4]. Indeed, if no constraints are imposed, one can set up a nonlinear system that provides independent components that are still mixed versions of the sources [4]. This result suggests that, instead of searching for a general framework, nonlinear BSS should be treated on a case-by-case basis by focusing on relevant classes of nonlinear models. Having this in mind, we tackle in this work the problem of BSS in the so-called linear-quadratic (LQ) model [5]. This class of models is appealing both in a practical

* L. T. Duarte would like to thank FAPESP for the financial support.

V. Vigneron et al. (Eds.): LVA/ICA 2010, LNCS 6365, pp. 263–270, 2010.

context —for instance, it is used in the design of gas sensor arrays [3] —and in a theoretical standpoint —it paves the way for dealing with polynomial mixtures.

A major issue in the development of BSS methods for LQ mixtures concerns the definition of the separating system structure. In a determined case (equal number of sources and mixtures), this problem is indeed tricky due to the difficulty in expressing the inverse of the mixing mapping in an analytical form. Possible solutions to this problem can be found in the nonlinear recurrent networks proposed in [5,6] or in the Bayesian approach of [7]. Moreover, in some particular cases —for instance when there are two sources and two mixtures — one can indeed find the inverse nonlinear mapping [5].

A second route for dealing with LQ mixtures relies on the following observation: when there are more mixtures than sources (overdetermined case), the inversion of the LQ mixing model becomes simpler as it can be performed using linear separating systems. Evidently, such a simplification opens the way for well-established ICA methods developed for the linear case. Furthermore, although we restrict our analysis to the LQ case, such a simplification is also interesting in the more general case of polynomial mixtures.

Even if the idea of separating LQ mixtures through linear ICA methods is not novel, the works that have exploited it focused on particular cases, such as sources in a finite alphabet [8] or circular sources [9]. In the present paper, however, we consider a more general framework, in which the only assumption made is that the sources are mutually statistically independent. The main difficulty here lies in the fact that, although overdetermined LQ models may admit a linear inverse, classical ICA strategies may not be able to separate LQ mixtures. Motivated by these difficulties, we develop an ICA method specially tailored for the considered problem.

2 Overdetermined Linear-Quadratic Mixing Model

Let us consider a problem with two sources[1] s_1 and s_2, which are assumed to be mutually statistically independent. In a LQ model, the i-th mixture is given by

$$x_i = a_{i1}s_1 + a_{i2}s_2 + a_{i3}s_1s_2, \ \forall i \in 1, \ldots, n_m, \tag{1}$$

where a_{ij} represents a mixing coefficient and n_m is the number of mixtures. The model (1) can be alternatively described through the following vector notation

$$\begin{bmatrix} x_1 \\ \vdots \\ x_{n_m} \end{bmatrix} = \begin{bmatrix} a_{11} & a_{12} & a_{13} \\ \vdots & \vdots & \vdots \\ a_{n_m1} & a_{n_m2} & a_{n_m3} \end{bmatrix} \begin{bmatrix} s_1 \\ s_2 \\ s_1s_2 \end{bmatrix}. \tag{2}$$

This representation suggests an insightful interpretation of the LQ model: it can be seen as a special case of a linear mixing model, in which the sources are given by s_1, s_2 and $s_3 = s_1s_2$ and, therefore, are no longer independent.

[1] This scenario is representative in the design of gas sensor arrays as one usually has binary mixtures of gases.

When the number of LQ mixtures is $n_m = 2$ (determined), there is no advantage in expressing the original LQ problem in a linear formulation. In fact, besides the presence of dependent sources, the resulting dual linear problem is underdetermined (less mixtures than sources). Performing BSS in such a scenario is quite difficult and requires the incorporation of further information.

Conversely, if, for instance[2], $n_m = 3$ (overdetermined LQ model), the resulting mixing matrix in (2) becomes square and, thus, can be inverted as follows

$$\begin{bmatrix} y_1 \\ y_2 \\ y_3 \end{bmatrix} = \begin{bmatrix} w_{11} & w_{12} & w_{13} \\ w_{21} & w_{22} & w_{23} \\ w_{31} & w_{32} & w_{33} \end{bmatrix} \begin{bmatrix} x_1 \\ x_2 \\ x_3 \end{bmatrix}, \tag{3}$$

where $\mathbf{y} = [y_1\ y_2\ y_3]^T$ represents the retrieved sources. That is, one can overcome the problem of how to define an LQ separating system by simply adding sensors into the array. Even better, the solution in this case is given by a matrix.

Of course, there remains the problem of how to find a separating matrix in the case of LQ mixtures. In a recent work, Castella [8] showed that, if the sources belong to a finite alphabet, cumulant-based ICA techniques, such as the JADE algorithm [10], can be used to adjust $\mathbf{W}$ in (3), despite the presence of mutually dependent sources in the linear formulation of Equation (2). The proposed approach in [8] is thus able to retrieve s_1, s_2 and $s_1 s_2$.

In the more general case of continuous sources, the presence of dependent sources in (2) does not allow one to apply ICA methods to adjust $\mathbf{W}$ in (3). Given that, instead of searching for the three sources s_1, s_2 and $s_3 = s_1 s_2$, we try to directly estimate s_1 and s_2 via a rectangular separating matrix, as follows

$$\begin{bmatrix} y_1 \\ y_2 \end{bmatrix} = \begin{bmatrix} w_{11} & w_{12} & w_{13} \\ w_{21} & w_{22} & w_{23} \end{bmatrix} \begin{bmatrix} x_1 \\ x_2 \\ x_3 \end{bmatrix}. \tag{4}$$

Structurally speaking, this separating system is also able to retrieve s_1 and s_2, possibly permuted and/or scaled. Indeed, this is achieved for all $\mathbf{A}$

$$\begin{bmatrix} w_{11} & w_{12} & w_{13} \\ w_{21} & w_{22} & w_{23} \end{bmatrix} = \mathbf{P} \begin{bmatrix} \alpha & 0 & 0 \\ 0 & \beta & 0 \end{bmatrix} \mathbf{A}^{-1}, \tag{5}$$

where $\mathbf{P}$ is a permutation matrix, and α and β are non-zero values representing a possible scaling of the retrieved signals. In the sequel, we discuss the use of ICA methods to adapt the rectangular matrix $\mathbf{W}$ in (4).

3 Toward a Linear ICA Algorithm for Overdetermined LQ Mixtures

ICA-based learning rules search for a matrix $\mathbf{W}$ that provides independent signals y_1 and y_2. At first glance, ICA techniques that are used in linear overdetermined models could be considered to separate LQ mixtures through (4). However, as it will be discussed in the sequel, the underlying nonlinear nature of the mixing process makes the application of some common ICA strategies difficult.

[2] In the rest of the paper, we restrict our analysis to the case of $n_m = 3$ mixtures.

3.1 Limitations of ICA Methods Based on Whitening as a Pre-processing Step

Often, ICA in overdetermined linear models is carried out in two steps. Firstly, the mixtures undergo a dimension reduction stage in order to obtain signals with dimension equal to the one of the sources —this is usually done via whitening[3] [10]. Then, ICA methods designed for determined models are applied.

For this two-step solution to work in the case of LQ mixtures, the process of dimension reduction should remove any trace of nonlinear mixing between the sources. Unfortunately, this cannot be achieved via whitening. To illustrate that, let us consider, as a working example, an LQ model ($n_m = 3$ mixtures and 2 uniformly distributed sources), where the mixing matrix (see the linear formulation of Equation (2)) is given by $\mathbf{A} = [1\ 0.7\ 0.3\ ;\ 0.6\ 1\ 0.5\ ;\ 0.5\ 0.5\ 0.6]$. We checked through simulations that the matrix $\mathbf{Q} = [8.67\ -1.21\ -9.28\ ;\ -1.21\ 8.47\ -8.39]$ provides a white two-dimensional signal. Yet, the combined system $\mathbf{QA}$ in this case is given by

$$\mathbf{QA} = \begin{bmatrix} 3.30 & 0.21 & -3.57 \\ -0.33 & 3.43 & -1.17 \end{bmatrix}, \tag{6}$$

that is, the nonlinear term $s_1 s_2$ remains in the whitened data.

3.2 Limitations of Natural Gradient Learning

From the last section, a more reasonable approach is to consider overdetermined ICA methods that do not require a whitening step. A possibility in this case can be found in the natural gradient algorithm. Although originally developed to operate in linear determined models, this method also works in the overdetermined case [11]. The learning rule in this case is given by

$$\mathbf{W} \leftarrow \mathbf{W} + \mu(\mathbf{I} - E\{f(\mathbf{y})\mathbf{y}^T\})\mathbf{W}, \tag{7}$$

where μ is the step size, $\mathbf{I}$ represents the identity matrix and $f(\cdot)$ is a nonlinear function that should be previously defined based on the source distributions[4] [10]. Given that (7) converges when $E\{f(\mathbf{y})\mathbf{y}^T\} = \mathbf{I}$, this learning rule is somehow trying to retrieve components that are nonlinearly decorrelated, a necessary but not sufficient condition for statistical independence (except if each component of $f(\mathbf{y})$ is the score function of the corresponding component of $\mathbf{y}$).

We tested (7) in the same working example as in the last section. The estimated matrix $\mathbf{W}$ in this case indeed provided nonlinearly decorrelated components satisfying $E\{f(\mathbf{y})\mathbf{y}^T\} = \mathbf{I}$ —we considered cubic functions $f(y_i) = y_i^3$,

---

[3] Whitening a vector $\mathbf{x}$ means finding a matrix $\mathbf{Q}$ that provides a vector $\mathbf{z} = \mathbf{Q}x$ whose covariance matrix is diagonal. Dimension reduction through whitening is based on the observation that the whitening matrix $\mathbf{Q}$ depends on the covariance matrix of $\mathbf{x}$, i.e. $\mathbf{R}_x$. Given that, one can have a lower dimensional vector $\mathbf{z}$ by only considering the eigenvectors associated with the largest eigenvalues of $\mathbf{R}_x$.

[4] Ideally, these functions should be as close as possible to the source score functions. However, even a rough approximation is enough to guarantee source separation in determined linear models.

which are typically used for sub-Gaussian sources [10]. However, the mixtures were not separated. This is shown in Figure 1, which depicts the joint distribution of the original sources and of the retrieved signals. It is interesting to note here that, although nonlinearly decorrelated, the retrieved signals are not statistically independent. That is, unlike in the linear case, the nonlinear decorrelation is not a safe route for independence in overdetermined LQ models.

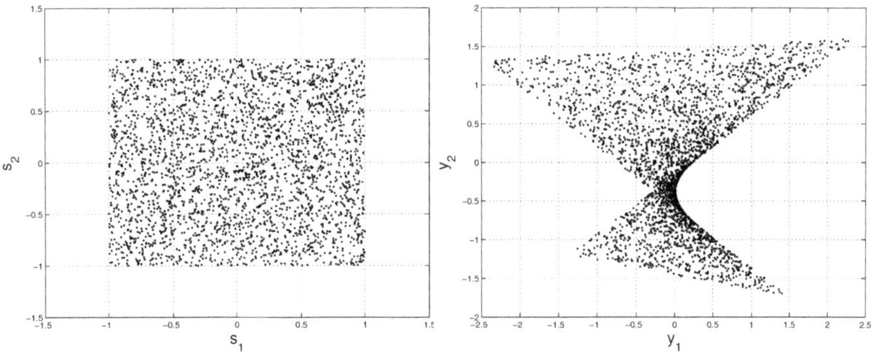

(a) Sources.

(b) Recovered signals.

Fig. 1. Application of natural gradient algorithm: scatter plots

3.3 Methods Based on the Minimization of the Mutual Information

We now consider a framework based on the minimization of the mutual information between the elements of $\mathbf{y}$, which is given by

$$I(\mathbf{y}) = H(y_1) + H(y_2) - H(\mathbf{y}), \tag{8}$$

where $H(\cdot)$ denotes the differential entropy [10]. Unlike in the nonlinear correlation, the mutual information offers a necessary but also sufficient condition for independence since it becomes null if and only if y_1 and y_2 are independent.

In [12], a framework to derive methods that minimize the mutual information[5] was introduced. Its application to linear models results in the following learning rule

$$\mathbf{W} \leftarrow \mathbf{W} + \mu E\{\beta_{\mathbf{y}}(\mathbf{y})\mathbf{x}^T\}, \tag{9}$$

where the i-th element of $\beta_{\mathbf{y}}(\mathbf{y})$, the (opposite of the) so-called score function difference vector of $\mathbf{y}$, is given by $\beta_{y_i}(y_i) = (-\partial \log p(\mathbf{y})/\partial y_i) - (-d \log p(y_i)/dy_i)$. We applied the method proposed in [13] to estimate this vector.

[5] Usually, the derivation of methods based on the mutual information makes use of a common trick to avoid the estimation of the joint term $H(\mathbf{y})$. They express it in terms of $H(\mathbf{x})$ by using the entropy transformation law [10]. However, we cannot use this strategy because $\mathbf{W}$ is not invertible in our case.

After performing some tests, the algorithm (9) was able to recover the original sources in some runs. However, we also noticed that in many trials the method only provided poor estimation. One could give two reasons for such a bad performance: either the algorithm is getting trapped in spurious minima and, thus, it is an optimization issue, or the considered model is not *separable* in the sense of ICA, i.e. retrieving independent components does not assure source separation. Note that, while the first issue could be solved by developing algorithms robust to local convergence, the second one would pose a serious problem as any attempt to perform BSS through ICA would become questionable.

To gain more insight into that question, we performed a series of tests with (9). At the end of each run, we estimated the average signal-to-interference (SIR) ratio[6] and the mutual information between the retrieved signals y_1 and y_2 —we considered the estimator proposed in [14]. The results obtained after 20 realizations —with uniformly distributed sources, mixing coefficients drawn from a normal distribution and random initialization of the separating matrix $\mathbf{W}$ —are plotted in Figure 2(a), in which each mark represents one realization. Note that when a low SIR was observed, the retrieved signals were still dependent as their mutual information was not null. This is an indicator that bad convergence here comes from the optimization itself and not from a separability problem.

4 A Robust ICA Method for Overdetermined LQ Mixtures

The results shown in Figure 2(a) revealed that the gradient-based learning rule of (9) may converge to local minima. A first possibility to deal with this problem is to consider global optimization methods such as evolutionary algorithms (EA). These methods are based on the notion of population, i.e. a set of possible candidate solutions (individuals) for the problem. At each iteration, new individuals are created from this population and, typically, the set of individuals that provides a better solution to the optimization problem is kept to the next iterations (selection). This population-based search gives EAs the ability of finding the global solution even when applied to multimodal cost functions.

The robustness to sub-optimal convergence in EAs comes at heavy computational burden. This is particularly problematic in the definition of an EA to perform ICA according to the minimum mutual information principle. Indeed, estimating the mutual information via accurate methods, such as the one presented in [14], is time demanding, and, since an EA performs many evaluations of the cost function during its execution, one may end up with a too slow method.

As an alternative to a direct application of an EA in our problem, we propose a hybrid scheme composed of two successive steps. Firstly, we indeed make use of an EA technique, the opt-aiNet algorithm (see [15] for details), to minimize the mutual information. However, instead of relying on a precise estimation of

[6] The SIR associated with a source and its estimate is given by: $\text{SIR}_i = 10\log\left(E\{\hat{s}_i^2\}/E\{(\hat{s}_i - \hat{y}_i)^2\}\right)$, where $\hat{s}_i$ and $\hat{y}_i$ denote, respectively, the actual source and its corresponding estimate after mean, variance and sign normalization.

the cost function, we consider the rougher and thus simpler mutual information estimator proposed in [16]. Therefore, this first step provides us with a coarse solution but near the global minimum. Then, in a second step, this coarse solution is refined by applying the learning rule (9).

In order to assess the performance of the proposed hybrid scheme, we conducted a set of simulations in the same scenario as considered in Section 3.3. In Figure 2(b), we show the results obtained after 20 runs. Whereas the simple application of (9) converged to a sub-optimal minimum in 9 out of 20 runs realizations, the proposed hybrid scheme was able to provide good estimates of the sources in 19 our of 20 realizations.

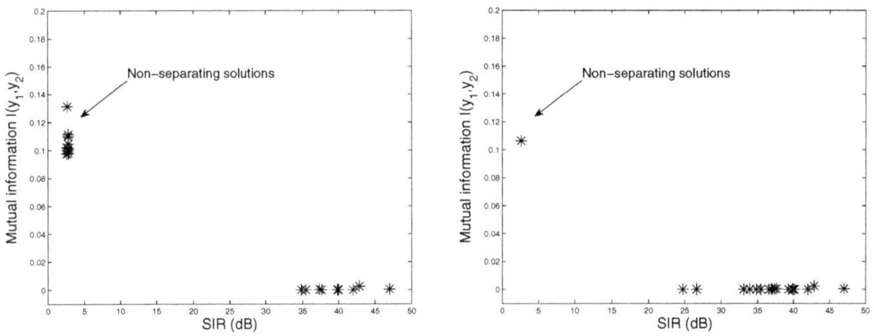

(a) Learning rule (9). (b) Proposed hybrid scheme.

Fig. 2. Analysis of the retrieved signals. Each mark corresponds to one realization.

5 Conclusions

In this work, we addressed the problem of BSS in overdetermined LQ mixtures. In this case, the mixing process can be inverted through linear structures. However, as illustrated by some examples, the application of common ICA strategies is not enough to perform source separation in the studied case. In view of this limitation, we introduced a hybrid scheme composed of a global optimization tool and of a gradient-based method for minimizing the mutual information between the retrieved signals. As checked via simulations, the proposed method is able to almost always avoid convergence to sub-optimal minima

In this first study, separability of overdetermined LQ models was only verified through simulations. As this approach is useful only for gaining some insight into this issue, a first perspective for future works is to study separability on a theoretical basis. A second point that deserves further investigation is related to the transformation of the original overdetermined problem into a determined one. We saw that whitening cannot be used here. Nonetheless, it would be interesting to investigate whether it is possible to obtain a determined system by considering additional prior information. Finally, we intent to extend the results obtained in the present work to scenarios in which the number of sources is larger than two and also to the more general case of polynomial mixtures.

References

1. Jutten, C., Karhunen, J.: Advances in blind source separation (BSS) and independent component analysis (ICA) for nonlinear mixtures. International Journal of Neural Systems 14, 267–292 (2004)
2. Duarte, L.T., Jutten, C., Moussaoui, S.: A Bayesian nonlinear source separation method for smart ion-selective electrode arrays. IEEE Sensors Journal 9(12), 1763–1771 (2009)
3. Bedoya, G.: Nonlinear blind signal separation for chemical solid-state sensor arrays. PhD thesis, Universitat Politecnica de Catalunya (2006)
4. Hyvärinen, A., Pajunen, P.: Nonlinear independent component analysis: existence and uniqueness results. Neural Networks 12, 429–439 (1999)
5. Hosseini, S., Deville, Y.: Blind separation of linear-quadratic mixtures of real sources using a recurrent structure. In: Mira, J., Álvarez, J.R. (eds.) IWANN 2003. LNCS, vol. 2686, Springer, Heidelberg (2003)
6. Deville, Y., Hosseini, S.: Recurrent networks for separating extractable-target nonlinear mixtures. part i: Non-blind configurations. Signal Processing 89, 378–393 (2009)
7. Duarte, L.T., Jutten, C., Moussaoui, S.: Bayesian source separation of linear-quadratic and linear mixtures through a MCMC method. In: Proc. of the IEEE MLSP (2009)
8. Castella, M.: Inversion of polynomial systems and separation of nonlinear mixtures of finite-alphabet sources. IEEE Trans. on Sig. Proc. 56(8), 3905–3917 (2008)
9. Abed-Meraim, K., Belouchiani, A., Hua, Y.: Blind identification of a linear-quadratic mixture of independent components based on joint diagonalization procedure. In: Proceedings of the IEEE ICASSP 1996, vol. 5, pp. 2718–2272 (1996)
10. Comon, P., Jutten, C. (eds.): Handbook of blind source separation, independent component analysis and applications. Academic Press, Elsevier (2010)
11. Zhang, L.Q., Cichocki, A., Amari, S.: Natural gradient algorithm for blind separation of overdetermined mixture with additive noise. IEEE Signal Processing Letters 6(11), 293–295 (2009)
12. Babaie-Zadeh, M., Jutten, C., Nayebi, K.: Differential of the mutual information. IEEE Signal Processing Letters 11(1), 48–51 (2004)
13. Pham, D.T.: Fast algorithm for estimating mutual information, entropies and score functions. In: Proceedings of the ICA, pp. 17–22 (2003)
14. Darbellay, G.A., Vajda, I.: Estimation of the information by an adaptive partitioning of the observation space. IEEE Trans. on Inf. Theory 45(4), 1315–1321 (1999)
15. de Castro, L.N., Timmis, J.: Artificial Immune Systems: A New Computational Approach. Springer, Heidelberg (2002)
16. Moddemeijer, R.: On estimation of entropy and mutual information of continuous distributions. Signal Processing 16(3), 233–248 (1989)

Constrained Complex-Valued ICA without Permutation Ambiguity Based on Negentropy Maximization

Qiu-Hua Lin, Li-Dan Wang, Jian-Gang Lin, and Xiao-Feng Gong

School of Information and Communication Engineering,
Dalian University of Technology, Dalian 116024, China
qhlin@dlut.edu.cn

Abstract. Complex independent component analysis (ICA) has found utility in separation of complex-valued signals such as communications, functional magnetic resonance imaging, and frequency-domain speeches. However, permutation ambiguity is a main problem of complex ICA for order-sensitive applications, e.g., frequency-domain speech separation. This paper proposes a semi-blind complex ICA algorithm based on negentropy maximization. The magnitude correlation of a source signal is utilized to constrain the separation process. As a result, the complex-valued signals are separated without permutation. Experiments with synthetic complex-valued signals, synthetic speech signals, and recorded speech signals are performed. The results demonstrate that the proposed algorithm can not only solve the permutation problem, but also achieve slightly improved separation compared to the standard blind algorithm.

Keywords: Independent component analysis, Complex ICA, Constrained ICA, Magnitude correlation, Negentropy maximization.

1 Introduction

Complex-valued independent component analysis (ICA) has found utility in many applications such as communications [1], analysis of functional magnetic resonance imaging (fMRI) [2], and convolutive source separation in the frequency domain [3].

Many blind complex ICA algorithms have been developed. These algorithms can be roughly divided into two categories: (1) matrix diagonalization based algorithms such as joint approximate diagonalization of eigenmatrices (JADE) [4] and the strongly uncorrelating transform (SUT) [5]. (2) nonlinear functions based algorithms such as the complex fastICA algorithm (CfastICA) [6], the complex kurtosis maximization algorithm [7], and the complex negentropy maximization (CMN) algorithm [8].

However, the blind complex ICA for separating complex-valued signals suffers from the same drawback of permutation ambiguity as blind ICA for separating real-valued signals. The permutation problem needs to be solved for order-sensitive

V. Vigneron et al. (Eds.): LVA/ICA 2010, LNCS 6365, pp. 271–278, 2010.

applications such as frequency-domain speech separation and high efficiency extraction of some desired components from fMRI data [9].

Recent work has suggested that semi-blind ICA can improve the potential of ICA by incorporating prior information into the estimation process [9]. This paper aims to propose a semi-blind complex ICA algorithm to solve the permutation ambiguity by utilizing prior information about the sources. Since the CMN algorithm can perform good and stable separation of both circular and noncircular sources with a wide range of statistics [8], we select CMN as the standard blind ICA to incorporate prior information. Considering that frequency-domain speech separation is especially sensitive to permutation ambiguity, and the amplitudes of the speech signals in neighboring frequency bins are strongly correlated, we here utilize the magnitude correlation of a source signal as a constraint to the CMN algorithm. Experiments with synthetic complex-valued signals, synthetic speech signals, and recorded speech signals are performed to demonstrate the efficacy of the proposed algorithm.

2 The CMN Algorithm

The CMN cost function is defined as [8]:

$$J(\mathbf{w}) = \max_{\|\mathbf{w}\|^2=1} E\left\{\left|G\left(\mathbf{w}^H\mathbf{x}\right)\right|^2\right\} \tag{1}$$

where $\mathbf{x} = [x_1,\ldots,x_n]^T$ are n observed mixtures of n unknown complex-valued independent source signals $\mathbf{s} = [s_1,\ldots,s_n]^T$, and $\mathbf{x}$ is prewhitened, i.e., $E\{\mathbf{x}\mathbf{x}^H\} = \mathbf{I}$; "$H$" denotes the conjugate transpose; $\mathbf{w}$ is a column of the unmixing matrix $\mathbf{W}$; $G(\cdot)$ is a nonlinear complex analytic function such as polynomials or transcendental functions. The optimal nonlinearity is the log of the joint distribution of the source being estimated. Compared to the CfastICA contrast function using only the amplitude of the complex-valued signals [6], CMN utilizes the full complex nature of signals including amplitude and phase. This property provides the advantage of generating an asymmetric class of functions that cannot be realized with CfastICA [8].

The complex Newton update used in [8] is:

$$\Delta\tilde{\mathbf{w}} = -\left(\left.\frac{\partial^2 f}{\partial\tilde{\mathbf{w}}^*\partial\tilde{\mathbf{w}}^T}\right|_{\mathbf{w}=\mathbf{w}_n}\right)^{-1}\left.\frac{\partial f}{\partial\tilde{\mathbf{w}}^*}\right|_{\mathbf{w}=\mathbf{w}_n} = -\tilde{\mathbf{H}}_f^{-1}\tilde{\nabla}_f^* \tag{2}$$

where $\Delta\tilde{\mathbf{w}} = \tilde{\mathbf{w}}_{n+1} - \tilde{\mathbf{w}}_n$, $\tilde{\mathbf{H}}_f$ is the complex Hessian, $\tilde{\nabla}_f$ is the complex gradient, "*" denotes the conjugate, and vectors are in the augmented form with the following definition:

$$\begin{aligned}\mathbf{z} &= [z_1,\ldots,z_N]^T \in C^N \\ \tilde{\mathbf{z}} &= [z_1, z_1^*,\ldots,z_N, z_N^*]^T \in C^{2N}\end{aligned} \tag{3}$$

The quasi-Newton algorithm for the optimization problem in (1) is derived as follows ($\mathbf{w}^+$ and $\mathbf{w}$ are used to replace $\mathbf{w}_{n+1}$ and $\mathbf{w}_n$ for brief) [8]:

$$\mathbf{w}^+ = -E\{G^*(y)G'(y)\mathbf{x}\} + E\{G'(y)G'^*(y)\}\mathbf{w} + E\{\mathbf{x}\mathbf{x}^T G^*(y)G''(y)\}\mathbf{w}^*$$
$$\mathbf{w}_{new} = \frac{\mathbf{w}^+}{\|\mathbf{w}^+\|} \tag{4}$$

where $y = \mathbf{w}^H\mathbf{x}$ and " $'$ " denotes the derivative.

3 The Proposed Algorithm

In the following, we derive a semi-blind CMN algorithm by utilizing the magnitude correlation of the source signals within the framework of constrained ICA [10]. The magnitude correlation is specifically used as an inequality constraint of the CMN cost function in (1):

$$\begin{aligned} &\underset{\|\mathbf{w}\|^2=1}{maximize} \quad J(\mathbf{w}) = E\left\{\left|G\left(\mathbf{w}^H\mathbf{x}\right)\right|^2\right\} \\ &s.t. \qquad g\left(\left|\mathbf{w}^H\mathbf{x}\right|\right) \le 0 \end{aligned} \tag{5}$$

where $g(|\,\mathbf{w}^H\mathbf{x}\,|)$ is the inequality constraint defined as:

$$g(|\,\mathbf{w}^H\mathbf{x}\,|) = \varepsilon(|\,\mathbf{w}^H\mathbf{x}\,|,|\,r\,|) - \xi \le 0 \tag{6}$$

and

$$\varepsilon(|\,\mathbf{w}^H\mathbf{x}\,|,|\,r\,|) = -E\{|\,\mathbf{w}^H\mathbf{x}\,|^2 \cdot |\,r\,|^2\} \tag{7}$$

where r is a reference signal, the magnitude of which has maximal correlation with that of an estimate y; ξ is a threshold to distinguish the estimate y from the others.

Based on the constrained optimization theory, we can optimize (5) using the following augmented Lagrangian function:

$$\overline{L}(\mathbf{w},\gamma,\mu) = J(\mathbf{w}) + \lambda\left(\mathbf{w}^H\mathbf{w} - 1\right) - \frac{1}{2\gamma}\left\{\max{}^2\left\{\gamma g\left(\left|\mathbf{w}^H\mathbf{x}\right|\right) + \mu, 0\right\} - \mu^2\right\} \tag{8}$$

where γ is a penalty parameter, μ is a Lagrange multiplier.

Rewrite (8) as:

$$\overline{L}(\mathbf{w},\gamma,\mu) = \overline{L}'(\mathbf{w},\gamma,\mu) + \lambda\left(\mathbf{w}^H\mathbf{w} - 1\right) \tag{9}$$

and refer to the complex Newton update in (2), we have $\tilde{\mathbf{H}}_{\overline{L}} = \tilde{\mathbf{H}}_{\overline{L}'} + \lambda\tilde{\mathbf{I}}$, $\tilde{\nabla}^*_{\overline{L}} = \tilde{\nabla}^*_{\overline{L}'} + \lambda\tilde{\mathbf{w}}$, and $\Delta\tilde{\mathbf{w}} = -\left(\tilde{\mathbf{H}}_{\overline{L}'} + \lambda\tilde{\mathbf{I}}\right)^{-1}\left(\tilde{\nabla}^*_{\overline{L}'} + \lambda\tilde{\mathbf{w}}\right)$, and then we derive

$$\left(\tilde{\mathbf{H}}_{\overline{L}'} + \lambda \tilde{\mathbf{I}}\right)\tilde{\mathbf{w}}^{+} = -\tilde{\nabla}^{*}_{\overline{L}'} + \tilde{\mathbf{H}}_{\overline{L}'}\tilde{\mathbf{w}} \tag{10}$$

and

$$\begin{aligned} \mathbf{H}_{\overline{L}'}\mathbf{w} = {} & E\left\{G'\left(\mathbf{w}^{H}\mathbf{x}\right)G''^{*}\left(\mathbf{w}^{H}\mathbf{x}\right)\right\}\mathbf{w} + E\left\{\mathbf{x}\mathbf{x}^{T}G^{*}\left(\mathbf{w}^{H}\mathbf{x}\right)G''\left(\mathbf{w}^{H}\mathbf{x}\right)\right\}\mathbf{w}^{*} \\ & - \mu E\left\{g'\left(\left|\mathbf{w}^{H}\mathbf{x}\right|\right)\right\}\mathbf{w} \\ \tilde{\nabla}^{*}_{\overline{L}'} = {} & E\left\{G^{*}\left(\mathbf{w}^{H}\mathbf{x}\right)G'\left(\mathbf{w}^{H}\mathbf{x}\right)\mathbf{x}\right\} - \mu E\left\{\mathbf{x}\left(\mathbf{w}^{H}\mathbf{x}\right)^{*} g'\left(\left|\mathbf{w}^{H}\mathbf{x}\right|\right)\right\} \end{aligned} \tag{11}$$

Combining (10) and (11), we obtain:

$$\begin{aligned} K\mathbf{w}^{+} = {} & -\left(E\left\{G^{*}\left(\mathbf{w}^{H}\mathbf{x}\right)G'\left(\mathbf{w}^{H}\mathbf{x}\right)\mathbf{x}\right\} - \mu E\left\{\mathbf{x}\left(\mathbf{w}^{H}\mathbf{x}\right)^{*} g'\left(\left|\mathbf{w}^{H}\mathbf{x}\right|\right)\right\}\right) \\ & + E\left\{G'\left(\mathbf{w}^{H}\mathbf{x}\right)G''^{*}\left(\mathbf{w}^{H}\mathbf{x}\right)\right\}\mathbf{w} + E\left\{\mathbf{x}\mathbf{x}^{T}G^{*}\left(\mathbf{w}^{H}\mathbf{x}\right)G''\left(\mathbf{w}^{H}\mathbf{x}\right)\right\}\mathbf{w}^{*} \\ & - \mu E\left\{g'\left(\left|\mathbf{w}^{H}\mathbf{x}\right|\right)\right\}\mathbf{w} \end{aligned} \tag{12}$$

where $K = \left(\mathbf{H}_{\overline{L}'} + \lambda\mathbf{I}\right)$ and can be removed due to the subsequent normalization of $\mathbf{w}$ to unit norm. Thus the learning rule for the proposed algorithm becomes

$$\begin{aligned} \mathbf{w}^{+} = {} & -\left(E\left\{G^{*}\left(\mathbf{w}^{H}\mathbf{x}\right)G'\left(\mathbf{w}^{H}\mathbf{x}\right)\mathbf{x}\right\} - \mu E\left\{\mathbf{x}\left(\mathbf{w}^{H}\mathbf{x}\right)^{*} g'\left(\left|\mathbf{w}^{H}\mathbf{x}\right|\right)\right\}\right) \\ & + E\left\{G'\left(\mathbf{w}^{H}\mathbf{x}\right)G''^{*}\left(\mathbf{w}^{H}\mathbf{x}\right)\right\}\mathbf{w} + E\left\{\mathbf{x}\mathbf{x}^{T}G^{*}\left(\mathbf{w}^{H}\mathbf{x}\right)G''\left(\mathbf{w}^{H}\mathbf{x}\right)\right\}\mathbf{w}^{*} \\ & - \mu E\left\{g'\left(\left|\mathbf{w}^{H}\mathbf{x}\right|\right)\right\}\mathbf{w} \\ \mathbf{w}_{new} = {} & \frac{\mathbf{w}^{+}}{\left\|\mathbf{w}^{+}\right\|} \end{aligned} \tag{13}$$

where the Lagrange multiplier μ is updated as:

$$\mu^{+} = \max\left\{\gamma g\left(\left|\mathbf{w}^{H}\mathbf{x}\right|\right) + \mu, 0\right\}. \tag{14}$$

4 Experiments and Results

To verify the efficacy of the proposed algorithm, we carry out experiments with synthetic complex-valued signals, synthetic speech signals, and recorded speech signals. The proposed algorithm is compared with the standard blind CMN algorithm. Specifically, $G(y) = \log(0.1 + y)$ is used for the nonlinear function in (13). For quantitative comparison of the separation performance, we mainly use the following two criteria:

(1) Signal to Noise Ratio (*SNR*)

$$SNR(\text{dB}) = 10\log_{10}\left(\frac{\sigma^2}{mse}\right)$$

where σ^2 is the variance of a source signal, *mse* denotes the mean square error between the source signal and the estimated signal. This is a widely used evaluation criterion measuring the approximation level between the recovered signal and the source signal.

(2) Correlation Coefficient ρ

$$\rho(y_i, s_i) = \frac{\left|\sum_t y_i(t)s_i(t)\right|}{\sqrt{\sum_t y_i^2(t)\sum_t s_i^2(t)}}$$

where $s_i(t)$ is the i-th source signal, $y_i(t)$ is its estimate. It is known that $\rho \in [0,1]$, and $\rho = 1$ means a perfect separation.

4.1 Experiments with Synthetic Complex-Valued Signals

The complex-valued source signals are formed with random numbers from different distributions. The mixing matrix is also randomly generated, the real and the imaginary of which are with entries from uniform distribution between 0 and 1. The reference signals were selected as the neighboring segments of the generated sources (e.g., a source and its reference are two adjacent segments from a synthetic signal), the magnitudes of which thus have maximal correlations with those of the corresponding sources compared to those of the others.

Fig. 1 shows one example for separating mixtures of two noncircular sources, the magnitudes for each complex-valued signals are shown. Specifically, s_1 is formed with random numbers from the binomial distribution and s_2 from the exponential distribution. r_1 is the reference signal corresponding to s_2, i.e., r_1 and s_2 are two neighboring segments from the same signals with exponential distribution; r_2 is the reference for s_1. Similarly, r_2 and s_1 are two adjacent segments with binomial distribution.

The separation results of the proposed algorithm are shown in Fig. 1(c). We can see that y_1 and y_2 are exactly the estimates of s_2 and s_1, that is, the two sources are recovered in the same order as r_1 and r_2. When we change the order of r_1 and r_2, the order of the two estimates changes correspondingly. Note that the estimates by CMN are very similar to those in Fig. 1(c) but with random order. The average *SNR*s and ρ for the two estimates by the two algorithms are computed, which are 13.70dB/0.96 for the proposed algorithm and 11.51dB/0.92 for CMN, respectively. The results of this example together with other simulations using different sources consistently show that, compared to CMN, the proposed algorithm can solve the permutation problem with slightly improved *SNR*/ρ by utilizing prior information about the sources.

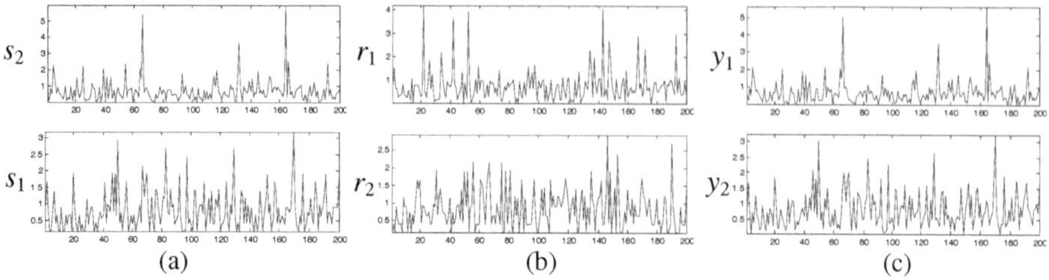

Fig. 1. One experiment with synthetic complex-valued signals (X-axis denotes the number of samples and Y-axis the magnitude). (a) Two source signals. (b) Two reference signals. (c) Two separated signals by the proposed algorithm.

4.2 Experiments with Synthetic Speech Signals

The example of experiments presented below is to mix the two speeches in Fig. 2(a) with three different orders of filters, which are 8, 128, and 256, respectively. The convolutive mixtures are first transformed to frequency domain by short time Fourier transform, and then separated with the proposed algorithm and CMN at each frequency bin. The reference signals for a frequency bin (e.g., *f*) are selected as the estimates at its neighboring frequency bin (e.g., *f*-1) since the amplitudes of the speech signals in neighboring frequency bins are strongly correlated.

The results for 8-order filter are displayed in Fig. 2(c) and (d) as an example. The separated signals by CMN are reordered by utilizing the amplitude correlation method at post-processing stage. Table 1 shows the comparison of average *SNR*s (dB) $/\rho$ for the two algorithms. We can see that the proposed algorithm can achieve frequency-domain speech separation as expected. It performs separation and permutation correction simultaneously with slight improvement in performance compared to the standard CMN utilizing no prior information.

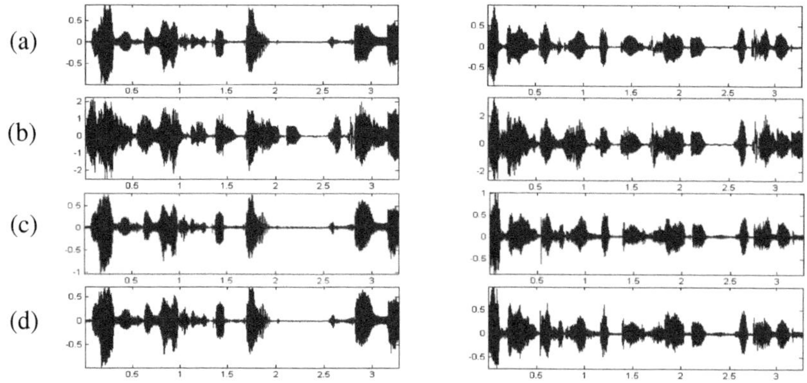

Fig. 2. One experiment with synthetic speech signals (X-axis denotes the number of samples ($\times 10^4$) and Y-axis the amplitude). (a) Two speech signals. (b) Two mixed speech signals with 8-order filter. (c) Two separated speech signals by the proposed algorithm. (d) Two separated signals by CMN with post-processing permutation correction using amplitude correlation.

Table 1. Comparison of average *SNR*s (dB) /ρ of the two estimates by the proposed algorithms and CMN

	8-order	128-order	256-order
Proposed	7.89/0.91	6.68/0.84	5.80/0.76
CMN	6.87/0.83	6.37/0.82	5.48/0.71

4.3 Experiments with Recorded Speech Signals

For a more comprehensive comparison, the actual noisy speeches from http://www.cnl.salk.edu/~tewon are used in this experiment. The speeches are recorded in a normal office room (3m×4m). The sources and sensors are placed in a rectangular (60cm×40cm) order with 60cm distance between the sources and the sensors. The sampling frequency is 16 kHz, the length of two speech signals is 7.4 s. Fig. 3(a) shows the recorded signals with one speaker saying the digits one to ten in English (*one two*$\cdots$*ten*), the other in Spanish (*uno dos*$\cdots$*diez*).

Similar to the experiments above, we still use the proposed algorithm and CMN for separation, the amplitude correlation is used by CMN for reordering post-process. The order of separation filer is selected as 256 which covers a delay of 16ms corresponding to 5 m. The separated signals are displayed in Fig. 3(b) and (c).

(a)
(b)
(c)

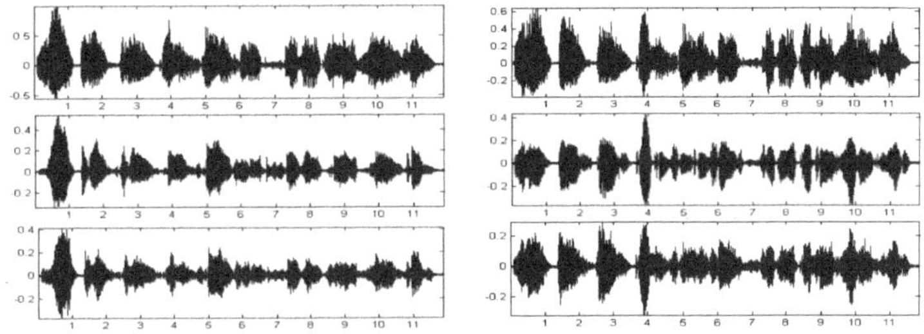

Fig. 3. One experiment with recorded speech signals (X-axis denotes the number of samples ($\times 10^4$) and Y-axis the amplitude). (a) Two recordings of noisy speeches. (b) Two separated speech signals by the proposed algorithm. (c) Two separated signals by CMN with post-processing permutation correction using amplitude correlation.

Because no source signals are available, it is impossible to quantitatively calculate *SNR* and correlation coefficient. Hence we only make subjective quality evaluation using average MOS (Mean Opinion Score). Ten people (5 males and 5 females, aged 20-25) are selected for experiment. They wear high-fidelity headphones, mark the quality of the separated speeches by the two algorithms in a quiet environment. As a result, the average MOS scores are 4.6 for the proposed algorithm and 4.1 for CMN. Therefore, the results from the recorded speech signals are consistent with those from the synthetic speech signals and the synthetic complex-valued signals. Compared to the blind CMN algorithm, the semi-blind CMN algorithm can correct the permutation during the separation process and achieve slight improvement as well.

5 Conclusion

In this paper, we present a constrained complex ICA algorithm based on negentropy maximization. The magnitude correlation of the source signal is utilized as a constraint of standard complex ICA to solve the permutation ambiguity. Experiments with synthetic complex-valued signals, synthetic speech signals, and recorded speech signals are performed. The results demonstrate that, by utilizing the prior information about the source signals, the proposed semi-blind algorithm can not only solve the permutation problem, but also yield slightly improved separation performance compared to the standard ICA algorithm.

Note that the proposed algorithm needs properly constructed reference signals. The references should have maximal correlations with the corresponding sources than with the other sources. In practice, the references can be constructed in different ways. Our future work includes exploring more efficient references and the constraints, presenting extensive comparison with other frequency-domain algorithms for speech separation, and examining its application to analysis of the original fMRI data with complex-values.

Acknowledgment. The authors would like to thank the reviewers for valuable comments to improve the paper. This work was supported by the National Natural Science Foundation of China under Grant No. 60971097.

References

1. Waheed, K., Salem, F.M.: Blind Information-Theoretic Multiuser Detection Algorithms for DS-CDMA and WCDMA Downlink Systems. IEEE Transactions on Neural Networks 16(4), 937–948 (2005)
2. Calhoun, V.D., Adali, T.: Unmixing fMRI with Independent Component Analysis. IEEE Engineering in Medicine and Biology Magazine 25, 79–90 (2006)
3. Sawada, H., Mukai, R., Araki, S., Makino, S.: Frequency-domain Blind Source Separation, Speech Enhancement. Springer, New York (2005)
4. Cardoso, J.F., Souloumiac, A.: Blind Beamforming for Non Gaussian Signals. IEEE Proc. Radar Signal Proc. 140, 362–370 (1993)
5. Eriksson, J., Koivunen, V.: Complex-valued ICA Using Second Order Statistics. In: 14th IEEE Signal Processing Society Workshop on Machine Learning for Signal Processing, Sao Luis, Brazil, pp. 183–191 (2004)
6. Bingham, E., Hyvärinen, A.: A Fast Fixed-point Algorithm for Independent Component Analysis of Complex Valued Signals. Int. J. Neural Systems 10, 1–8 (2000)
7. Li, H., Adali, T.: A Class of Complex ICA Algorithms Based on Kurtosis Maximization. IEEE Transactions on Neural Networks 19, 408–420 (2008)
8. Novey, M., Adali, T.: Complex ICA by Negentropy Maximization. IEEE Transactions on Neural Networks 19, 596–609 (2008)
9. Lin, Q.-H., Liu, J., Zheng, Y.R., Liang, H., Calhoun, V.D.: Semiblind Spatial ICA of fMRI Using Spatial Constraints. Human Brain Mapping 31, 1076–1088 (2010)
10. Lu, W., Rajapakse, J.C.: Approach and Applications of Constrained ICA. IEEE Transactions on Neural Networks 16, 203–212 (2005)

Time Series Causality Inference Using Echo State Networks

N. Michael Mayer[1], Oliver Obst[2], and Chang Yu-Chen[1]

[1] Department of Electrical Engineering
National Chung Cheng University
Min Hsiung, Chia-Yi, Taiwan
nmmayer@gmail.com
[2] CSIRO ICT Centre
Adaptive Systems Team
P.O. Box 76, Epping NSW 1710, Australia
oliver.obst@csiro.au

Abstract. One potential strength of recurrent neural networks (RNNs) is their – theoretical – ability to find a connection between cause and consequence in time series in an constraint-free manner, that is without the use of explicit probability theory. In this work we present a solution which uses the echo state approach for this purpose. Our approach learns probabilities explicitly using an online learning procedure and echo state networks. We also demonstrate the approach using a test model.

1 Introduction

A typical application of recurrent neural networks is the prediction of time series and modelling dynamical systems. Time series prediction is important to forecast e.g. economical data, and used to make decisions which, in turn, change economy. The term "causality" is used when past values of a time series provide significant information about future values of another time series [9]. One of the possible methods for causality inference is transfer entropy [8], however, it has the disadvantage of requiring a fair amount of data. Granger causality [2], on the other hand, is based on regression and uses less data, but is a linear method (non-linear extensions exist, see also [7] for a comparison between different methods). In this work, we propose a (non-linear) approach based on regression and a recent recurrent neural network learning method, which we briefly revisit in Sect. 2. Recent advances in this area have shown to be successful in time series prediction [4,6,3]. However, instead of using predictions of the neural networks directly, we take a different route and describe an approach using the prediction error to detect causal links in time series (Sect. 3). A related approach is presented in [5]; in our work, however, we present an analytical derivation. Our approach is demonstrated using simulation data (Sect. 4), and finally, we discuss our results in Sect. 6.

V. Vigneron et al. (Eds.): LVA/ICA 2010, LNCS 6365, pp. 279–286, 2010.

2 Echo State Networks

Echo State Networks (ESN) are an approach to address the problem of slow convergence in recurrent neural network learning. ESN consist of three layers (see Fig. 1): a) an input layer, where the stimulus is presented to the network; b) a randomly connected recurrent hidden layer; and c) the output layer. Connections in the output layer are trained to reproduce the training signal. The network dynamics is defined for discrete time-steps t, with the following equations:

$$\mathbf{x}_{\mathsf{lin},t+1} = \mathbf{W}\mathbf{x}_t + \mathbf{w}^{in}\mathbf{u}_t \tag{1}$$

$$\mathbf{x}_{t+1} = \mathsf{tanh}\left(\mathbf{x}_{\mathsf{lin},t+1}\right) \tag{2}$$

$$\mathbf{o}_t = \mathbf{w}^{out}\mathbf{x}_t \tag{3}$$

where the vectors $\mathbf{u}_t$, $\mathbf{x}_t$, $\mathbf{o}_t$ are the input and the neurons of the hidden layer and output layer respectively, and $\mathbf{w}^{in}$, $\mathbf{W}$, $\mathbf{w}^{out}$ are the matrices of the respective synaptic weight factors.

Connections in the hidden layer are random but the system needs to fulfil the so-called echo state condition. Jaeger [4] gives a definition; in the following a slightly more compact form of the echo state condition:

Consider a time-discrete recursive function $\mathbf{x}_{t+1} = F(\mathbf{x}_t, \mathbf{u}_t)$ that is defined at least on a compact sub-area of the vector-space $\mathbf{x} \in R^n$. and where $\mathbf{x}_t$ are to be interpreted as internal states and $\mathbf{u}_t$ is some external input sequence, i.e. the stimulus.

The definition of the echo-state condition is the following: Assume an infinite stimulus sequence: $\bar{\mathbf{u}}^\infty = \mathbf{u}_0, \mathbf{u}_1, \ldots$ and two random initial internal states of the system $\mathbf{x}_0$ and $\mathbf{y}_0$. To both initial states $\mathbf{x}_0$ and $\mathbf{y}_0$ the sequences $\bar{\mathbf{x}}^\infty = \mathbf{x}_0, \mathbf{x}_1, \ldots$ and $\bar{\mathbf{y}}^\infty = \mathbf{y}_0, \mathbf{y}_1, \ldots$ can be assigned.

$$\mathbf{x}_{t+1} = F(\mathbf{x}_t, \mathbf{u}_t) \tag{4}$$

$$\mathbf{y}_{t+1} = F(\mathbf{y}_t, \mathbf{u}_t) \tag{5}$$

Then the system $F(.)$ fulfils the echo-state condition if independent from the set $\mathbf{u}_t$ and for any $(\mathbf{x}_0,\mathbf{y}_0)$ and all real values $\epsilon > 0$ there exists a $\delta(\epsilon)$ for which

$$d(\mathbf{x}_t, \mathbf{y}_t) \leq \epsilon \tag{6}$$

for all $t \geq \delta$. The ESN is designed to fulfil the echo state condition.

2.1 Online Learning Using Recursive Least Squares

ESN can be trained using either an offline or an online learning procedure. For our approach, we are online learning the output layer using the recursive least square method (RLS). The combination of ESN and RLS has first been published by Jaeger [6]. The following update rule was used:

$$\alpha_t = s_t^{teach} - \mathbf{w}_{t-1}^{out} \cdot \mathbf{o}_t, \tag{7}$$

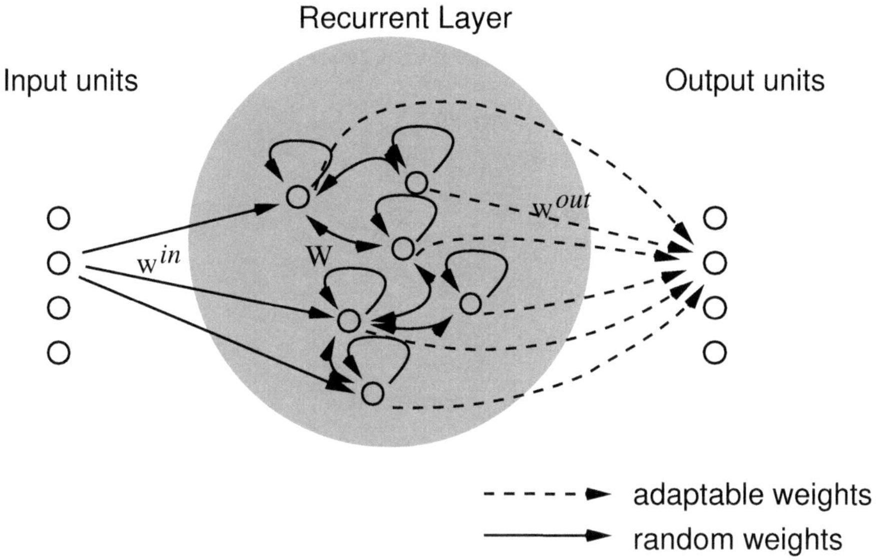

Fig. 1. ESN networks: Principle setup

$$g_t = p_t \cdot \mathbf{o}_t/(\lambda + \mathbf{o}_t{}^T \cdot p \cdot \mathbf{o}_{t)}, \tag{8}$$

$$p_t = 1./\lambda \cdot p_t - g_t \cdot \mathbf{o}_t{}^T \cdot p_t/\lambda, \tag{9}$$

$$\mathbf{w}_t^{out} = \mathbf{w}_t^{out} + (\alpha_t \cdot g_t^T), \tag{10}$$

where α_t represents the linear error vector and p_t the inverse of the autocorrelation, λ is close to 1 and is used as forgetting factor.

3 Modelling Probability Distributions by Using the Mean Square Error

Instead of training the output $\omega_t \in \Omega$ directly, we model a probability that a specific event has occurred with regard to the output. In other words: the aim is to train the network in that way that each of the output units represents the probability of an event. As one simplest way to do this we teach the output o_r of the network to reproduce the probability that the as a range of the – statistical – output variable $\Omega_r \subset \Omega$ that is of interest for the given task. The task of network is to find $p(\Omega_r|\mathbf{x}_t(\bar{\mathbf{u}}^\infty))$ in the following written short $p(\Omega_r)$. We define the teaching signal d_r as:

$$\begin{aligned} if\ (\omega_t \in \Omega_r)\ \ d_r &= 1 \\ else\ \ d_r &= 0 \end{aligned}$$

The mean square error (MSE) is the

$$E_{mse} =< (d_{r,t} - o_r)^2 >= p(\Omega_r)(1 - o_r)^2 + (1 - p(\Omega_r))o_r^2 \tag{11}$$

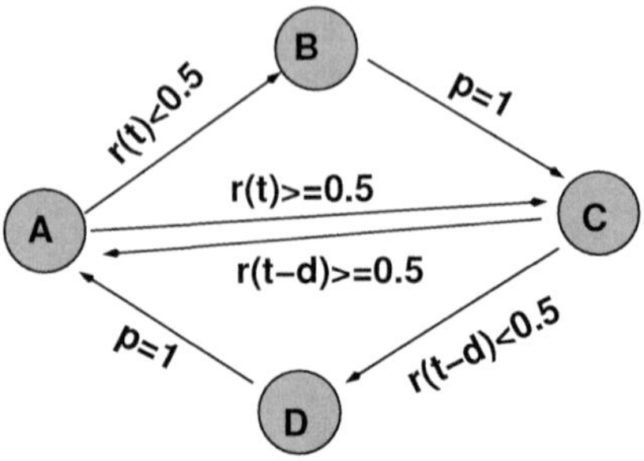

Fig. 2. Test model set-up

The derivative $\partial E_{mse}/\partial o_r$ set equal to zero yields the point at which E_{mse} is minimal:

$$o_r - p(\Omega_r) = 0 \tag{12}$$

Thus, the MSE is reached when $o_r = p(\Omega_r)$; we can assume

$$o_r \to p(\Omega_r), \tag{13}$$

for sufficiently long learning sequences. Since – with the common restrictions of reservoir computing– the full information of the input history is encoded in the activity state of the reservoir. Thus, –without additional efforts in the hidden layer– more information about statistical variables can be retrieved from additional output units: Because the optimal solution (absolute minimum of the MSE) can be derived, the network is going to find the true probability **as far as it is detectable by linear regression from the current state of the reservoir**. Usually, the quality of the network performance and the learning progress can be checked by measuring E_{mse}, where values close to zero represent a good network performance. It should be noted that for the learning rule outlined above the theoretical limit is above zero. Under the assumption that the $p(\Omega_r)$ is the true probability we get:

$$E_{min}(\Omega_r) = min_{o_r}(E_{mse}) = p(\Omega_r)(1 - p(\Omega_r)). \tag{14}$$

However, since in fact the true value $p(\Omega_r)$ is unknown, it is not a good idea to use $E_{mse} - E_{min}$ as a measure. However, it can be used to find out if the output node is deterministic (i.e. the output node takes either 0 or 1). In this case the minimal error is in fact 0 again.

Instead one could go the following way in that we can get a set of outputs that covers a complete range of the random variable in the way that for a range $r \in R$:

$$\cup_{r \in R} \Omega_r = \Omega, \tag{15}$$

$$\Omega_i \cap \Omega_j = \emptyset, \tag{16}$$

for all $i \neq j$. Obviously, we have $\sum_{r \in R} p(\Omega_r) = 1$. We can test the constraint in the network. We test the quality of the network output by testing measuring $\sum_r o_r$ which should be close to one if the network has adapted sufficiently.

Basing on the plausibility constraint it is very easy to define an error function for the network. In the simplest case, it can be assumed to train 2 outputs, of which the first represents the occurrence of an event $e \in \Omega_x$, whereas the second output is trained to record the non-occurrence $e \notin \Omega_x$. Thus, I train $(d_1 = 1, d_2 = 0)$ in the case $e \in \Omega_x$ and else $(d_1 = 0, d_2 = 1)$. In this case it can be assumed that the cost function

$$E_{total} = (o_1 + o_2 - 1)^2 \tag{17}$$

approaches zero after a sufficient long learning process since from Eq. 13 we get

$$E_{total} \to (p(\Omega_x) + p(\bar{\Omega}_x) - 1)^2 = (p(\Omega_x) + (1 - p(\Omega_x)) - 1)^2 = 0. \tag{18}$$

Thus, this energy function may serve better as an estimator how well the network has adapted to the particular current input history.

4 Simulation Details

We demonstrate the approach on a prediction task. Our test model cycles between four states (A, B, C, D). Figure 2 outlines the transition probabilities. Every time in which the state A is reached a random number $0 \leq r(t) < 1$ is drawn. In every time step the model transfers from one state to the next state. Is the current state the state A and the random value $r(t)$ smaller then 0.5 the model goes to state B else the model transfers to state C. From state B the model transfers to C. From C a the next step is either D if $r(t-d) < 0.5$ else the model returns to state A, where d is a positive or zero delay constant (0 delay indicates that the transition C-A happens in the same cycle as A-C, i.e. the complete cycle becomes A-C-A. Is the model in state D it always transits back to A.

The task of the network is to predict the next state from the previous. Had the states A, B, C, D been interpreted directly as Markov states the transitions from C to the next state would appear random with equal probability either to state A or D, i.e. in that interpretation the model is non-Markov. However, a perfect hidden Markov model (HMM) with 2^d hidden states would –presumably– be able to detect the fact that the choices of the transitions from A and those from C are linked.

The model is presented as an 8 dimensional vector in the following way to the network:

$$\begin{aligned} A &= [1, 0, 0, 0, 0, 1, 1, 1] \\ B &= [0, 1, 0, 0, 1, 0, 1, 1] \\ C &= [0, 0, 1, 0, 1, 1, 0, 1] \\ D &= [0, 0, 0, 1, 1, 1, 1, 0] \end{aligned}$$

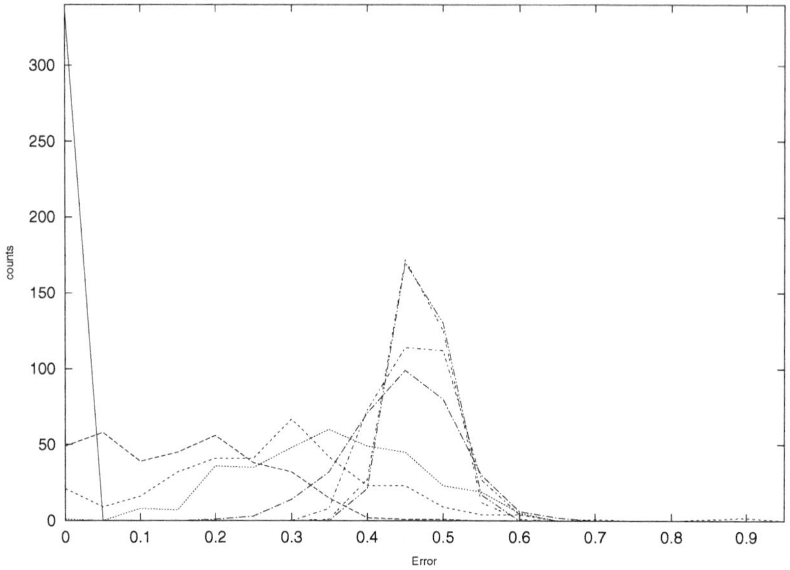

Fig. 3. Histograms of errors for different delays. Depicted is the MSE of probability of the transition to state D in the event of initial state C. If the network cannot detect the causality of transitions from C the error is at best 0.5. Full lines depicts the error at delay 0, dashes delay 1, small dashes delay 2, dotted delay 3, dash-dotted delay 4, double dashed delay 5, small dash-dotted delay 6.

The second half of each vector represents the inverse of the first. Thus, it can serve to find the cost function according to Eq. 18.

Obviously, the task becomes more complex as the delay constant increases. The inverse correlation matrix was set to $p = 0.0001 \cdot I$ where I is the identity matrix. The forgetting factor λ was set to 1.0, which sets the RLS into the non-forgetting mode. The recurrent matrix was set to random ortho-normal matrix which was multiplied by 0.98 (Fig. 3) and 1.14 (Fig. 4), which gives a slightly over-critical network. Since the input practically is never close to 0, the network stays non critical. The online learning was performed from the 1000th step onwards up to 18000th step. The different MSEs were recorded in the last 100 steps of each simulation.

5 Results

We tested the network for several delays and network sizes. To estimate the network performance we used the MSE error at the transition from node C, in following $E_{MSE,C}$. A network that is able to detect the causality between the history and the transition from node C can reach zero MSE, whereas for network that cannot detect the causality the transition appears to be stochastic

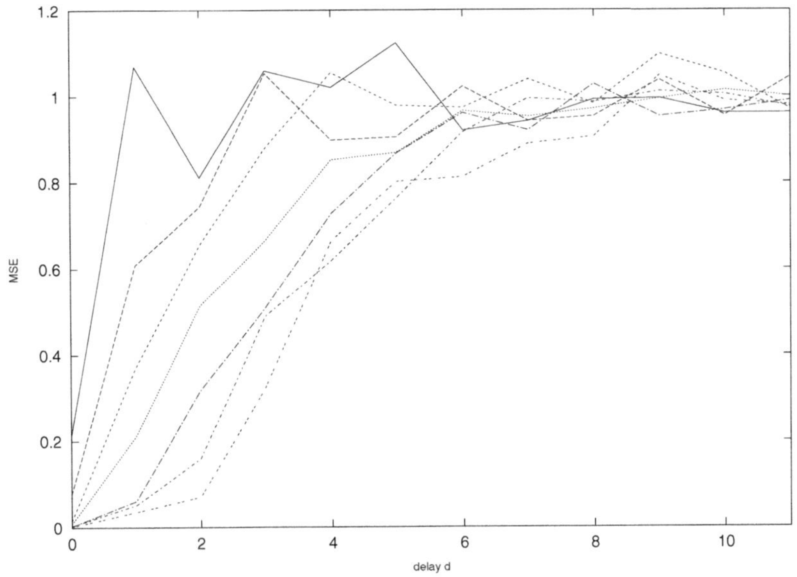

Fig. 4. Normalised errors for different network sizes at the hidden layer as a function of the delay. Full line 10 neurons, dashed 20 neurons, small dashes 30 neurons, dotted 50 neurons, dash-dotted 100 neurons, double dashed 150 neurons, and small dash-dotted 200 neurons.

with equal probability to state D and A. The MSE in this case can be determined by Eq. 14.

A histogram of errors for different delays is depicted in Fig. 3.

For sake of simplicity we used the MSE of transitions from A (stochastic, $E_{MSE,A}$) and B (deterministic, $E_{MSE,B}$) in Fig. 4. The plot depicts the values of

$$E_{norm} = (E_{MSE,C} - E_{MSE,B})/(E_{MSE,A} - E_{MSE,B}). \qquad (19)$$

Thus, values of E_{norm} around 0 can be interpreted in that way that the network is able to detect the causality relation between transitions from A and C. Fig. 3 results for different network sizes and delays. Each line represents different the performance of one network and different delays. The network sizes are 10,20,30,50,100,150, and 200 neurons in the hidden layer.

The value of E_{total} (cf. Eq. 18)) shows a very fast convergence to the final range almost immediately after the learning starts.

6 Discussion

We demonstrated an approach able to detect causalities in time series by using the mean square error of an ESN online learning procedure. Our current results indicate that the ability of this approach is limited to a few cycles – equivalent some dozens of steps. It can be expected that our results can be further improved by adapting the reservoir to the stimulus statistics. Some steps into this direction have been undertaken in [1]. It may also be possible to recursively improve the reservoir as soon as some causalities are detected, but this is subject to further investigation.

References

1. Boedecker, J., Obst, O., Mayer, N.M., Asada, M.: Initialization and self-organized optimization of recurrent neural network connectivity. HFSP J. 3(5), 340–349 (2009)
2. Granger, C.W.J.: Investigating causal relations by econometric models and cross-spectral methods. Econometrica 37(3), 424–438 (1969)
3. Hammer, B., Schrauwen, B., Steil, J.J.: Recent advances in efficient learning of recurrent networks. In: ESANN 2009 proceedings, European Symposium on Artificial Neural Networks, pp. 213–226 (2009)
4. Jaeger, H.: The 'echo state' approach to analysing and training recurrent neural networks. GMD Report 148, GMD German National Research Insitute for Computer Science (2001)
5. Jaeger, H.: Short term memory in echo state networks. GMD Report 152, GMD - German National Research Institute for Computer Science (2002)
6. Jaeger, H.: Adaptive nonlineaer systems identification with echo state networks. In: Advances in Neural Information Processing Systems, pp. 609–615 (2003)
7. Pereda, E., Quiroga, R.Q., Bhattacharya, J.: Nonlinear multivariate analysis of neurophysiological signals. Progress in Neurobiology (77), 1–37 (2005)
8. Schreiber, T.: Measuring information transfer. Physical Review Letters 85(2), 461–464 (2000)
9. Shibuya, T., Harada, T., Kuniyoshi, Y.: Causality quantification and its applications: structuring and modeling of multivariate time series. In: KDD 2009: Proceedings of the 15th ACM SIGKDD international conference on Knowledge discovery and data mining, pp. 787–796. ACM, New York (2009)

Complex Blind Source Separation via Simultaneous Strong Uncorrelating Transform

Hao Shen and Martin Kleinsteuber

Institute for Data Processing, Technische Universität München, Germany
{hao.shen,kleinsteuber}@tum.de

Abstract. In this paper, we address the problem of complex blind source separation (BSS), in particular, separation of nonstationary complex signals. It is known that, under certain conditions, complex BSS can be solved effectively by the so-called Strong Uncorrelating Transform (SUT), which simultaneously diagonalizes one Hermitian positive definite and one complex symmetric matrix. Our current work generalizes SUT to simultaneously diagonalize more than two matrices. A Conjugate Gradient (CG) algorithm for computing simultaneous SUT is developed on an appropriate manifold setting of the problem, namely complex oblique projective manifold. Performance of our method, in terms of separation quality, is investigated by several numerical experiments.

Keywords: Complex blind source separation, nonstationary signal, simultaneous strong uncorrelating transform, complex oblique projective manifold, conjugate gradient algorithm.

1 Introduction

In recent years, complex Independent Component Analysis (ICA) has become a prominent method for solving the problem of complex Blind Source Separation (BSS). Its applications can be found in convolutive blind source separation, wireless communication, and magnetic resonance imaging analysis. Although complex ICA and its real counterpart were born as twins [1], surprisingly, it was less understood from both theoretical and practical perspectives than real ICA.

Such a difference is mainly due to effects of the so-called *circularity* or *non-circularity* of complex signals on ICA models. Generally speaking, (non)-circularity describes statistical characteristics of real and imaginary parts of complex signals. A recent work by Eriksson and Koivunen [2] shows that, under certain scenario, where the source signals are non-circular and the values of their *circularity coefficients* are distinct, complex ICA can be solved effectively by the so-called Strong Uncorrelating Transform (SUT) [3], which utilizes only second-order statistics of observed mixtures. A robust extension of SUT, namely Generalized Uncorrelating Transform (GUT), has been developed as well in [4]. It is worth noticing that both approaches require a whitening of observations, which unfortunately is statistically inefficient in some applications, especially, when additive noise is present

V. Vigneron et al. (Eds.): LVA/ICA 2010, LNCS 6365, pp. 287–294, 2010.

[5]. A fixed-point algorithm for computing a single SUT without whitening has been developed in [6].

From an algorithmic point of view, both SUT and GUT require a simultaneous diagonalization of one Hermitian positive definite and one complex symmetric matrix. In this work, we are interested in blind separation of nonstationary complex signals. By exploiting the fact that second-order statistics of the nonstationary signals are time-varying in general, we develop a conjugate gradient (CG) algorithm to simultaneously diagonalize several Hermitian positive definite and complex symmetric matrices.

The paper is organized as follows. Section 2 introduces briefly the linear complex ICA problem and motivates a simultaneous SUT solution. In Section 3, we construct an appropriate manifold setting for the problem, namely, the complex oblique projective manifold. Section 4 develops an intrinsic conjugate gradient algorithm for computing a simultaneous SUT. Finally in Section 5, performance of our proposed approach in terms of separation quality is investigated by several experiments.

2 Complex Blind Source Separation

In this work, we denote by $(\cdot)^{\mathsf{T}}$ the matrix transpose, $\overline{(\cdot)}$ the complex conjugate, and $(\cdot)^{\mathsf{H}}$ the Hermitian transpose.

Let $s(t) = [s_1(t), \ldots, s_m(t)]^{\mathsf{T}} \in \mathbb{C}^m$ be an m-dimensional complex vector representing the time series of m statistically independent complex signals. The instantaneous linear complex BSS model is given by

$$w(t) = As(t), \tag{1}$$

where $A \in \mathbb{C}^{m\times m}$ is the mixing matrix of full rank and $w(t) = [w_1(t), \ldots, w_m(t)]^{\mathsf{T}} \in \mathbb{C}^m$ presents m observed linear mixtures of $s(t)$. Without loss of generality, we assume sources $s(t)$ have zero mean and unit variance, i.e.,

$$\mathbb{E}[s(t)] = 0, \qquad \text{and} \qquad \operatorname{cov}(s) := \mathbb{E}[s(t)s^{\mathsf{H}}(t)] = I_m, \tag{2}$$

where $\mathbb{E}[\cdot]$ denotes the expectation over time index t, and I_m is the $m\times m$ identity matrix. The expression $\operatorname{cov}(s)$ is referred to as the *complex covariance matrix* of $s(t)$. Furthermore, we assume that the so-called *pseudo-covariance matrix* of $s(t)$ has a real diagonal structure, i.e.

$$\operatorname{pcov}(s) := \mathbb{E}[s(t)s^{\mathsf{T}}(t)] = \Lambda \in \mathbb{R}^{m\times m}, \tag{3}$$

where $\Lambda := \operatorname{diag}(\lambda_1, \ldots, \lambda_m)$ and diagonal entries $\lambda_i \geq 0$, for all $i = 1, \ldots, m$, are called *circularity coefficient* of the corresponding signal $s_i(t)$. If all λ_i's are zero, then the sources $s(t)$ is called *second-order circular*. We refer to [2,6] and references therein for further discussions.

The task of the linear complex BSS problem (1) is to recover the source signals $s(t)$ by estimating the mixing matrix A or its inverse A^{-1} based only on the observations $w(t)$ via the demixing model

$$y(t) = X^{\mathsf{H}}w(t), \tag{4}$$

where $X^{\mathsf{H}} \in \mathbb{C}^{m \times m}$ is the demixing matrix, an estimation of A^{-1}, and $y(t) \in \mathbb{C}^m$ represents the corresponding extracted signals. According to theorem 11 in [1], a correct demixing matrix $X^* \in \mathbb{C}^{m \times m}$ can only be identified up to column-wise permutation and complex scaling, i.e., X^* is the inverse of A^{H} up to an $m \times m$ permutation matrix P and an $m \times m$ complex diagonal matrix D

$$X^* = A^{-\mathsf{H}} DP. \tag{5}$$

In the rest of this section, we quickly review the SUT approach and then generalize it to simultaneously diagonalize more than two matrices. Given the observation $w(t)$ from the ICA model (1), second-order statistics of $w(t)$ is computed as

$$\operatorname{cov}(w) := \mathbb{E}[w(t)w^{\mathsf{H}}(t)] = AA^{\mathsf{H}} \tag{6}$$

and

$$\operatorname{pcov}(w) := \mathbb{E}[w(t)w^{\mathsf{T}}(t)] = A\Lambda A^{\mathsf{T}}. \tag{7}$$

A matrix $X \in \mathbb{C}^{m \times m}$, which transforms both complex covariance (6) and pseudo-covariance matrix (7) into the following diagonal forms

$$X^{\mathsf{H}} \operatorname{cov}(w) X = I_m \qquad \text{and} \qquad X^{\mathsf{H}} \operatorname{pcov}(w) \overline{X} = \Lambda, \tag{8}$$

is called a *strong uncorrelating transform* of $w(t)$. According to theorem 2 in [3], if the values of circularity coefficients λ_i are distinct, then any SUT of $w(t)$ is a correct demixing matrix of the problem (1). It is important to notice that transforming the pseudo-covariance matrix into a real diagonal structure as shown in (8) is *not necessary* for solving the complex BSS problem (1).

Now let us consider the situation where sources $s(t)$ are nonstationary. Then in general, both covariance and pseudo-covariance matrices of $s(t)$, and consequently, $w(t)$ as well, are time-varying. One well-known technique to deal with this situation is to construct a set of covariance matrices of $w(t)$ at different time instances, and then simultaneously diagonalize this set [7]. Due to the fact that complex covariance does not necessarily provide complete second-order statistics of complex signals [8], we propose to construct additionally a set of pseudo-covariance matrices, then to simultaneously diagonalize these two sets in the similar manner as SUT (8), which is referred to here as *simultaneous SUT*.

To summarize, in what follows, we are interested in solving the following problem. Given a set of Hermitian positive matrices $\{C_i\}_{i=1}^N$ and a set of complex symmetric matrices $\{R_i\}_{i=1}^N$, the task is to find a nonsingular matrix X such that

$$X^{\mathsf{H}} C_i X \qquad \text{and} \qquad X^{\mathsf{H}} R_i \overline{X}, \tag{9}$$

for all $i = 1, \ldots, N$, are simultaneously diagonalized, or approximately simultaneously diagonalized subject to certain diagonality measure.

3 Complex Oblique Projective Manifold

In this section, we construct an appropriate manifold setting for the problem (1), such that the ambiguity due to column-wise complex scaling as shown in

(5) is eliminated. The optimal solutions of complex BSS on this search space are isolated in the generic case. This allows us to: (*i*) reduce the number of parameters, over which one needs to optimize, and, (*ii*) simplify the development of our proposed algorithm in the next section. We refer to [9] for deeper insights of the topic.

Let us denote by $GL(m,\mathbb{C})$ the set of all $m \times m$ invertible complex matrices. Starting point of our construction is the so-called *complex oblique manifold*, whose real counterpart was developed for real ICA [10], i.e.,

$$\mathcal{O}(m,\mathbb{C}) := \left\{X \in GL(m,\mathbb{C}) \,\middle|\, \operatorname{ddiag}(X^{\mathsf{H}}X) = I_m \right\}, \tag{10}$$

where $\operatorname{ddiag}(Z)$ forms a diagonal matrix, whose diagonal entries are those of Z. By the regular value theorem, the set $\mathcal{O}(m,\mathbb{C})$ is an $m(2m-1)$-real-dimensional differentiable manifold.

For a given $X = [x_1,\ldots,x_m] \in \mathcal{O}(m,\mathbb{C})$, it is clear that each column x_i identifies a one-complex-dimensional linear subspaces of $\mathbb{C}^m$. It is known that the set of all one-complex-dimensional linear subspaces of $\mathbb{C}^m$ forms a differentiable manifold, namely, the $(m-1)$-dimensional complex projective space $\mathbb{CP}^{m-1}$. In this work, we identify it as the set of all rank-one Hermitian projectors, i.e.

$$\mathbb{CP}^{m-1} := \left\{P \in \mathbb{C}^{m\times m} \,\middle|\, P^{\mathsf{H}} = P, P^2 = P, \operatorname{tr}(P) = 1\right\}. \tag{11}$$

Let us denote by

$$\mathfrak{u}(m) := \left\{\Omega \in \mathbb{C}^{m\times m} \,\middle|\, \Omega = -\Omega^{\mathsf{H}}\right\} \tag{12}$$

the set of skew-Hermitian matrices. Then, the tangent space at $P \in \mathbb{CP}^{m-1}$ is given by

$$T_P\mathbb{CP}^{m-1} := \{[P,\Omega] \mid \Omega \in \mathfrak{u}(m)\} \tag{13}$$

with matrix commutator $[A,B] := AB - BA$. Endowing $T_P\mathbb{CP}^{m-1}$ with the inner product

$$g\colon T_P\mathbb{CP}^{m-1} \times T_P\mathbb{CP}^{m-1} \to \mathbb{R}, \qquad g(\phi,\psi) := \Re\operatorname{tr}(\phi\cdot\psi) \tag{14}$$

turns $\mathbb{CP}^{m-1}$ into a Riemannian manifold. Here, $\Re Z$ is the real part of a complex number Z. Then, the geodesic through $P \in \mathbb{CP}^{m-1}$ in direction $\phi \in T_P\mathbb{CP}^{m-1}$ is given by

$$\gamma_{P,\phi}\colon \mathbb{R} \to \mathbb{CP}^{m-1}, \qquad \gamma_{P,\phi}(t) := \mathsf{e}^{t[\phi,P]}P\mathsf{e}^{-t[\phi,P]}. \tag{15}$$

Here, $\mathsf{e}^{(\cdot)}$ denotes the matrix exponential. Finally, the parallel transport of $\psi \in T_P\mathbb{CP}^{m-1}$ with respect to the Levi-Civita connection along the geodesic $\gamma_{P,\phi}(t)$ is given by

$$\tau_{P,\phi}(\psi) = \mathsf{e}^{[\phi,P]}\psi\mathsf{e}^{-[\phi,P]}. \tag{16}$$

By exploiting the fact that, for a given $X \in \mathcal{O}(m,\mathbb{C})$, the following matrix

$$XX^{\mathsf{H}} = \sum_{i=1}^{m} x_i x_i^{\mathsf{H}}, \tag{17}$$

where $x_i x_i^{\mathsf{H}}$, for all $i = 1, \ldots, m$, is a rank-one Hermitian projector, is positive definite, we construct a set of constrained collections of m rank-one Hermitian projectors as

$$\mathcal{Q}(m, \mathbb{C}) := \left\{ (P_1, \ldots, P_m) \,\middle|\, P_i \in \mathbb{CP}^{m-1}, \det\left(\sum_{i=1}^{m} P_i \right) > 0 \right\}. \tag{18}$$

It is worth noticing that $\mathcal{Q}(m, \mathbb{C})$ is an open and dense Riemannian submanifold of the m-times product of $\mathbb{CP}^{m-1}$ with the Euclidean product metric, i.e.

$$\overline{\mathcal{Q}(m, \mathbb{C})} = \underbrace{\mathbb{CP}^{m-1} \times \ldots \times \mathbb{CP}^{m-1}}_{m-\text{times}} =: \left(\mathbb{CP}^{m-1}\right)^m. \tag{19}$$

Here, $\overline{\mathcal{Q}(m, \mathbb{C})}$ denotes the closure of $\mathcal{Q}(m, \mathbb{C})$. It then follows, that the dimension of $\mathcal{Q}(m, \mathbb{C})$ is equal to the dimension of $\left(\mathbb{CP}^{m-1}\right)^m$, i.e.

$$\dim \mathcal{Q}(m, \mathbb{C}) = m \dim \mathbb{CP}^{m-1} = 2m(m-1), \tag{20}$$

and that the tangent spaces, the geodesics, and the parallel transport for $\mathcal{Q}(m, \mathbb{C})$ and $(\mathbb{CP}^{m-1})^m$ coincide locally. In other words, $\mathcal{Q}(m, \mathbb{C})$ is not a geodesically complete manifold.

Finally, we present the following results about $\mathcal{Q}(m, \mathbb{C})$ without further explanations. Given any $\Upsilon = (P_1, \ldots, P_m) \in \mathcal{Q}(m, \mathbb{C})$, the tangent space of $\mathcal{Q}(m, \mathbb{C})$ at Υ is

$$T_\Upsilon \mathcal{Q}(m, \mathbb{C}) \cong T_{P_1} \mathbb{CP}^{m-1} \times \ldots \times T_{P_m} \mathbb{CP}^{m-1}. \tag{21}$$

Let $\Phi = (\phi_1, \ldots, \phi_m) \in T_\Upsilon \mathcal{Q}(m, \mathbb{C})$ with $\phi_i \in T_{P_i} \mathbb{CP}^{m-1}$ for all $i = 1, \ldots, m$, the product metric on $T_\Upsilon \mathcal{Q}(m, \mathbb{C})$ is constructed as

$$G \colon T_\Upsilon \mathcal{Q}(m, \mathbb{C}) \times T_\Upsilon \mathcal{Q}(m, \mathbb{C}) \to \mathbb{R}, \qquad G(\Phi, \Psi) := \sum_{i=1}^{m} \Re \operatorname{tr}(\phi_i \cdot \psi_i). \tag{22}$$

The geodesic through $\Upsilon \in \mathcal{Q}(m, \mathbb{C})$ in direction $\Phi \in T_\Upsilon \mathcal{Q}(m, \mathbb{C})$ is given by

$$\gamma_{\Upsilon,\Phi} \colon \mathbb{R} \to \mathcal{Q}(m, \mathbb{C}), \qquad \gamma_{\Upsilon,\Phi}(t) := \left(\gamma_{P_1,\phi_1}(t), \ldots, \gamma_{P_m,\phi_m}(t) \right), \tag{23}$$

and the parallel transport of $\Psi \in T_\Upsilon \mathcal{Q}(m, \mathbb{C})$ with respect to the Levi-Civita connection along the geodesic $\gamma_{\Upsilon,\Phi}(t)$ is

$$\tau_{\Upsilon,\Phi}(\Psi) := \left(\tau_{P_1,\phi_1}(\psi_1), \ldots, \tau_{P_m,\phi_m}(\psi_m) \right). \tag{24}$$

4 A CG Algorithm for Simultaneous SUT

In this section we develop a CG algorithm for computing simultaneous SUT. First of all, we adapt a popular diagonality measure of matrices, namely the off-norm cost function, to our problem, i.e.

$$f \colon \mathcal{O}(m, \mathbb{C}) \to \mathbb{R}, \quad f(X) := \sum_{k=1}^{N} \tfrac{1}{2} \left\| \operatorname{off}(X^{\mathsf{H}} C_k X) \right\|_F^2 + \tfrac{1}{2} \left\| \operatorname{off}(X^{\mathsf{H}} R_k \overline{X}) \right\|_F^2, \tag{25}$$

where $\|\cdot\|_F$ denotes the Frobenius norm of matrices. A direct calculation gives

$$\begin{aligned} f(X) &= \sum_{i<j}^{m}\sum_{k=1}^{N} x_i^{\mathsf{H}} C_k x_j (x_i^{\mathsf{H}} C_k x_j)^{\mathsf{H}} + x_i^{\mathsf{H}} R_k \overline{x}_j (x_i^{\mathsf{H}} R_k \overline{x}_j)^{\mathsf{H}} \\ &= \sum_{i<j}^{m}\sum_{k=1}^{N} \operatorname{tr} x_i x_i^{\mathsf{H}} C_k x_j x_j^{\mathsf{H}} C_k + \operatorname{tr} x_i x_i^{\mathsf{H}} R_k \left(x_j x_j^{\mathsf{H}}\right)^{\mathsf{T}} R_k^{\mathsf{H}}. \end{aligned} \tag{26}$$

Clearly, the function f induces the following function $\widetilde{f}$ on $\mathcal{Q}(m, \mathbb{C})$

$$\widetilde{f}\colon \mathcal{Q}(m,\mathbb{C}) \to \mathbb{R}, \qquad \widetilde{f}(\Upsilon) := \sum_{i<j}^{m}\sum_{k=1}^{N} \operatorname{tr} P_i C_k P_j C_k + \operatorname{tr} P_i R_k P_j^{\mathsf{T}} R_k^{\mathsf{H}}. \tag{27}$$

We now compute the first derivative of $\widetilde{f}$ at $\Upsilon \in \mathcal{Q}(m,\mathbb{C})$ in direction $\Phi \in T_\Upsilon \mathcal{Q}(m,\mathbb{C})$ as

$$\mathrm{D}\,\widetilde{f}(\Upsilon)\Phi = \sum_{i\neq j} \operatorname{tr} \phi_i A P_j A + \sum_{i=1}^{m} \operatorname{tr} \phi_i \left(\sum_{i<j}^{m} B P_j^{\mathsf{T}} B^{\mathsf{H}} + \sum_{i>j}^{m} B^{\mathsf{T}} P_j^{\mathsf{T}} \overline{B} \right). \tag{28}$$

Then, the Riemannian gradient of $\widetilde{f}$ at $\Upsilon \in \mathcal{Q}(m,\mathbb{C})$, i.e. $\Phi := \nabla_{\widetilde{f}}(\Upsilon) \in T_\Upsilon \mathcal{Q}(m,\mathbb{C})$, is computed, for each element $\phi_i \in T_{P_i}\mathbb{CP}^{m-1}$, as follows

$$\phi_i = \left[P_i, \left[P_i, \sum_{i\neq j} A P_j A + \sum_{i<j}^{m} B P_j^{\mathsf{T}} B^{\mathsf{H}} + \sum_{i>j}^{m} B^{\mathsf{T}} P_j^{\mathsf{T}} \overline{B} \right]\right]. \tag{29}$$

A conjugate gradient algorithm for minimizing the function $\widetilde{f}$ as defined in (27) is summarized as follows.

Algorithm 1. *A conjugate gradient algorithm*

Step 1: Given an initial guess $\Upsilon^{(0)} \in \mathcal{Q}(m,\mathbb{C})$ and set $i = 0$.

Step 2: Set $i = i + 1$, let $\Upsilon^{(i)} = \Upsilon^{(i-1)}$, and compute

$$\Phi^{(1)} = \Psi^{(1)} = -\nabla_{\widetilde{f}}\left(\Upsilon^{(i)}\right).$$

Step 3: For $j = 1, \ldots, 2m^2 - 2m - 1$:

(*i*) Update $\Upsilon^{(i)} \leftarrow \gamma_{\Upsilon^{(i)},\Phi^{(i)}}(\lambda^*)$, where

$$\lambda^* = \operatorname*{argmin}_{\lambda\in\mathbb{R}} \widetilde{f} \circ \gamma_{\Upsilon^{(i)},\Phi^{(i)}}(\lambda);$$

(*ii*) Compute $\Psi^{(j+1)} = -\nabla_{\widetilde{f}}\left(\Upsilon^{(i)}\right)$;

(*iii*) Update $\Phi^{(j+1)} \leftarrow \Psi^{(j+1)} + \mu\, \tau_{\Upsilon^{(i)},\Phi^{(i)}}(\Phi^{(j)})$, where μ is chosen such that $\tau_{\Upsilon^{(i)},\Phi^{(i)}}(\Phi^{(j)})$ and $\Phi^{(j+1)}$ conjugate with respect to the Hessian of $\widetilde{f}$ at $\Upsilon^{(i)}$.

Step 4: If $\left\|\Upsilon^{(i+1)} - \Upsilon^{(i)}\right\|$ is small enough, stop.

Otherwise, go to Step 2.

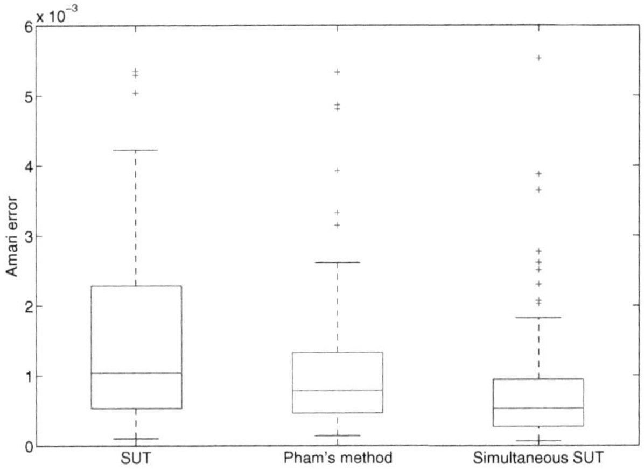

Fig. 1. Separation performance of the proposed CG algorithm

Note that, an iteration from Step 2 to Step 4 is referred to as a CG-sweep. It is known that, finding the local or global minimum of a restricted cost function $\widetilde{f} \circ \gamma_{\Upsilon^{(i)}, \Phi^{(i)}}(\lambda)$ in Step 3-(i) is often unfeasible in practice. In this work, we follow an approach of selecting the step-size λ proposed in [11], which is based on a one-dimensional Newton step, i.e.

$$\lambda^* = -\frac{\frac{\mathrm{d}}{\mathrm{d}\,t} \widetilde{f} \circ \gamma_{\Upsilon^{(i)}, \Phi^{(i)}}(t)\Big|_{t=0}}{\left|\frac{\mathrm{d}^2}{\mathrm{d}\,t^2} \widetilde{f} \circ \gamma_{\Upsilon^{(i)}, \Phi^{(i)}}(t)\Big|_{t=0}\right|}. \tag{30}$$

For updating the direction parameter μ in Step 3-(iii), we confine ourselves to a formula proposed recently in [11]

$$\mu = \frac{G\big(\Psi^{(j+1)}, \Psi^{(j+1)} - \tau\Psi^{(j)}\big)}{G\big(\Psi^{(j)}, \Phi^{(j)}\big)}. \tag{31}$$

5 Numerical Experiments

In our experiment, we investigate performance of our method in terms of separation quality, compared with the single SUT approach and a method developed by Pham in [7], which simultaneously diagonalizes a set of Hermitian positive definite matrices. Separation performance is measured by the normalized Amari error proposed in [12]. Generally, the smaller the Amari error, the better the separation.

The task of our experiment is to jointly diagonalize a set of Hermitian positive definite matrices $\{C_i\}_{i=1}^N$ and a set of complex symmetric matrices $\{R_i\}_{i=1}^N$, which are constructed by

$$C_i = A\Lambda_i A^{\mathsf{H}} + \varepsilon E_H \qquad \text{and} \qquad R_i = A\widehat{\Lambda}_i A^{\mathsf{T}} + \varepsilon E_S \tag{32}$$

where $A \in GL(m, \mathbb{C})$ is randomly picked, real diagonal entries of Λ_i and $\widehat{\Lambda}_i$ are drawn from a uniform distribution on the interval $(0, 10)$, matrices $E_H \in \mathbb{C}^{m\times m}$ and $E_S \in \mathbb{C}^{m\times m}$ are a Hermitian and a complex symmetric matrix, respectively, whose real and imaginary parts are generated from a uniform distribution on the unit interval $(-0.5, 0.5)$, representing additive stationary noises, and $\varepsilon \in \mathbb{R}$ is the noise level. In particular, to make a fair comparison between Pham's approach and ours, we set the number of Hermitian positive definite matrices for Pham's method to diagonalize to be equal to the total number of matrices for our algorithm to diagonalize, i.e. $2N$.

We set $m = 5$, $N = 5, \varepsilon = 0.01$, and run 100 tests. The quartile based boxplot of Amari errors for each method are drawn in Figure 1. Our proposed approach, simultaneous SUT, significantly outperforms the other two methods in terms of Amari error consistently.

Acknowledgments. This work has been supported in parts by the German DFG funded Cluster of Excellence CoTeSys - Cognition for Technical Systems.

References

1. Comon, P.: Independent component analysis, a new concept? Signal Processing 36(3), 287–314 (1994)
2. Eriksson, J., Koivunen, V.: Complex random vectors and ICA models: Identifiability, uniqueness, and separability. IEEE Transactions on Information Theory 52(3), 1017–1029 (2006)
3. Eriksson, J., Koivunen, V.: Complex-valued ICA using second order statistics. In: Proceedings of the IEEE-MLSP 2004, pp. 183–191 (2004)
4. Ollila, E., Koivunen, V.: Complex ICA using generalized uncorrelating transform. Signal Processing 89, 365–377 (2009)
5. Cardoso, J.F.: Blind signal separation: Statistical principles. Proceedings of the IEEE 86(10), 2009–2025 (1998)
6. Douglas, S.C., Eriksson, J., Koivunen, V.: Equivariant algorithms for estimating the strong-uncorrelating transform in complex independent component analysis. In: Rosca, J.P., Erdogmus, D., Príncipe, J.C., Haykin, S. (eds.) ICA 2006. LNCS, vol. 3889, pp. 57–65. Springer, Heidelberg (2006)
7. Pham, D.T.: Joint approximate diagonalization of positive definite Hermitian matrices. SIAM Journal on Matrix Analysis and Applications 22, 1136–1152 (2001)
8. Neeser, F., Massey, J.: Proper complex random processes with applications to information theory. IEEE Transactions on Information Theory 39(4), 1293–1302 (1993)
9. Spivak, M.: A Comprehensive Introduction to Differential Geometry, 3rd edn., vol. 1 - 5. Publish or Perish, Inc. (1999)
10. Afsari, B.: Sensitivity analysis for the problem of matrix joint diagonalization. SIAM Journal on Matrix Analysis and Applications 20, 1148–1171 (2008)
11. Kleinsteuber, M., Hüper, K.: An intrinsic CG algorithm for computing dominant subspaces. In: Proceedings of IEEE-ICASSP 2007, pp. IV1405–IV1408 (2007)
12. Amari, S., Cichocki, A., Yang, H.H.: A new learning algorithm for blind signal separation. In: Advances in Neural Information Processing Systems, vol. 8, pp. 757–763 (1996)

A General Approach for Robustification of ICA Algorithms*

Matthew Anderson and Tülay Adalı

Machine Learning for Signal Processing Laboratory
University of Maryland Baltimore County, Baltimore, MD 21250
matt.anderson@ieee.org, adali@umbc.edu
http://mlsp.umbc.edu/

Abstract. This paper presents a general and robust approach to mitigating impact of outliers in independent component analysis applications. The approach detects and removes outlier samples from the dataset and has minimal impact on the overall performance when the dataset is free of outliers. It also has minimal computational burdens, is simply parameterized, and readily implemented. Significant gains in performance is shown for algorithms when outliers are present.

Keywords: Independent component analysis, non-circular, outliers.

1 Introduction

A common assumption in the design of most independent component analysis (ICA) algorithms is that no outliers are present. As we show with simple examples, the presence of even a single outlier sample can significantly degrade the performance of ICA. In this paper, we develop an approach that can robustify any ICA algorithm when used as a preprocessing step. The algorithm is given for complex-valued data and is readily reduced to handle real-valued data.

We motivate the importance of outlier treatment with an illustration of their impact on performance via simulation in the next section. In Section 3, the genesis of this work, a paper by Ollila and Koivunen [1], is briefly summarized as it motivates the approach of Section 4, namely outlier editing based on a second-order statistic. Section 5 presents several experiments that analyze the sensitivity and performance of the proposed approach. Conclusions are provided in Section 6.

2 Motivation for Outlier Treatment

It is well known and desired that ICA algorithms should have inherent robustness to outliers [2]. Lack of robustness to outliers is a major critique of measures of non-Gaussianity such as kurtosis which quickly degrade when outliers are

* This work is supported by the NSF grants NSF-CCF 0635129 and NSF-IIS 0612076.

V. Vigneron et al. (Eds.): LVA/ICA 2010, LNCS 6365, pp. 295–302, 2010.

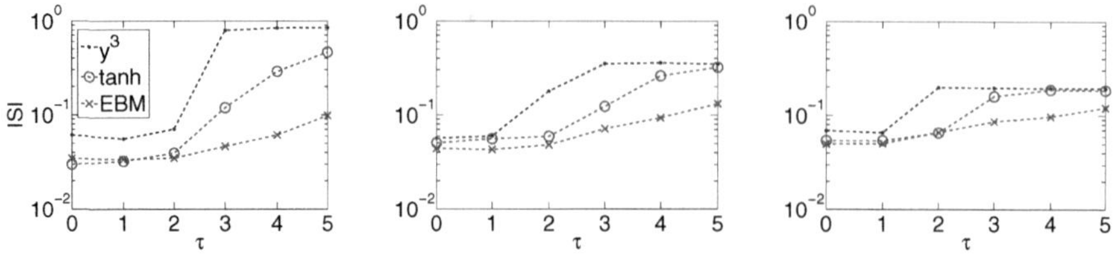

(a) $N = 2$ (b) $N = 5$ (c) $N = 10$

Fig. 1. Median performance measure for uniform sources versus the outlier extent, τ, of 100 trials for FastICA (with y^3 and tanh nonlinearities) and EBM algorithms. Performance degrades with increasing outlier extent.

present. To improve the robustness, other measures of non-Gaussianity are used in ICA algorithms that possess saturating type characteristics. Such behavior restricts the influence of the outliers on algorithm performance. An example of such a saturating function is the tangent hyperbolic function used in FastICA [2].

Here, we illustrate with a simple example that even though *some* robustness is introduced by using these "dampened" nonlinear functions their effectiveness is quite limited. To evaluate the effect on ICA performance, three algorithms are examined. The first algorithm is the FastICA algorithm[1] using the kurtosis cost function. The second algorithm is again FastICA but with the saturating tangent hyperbolic cost. The last algorithm is the real-valued version of entropy bound minimization (EBM) [3]. It makes no explicit claim about robustness but by selecting the "best" solution among a number of nonlinear functions for each source, the algorithm implicitly considers robustness. We simulate two independent uniformly distributed sources, Unif$(-1, 1)$, mixed with a random mixing matrix. One of the 5000 source samples is replaced by an outlier. The magnitude of the outlier is set to have values of $\tau \geq 0$ and its sign is randomly chosen. We use $\tau = 0$ to indicate no outliers injected and note for this formulation $\tau \leq 1$ is not an outlier. The performance metric, median inter-symbol-interference (ISI), is shown in Fig. 1 for $N = 2$, 5, and 10 independent uniform sources. The results indicate that EBM is the most robust and indeed FastICA using the tanh score function is more robust to outliers than FastICA using the kurtosis score function. As expected, the baseline performance, $\tau = 0$, degrades as the number of sources increases, since the number of samples is fixed. The benefit of using the tanh score function is nominally present when $\tau \leq 3$ and is essentially non-existent for more extreme outliers. Similar trends are observed when the sources are from the generalized Gaussian distribution (GGD) with randomly drawn shape parameters.

The simple example clearly illustrates the potential dramatic effect of outliers. Current ICA algorithms address "mild" outliers with some success.

[1] FastICA algorithm available at `http://www.cis.hut.fi/projects/ica/fastica/`

The remainder of the paper is dedicated to improving the performance of ICA algorithms in the presence of outliers.

3 Background

Generally, ICA is used to identify latent independent sources observed via an unknown instantaneous linear mixing. The N independent latent sources are denoted by the column vector $\mathbf{s}(t) = [s_1(t), \ldots, s_N(t)]^T \in \mathbb{C}^N$, where t is the sample index and the superscript T denotes the transpose. The sources are observed indirectly via $\mathbf{x}(t) = \mathbf{A}\mathbf{s}(t)$, where $\mathbf{A} \in \mathbb{C}^{N \times N}$ is an unknown invertible linear mixture matrix. There are T observations of the linear mixture available. The demixing matrix, $\mathbf{W}$, is estimated to form an estimate of the latent sources, $\mathbf{y}(t)$, via $\mathbf{y}(t) = \mathbf{W}\mathbf{x}(t)$. When the sample index, t, is omitted then all the samples of the vector are concatenated to form an $N \times T$ matrix.

The motivation for this paper is derived from the generalized uncorrelating transform (GUT) [1] which is a robustification of the strongly uncorrelating transform (SUT) algorithm introduced in [4]. SUT is an algorithm for blindly identifying complex non-circular sources that have distinct circularity coefficients. By jointly diagonalizing the covariance and the pseudo-covariance matrices, SUT achieves source separation in an efficient manner using only second-order statistics. Here we simply denote the SUT estimate of the demixing matrix as $\mathbf{W}_{\mathrm{SUT}} = \mathrm{SUT}(\mathbf{x})$.

In [1] the SUT approach is extended to address outliers. The critique about the sensitivity of kurtosis to outliers can be extended to SUT and motivates the GUT formulation which can be simply summarized as a weighted SUT, with more weighting on "smaller" values and less weighting on "larger" values. To be clear, in this paper the term 'outlier' is a small set of samples that have a large magnitude relative to all other samples.

The GUT approach is shown in [1] to be a separating matrix estimator that is still based on second-order statistics. The general concept is to apply the SUT algorithm to estimates of the covariance and pseudo-covariance matrices that have been estimated in robust manners. An essential idea is presented for measuring how far each sample is in a normalized statistical measure from other samples. The measure is denoted as $\gamma_{\mathbf{x}}(t) = \mathbf{x}^H(t)\mathbf{C}_{\mathbf{x}}^{-1}\mathbf{x}(t)$, where $\mathbf{C}_{\mathbf{x}} = E\{\mathbf{x}(t)\mathbf{x}^H(t)\}$, $E\{\cdot\}$ denotes the expectation operator, and superscript H is the complex conjugate transpose. The quantity $\gamma_{\mathbf{x}}(t)$ is termed the generalized-inner-product (GIP) sample [5]. One possible approach for robust estimation of the covariance and pseudo-covariance matrices is to first estimate the GIP values for each sample using the mixture outer product average, $\mathbf{x}\mathbf{x}^H/T$, to estimate the covariance matrix, $\mathbf{C}$. Various weighting functions can be formulated with the estimated GIP values so that samples with large GIP values are deweighted relative to samples with small GIP values. In [1], several weighting functions, $\varphi_t(\gamma_{\mathbf{x}})$, are examined and based on the analysis, a Huber-based weighting function that varies with N is used as shown in Fig. 2. Finally, this

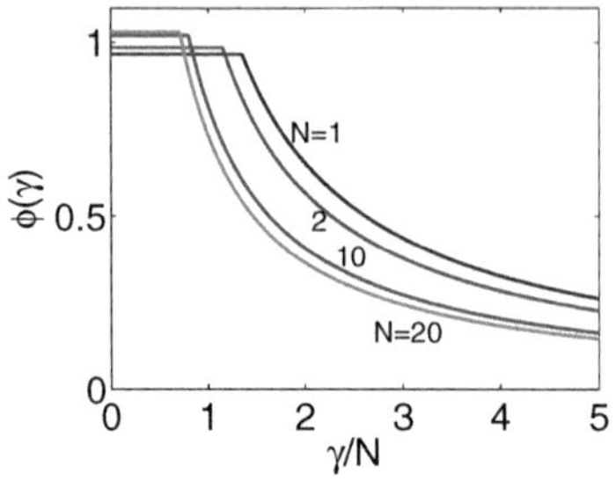

Fig. 2. Example showing Huber-type weighting function suggested in [1] for various dimensions. Shows deweighting of large GIP normalized, γ/N, values.

particular GUT implementation is summarized as SUT with weighted samples:

$$\mathbf{x}_{\mathrm{GUT}}(t) = \sqrt{\varphi_t(\gamma_{\mathbf{x}})}\mathbf{x}(t) \qquad t = 1, \ldots, T \tag{1}$$

$$\mathbf{W}_{\mathrm{GUT}} = \mathrm{SUT}(\mathbf{x}_{\mathrm{GUT}}). \tag{2}$$

4 Outlier Detection

In this section, a method for robust outlier detection is formulated based on the GIP statistic. The goal is to find a method that can reliably detect and edit outliers without discarding non-outliers. Then the ICA approaches based on the assumption of no outliers can be utilized without modification.

The GIP statistic is characterized in [6] under the assumption that samples are normally distributed. It can be argued that mixtures observed in ICA applications tend toward the normal as the number of sources increases using a central limit theorem type argument. Thus in ICA applications, the GIP statistic is well characterized and can be used to detect outliers.

There are number of potential approaches for using the GIP statistic to detect outlier samples. We use a *non-recursive* approach such that the GIP statistic is only estimated once for each sample using a covariance matrix estimate that is fixed for all samples and is potentially contaminated by outliers. The approach exploits the known distribution of the GIP [6] to set a threshold for valid GIP values based on an acceptable probability for falsely declaring an outlier, p_{fa}. More explicitly the distribution of the GIP under the Gaussian assumption is

$$p(\gamma) = \frac{\gamma^{N-1}}{\Gamma(N)} \exp(-\gamma) \quad 0 \leq \gamma < \infty. \tag{3}$$

This is the standard chi-square distribution with $2N$ degrees of freedom scaled by $1/2$ so that if $Z \sim \chi^2(2N)$, then $\gamma = z/2$. From this characterization, the threshold for the maximum acceptable GIP value for some number of components, N, and a desired p_{fa} is then,

$$\gamma_{\max} = F_{\chi^2}^{-1}(1 - p_{\mathrm{fa}}, 2N), \tag{4}$$

where $F_{\chi^2}(x, v)$ is the cumulative probability distribution function for the chi-square distribution with v degrees of freedom.

Since the proposed approach is non-recursive, the presence of just a few outliers can inflate the covariance matrix estimate. This inflation is advantageous as it allows us to increase the GIP threshold, i.e., use a lower p_{fa}, and still retain a high probability of outlier detection. Furthermore, the robustness of ICA algorithms to mild outliers illustrated previously, further justifies using a high threshold to detect and edit large outliers.

5 Application Examples

In this section, the performance of the proposed outlier detect and edit approach is demonstrated using three examples; GGD sources, heavy-tailed sources, and a communications example.

To examine outlier performance we use the model for injecting outliers defined in [1], where a fraction, f_{out}, of the samples are randomly selected to be replaced by outliers. The outliers have the ith component given by $\mathbf{x}_{\mathrm{out},i} = b_i u_i z_i$, where b_i is equally likely ± 1, u_i is $\mathrm{U}(1, \tau)$, z_i is the maximum modulus of the ith mixture component. The larger τ is, the more extreme the modeled outliers.

To examine performance of ICA algorithms, a common performance index sometimes referred to as the inter-symbol-interference (ISI) metric is used,

$$\mathrm{ISI}(\mathbf{G}) \triangleq \frac{\sum_{n=1}^{N}\left(\sum_{m=1}^{N} \frac{g_{n,m}}{\max_p g_{n,p}} - 1\right) + \sum_{m=1}^{N}\left(\sum_{n=1}^{N} \frac{g_{n,m}}{\max_p g_{p,m}} - 1\right)}{2N(N-1)}, \quad (5)$$

where $\mathbf{G} = |\mathbf{WA}|$. The ISI is a normalized measure in that $0 \leq \mathrm{ISI} \leq 1$, where 0 is ideal. Additionally, we note that the ISI performance index is sensitive to the scaling of $\mathbf{W}$ and $\mathbf{A}$. We address this scaling ambiguity by normalizing $\mathbf{W}$ and $\mathbf{A}$ so that the original latent sources and their estimations have the same variance.

The results of SUT and GUT are compared to the complex entropy bound minimization (C-EBM) algorithm which has been shown to provide superior and robust performance for a wide range of complex source types [7].

5.1 Generalized Gaussian Sources

The sources are generated using a complex non-circular GGD source generator[2] [8] with unitary power for both the real and imaginary components, uniformly distributed shape parameter, Unif(0.1, 3.1), and a uniformly distributed correlation coefficient, Unif(0, 1). We note that the generated sources are both super and sub-Gaussian, each with distinct circularity coefficients as required by SUT and GUT theory. The mixing matrix, $\mathbf{A}$, elements, $a_{i,j}$, are complex and normally distributed.

[2] Complex non-circular GGD code at: `http://mlsp.umbc.edu/codes/getComplexGGD.m`

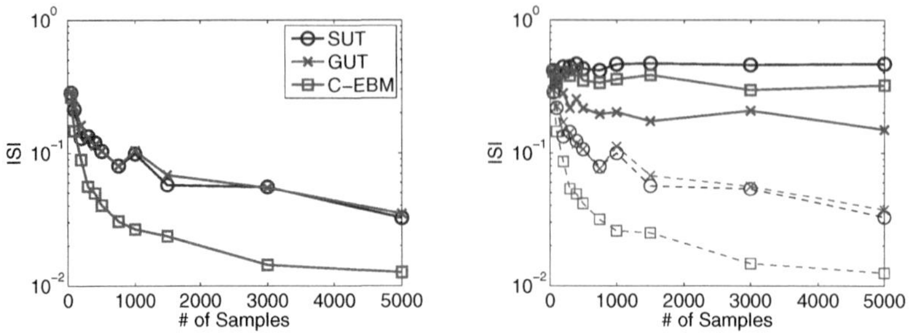

(a) No Outliers (b) 1% Outliers

Fig. 3. Performance of three ICA algorithms, the solid lines have no editing and the dashed lines are using GIP-based editing ($p_{\text{fa}} = 0.005$)

The results for $N = 4$ sources are shown in Fig. 3. The potential negative effect associated with false or over-editing (Fig. 3a) is not discernible as less than 0.5% of the samples are erroneously discarded when no outliers are present. The benefit of editing is illustrated when outliers are introduced with $f_{\text{out}} = 0.01$ and $\tau = 5$ in Fig. 3b. Without editing, GUT provides the most robust performance in presence of outliers, yet still its performance is degraded compared to outlier free performance. With editing, the performance of all three approaches is nearly restored to that achieved when no outliers are present and no editing is performed. Similar results are achieved for differing number of sources and a wide range of reasonable user p_{fa} settings. Simple simulations can guide the user to the appropriate choice for the p_{fa} parameter according to the anticipated source, mixture, and outlier characteristics. The benefit of outlier editing and the GUT approach reduces as the outlier extent, τ, decreases.

5.2 Heavy-Tailed Sources

The following example illustrates how the GIP-based editing can actually improve performance even when there are no outliers in the dataset. In the simulation example of [1] titled "Setting C", $N = 4$ sources are simulated with $\text{Re}(s_1) \sim \text{Logist}(1)$, $\text{Im}(s_1) \sim \text{Logist}(2)$, $\text{Re}(s_2) \sim \text{Cau}(1)$, $\text{Im}(s_2) \sim \text{Cau}(2)$, $s_3 \sim \text{Ell}(1,0)$, $s_4 \sim \text{CN}(\epsilon, \vartheta)$, with $\epsilon \sim \text{Unif}(0,1)$, $\vartheta \sim \text{Unif}(-\pi/2, \pi/2)$, where $\text{Logist}(k)$ and $\text{Cau}(k)$ are the logistic and Cauchy distributions with scale parameter k, $\text{CN}(\epsilon, \vartheta)$ is the unit variance complex normal distribution with eccentricity ϵ and orientation ϑ, and $\text{Ell}(1,0)$ has a uniform distribution on the unit complex circle. The performance with and without GIP editing is shown in Fig. 4. For the heavy-tailed distributions, since we would need a significantly large number of samples to approximate the actual distribution and hence the true statistics, in this case, the performances of both GUT and C-EBM improve after editing the occasionally large samples generated in these cases.

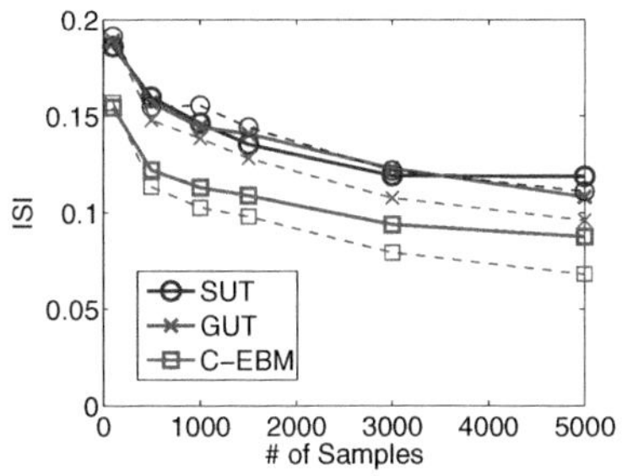

Fig. 4. Performance of three ICA algorithms with no outliers present, the solid lines have no editing and the dashed lines use GIP-based editing ($p_{\text{fa}} = 0.005$)

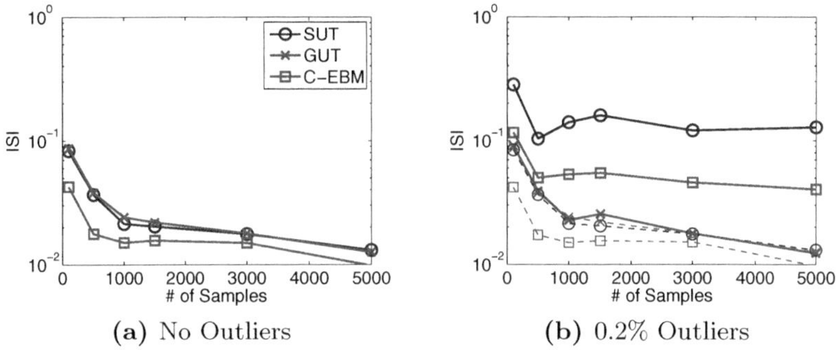

Fig. 5. Communication example performance, without editing (solid) and with GIP editing (dashed) using $p_{\text{fa}} = 0.005$

5.3 Communications Example

We now use the simple communication example of [1] to further illustrate performance. The communication channel is a three element uniform linear array with half-wavelength interelement spacing. Three independent signals impinge on the array — a BPSK, a 8-QAM, and a circular Gaussian signal, with equal power from angles of $-20°$, $5°$, and $35°$, respectively, relative to the array normal. Each signal is corrupted by additive Gaussian noise, i.e., $\mathbf{x} = \mathbf{A}\mathbf{s} + \mathbf{n}$, such that the signal to noise ratio is 20 dB. With the additive noise, none of the examined ICA approaches can claim optimality since all are derived assuming noiseless data. In Fig. 5a, complex EBM is better than SUT, which is only slightly better than GUT. The effect of editing cannot be discerned from the plots since the number of outliers removed is minimal ($< 0.5\%$) when no outliers are present.

Outliers are added using the same procedure described previously with $f_{\text{out}} = 0.002$ and $\tau = 5$. The resulting performance with and without editing is shown in Fig. 5b. This example shows that C-EBM and SUT nearly recover outlier-free performance after outlier editing. We also notice that the GUT implementation is robust in this example as the performance with outliers present is nearly identical to the outlier-free performance.

6 Conclusions

The effect of outliers on ICA algorithms is a significant problem in many "real-world" applications. Current approaches incorporate robustness to outliers within the ICA algorithm itself. These approaches can be sufficient for mild outliers. In this paper, we present a robust and flexible approach to handling outliers based on the GIP statistic. The procedure is not computationally burdensome, as it is based on a second-order statistic. The approach becomes agnostic to source distributions as the number of sources is increased based on central limit theorem type arguments.

The suggested approach is to use the GIP-based editing to remove large outliers in a preprocessing step. Any remaining mild outliers after editing can be handled by using a robust ICA algorithm. We note that the outlier editing should be performed prior to the whitening procedure used by many ICA algorithms. The argument for using this outlier preprocessing step is demonstrated via examples. Specifically, the degradation in performance due to editing when no outliers are present is minimal and the performance gain when there are outliers present is potentially substantial.

References

1. Ollila, E., Koivunen, V.: Complex ICA using generalized uncorrelating transform. Signal Processing 89(4), 365–377 (2009)
2. Hyvarinen, A., Karhunen, J., Oja, E.: Independent Component Analysis. Wiley Interscience, Hoboken (2001)
3. Li, X.L., Adalı, T.: A novel entropy estimator and its application to ICA. In: IEEE International Workshop on Machine Learning for Signal Processing, MLSP 2009, pp. 1–6 (September 2009)
4. Eriksson, J., Koivunen, V.: Complex-valued ICA using second order statisitcs. In: Proceedings of the 2004 14th IEEE Signal Processing Society Workshop, Machine Learning for Signal Processing, pp. 183–192 (2004)
5. Teixeira, C.M., Bergin, J.S., Techau, P.M.: Adaptive thresholding of non-homogeneity detection for STAP applications. In: Proceedings of the IEEE Radar Conference, pp. 355–360 (April 2004)
6. Rangaswamy, M., Michels, J.H., Himed, B.: Statistical analysis of the non-homogeneity detector for STAP applications. Digital Signal Processing 14(3), 253–267 (2004)
7. Li, X.L., Adalı, T.: Complex independent component analysis by entropy bound minimization. IEEE Trans. Circuits Syst. I, Reg. Papers (in press)
8. Novey, M., Adalı, T., Roy, A.: A complex generalized gaussian distribution; characterization, generation, and estimation. IEEE Transactions on Signal Processing 58, 1427–1433 (2010)

Strong Sub- and Super-Gaussianity

Jason A. Palmer[1]*, Ken Kreutz-Delgado[2], and Scott Makeig[1]

[1] Swartz Center for Computational Neuroscience
[2] Department of Electrical and Computer Engineering
University of California San Diego, La Jolla, CA 92093
{jason,scott}@sccn.ucsd.edu, kreutz@ece.ucsd.edu

Abstract. We introduce the terms *strong sub- and super-Gaussianity* to refer to the previously introduced class of densities log-concave is x^2 and log-convex in x^2 respectively. We derive relationships among the various definitions of sub- and super-Gaussianity, and show that strong sub- and super-Gaussianity are related to the score function being star-shaped upward or downward with respect to the origin. We illustrate the definitions and results by extending a theorem of Benveniste, Goursat, and Ruget on uniqueness of separating local optima in ICA.

1 Introduction

In their seminal work on blind deconvolution, Benveniste, Goursat, and Ruget [3] proposed a definition of sub- and super-Gaussianity, and used it to derive conditions for blind identifiability of the unmixing deconvolutive system by minimization of the expected value of certain classes of functions. Unlike the Cardoso-Amari stability conditions [1,7], which are local stability conditions for separating solutions based on second derivatives, the BGR conditions based on super- and sub-Gaussianity give conditions not only for stability, but also for uniqueness of separating local optima, and are based on first derivatives, not requiring finite curvature. While their work is credited with being fundamental [10,8], the BGR definition does not seem to have been employed much in subsequent development in the ICA/BSS and latent variable communities, though the definition of super-Gaussianity turns out to coincide with that used to guarantee monotonicity in the sparse solution of underdetermined systems [15]. In this paper, we attempt to illuminate the BGR concept of sub- and super-Gaussianity, which we refer to as *strong sub-* and *super-Gaussianity*, by situating the definition in a nested hierarchy that includes the more commonly used characterizations such as kurtosis, and Gaussian density crossings. We also extend the class of functions that lead to unique globally optimal separating solutions for super-Gaussian sources to the more naturally related class of star-shaped functions.

2 Definitions and Set Inclusion Relationships

Qualitatively, super-Gaussianity may be thought of as implying *both* a sharper peak and heavier tail than the corresponding Gaussian, while sub-Gaussian densities have

* This research was partially supported by NSF grants ISS-0613595 and CCF-0830612.

V. Vigneron et al. (Eds.): LVA/ICA 2010, LNCS 6365, pp. 303–310, 2010.

flatter peaks, "heavier shoulders", and lighter tails. Quantitatively, we shall consider in particular the following four definitions of the classes of sub- and super-Gaussian densities.

1. *Fourth-order cumulant.* Let K_+ to be the set of symmetric densities with positive or infinite fourth-order cumulant, or excess kurtosis. Let K_- be the set of densities with negative fourth-order cumulant.
2. *Density crossing inequalities.* Let DC_+ be the set of symmetric, finite variance densities that cross a Gaussian density of equal variance exactly four times, with higher density at the origin and on the tails. Let DC_- be defined similarly, with lower density at the origin and on the tails.
3. *Strong sub-/super-Gaussianity.* Let SS_+ be the set of symmetric densities, $p(x)$, such that $\log p(\sqrt{x})$ is convex on $(0, \infty)$. or equivalently densities of the form $\exp(-f(x))$ such that $f'(x)/x$ is non-increasing on $(0, \infty)$. Let SS_- be the set of symmetric densities with $\log p(\sqrt{x})$ concave on $(0, \infty)$, or equivalently $p(x) = \exp(-f(x))$ with $f'(x)/x$ non-decreasing on $(0, \infty)$.
4. *Convexity of score function.* Let SC_+ be the set of symmetric densities $p(x) = \exp(-f(x))$ such that $f'(x)$ is concave on $(0, \infty)$. Let SC_- be the set of densities with $f'(x)$ convex on $(0, \infty)$.

Consider a random variable X with probability density $p(x)$, mean $\mu = E\{X\}$, and variance $\sigma^2 = E\{(X - \mu)^2\}$. We can define a measure based on the fourth moment that is invariant to changes in mean and scale, sometimes called the normalized or standardized fourth moment, or kurtosis, $\kappa = E\{(X-\mu)^4\}/\sigma^4$. Employing cumulants, we find that the first and second cumulants equal the mean and variance, and the fourth cumulant, γ, is equal to $\kappa - 3$. The fourth cumulant is sometimes called the excess kurtosis, since the kurtosis of the standard Normal density is 3. The most commonly used definition of sub- and super-Gaussianity involves the sign of the fourth cumulant, i.e. the kurtosis relative to a Gaussian of equal variance. If the kurtosis exceeds that of the Gaussian, then X, or its density $p(x)$, is said to be super-Gaussian. Likewise if γ is negative, or $\kappa < 3$, then X is said to be sub-Gaussian.

2.1 Density Crossings and Karlin's Theory

Let $q(x)$ be a Gaussian density with variance equal to that of the symmetric density $p(x)$. According to [9], the proposition was known since R. A. Fisher that any symmetric density which has sharper peak and heavier tails at unit variance, defined by $p(x)$ crossing $q(x)$ four times, with $p(x) > q(x)$ near $x = 0$, and $p(x) > q(x)$ as $|x| \to \infty$, will have positive fourth cumulant. Finucan [9] proves this proposition.[1] The relationship between density crossings and moments in clarified by the work of S. Karlin, which we shall briefly describe below.

Let $p(x)$ and $q(x)$ be symmetric probability densities. We first note that any symmetric function $\psi(x)$ that is increasing on $(0, \infty)$ can be used as a normalizing condition, $\int \psi(x)p(x)dx = \int \psi(x)q(x)dx$ in the definition of a class of moment-based sub- and super-Gaussianity. We employ the two sign change form of Lemma A of [11].

[1] In [13] the authors prove a similar theorem, where a concept of "over-Gaussianity" is defined using the criterion of heavier tail only.

Lemma 1. *If $\int \psi(x)p(x)dx = \int \psi(x)q(x)dx$, with $\psi(x)$ symmetric and increasing on $(0,\infty)$, then $p(x)$ and $q(x)$ cross each other at least four times, i.e. $p(x) - q(x)$ has at least two sign changes on $(0,\infty)$.*

We say that $\varphi(x)$ is *convex with respect to* $\psi(x)$ on (a,b) if,

$$\begin{vmatrix} 1 & \psi(x_1) & \varphi(x_1) \\ 1 & \psi(x_2) & \varphi(x_2) \\ 1 & \psi(x_3) & \varphi(x_3) \end{vmatrix} > 0, \quad a \le x_1 < x_2 < x_3 \le b$$

Ordinary convexity is obtained for $\psi(x) = x$. Just as a convex function can be intersected by a linear function at most two times, if $\varphi(x)$ is convex with respect to $\psi(x)$, then $\varphi_{\alpha,\beta}(x) = \alpha\psi(x) + \beta$ and $\psi_{\gamma,\delta}(x) = \gamma\varphi(x) + \delta$ can intersect at most two times for all $\alpha, \beta, \gamma, \delta$. And when $\alpha, \gamma > 0$, and $\varphi_{\alpha,\beta}(x)$ and $\psi_{\gamma,\delta}(x)$ intersect two times on (a,b), then $\varphi_{\alpha,\beta}(x) > \psi_{\gamma,\delta}(x)$ in a neighborhood of b. If $\varphi(x)$ is convex with respect to $\psi(x)$, we also say that $\psi(x)$ is *concave with respect to* $\varphi(x)$. It is clear from the definition that $\varphi(x)$ is convex with respect to $\psi(x)$ on (a,b) if and only if $\varphi(\psi^{-1}(x))$ is convex on $(\psi(a), \psi(b))$.

Karlin and Novikoff's Lemma B states.

Lemma 2. *If $\int \psi(x)p(x)dx = \int \psi(x)q(x)dx$, with $\psi(x)$ symmetric and increasing on $(0,\infty)$, and $p(x)$ and $q(x)$ intersect exactly two times on $[0,\infty)$, then $\int \varphi(x)p(x)dx \ge \int \varphi(x)q(x)dx$ for all $\varphi(x)$ convex with respect to $\psi(x)$.*

Taking $q(x)$ to be Gaussian, $\psi(x) = x^2$, and $\varphi(x) = x^4$ yields the Finucan-Fisher theorem. Taking $q(x)$ Gaussian and $\psi(x) = |x|$ in Thm. 1 implies that $E\{\varphi(X)\} \ge E\{\varphi(Z)\}$ for Gaussian Z and strongly super-Gaussian X satisfying $E\{|X|\} = E\{|Z|\}$, allowing the moment-based definition to extend to random variables which do not have finite variance.

2.2 Strong Sub- and Super-Gaussianity

If we assume that $\log p(x)$ is convex with respect to $\log q(x)$, or vice-versa, then the normalized versions of $p(x)$ and $q(x)$, which intersect at least two times on $(0,\infty)$, must in fact intersect exactly two times, since otherwise $\log p(x)$ and $\log q(x)$ would also intersect more than two times contradicting the relative convexity of $\log p(x)$ and $\log q(x)$. We thus have the following theorem.

Theorem 1. *Let $p(x)$ and $q(x)$ be unimodal, symmetric probability densities, and let $\int \psi(x)p(x)dx = \int \psi(x)q(x)dx$, with $\psi(x)$ increasing on $[0,\infty)$. Let $\varphi(x)$ be convex with respect to $\psi(x)$ on $[0,\infty)$. Then if $\log p(x)$ is convex with respect to $\log q(x)$, then $\int \varphi(x)p(x)dx \ge \int \varphi(x)q(x)dx$.*

If we take $q(x)$ to be Gaussian, then the condition that $\log p(x)$ be convex (concave) with respect to $\log q(x)$ on $[0,\infty)$ is equivalent to saying that $-\log p(x)$ is convex (concave) with respect to x^2, or $-\log p(\sqrt{x})$ is convex (concave) on $(0,\infty)$. This condition forms the definition of strong sub- and super-Gaussianity proposed by [4].

Definition 1. *A symmetric probability density $p(x)$ is **strongly super-gaussian** if $p(\sqrt{x})$ is log-convex on $(0,\infty)$, and **strongly sub-gaussian** if $p(\sqrt{x})$ is log-concave on $(0,\infty)$.*

The requirement that $f(x) \triangleq -\log p(x)$ be convex or concave with respect to x^2 can be expressed differentially in the same way that ordinary convexity can. Letting x_1 tend to x_2 in the determinantal definition of relative convexity yields the first order condition that $\varphi(x)$ is convex with respect to $\psi(x)$ if $\varphi'(x)/\psi'(x)$ is non-decreasing. Thus the differential condition for $f(x)$ convex (concave) with respect to x^2 on (a, b) is that,

$$f'(x)/x \quad \text{non-decreasing (non-increasing) on } (a, b)$$

This condition on the first derivative of $f(x)$ can be expressed more intuitively using the concept of *star-shaped functions* [6]. Recall that a set S is star-shaped with respect to a point $x \in S$ if every point $z \in S$ can be "seen" by x, i.e. for every point y on the line joining x and z, $y \in S$. Classically, a function is said to be star-shaped if the epigraph of $f(x)$ (i.e. points (x, y) such that $y \geq f(x)$) is star-shaped with respect to the origin. This definition seems to be overly restrictive. We extend the classical definition of star-shaped functions to include *star-shaped upward* and *star-shaped downward* functions.

Definition 2. *A function $f(x)$ is* ***star-shaped upward*** *if the epigraph of $f(x)$ is star-shaped with respect to the origin. $f(x)$ is* ***star-shaped downward*** *if the hypograph (i.e. (x, y) such that $y \leq f(x)$) is star-shaped with respect to the origin.*

Geometrically, if a line segment is drawn joining the origin to the point $(x, f(x))$, the slope is given by $f(x)/x$. If $f(x)$ is star-shaped upward, then by definition for $0 \leq \alpha \leq 1$, we have $f(\alpha x) \leq \alpha f(x)$, or $f(\alpha x)/(\alpha x) \leq f(x)/x$. Thus an equivalent defining criterion of star-shaped upward (downward) functions is that $f(x)/x$ be non-decreasing (non-increasing). We may thus formulate a natural geometric definition of strong sub- and super-Gaussianity in terms of the (location) score function $\psi(x) \triangleq (d/dx) \log p(x)$: *A symmetric density $p(x)$ is strongly super-Gaussian (sub-Gaussian) if and only if the score function $\psi(x)$ is star-shaped upward (downward).*

We note from the functions $\exp(-|x|^p)$, which are super-Gaussian for $p < 2$ and sub-Gaussian for $p > 2$ (by all definitions) that the score function in this case is convex in the super-Gaussian case, and concave in the sub-Gaussian case. We might thus define a (somewhat strict) form of sub- and super-Gaussianity accordingly as the score function is convex or concave on $(0, \infty)$. Since functions that are non-positive and convex are also star-shaped upward, and functions that are non-positive and concave are star-shaped downward, we see that the score convex and concave classes SC_+ and SC_- are strict subsets of strong super- and sub-Gaussian classes SS_+ and SS_- respectively.

Putting these relationships together, we have the following.

Theorem 2. *We have,*

$$K_+ \supset DC_+ \supset SS_+ \supset SC_+$$

and similarly for the sub-Gaussian sets, with the unique intersection of all sub- and super-Gaussian sets being the Gaussian density, as illustrated in the Venn diagram in Figure 2a.

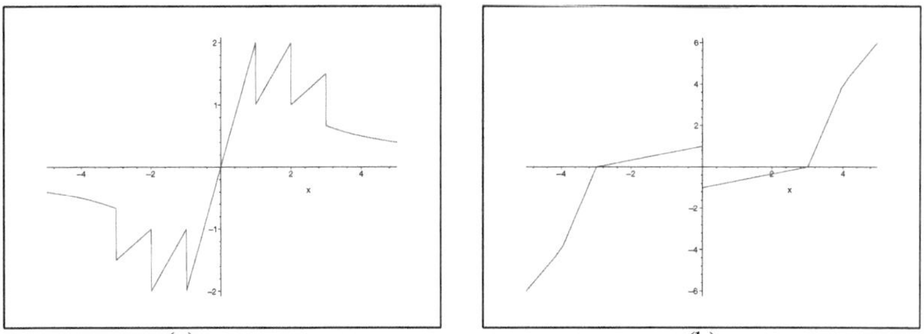

(a) (b)

Fig. 1. Star-shaped functions. (a) Star-shaped downward function corresponding to a strong super-Gaussian density. (b) Star-shaped upward function corresponding to a (non-unimodal) strongly sub-Gaussian density. In each function, the points on the graph are "visible" to the origin (joined by a line segment without intersecting the graph). Star-shaped downward functions must be odd and non-negative on $(0, \infty)$, so that strong super-Gaussians must be unimodal. Star-shaped downward functions may be negative near the origin, and thus strong sub-Gaussians can be bimodal. Neither require continuity, monotonicity, differentiability, or convexity or concavity.

2.3 Scale Mixture Representations

Gaussian scale mixtures constitute a large class of super-Gaussian densities, closed under convolution, and in fact are the uniformly convergent limit of a sequence functions that may be called n-times monotone [17].

Definition 3. *$p(x)$ is **n-times monotone** on (a, b) if $(-1)^k f^{(k)}(x)$ is non-negative, non-increasing and convex on (a, b) for $k = 0, 1, 2, \ldots, n-2$.*

Thus n-times monotone functions have derivatives of alternating sign up to order n when sufficiently differentiable. If this holds for all n, then we have *complete monotonicity* [16]: A function $f(x)$ is *completely monotonic* on (a, b) if $(-1)^n f^{(n)}(x) \geq 0, \quad n = 0, 1, \ldots$ for every $x \in (a, b)$. Bernstein's theorem [16, Thm. 12b] states: A necessary and sufficient condition that $p(x)$ should be completely monotonic on $(0, \infty)$ is that $p(x) = \int_0^\infty e^{-tx} d\alpha(t)$, where $\alpha(t)$ is non-decreasing on $(0, \infty)$. Similarly, the following theorem of Williamson [17] states the conditions for n-times monotonicity.

Theorem 3. *A necessary and sufficient condition that $p(x)$ should be n-times monotonic on $(0, \infty)$ is that $p(x) = \int_0^\infty (1 - tx)_+^{n-1} d\alpha(t)$, where $\alpha(t)$ is non-decreasing and bounded below on $(0, \infty)$.*

We define the class $M^2(n)$ to be the set of functions n-times monotone in x^2:

Definition 4. *The class of functions $M^2(n)$ consists of all functions of the form,*

$$p(x) = \int_0^\infty (1 - tx^2)_+^{n-1} d\alpha(t)$$

where $\alpha(t)$ is non-decreasing and bounded below on $(0, \infty)$.

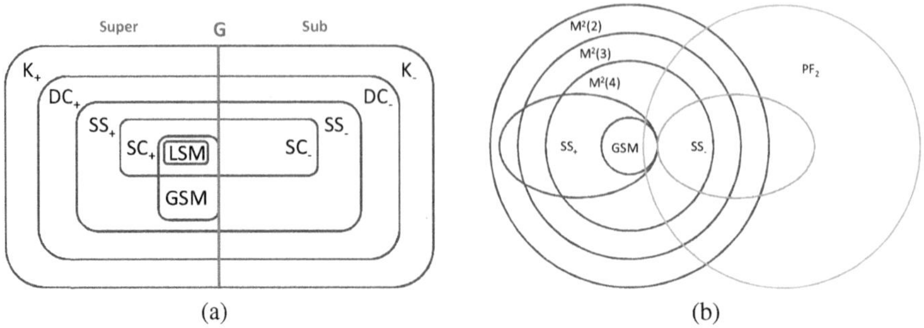

Fig. 2. (a) Venn diagram showing set inclusions among sub- and super-Gaussian desnties. Gaussian (G) is represented by the central line, and lies in the intersection of all sets except LSM. K_+ and K_- are defined by moment-based criteria, e.g. excess kurtosis, and form the outermost shell. The set of densities satisfying crossing properties with respect to a normalized Gaussian density, DC_+ and DC_- form subsets of K_+ and K_- respectively, as shown by Karlin's theory. The set of strong sub- and super-Gaussians, SS_- and SS_+ are again strict subsets of the density crossing classes, and the score concave and convex classes, SC_+ and SC_- are strict subsets of SS_+ and $SS_=$. The class of Gaussian scale mixtures, GSM, is a strict subset of SS_+, and strictly contains the set of Laplacian scale mixtures, LSM, which is itself strictly contained in the class of score convex densities, SC_+. (b) Venn diagram showing set inclusions among strong sub- and super-Gaussian classes, various scale mixtures, and the log-concave class. The class of densities n-times monotone in x^2, $M^2(n)$, tends to GSM as $n \to \infty$. The class of strong super-Gaussians, SS_+ is strictly contained in $M^2(2)$, but not in $M^2(n)$ for $n \geq 3$. The set of unimodal strong sub-Gaussians is strictly contained in the log-concave class, PF_2 (Polya frequency functions of order 2), which however also contains densities that are in SS_+.

By Williamson's theorem, this is equivalent to $(-1)^k(d/dx)^k p(\sqrt{x})$ non-negative, non-increasing, and convex on $(0,\infty)$, for $k = 1,2,\ldots,n-2$. It is obvious from this that $M^2(n) \subset M^2(m)$ for $m < n$. By the Bernstein-Widder theorem, we have the result that [12,2], a function $p(x)$ can be represented as a Gaussian scale mixture if and only if $p(\sqrt{x})$ is completely monotonic on $(0,\infty)$. Concavity of $-\log p(\sqrt{x})$ follows from the complete monotonicity of $p(\sqrt{x})$ since sums of log-convex functions are again log-convex [5, §3.5.2]. Thus, completely monotonic functions, being scale mixtures of the log-convex on $(0,\infty)$ function, $\exp(-x)$, are also log-convex on $(0,\infty)$. We thus have the following [14], *All Gaussian scale mixtures are strongly super-Gaussian, i.e.* $GSM \subset SS_+$. The inclusion is strict.

We also note that since strong super-Gaussianity of $p(x)$ implies that $p(\sqrt{x})$ is convex (and non-negative and non-increasing) on $(0,\infty)$, $p(x) \in M^2(2)$ by Williamson's theorem. Hence $SS_+ \subset M^2(2)$. $M^2(n)$ in fact includes sub- as well as super-Gaussian densities (the functions $(1-x^2)_+^n$ are all sub-Gaussian,) but the scale mixture representation using $M^2(2)$ will be useful when we consider the blind deconvolution theorem. It is unclear whether the class SS_+ admits an equivalent scale mixture representation, but we shall show in the next section that for $p(x) \in SS_+$, $\log p(x)$ can be represented by a scale mixture, and thus $p(x)$ can be represented by a type of product mixture.

We finally note that the classes $M^2(n)$, along with the limiting class GSM, are closed under convolution. The closure of GSM is straightforward, owing to the well-known closure of the Gaussian density under convolution. In fact, for X and Y Gaussian scale mixtures, $X = \xi_1^{1/2} Z_1$, $Y = \xi_2^{1/2} Z_2$ where ξ_1, ξ_2 are non-negative i.i.d, and Z_1 and Z_2 are i.i.d. Gaussian, we have $X + Y \stackrel{d}{=} (\xi_1 + \xi_2)^{1/2} Z$ [12]. The following are used in the sequel is stated without proof due to space constraints.

Theorem 4. *The classes $M^2(n)$ are closed under convolution. Furthermore, if $p(x) \in M^2(m)$ and $q(x) \in M^2(n)$, then $\int p(x-t)q(t)dt \in M^2(\min(m,n))$.*

Theorem 5. *If $\psi(x)$ is star-shaped downward, and $p(x) \in M^2(2)$, then $\int_{-\infty}^{\infty} p(x - t)\psi(t)dt$ is star-shaped downward.*

3 Uniqueness of Separating Local Optima in ICA

To illustrate the application of the definition of strong sub- and super-Gaussianity (the BGR definition) we prove the following theorem.

Theorem 6. *Let $\mathbf{x} = \mathbf{A}\mathbf{s}$ where s_i are i.i.d., zero mean, and strongly super-Gaussian. If $\varphi'(y)$ is star-shaped downward, then,*

$$L(\mathbf{w}) = E\{\varphi(\mathbf{w}^T\mathbf{x})\}$$

has a local minimum $\mathbf{w}^*$ with $\|\mathbf{w}^*\| = 1$, if and only if $\mathbf{w}^{*T}\mathbf{A} = c\,\mathbf{e}_j^T$ for some j, i.e. $\mathbf{w}^{*T}$ is a row of the inverse of $\mathbf{A}$ up to scaling and permutation.

Proof. Define $\mathbf{c}^T = \mathbf{w}^T\mathbf{A}$. Suppose $c_i \neq 0$ and $c_j \neq 0$. Define $\tilde{\varphi}(y) \triangleq \int \varphi(u)h(y - u)du$, where $h(u)$ is the symmetric probability density function of $u = \sum_{k \neq i,j} c_k s_k$. We have $h(u) \in M^2(2)$ by Theorem 4, and,

$$E\{\varphi(\mathbf{w}^T\mathbf{x})\} = E\{\tilde{\varphi}(c_i s_i + c_j s_j)\}$$

with $\tilde{\varphi}$ star-shaped downward by Theorem 5. Let $y = R\cos(\theta)s_i + R\sin(\theta)s_j$, where $R = 1 - \sum_{k \neq i,j} w_k^2$. Consider the function, $(d/d\theta)\, E\{\varphi(y)\,;\,\theta\}$. From [3] we have,

$$\begin{aligned}(d/d\theta)\, E\{\varphi(y)\,;\,\theta\} &= \int_0^\infty r^3 \int_0^{\pi/4} \exp\big(-g(r\cos(\phi)) - g(r\sin(\phi))\big)\sin(2\phi) \\ &\quad \times \left(\frac{g'(r\cos(\phi))}{r\cos(\phi)} - \frac{g'(r\sin(\phi))}{r\sin(\phi)}\right)\Big(b(\phi - \theta) - b(\phi + \theta)\Big)d\phi\, dr \qquad (1)\end{aligned}$$

where $b(\phi) \triangleq \varphi(r\cos(\phi)) + \varphi(r\sin(\phi))$. Note that the function $b(\phi)$ satisfies the symmetries, $b(\phi) = b(-\phi) = b(\pi/2 - \phi)$ by the symmetry in φ. Also, we have,

$$b'(\phi) = \tfrac{1}{2}r^2\sin(2\phi)\left(\frac{\varphi'(r\sin(\phi))}{r\sin(\phi)} - \frac{\varphi'(r\cos(\phi))}{r\cos(\phi)}\right)$$

Since $g'(x)/x$ is non-decreasing (sources are strongly super-Gaussian), and $\varphi'(x)/x$ is non-decreasing ($\varphi'(x)$ is star-shaped downward) on $(0, \infty)$, and both are increasing on a common interval of nonzero measure, we have that the integrand of (1) is positive, and thus,

$$(d/d\theta)\, E\{\varphi(y)\,;\,\theta\} > 0, \quad \theta \in (0, \pi/4)$$

and $\mathbf{w}$ cannot be a local optimum, and one of c_i or c_j must be zero at all local optima.

4 Conclusion

We defined strong sub- and super-Gaussianity, and derived relationships among various definition of sub- and super-Gaussianity. We extended a result of Benveniste, Goursat, and Ruget [3] to include star-shaped downward non-linearities. Similar results can be derived for sub-Gaussians and star-shaped upward functions, but under more restrictive conditions due to limitations on closure under convolution.

References

1. Amari, S.-I., Chen, T.-P., Cichocki, A.: Stability analysis of learning algorithms for blind source separation. Neural Networks 10(8), 1345–1351 (1997)
2. Andrews, D.F., Mallows, C.L.: Scale mixtures of normal distributions. J. Roy. Statist. Soc. Ser. B 36, 99–102 (1974)
3. Benveniste, A., Goursat, M., Ruget, G.: Robust identification of a nonminimum phase system. IEEE Transactions on Automatic Control 25(3), 385–399 (1980)
4. Benveniste, A., Métivier, M., Priouret, P.: Adaptive algorithms and stochastic approximations. Springer, Heidelberg (1990)
5. Boyd, S., Vandenberghe, L.: Convex Optimization. Cambridge University Press, Cambridge (2004)
6. Bruckner, A.M., Ostrow, E.: Some function classes related to the class of convex functions. Pacific J. Math. 12(4), 1203–1215 (1962)
7. Cardoso, J.-F., Laheld, B.H.: Equivariant adaptive source separation. IEEE. Trans. Sig. Proc. 44(12), 3017–3030 (1996)
8. Comon, P., Jutten, C. (eds.): Handbook of Blind Source Separation: Independent Component Analysis and Applications. Elsevier, Amsterdam (2010)
9. Finucan, H.M.: A note on kurtosis. Journal of the Royal Statistical Society, Series B (Methodological) 26(1), 111–112 (1964)
10. Haykin, S.: Neural Networks: a comprehensive foundation. Prentice-Hall, Englewood Cliffs (1999)
11. Karlin, S., Novikoff, A.: Generalized convex inequalities. Pacific J. Math. 13, 1251–1279 (1963)
12. Keilson, J., Steutel, F.W.: Mixtures of distributions, moment inequalities, and measures of exponentiality and Normality. The Annals of Probability 2, 112–130 (1974)
13. Mansour, A., Jutten, C.: What should we say about the kurtosis? IEEE Signal Processing Letters 6(12), 321–322 (1999)
14. Palmer, J.A., Kreutz-Delgado, K., Wipf, D.P., Rao, B.D.: Variational EM algorithms for non-gaussian latent variable models. In: Advances in Neural Information Processing Systems. MIT Press, Cambridge (2006)
15. Rao, B.D., Engan, K., Cotter, S.F., Palmer, J., Kreutz-Delgado, K.: Subset selection in noise based on diversity measure minimization. IEEE Trans. Signal Processing 51(3) (2003)
16. Widder, D.V.: The Laplace Transform. Princeton University Press, Princeton (1946)
17. Williamson, R.: Multiply monotone functions and their Laplace transforms. Duke Math. J. 23, 189–207 (1956)

Hybrid Channel Estimation Strategy for MIMO Systems with Decision Feedback Equalizer

Héctor José Pérez-Iglesias, Adriana Dapena,
Paula M. Castro, and José A. García-Naya

Department of Electronics and Systems. University of A Coruña
Campus de Elviña s/n, 15071 A Coruña, Spain
{hperez,adriana,pcastro,jagarcia}@udc.es
http://gtec.udc.es

Abstract. We propose combining supervised and unsupervised algorithms in order to improve the performance of multiple-input multiple-output digital communication systems which make use of decision-feedback equalizers at the receiver. The basic idea is to avoid the periodical transmission of pilot symbols by using a simple criterion to determine the time instants when the performance obtained with an unsupervised algorithm is poor or, equivalently, those instants when pilot symbols must be transmitted. Simulation results show how the novel approach provides an adequate BER with a low overhead produced by the transmission of pilot symbols.

1 Introduction

Decision-Feedback Equalizer (DFE) was initially proposed to reduce the effect of multiple-delayed copies of a signal transmitted over *Single-Input Single-Output* (SISO) systems [1], i.e., to equalize the channel. It consists of two linear filters: the feedforward filter, whose input is the received sequence, and the feedback filter, whose input is the detected sequence. The basic idea is to use a feedback from past decisions to cancel the interference of the symbols already detected. Extensions of this idea to *Multiple-Input Multiple-Output* (MIMO) systems have been proposed by several authors (see, for instance, [2–4]). In fact, DFE has been included in several standards like *Digital Terrestrial Multimedia Broadcast* (DTMB) and 802.16 (WiMAX).

In this work, we propose a novel way to combine supervised and unsupervised algorithms in order to improve the performance of MIMO DFE systems. The basic idea is to use an unsupervised algorithm joined with a simple detection criterion to determine the instant where the channel has suffered a considerable variation. When this event occurs, the receiver sends an "alarm" to the transmitter by means of a feedback or reverse channel indicating that some pilot symbols must be transmitted. Such reverse channel is actually implemented in most of the standards [5]. In the rest of time, an unsupervised adaptive algorithm is used to track channel variations. Simulation results show that the proposed scheme leads to good performance in terms of *Bit Error Rate* (BER) and avoids periodical transmissions of pilot symbols.

V. Vigneron et al. (Eds.): LVA/ICA 2010, LNCS 6365, pp. 311–318, 2010.

2 System Model

We consider a MIMO system with N_t transmit antennas and N_r receive antennas. The data symbols $\mathbf{u} = [u_1, ..., u_{N_t}]^T$ are transmitted to the different receive antennas such as, for flat fading channels, the received signals have the following form

$$y_j(t) = \sum_{i=1}^{N_t} h_{j,i}(t)u_i(t) + n_j(t)$$

In a compact form, we can express the vector of received signals, $\mathbf{y}(t)$, as follows

$$\mathbf{y}(t) = \mathbf{H}(t)\mathbf{u}(t) + \mathbf{n}(t) \tag{1}$$

where $\mathbf{u}(t)$ is the source vector, $\mathbf{n}(t)$ is the noise vector, and $\mathbf{H}(t)$ contains the channel coefficients $h_{j,i}(t)$ from the i-th transmit antenna to the j-th receive antenna. In general, if we let $f[n] = f(nT_\mathrm{s} + \Delta)$ denote samples of $f(t)$ every T_s seconds with Δ being the sampling delay and T_s the symbol time, then sampling $\mathbf{y}(t)$ every T_s seconds yields the discrete time signal $\mathbf{y}[n] = \mathbf{y}(nT_\mathrm{s} + \Delta)$ given by

$$\mathbf{y}[n] = \mathbf{H}[q]\mathbf{u}[n] + \mathbf{n}[n] \tag{2}$$

where $n = 0, 1, 2, \ldots$ corresponds to samples spaced with T_s and q denotes the slot time. For brevity, henceforth we omit the slot index q. We will consider that the channel remains unchanged during a block of N_B symbols (i.e over the data frame) and that the transmit sources are independent and identically distributed with unit power, i.e. $\mathbf{C_u} = E[\mathbf{u}[n]\mathbf{u}^H[n]] = \mathbf{I}$, where $\mathbf{I}_{N_t}$ denotes the identity matrix.

Note that the discrete time model in Equation (2) is equivalent to the continuous time model in Equation (1) only if the *Inter-Symbol Interference* (ISI) between samples is avoided, i.e. if the Nyquist criterion is satisfied. In that case, we will be able to reconstruct the original continuous signal from samples by means of interpolation. This channel model is known as time-varying flat block fading channel and this assumption is made in the rest of this paper.

2.1 Decision Feedback Equalizer

Channel equalization is often used at the receiver to combat the distortion introduced by the channel. As it can be seen in Figure 1, the estimated signal $\hat{\mathbf{u}}[n]$ can be expressed as

$$\hat{\mathbf{u}}[n] = \mathbf{F}\mathbf{y}[n] + (\mathbf{I}_{N_t} - \mathbf{B})\,\tilde{\mathbf{u}}[n] \tag{3}$$

where $\mathbf{F}$, $\mathbf{B}$, and $\mathbf{I}_{N_t}$ represent the feedfordward filter, the feedback filter, and the identity matrix, respectively. The vector $\tilde{\mathbf{u}}[n]$ denotes the quantized symbols. Since the estimated signals can be recovered in a different order than the transmitted ones, we introduce the permutation matrix $\mathbf{P}$ such as

$$\hat{\mathbf{u}}[n] = \mathbf{F}\mathbf{y}[n] + (\mathbf{I}_{N_t} - \mathbf{B})\,\mathbf{P}\tilde{\mathbf{u}}_p[n].$$

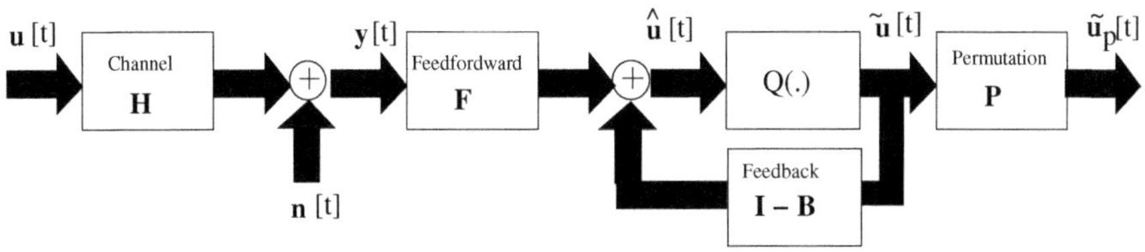

Fig. 1. MIMO system with DFE

The DFE feedforward and feedback filters can be found by minimizing the *Mean Square Error* (MSE) between a permutation of the transmit signals and their respective estimates by restricting the feedback filter $\mathbf{B}$ to being lower triangular [2, 3], i.e.

$$\{\mathbf{P}_{MSE}^{DFE}, \mathbf{F}_{MSE}^{DFE}, \mathbf{B}_{MSE}^{DFE}\} = E\left[\|\mathbf{Pu}[n] - \hat{\mathbf{u}}[n]\|_2^2\right] \quad \text{s.t.: } \mathbf{B} \text{ is unit lower triangular.}$$

The procedure described in [2] to find the matrices $\mathbf{P}, \mathbf{B}$, and $\mathbf{F}$ can be summarized as follows:

- Step 1: Compute $\mathbf{\Phi} = \left(\mathbf{H}^H\mathbf{C}_\mathbf{n}^{-1}\mathbf{H} + \mathbf{C}_\mathbf{u}^{-1}\right)^{-1}$
- Step 2: Initialize $\mathbf{P} = \mathbf{I}_{N_t}$ and $\mathbf{D} = \mathbf{O}_{N_t \times N_t}$
- Step 3: **for** $i = 1, \dots, N_t$
  - Find $l = \arg \min_{l'=1,\dots,N_t} \mathbf{\Phi}(l', l')$
  - Set $\mathbf{P}_i = \mathbf{I}_{N_t}$ whose i-th and l-th rows are exchanged
  - Compute $\mathbf{P} = \mathbf{P}_i\mathbf{P}$ and $\mathbf{\Phi} = \mathbf{P}_i\mathbf{\Phi}\mathbf{P}_i^T$
  - Let $\mathbf{D}(i, i) = \mathbf{\Phi}(i, i)$
  - Compute $\mathbf{\Phi}(i : N_t, i) = \mathbf{\Phi}(i : N_t, i)/\mathbf{D}(i, i)$
  - Compute $\mathbf{\Phi}(i + 1 : N_t, i + 1 : N_t) = \mathbf{\Phi}(i + 1 : N_t, i + 1 : N_t)$ $- \mathbf{\Phi}(i + 1 : N_t, i)\mathbf{\Phi}(i + 1 : N_t, i)^H\mathbf{D}(i, i)$
  - Let $\mathbf{L}$ be the lower triangular part of $\mathbf{\Phi}$
- Step 4: Compute $\mathbf{B} = \mathbf{L}^{-1}$ and $\mathbf{F} = \mathbf{D}\mathbf{L}^H\mathbf{P}\mathbf{H}^H{\mathbf{C}_\mathbf{n}}^{-1}$

Some remarks must be made about this implementation. First, the estimation of channel matrix is used in the first step to compute $\mathbf{\Phi}$ and in the last one to compute $\mathbf{F}$. Second, with the aim of minimizing the effect derived from feeding back erroneous decisions, the iterative procedure allows to be extracted the signals in a descending order according to the diagonal elements in the MSE matrix $\mathbf{\Phi}$. Finally, note that $\mathbf{C}_\mathbf{n} = \sigma_n^2\mathbf{I}_N$ and $\mathbf{C}_\mathbf{u} = \mathbf{I}_N$.

In order to implement the DFE procedure described above (Step 1 and 4), the receiver must acquire the channel matrix $\mathbf{H}$. Since the transmission model in Equation (2) corresponds to a linear combination of the transmit signals, we will consider a linear recovering system whose outputs are computed as follows

$$\mathbf{z}[n] = \mathbf{W}^H[n]\mathbf{y}[n] \tag{4}$$

where $\mathbf{W}[n]$ is a $N_r \times N_t$ matrix that can be found using different supervised and unsupervised algorithms.

The classical way of estimating $\mathbf{H}$ uses pilot symbols periodically sent from the transmit to the receive antennas. An important family of unsupervised filtering algorithms arises from considering the minimization of the MSE between the outputs $\mathbf{z}[n]$ and the desired signals $\mathbf{u}[n]$ [6]. Mathematically, the cost function can be written as

$$J_{\mathrm{MSE}} = \sum_{i=1}^{N_B} E\left[|z_i[n] - u_i[n]|^2\right] \tag{5}$$

$$= E\left[tr\left((\mathbf{W}^H[n]\mathbf{y}[n] - \mathbf{u}[n])(\mathbf{W}^H[n]\mathbf{y}[n] - \mathbf{u}[n])^H\right)\right]. \tag{6}$$

In this case, the optimum separating matrix can be obtained by determining the points where the gradient of J vanishes, i.e.,

$$\nabla_{\mathbf{W}} J_{\mathrm{MSE}} = 0 \Rightarrow \mathbf{W}_{opt} = \mathbf{C_y}^{-1}\mathbf{C_{yu}} \tag{7}$$

where $\mathbf{C_y} = E[\mathbf{y}[n]\mathbf{y}^H[n]]$ is the autocorrelation of the observations and $\mathbf{C_{yu}} = E[\mathbf{y}[n]\mathbf{u}^H[n]]$ is the cross-correlation between the observations and the desired signals.

The transmission of pilot symbols can be avoided by means of using *Blind Source Separation* (BSS) algorithms [7], which simultaneously estimates the channel matrix from the corresponding realizations of the observed vector $\mathbf{y}[n]$. One of the best known BSS algorithms has been approached by Bell and Sejnowski in [8]. Given an activation function $h(.)$, the idea proposed by these authors is to obtain the weighted coefficients of a Neural Network, $\mathbf{W}[n]$, in order to maximize the mutual information between the outputs before the activation function, $\mathbf{h}(\mathbf{z}[n]) = \mathbf{h}(\mathbf{W}^H[n]\mathbf{y}[n])$, and its inputs $\mathbf{y}[n]$. The learning rule of Infomax [8] is given by

$$\mathbf{W}[n+1] = \mathbf{W}[n] + \mu\mathbf{W}[n]\left(\mathbf{z}[n]\mathbf{g}^H(\mathbf{z}[n]) - \mathbf{I}_{N_t}\right) \tag{8}$$

where $\mathbf{g}(\mathbf{z}[n]) = [-h_1''(z_1[n])/h_1'(z_1[n]), \cdots, -h_{N_t}''(z_{N_t}[n])/h_{N_t}'(z_{N_t}[n])]^T$ depends on the activation function ($h'(.)$ and $h''(.)$ represent the first and the second derivative of $h(.)$, respectively). The expression in Equation (8) admits an interesting interpretation by means of the use of the non–linear function $g(z) = z^*(1-|z|^2)$. In this case, Castedo and Macchi in [9] have shown that the Bell and Sejnowski rule can be interpreted as a generalization of the *Constant Modulus Algorithm* (CMA).

3 Hybrid Approach

We propose to combine the two channel estimation methods above shown in order to obtain a performance close to the supervised approach, but by means of using a lower number of pilot symbols. Let $\mathbf{W}_u[n]$ and $\mathbf{W}_s[n]$ be the matrices of coefficients for the unsupervised and supervised modules, respectively. We start with an initial estimation of the channel matrix obtained using the supervised method given by Equation (7), $\hat{\mathbf{H}} = \mathbf{W}_s^{-H}[n]$. This estimation is used to initialize the unsupervised algorithm to $\mathbf{W}_u[n] = \hat{\mathbf{H}}^{-H}$.

Each time a new frame is received, the unsupervised algorithm updates sample by sample the separating matrix using the rule (8) and the channel matrix needed by DFE is estimated as $\hat{\mathbf{H}} = \mathbf{W}_u^{-H}[n]$. A "decision module" determines if the estimation obtained with the unsupervised algorithm is poor due, for example, to a large variation in the channel. When this event occurs, the receiver sends an "alarm" to the transmitter. At this instant, a frame of pilot symbols must be sent by the transmitter. At the receiver, the supervised algorithm estimates the channel from pilot symbols using Equation (7). This solution is used to initialize the unsupervised algorithm.

The important question is how to determine the instants where the unsupervised algorithm presents a poor performance. By combining both Equations (2) and (4), the output $\mathbf{z}[n]$ can be rewritten as a linear combination of the sources

$$\mathbf{z}[n] = \mathbf{\Gamma}[n]\mathbf{u}[n] + \mathbf{W}^H[n]\mathbf{n}[n] \tag{9}$$

where $\mathbf{\Gamma}[n] = \mathbf{W}^H[n]\mathbf{H}$ represents the overall mixing/separating system (or gain matrix). This means that each output contains a term corresponding to the desired source and another one due to the *MultiUser Interference* (MUI). It is interesting to note that the initialization of the unsupervised algorithm removes the permutation ambiguity inherent in this class of learning rules. Thus, each output will have the form

$$z_i[n] = \gamma_{ii}[n]u_i[n] + \sum_{j=1, i\neq j}^{N_t} \gamma_{ij}[n]u_j[n] + \mathbf{w}_i^H\mathbf{n}[n]. \tag{10}$$

By dividing this equation by $\gamma_{ii}[n]$ and considering that the noisy component is small compared to the another terms, we obtain that the power of each output is given by

$$\frac{E[|z_i[n]|^2]}{|\gamma_{ii}[n]|^2} = E[|u_i[n]|^2] + \sum_{j=1, i\neq j}^{N_t} \frac{|\gamma_{ij}[n]|^2}{|\gamma_{ii}[n]|^2} E[|u_j[n]|^2] = E[|u_i[n]|^2] + MUI_i$$

where the MUI term is implicitly defined. When MUI is high, the channel matrix estimate is poor. In that case, a pilot frame must be transmitted, i.e.,

$$MUI = \sum_{i=1}^{N_t} MUI_i = \sum_{i=1}^{N_t}\sum_{\substack{j=1\\ i\neq j}}^{N_t} \frac{|\gamma_{ij}[n]|^2}{|\gamma_{ii}[n]|^2} > \beta \rightarrow \text{Send an``alarm''} \tag{11}$$

where β is a real positive number (threshold). The gains $\gamma_{ij}[n]$ can be computed using $\mathbf{\Gamma}[n] = \mathbf{W}_u^H[n]\hat{\mathbf{H}}$, where $\hat{\mathbf{H}}$ is an initial estimation of the channel matrix obtained with the supervised approach. Obviously, a small value for β reduces the error, but it also increases the number of pilot symbols.

4 Experimental Results

In order to show the performance achieved with the proposed hybrid scheme, we present the results for several computer simulations performed considering that 4900 symbols are transmitted in frames of size $N_B = 100$ using a QPSK modulation. The system considers four transmit and receive antennas. The channel matrix is updated using the model $\mathbf{H} = (1-\alpha)\mathbf{H} + \alpha\mathbf{H}_{\text{new}}$, where $\mathbf{H}_{\text{new}}$ is a 4×4 matrix randomly generated according to a Rayleigh distribution.

We have evaluated the performance of the following proposed schemes:

- *Perfect Channel State Information* (CSI) at the receiver, i.e., the DFE uses the true channel matrix.
- The supervised approach in Equation (7) computed using a frame of N_B pilot symbols transmitted each 10 frames.
- The generalized Infomax algorithm initialized to the matrix obtained with the supervised approach. The step-size parameter and the non-linearity was, respectively, $\mu = 0.001$ and $g(z) = z^*(1 - |z|^2)$.
- The hybrid approach using the generalized Infomax algorithm with different thresholds, step-size parameter given by $\mu = 0.001$, non-linearity $g(z) = z^*(1 - |z|^2)$, and $\beta = 0.1$. A frame of N_B pilot symbols is used when the error is higher than the corresponding threshold.

The results have been obtained by averaging 1000 independent experiments. Pilot frames have not been considered to compute the Bit Error Rate.

In the first experiment we have considered that the channel remains constant during 10 frames. This assumption is an ideal situation for the supervised approach because it corresponds to the same instants where pilot frames are transmitted. Figure 2 (a) shows the BER and the number of transmit pilot symbols in terms of the channel updating parameter α for a SNR of 15 dB. Note the considerably improvement in BER obtained with the hybrid approach compared to the Infomax algorithm. Note also that the value for BER is close to that obtained with the supervised approach but now the number of pilots is considerably smaller.

Figure 3 (a) shows the BER and the number of pilot symbols for several values of SNR given $\alpha = 0.05$. Again, the hybrid approach improves on the results obtained with the unsupervised algorithm and it achieves a BER close to that of the supervised approach with fewer pilot symbols. Note that the Infomax algorithm shows a floor effect at a SNR of 8 dB while this effect appears for the hybrid approach at about 12 dB where it is clear that the BER is quite low.

In the second experiment, we have considered that the channel remains constant during a number of frames randomly generated in each computer simulation from the interval $[10, 15]$. Figure 2 (b) shows the performance at a SNR of 15 dB for different values of the updating channel parameter α. It can be seen that the supervised BER obtained with the hybrid approach matches the BER obtained with the supervised algorithm. Comparing to part (a), we observe that the number of pilot symbols is reduced. The same conclusion is obtained from

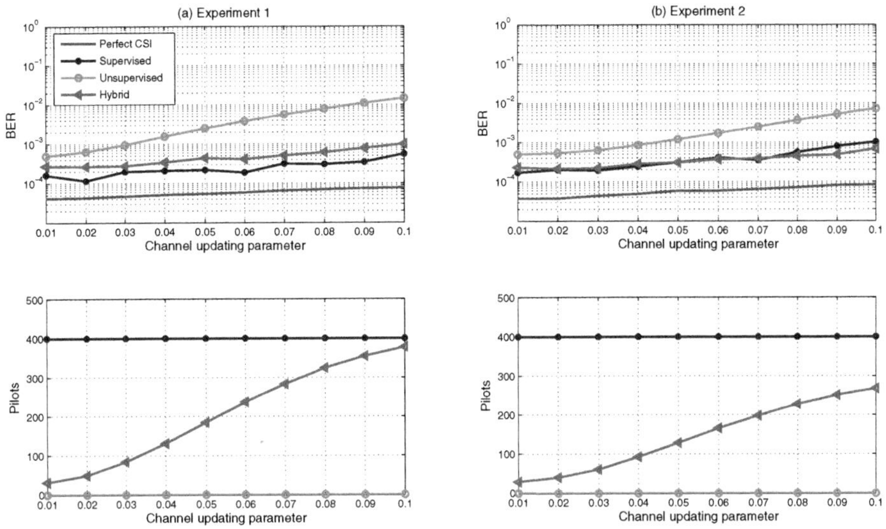

Fig. 2. BER and number of pilots in terms of the channel updating parameter for a SNR of 15 dB. The channel remains constant during 10 frames (Experiment 1) or during a value between 10 and 15 frames (Experiment 2).

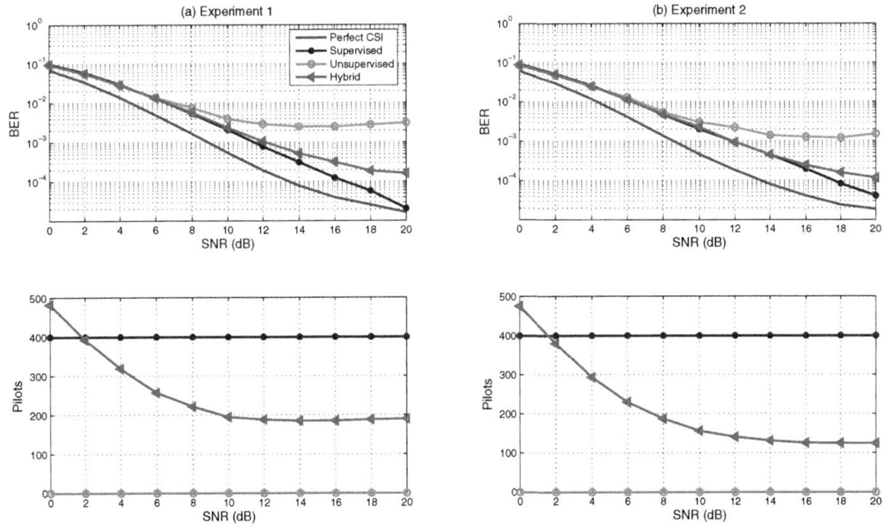

Fig. 3. BER and number of pilots in terms of SNR for $\alpha = 0.05$

Figure 3 (b), which corresponds to the performance for different SNRs considering $\alpha = 0.05$. Note that now the floor effect of the hybrid approach appears at a SNR of 15 dB corresponding to a BER of 2×10^{-4}. For this value only 130 pilot symbols are transmitted, which represents a considerable reduction in comparison to the 400 symbols needed by the supervised approach.

5 Conclusions

This paper deals with the utilization of supervised and unsupervised algorithms for estimating the channel matrix in systems with DFE at the receiver. Given a communication model where the channel is block flat fading, we have proposed a simple way to dynamically determine the instants when pilot symbols must be transmitted. Simulation results show that the novel approach provides an adequate BER with a low overhead produced by the transmission of these pilot symbols.

The method to determine the instants where pilots are required depends on a threshold β that must be selected to obtain a compromise between SER and number of pilot frames. Further work deals with determine the analytical expression to obtain this parameter.

Acknowledgment

This work was supported by Xunta de Galicia, Ministerio de Ciencia e Innovación of Spain, and FEDER funds of the European Union under grants number 09TIC008105PR, TEC2007-68020-C04-01 and CSD2008-00010.

References

1. Austin, M.E.: Decision feedback equalization for digial communication over dispersive channels. Technical report 437. Lincoln Laboratory (1967)
2. Kusume, K., Joham, M., Utschick, W.: MMSE block decision-feedback equalizer for spatial multiplexing with reduced complexity. In: Proc. IEEE Global Telecommunications Conference, vol. 4, pp. 2540–2544 (2004)
3. Fischer, R.F.H.: Precoding and Signal Shaping for Digital Transmission. John Wiley & Sons, Chichester (2002)
4. Joham, M.: Optimization of Linear and Nonlinear Transmit Signal Processing. PhD dissertation. Munich University of Technology (2004)
5. Philips: Comparison Between MU-MISO Codebook-based Channel Reporting Techniques for LTE Downlink. 3GPP TSG RAM WG1, Tech. Rep. R1-062483 (2006)
6. Haykin, S.: Neural Networks A Comprehensive Foundation. Macmillan College Publishing Company, New York (1994)
7. Comon, P., Jutten, C.: Handbook of Blind Source Separation, Independent Component Analysis and Applications. Academic Press, London (2010)
8. Bell, A., Sejnowski, T.: An information-maximization approach to blind separation and blind deconvolution. Neural Computation 7(6), 1129–1159 (1995)
9. Castedo, L., Macchi, O.: Maximizing the information transfer for adaptive unsupervised source separation. In: Proc. SPAWC 1997, Paris, France, pp. 65–68 (1997)

An Alternating Minimization Method for Sparse Channel Estimation

Rad Niazadeh[1], Massoud Babaie-Zadeh[1,⋆], and Christian Jutten[2]

[1] Department Of Electrical Engineering, Sharif University of Technology, Tehran, Iran
[2] GIPSA-Lab, Grenoble, and Institut Universitaire de France, France
radniazadeh@gmail.com,
mbzadeh@yahoo.com, Christian.Jutten@inpg.fr

Abstract. The problem of estimating a sparse channel, i.e. a channel with a few non-zero taps, appears in many fields of communication including acoustic underwater or wireless transmissions. In this paper, we have developed an algorithm based on Iterative Alternating Minimization technique which iteratively detects the location and the value of the channel taps. In fact, at each iteration we use an approximate Maximum A posteriori Probability (MAP) scheme for detection of the taps, while a least square method is used for estimating the values of the taps at each iteration. For approximate MAP detection, we have proposed three different methods leading to three variants for our algorithm. Finally, we experimentally compared the new algorithms to the Cramér-Rao lower bound of the estimation based on knowing the locations of the taps. We experimentally show that by selecting appropriate preliminaries for our algorithm, one of its variants almost reaches the Cramér-Rao bound for high SNR, while the others always achieve good performance.

1 Introduction

In this paper, we are going to investigate the problem of sparse channel estimation which appears in acoustic underwater or wireless transmissions [1]. In this problem, we want to estimate channel taps by observing the output of the channel while the channel is sparse, i.e. it has a few non-zero taps. Mathematically, we have the following model:

$$y(k) = u(k) * h(k) + n(k), \tag{1}$$

in which $y(k)$ is the output signal of the channel, $h(k)$ is the K-sparse[1] channel with N taps, $u(k)$ is L-length training sequence at the input of the channel and

⋆ This work has been partially funded by Iran Telecom Research Center (ITRC), and also by center for International Research and Collaboration (ISMO) and French embassy in Tehran in the framework of a GundiShapour collaboration program.

[1] By K-sparse we mean that there are at most K non-zero elements in $\mathbf{h} = [h(0), h(1), \ldots, h(N)]^T$.

V. Vigneron et al. (Eds.): LVA/ICA 2010, LNCS 6365, pp. 319–327, 2010.

$n(k)$ is a Gaussian noise. By observing all of the $M = N + L - 1$ output symbols of the channel, we have the following model in matrix form:

$$\mathbf{y} = \mathbf{U}\mathbf{h} + \mathbf{n} = \mathbf{U}_{\mathbf{h}}\mathbf{b} + \mathbf{n}. \tag{2}$$

in which $\mathbf{y}$ is the $M \times 1$ vector of output symbols of the channel, $\mathbf{n} \sim N(\sigma^2\mathbf{I}_M, 0)$ is an $M \times 1$ Gaussian noise vector and $\mathbf{U}$ is the $M \times N$ full-rank training matrix (as in [2]). Additionally, $\mathbf{U}_{\mathbf{h}} \triangleq \mathbf{U}\mathrm{diag}(\mathbf{h})$ and $\mathbf{b} \in \{0, 1\}^N$ is a binary vector that indicates the locations of non-zero taps i.e. :

$$\forall i \in \{1, 2, \ldots N\} : \quad b_i = \begin{cases} 1 \text{ if } h_i \neq 0, \\ 0 \text{ if } h_i = 0. \end{cases} \tag{3}$$

The goal of our problem is to find an appropriate estimation for $\mathbf{h}$ based on the observation vector, $\mathbf{y}$. If the estimator has no prior information about $\mathbf{h}$ i.e. it is unstructured, then it is well known [2] that the Least Square (LS) estimator will attain the best estimate for $\mathbf{h}$ in the sense of Mean Square Error (MSE). This estimator finds the solution of the following problem:

$$\hat{\mathbf{h}} = \underset{\mathbf{h}}{\mathrm{argmin}} \|\mathbf{y} - \mathbf{U}\mathbf{h}\|_2^2. \tag{4}$$

It is possible to show [2] that the above LS estimator achieves the Cramér-Rao lower bound of the problem. For the case when the estimators have prior knowledge about the location or number of taps of $\mathbf{h}$, it is possible that we obtain better estimations compared to the least square solution. Indeed, if a genie aids us with the location of the non-zero taps of $\mathbf{h}$, then the Structural Least Square (SLS) estimator, which will find the solution of the following problem:

$$\hat{\mathbf{h}}_\tau = \underset{\mathbf{h}_\tau}{\mathrm{argmin}} \|\mathbf{y} - \mathbf{U}_\tau \mathbf{h}_\tau\|_2^2. \tag{5}$$

will be an efficient estimator [1]. In (5), $\tau \subset \{1, 2, \ldots N\}$ is the support of $\mathbf{h}$, $\mathbf{U}_\tau$ is a sub-matrix of $\mathbf{U}$ that includes columns corresponding to the indices in τ and $\mathbf{h}_\tau$ is a $K \times 1$ vector that contains the non-zero taps of $\mathbf{h}$. Here, by 'efficient' we mean that this estimator will reach CRB of the estimation problem in which, the estimator will estimate $\mathbf{h}$ based on $\mathbf{y}$ and τ. In literature, this bound is known as CRB-S[2].

By comparing the Cramér-Rao bound of the unstructured estimation problem, known as CRB-US, with CRB-S, it is shown that CRB-S is much less than CRB-US [2] and their difference will increase as the $\mathbf{h}$ becomes sparser. Consequently, it is conceivable that we design an estimation technique based on $\mathbf{y}$ and some prior information about the sparsity degree of $\mathbf{h}$ so that its MSE will be close to CRB-S. Unfortunately, structured LS estimator which achieves CRB-S is not practical, because its estimation is based on knowing the exact location of the taps. The design of a practical estimator with MSE close to that of SLS estimator is of interest. So, many efforts have been done to find a practical solution for this estimation problem based on minimum knowledge about the sparsity of the original vector. Candès *et al.* [3] and Haupt *et al.* [4] proposed estimators that can achieve CRB-S to a factor of $\log M$. Similarly, Babadi *et al.* [5] showed that with the use of an estimator known as 'Typical' Estimator we can asymptotically achieve the Cramér-Rao bound of the genie aided estimation. Additionally, the work done by Carbonelli *et al.* [2] proposed practical algorithms for this problem which can reach close to CRB-S in the MSE sense.

In this paper, we try to find practical solutions for the problem of estimating the channel taps from noisy outputs of the channel. Our proposed algorithm is based on Alternating Minimization [6] technique for joint estimation of $\mathbf{b}$ and $\mathbf{h}$ at each iteration step. More precisely, our algorithm detects non-zero taps locations i.e. $\mathbf{b}$, iteratively based on a MAP scheme. Simultaneously, we will use the detected locations for finding the original sparse vector using an structured least square estimation at each iteration. For approximate MAP detection, we propose three methods. We will compare the MSE curve of all of the variations of the proposed algorithm to that of CRB-S, CRB-US and also with MSE curve of ITD-SE algorithm introduced in [2] and we will discuss our results.

So, this paper is organized as follows. In the next section, we will investigate our proposed MAP estimation of the taps based on the Bernoulli model for the channel taps. We will introduce an alternating minimization technique that we used at each iteration for joint estimation of the location and the value of the taps. Then, we will develop our algorithms to find an approximate MAP solution for $\mathbf{b}$ in each iteration. We will discuss the theory behind each one and the steps of the proposed variants of the main algorithm. Finally, we have an experimental results section in which we compare our algorithm with CRB-S, CRB-US and ITD-SE and we will discuss the results. Note that experimental results are computed on simulated data.

2 Iterative Approximate MAP-SLS Algorithms

2.1 MAP Detection and Iterative Alternating Minimization

In this section we introduce our main strategy for jointly estimating of the channel taps and the location of the taps i.e. $\mathbf{h}$ and $\mathbf{b}$ in the model described in (2). We use an iterative procedure based on alternating minimization for jointly finding both of the $\mathbf{h}$ and $\mathbf{b}$. To develop our algorithm, first assume that we have an appropriate estimate for $\mathbf{b}$ i.e. $\hat{\mathbf{b}}$ at a given iteration step. So, we estimate $\mathbf{h}$ at this iteration by finding the solution of the following problem:

$$\hat{\mathbf{h}} = \underset{\mathbf{h}}{\operatorname{argmin}} \|\mathbf{y} - \mathbf{U}\operatorname{diag}(\hat{\mathbf{b}})\mathbf{h}\|_2 = \underset{\mathbf{h}}{\operatorname{argmin}} \|\mathbf{y} - \mathbf{U}_{\hat{\mathbf{b}}}\mathbf{h}\|_2 = ({\mathbf{U}_{\hat{\mathbf{b}}}}^T\mathbf{U}_{\hat{\mathbf{b}}})^{\dagger}{\mathbf{U}_{\hat{\mathbf{b}}}}^T\mathbf{y}. \quad (6)$$

in which $(.)^{\dagger}$ denotes the pseudo-inverse operator. Note that this estimate is equal to the structured least square solution based on the location vector $\mathbf{b}$. On the other hand, if we have an appropriate estimation for $\mathbf{h}$ at a given iteration step and we want to obtain a more accurate estimate for $\mathbf{b}$, then we can make a MAP estimation for $\mathbf{b}$ based on the following observation:

$$\mathbf{y} \approx \mathbf{U}_{\hat{\mathbf{h}}}\mathbf{b} + \mathbf{n}. \quad (7)$$

for the MAP estimation, we need prior knowledge on the probability distribution of $\mathbf{b}$. In this paper, we assume an i.i.d Bernoulli distribution for the channel location vector, $\mathbf{b}$, based on the sparsity degree of the channel. Mathematically, if we define $P_a = \frac{K}{N} < \frac{1}{2}$ then we assume that $\mathbb{P}\{b_i = 1\} = P_a$ and so:

$$\mathbb{P}\{\mathbf{b}\} = \prod_{i=1}^{N} \mathbb{P}\{b_i\} = (1 - P_a)^{(N - \|\mathbf{b}\|_0)} P_a^{\|\mathbf{b}\|_0}. \tag{8}$$

the probability of vector $\mathbf{b}$ is computed for a given vector with exactly K components and not any vector with exactly K components. Now, we can find the MAP solution as:

$$\begin{aligned}\hat{\mathbf{b}}_{\text{MAP}} &= \underset{\mathbf{b}\in\{0,1\}^N}{\text{argmax}}\ \mathbb{P}\{\mathbf{y}|\hat{\mathbf{h}}, \mathbf{b}\}\mathbb{P}\{\mathbf{b}\} = \underset{\mathbf{b}\in\{0,1\}^N}{\text{argmax}}\ \exp\left(-\frac{1}{2\sigma^2}\|\mathbf{y} - \mathbf{U}_{\hat{\mathbf{h}}}\mathbf{b}\|_2^2\right)\left(\frac{P_a}{1 - P_a}\right)^{\|\mathbf{b}\|_0} \\ &= \underset{\mathbf{b}\in\{0,1\}^N}{\text{argmin}}\ \|\mathbf{y} - \mathbf{U}_{\hat{\mathbf{h}}}\mathbf{b}\|_2^2 + \lambda\|\mathbf{b}\|_0. \end{aligned} \tag{9}$$

in which $\lambda = 2\sigma^2 \ln(\frac{1-P_a}{P_a}) > 0$, according to the fact that $P_a < \frac{1}{2}$. Note that in (9) we minimize a function which is a combination of sparsity degree of $\mathbf{b}$ and l_2-norm of error $\mathbf{y} - \mathbf{U}_{\hat{\mathbf{h}}}\mathbf{b}$. Indeed, λ is an increasing function of σ which intuitively seems correct: as the noise increases, the importance of sparsity becomes greater in the estimation according to the fact that $\mathbf{y}$ is more noisy.

It is important to mention that the problem in (9), is exponentially complex with respect to N according to the search over $\{0,1\}^N$ for finding $\hat{\mathbf{b}}_{\text{MAP}}$. Although this problem seems to be impossible to solve, one may use an approximation method to find the MAP solution, as we will do in this paper.

To find the solutions of (9) and (6) simultaneously, we use an iterative alternating minimization approach. In fact, at each iteration step, by the use of the approximation for $\mathbf{b}$ in the previous iteration, we will find the solution to (6), then by using this solution we will find an approximate solution for (9). We have proposed 3 methods for finding an approximate solution to (9) i.e. MAP detection of the location vector at each iteration step of the proposed alternating minimization method. These methods are based on the algorithms introduced in [7] for finding the the user activity vector in CDMA[2], although they have been changed for being applied to our problem.

Finally, for the initialization of the algorithm, we use $\hat{\mathbf{b}}_0 = [1, \ldots, 1]^T$. This algorithm, which we call Approximate MAP-SLS, is summarized in Algorithm 1 with all its variants. Each of its variants, uses one of the the proposed methods for approximate MAP detection, which are described in the following sections.

2.2 Approximate l_2-MAP with Threshold

Based on the idea of Ridge Detector introduced in [7], we replace $\|\mathbf{b}\|_0$ by $\|\mathbf{b}\|_2^2$ and then, we will find the solution of this new problem in the real field. In fact, we find the solution of the following problem:

$$\hat{\mathbf{b}}_{\mathbf{r}} = \underset{\mathbf{b}\in\mathbb{R}^N}{\text{argmin}}\ \|\mathbf{y} - \mathbf{U}_{\hat{\mathbf{h}}}\mathbf{b}\|_2^2 + \lambda\|\mathbf{b}\|_2^2. \tag{10}$$

[2] Code Division Multiple Access.

By taking the gradient of the target function for optimization with respect to $\mathbf{b}$, we can find a closed form solution for (10). So, we have[3]:

$$\hat{\mathbf{b}}_{\mathbf{r}} = (\mathbf{U}_{\hat{\mathbf{h}}}{}^T \mathbf{U}_{\hat{\mathbf{h}}} + \lambda I)^{-1} \mathbf{U}_{\hat{\mathbf{h}}}{}^T \mathbf{y}. \tag{11}$$

After that, we will obtain $\hat{\mathbf{b}}$ by quantizing the solution of (11). More accurately:

$$\forall i \in \{1, 2, \ldots N\}: \quad \hat{b}_i = \begin{cases} 1 & \text{if} \quad \hat{b}_r(i) \geq \gamma(\sigma), \\ 0 & \text{if} \quad \hat{b}_r(i) < \gamma(\sigma). \end{cases} \tag{12}$$

in which, $\gamma(\sigma)$ is a threshold. By the use of simulation, we see that by choosing a suitable function for $\gamma(\sigma)$, we can find an approximately accurate estimate for $\mathbf{b}$, although there is no mathematical convenient way to show the accuracy of our proposed algorithm. This algorithm is so simple, although it needs a suitable function for $\gamma(\sigma)$. It is almost obvious from (11) that $\gamma(\sigma)$ needs to be increasing with respect to σ. So, in our algorithm we choose a simple increasing function for $\gamma(\sigma)$ i.e. $\gamma(\sigma) = \alpha\sigma^2$. Note that the parameter σ is assumed to be known to the detector.

Based on this method, we propose a variant for the main algorithm discussed in Sect. 2.1. The results of this variant of the main algorithm is examined for two values of α and it is experimentally seen that by choosing an appropriate α, we can have an acceptable amount of accuracy. The exact steps of the algorithm is summarized in Algorithm 1.

2.3 Approximate LASSO-MAP with Threshold

Inspired by LASSO Detector introduced in [7] and similar to the algorithm introduced in Sect. 2.2, we replace $\|\mathbf{b}\|_0$ by $\|\mathbf{b}\|_1$ in (9) and find the solution in the real field. In fact we have:

$$\hat{\mathbf{b}}_{\mathbf{r}} = \underset{\mathbf{b} \in \mathbb{R}^N}{\operatorname{argmin}} \|\mathbf{y} - \mathbf{U}_{\hat{\mathbf{h}}} \mathbf{b}\|_2^2 + \lambda \|\mathbf{b}\|_1. \tag{13}$$

The problem in (13) can be considered as LASSO problem introduced in [8]. LASSO is a shrinkage and selection method for linear regression. It minimizes the usual sum of squared errors, with a bound on the sum of the absolute values of the coefficients. It has connections to soft-thresholding of wavelet coefficients, forward stage wise regression, and boosting methods [8]. We use block coordinate [9] algorithm for LASSO to find the solution of (13). Afterwards, we use the quantization method of (12) to obtain $\hat{\mathbf{b}}$. This algorithm is summarized in Algorithm 1, and the result of this algorithm is examined experimentally in Sect. 3.

[3] Note that $\mathbf{U}_{\hat{\mathbf{h}}}{}^T \mathbf{U}_{\hat{\mathbf{h}}} + \lambda I$ is non-singular, because all of its eigenvalues are greater or equal to $\lambda > 0$ and so, we can use inverse operation in (11).

2.4 Approximate Backward-Detection MAP

In this section, we try to find an approximate search method for finding the solution of (9). Similar to [7], assume that we have a QR decomposition of $\mathbf{U}_{\hat{\mathbf{h}}}$ i.e. $\mathbf{U}_{\hat{\mathbf{h}}} = \mathbf{Q}\mathbf{R}_{\hat{\mathbf{h}}}$ in which $\mathbf{Q}$ is a unitary matrix and $\mathbf{R}_{\hat{\mathbf{h}}}$ is an upper triangular matrix. Now, we have:

$$\begin{aligned}\hat{\mathbf{b}} &= \underset{\mathbf{b}\in\{0,1\}^N}{\operatorname{argmin}} \|\mathbf{y} - \mathbf{U}_{\hat{\mathbf{h}}}\mathbf{b}\|_2^2 + \lambda\|\mathbf{b}\|_0 = \underset{\mathbf{b}\in\{0,1\}^N}{\operatorname{argmin}} \|\mathbf{Q}^T\mathbf{y} - \mathbf{Q}^T\mathbf{Q}\mathbf{R}_{\hat{\mathbf{h}}}\mathbf{b}\|_2^2 + \lambda\|\mathbf{b}\|_0 \\ &= \underset{\mathbf{b}\in\{0,1\}^N}{\operatorname{argmin}} \sum_{i=1}^{M} \Big(y'_i - \sum_{j=i}^{N} R_{\hat{h}}(i,j)b_j\Big)^2 + \lambda \sum_{1}^{N} b_i \\ &= \underset{\mathbf{b}\in\{0,1\}^N}{\operatorname{argmin}} \sum_{i=1}^{N} \Big[\Big(y'_i - \sum_{j=i}^{N} R_{\hat{h}}(i,j)b_j\Big)^2 + \lambda b_i\Big] + \sum_{i=N+1}^{M} {y'_i}^2 \\ &= \underset{\mathbf{b}\in\{0,1\}^N}{\operatorname{argmin}} \sum_{i=1}^{N} \Big[\Big(y'_i - \sum_{j=i}^{N} R_{\hat{h}}(i,j)b_j\Big)^2 + \lambda b_i\Big]. \end{aligned} \tag{14}$$

It is seen from (14) that once the estimates $\{\hat{b}_j\}_{j=i+1}^N$ are available, the optimal solution for $\hat{b}_i$ can be obtained by the following equation, regardless of the values for $\{\hat{b}_j\}_{j=1}^{i-1}$:

$$\hat{b_i}: \quad (y'_i - R_{\hat{h}}(i,i) - \sum_{j=i+1}^{N} R_{\hat{h}}(i,j)\hat{b}_j)^2 + \lambda \underset{1}{\overset{0}{\gtrless}} (y'_i - \sum_{j=i+1}^{N} R_{\hat{h}}(i,j)\hat{b}_j)^2. \tag{15}$$

by doing some manipulations, (15) can be simplified as :

$$\hat{b_i}: \quad y'_i \underset{0}{\overset{1}{\gtrless}} \frac{{R_{\hat{h}}}^2(i,i) + 2R_{\hat{h}}(i,i)\sum_{j=i+1}^{N} R_{\hat{h}}(i,j)\hat{b}_j + \lambda}{2R_{\hat{h}}(i,i)}. \tag{16}$$

Accordingly, by backward detection of b_i for $i = N, N-1, \ldots, 1$ using (16), we can find a near exact solution for (9). Based on this method, we develop a variation for our main algorithm which is summarized in Algorithm 1. Additionally, this version of our main algorithm is examined experimentally in the Sect. 3.

3 Experimental Results

In this part, we examine the efficiency of the proposed Iterative Approximate MAP-SLS algorithm and its variants, summarized in Algorithm 1 and we will compare it to ITD-SE [2]. For the sake of showing the effect of α, we develop two experiments. In both of the experiments, we choose $M = 50$, $N = 20$ and $K = 5$. We generate a separate K-sparse random signal as the sparse channel for each experiment. Indeed, elements of $\mathbf{u}$ are generated independently random according to $N(0,1)$ as the training data[4] and $\mathbf{u}$ is generated from $\mathbf{U}$, as in [2]. For finding

[4] It is better to use actual symbols as the training sequence, but, as we have seen in simulations not presented in this article, there is no major difference in the results.

Algorithm 1. Main algorithm

init: $k \leftarrow$ max-iterations
$\hat{\mathbf{b}}_0 = [1, 1, \ldots, 1]^T$
for $i \leq k$ **do**
$\hat{\mathbf{h}}_i = (\mathbf{U}_{\hat{\mathbf{b}}_{i-1}}^T \mathbf{U}_{\hat{\mathbf{b}}_{i-1}})^{\dagger} \mathbf{U}_{\hat{\mathbf{b}}_{i-1}}^T \mathbf{y}$ ▷ Structural least square estimator

if MAP-algorithm=l_2-MAP with Thresholding **then**
$\hat{\mathbf{b}}_\mathbf{r} = (\mathbf{U}_{\hat{\mathbf{h}}}{}^T \mathbf{U}_{\hat{\mathbf{h}}} + \lambda I)^{\dagger} \mathbf{U}_{\hat{\mathbf{h}}}{}^T \mathbf{y}$

else if MAP-algorithm=LASSO-MAP with Thresholding **then**
$\hat{\mathbf{b}}_\mathbf{r} = \underset{\mathbf{b} \in \mathbb{R}^N}{\text{argmin}} \; \|\mathbf{y} - \mathbf{U}_{\hat{\mathbf{h}}} \mathbf{b}\|_2^2 + \lambda \|\mathbf{b}\|_1$
▷ Solution is found using block-coordinate algorithm for LASSO
end if

if MAP-algorithm=Backward-Detection MAP **then**
$\mathbf{U}_{\hat{\mathbf{h}}} = \mathbf{Q}\mathbf{R}_{\hat{\mathbf{h}}}$
$\mathbf{y}' \;\; = \mathbf{Q}^T \mathbf{y}$
for $i = N, N-1, \ldots 1$ **do**
Given $\{\hat{b}_j\}_{i+1}^N$
$\hat{b}_i : \;\; y_i' \underset{0}{\overset{1}{\gtrless}} \frac{R_{\hat{h}}{}^2(i,i) + 2R_{\hat{h}}(i,i) \sum_{j=i+1}^N R_{\hat{h}}(i,j)\hat{b}_j + \lambda}{2R_{\hat{h}}(i,i)}$

end for
else
$\forall i \in \{1, 2, \ldots N\} : \quad \hat{b}_i = \begin{cases} 1 & \text{if } \hat{b}_r(i) \geq \gamma(\sigma), \\ 0 & \text{if } \hat{b}_r(i) < \gamma(\sigma). \end{cases} \qquad \gamma(\sigma) = \alpha\sigma^2$
end if
end for
$\hat{\mathbf{h}}_{\text{final}} = (\mathbf{U}_{\hat{\mathbf{b}}_k}^T \mathbf{U}_{\hat{\mathbf{b}}_k})^{\dagger} \mathbf{U}_{\hat{\mathbf{b}}_k}^T \mathbf{y}$

MSE, each experiment is repeated 100 times and the averaging over all these runs is taken. We choose $\alpha = 1.6$ for experiment 1 and $\alpha = 1$ for experiment 2. We run the simulations on an 2.8GHZ Intel Core2Duo CPU. The MSE vs SNR curves for both of the experiments are shown in Fig. 3. CRB-S and CRB-US for MSE are equal to $\sigma^2 \text{Tr}\{(\mathbf{U}_\mathbf{h}{}^T \mathbf{U}_\mathbf{h})^{-1}\}$ and $\sigma^2 \text{Tr}\{(\mathbf{U}^T \mathbf{U})^{-1}\}$ respectively as in [1]. Computational complexity of different algorithms are compared by the use of CPU time, shown in Tab. 3. We use CPU time as a metric for roughly comparison of computational complexity. Note that for the validity of our comparison, we use 10 iterations for all of the algorithms in the main loop. As it is seen from Fig. 1, by choosing $\alpha = 1.6$ the accuracy of Iterative l_2-MAP with Thresholding is better than that of ITD-SE and Approximate Backward-Detection MAP, although Approximate Backward-Detection MAP has the advantage that it does not need any thresholding and any pre-settings for $\gamma(\sigma)$. But by choosing $\alpha = 1$, all of these three algorithms will have the same accuracy. In both of the experiments, Iterative LASSO-MAP with Thresholding almost reaches to the CRB-S at high SNR, while it suffers from poor behaviour at low SNR. In fact, this

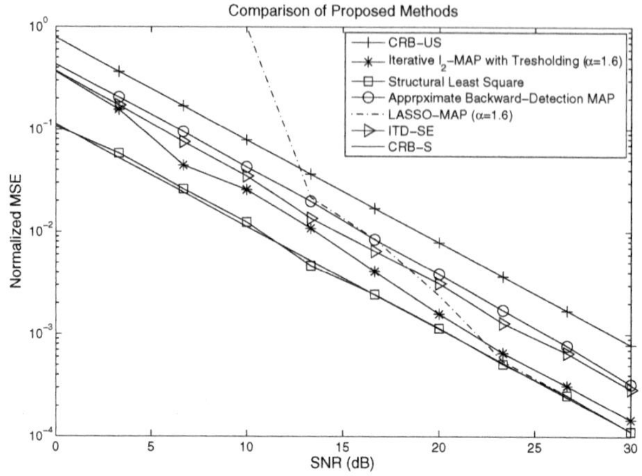

(a) experiment 1., $\alpha = 1.6$

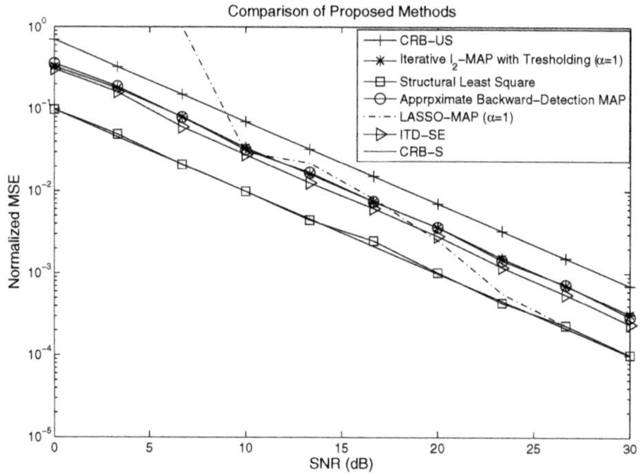

(b) experiment 2., $\alpha = 1$

Fig. 1. Normalized MSE vs SNR curves for proposed algorithms and ITD-SE

Table 1. Comparison of CPU time for various algorithms

CPU time in seconds	
Iterative l_2-MAP with Thresholding	0.0055
Iterative LASSO-MAP with Thresholding	0.1109
Approximate Backward-Detection MAP	0.0204
ITD-SE	0.0890

algorithm is near-optimal at high SNR, but its complexity is much more than the others by Tab. 1.

4 Conclusion

In this paper, we have proposed a new method for solving the problem of channel estimation. Our method is based on an alternating minimization approach. In fact, at each iteration, MAP detection of the location of the taps and structured least square estimation are applied simultaneously. For the MAP detection part, we proposed three methods and so we proposed three variants for our algorithm. We compared our proposed variants with ITD-SE introduced in [2] which in known as an efficient method for our problem. All the proposed methods have better or at worst case equal accuracy (in the sense of MSE) compared to ITD-SE, while having less complexity except for LASSO-MAP with Thresholding. However, we experimentally see that LASSO-MAP with Thresholding almost reaches the CRB-S at high SNR and so it is near-optimal. For further work around this subject, one can perform further experiments on actual signals, coming form communication or seismic and examine the performance of our proposed methods.

References

1. Sharp, M., Scaglione, A.: Estimation of sparse multipath channels. In: IEEE Military Communications Conference, MILCOM 2008, pp. 1–7 (2008)
2. Carbonelli, C., Vedantam, S., Mitra, U.: Sparse channel estimation with zero tap detection. IEEE Transactions on Wireless Communications 6(5), 1743–1763 (2007)
3. Candès, E., Tao, T.: The Dantzig selector: statistical estimation when p is much larger than n. Annals of Statistics 35(6), 2313–2351 (2007)
4. Haupt, J., Nowak, R.: Signal reconstruction from noisy random projections. IEEE Transactions on Information Theory 52(9), 4036–4048 (2006)
5. Babadi, B., Kalouptsidis, N., Tarokh, V.: Asymptotic achievability of the Cramér–Rao bound for noisy compressive sampling. IEEE Transactions on Signal Processing 57(3), 1233–1236 (2009)
6. Nocedal, J., Wright, S.: Numerical optimization. Springer, Heidelberg (1999)
7. Zhu, H., Giannakis, G.: Sparsity-embracing multiuser detection for CDMA systems with low activity factor. In: Proceedings of ISIT 2009, vol. 1, pp. 164–168 (2009)
8. Tibshirani, R.: Regression shrinkage and selection via the LASSO. Journal of the Royal Statistical Society, Series B (Methodological) 58(1), 267–288 (1996)
9. Meier, L., Van de Geer, S., Buhlmann, P.: The group LASSO for logistic regression. Journal-Royal Statistical Society, Series B Statistical Methodology 70(1), 53 (2008)

A Method for Filter Equalization in Convolutive Blind Source Separation

Radoslaw Mazur and Alfred Mertins

Institute for Signal Processing, University of Lübeck, 23538 Lübeck, Germany
{mazur,mertins}@isip.uni-luebeck.de

Abstract. The separation of convolutive mixed signals can be carried out in the time-frequency domain, where the task is reduced to multiple instantaneous problems. This direct approach leads to the permutation and scaling problems, but it is possible to introduce an objective function in the time-frequency domain and minimize it with respect to the time domain coefficients. While this approach allows for the elimination of the permutation problem, the unmixing filters can be quite distorted due the unsolved scaling problem. In this paper we propose a method for equalization of these filters by using the scaling ambiguity. The resulting filters have a characteristic of a Dirac pulse and introduce less distortion to the separated signals. The results are shown on a real-world example.

1 Introduction

The blind source separation method (BSS) is used to recover signals from observed mixtures. It is called blind as neither the original signals nor the mixing system is known. For the instantaneous case, different methods have been proposed [2,6,4]. In the case of real world acoustic mixtures of speech the situation is more complicated, as the signals arrive multiple times with different lags. This behavior can be modeled using FIR filters, but for realistic scenarios the length can reach several thousand taps. In this case, the separation can be performed only using filters with similar lengths.

An often used approach is the transformation to the time-frequency domain where the convolution becomes a multiplication [18]. This allows the use of instantaneous methods in each frequency bin independently. The major drawback is the arbitrary permutation in each frequency bin which has to be corrected, or the whole process fails. Although different methods have been proposed [15,3,10,16,19,14], the correction can not be calculated reliable in all cases.

The unmixing filters can be calculated directly in the time domain [5,1] but these algorithms suffer from high computational costs. In [13] an alternative method has been proposed, where the objective function is formulated in the frequency domain and minimized with respect to the time coefficients. This method combines the effectiveness of the frequency domain approaches with the absence of the permutation problem of the time domain methods. However, the scaling in the different frequencies is not addressed and therefore quite arbitrarily. This leads to coloration and added reverberation in the separated signals.

V. Vigneron et al. (Eds.): LVA/ICA 2010, LNCS 6365, pp. 328–336, 2010.

The postfilter-method in [7] tries to recover the signals as they have been recorded at the microphones and thus accepts all filtering done by the mixing system without adding new distortions. In [9], with the minimal distortion principle, a similar technique has been proposed.

New approaches as proposed in [11] and [12] solve the scaling problem with the aim of filter shortening or shaping. The methods of [11] and [12] use the instantaneous separation in the time-frequency domain as in [18] and allow for a simple calculation of the scaling coefficients. As these approaches are able to enhance the separation performance and reduce the reverberation at the same time, we propose to use these methods in combination with the algorithm from [13]. As this algorithm calculates only time domain unmixing coefficients the calculation of the scaling coefficients has to be modified. In this paper we will show how to calculate the scaling coefficients in this setup and apply an equalization method. The results will be shown in a real world example.

2 Problem Statement

The mixing system of real-world acoustic scenarios is convolutive and can be described using FIR filters of length L where L can reach several thousand. With N sources and N mixtures, the source vector $\boldsymbol{s}(n) = [s_1(n), \ldots, s_N(n)]^T$, and negligible measurement noise, the observation signals are given by

$$\boldsymbol{x}(n) = \mathbf{H}(n) * \mathbf{s}(n) = \sum_{l=0}^{L-1} \mathbf{H}(l)\boldsymbol{s}(n-l) \tag{1}$$

where $\mathbf{H}(n)$ is a sequence of $N \times N$ matrices containing the impulse responses of the mixing channels. For the separation, we use FIR filters of length $M \geq L-1$ and obtain

$$\boldsymbol{y}(n) = \mathbf{W}(n) * \boldsymbol{x}(n) = \sum_{l=0}^{M-1} \mathbf{W}(l)\boldsymbol{x}(n-l) \tag{2}$$

with $\boldsymbol{y}(n) = [y_1(n), \ldots, y_N(n)]^T$ being the vector of separated outputs and $\mathbf{W}(n)$ containing the time domain unmixing coefficients. Fig. 1 shows the scenario for two sources and sensors. The overall system is given by

$$\boldsymbol{y}(n) = \mathbf{W}(n) * \mathbf{H}(n) * \boldsymbol{s}(n) = \mathbf{G}(n) * \boldsymbol{s}(n), \tag{3}$$

which reduces to a multiplication in the time-frequency domain:

$$\boldsymbol{Y}(\omega, \tau) \approx \boldsymbol{W}(\omega)\boldsymbol{H}(\omega)\boldsymbol{S}(\omega, \tau) = \boldsymbol{G}(\omega)\boldsymbol{S}(\omega, \tau). \tag{4}$$

The only sources of information for estimating $\mathbf{W}(n)$ are the statistical properties of the observed signals $\boldsymbol{x}(n)$. Using the time-frequency approaches the overall system can only be estimated up to an arbitrary order and scaling:

$$\boldsymbol{G}(\omega) = \boldsymbol{P}(\omega)\boldsymbol{D}(\omega) \tag{5}$$

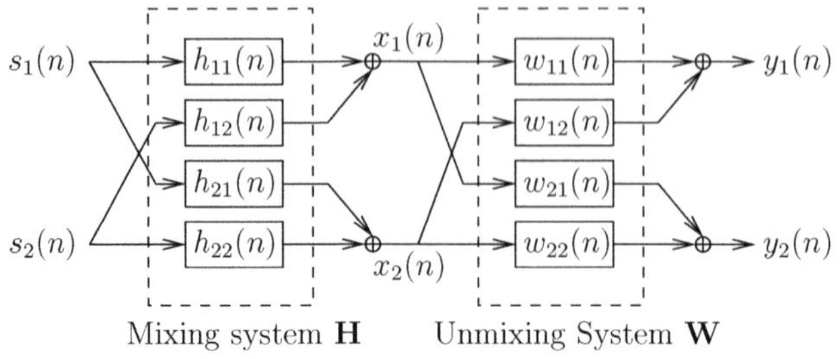

Fig. 1. BSS model with two sources and sensors

with $\boldsymbol{P}(\omega)$ being a permutation matrix and $\boldsymbol{D}(\omega)$ an arbitrary diagonal matrix. For a successful separation the permutation matrices $\boldsymbol{P}(\omega)$ have to be the same at all frequencies. All matrices $\boldsymbol{D}(\omega)$ which are not the identity introduce filtering to the separated signals.

3 Blind Separation Algorithm

The method from [13] does not suffer from the previously addressed permutation problem. This is achieved by using the integrated Kullback-Leibler divergence in the frequency domain as the objective function and minimizing it with respect to the time-domain matrices $\mathbf{W}(n)$. The update rule reads

$$\mathbf{W}^{l+1}(n) = \mathbf{W}^{l}(n) - \mu \frac{\partial f(\mathcal{W})}{\partial \mathbf{W}^{l}(n)} \tag{6}$$

with l being the iteration index, $f(\cdot)$ the integrated Kullback-Leibler divergence and $\mathcal{W} = [\mathbf{W}(0), \mathbf{W}(1), \dots, \mathbf{W}(M-1)]$. The gradient is calculated in [13] as

$$\frac{\partial f(\mathcal{W}^{l})}{\partial \mathbf{W}^{l}(n)} = \int_{-\pi}^{\pi} [\boldsymbol{I} - \boldsymbol{D}^{-1}(l,\omega)\boldsymbol{P}(l,\omega)]\boldsymbol{W}^{l}(e^{j\omega})e^{j\omega n} d\omega \tag{7}$$

where

$$\boldsymbol{D}(l,\omega) = \mathrm{diag}\left([\sigma_1^{r_1}(l,\omega), \dots, \sigma_N^{r_N}(l,\omega)]^{\mathrm{T}}\right) \tag{8}$$

and

$$\boldsymbol{P}(l,\omega) = \boldsymbol{Y}^{r-1}(l,e^{j\omega})\boldsymbol{Y}^{\mathrm{H}}(l,e^{j\omega}), \tag{9}$$

$$\begin{aligned} \mathbf{Y}^{r-1}(l,e^{j\omega}) = \Big[& \left|Y_1(l,e^{j\omega})\right|^{r_1-1} e^{j\theta(Y_1(l,e^{j\omega}))}, \dots \\ & \dots, \left|Y_N(l,e^{j\omega})\right|^{r_N-1} e^{j\theta(Y_N(l,e^{j\omega}))}\Big]^{\mathrm{T}} \end{aligned} \tag{10}$$

with $Y_i(e^{j\omega})$ being the short-time Fourier transforms of $y_i(n)$, $i = 1, 2, \dots, N$ and

$$\sigma_p^{r_p}(l,\omega) = \beta\sigma_p^{r_p}(l,\omega) + (1-\beta)\left|y_p(l,e^{j\omega})\right|^{r_p}. \tag{11}$$

The parameter β with $0 < \beta < 1$ is a moving-average parameter, and r_p is the order of an assumed generalized Gaussian source model.

This method is able to separate real-room recordings as it is capable of dealing with long filters, but it suffers from linear distortions which are introduced by the unmixing filters. A new method for resolving this problem will be presented in the next section.

4 Resolving the Scaling Ambiguity

A commonly used method for solving the scaling ambiguity is the minimal distortion principle (MDP) as proposed in [9]. The frequency unmixing matrices are calculated as

$$\boldsymbol{W}'(\omega) = \text{dg}(\boldsymbol{W}^{-1}(\omega)) \cdot \boldsymbol{W}(\omega) \tag{12}$$

with dg(·) returning the argument with all off-diagonal elements set to zero. In conjunction with the BSS algorithm presented in the last section there are two possibilities of employing it. The simple way is to carry out the iteration as in equation (6) and after convergence, transform the filters to the frequency domain by the Discrete Fourier Transform (DFT) where (12) can be carried out for all frequencies. The time-domain filters are then obtained by the inverse DFT. A better approach is to apply the MDP after every step of (6). As it will be shown in the simulations section, this method is able to greatly enhance the separation performance.

Both methods yield filters that have quite arbitrary form. Besides the main peak the filters have lots of large coefficients, which leads to coloration and reverberation in the separated signals. For reducing this coloration we propose to adapt the method from [12]. For this, it needs to be changed from a frequency by frequency method to an algorithm in which all frequencies are processed jointly. The time domain unmixing filters $\boldsymbol{w}_{ij}$ can be calculated in the dependency of the scaling coefficients $\boldsymbol{c}_j = [c_j(\omega_0), \ldots, c_j(\omega_{K-1})]^T$ as

$$\begin{aligned} \boldsymbol{w}_{ij} &= \bar{\mathcal{F}} \cdot \boldsymbol{E}_{ij} \cdot \boldsymbol{B} \cdot \boldsymbol{c}_j \\ &= \boldsymbol{V}_{ij} \cdot \boldsymbol{c}_j \end{aligned} \tag{13}$$

with

$$\boldsymbol{E}_{ij} = \text{diag}([\mathcal{R}(\mathbf{W}_{ij})\mathcal{I}(\mathbf{W}_{ij})] \tag{14}$$

being a diagonal matrix of the frequency domain unmixing coefficients with separated real and imaginary parts. As the resulting filters $\boldsymbol{w}_{ij}$ and the scaling coefficients $\boldsymbol{c}_j$ are real, it is possible to take advantage of the symmetry properties of the DFT. With M being the length of $\boldsymbol{w}_{ij}$ there are only $K = M/2+1$ scaling coefficients. $\bar{\mathcal{F}}$ is obtained from the $(M \times M)$-IDFT matrix $\mathcal{F}$ by concatenation of the real and imaginary parts such that the multiplication with $\boldsymbol{E}_{ij}$ is real again. Finally, $\boldsymbol{B}$ consists of concatenated identity matrices and is responsible

Table 1. Comparison of the signal-to-interference ratios in dB and the distortions measured by the SFM

	Left	Right	Overall	SFM
MDP (1)	2.85	7.37	5.77	0.32
MDP (2)	8.26	8.88	8.75	0.33
New Alg.	9.91	12.10	11.46	0.80

for aligning the scaling coefficients to both the real and imaginary parts of the transformation matrices.

The scaling factors $c_j(\omega)$ have to be calculated for all filters belonging to the same output j simultaneously. This can be achieved by stacking $\boldsymbol{V}_{ij}$ into $\bar{\boldsymbol{V}}_j$ and minimizing

$$||\bar{\mathbf{w}}_j \bar{\boldsymbol{V}}_j - \boldsymbol{c}_j||_{\ell_2} \tag{15}$$

where $\bar{\mathbf{w}}_j$ is the vertical concatenation of some desired filters. For the proposed equalization, these desired filters $\mathbf{w}_{ij}$ are defined to consist of zeros and have a single one at the position where the corresponding $\boldsymbol{w}_{ij}$ have the main peak when calculated using the MDP. The solution is given by $\boldsymbol{c}_j = \bar{\boldsymbol{V}}_j^+ \bar{\mathbf{w}}_j$, with $\bar{\boldsymbol{V}}_j^+$ being the pseudoinverse of $\bar{\boldsymbol{V}}_j$.

5 Simulations

Simulations have been done on real-world recordings of eight seconds of speech sampled at 16 kHz. The length of the unmixing filters was $M = 1024$. As single contributions of signals at the microphones are available, the separation performance can be calculated as in [17]. The coloration done by the unmixing system is measured in the terms of spectral flatness measure (SFM) [8]. With SFM being one a filter is an all-pass and does not color the signals. A value near zero indicates very strong distortions.

The separation was successful and no permutation occurred. The results after applying the normalization after convergence are shown in the first line of Table 1. The separation performance is quite poor and with an average SFM $= 0.32$ the signals are colorated. Applying the normalization in every step enhances the separation performance, but the coloration is still the same. In the last line of Table 1 the results for the equalized filters are shown. The new algorithm is able to enhance the separation performance even more and the coloration is reduced. With an average SFM $= 0.80$ the filters have a lot more all-pass characteristic which also can be seen in the Figs. 2 and 3 where the filter set before and after equalization is compared. The main peak has been enhanced, while the other coefficients are scaled down. The energy of these coefficients has been reduced by approximately 10 dB.

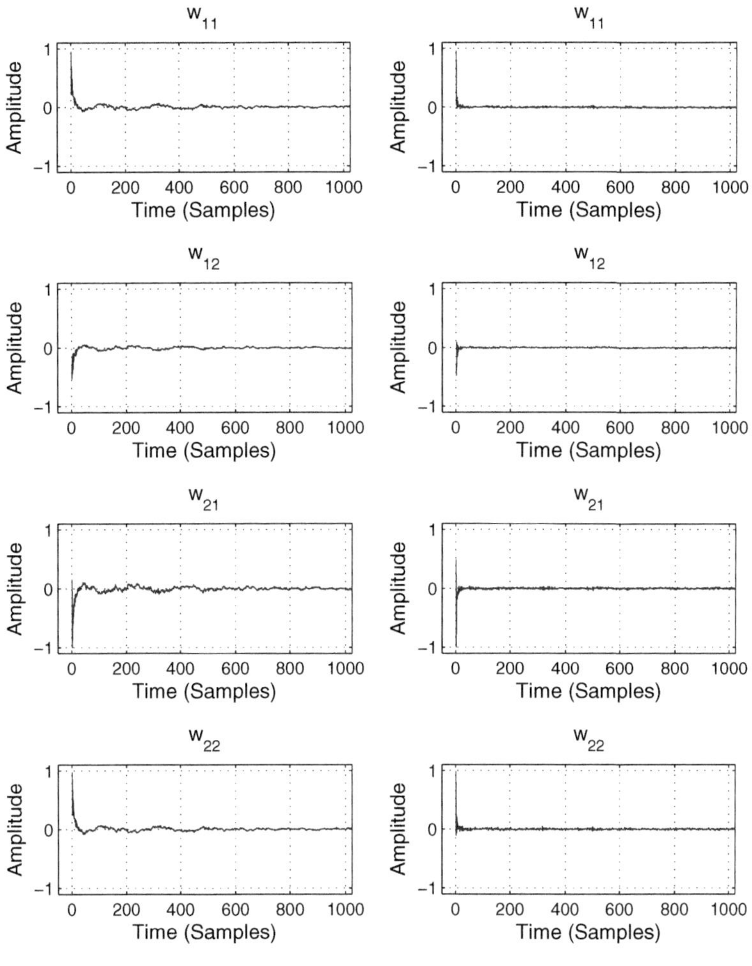

Fig. 2. Comparison of filter sets using the minimal distortion principle (left) and the new method (right)

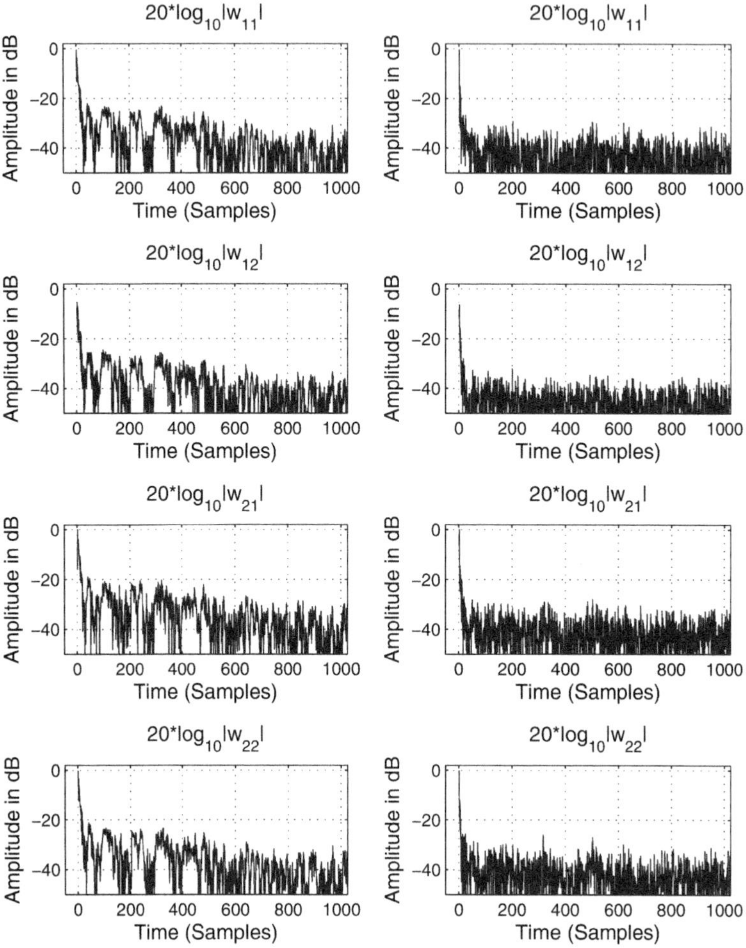

Fig. 3. Magnitudes of filters designed via the minimal distortion principle (left) and the new method (right)

6 Summary

In this paper, we have proposed to use the scaling ambiguity of convolutive blind source separation for equalization of the unmixing filters. We calculated a set of scaling factors that lead to unmixing filters with a more all-pass characteristic. This leads to less coloration of the separated signals and enhanced separation performance. The algorithm has been tested on a real-world example.

References

1. Aichner, R., Buchner, H., Araki, S., Makino, S.: On-line time-domain blind source separation of nonstationary convolved signals. In: Proc. 4th Int. Symp. on Independent Component Analysis and Blind Signal Separation (ICA 2003), Nara, Japan, pp. 987–992 (April 2003)
2. Amari, S.I., Cichocki, A., Yang, H.H.: A new learning algorithm for blind signal separation. In: Advances in Neural Information Processing Systems, vol. 8. MIT Press, Cambridge (1996)
3. Anemüller, J., Kollmeier, B.: Amplitude modulation decorrelation for convolutive blind source separation. In: Proceddings of the second international workshop on independent component analysis and blind signal separation, pp. 215–220 (2000)
4. Cardoso, J.F., Soulomiac, A.: Blind beamforming for non-Gaussian signals. Proc. Inst. Elec. Eng., pt. F. 140(6), 362–370 (1993)
5. Douglas, S.C., Sawada, H., Makino, S.: Natural gradient multichannel blind deconvolution and speech separation using causal FIR filters. IEEE Trans. Speech and Audio Processing 13(1), 92–104 (2005)
6. Hyvärinen, A., Oja, E.: A fast fixed-point algorithm for independent component analysis. Neural Computation 9, 1483–1492 (1997)
7. Ikeda, S., Murata, N.: A method of blind separation based on temporal structure of signals. In: Proc. Int. Conf. on Neural Information Processing, pp. 737–742 (1998)
8. Johnston, J.D.: Transform coding of audio signals using perceptual noise criteria. IEEE Journal on Selected Areas in Communication 6(2), 232–314 (1988)
9. Matsuoka, K.: Minimal distortion principle for blind source separation. In: Proceedings of the 41st SICE Annual Conference, August 5-7, vol. 4, pp. 2138–2143 (2002)
10. Mazur, R., Mertins, A.: An approach for solving the permutation problem of convolutive blind source separation based on statistical signal models. IEEE Trans. Audio, Speech, and Language Processing 17(1), 117–126 (2009)
11. Mazur, R., Mertins, A.: A method for filter shaping in convolutive blind source separation. In: Adali, T., Jutten, C., Romano, J.M.T., Barros, A.K. (eds.) ICA 2009. LNCS, vol. 5441, pp. 282–289. Springer, Heidelberg (2009)
12. Mazur, R., Mertins, A.: Using the scaling ambiguity for filter shortening in convolutive blind source separation. In: Proc. IEEE Int. Conf. Acoust, Taipei, Taiwan, pp. 1709–1712 (April 2009)
13. Mei, T., Xi, J., Yin, F., Mertins, A., Chicharo, J.F.: Blind source separation based on time-domain optimizations of a frequency-domain independence criterion. IEEE Trans. Audio, Speech, and Language Processing 14(6), 2075–2085 (2006)
14. Mukai, R., Sawada, H., Araki, S., Makino, S.: Blind source separation of 3-d located many speech signals. In: 2005 IEEE Workshop on Applications of Signal Processing to Audio and Acoustics, pp. 9–12 (October 2005)

15. Rahbar, K., Reilly, J.P.: A frequency domain method for blind source separation of convolutive audio mixtures. IEEE Trans. Speech and Audio Processing 13(5), 832–844 (2005)
16. Sawada, H., Mukai, R., Araki, S., Makino, S.: A robust and precise method for solving the permutation problem of frequency-domain blind source separation. IEEE Trans. Speech and Audio Processing 12(5), 530–538 (2004)
17. Schobben, D., Torkkola, K., Smaragdis, P.: Evaluation of blind signal separation methods. In: Proc. Int. Workshop Independent Component Analysis and Blind Signal Separation, Aussois, France (January 1999)
18. Smaragdis, P.: Blind separation of convolved mixtures in the frequency domain. Neurocomputing 22(1-3), 21–34 (1998)
19. Wang, W., Chambers, J.A., Sanei, S.: A novel hybrid approach to the permutation problem of frequency domain blind source separation. In: Puntonet, C.G., Prieto, A.G. (eds.) ICA 2004. LNCS, vol. 3195, pp. 532–539. Springer, Heidelberg (2004)

Cancellation of Nonlinear Inter-Carrier Interference in OFDM Systems with Nonlinear Power-Amplifiers

C. Alexandre R. Fernandes[1], João Cesar M. Mota[2], and Gérard Favier[3]

[1] Federal University of Ceará, Computer Engineering
rua Anahid Andrade 472, 62011-000, Sobral, Brazil
alexandrefernandes@ufc.br

[2] Federal University of Ceará, Dept. of Teleinformatics Engineering
Campus do Pici, 60.755-640, 6007 Fortaleza, Brazil
mota@gtel.ufc.br

[3] I3S Laboratory, University of Nice-Sophia Antipolis, CNRS
2000 route des Lucioles, BP 121, 06903, Sophia-Antipolis Cedex, France
favier@i3s.unice.fr

Abstract. Due to a high peak-to-average power ratio (PAPR), orthogonal frequency division multiplexing (OFDM) signals are often driven at the nonlinear region of power amplifiers (PAs). As a consequence, the orthogonality between the subcarriers is broken and nonlinear inter-carrier interference (ICI) is introduced. In this paper, we proposed two techniques for canceling nonlinear ICI in wireless OFDM communication systems with nonlinear PAs. The proposed techniques are based on the concept of power diversity, which consists in a transmission scheme that re-transmits the symbols several times with a different transmission power each time. The main advantage of using the power diversity is that the problem of canceling the nonlinear ICI can be viewed as a source separation problem, where the ICI terms correspond to "virtual" sources. The proposed techniques are able to provide a more robust transmission at the cost of a lower transmission rate.

Keywords: OFDM, nonlinear power amplifier, inter-carrier interference, power diversity, source separation.

1 Introduction

In this paper, two techniques for canceling nonlinear inter-carrier interference (ICI) in orthogonal frequency division multiplexing (OFDM) wireless communication systems with nonlinear radio frequency power amplifiers (PAs) are proposed. An important drawback in OFDM systems is that the transmitted signals are characterized by a high peak-to-average power ratio (PAPR) [1, 10] and, as a consequence, in some situations, the OFDM signal is driven at the nonlinear region of the PA. Nonlinear inter-carrier interference (ICI) is then introduced, which may significantly deteriorate the recovery of the information symbols.

V. Vigneron et al. (Eds.): LVA/ICA 2010, LNCS 6365, pp. 337–345, 2010.

Theoretical analysis and performance of OFDM signals in nonlinear wireless channels have been widely studied in the literature [2, 10]. In this paper, the PA is modeled as a polynomial with complex-valued coefficients. This model, also called quasi-memoryless polynomial model, is widely used in the literature to represent PAs [3].

The proposed techniques are based on the concept of power diversity, which consists in a transmission scheme that re-transmits the symbols several times with a different transmission power each time. The power diversity induces a multi-channel representation, allowing a perfect recovery of the information symbols in the noiseless case. As it will be viewed, the key point of this approach is that the problem of canceling the nonlinear ICI can be viewed as a source separation problem, where the information symbols correspond to the source of interest and the ICI can be viewed as "virtual" sources. The main drawback of this approach is the fact that the transmission rate is divided by the repetition factor, i.e. the number of times that every symbol is transmitted. However, in many cases, it is possible to use a repetition factor equal to 2. Moreover, the techniques are proposed for two different scenarios. The first one assumes that the wireless channel frequency response and the PA coefficients are known, while the second one assumes that only the PA coefficients are known. In the second scenario, it is assumed that some subcarriers are dedicated to pilot symbols that are used to estimate the wireless channel frequency response.

Input backoff and signal predistortion [1, 3–5] are popular approaches used to combat PA nonlinearities in communication systems. It should be mentioned that the backoff approach reduces the power efficiency, as the transmitter uses only a small portion of its allowed input range. Besides, our approach for compensating the nonlinear distortions at the receiver provides some advantages over predistortion schemes. One of these advantages is that few modifications in the portable units are necessary to accommodate the nonlinearity compensation in the uplink. Moreover, our approach may take other channel nonlinearities into account, contrarily to predistortion schemes that generally compensate the nonlinear distortions of a single nonlinear block. Other techniques for nonlinear interference rejection at the receiver side of OFDM systems have been proposed, most of them based on iterative methods as, for instance, [6, 7].

2 System Model

A simplified scheme of the discrete-time equivalent baseband OFDM system used in this work is shown in Fig. 1. Let us denote by N the number of subcarriers, $\bar{s}_{i,n}$ the frequency-domain data symbol at the nth subcarrier and i^{th} symbol period, and $\bar{\mathbf{s}}(i) = [\bar{s}_{i,1} \cdots \bar{s}_{i,N}]^T \in \mathbb{C}^{N\times 1}$ the vector containing all the N data symbols of the ith symbol period. The frequency-domain data symbols $\bar{s}_{i,n}$, for $1 \leq n \leq N$, are assumed to be independent and identically distributed (i.i.d.), with a uniform distribution over a QAM alphabet. All the variables with an overline correspond to frequency-domain signals.

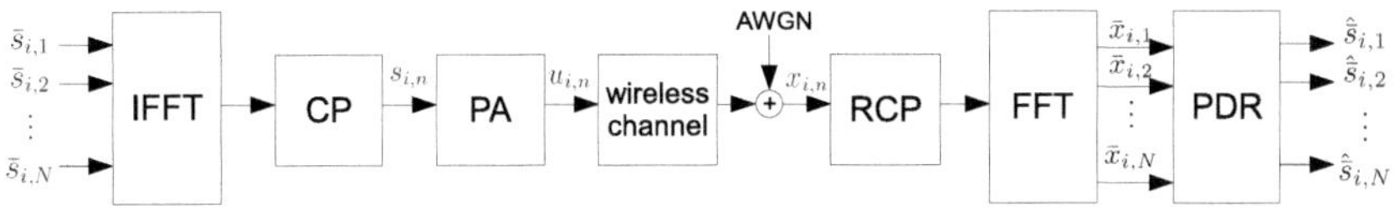

Fig. 1. Discrete-time equivalent baseband SISO-OFDM system

The time-domain OFDM symbol $s_{i,n}$, for $1 \leq n \leq N$, is obtained by taking the Inverse Fast Fourier Transform (IFFT) of the frequency-domain data symbols and, then, a cyclic prefix (CP) is inserted in order to avoid intersymbol interference (ISI) and ICI. Indeed, when the PA is linear, the cyclic prefix ensures the orthogonality between the subcarriers. However, as it will be shown in the sequel, this is not the case for a nonlinear PA.

The time-domain symbol with the CP is then amplified by a PA that is modeled as a polynomial of order $2K+1$. Denoting by $u_{i,n}$ $(1 \leq n \leq N)$ the output of the PA, we have:

$$u_{i,n} = \sum_{k=0}^{K} f_{2k+1} |s_{i,n}|^{2k} s_{i,n} = \sum_{k=0}^{K} f_{2k+1} \psi_{2k+1}(s_{i,n}), \tag{1}$$

where f_{2k+1}, for $0 \leq k \leq K$, are the equivalent baseband coefficients of the polynomial that models the PA and the operator $\psi_{2k+1}(\cdot)$ is defined as $\psi_{2k+1}(a) = |a|^{2k} a$. The equivalent baseband polynomial model (1) includes only the odd-order power terms with one more non-conjugated term than conjugated terms because the other nonlinear products of input signals correspond to spectral components lying outside the channel bandwidth, and can therefore be eliminated by bandpass filtering [8].

The signal $u_{i,n}$ is transmitted through a frequency-selective fading wireless channel and, at the receiver, the CP is removed (RCP box in Fig. 1) from the time-domain received signal and the FFT is then calculated. Assuming that the length of the cyclic prefix is higher than or equal to the channel delay spread, it can be shown that the frequency-domain received signals can be written as [8]:

$$\bar{\mathbf{x}}(i) = \Lambda \sum_{k=0}^{K} f_{2k+1} \bar{\boldsymbol{\psi}}_{2k+1}(\mathbf{V}^H \bar{\mathbf{s}}(i)) + \bar{\mathbf{n}}(i). \tag{2}$$

where $\bar{\mathbf{x}}(i) \in \mathbb{C}^{N\times 1}$ is the vector of frequency-domain received signals at the ith symbol period, $\Lambda \in \mathbb{C}^{N\times N}$ is a diagonal matrix containing samples of the channel frequency response λ_n $(1 \leq n \leq N)$, $\mathbf{V} \in \mathbb{C}^{N\times N}$ is the FFT matrix of dimension N, with $[\mathbf{V}]_{p,q} = e^{-\jmath 2\pi(p-1)(q-1)/N}$ $(1 \leq p, q \leq N)$, and $\bar{\mathbf{n}}(i) \in \mathbb{C}^{N\times 1}$ is a vector containing additive white Gaussian noise (AWGN) components of variance σ^2. Besides, the operators $\boldsymbol{\psi}_{2k+1}(\cdot)$ and $\bar{\boldsymbol{\psi}}_{2k+1}(\cdot)$ are defined as $\boldsymbol{\psi}_{2k+1}(\mathbf{a}) = [\psi_{2k+1}(a_1) \cdots \psi_{2k+1}(a_N)]^T \in \mathbb{C}^{N\times 1}$, for $\mathbf{a} = [a_1 \cdots a_N] \in \mathbb{C}^{N\times 1}$, and $\bar{\boldsymbol{\psi}}_{2k+1}(\mathbf{a}) = \mathbf{V}\boldsymbol{\psi}_{2k+1}(\mathbf{a}) \in \mathbb{C}^{N\times 1}$.

Equation (2) shows that the frequency-domain received signal $\bar{x}_{i,n}$ equals a scaled version of the frequency-domain data symbol $\lambda_n f_1 \bar{s}_{i,n}$ plus nonlinear ICI terms $\sum_{k=1}^{K} \lambda_n f_{2k+1} [\bar{\boldsymbol{\psi}}_{2k+1}(\mathbf{s}(i))]_n$, where $[\bar{\boldsymbol{\psi}}_{2k+1}(\mathbf{s}(i))]_n$ denotes the

n^{th} element of $\bar{\psi}_{2k+1}(\mathbf{s}(i))$. It should be remarked that the nonlinear ICI term $[\bar{\psi}_{2k+1}(\mathbf{s}(i))]_n$ depends on the information symbols of all the other subcarriers at symbol period i, which means that each subcarrier interferes on all the other subcarriers. The techniques proposed in this paper, called Power Diversity based Receivers (PDRs), are placed after the FFT block and their purpose is to eliminate the nonlinear ICI and to remove the scalar factor $\lambda_n f_1$.

The frequency-domain received signal $\bar{x}_{i,n}$ can be rewritten in a vector form:

$$\bar{x}_{i,n} = \lambda_n \mathbf{f}^T \bar{\boldsymbol{\phi}}_{i,n} + \bar{n}_{i,n}, \tag{3}$$

where $\mathbf{f} = [f_1 \quad f_3 \cdots f_{2K+1}]^T \in \mathbb{C}^{(K+1)\times 1}$, $\bar{\boldsymbol{\phi}}_{i,n} = [\bar{s}_{i,n} \quad [\bar{\psi}_3(\mathbf{V}^H\bar{\mathbf{s}}(i))]_n \cdots [\bar{\psi}_{2K+1}(\mathbf{V}^H\bar{\mathbf{s}}(i))]_n]^T \in \mathbb{C}^{(K+1)\times 1}$ is a vector containing the information symbol and the nonlinear ICI components of the nth subcarrier and i^{th} symbol period, and $\bar{n}_{i,n}$ is the corresponding noise component in the frequency domain.

3 Power Diversity

In this section, we introduce the concept of power diversity by presenting a transmission scheme that consists of re-transmitting the information symbols several times with different transmission powers. Then, we briefly discuss how the problem of estimating the information symbols can be viewed as a source separation problem.

3.1 Transmission Scheme

The power diversity transmission scheme can be summarized as follows. The frequency-domain information symbol $\bar{s}_{i,n}$ at the nth subcarrier $(1 \leq n \leq N)$ and i^{th} symbol period $(1 \leq i \leq I_B)$ is transmitted L times with transmission powers that are multiplied by the factors $P_1, ..., P_L$, as follows:

$$\bar{s}^{(pd)}_{((i-1)L+l),n} = \sqrt{P_l}\, \bar{s}_{i,n}, \quad \text{for} \quad 1 \leq l \leq L, \tag{4}$$

where $\bar{s}^{(pd)}_{((i-1)L+l),n}$ is the "weighted" frequency-domain symbol associated with the nth subcarrier and $((i-1)L+l)^{th}$ symbol period. Note that the transmission power factors $P_1, ..., P_L$ are the same for all the subcarriers.

Denoting by $\bar{x}^{(pd)}_{((i-1)L+l),n}$, $1 \leq l \leq L$, the L frequency-domain received signal samples associated with the frequency-domain information symbol $\bar{s}_{i,n}$, we can define the following vector $\bar{\mathbf{x}}^{(pd)}(i) = [\bar{x}^{(pd)}_{((i-1)L+1),n} \quad \bar{x}^{(pd)}_{((i-1)L+2),n} \cdots \bar{x}^{(pd)}_{iL,n}]^T \in \mathbb{C}^{L\times 1}$. Assuming that the samples of the channel frequency response channel λ_n $(1 \leq n \leq N)$ and PA coefficients f_{2k+1} $(0 \leq k \leq K)$ are time-invariant over L symbol periods, we can write from (3) and (4):

$$\bar{\mathbf{x}}^{(pd)}_{i,n} = \lambda_n \mathbf{H} \bar{\boldsymbol{\phi}}_{i,n} + \bar{\mathbf{n}}^{(pd)}_{i,n}, \tag{5}$$

where

$$\mathbf{H} = \begin{bmatrix} P_1^{\frac{1}{2}} & \cdots & P_1^{\frac{2K+1}{2}} \\ \vdots & \ddots & \vdots \\ P_L^{\frac{1}{2}} & \cdots & P_L^{\frac{2K+1}{2}} \end{bmatrix} \begin{bmatrix} f_1 & \cdots & 0 \\ \vdots & \ddots & \vdots \\ 0 & \cdots & f_{2K+1} \end{bmatrix} \in \mathbb{C}^{L\times(K+1)}, \tag{6}$$

and $\bar{\mathbf{n}}_{i,n}^{(pd)} \in \mathbb{C}^{L\times 1}$ is a noise vector.

3.2 Source Separation Interpretation

The main motivation for using power diversity is that the problem of estimating the information symbols can be viewed as a source separation problem, where the nonlinear ICI terms are "virtual" sources and the information symbols correspond to the source of interest. The proposed transmission scheme induces L subchannels, where each re-transmission period corresponds to a subchannel.

It is interesting to remark that a classical antenna/sensor array would not succeed to induce a useful multi-channel representation, as all sources (real and virtual) are located in the same point in the space domain. In this case, the array response matrix would have unit rank, exhibiting degenerate discrimination. On the other hand, when diversity is used in power domain, it can be demonstrated that the matrix $\lambda_n \mathbf{H}$, which plays the role of "array response" of the nth subcarrier, is full-rank when $P_i \neq P_j$, for $1 \leq i \neq j \leq L$. Under that condition, if $L \geq (K+1)$, this multi-channel representation allows the perfect recovery of $\bar{s}_{i,n}$ in the noiseless case.

The main drawback of this approach is the fact that the transmission rate is divided by L. However, based on a realistic assumption that the PA can be modeled using a third-order polynomial ($K+1=2$) [3–5, 10], we can use $L=2$, which minimizes the transmission rate loss.

4 Power Diversity-Based Receivers (PDRs)

The receivers based on the power diversity transmission scheme are presented in this section. The first one assumes the knowledge of the channel frequency response and PA coefficients. In practice, the PA parameters have to be estimated at the transmitter and this information has to be sent to the receiver. The transmission of these parameters must be included in the system initialization. This scheme was used previously by other authors in the context of OFDM systems with nonlinear PA [7]. In what concerns the channel frequency response, there are several methods in the literature to estimate these parameters in nonlinear OFDM systems [7, 8]. The second proposed receiver also assumes the knowledge of the PA coefficients, but it does not assume that the channel frequency response is known a priori. Instead, it jointly estimates the information symbols and the wireless channel coefficients iteratively, by using some subcarriers dedicated to pilot symbols.

4.1 PDR with Channel Knowledge (PDR-CK)

As mentioned in Section 3.2, the PDR considers the estimation of the frequency-domain information symbols as a source separation problem. The information symbols are estimated as:

$$\hat{\bar{s}}_{i,n} = \mathbf{w}_n \bar{\mathbf{x}}_{i,n}^{(pd)}, \tag{7}$$

for $1 \le n \le N$, where $\mathbf{w}_n \in \mathbb{C}^{1\times L}$ contains the coefficients of the separator, calculated by minimizing the following mean square error (MSE) cost function: $J_n = \mathbb{E}[|\bar{s}_{i,n} - \mathbf{w}_n \bar{\mathbf{x}}_{i,n}^{(pd)}|^2]$, whose solution is given by:

$$\mathbf{w}_n = \lambda_n^* \mathbf{r}_{\bar{\phi}} \mathbf{H}^H \left(\mathbf{H}\mathbf{R}_{\bar{\phi}}\mathbf{H}^H |\lambda_n|^2 + \mathbf{I}_L \sigma^2 \right)^{-1} \in \mathbb{C}^{1\times L}, \tag{8}$$

where $\mathbf{R}_{\bar{\phi}} = \mathbb{E}[\bar{\boldsymbol{\phi}}_{i,n}\bar{\boldsymbol{\phi}}_{i,n}^H] \in \mathbb{C}^{(K+1)\times(K+1)}$, $\mathbf{r}_{\bar{\phi}} = \mathbb{E}[\bar{s}_{i,n}\bar{\boldsymbol{\phi}}_{i,n}^H] \in \mathbb{C}^{1\times(K+1)}$ and $\mathbf{I}_L$ is the identity matrix of order L. The expressions for the values of $\mathbf{R}_{\bar{\phi}}$ and $\mathbf{r}_{\bar{\phi}}$ are omitted due to a lack of space.

4.2 PDR with Channel Estimation (PDR-CE)

The PDR-CE assumes that pilot symbols are allocated in subcarriers regularly spaced in the channel bandpass. These subcarriers are denoted by the set $\mathcal{N} = \{n_1, \cdots, n_D\}$, where D is the number of subcarriers dedicated to pilot symbols. This technique provides an initial estimate of the channel coefficients assuming that the PA is linear, then, it iteratively re-estimates the channel coefficients and information symbols. The initial estimate of channel frequency response on the pilot subcarriers is obtained assuming that the PA is linear. In that case, the optimal estimate of λ_n is found by using the maximum ratio combining (MRC) method:

$$\hat{\lambda}_n^{(0)} = \left[\frac{[P_1^{\frac{1}{2}} \cdots P_L^{\frac{1}{2}}]}{\sum_{l=1}^{L} P_l^2} \bar{\mathbf{x}}_{i,n}^{(pd)} \right] \frac{1}{\bar{s}_{i,n}} \tag{9}$$

for $n \in \mathcal{N}$. The initial channel frequency response on the other subcarriers are estimated from the coefficients obtained in Step 1 by interpolation, as in Step 4 of the algorithm.

At each symbol period i, the proposed receiver carries out the following steps:

<u>For $it = it + 1$:</u>

1. Estimate $\hat{\bar{s}}_{i,n}^{(it)}$ $(1 \le n \le N)$ from (7) and (8), using $\hat{\lambda}_n^{(it-1)}$.
2. Project $\hat{\bar{s}}_{i,n}^{(it)}$ $(1 \le n \le N)$ onto the QAM alphabet and construct $\hat{\bar{\boldsymbol{\phi}}}_{i,n}^{(it)} = [\hat{\bar{s}}_{i,n}^{(it)} \ \ [\bar{\psi}_3(\mathbf{V}^H\hat{\bar{\mathbf{s}}}^{(it)}(i))]_n \cdots [\bar{\psi}_{2K+1}(\mathbf{V}^H\hat{\bar{\mathbf{s}}}^{(it)}(i))]_n]^T$ from $\hat{\bar{s}}_{i,n}^{(it)}$, for $1 \le n \le N$.
3. Estimate the channel frequency response on the pilot subcarriers as:
$$\hat{\lambda}_n^{(it)} = \left[[\mathbf{u}_{i,n}^{(it)}]^H \bar{\mathbf{x}}_{i,n}^{(pd)} \right] \frac{1}{[\mathbf{u}_{i,n}^{(it)}]^H \mathbf{u}_{i,n}^{(it)}}, \tag{10}$$
for $n \in \mathcal{N}$, where $\mathbf{u}_{i,n}^{(it)} = \mathbf{H}\hat{\bar{\boldsymbol{\phi}}}_{i,n}^{(it)}$. That corresponds to the optimal estimate of λ_n given $\mathbf{H}$ and $\hat{\bar{\boldsymbol{\phi}}}_{i,n}^{(it)}$.

4. The channel frequency response on the other subcarriers ($\hat{\lambda}_n^{(it)}$, for $1 \leq n \leq N$) are calculated by interpolation from the channel frequency response on the pilot subcarriers ($\hat{\lambda}_n^{(it)}$, for $n \in \mathcal{N}$) obtained in Step 3. See [7, 11], for more details about the interpolation procedure. In the simulation results section, the interpolation is done by using truncated FFT matrices [7, 11].
5. If $\sum_{n=1}^{N} |\hat{\lambda}_n^{(it)} - \hat{\lambda}_n^{(it-1)}|^2 / \sum_{n=1}^{N} |\hat{\lambda}_n^{(it-1)}|^2 < \epsilon$, stop. Otherwise, go to Step 1.

5 Simulation Results

In this section, the proposed techniques are evaluated by means of simulations. A OFDM system with a third-order polynomial PA with coefficients equal to $f_1 = 0.9798 - 0.2887\jmath$ and $f_3 = -0.2901 + 0.4350\jmath$ [9], and a wireless link with frequency selective fading due to multipath propagation has been considered for the simulations. The results were obtained with $N = 64$ subcarriers and BPSK (binary phase shift keying) transmitted signals, via Monte Carlo simulations using 500 independent data realizations. In all the simulations, the PDRs use a repetition factor $L = 2$, with $P_1 = 1.2$ and $P_2 = 0.6$.

Fig. 2 shows the bit error rate (BER) versus the noise variance for 6 different techniques. The PDR-CK and PDR-CE correspond to the proposed techniques, while 1-tap eqz-CK and 1-tap eqz-CE correspond to 1-tap equalizers (complex automatic gain controls) that simply divide the received signal $\bar{x}_{i,n}$ by the corresponding channel coefficient, with channel knowledge and channel estimation, respectively. For the 1-tap equalizers, we have used the transmission power that maximizes signal to noise plus interference (SNIR) ratio. The derivation of this optimal power is omitted due to a lack of space. Moreover, we also show the BER provided by 1-tap equalizers in the case of a linear PA, with channel knowledge and channel estimation. All the techniques that use channel estimation assume that 8 subcarries regularly spaced in the channel bandpass are dedicated to pilot symbols. The channel output signal to noise ratio (SNR) is not used as the

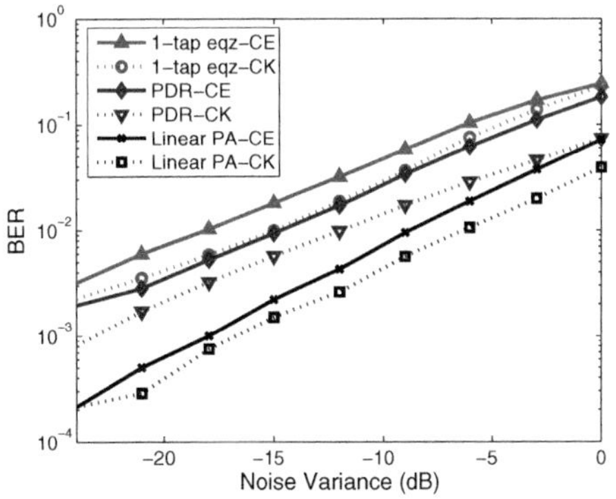

Fig. 2. BER versus Noise Variance

independent variable in Fig. 2 because it is a nonlinear non-monotonic function of the transmission power.

Remark, for the nonlinear PAs, the proposed PDRs provide BERs significantly lower than the 1-tap equalizers, with CK as well as with CE. This shows that the proposed techniques improve the transmission robustness with respect to a standard OFDM receiver. Besides, PDR-CK provides a gain of 2.5 to 5dB in noise variance with respect to the PDR-CE. Note also that, as expected, when the PA is linear, the BERs are much lowers than the ones obtained with the nonlinear PA.

6 Conclusion

Two techniques for canceling nonlinear ICI in OFDM systems with nonlinear PAs have been proposed in this paper. These techniques are based on the power diversity transmission scheme that re-transmits all the symbols several times with a different transmission power each time. We have tested the proposed receivers by means of simulations and they have significantly outperformed the 1-tap equalizers, which correspond to a standard approach to recover OFDM symbols. The main drawback of the PDRs is the fact that the transmission rate is divided by the repetition factor. However, based on a realistic assumption that the PA can be modeled using a third-order polynomial, we can use a repetition factor equal to 2. That means that the PDRs may provide a more robust transmission at the cost of a lower transmission rate. In a future work, we will compare the performance of the proposed techniques with other methods for interference rejection in nonlinear OFDM systems and the PDRs will be extended to the case of a nonlinear PA with memory.

References

1. D'Andrea, A.N., Lottici, V., Reggiannini, R.: Nonlinear predistortion of OFDM signals over frequency-selective fading channels. IEEE Transactions on Communications 49(5), 837–843 (2001)
2. Costa, E., Pupolin, S.: M-QAM-OFDM System Performance in the Presence of a Nonlinear Amplifier and Phase Noise. IEEE Transactions on Communications 50(3), 462–472 (2002)
3. Ding, L.: Digital Predistortion of Power Amplifiers for Wireless Applications. School of Electrical and Computer Engineering, Georgia Institute of Technology, Georgia, USA (2004)
4. Aschbacher, E.: Digital Pre-distortion of Microwave Power Amplifiers. Vienna University of Technology, Austria (2005)
5. Ding, L., Zhou, G.T., Morgan, D.R., Ma, Z., Kenney, J.S., Kim, J., Giardina, C.R.: A Robust Digital Baseband Predistorter Constructed Using Memory Polynomials. IEEE Transactions on Communications 52(1), 159–165 (2004)
6. Ermolova, N.Y., Nefedov, N., Haggman, S.: An iterative method for non-linear channel equalization in OFDM systems. In: IEEE International Symposium on Personal, Indoor and Mobile Radio Communications, Barcelona, Spain , vol. 1, pp. 484–488 (2004)

7. Gregorio, F., Werner, S., Laakso, T.I., Cousseau, J.: Receiver Cancellation Technique for Nonlinear Power Amplifier Distortion in SDMA–OFDM Systems. IEEE Transactions on Vehicular Technology 56(5), 2499–2516 (2007)
8. Fernandes, C.A.: Nonlinear MIMO Communication Systems: Channel Estimation and Information Recovery using Volterra Models. University of Nice - Sophia Antipolis, France & Federal University of Ceará, Brazil (2009)
9. Bohara, V.A., Ting, S.H.: Theoretical analysis of OFDM signals in nonlinear polynomial models. In: International Conference on Information, Communications and Signal Processing, Singapore City, Singapore, pp. 10–13 (2007)
10. Bohara, V.A., Ting, S.H.: Analysis of OFDM Signals in Nonlinear High Power Amplifier with Memory. In: IEEE International Conference on Communications, Beijing, China, pp. 3653–3657 (2008)
11. Tang., H., Lau, K.Y., Brodersen, R.W.: Interpolation-based maximum likelihood channel estimation using OFDM pilot symbols. In: Global Telecommunications Conference, Taipei, Taiwan, vol. 2, pp. 1860–1864 (2002)

Probabilistic Latent Tensor Factorization

Y. Kenan Yılmaz and A. Taylan Cemgil*

Department of Computer Engineering,
Boğaziçi University, 34342 Bebek, Istanbul, Turkey
kenan@sibnet.com.tr, taylan.cemgil@boun.edu.tr

Abstract. We develop a probabilistic framework for multiway analysis of high dimensional datasets. By exploiting a link between graphical models and tensor factorization models we can realize any arbitrary tensor factorization structure, and many popular models such as CP or TUCKER models with Euclidean error and their non-negative variants with KL error appear as special cases. Due to the duality between exponential families and Bregman divergences, we can cast the problem as inference in a model with Gaussian or Poisson components, where tensor factorisation reduces to a parameter estimation problem. We derive the generic form of update equations for multiplicative and alternating least squares. We also propose a straightforward matricisation procedure to convert element-wise equations into the matrix forms to ease implementation and parallelisation.

Keywords: Tensor factorisation, Non-negative decompositions, NMF, NTF, CP, TUCKER, EM algorithm, Graphical models.

1 Introduction

Advances in computing power, data acquisition, storage technologies made it possible to collect and process huge amounts of data in many disciplines. Yet, in order to extract useful information effective and efficient computational tools are needed. In this context, matrix factorisation techniques have emerged as a useful paradigm [10,15]. Clustering, ICA, NMF, LSI, collaborative filtering and many such methods can be expressed and understood as matrix factorisation problems. Thinking of a matrix as the basic data structure maps well onto special purpose hardware (such as a GPU unit) to make algorithms run faster via parallelisation. Moreover, matrix computations come with a toolbox of well understood algorithms with precise error analysis, performance guarantees and extremely efficient standard implementations (e.g., SVD).

A useful method in multiway analysis is *tensor factorization* (TF) to extract hidden structure in data that consists of more than two entities. However, since there are many more natural ways to factorise a multiway array, there exists a plethora of related models with distinct names discussed in detail in recent

* This research is funded by the Bogazici University Research Fund under grant BAP 09A105P.

V. Vigneron et al. (Eds.): LVA/ICA 2010, LNCS 6365, pp. 346–353, 2010.

tutorial reviews [9,1]. A recent book [5] outlined various optimization algorithms for non-negative TF for alpha and beta divergences. The idea of sparse non-negative TUCKER was discussed in [12]. Use of the probabilistic approach, then, for the matrix factorization (PMF) was presented by [13] while probabilistic non-negative TF came out in [14]. However, all these works focus on isolated models.

The motivation behind this paper is pave the way to a unifying framework in which any arbitrary TF structure can be realized and the associated inference algorithm can be derived automatically using matrix computation primitives. For this, we introduce a notation for TF models that closely resembles probabilistic graphical models [7]. We also propose a probabilistic approach to multiway analysis as this provides a natural framework for model selection and handling missing values. We focus on using the KL divergence and the Euclidean metric to cover both unconstrained and non-negative decompositions. Our probabilistic treatment generalises the statistical treatment of NMF models described in [4,6].

2 Tensor Factorization (TF) Model

Following the established jargon, we call a N-way array $X \in \mathcal{X}^{I_1 \times I_2 \times \cdots \times I_N}$ simply a 'tensor'. Here, I_n are finite index sets, where i_n is the corresponding index. We denote an element of the tensor $X(i_1, i_2, \ldots, i_N) \in \mathcal{X}$ as $X^{i_1, i_2, \ldots, i_N}$. Similarly, given the index set $W = \{i_1, \ldots, i_N\}$ we use the notation $X(w)$ to denote an element of $X^{i_1, i_2, \ldots, i_N}$.

We associate with each TF model an undirected graph, where each vertex corresponds to an index. We let V be the set of vertices $V = \{v_1, \ldots v_n, \ldots, v_N\}$. Our objective is to estimate a set of tensors $\mathcal{Z} = \{Z_\alpha | \alpha = 1 \ldots N\}$ such that

$$\text{minimize } d(X||\hat{X}) \text{ s.t. } \hat{X}(w) = \sum_{\bar{w} \in \bar{W}} \prod_{\alpha} Z_\alpha(v_\alpha) \tag{1}$$

where the function d is a suitable error measure, which we define later. Each Z_α is associated with an index set V_α such that $V = \cup_\alpha V_\alpha$. Two distinct sets V_α and $V_{\alpha'}$ can have nonempty intersection but they don't contain each other. We define a set of 'visible' indices $W \subseteq V$ and 'invisible' indices $\bar{W} \subseteq V$ such that $W \cup \bar{W} = V$ and $W \cap \bar{W} = \emptyset$.

Example 1 (TUCKER3 Factorization). The TUCKER3 factorization [8,9] aims to find Z_α for $\alpha = 1 \ldots 4$ that solve the following optimization problem where in our notation, the TUCKER3 model is given by $N = 4$, $V = \{p, q, r, i, j, k\}$, $V_1 = \{i, p\}$, $V_2 = \{j, q\}$, $V_3 = \{k, r\}$, $V_4 = \{p, q, r\}$ and $W = \{i, j, k\}$, $\bar{W} = \{p, q, r\}$.

$$\text{minimize } d(X||\hat{X}) \text{ s.t. } \hat{X}^{i,j,k} = \sum_{p,q,r} Z_1^{i,p} Z_2^{j,q} Z_3^{k,r} Z_4^{p,q,r} \qquad \forall i, j, k \tag{2}$$

In this paper for the error measure d, we use KL divergence and Euclidean distance that give rise to two variants that we call as PLTF_{KL} (*Probabilistic Latent Tensor Factorization*) and PLTF_{EU} respectively. Alternatives such as NMF with

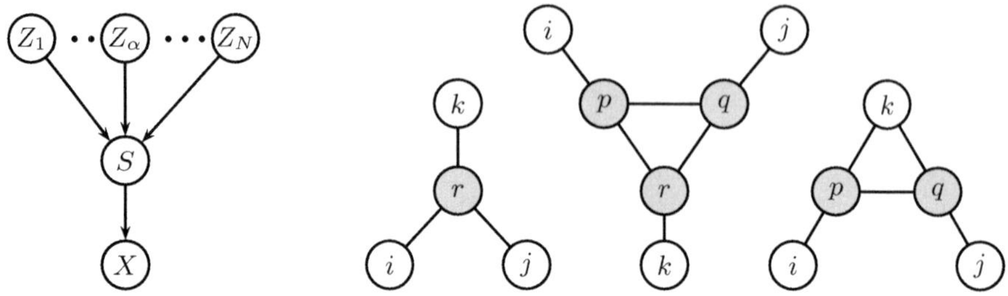

Fig. 1. The DAG on the left is the graphical model of PLTF. X is the observed multiway data and Z_α's are the parameters. The latent tensor S allows us to treat the problem in a data augmentation setting and apply the EM algorithm. On the other hand, the factorisation implied by TF models can be visualised using the semantics of undirected graphical models where cliques (fully connected subgraphs) correspond to individual factors. The undirected graphs on the right represent CP, TUCKER3 and PARATUCK2 models in the order. The shaded indices are hidden, i.e., correspond to the dimensions that are not part of X.

IS divergence also exist in [6]. Due to the duality between the Poisson likelihood and KL divergence, and between the Gaussian likelihood and Euclidean distance [3], solving the TF problem in (1) is equivalent to finding the ML solution of $p(X|Z_{1:N})$ [6].

2.1 Probability Model

For *PLTF*, we write the following generative model such that $W \cup \bar{W} = \cup_\alpha V_\alpha = V$ and for their instantiations $(w, \bar{w}) = \cup_\alpha v_\alpha = v$

$$\Lambda(v) = \prod_{\alpha}^{N} Z_\alpha(v_\alpha) \quad \text{model paramaters to estimate} \tag{3}$$

$$S(w, \bar{w}) \sim \mathcal{PO}(S; \Lambda(v)) \quad \text{element of latent tensor for } PLTF_{KL} \tag{4}$$

$$S(w, \bar{w}) \sim \mathcal{N}(S; \Lambda(v), 1) \quad \text{element of latent tensor for } PLTF_{EU} \tag{5}$$

$$X(w) = \sum_{\bar{w} \in \bar{W}} S(w, \bar{w}) \quad \text{model estimate after augmentation} \tag{6}$$

$$M(w) = \begin{cases} 0 & X(w) \text{ is missing} \\ 1 & \text{otherwise} \end{cases} \quad \text{mask array} \tag{7}$$

Note that due to reproductivity property of Possion and Gaussian distributions [11] the observation $X(w)$ has the same type of distribution as $S(w, \bar{w})$.

Next, *PLTF* handles the missing data smoothly by the following observation model [13,4]

$$p(X|S)p(S|Z_{1:N}) = \prod_{w \in W} \prod_{\bar{w} \in \bar{W}} \left(p(X(w)|S(w, \bar{w}))\, p(S(w, \bar{w})|Z_{1:N})\right)^{M(w)} \tag{8}$$

2.2 PLTF$_{KL}$ Fixed Point Update Equation

We can easily optimise for Z_α by an EM algorithm. The loglikelihood $\mathcal{L}_{KL}$ is

$$\sum_{w \in W} \sum_{\bar{w} \in \bar{W}} M(w) \left(S(w,\bar{w}) \log \left(\prod_\alpha Z_\alpha(v_\alpha) \right) - \left(\prod_\alpha Z_\alpha(v_\alpha) \right) - \log S(w,\bar{w})! \right)$$

subject to the constraint $X(w) = \sum_{\bar{w}} S(w,\bar{w})$ whenever $M(w) = 1$. The E-step is calculated by identifying the posterior of S as a multinomial distribution [11] with the following sufficient statistics

$$\langle S(w,\bar{w}) \rangle = \frac{X(w) \prod_\alpha Z_\alpha(v_\alpha)}{\sum_{\bar{w} \in \bar{W}} \prod_\alpha Z_\alpha(v_\alpha)} = \frac{X(w)}{\hat{X}(w)} \prod_\alpha Z_\alpha(v_\alpha) \tag{9}$$

where $\hat{X}(w)$ is the model estimate as $\hat{X}(w) = \sum_{\bar{w} \in \bar{W}} \prod_\alpha Z_\alpha(v_\alpha)$. The M-step is

$$\frac{\partial \mathcal{L}_{KL}}{\partial Z_\alpha(v_\alpha)} = 0 \qquad \Rightarrow \qquad Z_\alpha(v_\alpha) = \frac{\sum_{v \notin V_\alpha} M(w) \langle S(w,\bar{w}) \rangle}{\sum_{v \notin V_\alpha} M(w) \prod_{\alpha' \neq \alpha} Z_{\alpha'}(v_{\alpha'})} \tag{10}$$

After substituting (9) in (10) we obtain the following fixed point update for Z_α

$$Z_\alpha(v_\alpha) \leftarrow Z_\alpha(v_\alpha) \frac{\sum_{v \notin V_\alpha} M(w) \frac{X(w)}{\hat{X}(w)} \prod_{\alpha' \neq \alpha} Z_{\alpha'}(v_{\alpha'})}{\sum_{v \notin V_\alpha} M(w) \prod_{\alpha' \neq \alpha} Z_{\alpha'}(v_{\alpha'})} \tag{11}$$

Definition 1. *We define the tensor valued function $\Delta_\alpha(A) : \mathbb{R}^{|A|} \to \mathbb{R}^{|Z_\alpha|}$ (associated with Z_α) as*

$$\Delta_\alpha(A) \equiv \left[\sum_{v \notin V_\alpha} \left(A(w) \prod_{\alpha' \neq \alpha} Z_{\alpha'}(v_{\alpha'}) \right) \right] \tag{12}$$

$\Delta_\alpha(A)$ is an object the same size of Z_α. We also use the notation $\Delta_{Z_\alpha}(A)$ especially when Z_α are assigned distinct letters. $\Delta_\alpha(A)(v_\alpha)$ refers to a particular element of $\Delta_\alpha(A)$. Using this new definition, we rewrite (11) more compactly as

$$Z_\alpha \leftarrow Z_\alpha \circ \Delta_\alpha(M \circ X/\hat{X}) / \Delta_\alpha(M) \tag{13}$$

where $\circ$ and $/$ stand for element wise multiplication (Hadamard product) and division respectively. Later we develop the explicit matrix forms of these updates.

2.3 PLTF$_{EU}$ Fixed Point Update Equation

The derivation follows closely Section 2.2 where we merely replace the Poisson likelihood with that of a Gauissian. The complete data loglikelihood becomes

$$\mathcal{L}_{EU} = \sum_{w \in W} \sum_{\bar{w} \in \bar{W}} M(w) \left(-\frac{1}{2} \log(2\pi) - \frac{1}{2} \left(S(w,\bar{w}) - \prod_\alpha Z_\alpha(v_\alpha) \right)^2 \right) \tag{14}$$

subject to the constraint $X(w) = \sum_{\bar{w}} S(w, \bar{w})$ for $M(w) = 1$. The sufficient statistics of the Gaussian posterior $p(S|Z, X)$ are available in closed form as

$$\langle S(w, \bar{w}) \rangle = \prod_{\alpha}^{N} Z_\alpha(v_\alpha) - \frac{1}{K}(X(w) - \hat{X}(w)) \tag{15}$$

where K is the cardinality i.e. $K = |\bar{W}|$. Then, the solution of the M step after plugging (15) in $\frac{\partial \mathcal{L}_{EU}}{\partial Z_\alpha(v_\alpha)}$ and by setting it to zero

$$\begin{aligned} \frac{\partial \mathcal{L}_{EU}}{\partial Z_\alpha(v_\alpha)} &= \sum_{v \notin V_\alpha} M(w) \left(\left(X(w) - \hat{X}(w) \right) \prod_{\alpha' \neq \alpha} Z_{\alpha'}(v_{\alpha'}) \right) \\ &= \Delta_\alpha(M \circ X) - \Delta_\alpha(M \circ \hat{X}) = 0 \end{aligned} \tag{16}$$

The solution of this fixed point equation leads to two related but different iterative schemata: multiplicative updates (MUR) and alternating least squares (ALS).

PLTF$_{EU}$ Multiplicative Update Rules (MUR). This method is indeed gradient ascent similar to [10] by setting $\eta(v_\alpha) = Z_\alpha(v_\alpha)/\Delta_\alpha(M \circ \hat{X})(v_\alpha)$ as

$$Z_\alpha(v_\alpha) \leftarrow Z_\alpha(v_\alpha) + \eta(v_\alpha) \frac{\partial \mathcal{L}_{EU}}{\partial Z_\alpha(v_\alpha)} \tag{17}$$

Then the update rule becomes simply

$$Z_\alpha \leftarrow Z_\alpha \circ \Delta_\alpha(M \circ X) / \Delta_\alpha(M \circ \hat{X}) \tag{18}$$

PLTF$_{EU}$ Alternating Least Squares (ALS). The idea behind ALS is to obtain a closed form solution for Z_α directly from (16)

$$\Delta_\alpha(M \circ X) = \Delta_\alpha(M \circ \hat{X}) \tag{19}$$

Note that $\hat{X}$ depends on Z_α, see (1). This equation can be solved for Z_α by least squares, as it is linear in Z_α. If there is no missing data ($M(w) = 1$ for all w), the result is available in closed form. To see this, we write all the tensors in matrix form and write the solution explicitly using standard matrix algebra.

3 Matricization

Matricization as defined in [8,9] is the operation of converting a multiway array into a matrix by reordering the column fibers. In this paper we refer to this definition as 'unfolding' and refer to *matricization* as the procedure to convert an element-wise equation (such as (19)) into a corresponding matrix form. We use Einstein's summation convention where repeated indices are added over.

The conversion rules are given in Table 1. Our notation is best illustrated with an example: consider a matrix $X^{i,j}$ with row index i and column index j. If we assume a column by column memory layout, we refer to the vectorisation of $\mathbf{vec}\, X$ (vertical concatenation of columns) as X_{ji}; adopting a 'faster index last' convention and we drop the commas. Here i is the faster index since when traversing the elements of the matrix X in sequence i changes more rapidly. With this, we arrive at the following definition:

Definition 2. *Consider a multiway array $X \in \mathbb{R}^{I_1 \times \ldots \times I_L}$ with a generic element denoted by $X^{i_1,i_2,\ldots,i_L}$. The mode-n unfolding of X is the matrix $X_{(n)} \in \mathbb{R}^{I_n \times \prod_{k \neq n} I_k}$ with row index i_n where*

$$X_{(n)} \equiv X_{i_n}^{i_L \ldots i_{n-1} i_{n+1} \ldots i_2 i_1} \tag{20}$$

Table 1. Index notation used to unfold a multiway array into the matrix form. Following Einstein convention, duplicate indices are summed over. Khatri-Rao product and mode-n unfolding are implemented in N-way Toolbox [2] as *krb*() and *nshape*().

Equivalance	Matlab	Remark
$X_i^j \equiv X$	X	Matrix notation
$X_i^{kj} \equiv X_{(1)}$	$nshape(X,1)$	Array (mode-1 unfolding)
$X_i^j \equiv (X^T)_j^i$	X'	Transpose
$\mathbf{vec}\, X_i^j \equiv (X)_{ji}$	$X(:)$	Vectorize
$X_i^j Y_j^p \equiv (XY)_i^p$	$X * Y$	Matrix product
$X_i^p Y_j^p \equiv (X \odot Y)_{ij}^p$	$krb(X,Y)$	Khatri-Rao product
$X_i^p Y_j^q \equiv (X \otimes Y)_{ij}^{pq}$	$kron(X,Y)$	Kronecker product

In the following, we illustrate the derivation of the well known TUCKER3 factorization. Alternative models can be derived and implemented similarly. To save the space we only give derivations for factor A and core tensor G and omit the others. Further details, model examples and reference implementations in Matlab can be found from `http://www.sibnet.com.tr/pltf`.

Example 2 (Derivation of matrix form update rules for the TUCKER3 decomposition). We compute first the prediction in matrix form

$$\hat{X}^{i,j,k} = \sum_{pqr} G^{p,q,r} A^{i,p} B^{j,q} C^{k,r} \tag{21}$$

$$(\hat{X}_{(1)})_i^{kj} = (G_{(1)})_p^{rq} A_i^p B_j^q C_k^r = \left((AG_{(1)})((C \otimes B)^T)\right)_i^{kj} \tag{22}$$

$$\hat{X}_{(1)} = AG_{(1)}(C \otimes B)^T \tag{23}$$

Algorithm 1. *Probabilistic Latent Tensor Factorisation*

for epoch = 1 ... MAXITER **do**
  for $\alpha = 1 \ldots N$ **do**
    $\hat{X} \leftarrow \sum_{\bar{w} \in \bar{W}} \prod_{\alpha} Z_{\alpha}(V_{\alpha})$
    if KL: $Z_{\alpha} \leftarrow Z_{\alpha} \circ \Delta_{\alpha}(M \circ X/\hat{X})/\Delta_{\alpha}(M)$
    if EUC-MUR: $Z_{\alpha} \leftarrow Z_{\alpha} \circ \Delta_{\alpha}(M \circ X)/\Delta_{\alpha}(M \circ \hat{X})$
    if EUC-ALS: Solve $\Delta_{\alpha}(M \circ X) = \Delta_{\alpha}(M \circ \sum_{\bar{w} \in \bar{W}} \prod_{\alpha} Z_{\alpha}(V_{\alpha}))$ for Z_{α}
  end for
end for

Now, $\Delta_{Z_{\alpha}}$ for all α can also be represented in matrix form. The functions Δ_A and Δ_G are

$$\Delta_A(X) \equiv (X_{(1)})_i^{kj} B_j^q C_k^r G_p^{rq} \equiv X_{(1)}(C \otimes B) G_{(1)}^T \tag{24}$$

$$\Delta_G(X) \equiv (X_{(1)})_i^{kj} A_i^p B_j^q C_k^r \equiv A^T X_{(1)}(C \otimes B) \tag{25}$$

– if KL ((13)) we evaluate $\Delta_{\alpha}(Q)$ and $\Delta_{\alpha}(M)$ where $Q = M \circ (X/\hat{X})$

$$A \leftarrow A \circ \frac{Q_{(1)}(C \otimes B) G_{(1)}^T}{M_{(1)}(C \otimes B) G_{(1)}^T} \qquad G_{(1)} \leftarrow G_{(1)} \circ \frac{(A^T Q_{(1)})(C \otimes B)}{(A^T M_{(1)})(C \otimes B)} \tag{26}$$

– if EUC-ALS. We solve $\Delta_{\alpha}(X) = \Delta_{\alpha}(\hat{X})$ (19) when there are no missing observations, i.e., $M(w) = 1$ for all w. We show only the updates for the core tensor G. The pseudo-inverse of A is denoted by $A^{\dagger}$. From (25) we have

$$A^T X_{(1)}(C \otimes B) = A^T \left(A G_{(1)} (C \otimes B)^T \right) (C \otimes B) \tag{27}$$

$$G_{(1)} \leftarrow A^{\dagger} X_{(1)} \left((C \otimes B)^T \right)^{\dagger}. \tag{28}$$

4 Discussion

The main saving in our framework appears in the computation of Δ_{α}, that is computationally equivalent to computing expectations under probability distributions that factorise according to a given graph structure. As is the case with graphical models, this quantity can be computed a-la belief propagation: algebraically we distribute the summation over all $v \notin V_{\alpha}$ and compute the sum in stages. For MUR, the intermediate computations carried out when computing the denominator and numerator can be reused, which leads to further savings (e.g., see (26)). Perhaps more importantly, PLTF encourages the researchers to 'invent' new factorization models appropriate to their applications. Pedagogically, the framework guides building new models as well as deriving update equations for KL and Euclidean cost functions. Indeed, results scattered in the literature can be derived in a straightforward manner.

Due to space constraints, in this paper we could not detail on model selection issues, i.e., questions regarding to the dimensions of latent indices and the selection of an optimal factorisation structure, guided by data. It turns out, exploiting the probabilistic interpretation and choosing an appropriate prior, it is indeed possible to approximate the marginal likelihood $p(X) = \int dZp(X, Z)$ for doing Bayesian model selection, using techniques described in [4].

References

1. Acar, E., Yener, B.: Unsupervised multiway data analysis: A literature survey. IEEE Transactions on Knowledge and Data Engineering 21, 6–20 (2009)
2. Andersson, C.A., Bro, R.: The N-way toolbox for MATLAB. Chemometrics and Intelligent Laboratory Systems 52, 1–4 (2000)
3. Banerjee, A., Merugu, S., Dhillon, I.S., Ghosh, J.: Clustering with Bregman divergences. Journal of Machine Learning Research 6, 1705–1749 (2005)
4. Cemgil, A.T.: Bayesian inference for nonnegative matrix factorisation models. Computational Intelligence and Neuroscience 2009, 1–17 (2009)
5. Cichocki, A., Zdunek, R., Phan, A.H., Amari, S.: Nonnegative Matrix and Tensor Factorization. Wiley, Chichester (2009)
6. Fevotte, C., Cemgil, A.T.: Nonnegative matrix factorisations as probabilistic inference in composite models. In: Proc. 17th EUSIPCO (2009)
7. Jordan, M.I. (ed.): Learning in Graphical Models. MIT Press, Cambridge (1999)
8. Kiers, H.A.L.: Towards a standardized notation and terminology in multiway analysis. Journal of Chemometrics 14, 105–122 (2000)
9. Kolda, T.G., Bader, B.W.: Tensor decompositions and applications. SIAM Review 51(3), 455–500 (2009)
10. Lee, D.D., Seung, H.S.: Algorithms for non-negative matrix factorization. In: NIPS, vol. 13, pp. 556–562 (2001)
11. Meyer, P.L.: Introductory Probability and Statistical Applications, 2nd edn. Addison-Wesley, Reading (1970)
12. Mørup, M., Hansen, L.K., Arnfred, S.M.: Algorithms for sparse non-negative TUCKER. Neural Computation 20(8), 2112–2131 (2008)
13. Salakhutdinov, R., Mnih, A.: Probabilistic matrix factorization. Advances in Neural Information Processing Systems 20 (2008)
14. Schmidt, M., Mohamed, S.: Probabilistic non-negative tensor factorisation using Markov Chain Monte Carlo. In: 17th European Signal Processing Conference (2009)
15. Singh, A.P., Gordon, G.J.: A unified view of matrix factorization models. In: Daelemans, W., Goethals, B., Morik, K. (eds.) ECML PKDD 2008, Part II. LNCS (LNAI), vol. 5212, pp. 358–373. Springer, Heidelberg (2008)

Nonorthogonal Independent Vector Analysis Using Multivariate Gaussian Model*

Matthew Anderson, Xi-Lin Li, and Tülay Adalı

Machine Learning for Signal Processing Laboratory
University of Maryland Baltimore County, Baltimore, MD 21250
matt.anderson@ieee.org, lixilin@umbc.edu, adali@umbc.edu
http://mlsp.umbc.edu/

Abstract. We consider the problem of joint blind source separation of multiple datasets and introduce an effective solution to the problem. We pose the problem in an independent vector analysis (IVA) framework utilizing the multivariate Gaussian source vector distribution. We provide a new general IVA implementation using a decoupled nonorthogonal optimization algorithm and establish the connection between the new approach and another approach using second-order statistics, multiset canonical correlation analysis. Experimental results are given to demonstrate the success of the new algorithm in achieving reliable source separation for both Gaussian and non-Gaussian sources.

Keywords: Canonical correlation analysis (CCA); independent vector analysis (IVA).

1 Introduction

The problem of joint blind source separation (BSS) of multiple datasets arises in many applications such as joint analysis of multi-subject data in medical studies or when working with transformed data in multiple bins. An example joint BSS application is the identification of meaningful brain activations in functional magnetic resonance imaging (fMRI) datasets from multiple subjects concurrently [1,2]. In such problems, it is natural to assume that each latent source within a dataset is both related to a single latent source within each of the other datasets *and* independent of all the other sources within the dataset. Joint BSS solutions can exploit the dependencies of sources across datasets potentially resulting in performance beyond what is achievable with BSS algorithms applied to each dataset individually. Additionally, joint BSS algorithms can "align" dependent sources across the datasets. Using BSS algorithms individually does not automatically provide such an alignment. Rather, an additional algorithm is needed to resolve this so-called permutation ambiguity.

An extension of canonical correlation analysis from two datasets to multiple datasets, termed multiset canonical correlation analysis (MCCA) [3], has been shown to achieve joint BSS under certain conditions [2] and has been noted to

* This work is supported by the NSF grants NSF-CCF 0635129 and NSF-IIS 0612076.

V. Vigneron et al. (Eds.): LVA/ICA 2010, LNCS 6365, pp. 354–361, 2010.

be quite promising for analysis of fMRI data from multiple subjects. MCCA as introduced in [3] utilizes a number of cost functions that are all ad hoc in nature and are based on second-order statistics. After pre-whitening the datasets, a deflationary procedure is used to achieve the decomposition that results in orthogonal demixing matrices. The use of second-order statistics results in an efficient joint BSS solution that can be widely applied. However, the ad hoc nature of the cost functions, the error accumulation of the deflationary procedure, and the limiting nature of the orthogonality constraint on the demixing matrix limit the performance of this MCCA implementation.

A more general and principled approach for joint BSS is to minimize the mutual information across the sources to achieve independent vector analysis (IVA). IVA as originally derived in [4] was designed to address another joint BSS problem, namely BSS of data transformed from the temporal domain to the frequency domain. The implementation assumes a multivariate Laplace as the form of the *source component vector* (SCV) distribution and no second-order correlation within the SCVs. Both assumptions are strong and limit the potential applications of this IVA implementation.

In this paper, we pose the problem as mutual information minimization as in IVA, but propose to exploit second-order statistical information across datasets by assuming the SCV distributions are multivariate Gaussians. The resulting cost function is shown to be equivalent to a particular MCCA cost function. We then minimize the IVA cost function with a set of nonorthogonal demixing matrices found via a novel decoupled gradient-based optimization scheme. We show using simulations that this algorithm solves joint BSS for linearly dependent Gaussian and non-Gaussian sources. The new algorithm extends the application domain of IVA to joint BSS when sources across datasets possess linear dependence.

2 Joint BSS Problem Formulation

We first formulate the joint BSS problem. There are K datasets each formed from linear mixtures of N independent sources,

$$\mathbf{x}^{[k]} = \mathbf{A}^{[k]}\mathbf{s}^{[k]}, \quad 1 \leq k \leq K.$$

The zero-mean source vector, $\mathbf{s}^{[k]} = \left[s_1^{[k]}, \ldots, s_N^{[k]}\right]^T$, and the invertible mixing matrix, $\mathbf{A}^{[k]}$, are unknown real-valued quantities to be estimated. The nth SCV, $\mathbf{s}_n = \left[s_n^{[1]}, \ldots, s_n^{[K]}\right]^T$, is independent of all other SCVs. Thus, the joint distribution for all the sources is the product of the SCV distributions, or $p(\mathbf{s}_1, \ldots, \mathbf{s}_N) = \prod_{n=1}^{N} p(\mathbf{s}_n)$. The BSS solution finds K demixing matrices and source vector estimates for each dataset, with the kth ones denoted as $\mathbf{W}^{[k]}$ and $\mathbf{y}^{[k]} = \mathbf{W}^{[k]}\mathbf{x}^{[k]}$, respectively. The estimate of the nth component from the kth dataset is given by $y_n^{[k]} = \left(\mathbf{w}_n^{[k]}\right)^T \mathbf{x}^{[k]} = \sum_{l=1}^{N} w_{n,l}^{[k]} x_l^{[k]}$, where $\left(\mathbf{w}_n^{[k]}\right)^T$ is the nth row of $\mathbf{W}^{[k]}$. The estimate of the nth SCV is given as $\mathbf{y}_n = \left[y_n^{[1]}, \ldots, y_n^{[K]}\right]^T$, thus requiring the estimated demixing matrices to result in aligned sources. Lastly, the mixing matrices are distinct for each dataset and are not necessarily related.

3 IVA Cost Function

The goal of IVA, the identification of the independent SCVs, can be achieved by minimizing the mutual information among the source component vectors:

$$\begin{aligned}\mathcal{I}_{\text{IVA}} \triangleq I\left[\mathbf{y}_1;\ldots;\mathbf{y}_N\right] &= \sum_{n=1}^{N} H\left[\mathbf{y}_n\right] - H\left[\mathbf{y}_1,\ldots,\mathbf{y}_N\right] \\ &= \sum_{n=1}^{N} H\left[\mathbf{y}_n\right] - H\left[\mathbf{W}^{[1]}\mathbf{x}^{[1]},\ldots,\mathbf{W}^{[K]}\mathbf{x}^{[K]}\right] \\ &= \sum_{n=1}^{N} H\left[\mathbf{y}_n\right] - \sum_{k=1}^{K} \log\left|\det\left(\mathbf{W}^{[k]}\right)\right| - C_1. \end{aligned} \quad (1)$$

We note that the entropy[1] of a linear invertible transformation, $\mathbf{Wx}$, is given by $\log|\det(\mathbf{W})| + H[\mathbf{x}]$ and the determinant of a block diagonal matrix is the product of the determinants of the individual blocks. The last term of (1), C_1, denotes the constant term $H\left[\mathbf{x}^{[1]},\ldots,\mathbf{x}^{[K]}\right]$. The regularization term, $\log\left|\det\left(\mathbf{W}^{[k]}\right)\right|$, penalizes demixing matrices having small determinants and by restricting demixing matrices to be orthogonal the penalty term becomes fixed. Thus, nonorthogonal demixing matrices enable lower cost functions to be found.

The cost function is the same as in [1,4], however the derivation differs because rather than using the Kullback-Leibler distance, here the relationship between mutual information and the difference between the sum of marginal entropies and the joint entropy is exploited.

By noting that $H[\mathbf{y}_n] = \sum_k H\left[y_n^{[k]}\right] - I[\mathbf{y}_n]$, then (1) is

$$\mathcal{I}_{\text{IVA}} = \sum_{n=1}^{N}\left(\sum_{k=1}^{K} H\left[y_n^{[k]}\right] - I\left[\mathbf{y}_n\right]\right) - \sum_{k=1}^{K}\log\left|\det\left(\mathbf{W}^{[k]}\right)\right| - C_1.$$

This representation shows that minimizing the cost function simultaneously minimizes the entropy of all components *and* maximizes the mutual information within each SCV.

4 Decoupled Nonorthogonal Optimization for IVA

After identifying the cost function, an approach to its minimization is devised. The approach of [1,4] is to compute the derivative of the cost function with respect to each demixing matrix,

$$\frac{\partial \mathcal{I}_{\text{IVA}}}{\partial \mathbf{W}^{[k]}} = E\left\{\phi^{[k]}\left(\mathbf{y}\right)\left(\mathbf{x}^{[k]}\right)^{T}\right\} - \left(\mathbf{W}^{[k]}\right)^{-T}, \quad (2)$$

[1] Since discrete-valued variables are not considered in this paper, we refer to differential entropy as simply entropy. Furthermore, $H[\mathbf{x}]$ is the entropy of $\mathbf{x}$, and $I[\mathbf{x};\mathbf{y}]$ is the mutual information between $\mathbf{x}$ and $\mathbf{y}$.

where,

$$\phi^{[k]}(\mathbf{y}) = \left[\phi^{[k]}(\mathbf{y}_1), \ldots, \phi^{[k]}(\mathbf{y}_N)\right]^T = -\left[\frac{\partial \log\{p(\mathbf{y}_1)\}}{\partial y_1^{[k]}}, \ldots, \frac{\partial \log\{p(\mathbf{y}_N)\}}{\partial y_N^{[k]}}\right]^T,$$

is termed the multivariate score function. Then, each of the K matrices are updated sequentially by

$$\mathbf{W}^{[k]} \leftarrow \mathbf{W}^{[k]} - \mu \frac{\partial \mathcal{I}_{\text{IVA}}}{\partial \mathbf{W}^{[k]}}, \tag{3}$$

where μ is the scalar positive step size. Additionally, it is suggested in [1,4] to use natural gradient updates for improved convergence.

Next we note that each row of a demixing matrix corresponds to an estimated source, with a direction for reducing (1) specified by the directions of the rows of (2). Thus, using a single step size to update an entire demixing matrix gives equal weight to each direction when in fact this might be undesirable depending on the cost function surface. We propose to sequentially update each row of each demixing matrix. This enables the tailoring of the step size for each direction, resulting in faster convergence per iteration than that achieved heretofore.

Following [5], we let $\mathbf{h}_n^{[k]}$ be any unit length vector such that $\widetilde{\mathbf{W}}_n^{[k]}\mathbf{h}_n^{[k]} = \mathbf{0}$, where $\widetilde{\mathbf{W}}_n^{[k]}$ is the $(N-1) \times N$ matrix formed by removing the nth row of the demixing matrix $\mathbf{W}^{[k]}$. Then, it can be shown that [5]:

$$\left|\det\left(\mathbf{W}^{[k]}\right)\right| = \left|\left(\mathbf{h}_n^{[k]}\right)^T \mathbf{w}_n^{[k]}\right| S_n^{[k]}, \tag{4}$$

where $S_n^{[k]} = \sqrt{\left|\det\left(\widetilde{\mathbf{W}}_n^{[k]}\left(\widetilde{\mathbf{W}}_n^{[k]}\right)^T\right)\right|}$, is independent of $\mathbf{w}_n^{[k]}$. Substituting (4) into (1) we have,

$$\begin{aligned}\mathcal{I}_{\text{IVA}} &= \sum_{m=1}^{N} H[\mathbf{y}_m] - \sum_{l=1}^{K}\left[\log\left(\left|\left(\mathbf{h}_n^{[l]}\right)^T \mathbf{w}_n^{[l]}\right|\right) + \log\left(S_n^{[l]}\right)\right] - C_1 \\ &= H[\mathbf{y}_n] - \log\left(\left|\left(\mathbf{h}_n^{[k]}\right)^T \mathbf{w}_n^{[k]}\right|\right) - C_2,\end{aligned}$$

where we note that $H[\mathbf{y}_m]$ is independent of $\mathbf{w}_n^{[k]}$ for $m \neq n$ and we let C_2 be the new constant containing all the terms that are fixed with respect to changes in $\mathbf{w}_n^{[k]}$. Then, the IVA cost function derivative with respect to $\mathbf{w}_n^{[k]}$ is

$$\begin{aligned}\frac{\partial \mathcal{I}_{\text{IVA}}}{\partial \mathbf{w}_n^{[k]}} &= -E\left\{\frac{\partial \log p(\mathbf{y}_n)}{\partial y_n^{[k]}} \frac{\partial y_n^{[k]}}{\partial \mathbf{w}_n^{[k]}}\right\} - \frac{\partial \log\left|\left(\mathbf{h}_n^{[k]}\right)^T \mathbf{w}_n^{[k]}\right|}{\partial \mathbf{w}_n^{[k]}} \\ &= E\left\{\phi^{[k]}(\mathbf{y}_n)\,\mathbf{x}^{[k]}\right\} - \frac{\mathbf{h}_n^{[k]}}{\left(\mathbf{h}_n^{[k]}\right)^T \mathbf{w}_n^{[k]}}.\end{aligned} \tag{5}$$

The Gaussian assumption has found wide use in many signal processing and data analysis applications, and in [2] it has been shown that second-order statistics are successfully able to solve the joint BSS problem. Hence, we use the zero-mean and real-valued multivariate Gaussian distribution,

$$p(\mathbf{y}_n|\boldsymbol{\Sigma}_n) = \frac{1}{(2\pi)^{K/2}\det(\boldsymbol{\Sigma}_n)^{1/2}}\exp\left(-\frac{1}{2}\mathbf{y}_n^T\boldsymbol{\Sigma}_n^{-1}\mathbf{y}_n\right),$$

as the assumed form of the K dimensional SCV distribution with covariance $\boldsymbol{\Sigma}$. We calculate $\boldsymbol{\phi}(\mathbf{y}_n)$ required for the update rules as,

$$\boldsymbol{\phi}(\mathbf{y}_n) = -\frac{\partial \log\{p(\mathbf{y}_n)\}}{\partial \mathbf{y}_n} = \boldsymbol{\Sigma}_n^{-1}\mathbf{y}_n. \tag{6}$$

The new iterative IVA solution using the multivariate Gaussian distribution, named IVA-MG, is now summarized. After initializing the demixing matrices, $\mathbf{W}^{[k]}$, for each demixing matrix row, $\mathbf{w}_n^{[k]}$, we construct $\mathbf{h}_n^{[k]}$ using one of several possible methods. We suggest using the efficient recursive approach given in the appendix of [6]. Then we use the maximum likelihood estimate for the covariance matrix of the nth SCV, $\hat{\boldsymbol{\Sigma}}_n = \left(\sum_{t=1}^{T}\mathbf{y}_{n,t}\mathbf{y}_{n,t}^T\right)/T$, in (6), which is used to compute (5). Lastly, (5) is used to iteratively update $\mathbf{w}_n^{[k]}$ via

$$\mathbf{w}_n^{[k]} \leftarrow \mathbf{w}_n^{[k]} - \mu\frac{\partial \mathcal{I}_{\text{IVA}}}{\partial \mathbf{w}_n^{[k]}}. \tag{7}$$

5 IVA with Multivariate Gaussian Model and MCCA

We provide further insight by examining the IVA-MG cost function more closely. Substituting into (1) the entropy of a K dimensional multivariate Gaussian vector, $H(\mathbf{y}) = 1/2\log\left[(2\pi e)^K\prod_{k=1}^{K}\lambda_k\right]$, where λ_k is the kth eigenvalue of the covariance matrix $\boldsymbol{\Sigma}$, gives the IVA-MG cost function,

$$\mathcal{I}_{\text{IVA-MG}} = \frac{NK\log(2\pi e)}{2} + \frac{1}{2}\log\left(\prod_{n=1}^{N}\prod_{k=1}^{K}\lambda_{k,n}\right) - \sum_{k=1}^{K}\log\left|\det\left(\mathbf{W}^{[k]}\right)\right| - C_1.$$

The kth eigenvalue of the covariance matrix associated with the nth SCV is denoted as $\lambda_{k,n}$. The cost function indicates that the product of the SCV covariance eigenvalues should be minimized. Under the constraint that the sum of the eigenvalues is fixed, then the minimal cost is when each covariance matrix is as ill-conditioned as possible, i.e., the eigenvalues are maximally spread apart. This condition is equivalent to finding SCVs with maximally correlated components.

In MCCA, five cost functions are proposed to indirectly achieve this condition. The GENVAR cost function is one of these five ad hoc cost functions. Since MCCA adopts a deflationary approach, the GENVAR cost function is the product of eigenvalues associated with a single SCV covariance matrix. If the GENVAR cost function is naturally extended to be the product of the covariance eigenvalues for all N estimated SCVs and the demixing matrices are orthogonal, then the MCCA and IVA-MG cost functions are identical.

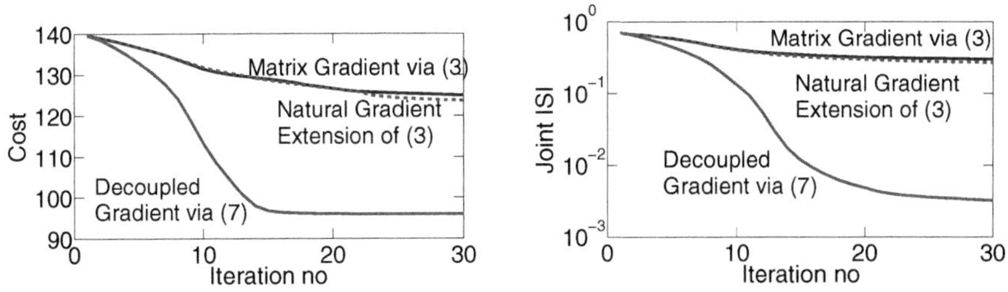

(a) IVA-MG cost versus iteration **(b)** Joint ISI versus iteration

Fig. 1. Example IVA-MG cost and performance measure versus iteration for $N = K = 10$ multivariate Gaussian sources, using the matrix gradient, natural gradient extension (dashed), and decoupled gradient optimization approaches

6 Simulations

The performance of the proposed IVA-MG algorithm is shown via simulations. A zero-mean unit-variance generalized Gaussian distribution (GGD) source is parameterized by a shape parameter, $\alpha > 0$, with $\alpha < 2$ super-Gaussian, $\alpha > 2$ sub-Gaussian, and $\alpha = 2$ Gaussian. Following [2], a linear transformation is used to generate each SCV, $\mathbf{s}_n = \mathbf{V}_n \mathbf{\Lambda}_n^{1/2} \mathbf{z}_n$, where $\mathbf{V}_n$ and $\mathbf{\Lambda}_n$ are eigenvector and eigenvalue matrices of $\mathbf{\Sigma}_n$ and $\mathbf{z}_n$ is T samples of K independent GGD sources, so that $E\left\{\mathbf{s}_n \mathbf{s}_n^T\right\} = \mathbf{\Sigma}_n$. The kth entry of each SCV is used as a latent source for the kth dataset. Random mixing matrices are used to generate each dataset. The results are comparatively insensitive to sample size, so we only consider one sample size, $T = 10\,000$.

Performance is assessed using an extension of the normalized inter-symbol-interference (ISI), termed here joint ISI, a measure similar to that used in [2]. Letting $\mathbf{G} = \sum_{k=1}^{K} \left|\mathbf{G}^{[k]}\right|$, where $\mathbf{G}^{[k]} = \mathbf{W}^{[k]} \mathbf{A}^{[k]}$ then,

$$\mathrm{ISI}_{\mathrm{JNT}}\left(\mathbf{G}\right) \triangleq \frac{\sum_{n=1}^{N}\left(\sum_{m=1}^{N} \frac{g_{n,m}}{\max_p g_{n,p}} - 1\right) + \sum_{m=1}^{N}\left(\sum_{n=1}^{N} \frac{g_{n,m}}{\max_p g_{p,m}} - 1\right)}{2N\left(N-1\right)}.$$

The joint ISI metric penalizes SCV estimates which are not consistently aligned across datasets and it is normalized so that $0 \leq \mathrm{ISI}_{\mathrm{JNT}} \leq 1$ with 0 being ideal. We note that the sensitivity of the joint ISI metric to scaling of $\mathbf{G}^{[k]}$ is addressed by scaling the original sources and their estimates to have unit variance.

An illustration for a single trial is given in Fig. 1a of the cost function at each iteration provided by decoupling the updates of the rows of a demixing matrix using (5)-(7), where we define an iteration as an update of all K demixing matrices. We compare this convergence performance with using (2) and (3) to update each demixing matrix with a single step size and its natural gradient extension. There is negligible difference between the matrix update optimizations and both approaches are inferior to the decoupled gradient updates. Fig. 1b

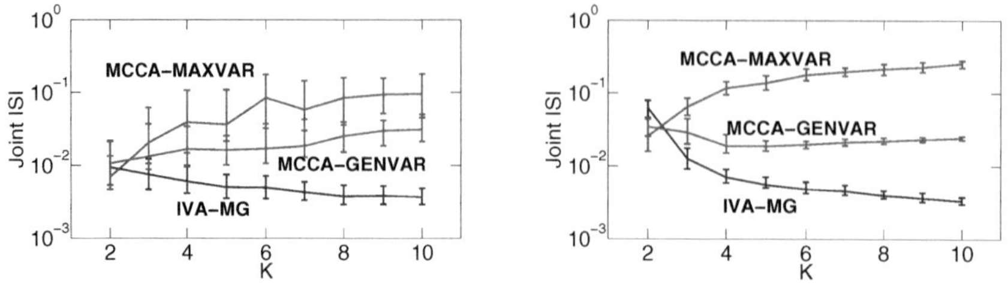

(a) $N = 2$ sources (b) $N = 10$ sources

Fig. 2. Median joint ISI for 100 trials versus the number of datasets K using multivariate Gaussian sources. Error bars indicate 25th and 75th percentile.

indicates that the joint ISI performance metric decreases as the cost function decreases.

In Fig. 2, purely Gaussian sources are considered and the joint ISI metric is shown versus the number of datasets. The performance for IVA-MG is compared to MCCA using the MAXVAR and GENVAR cost functions defined in [3]. We see IVA-MG provides superior performance. Since, the GENVAR cost function of MCCA is the closest to the IVA-MG cost function it outperforms the MAXVAR cost function. Furthermore, for IVA-MG the joint ISI decreases as the number of datasets increases, with the largest rate of decrease occurring for the first few datasets. The benefit of exploiting dependencies across datasets increases with the number of sources.

In many applications the sources of interest are non-Gaussian, as when using independent component analysis (ICA) for a single dataset. Therefore we consider GGD sources to demonstrate that for non-Gaussian sources, second-order methods are able to perform joint BSS when each source is correlated with a source in each dataset. Specifically, the elements of $\mathbf{z}_n$ are GGD sources with shape parameters selected randomly between $[1, 1.5]$ or $[2.5, 3]$. Performance in Fig. 3 is similar to the performance in Fig. 2, indicating reliable performance for IVA-MG with second-order correlated non-Gaussian SCV. The same robust behavior holds when the range of the GGD shape parameter is extended to generate either more super or sub-Gaussian sources. For this dataset, the original implementation of IVA [1,4] using the uncorrelated multivariate Laplace (IVA-ML) is also evaluated. The performance is poor due to the sources being second-order correlated and the presence of sub-Gaussian sources.

We also examine an alternative approach to joint BSS, using ICA separately K times followed by a source alignment algorithm. We approximate the performance bound of such an alternative using the EFICA algorithm of [7] on each dataset, followed by a permutation algorithm to optimally align estimated sources using *clairvoyant knowledge*. We denote this unrealizable approach, OA-EFICA, for optimally aligned EFICA. We use the EFICA algorithm since it is specifically designed to address GGD sources. Comparing OA-EFICA and IVA-MG in Fig. 3 illustrates the benefit of the latter's joint BSS formulation.

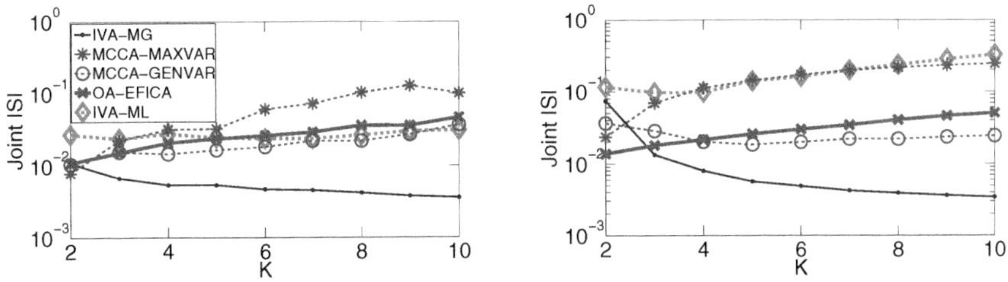

(a) $N = 2$ sources (b) $N = 10$ sources

Fig. 3. Median joint ISI for 100 trials versus the number of datasets K using second-order correlated GGD sources for each SCV

7 Conclusions

An efficient nonorthogonal implementation of IVA, IVA-MG, using second-order statistics has been given to solve the joint BSS problem. The relationship between the IVA-MG cost function and the GENVAR cost function used in MCCA is shown. We illustrate that IVA-MG is able to exploit second-order correlation of non-Gaussian sources. Additionally in the considered simulations, it provides superior performance compared to using a state-of-the art ICA algorithm on each dataset individually that is then optimized using clairvoyant knowledge to resolve the permutation ambiguity.

For future work we intend to investigate expanding the assumed form of the SCV distribution. In particular, the more general class of elliptical multivariate distributions are promising but the use of these requires the development of an efficient estimation approach for the distribution parameters.

References

1. Lee, J.H., Lee, T.W., Jolesz, F.A., Yoo, S.S.: Independent vector analysis (IVA): Multivariate approach for fMRI group study. NeuroImage 40, 86–109 (2008)
2. Li, Y.O., Adalı, T., Wang, W., Calhoun, V.D.: Joint blind source separation by multiset canonical correlation analysis. IEEE Trans. Signal Process. 57, 3918–3929 (2009)
3. Kettenring, J.R.: Canonical analysis of several sets of variables. Biometrika 58, 433–451 (1971)
4. Kim, T., Eltoft, T., Lee, T.W.: Independent vector analysis: An extension of ICA to multivariate components, 165–172 (2006)
5. Li, X.L., Zhang, X.D.: Nonorthogonal joint diagonalization free of degenerate solution. IEEE Trans. Signal Process. 55, 1803–1814 (2007)
6. Li, X.L., Adalı, T.: Independent component analysis by entropy bound minimization. IEEE Transaction on Signal Processing (to be published)
7. Koldovský, Z., Tichavský, P., Oja, E.: Efficient variant of algorithm FastICA for independent component analysis attaining the Cramér-Rao lower bound. IEEE Trans. Neural Netw. 17 (September 2006)

Deterministic Blind Separation of Sources Having Different Symbol Rates Using Tensor-Based Parallel Deflation

André L.F. de Almeida[1], Pierre Comon[2], and Xavier Luciani[3]

[1] Wireless Telecom Research Group, Department of Teleinformatics Engineering, Federal University of Ceará, CP 6005, Campus do Pici, 60455-760, Fortaleza, Brazil
[2] Laboratoire d'Informatique, Signaux et Systèmes de Sophia-Antipolis, UMR6070, UNSA CNRS, 2000, route des Lucioles, Les Algorithmes, bât. Euclide B, BP 121, 06903 Sophia Antipolis Cedex, France
[3] Laboratoire Traitement du Signal et de l'Images, Université de Rennes 1, U642 INSERM, Campus de Beaulieu, bât. 22, 35042 Cedex, Rennes, France
andre@gtel.ufc.br, pcomon@i3s.unice.fr, lucianix@gmail.com

Abstract. In this work, we address the problem of blind separation of non-synchronous statistically independent sources from underdetermined mixtures. A deterministic tensor-based receiver exploiting symbol rate diversity by means of parallel deflation is proposed. By resorting to bank of samplers at each sensor output, a set of third-order tensors is built, each one associated with a different source symbol period. By applying multiple Canonical Decompositions (CanD) on these tensors, we can obtain parallel estimates of the related sources along with an estimate of the mixture matrix. Numerical results illustrate the bit-error-rate performance of the proposed approach for some system configurations.

Keywords: Blind source separation, underdetermined mixtures, non-synchronous sources, tensor decomposition, parallel deflation.

1 Introduction

Blind Source Separation (BSS) and Blind Identification (BI) of underdetermined mixtures have now become classical problems in signal processing and telecommunications. A large part of the related algorithms resorts to Higher Order statistics, notably by exploiting the multilinear properties of cumulant tensors. For instance, FOOBI-1 and FOOBI-2 [1] algorithms are based on fourth order statistics whereas the 6-BIOME algorithm (also referred to as BIRTH) [2] relies on sixth order cumulants.

In many applications, the sources are known to be cyclostationnary. Indeed, this property appears as soon as digital communication signals are oversampled. In particular, the behavior of second- and fourth-order BSS algorithms in a cyclostationary context has been addressed in [3]. A partially unbiased estimator of the sixth-order cumulant tensor has been later proposed in [4] by taking

V. Vigneron et al. (Eds.): LVA/ICA 2010, LNCS 6365, pp. 362–369, 2010.

into account the knowledge of the source cyclic frequencies. More generally, algorithms exploiting the cyclostationarity property resort to statistical tools.

In this study, we address the problem of underdetermined BSS and BI in the cyclostationary context by using a deterministic tensor-based receiver. Several methods have been proposed in the literature in order to solve BI and BSS problem using different multilinear models [5,7,8]. All of these works assume synchronous sources. This assumption is often unrealistic in non-cooperative systems (e.g. interception). The proposed receiver assumes non-synchronous sources. It consists of using a bank of sampling chains at each receive sensor, each one being adapted to the symbol rate of a given source, which are assumed to be known or estimated. At the output of the sampling stage, a set of third-order tensors is built, each one associated with a different sampling rate. As a consequence, we can obtain as many tensors as pre-detected sources. By applying successive canonical decompositions (CanD) [9] of these tensors, we obtain successive estimates of the related sources along with a mixture estimate. This approach can be connected to the parallel deflation method proposed in [10,11].

This paper is organized as follows. In Section 2, we introduce the system model. In Section 3, a tensor-based formulation of the received data model is given and the proposed parallel deflation receiver is presented. Section 4 presents our simulation results and the paper is concluded in Section 5.

2 System Model

Let us consider a uniform linear array of I sensors receiving signals from P sources. Let $\{s_p(t)\}$ be the information-bearing signal of the p-th source, $a_{i,p}$ be the sensor response of the i-th sensor to the p-th source, and $h_p(t)$ the time-domain signature of the p-th source, which comprises channel fading, transmit and receive filter responses. The baseband-equivalent signal received at the i-th sensor can be written, in absence of noise, as:

$$y_i(t) = \sum_{p=1}^{P} a_{i,p} x_p(t), \tag{1}$$

where $x_p(t) = \sum_{n=0}^{N-1} h_p(t - nT_p) s_p(n)$. Suppose that the P sources have different symbol rates $1/T_p$, where T_p is the symbol period of the p-th source. By sampling the received signal $\{y_p(t)\}$ at a reference sampling rate $1/T_s$, we obtain:

$$y_i(mT_s) = \sum_{p=1}^{P} a_{i,p} \left(\sum_{n=0}^{N-1} h_p(mT_s - nT_p) s_p(nT_s) \right), \tag{2}$$

where $m = 0, \ldots, M-1$, and M corresponds to the number of collected samples of the received signal. We assume that $h_p(t)$ is zero outside an interval $[0, K_p)$, $K_p < T_p$, $p = 1, \ldots, P$, which means that the channel of each of the sources has no temporal dispersion, i.e. we have an instantaneous mixture of sources, or a flat-fading channel in communications terminology. Let us assume that the source symbol periods are organized in increasing order, i.e. $T_s = T_1 < T_2 < \ldots < T_P$, where the symbol period of the first source is assumed to be equal to the reference sampling period. The received signal samples are collected during a fixed observation window of MT_s seconds, regardless of the used sampling period. Therefore, when sampling at a rate $1/T_j$ associated with the j-th source, the number M_j of collected received signal samples is given by:

$$M_j = \left\lceil M\left(\frac{T_s}{T_j}\right)\right\rceil = \left\lceil \frac{M}{\alpha_j}\right\rceil, \quad \alpha_j \in \mathbb{R}^+. \tag{3}$$

Now, suppose that the sampling rate $1/T_s$ at the receiver is L times higher than the p-th source symbol rate, i.e. $T_s = T_p/L$. This means that the received data is oversampled by a factor of L w.r.t. the p-th source. Then, (2) is equivalent to the following model:

$$y_i\big((m_j + l/L)T_j\big) = \sum_{p=1}^{P} a_{i,p} h_p(lT_j/L) s_p(m_j T_j), \tag{4}$$

$m_j = 0, \ldots, M_j - 1$, $l = 0, \ldots, L-1$, $j = 1, \ldots, P$. Suppressing the term T_j, (4) can be simplified to:

$$y_i^{(j)}(m_j + l/L) = \sum_{p=1}^{P} a_{i,p} h_p^{(j)}(l/L) s_p^{(j)}(m_j), \tag{5}$$

where $y_i^{(j)}(m_j+l/L) \doteq y_i\big((m_j+l/L)T_j\big)$, $h_p^{(j)}(l/L) \doteq h_p(lT_j/L)$, and $s_p^{(j)}(m_j) \doteq s_p(mT_j)$. Note that $h_p^{(j)}(l/L)$ is the l-th polyphase component of the p-th source channel impulse response obtained by resampling the original channel impulse response at a factor of L times the j-th source symbol rate. Likewise, $s_p^{(j)}(m_j)$ is obtained by resampling the p-th source symbol sequence at the j-th source symbol rate.

3 Parallel Deflation Receiver

In this section, we show that the received signal model given by (5) can be formulated as a third-order CanD model. Let $a_{i,p}$, $\bar{h}_{l,p}^{(j)} \doteq h_p^{(j)}((l-1)/L)$ and $\bar{s}_{m,p}^{(j)} \doteq s_p^{(j)}(m_j - 1)$ be, respectively, the entries of the matrices $\mathbf{A} \in \mathbb{C}^{I \times P}$, $\bar{\mathbf{H}}^{(j)} \in \mathbb{C}^{L \times P}$, and $\bar{\mathbf{S}}^{(j)} \in \mathbb{C}^{M_j \times P}$ collecting sensor responses, channel coefficients and transmitted symbols. Define $\bar{y}_{i,m_j,l}^{(j)} \doteq y_i\big((m_j+(l-1)/L)T_j\big)$ as the (i, m_j, l)-th entry of a third-order tensor $\bar{\mathcal{Y}}^{(j)} \in \mathbb{C}^{I \times M_j \times L}$ collecting the overall received signal sampled at a rate of T_j/L, $i \in [1, I], m_j \in [1, M_j], l \in [1, L]$. With these definitions, we can rewrite (4) as a CanD model:

$$\bar{y}_{i,m_j,l}^{(j)} = \sum_{p=1}^{P} a_{i,p} \bar{s}_{m_j,p}^{(j)} \bar{h}_{l,p}^{(j)}, \quad j = 1, \ldots, P, \tag{6}$$

or, alternatively, in unfolded matrix forms:

$$\bar{\mathbf{Y}}_1^{(j)} = (\bar{\mathbf{H}}^{(j)} \odot \bar{\mathbf{S}}^{(j)})\mathbf{A}^T \in \mathbb{C}^{LM_j \times I} \tag{7}$$

$$\bar{\mathbf{Y}}_2^{(j)} = (\mathbf{A} \odot \bar{\mathbf{H}}^{(j)})\bar{\mathbf{S}}^{(j)T} \in \mathbb{C}^{IL \times M_j} \tag{8}$$

$$\bar{\mathbf{Y}}_3^{(j)} = (\bar{\mathbf{S}}^{(j)} \odot \mathbf{A})\bar{\mathbf{H}}^{(j)T} \in \mathbb{C}^{M_j I \times L}, \tag{9}$$

where $\odot$ denotes the Khatri-Rao (columnwise Kronecker) product.

Therefore, by performing P sampling operations over the received signal, each one w.r.t. a different source symbol rate, we can construct P different CanD tensors $\bar{\mathcal{Y}}^{(1)}, \ldots, \bar{\mathcal{Y}}^{(P)}$, each one of which will be associated with the detection of a given source. Note that only the j-th column of $\bar{\mathbf{S}}^{(j)}$ contains useful information (in this case it contains the j-th source symbol sequence). Due to the symbol rate mismatch between sources, the remaining $P-1$ columns of $\bar{\mathbf{S}}^{(j)}$ only correspond to structured noise generated by the corresponding sources.

The proposed method to blindly extract the source signals (as well as to estimate the sensor and channel responses), is to fit each of the received signal tensors $\bar{\mathcal{Y}}^{(1)}, \ldots, \bar{\mathcal{Y}}^{(P)}$ to its associated third-order CanD tensor model in the least squares (LS) sense using the unfolded factorizations (7)-(9). In this case, the tensor-based detection has P processing stages. At each stage, we can use, for instance, the alternating least squares (ALS) algorithm, which consists of alternated estimations of the sensor, symbol and channel matrices according to the following criterion [5]:

$$\min_{\{\widehat{\mathbf{A}}, \widehat{\bar{\mathbf{S}}}^{(j)}, \widehat{\bar{\mathbf{H}}}^{(j)}\}} \|\bar{\mathcal{Y}}^{(j)} - \sum_{p=1}^{P} \mathbf{a}_p \circ \bar{\mathbf{s}}_p^{(j)} \circ \bar{\mathbf{h}}_p^{(j)}\|^2, \quad j = 1, \ldots, P. \tag{10}$$

Note that we need to process all the P tensors $\bar{\mathcal{Y}}^{(1)}, \ldots, \bar{\mathcal{Y}}^{(P)}$ in order to extract the symbol sequence of all the sources. Of course, this is due to the fact that only the j-th source symbol sequence can be extracted from $\bar{\mathcal{Y}}^{(j)}$, the remaining symbol sequences appear only as structure interference due to source symbol rate mismatch.

Uniqueness Issue

The CanD model (6) representing the received signal tensor $\bar{\mathcal{Y}}^{(j)} \in \mathbb{C}^{I \times M_j \times L}$ enjoys the essential uniqueness property, which means that $\{\mathbf{A}, \bar{\mathbf{S}}^{(j)}, \bar{\mathbf{H}}^{(j)}\}$ and $\{\widehat{\mathbf{A}}, \widehat{\bar{\mathbf{S}}}^{(j)}, \widehat{\bar{\mathbf{H}}}^{(j)}\}$ giving rise to the same tensor $\bar{\mathcal{Y}}^{(j)}$ are linked by the following relations $\widehat{\mathbf{A}} = \mathbf{A}\boldsymbol{\Pi}\boldsymbol{\Delta}_A$, $\widehat{\bar{\mathbf{S}}}^{(j)} = \mathbf{C}\boldsymbol{\Pi}\boldsymbol{\Delta}_S$, $\widehat{\bar{\mathbf{H}}}^{(j)} = \bar{\mathbf{H}}^{(j)}\boldsymbol{\Pi}\boldsymbol{\Delta}_H$, with $\boldsymbol{\Delta}_A\boldsymbol{\Delta}_S\boldsymbol{\Delta}_H = \mathbf{I}_P$, where $\boldsymbol{\Pi}$ is a permutation matrix whereas $\boldsymbol{\Delta}_A$,$\boldsymbol{\Delta}_S$, and $\boldsymbol{\Delta}_H$ are diagonal matrices. A sufficient condition for such an uniqueness was first established in [6] and generalized in [5] to the complex case. This condition states that the CanD model (6) is essentially unique if $k_{\mathbf{A}} + k_{\bar{\mathbf{S}}^{(j)}} + k_{\bar{\mathbf{H}}^{(j)}} \geq 2(P+1)$, where $k_{\mathbf{A}}$ denotes the Kruskal-rank, also called k-rank, of $\mathbf{A}$, i.e. the greatest integer $k_{\mathbf{A}}$ such that any set of $k_{\mathbf{A}}$ columns of $\mathbf{A}$ is linearly independent. The rank and the Kruskal-rank of $\mathbf{A}$ are related by the inequality $k_{\mathbf{A}} \leq rank(\mathbf{A})$. Assuming that

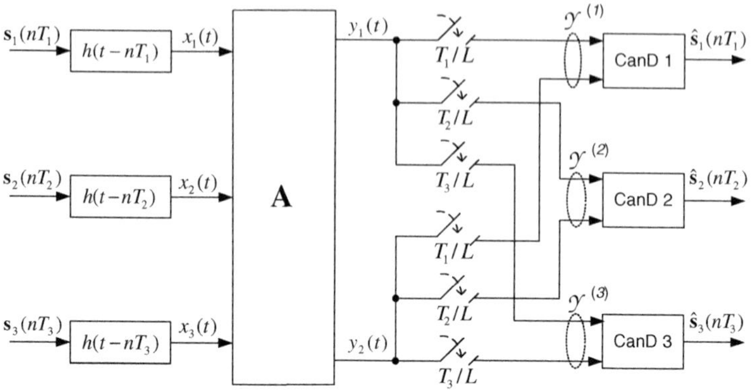

Fig. 1. Block-diagram of the proposed tensor-based parallel deflation receiver

$\mathbf{A}$, $\bar{\mathbf{S}}^{(j)}$ and $\bar{\mathbf{H}}^{(j)}$ are full rank matrices, the Kruskal-rank and the rank coincide, so that

$$min(I, P) + min(M_j, P) + min(L, P) \geq 2(P + 1) \tag{11}$$

ensures the essential uniqueness of the CanD model (6) for the received signal tensor sampled at a rate of T_j/L.

In this work, we are interested in the case of underdetermined mixtures (i.e. more sources than sensors). According to condition (11), if $M_j \geq P$ and $L \geq P$, then $I = 2$ sensors are enough to blindly separate P sources.

Tensor-based algorithm for parallel deflation

* *Here, we describe the j-th step of the algorithm which corresponds to the extraction of the j-th source. This process has to be repeated for $j = 1, \ldots, P$:*

1. Resample the output signal of each sensor at the frequency L/T_j;
2. Build the three-way tensor$\bar{\mathcal{Y}}^{(j)} \in \mathbb{C}^{L \times M_j \times I}$ suitable for the j-th source;
3. Estimate loadings matrices $\widehat{\mathbf{A}}$, $\widehat{\bar{\mathbf{S}}}^{(j)}$, $\widehat{\bar{\mathbf{H}}}^{(j)}$ describing the CanD of tensor $\bar{\mathcal{Y}}^{(j)}$ by minimizing criterion (10) in the LS sense (e.g. using the ALS algorithm [5]);
4. Evaluate and compare the "discrete" structure of each column of $\widehat{\bar{\mathbf{S}}}^{(j)} = [\widehat{\bar{\mathbf{s}}}_1^{(j)}, \ldots, \widehat{\bar{\mathbf{s}}}_P^{(j)}]$ in order to choose the suitable vector corresponding to the j-th source;

Remark 1: In addition to blind source separation, this algorithm also provides an estimation of the mixing matrix. A natural approach consists of estimating one column of matrix $\mathbf{A}$ at each deflation layer. Since only a single source is extracted at each detection layer, we eliminate the inherent column permutation ambiguity that would exist in a joint CanD estimation of equal symbol rate sources. The mixture identification problem will be addressed in a future work.

4 Performance Evaluation

In this section, computer simulation results are provided for the performance evaluation of the proposed tensor-based deflation receiver in a wireless communication system. The receiver is equipped with a uniform linear array of half-wavelength spaced sensors. The propagation channel associated with each source is characterized by a complex envelope, an angle of arrival and a delay of arrival (assumed to be negligible with respect to the symbol source period). All these parameters are assumed to be constant during a data block of duration MT_s seconds. At each sensor, the received signal is sampled at the Nyquist rate at the input of each deflation layer, i.e. $T_e = T_j/2$, $j = 1, \ldots, P$. The transmit/receive filters are raised cosines with roll-off factor 0.3. We assume binary phase shift keying (BPSK) modulation for all the sources. The sources carrier residues are assumed to be negligible. Additive noise samples are modeled as complex Gaussian random variables with equal variance for all the sensors.

We compute the bit error rate (BER) for several signal-to-noise ratios (SNR) and for different system parameters (number of sources, number of sensors, sampling factor and data block size). The results are validated from 1000 Monte-Carlo runs. Each run is associated with a different realization of the source symbol vectors, mixing matrix, noise tensor, angles of arrival, complex envelopes and the ratio of source symbol periods. The angles of arrival of the sources are randomly drawn between 0 and 80° according to a uniform distribution. The symbol period T_1 of source 1 is taken as the reference sampling period, while those of the remaining sources $(T_2, \ldots, T_P)$ are randomly varied at each run are given by $T_p = T_p^0 \pm \Delta_p$, $p = 2, \ldots, P$, where T_p^0 is the median value of the p-th source symbol period, and Δ_p is a random variable uniformly distributed between $[0, 1/2]$, which is assumed to be the same for all the sources. Our simulations suppose that $T_1, \ldots, T_P$ are known or have been estimated in a previous processing stage. The estimation of the symbol period is outside the scope of this paper. However, the methods proposed in [12,13] can be applied in this context. We consider two underdetermined mixture setups: i) 3 sources and 2 sensors and ii) 4 sources et 3 sensors. We define $\bar{\mathcal{Y}}^{(j)} \in \mathbb{C}^{L \times M_j \times I}$ as the noise free data tensor built from the j-th sampler output L/T_j. The noise samples are stored in a tensor $\mathcal{N}^{(j)} \in \mathbb{C}^{L \times M_j \times I}$, so that the received data tensor $\tilde{\bar{\mathcal{Y}}}^{(j)}$ is given by $\tilde{\bar{\mathcal{Y}}}^{(j)} = \bar{\mathcal{Y}}^{(j)} + \mathcal{N}^{(j)}$. The SNR is defined as $\text{SNR} = 10\log_{10} \frac{\|\bar{\mathcal{Y}}^{(j)}\|_F^2}{\|\mathcal{N}^{(j)}\|_F^2}$ (dB), where the variance of $\mathcal{N}^{(j)}$ is computed in order to obtained the desired SNR level.

We first show the BER performance of the proposed receiver in the underdetermined case of 3 sources and 2 sensors ($P = 3, I = 2$). The data block size is equal to 100 samples for source 1 (deflation layer 1). Since the length of the considered data block is fixed to $M = M_1 = 100$ for source 1, the number of samples $M_2, \ldots, M_P$ processed by layers 2 to P is lower than 100 and depends on the source symbol period ratios each run. The average BER values are computed over the 1000 runs and the P sources. These values are reported in Table 1 for different SNR's and sampling factors. Herein, the source symbol periods are given by $T_1 = T_e$, $T_2 = (1.5 \pm \Delta_2)T_e$, $T_3 = (2.5 \pm \Delta_3)T_e$, where Δ_2 et Δ_3 follow

Table 1. Effect of sampling factor and SNR on the average BER ($P = 3$, $I = 2$)

SNR (dB)	Oversampling factor			
	$L = 3$	$L = 4$	$L = 6$	$L = 8$
0	0.2501	0.1868	0.1491	0.0688
5	0.1633	0.0852	0.0855	0.0398
10	0.1125	0.0553	0.0541	0.0334
15	0.0957	0.0418	0.0409	0.0322
20	0.0809	0.0390	0.0410	0.0294

Table 2. Average BER or the best estimated source (BER_{max}) and for the worst estimated sources (BER_{min}) computed over 1000 Monte-Carlo runs ($N = 50$, $L = 4$)

SNR (dB)	System configuration			
	$P = 3, I = 2$		$P = 4, I = 3$	
	BER_{max}	BER_{min}	BER_{max}	BER_{min}
0	0.2222	0.1253	0.2637	0.1495
5	0.1179	0.0427	0.1670	0.0620
10	0.0837	0.0161	0.1383	0.0345
15	0.0621	0.0121	0.1328	0.0332
20	0.0587	0.0092	0.1421	0.0384

Table 3. Comparison with the MAP estimator ($P = 3$, $I = 2$, SNR=25dB)

	Configuration		
	$P = 6, N = 50$	$P = 6, N = 100$	$P = 8, N = 100$
CanD deflation	0.023	0.025	0.013
MAP estimator	0.014	0.020	0.005

an uniform distribution in the range [0, 0.5]. One can first note a significant degradation of the BER when the sampling factor is equal to 3. Better results can be obtained by increasing the sample factor. Only a slight improvement is observed for $L = 4$ and $L = 6$, although satisfying results are obtained for $L = 8$. Table 2 compares the average BER values of the best estimated source with the worst estimated one. The data block size and the sampling factor are fixed to $N = 100$ and $L = 4$, respectively. We now consider two cases: $(P = 3, I = 2)$ and $(P = 4, I = 3)$. In the first case, the source symbol periods are $T_1 = T_e$, $T_2 = (1.5 \pm \Delta_2)T_e$, $T_3 = (2.5 \pm \Delta_3)T_e$. In the second case, we have $T_1 = T_e$, $T_2 = (1.4 \pm \Delta_2)T_e$, $T_3 = (1.8 \pm \Delta_3)T_e$, $T_4 = (2.2 \pm \Delta_4)T_e$. We can note significant performance deviations between different deflation layers. One should note that the symbol period ratios between different sources are changed at each Monte Carlo run. Thus, when any two sources have very close symbol periods, their parallel extraction becomes more difficult and performance will be affected. With an eye to a reference comparison, we have also simulated the maximum-a-posteriori (MAP) estimator based on the estimated channel, for $L = 3$, $I = 2$ and SNR=20dB. According to Table 3 the performance gap between both receivers is not significant for $(P = 6, N = 50)$ and $(P = 6, N = 100)$.

5 Conclusions and Perspectives

We have proposed a new deterministic tensor-based receiver for blind source separation and channel identification in the case of underdetermined mixtures. The proposed receiver efficiently exploits symbol rate diversity and uses a CanD-based parallel deflation approach to individually extract each source. According to our results, satisfactory performances can be obtained, especially when the source symbol rates are not too close. In addition to source separation, our approach allows to blindly estimate the full mixture matrix either jointly or separately for each source. Perspectives include comparisons with existing statistical BSS/BI approaches and a generalization to convolutive mixtures.

References

1. De Lathauwer, L., Castaing, J., Cardoso, J.-F.: Fourth-order cumulant-based blind identification of underdetermined mixtures. IEEE Trans. Signal Processing 55(2), 2965–2973 (2007)
2. Albera, L., Ferreol, A., Comon, P., Chevalier, P.: Blind identification of Overcomplete MixturEs of sources (BIOME). Linear Algebra Applications, Special Issue on Linear Algebra in Signal and Image Processing 391C, 3–30 (2004)
3. Ferreol, A., Chevalier, P.: On the behavior of current second and higher order blind source separation methods for cyclostationary sources. IEEE Trans. Sig. Proc. 48, 1712–1725 (2002)
4. De Almeida, A.L.F., Luciani, X., Comon, P.: Blind identification of underdetermined mixtures based on the hexacovariance and higher-order cyclostationarity. In: SSP 2009, Cardiff, Wales, UK, August 31 - September 3 (2009)
5. Sidiropoulos, N.D., Giannakis, G.B., Bro, R.: Blind PARAFAC receivers for DS-CDMA systems. IEEE Trans. Signal Process. 48(3), 810–822 (2000)
6. Kruskal, J.B.: Three-way arrays: rank and uniqueness of trilinear decompositions. with application to arithmetic complexity and statistics. Lin. Algebra Appl. 18, 95–138 (1977)
7. De Almeida, A.L.F., Favier, G., Mota, J.C.M.: PARAFAC-based unified tensor modeling for wireless communication systems with application to blind multiuser equalization. Signal Processing 87(2), 337–351 (2007)
8. Nion, D., De Lathauwer, L.: Block component model-based blind DS-CDMA receiver. IEEE Trans. Signal Process. 56(11), 5567–5579 (2008)
9. Comon, P.: Tensor Decompositions, State of the Art and Applications. In: McWhirter, J.G., Proudler, I.K. (eds.) Mathematics in Signal Processing V, pp. 1–24. Clarendon Press, Oxford (2002)
10. Zarzoso, V., Rota, L., Comon, P.: Deflation parallele avec des contrastes APF pour l'extraction aveugle de sources. In: Gretsi 2005, Louvain La Neuve, Belgique, September 5-10 (2005)
11. Rota, L., Zarzoso, V., Comon, P.: Parallel deflation with alphabet-based criteria for blind source extraction. In: IEEE SSP 2005, Bordeaux, France, July 2005, pp. 17–20 (2005)
12. Houcke, S., Chevreuil, A., Loubaton, P.: Blind equalization: case of an unknown symbol period. IEEE Trans. on Signal Processing 51(3), 781–793 (2003)
13. Jallon, P., Chevreuil, A., Loubaton, P., Chevalier, P.: Separation of convolutive mixtures of cyclostationary sources: a contrast function based approach. In: Puntonet, C.G., Prieto, A.G. (eds.) ICA 2004. LNCS, vol. 3195, pp. 508–515. Springer, Heidelberg (2004)

Second Order Subspace Analysis and Simple Decompositions

Harold W. Gutch[1,2], Takanori Maehara[3], and Fabian J. Theis[2,4]

[1] Max Planck Institute for Dynamics and Self-Organization, Department of Nonlinear Dynamics, Germany
[2] Technical University of Munich, Germany
[3] University of Tokyo, Japan
[4] Helmholtz-Institute Neuherberg, Germany
harold.gutch@ds.mpg.de, maehara@misojiro.t.u-tokyo.ac.jp, fabian.theis@helmholtz-muenchen.de

Abstract. The recovery of the mixture of an N-dimensional signal generated by N independent processes is a well studied problem (see e.g. [1,10]) and robust algorithms that solve this problem by Joint Diagonalization exist. While there is a lot of empirical evidence suggesting that these algorithms are also capable of solving the case where the source signals have block structure (apart from a final permutation recovery step), this claim could not be shown yet - even more, it previously was not known if this model separable at all. We present a precise definition of the subspace model, introducing the notion of simple components, show that the decomposition into simple components is unique and present an algorithm handling the decomposition task.

The general task of Blind Source Separation (BSS) can be formulated as follows: Given a number of source signals $(\mathbf{S}_1, \ldots, \mathbf{S}_N) = \mathbf{S}$ unknown to the observer and some kind of mixture $f(\mathbf{S}_1, \ldots, \mathbf{S}_N)$, recover the sources given only the mixture $\mathbf{X} = f(\mathbf{S})$. In general, f has to be assumed to be invertible (otherwise full recovery would not be possible), but explicit knowledge is not assumed. A common assumption on f is linearity, i.e. it can be expressed as a matrix $\mathbf{A}$.

If the sources can be assumed to have some time-like structure, e.g. audio signals, financial data or fMRI recordings from EEG signals, an approach incorporating the time structure may be taken. Here the time structure is given mathematical meaning by modeling the sources as stochastic processes, and they are usually assumed to be wide sense stationary (w.s.s.). In this case, a computationally simple approach may be taken by looking at the cross-autocorrelation matrices of the sources – that is, taking only into account first and second order moments. In case all sources are one-dimensional, this means that the N stochastic source processes all are independent and then the auto-crosscorrelation matrices of the sources are block-diagonal. It is known that then any two sources can be separated as long their correlation coefficients differ under at least one time shift [1], and various algorithms solving this task numerically have been suggested, e.g. [1,5,8,10], however general results for the higher dimensional case so far were unknown.

V. Vigneron et al. (Eds.): LVA/ICA 2010, LNCS 6365, pp. 370–377, 2010.

In this contribution, we present a framework for the general Second Order BSS model, arguing that unlike in Independent Subspace Analysis (ISA), where the key assumption on the sources was irreducibility [4], rather what we call *simplicity* of the sources should be preferred, where irreducible sources with a common autocorrelation structure are combined to form *simple subspaces.* We show that any sources are uniquely linearly decomposable into simple subspaces and, reducing the problem to the problem of Joint Block Diagonalization, we present two algorithms performing this task for the noiseless case as well as for the noisy case, which have been successfully used for ICA [6]. We compare the performance of our algorithm to a standard heuristic algorithm for Joint Block Diagonalization [9] on toy data. Our analysis is restricted to real-valued signals, although extensions to the complex case are possible.

1 Second Order Subspace Analysis

1.1 Stochastic Processes and Second Order Structure

Assume an n-dimensional real-valued centered, wide sense stationary (w.s.s.) stochastic process $\mathbf{S}$ which we may write as $\mathbf{S}(t) = (\mathbf{S}_1(t), \ldots, \mathbf{S}_N(t))$ (where $\sum_{i=1}^{N} \dim(\mathbf{S}_i) = n$) such that for any pair $i \neq j$, the auto-crosscorrelation vanishes, i.e. the matrix

$$\mathbf{R}_{\mathbf{S}}(\tau) := E\{\mathbf{S}(t)\mathbf{S}(t-\tau)^\top\}$$

is block-diagonal for all $\tau \in \mathbb{R}$, where the k-th block has the same size as $\mathbf{S}_k$. Furthermore assume some invertible but otherwise unknown mixing matrix $\mathbf{A}$ and let $\mathbf{X}(t) := \mathbf{A}\mathbf{S}(t)$. We then have

$$\mathbf{R}_{\mathbf{X}}(\tau) = E\{\mathbf{A}\mathbf{S}(t)(\mathbf{A}\mathbf{S}(t-\tau))^\top\} = \mathbf{A}\mathbf{R}_{\mathbf{S}}(\tau)\mathbf{A}^\top \ .$$

The task now is to reconstruct $\mathbf{S}(t)$ given only $\mathbf{X}(t)$. Obviously this can be performed at most up to invertible mixings within the source blocks $\mathbf{S}_i(t)$: $\mathbf{R}_{\mathbf{LS}}(\tau)$ again is block-diagonal with the same block sizes for any invertible block-diagonal matrix $\mathbf{L}$. Additionally to independence obviously some kind of minimality condition is required, as we could otherwise group together any number of sources and get a new (coarser) decomposition fulfilling that any two sources are independent. The intuitive minimality condition that immediately comes to mind is the condition of irreducibility, i.e. we demand that for every source $\mathbf{S}_i(t)$ there is no invertible matrix $\mathbf{T}$ such that $\mathbf{T}\mathbf{S}_i(t) = (\mathbf{S}_i^1(t), \mathbf{S}_i^2(t))$ with auto-decorrelated $\mathbf{S}_i^1(t)$, $\mathbf{S}_i^2(t)$. However we argue that this notation is misleading. To visualize this, imagine the simple case of three 1-dimensional components with identical second order structure in the first and second component, i.e. $E\{S_1(t)S_1(t-\tau)^\top\} = E\{S_2(t)S_2(t-\tau)^\top\}$ for all delays $\tau \in \mathbb{R}$. It is easy to see that then any orthogonal mixing within these two components will still preserve auto-decorrelation of them, so irreducibility alone clearly can lead to misleading results. Previously it was simply assumed that the sources do not exhibit this behavior [1,10], and it indeed can be argued that in general among all higher dimensional stochastic processes those where two components have identical second order structure occur almost never, but for a rigorous analysis of second order processes this case has to be taken into account.

1.2 Whitening

Applying the common pre-whitening step to every source and to the observations, we may assume that both are decorrelated (i.e. $\mathbf{R_S}(0) = \mathbf{R_X}(0) = \mathbf{I}$), and hence that the mixing matrix $\mathbf{A}$ is not only invertible, but even orthogonal.

1.3 Irreducible and Simple Components

Having reduced the problem to the case of merely orthogonal mixings, we can give a concise definition of the mathematical terms we will use.

Definition 1. *Two w.s.s. processes* $\mathbf{S}_1$ *and* $\mathbf{S}_2$ *are called* second oder temporally independent *if*

$$E\{\mathbf{S}_1(t)\mathbf{S}_2(t-\tau)^\top\} = 0$$

for all $\tau \in \mathbb{R}$.

As $\mathbf{S}_1$ and $\mathbf{S}_2$ are w.s.s., this is equivalent to block-diagonality of

$$\mathbf{R_S}(\tau) = \begin{pmatrix} \mathbf{R}_{\mathbf{S}_1}(\tau) & 0 \\ 0 & \mathbf{R}_{\mathbf{S}_2}(\tau) \end{pmatrix}$$

for $\mathbf{S} := (\mathbf{S}_1, \mathbf{S}_2)$, and we will call two such processes from now on simply *independent.*

Definition 2. *An* n*-dimensional process* $\mathbf{S}$ *is called* reducible *if there is some* $\mathbf{A} \in O(n)$ *such that* $\mathbf{AS} = (\mathbf{S}_1, \mathbf{S}_2)$ *with independent* $\mathbf{S}_1$, $\mathbf{S}_2$. *A w.s.s. process that is not reducible is called* irreducible.

Obviously any finite-dimensional process $\mathbf{S}$ can be decomposed into irreducible components: Either it already is irreducible or we can find an $\mathbf{A}$ such that $\mathbf{AS} = (\mathbf{S}_1, \mathbf{S}_2)$ with independent $\mathbf{S}_1$, $\mathbf{S}_2$. If these are irreducible, we are finished, otherwise we decompose whichever of the two is reducible, until, after a finite number $k < \dim(\mathbf{S})$ of steps, we are left with irreducible components.

Definition 3. *Two w.s.s. processes* $\mathbf{S}_1$ *and* $\mathbf{S}_2$ *are said to have* common temporal structure *if there is some orthogonal* $\mathbf{A}$ *such that*

$$\mathbf{R}_{\mathbf{S}_1}(\tau) = \mathbf{R}_{\mathbf{AS}_2}(\tau) .$$

The following theorem shows the indeterminacies in decompositions into irreducible components, the proof of which we have omitted due to lack of space:

Theorem 1. *Assume a w.s.s. process* $\mathbf{S} = (\mathbf{S}_1, \ldots, \mathbf{S}_N)$ *with independent, irreducible* $\mathbf{S}_i$ *and an orthogonal* $\mathbf{A}$ *such that* $(\mathbf{X}_1, \ldots, \mathbf{X}_M) = \mathbf{AS}$, *again with independent, irreducible* $\mathbf{X}_j$. *Let* $\mathbf{A}_{ij}$ *be the submatrices of* $\mathbf{A}$ *of sizes* $\dim(\mathbf{X}_i) \times \dim(\mathbf{S}_j)$. *Then:*

1. $N = M$.
2. *If* $\mathbf{A}_{ij} \neq 0$, *then* $\dim(\mathbf{S}_i) = \dim(\mathbf{X}_j)$ *and* $\alpha_{ij}\mathbf{A}_{ij}$ *is orthogonal for some* $\alpha_{ij} \in \mathbb{R} \setminus \{0\}$.

Corollary 1. *It is possible to separate any two independent, irreducible components with different temporal structure.*

Proof. Assume $\mathbf{S} = (\mathbf{S}_1, \ldots, \mathbf{S}_N)$ and $(\mathbf{X}_1, \ldots, \mathbf{X}_N) = \mathbf{AS}$, two different independent, irreducible representations of a w.s.s. process with some orthogonal $\mathbf{A}$. Then $\mathbf{R}_{\mathbf{X}}(\tau) = \mathbf{R}_{\mathbf{AS}}(\tau) = \mathbf{A}\mathbf{R}_{\mathbf{S}}(\tau)\mathbf{A}^\top$ and as $\mathbf{A}$ is orthogonal this is equivalent to $\mathbf{R}_{\mathbf{X}}(\tau)\mathbf{A} = \mathbf{A}\mathbf{R}_{\mathbf{S}}(\tau)$. Then also $\mathbf{R}_{\mathbf{X}_i}(\tau)\mathbf{A}_{ij} = \mathbf{A}_{ij}\mathbf{R}_{\mathbf{S}_j}(\tau)$ for every i, j.

Our claim now says that for any $l \neq m$ where $\mathbf{S}_l$ and $\mathbf{S}_m$ have different temporal structure, there is no k such that both $\mathbf{A}_{kl}$ and $\mathbf{A}_{km}$ are non-zero. Let us assume indices $l \neq m$ and k such that both $\mathbf{A}_{kl}$ and $\mathbf{A}_{km}$ are non-zero. According to Theorem (1), any non-zero $\mathbf{A}_{ij}$ implies that $\alpha_{ij}\mathbf{A}_{ij}$ is orthogonal for some real non-zero α_{ij}, in which case $\mathbf{A}_{ij}^{-1} = \alpha_{ij}^2\mathbf{A}_{ij}^\top$. Then

$$\mathbf{R}_{\mathbf{X}_k}(\tau) = \mathbf{A}_{kl}\mathbf{R}_{\mathbf{S}_l}(\tau)\mathbf{A}_{kl}^{-1} = \mathbf{A}_{kl}\mathbf{R}_{\mathbf{S}_l}\alpha_{kl}^2\mathbf{A}_{kl}^\top = \mathbf{R}_{(\alpha_{kl}\mathbf{A}_{kl})\mathbf{S}_l}(\tau)$$

so $\mathbf{X}_k$ and $\mathbf{S}_l$ have common temporal structure. Using the same arguments, we see that $\mathbf{X}_k$ and $\mathbf{S}_m$ also have common temporal structure, and then so do $\mathbf{S}_l$ and $\mathbf{S}_m$.

Conversely, if two or more irreducible, independent components of a process have common temporal structure, it is possible to mix them non-trivially and again get an irreducible, independent representation. For the problem of demixing a set of independent processes therefore the following dilemma arises: If some of the source processes (say $\mathbf{S}_1$ and $\mathbf{S}_2$) have common temporal structure, it is impossible to separate them in our framework. Indeed, if we have found an irreducible, independent demixing of the mixtures $\mathbf{X}$, two of these recovered components, say $\mathbf{S}_1'$ and $\mathbf{S}_2'$ may be non-trivial mixings of $\mathbf{S}_1$ and $\mathbf{S}_2$. As this might be the case for all tuples of sources with common temporal structure, we gather all such sources with common temporal structure together to form a *simple component.* Two different simple components of a process then of course still are independent, although they not necessarily have to be irreducible. As we can separate any two irreducible components with different temporal structure, we can also separate any two different simple components.

Theorem 2 (Uniqueness of simple decompositions). *Assume a w.s.s. process* $\mathbf{S} = (\mathbf{S}_1, \ldots, \mathbf{S}_N)$ *with independent, simple* $\mathbf{S}_i$ *and an orthogonal* $\mathbf{A}$ *such that* $(\mathbf{X}_1, \ldots, \mathbf{X}_M) = \mathbf{AS}$, *again with independent, simple* $\mathbf{X}_j$. *Let* $\mathbf{A}_{ij}$ *be the submatrices of* $\mathbf{A}$ *of sizes* $\dim(\mathbf{X}_i) \times \dim(\mathbf{S}_j)$. *Then:*

1. $N = M$.
2. *For every* i *there is exactly one* j *such that* $\mathbf{A}_{ij} \neq 0$. *Furthermore then* $\mathbf{A}_{ij}$ *is orthogonal.*

We argue that a decomposition of $\mathbf{S}$ into simple components makes more sense than a decomposition into irreducible components: Unlike a decomposition into irreducible components, the decomposition into independent, simple components is unique (up to choice of basis within the components). The task of second order subspace analysis therefore is as follows:

Given an n-dimensional w.s.s. process $\mathbf{X}$, find an orthogonal $\mathbf{A}$ such that $\mathbf{AX} = (\mathbf{S}_1, \ldots, \mathbf{S}_N)$ with independent, simple processes $\mathbf{S}_k$.

2 Second Order Subspace Analysis by Joint Block Diagonalization

We propose solving the task of second order subspace analysis by Joint Block Diagonalization (JBD). Noiseless JBD can be described as follows: Given a set of matrices $\mathcal{M} = \{\mathbf{M}_1, \ldots, \mathbf{M}_T\}$, find an orthogonal $\mathbf{A}$ such that $\mathbf{A}^\top \mathbf{M}_t \mathbf{A}$ adheres to a common block structure, i.e. for every $1 \leq t \leq T$ we have

$$\mathbf{A}^\top \mathbf{M}_t \mathbf{A} = \mathrm{bdiag}(\mathbf{S}_t^1, \ldots, \mathbf{S}_t^n)$$

where the size of $\mathbf{S}_t^j$ does not depend on t. Here bdiag(.) produces a block diagonal matrix with the arguments as blocks. The block structure is assumed to be minimal, i.e. for no j there is an orthogonal $\mathbf{B}$ with $\mathbf{B}^\top \mathbf{S}_t^j \mathbf{B} = \mathrm{bdiag}(\mathbf{S}_t^{j1}, \mathbf{S}_t^{j2})$.

With this, we solve the problem of decomposing our observation $\mathbf{X}$ into simple blocks by performing JBD of matrices $\mathbf{R_X}(\tau_t)$ for some time lags τ_t, the choice of which is non-trivial. Observe that when decorrelating (preprocessing), we have made use of all information in the lag-less cross-correlation. This and the fact that $\mathbf{R_X}(-\tau) = \mathbf{R_X}(\tau)^\top$ for all τ allows us to restrict the choice of τ_t to only positive time lags.

We call the i-th and j-th block of our block-diagonal matrices *equivalent* if there is some orthogonal $\mathbf{A}$ such that $\mathbf{A}^\top \mathbf{S}_t^i \mathbf{A} = \mathbf{S}_t^j$ for every t. Gathering all equivalent blocks to a (reducible) *simple* block, JBD has the same indeterminacies as second order subspace analysis. Again, $\mathbf{A}$ is unique up to block-wise permutations of $\mathbf{A}$, where the block sizes are given by the sizes of the irreducible components $\mathbf{S}_t^i$, and orthogonal transformations within simple blocks.

2.1 Joint Block Diagonalization by Joint Diagonalization

If there is some $\mathbf{A}$ such that all matrices $\mathbf{A}^\top \mathbf{M}_t \mathbf{A}$ are diagonal, this task is called Joint Diagonalization, and it can be solved with iterative Givens rotations [2], and here the angles of the rotations can be calculated in closed form [3].

It was conjectured that application of this algorithm to the symmetrized matrices $\{\mathbf{M}_1 + \mathbf{M}_1^\top, \ldots, \mathbf{M}_T + \mathbf{M}_T^\top\}$ gives a permutation of a Joint Block Diagonalizer of $\mathcal{M}$, giving rise to a Joint Block Diagonalizing algorithm where all that is left to do is recovery of the final permutation matrix [9]. In case the data set is noisy, the recovery of the permutation matrix needs some thresholding, the choice of which is non-trivial.

However, this set of steps does not work in general: If every $\mathbf{M} \in \mathcal{M}$ is anti-symmetric (i.e. $\mathbf{M}+\mathbf{M}^\top$ is a diagonal matrix), this approach will consist of Joint Diagonalization of already diagonal matrices, making the Joint Diagonalization algorithm return simple the identity (or, in the noisy case, a matrix close to the identity), but in general a finer JBD of $\mathcal{M}$ might be possible.

2.2 Joint Block Diagonalization via the Matrix *-Algebra

In [7], from a theoretical point of view, the problem of JBD is approached by looking at the *matrix *-algebra* generated by the matrices $\mathbf{M}_1, \ldots, \mathbf{M}_T$. A matrix *-algebra is a set $\mathcal{T} \subseteq \mathrm{Mat}(n \times n)$ (the $n \times n$ matrices with real entries) where $\mathbf{I}_n \in \mathcal{T}$ and for all $\mathbf{A}, \mathbf{B} \in \mathcal{T}$ also $\mathbf{A}^\top, \mathbf{AB}, a\mathbf{A}+b\mathbf{B} \in \mathcal{T}$ for all real a, b. The set of all $n \times n$ matrices forms a matrix *-algebra, and the set of all matrices according to a given block diagonal form forms a matrix *-subalgebra thereof. A set $\mathcal{M}$ *generates* a matrix *-algebra $\mathcal{T}$ if $\mathcal{T}$ is the smallest matrix *-algebra containing $\mathcal{M}$. A set S is said to be a *basis* of the matrix *-algebra $\mathcal{T}$ if every $\mathbf{A} \in \mathcal{T}$ can be written as a linear combination of elements in S. The key reduction given in [7] now is as follows:

Theorem 3. *Let $\mathcal{T}$ be a matrix *-algebra with basis $\{\mathbf{M}_1, \ldots, \mathbf{M}_T\}$. Then with probability $p = 1$ any arbitrary normalized linear combination*

$$\mathbf{M} := \sum_{t=1}^{T} c_t \mathbf{M}_t \text{ (where} \sum_t c_t^2 = 1)$$

has the property that a diagonalizer $\mathbf{P}$ of the symmetric matrix $\mathbf{M} + \mathbf{M}^\top$ gives a decomposition of $\mathcal{T}$ into irreducible block form, that is, that

$$\mathbf{P}^\top \mathbf{M}_t \mathbf{P} = \mathrm{bdiag}(\mathbf{S}_t^1, \ldots, \mathbf{S}_t^n)$$

where every $\mathbf{S}_t^j$ is irreducible.

In order to use this theorem for JBD, a basis S of $\mathcal{T}$, the matrix *-algebra generated by $\mathcal{M}$, has to be calculated. After this, we merely need to diagonalize the symmetrized version of an arbitrary linear combination of the elements in S, e.g. by eigenvalue decomposition.

In [7] an algorithm was presented that calculates a basis of the matrix *-algebra $\mathcal{T}$ generated by the set $\mathcal{M}$. Furthermore, this algorithm is capable of error-control through a parameter ε, making it capable of being applicable in the presence of noise on the data. Essentially the matrices forming a basis of $\mathcal{T}$ correspond to eigenvectors of a specific matrix, and linear combinations of the matrices correspond to linear combinations of the eigenvectors. In the noiseless case eigenvectors corresponding to non-zero eigenvalues correspond to elements of the basis of $\mathcal{M}$, while in the noisy case we select the eigenvectors corresponding to eigenvalues whose absolute value is larger than ε.

2.3 Simple Decomposition by Irreducible Decomposition

Both approaches, the empirical JBD by JD and the exact JBD via the matrix *-algebra, give an irreducible decomposition. In order to find a simple decomposition, we have to group together some of the irreducible blocks, namely, for a given i, we want all j for which there is some orthogonal $\mathbf{A}$ such that $\mathbf{A}^\top \mathbf{S}_t^i \mathbf{A} = \mathbf{S}_t^j$ for all t. However, as the JBD algorithm based on the matrix *-algebra is probabilistic (and the specific diagonalizer $\mathbf{P}$ depends on the specific linear combination chosen), an alternative approach is possible: We can perform JBD once, giving us a diagonalizer $\mathbf{P}_1$ and then perform JBD on the resulting set $\mathbf{P}_1^\top \mathbf{M}_t \mathbf{P}_1 = \mathrm{bdiag}(\mathbf{S}_t^1, \ldots, \mathbf{S}_t^n)$, giving us a second diagonalizer $\mathbf{P}_2$. If $\mathbf{P}_2$ is a block diagonal matrix with blocks of the sizes of the $\mathbf{S}_t^i$ (or a block permutation thereof), we conclude that the irreducible components coincide with the simple components. However if $\mathbf{P}_2$ consists of non-trivial operations within blocks, we may safely assume that all such blocks belong to a larger simple component, and we gather all blocks according to this. Here non-trivial operations are detected by more than two blocks containing entries with absolute values larger than some threshold δ. In theory we may perform this multiple times, although a single such run already suffices.

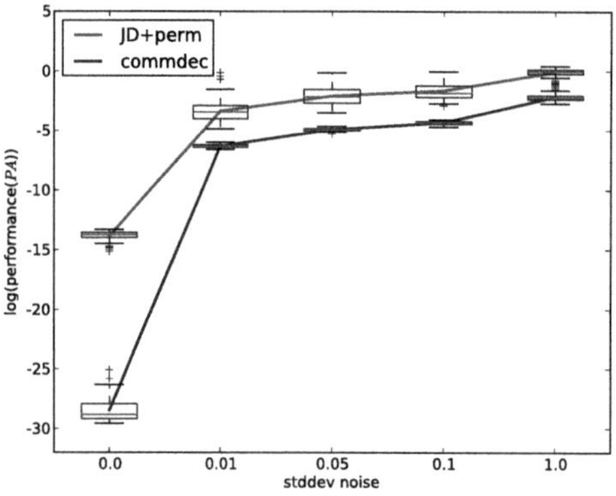

Fig. 1. Comparison of the two algorithms. Performance (y-axis) depicts the total Frobenius norm of the off-block-diagonal values of $\mathbf{WA}$, smaller values meaning better reconstruction. The suggested algorithm "commdec" outperforms the heuristic combination of JD with permutation recovery.

3 Simulations

In order to compare the validity of the proposed algorithm with the standard heuristic algorithm consisting of Joint Diagonalization and permutation recovery, we generated $N = 10$ block diagonal matrices with block sizes $3+3+3$, the entries of which were independently sampled from a Gaussian distribution (mean 0, stddev 1). We then mixed the matrices with a uniformly sampled orthogonal matrix and to each matrix added random noise sampled from a centered

Gaussian distribution with rising standard deviation $\delta = 10^{-2} \cdot \{0, 1, 5, 10, 100\}$. We compared the performance of the heuristic algorithm and the proposed joint diagonalization algorithm in dependence of the noise, depicting the averages over 100 runs, evaluating the squared sum of the off-block-diagonal values of the product of the suggested recovery matrix $\mathbf{W}$ and the mixing matrix $\mathbf{A}$.

4 Conclusion

We have presented a complete theoretical framework in which data sets with time structure can be analyzed and classified in terms of independence, using moments of at most second order. We have presented insight in the exact indeterminacies of the model, extending previously known results to both irreducible components of size larger than 1 and to the case of components with identical time structure, which so far was ignored. A provably exact and efficient probabilistic algorithm (in the noiseless case) and based on this, an efficient probabilistic algorithm for the noisy case were presented. Apart from the obvious application to real life data sets, future work could include convergence analysis of the algorithm in the noisy case and non-probabilistic algorithms performing this task.

References

1. Belouchrani, A., Abed-Meraim, K., Cardoso, J.-F., Moulines, E.: A blind source separation technique using second-order statistics. IEEE Transactions on Signal Processing 45(2), 434–444 (1997)
2. Bunse-Gerstner, A., Byers, R., Mehrmann, V.: Numerical methods for simultaneous diagonalization. SIAM J. Matrix Anal. Appl. 14(4), 927–949 (1993)
3. Cardoso, J.-F., Souloumiac, A.: Jacobi angles for simultaneous diagonalization. SIAM J. Matrix Anal. Appl. 17(1), 161–164 (1996)
4. Gutch, H.W., Theis, F.J.: Independent subspace analysis is unique, given irreducibility. In: Davies, M.E., James, C.J., Abdallah, S.A., Plumbley, M.D. (eds.) ICA 2007. LNCS, vol. 4666, pp. 49–56. Springer, Heidelberg (2007)
5. Liu, W., Mandic, D.P., Cichocki, A.: Blind source extraction based on a linear predictor. IET Signal Process. 1(1), 29–34 (2007)
6. Maehara, T., Murota, K.: Error-controlling algorithm for simultaneous block-diagonalization and its application to independent component analysis. JSIAM Letters (submitted)
7. Maehara, T., Murota, K.: Algorithm for error-controlled simultaneous block-diagonalization of matrices. Technical Report METR-2009-53 (December 2009)
8. Molgedey, L., Schuster, H.G.: Separation of a mixture of independent signals using time delayed correlations. Phys. Rev. Lett. 72(23), 3634–3637 (1994)
9. Theis, F.J.: Towards a general independent subspace analysis. In: Proc. NIPS, pp. 1361–1368 (January 2006)
10. Tong, L., Soon, V.C., Huang, Y.-F., Liu, R.: AMUSE: a new blind identification algorithm. In: IEEE International Symposium on Circuits and Systems, vol. 3, pp. 1784–1787 (1990)

Sensitivity of Joint Approximate Diagonalization in FD BSS

Savaskan Bulek and Nurgun Erdol

Department of Electrical Engineering,
Florida Atlantic University, Boca Raton, Florida, USA
sbulek@fau.edu

Abstract. This paper investigates the sensitivity of the joint approximate diagonalization of a set of time varying cross-spectral matrices for blind separation of convolutive mixtures of speech signals. We introduce the multitaper method of cross-spectrum estimation. Based on the work of [1] factors affecting the sensitivity of the joint approximate diagonalization problem were investigated. We studied the effect of the number of matrices in the set, and observed that there exists a link between the uniqueness of the joint diagonalizer measured by modulus of uniqueness parameter and the estimation of demixing system parameters.

1 Introduction

Joint approximate diagonalization (JAD) is a method frequently used to separate convolutively combined source signals in the frequency domain. Its practical success depends on the fulfillment of the assumed properties as measured on or estimated from finite length data. In [1] sensitivity of the JAD of a set of perturbed matrices is analyzed and it is shown that the factors affecting the bias of the demixing system parameters are (i) the modulus of uniqueness and (ii) the norm of the perturbation matrix. The influence of these factors is typically reduced by increasing the cardinal number D of the set of matrices to be jointly approximately diagonalized.

The perturbation model of [1] may be used to study the effects of inaccuracies in the source cross spectral matrix, which is assumed to be diagonal. Modeling errors, noise and spectral estimation errors may cause violation of the assumptions on the sources. The study of spectral estimation effects is particularly important in applications of JAD to blind speech separation in real-time. One often finds in practice most of the separation algorithms work equally well without the real-time constraints of data lengths over which statistical estimates are computed. In this paper we study the effects of perturbations of the source spectral matrix on the JAD of the cross spectral matrices of D contiguous frames of mixtures. We measure the off diagonal perturbations of the source cross-spectral matrix on the performance as the sum of a diagonal matrix and a non-diagonal symmetric matrix.

Auto and cross spectral estimation plays a critical role in the JAD of a set of cross-spectral matrices. This is typically done using the Welch method of

V. Vigneron et al. (Eds.): LVA/ICA 2010, LNCS 6365, pp. 378–385, 2010.

averaging direct spectral estimates obtained from short-time Fourier transforms (STFT) of segments of data [2], [3]. The data length must be large enough so that bias and variance errors due to the short segment lengths do not offset the variance reduction due to averaging and must be small enough to maintain stationarity. Long data lengths introduce large processing delays as well as a risk of averaging over nonstationary data segments whereby the idea of exploiting nonstationarity would be rendered ineffective. In this paper we propose to use the multitaper method to estimate the cross-spectral matrix [4]. As opposed to Welch's method of weighted overlapped segment averaging, the multitaper method is based on averaging direct estimates of a single segment of data over multiple windows or tapers. The averaging over tapers reduces the estimate variance effectively at the expense of a slight increase in bias.

The paper is organized as follows. In Sect. 2 the problem of convolutive BSS in both time and frequency domains is briefly reviewed along with an existing frequency domain demixing criterion in Sect. 2.1, which is employed in the numerical tests. Section 2.2 describes the estimation of the cross-spectral matrices by the multitaper method. The effects of the imperfect cross-spectral matrix estimation is analyzed in Sect. 2.3. The performance of the demixing method along with the factors affecting its sensitivity are examined by simulations in Sect. 3. Finally, the work is summarized in Sect. 4.

2 System Model

We consider the following convolutive mixing model: $\mathbf{X}(n) = \sum_l \mathbf{H}(l)\mathbf{S}(n-l)$, where $\mathbf{S}(n)$ denotes the real-valued column vector of sources $S_1(n), \cdots, S_{N_s}(n)$, $\mathbf{X}(n)$ denotes the real-valued column vector of mixtures $X_1(n), \cdots, X_{N_x}(n)$, and $\mathbf{H}(n)$ is the real-valued $(N_x \times N_s)$ multichannel mixing system whose (i,j)th element is the impulse response $H_{ij}(n)$. Accordingly, the demixing system is modeled as: $\mathbf{Y}(n) = \sum_l \mathbf{W}(l)\mathbf{X}(n-l)$, where $\mathbf{Y}(n)$ denotes the real-valued column vector of outputs $Y_1(n), \cdots, Y_{N_y}(n)$, and $\mathbf{W}(n)$ is the real-valued $(N_y \times N_x)$ multichannel demixing system whose (i,j)th element is the impulse response $W_{ij}(n)$.

The aim of blind source separation is to recover the source signals $\mathbf{S}(n)$ at the outputs $\mathbf{Y}(n)$ through the estimation of the true demixing system $\mathbf{W}(n)$ using the mixture signals $\mathbf{X}(n)$ and some assumptions on $\mathbf{S}(n)$ and $\mathbf{H}(n)$. The problem of estimating the demixing system parameters in time domain is high computational complexity and low convergence rate [5]. The convolutive mixtures in time domain can be decomposed into instantaneous mixtures in frequency domain via STFT as:

$$\mathbf{X}(f,i) = \mathbf{H}(f)\mathbf{S}(f,i) \ , \tag{1}$$

where $\mathbf{X}(f,i)$, $\mathbf{S}(f,i)$ are the STFT of the mixture and source signals at frequency f and segment i. Here, $\mathbf{H}(f)$ denotes the complex-valued $(N_x \times N_s)$ frequency response of the LTI mixing system. Accordingly, the demixing system can be written as:

$$\mathbf{Y}(f,i) = \mathbf{W}(f)\mathbf{X}(f,i) \ , \tag{2}$$

$\mathbf{Y}(f, i)$ being the STFT of the output signals, and $\mathbf{W}(f)$ denotes the complex-valued $(N_y \times N_x)$ frequency response of the demixing system. The STFT of the source and the output signals are related as: $\mathbf{Y}(f, i) = \mathbf{G}(f)\mathbf{S}(f, i)$, where $\mathbf{G}(f) = \mathbf{W}(f)\mathbf{H}(f)$ denotes the complex-valued $(N_y \times N_s)$ frequency response of the cascade system.

The idea of frequency domain separation is to estimate the demixing system parameters $\mathbf{W}(f)$ at each frequency independently. This results in lower computational complexity and higher convergence rate. Estimation of the demixing system is done by finding $\mathbf{Y}(f, i)$ that will restore some statistical features of the source signals $\mathbf{S}(f, i)$. Assumptions on the source signals are the following: (i) $\mathbf{S}(n)$ are zero-mean nonstationary processes, (ii) over each segment i they are second-order jointly stationary processes, (iii) they are uncorrelated, hence all the cross-covariances and cross-spectra between $S_k(.)$ and $S_l(.)$ vanish for $k \neq l = 1, \cdots, N_s$. One method based on second-order statistics is the JAD of a set of cross-spectral matrices. The underlying idea behind this approach is to exploit the time-varying vanishing source cross-spectra [5], [2], [3]. Once the demixing system is found for each frequency then the next step is to correct the permutation and scaling of the columns of the demixing system. Using this approach overall separation performance depends on the success of these three steps: demixing matrix estimation, permutation and scaling correction. Current state-of-the-art scaling correction method is [6], as for the permutation correction there is a wide variety of methods, however, it still seems to be a open problem, see [3], for instance. Errors introduced in the first step of demixing system estimation will have highly negative effects on the latter two. Therefore, it is desirable to perform the demixing matrix estimation at each frequency as accurate as possible.

2.1 Joint Diagonalization of a Set of Cross-Spectral Matrices

The source cross-spectral matrix $\mathbf{R}_s(f, i)$ at frequency f and segment i is defined as $\mathbf{R}_s(f, i) \equiv E[\mathbf{S}(f, i)\mathbf{S}^\dagger(f, i)]$, where the superscript † denotes the complex-conjugate transpose (Hermitian). Because of the assumption that the source signals are zero mean and uncorrelated, $\mathbf{R}_s(f, i)$ is a real-valued $(N_s \times N_s)$ diagonal matrix $\forall f,\ i$. The same definition applies to mixture and output cross-spectral matrices $\mathbf{R}_x(f, i)$, $\mathbf{R}_y(f, i)$. According to (1) the mixture cross-spectral matrix $\mathbf{R}_x(f, i)$ can be written as

$$\mathbf{R}_x(f, i) = \mathbf{H}(f)\mathbf{R}_s(f, i)\mathbf{H}^\dagger(f) \ . \tag{3}$$

Similarly, using (2) in the definition of $\mathbf{R}_y(f, i)$ one gets

$$\mathbf{R}_y(f, i) = \mathbf{W}(f)\mathbf{R}_x(f, i)\mathbf{W}^\dagger(f) \ . \tag{4}$$

The goal of JAD is to find a non-orthogonal non-singular complex-valued matrix $\mathbf{W}(f)$ that will jointly diagonalize the set of D matrices $\mathbf{R}_y(f, i),\ i = 1, \cdots, D$. Under certain uniqueness conditions it was shown in [1] that the joint diagonalizer matrix $\mathbf{W}(f)$ satisfies $\mathbf{W}(f)\mathbf{H}(f) = \mathbf{\Pi}(f)\mathbf{\Lambda}(f)$, where $\mathbf{\Pi}(f)$ and $\mathbf{\Lambda}(f)$

are permutation and diagonal scaling matrices representing the inherent permutation and scaling ambiguities that arise because of the estimation of $\mathbf{W}(f)$ based on $\mathbf{R}_x(f,i)$ without the knowledge of $\mathbf{H}(f)$ or $\mathbf{R}_s(f,i)$. In such a case the matrices $\mathbf{W}(f)$ and $\mathbf{H}^{-1}(f)$ are said to be essentially equal.

Uniqueness Conditions. In [1] the uniqueness of the solution for the JAD problem is quantified by the introduction of a parameter ρ called the modulus of uniqueness. It is derived from the set of diagonal source cross-spectral matrices $\mathbf{R}_s(f,i),\ i=1,\cdots,D$ in the following way; first a real-valued $(D\times N_s)$ matrix $\mathbf{\Psi}(f)$ is constructed from the vertical concatenation of the D diagonals of $\mathbf{R}_s(f,i)$

$$\mathbf{\Psi}(f)=\begin{bmatrix} R^s_{11}(f,1) & R^s_{22}(f,1) & \cdots & R^s_{N_sN_s}(f,1)\\ \vdots & \ddots & \ddots & \vdots\\ R^s_{11}(f,D) & R^s_{22}(f,D) & \cdots & R^s_{N_sN_s}(f,D)\end{bmatrix}=\left[\psi_1(f)\ \psi_2(f)\cdots\ \psi_{N_s}(f)\right]\ , \tag{5}$$

where the real-valued $(D\times 1)$ column vector $\psi_i(f)$ denotes the ith column of $\mathbf{\Psi}(f)$. The modulus of uniqueness $\rho(f)$ for $\mathbf{R}_s(f,i),\ i=1,\cdots,D$ is defined as the maximum of absolute value of the cosine of the angle $\rho_{ij}(f)$ between the columns $\psi_i(f),\psi_j(f)$ of $\mathbf{\Psi}(f)$. More specifically, $\rho_{ij}(f)=\psi_i^\dagger(f)\psi_j(f)/\left(\|\psi_i(f)\|\,\|\psi_j(f)\|\right)$ for $i\neq j=1,\cdots,N_s$ and $\rho(f)=\max_{i,j}|\rho_{ij}(f)|$. It basically measures the collinearity of the columns of $\mathbf{\Psi}(f)$ and the uniqueness condition can be formulated as $\rho(f)<1$. Unless the source signals have identical time-varying spectra it can be shown that this condition is satisfied.

2.2 Cross-Spectral Matrix Estimation Using Multitaper Method

We consider the following class of multitaper estimators [4]

$$\hat{R}^s_{lm}(f,i)=\frac{1}{K}\sum_{k=0}^{K-1}\left\{\sum_{p=1}^{N}\omega_k(p)S_l(p+(i-1)N)e^{-\jmath 2\pi fp}\right\}$$
$$\left\{\sum_{q=1}^{N}\omega_k(q)S_m(q+(i-1)N)e^{\jmath 2\pi fq}\right\},\quad l,m=1,\cdots N_s\ , \tag{6}$$

where each time domain signal $S_m(.)$ is divided into nonoverlapping segments $S_m(q+(i-1)N)$ each of size N, for example the ith segment of the mth signal utilizes the samples $S_m(1+(i-1)N),\cdots,S_m(iN)$. Here K denotes the total number of tapers $\omega_k(1),\cdots,\omega_k(N)$, $k=1,\cdots,K$. $\hat{R}^s_{lm}(f,i)$ is the (l,m)th element of the Hermitian $(N_s\times N_s)$ estimated source cross-spectral matrix $\hat{\mathbf{R}}_s(f,i)$. Basically the multitaper cross-spectrum estimator (6) is an average of K direct cross-spectrum estimators where the underlying pair of signal segments is the same, but different tapers are used. It is assumed that the tapers are real-valued and orthonormal, that is $\sum_{p=1}^{N}\omega_k(p)\omega_l(p)=\delta(k-l)$. The product of the segment size N and analysis bandwidth B is the time bandwidth product NB, and it plays a critical role in taper design. The Slepian tapers are the sequences of length N with maximum energy concentration in a frequency band $(f-B, f+B)$. The total number of tapers K is given by $K = 2NB - 1$.

2.3 Effects of Imperfect Cross-Spectral Matrix Estimation

For the finite sample case due to estimation errors $\hat{\mathbf{R}}_s(f,i)$ will no longer be diagonal. Let $\hat{\mathbf{R}}_s(f,i)$ be written as the following

$$\hat{\mathbf{R}}_s(f,i) = \hat{\mathbf{R}}_s^d(f,i) + \hat{\mathbf{R}}_s^o(f,i) \ , \tag{7}$$

where $\hat{\mathbf{R}}_s^d(f,i)$ is a $(N_s \times N_s)$ real-valued diagonal matrix with the same diagonal as $\hat{\mathbf{R}}_s(f,i)$, and $\hat{\mathbf{R}}_s^o(f,i)$ is Hermitian $(N_s \times N_s)$ matrix with zero diagonal. In other words, the estimation errors on the auto-cross spectra are absorbed into $\hat{\mathbf{R}}_s^d(f,i)$. In Sect. 3 the modulus of uniqueness parameter $\rho(f)$ will be estimated based on $\hat{\mathbf{R}}_s^d(f,i)$.

From (3) and (7) the estimator $\hat{\mathbf{R}}_x(f,i)$ of the $(N_x \times N_x)$ mixture cross-spectral matrix $\mathbf{R}_x(f,i)$ is given by

$$\begin{aligned} \hat{\mathbf{R}}_x(f,i) &= \mathbf{H}(f)\hat{\mathbf{R}}_s^d(f,i)\mathbf{H}^\dagger(f) + \mathbf{H}(f)\hat{\mathbf{R}}_s^o(f,i)\mathbf{H}^\dagger(f) \\ &= \hat{\mathbf{R}}_x^d(f,i) + \hat{\mathbf{R}}_x^o(f,i) \ . \end{aligned} \tag{8}$$

Equation (8) explicitly shows the perturbation matrix $\hat{\mathbf{R}}_x^o(f,i)$. Because of this perturbation matrix exact joint diagonalization of the set $\hat{\mathbf{R}}_x(f,i),\ i = 1,\cdots,D$ of D matrices via $\mathbf{W}(f)\hat{\mathbf{R}}_x(f,i)\mathbf{W}^\dagger(f)$ will not be possible. Rather an approximate diagonalization of the set would be achieved. As a measure of joint deviation of $\hat{\mathbf{R}}_x(f,i),\ i = 1,\cdots,D$ from diagonality the following measure proposed in [7] will be used

$$J\left(\mathbf{W}(f)\right) = \sum_{i=1}^{D} \log \frac{\det \operatorname{diag} \mathbf{W}(f)\hat{\mathbf{R}}_x(f,i)\mathbf{W}^\dagger(f)}{\det \mathbf{W}(f)\hat{\mathbf{R}}_x(f,i)\mathbf{W}^\dagger(f)} \ , \tag{9}$$

where det is the determinant operator, and diag($\mathbf{A}$) is a diagonal matrix with the same diagonal as $\mathbf{A}$. It can be shown that $J \geq 0$ with equality if and only if $\mathbf{W}(f)\hat{\mathbf{R}}_x(f,i)\mathbf{W}^\dagger(f)$ is diagonal $\forall i$. Furthermore, J is scale and permutation invariant, that is $J(\mathbf{\Pi\Lambda W}) = J(\mathbf{W})$. It is important to note that $J\left(\mathbf{W}(f)\right)$ is a suitable measure only for positive-definite matrices $\hat{\mathbf{R}}_x(f,i)$, which is the case for (6).

Equation (9) is used as a cost function in the search of its non-orthogonal minimizer $\mathbf{W}(f)$ [7]. The perturbation matrix $\hat{\mathbf{R}}_x^o(f,i)$ will introduce bias to the minimizer $\mathbf{W}(f)$ of (9). In [1] it was shown that the closer ρ gets to unity, the larger the bias of the demixing system parameters, and the larger the norm of the perturbation term the higher is the bias. Next we will study the effects of the cardinal number D on ρ and on the demixing performance using numerical simulations.

3 Numerical Simulations

We consider the two input two output scenario, that is $N_s = N_x = N_y = 2$. The source signals $S_1(n), S_2(n)$ are male and female speech utterances from the TIMIT database each of a total duration of 20 s with a sampling rate of $f_s = 16$ kHz. Since the speech signals are nonstationary, they are subdivided into 400 nonoverlapping segments each of duration 50 ms ($N = 800$ samples), over which the stationarity assumption approximately holds. The source cross-spectral matrices $\hat{\mathbf{R}}_s(f,i)$ are estimated at $N_{\text{fft}} = 2^{10}$ frequency points over each segment $i = 1, \cdots, 400$ using (6) with $K = 3$ Slepian tapers.The four mixing channels $H_{ij}(n)$ are chosen as 8-tap FIR filters. All the filter coefficients are drawn randomly from a zero-mean unit-variance uniform distribution. The mixture cross-spectral matrices $\hat{\mathbf{R}}_x(f,i)$ are generated using the frequency response functions of the mixing channels according to (3). The cardinal number D of the set $\hat{\mathbf{R}}_x(f,i)$, $i = k, \cdots, k+D-1$ is chosen as a variable. Here k denotes the start index of the kth set. For fixed D, $k = 1, \cdots, 400 - D + 1$ and we have a total of $400 - D + 1$ sets of D matrices. For each set k we estimate the demixing system $\mathbf{W}^{(k)}(f)$ and measure the performance by $\text{Index}\left(\mathbf{G}^{(k)}(f)\right)$ defined by

$$\text{Index}(\mathbf{G}(f)) \equiv \sum_{i=1}^{N_y}\left[\sum_{j=1}^{N_s}\frac{|G_{ij}(f)|^2}{\max_k |G_{ik}(f)|^2} - 1\right] + \sum_{j=1}^{N_s}\left[\sum_{i=1}^{N_y}\frac{|G_{ij}(f)|^2}{\max_k |G_{kj}(f)|^2} - 1\right] . \tag{10}$$

Alternatively, we can get an overall performance index by averaging over frequency $\text{Index}\left(\mathbf{G}^{(k)}\right)$, and by averaging over both k and f $\text{Index}(\mathbf{G})$. Similarly, the modulus of uniqueness parameter $\rho^{(k)}(f)$ for each frequency and set, $\rho^{(k)}$ for each set obtained by averaging across frequency, and ρ obtained by averaging both over k and f will be used in the simulations. Dependence of the $\text{Index}(\mathbf{G})$ with respect to the cardinal number D is shown in panel (a) of Fig. 1 in a dB scale. The blue line is obtained by averaging $\text{Index}\left(\mathbf{G}^{(k)}\right)$ over k. In other words each star corresponds to an average of $400 - D + 1$ $\text{Index}\left(\mathbf{G}^{(k)}\right)$ values. The red lines indicate the 95% confidence interval. It is clearly seen that increasing D improves the separation performance, which is in agreement with [1]. This improvement is related to decreasing values of the modulus of uniqueness parameter ρ with increasing D as is shown in panel (b) of Fig. 1. Figure 2 shows the histogram of the performance index and the modulus of uniqueness parameter both averaged over frequency for $D = 5,\ 50$. The top plots compare the histogram, in terms of the percentage of the $400 - D + 1$ outcomes in each bin, of $\text{Index}(\mathbf{G})$ using (a) $D = 5$ and (b) $D = 50$ cross-spectral matrices in each set. The bottom plots show the distribution of ρ for (c) $D = 5$ and (d) $D = 50$. These plots are detailed analysis of the first and last points (D) in Fig. 1. It is clearly seen that the $\text{Index}(\mathbf{G})$ take values below -20 dB for $D = 50$, whereas for $D = 5$ case it is between -20 and 0 dB. The ρ values are distributed close to 0 with small dispersion around the mean value 0.1131 for $D = 50$, on the other hand the dispersion is large around the mean value of 0.4283 with $D = 5$. Next, we examine the frequency distribution of the performance index

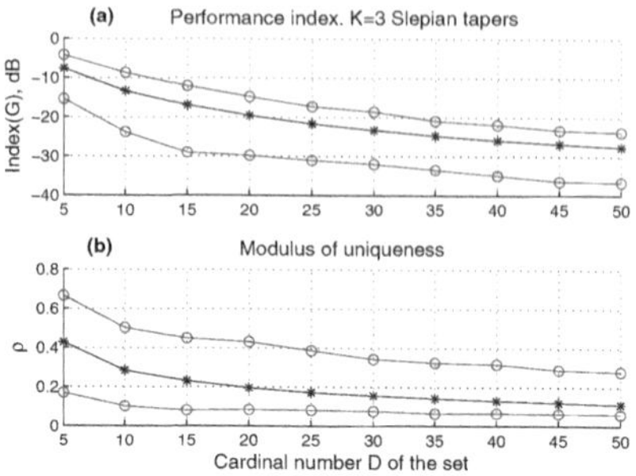

Fig. 1. Dependence of the mean (a) performance measure and (b) modulus of uniqueness wrt the cardinal number D along with 95% confidence intervals

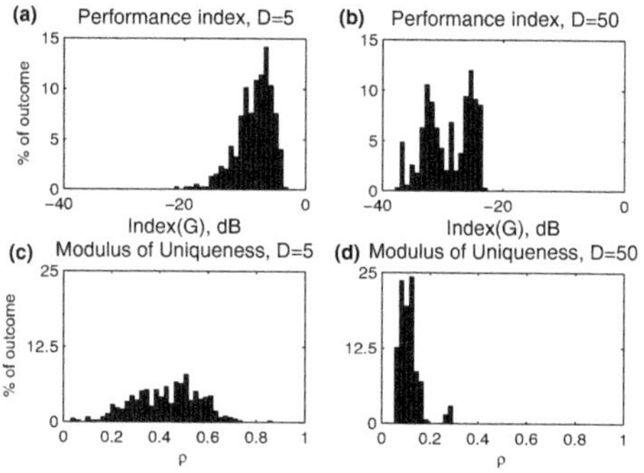

Fig. 2. Histogram of the performance index in panels (a),(b) and the modulus of uniqueness parameter in panels (c),(d) both averaged over frequency

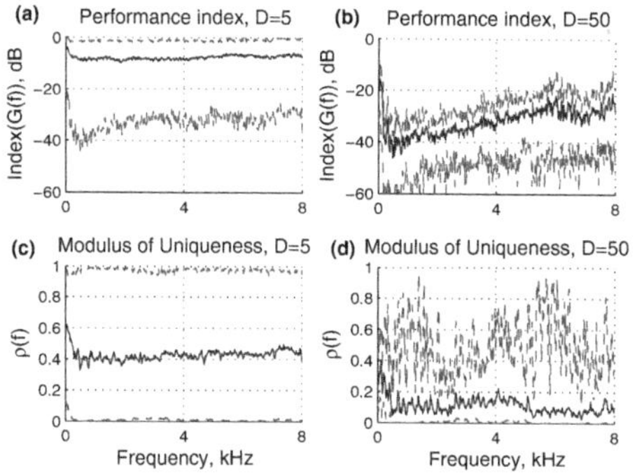

Fig. 3. Mean performance index (a),(b) and mean modulus of uniqueness (c),(d) as a function of frequency with 95% confidence intervals. (a),(c) $D = 5$, (b),(d) $D = 50$.

and the modulus of uniqueness parameter using $K = 3$ Slepian tapers, and for $D = 5,\ 50$. Figure 3 shows the mean and 95% confidence interval, obtained by using $400 - D + 1$ values Index $\left(\mathbf{G}^{(k)}(f)\right)$ and $\rho^{(k)}(f)$ for each f. The top plots compare the performance index (a) $D = 5$ and (b) $D = 50$, and the bottom plots compare the modulus of uniqueness for (c) $D = 5$ and (d) $D = 50$. The results are similar to Fig. 2 showing the correlation between the $\rho(f)$ and Index $(\mathbf{G}(f))$. The closer the ρ to zero, the better is the performance.

4 Conclusion

We introduced the multitaper method of cross-spectrum estimation in the joint approximate diagonalization (JAD) of the set of cross-spectral matrices. Based on the work of [1] factors affecting the sensitivity of the JAD problem were investigated in the application of the convolutive mixtures of speech signals. We studied the effect of the number of matrices in the set, and observed that there exists a link between the modulus of uniqueness parameter and the estimation of demixing system parameters.

References

1. Afsari, B.: Sensitivity analysis for the problem of matrix joint diagonalization. SIAM J. Matrix AA. 30, 1148–1171 (2008)
2. Parra, L., Spence, C.: Convolutive blind separation of non-stationary sources. IEEE Trans. Speech & Audio P. 8, 320–327 (2000)
3. Serviere, C., Pham, D.T.: Permutation correction in the frequency domain in blind separation of speech mixtures. EURASIP J. App. SP, 1–16 (2006)
4. Walden, A.T.: A unified view of multitaper multivariate spectral estimation. Biometrika 87, 767–788 (2000)
5. Wu, H.C., Principe, J.C.: Simultaneous diagonalization in the frequency domain (SDIF) for source separation. In: 1st Int. Workshop ICA & Signal Separation, pp. 245–250 (1999)
6. Matsuoka, K.: Minimal distortion principle for blind source separation. In: 41st SICE Conf., pp. 2138–2143 (2002)
7. Pham, D.T.: Joint approximate diagonalization of positive definite Hermitian matrices. SIAM J. Matrix AA. 22, 1136–1152 (2001)

Blind Compressed Sensing: Theory

Sivan Gleichman and Yonina C. Eldar*

Department of Electrical Engineering, Technion - Israel Institute of Technology

Abstract. Compressed sensing successfully recovers a signal, which is sparse under some basis representation, from a small number of linear measurements. However, prior knowledge of the sparsity basis is essential for the recovery process. In this work we define the blind compressed sensing problem, which aims to avoid the need for this prior knowledge, and discuss the uniqueness of its solution. We prove that this problem is ill possed in general unless further constraints are imposed. We then suggest three possible constraints on the sparsity basis that can be added to the problem in order to render its solution unique. This allows a general sampling and reconstruction system that does not require prior knowledge of the sparsity basis.

1 Introduction

Sparse signal representations have gained popularity in recent years in many theoretical and applied areas [2, 3, 11]. A finite dimensional vector is referred to as sparse under a basis if its representation under the basis transform contains a small number of nonzero entries. Compressed sensing (CS) [3, 11] focuses on the role of sparsity in reducing the number of measurements needed to represent a finite dimensional vector $x \in \mathbb{R}^m$. The vector x is measured by $b = Ax$, where $A \in \mathbb{R}^{n \times m}$ and $n \ll m$. In this formulation, determining x from b is ill possed in general. However, if x is known to be sparse in a given basis P, then under additional conditions on A [4], the measurements b determine x uniquely.

Many CS methods have been proposed to recover x from b [2, 3, 11]. However, all these approaches use the prior knowledge of the sparsity basis P. We introduce the concept of blind compressed sensing (BCS), which aims to avoid this prior knowledge. That is, the goal is to recover the vector x from a small number of measurements, where the only prior is that there exists some basis P in which x is sparse. BCS was also discussed in [1] in the context of bounding the estimation error.

Any vector is sparse in a basis which contains the vector itself. Therefore, without knowing P the recovery of a single vector out of a small number of measurements is ill possed. As we show, even if we have multiple signals that share the same (unknown) sparsity basis, BCS remains ill-posed. In order for the

* This work was supported in part by the Israel Science Foundation under Grant no. 1081/07 and by the European Commission in the framework of the FP7 Network of Excellence in Wireless COMmunications NEWCOM++ (contract no. 216715).

V. Vigneron et al. (Eds.): LVA/ICA 2010, LNCS 6365, pp. 386–393, 2010.

measurements to determine x uniquely we need an additional constraint. In this paper we discuss three possible constraints on the sparsity basis, which enable blind recovery. For each constraint we define the constrained BCS problem and prove conditions under which its solution is unique.

The first constraint we consider relies on the fact that over the years there have been several bases, such as wavelets [9], that have been considered "good" in the sense that they are known to sparsely represent many natural signals. We therefore treat the setting in which the unknown basis P is one of a finite and known set of bases. The next constraint allows P to contain any sparse enough combination of the columns of a given dictionary. For both these classes of constrains we show that a Gaussian random measurement matrix satisfies the uniqueness conditions we develop with probability 1.

The third constraint we present is inspired by multichannel systems, where the signals from each channel are sparse under separate bases. In our setting this translates to the requirement that P is block diagonal. For simplicity, and following several previous works [7,10], we impose in addition that P is orthogonal. While the first two constraints enable the recovery of a single vector x, here we require an ensemble of signals X, all sparse in the same basis. Using this structure we develop uniqueness conditions, and show that a suitable choice of random matrix A satisfies the uniqueness conditions with probability 1. This work focuses of the theory of the BCS problem. In [6] we introduce efficient methods to blindly reconstruct the signals from the measurements, under each of these constraints.

The remainder of the paper is organized as follows. In Section 2 we define the BCS problem and prove that it is ill posed. In Section 3 we consider the three constrained BCS problems.

2 BCS: Definition and Uniqueness

The goal of CS is to reconstruct a vector $x \in \mathbb{R}^m$ from its measurements $b = Ax$, where $A \in \mathbb{R}^{n \times m}$ and $n \ll m$, under the constraint that x is sparse in some known basis P. The CS problem can be formulated as:

$$\hat{s} = \arg\min ||b - APs||_2^2 \qquad \text{s.t.} \qquad ||s||_0 \leq k, \tag{1}$$

where $||\cdot||_0$ is the ℓ_0 semi-norm which counts the number of nonzero elements of the vector, and k is given. The reconstructed signal is then $\hat{x} = P\hat{s}$. In [4] the authors define the *spark* of a matrix, denoted by $\sigma(\cdot)$, which is the smallest possible number of linearly dependent columns. They prove that if $||s||_0 \leq k$, and $\sigma(AP) \geq 2k$, then the solution to (1) is unique.

It is easy to see that if A is an i.i.d. Gaussian random matrix, then $\sigma(AP) = n + 1$ with probability 1 for any fixed orthogonal basis P. This leads to the universality property of CS which offers a sampling process that does not require knowledge of P, as long as P is orthogonal and $||s||_0 \leq n/2$. However, even in this case all existing CS algorithms require the knowledge of P for the reconstruction process. The idea of BCS is to avoid entirely the need for this prior knowledge, and perform both the sampling and the reconstruction without knowing the sparsity basis.

2.1 BCS Problem Definition

The problem of BCS seems impossible at first, since every signal is sparse under a basis that contains the signal itself. This would imply that BCS allows reconstruction of any signal from a small number of measurements without any prior knowledge, which is clearly impossible. Our approach then, is to sample an ensemble of signals that are all sparse under the same basis. Later on we revisit problems with only one signal, but with additional constraints.

Let $X \in \mathbb{R}^{m \times N}$ denote a matrix whose columns are the original signals, and let $S \in \mathbb{R}^{m \times N}$ denote the corresponding sparse vectors, such that $X = PS$ for some basis $P \in \mathbb{R}^{m \times m}$. The signals are all sampled using a measurement matrix $A \in \mathbb{R}^{n \times m}$, producing $B = AX$. For the measurements to be compressed the dimensions must satisfy $n < m$, where the compression ratio is $L = m/n$. Following [5] we assume the maximal number of nonzero elements in each of the columns of S, is known to equal k. We refer to such a matrix S as a k-sparse matrix. The BCS problem can be formulated as follows.

Problem 1. Given the measurements B and the measurement matrix A find the signal matrix X such that $B = AX$ where $X = PS$ for some basis P and k-sparse matrix S.

An important question is the uniqueness of the BCS solution, namely the uniqueness of the signal matrix X which solves Problem 1. Unfortunately, although Problem 1 seems quite natural, its solution is not unique regardless of the choice of A or the value of N and k. Therefore, an additional constraint is needed for Problem 1 to be well defined. We prove this result by reducing the problem to an equivalent one, using the field of dictionary learning (DL), and proving that the solution to this equivalent problem is not unique.

2.2 Dictionary Learning

The field of DL [5, 8] focuses on finding a sparse matrix $S \in \mathbb{R}^{m \times N}$ and a dictionary $D \in \mathbb{R}^{n \times m}$ such that $B = DS$ where only $B \in \mathbb{R}^{n \times N}$ is given. Usually in DL the dimensions satisfy $n \ll m$. BCS can be viewed as a constrained DL problem, with $D = AP$ for a given A and an unknown basis P. However, there is an important difference in the output of DL and BCS. DL provides the dictionary $D = AP$ and the sparse matrix S, but in BCS we are interested in recovering the unknown signals $X = PS$. As we prove in the next subsection this difference renders the BCS an ill possed problem.

An important question is the uniqueness of the DL factorization. That is, given a matrix B we seek the uniqueness of the pair D, S such that $B = DS$ and S is k-sparse. Note that scaling and signed permutation of the columns of D and rows of S respectively do not change the product $B = DS$. Therefore, there cannot be a unique pair D, S, so that in the context of DL the term uniqueness refers to uniqueness up to scaling and signed permutation. Sufficient conditions on D and S for DL uniqueness are proven in [5]. The condition on D is $\sigma(D) \geq 2k$, which guarantees that given D there is a unique S. We refer to the conditions on S as the *richness conditions*, which require in general that the columns of S should be diverse with respect to both the locations and the values of the nonzero elements. The exact richness conditions can be found in [5]. One of the conclusions from the richness conditions is that N must be at least $\binom{m}{k}(k+1)$. Nevertheless, it was shown in [5] that in practice far fewer signals are needed. Heuristically, the number of signals should grow at least linearly with the length of the signals.

2.3 BCS Uniqueness

Under the conditions above the DL solution given B is unique up to scaling and signed permutations. Without loss of generality we can ignore this ambiguity, since we are only interested in the product PS and not in P or S themselves. Therefore, we assume that applying DL on B provides the exact $D = AP$ and S. If we can extract P out of D, then we can recover the correct signals $X = PS$. Therefore, under the DL uniqueness conditions Problem 1 is equivalent to the following problem.

Problem 2. Given $D \in \mathbb{R}^{n \times m}$ and $A \in \mathbb{R}^{n \times m}$, where $n < m$, find a basis P such that $D = AP$.

We therefore focus on Problem 2 and prove that it has no unique solution. Assume P_1 is a basis, i.e., has full rank, and satisfies $D = AP_1$. Decompose P_1 as $P_1 = P_{N^\perp} + P_N$ where the columns of P_N are in $N(A)$, the null space of A, and those of $P_{N^\perp}$ are in its orthogonal complement $N(A)^\perp$. Note that necessarily $P_N \neq 0$, otherwise the matrix $P_1 = P_{N^\perp}$ is in $N(A)^\perp$ and has full rank. However, since the dimension of $N(A)^\perp$ is at most $n < m$, there is no $m \times m$ full rank matrix whose columns are all in $N(A)^\perp$.

Next define the matrix $P_2 = P_{N^\perp} - P_N \neq P_1$, which satisfies $D = AP_2$. Moreover, $P_N^T P_{N^\perp} = 0$ so that $P_1^T P_1 = P_2^T P_2$. Therefore, since P_1 has full rank so does P_2, such that both of them are solutions to Problem 2. In fact there are many more solutions; some of them can be found by changing the signs of only part of the columns of P_N.

When the DL solution given B is unique, BCS is equivalent to Problem 2 which has no unique solution. Obviously if the DL solution given B is not unique, then the BCS solution is not unique. Therefore, in order to guarantee BCS uniqueness additional constraints are needed. We focus below on three such constraints, summarized in Table 1.

3 Constrained BCS

In this section we introduce three possible constraints on the BCS problem, and prove they lead to uniqueness under specific conditions.

Table 1. Summary of constraints on P

The constraint	Conditions for uniqueness
Finite Set (Section 3.1): P is in a given finite set of possible bases Ψ.	• $\sigma(AP) \leq 2k$ for any $P \in \Psi$. • A is k-rank preserving of Ψ (Definition 1).
Sparse Basis (Section 3.2): P is k_P-sparse under a given dictionary Φ.	• $\sigma(A\Phi) \geq 2k_P k$.
Structure (Section 3.3): P is orthogonal $2L$-block diagonal.	• The richness conditions on S. • A is a union of L orthogonal bases. • $\sigma(AP) = n + 1$. • A is not inter-block diagonal (Definition 2).

3.1 Finite Set of Bases

Over the years a variety of bases were proven to lead to sparse representations of many natural signals, such as wavelets [9]. Therefore, when the basis is unknown it is natural to try one of these choices. That is, the first constraint we consider limits the number of possible bases P to a finite and given set of bases. The new constrained BCS, instead of Problem 1, is then:

Problem 3. Given the measurements B, the measurement matrix A and a finite set of bases Ψ, find the signal matrix X such that $B = AX$ and $X = PS$ for some basis $P \in \Psi$ and k-sparse matrix S.

To consider the uniqueness in this setting we rely on the following definition.

Definition 1. *A is k-rank preserving of the bases set Ψ if any two index sets T, J of size k and any two bases $P, \bar{P} \in \Psi$ satisfy*

$$rank(A[P_T, \bar{P}_J]) = rank[P_T, \bar{P}_J]. \tag{2}$$

where $P_T, \bar{P}_J$ are the sub-matrices of $P, \bar{P}$ containing only the columns with indices in T, J respectively.

We now show that if $\sigma(AP) \geq 2k$ for any $P \in \Psi$ and if A is k-rank preserving of the bases set Ψ, then the solution to Problem 3 is unique even when there is only one signal, so that $N = 1$. In this case instead of the matrices X, S, B we deal with the vectors x, s, b respectively.

Assume $x = Ps$ satisfies $b = Ax$ where $||s||_0 \leq k$ and $P \in \Psi$. Uniqueness is achieved if there is no $\bar{x} \neq x$ and $\bar{P} \in \Psi$ such that $b = A\bar{x}$, and $\bar{x} = \bar{P}\bar{s}$ for $||\bar{s}||_0 \leq k$. Since $\sigma(AP) \geq 2k$ if $\bar{P} = P$ then $\bar{x} = x$. Otherwise, let T, J denote the index set of the nonzero elements in $s, \bar{s}$ respectively, so that $x = P_T s_T$ and $\bar{x} = \bar{P}_J \bar{s}_J$. Therefore, $b = A\bar{P}_J\bar{s}_J = AP_T s_T$, which implies that the matrix $A[P_T, \bar{P}_J]$ has a null space. Since A is k-rank preserving of Ψ the null space of $A[P_T, \bar{P}_J]$ equals the null space of $[P_T, \bar{P}_J]$. Therefore, $A\bar{P}_J\bar{s}_J = AP_T s_T$ if and only if $\bar{P}_J\bar{s}_J = P_T s_T$, which implies $\bar{x} = x$, so that the solution to Problem 3 is unique. For $N > 1$ the same conditions are sufficient for uniqueness, since we can look at every signal separately. However, for $N > 1$ sometimes the condition that A must be k-rank preserving can be relaxed.

As long as $k \leq n/2$, in order to satisfy $\sigma(AP) \geq 2k$ with probability 1 we can require all $P \in \Psi$ to be orthogonal and generate A from an i.i.d. Gaussian distribution. Alternatively, since the number of bases is finite, we can verify this condition for all the products AP. An i.i.d Gaussian A is also guaranteed with probability 1 to be k-rank preserving of any bases set Ψ; See [6].

3.2 Sparse Basis

The second constraint we consider is the sparsity of the basis P. We assume that the columns of P are sparse under some known dictionary Φ, so that there exists some unknown k_p-sparse matrix Z such that $P = \Phi Z$. In order for P to be a basis, Φ must have full row rank, and Z must have full column rank. The constrained BCS in this case is then:

Problem 4. Given the measurements B, the measurement matrix A and the dictionary Φ, which has full row rank, find the signal matrix X such that $B = AX$ where $X = \Phi Z S$ for some k-sparse matrix S and k_p-sparse and full column rank matrix Z.

This problem is similar to that studied in [12] in the context of sparse DL. The difference is that [12] finds the matrices Z, S, while we are only interested in their product. The motivation behind Problem 4 is to overcome the disadvantage of the previously discussed Problem 3 in which the bases are fixed. When using a sparse basis we enhance the adaptability of the dictionary Φ to different signals by allowing any sparse enough combination of its columns

Here too we start by proving the uniqueness conditions in case $N = 1$. Therefore, instead of matrices X, S, B we deal with vectors x, s, b respectively. Since $||s||_0 \leq k$ and Z is k_p-sparse, the vector $c = Zs$ necessarily satisfies $||c||_0 \leq k_p k$. Therefore, Problem 4 can be formulated as:

$$\hat{c} = \arg\min_{c} ||b - A\Phi c||_2^2 \qquad \text{s.t. } ||c||_0 \leq k_p k, \tag{3}$$

where the recovery is $x = \Phi\hat{c}$. According to [4] the solution to (3) is unique if $\sigma(A\Phi) \geq 2k_p k$. The same condition is sufficient even when $N > 1$, since we can look at every signal separately.

Note that in Problem 4 the matrix Z necessarily has full column rank, while this constraint is dropped in (3). However, if the solution without this constraint is unique then obviously the solution with it is also unique.

3.3 Structural Constraint

Another possible constraint is a structural constraint on the basis P. Motivated by multichannel systems we require P to be block diagonal, and following several previous works [7, 10], we impose in addition that P is orthogonal. The constrained BCS problem is then:

Problem 5. Given the measurements B and the measurement matrix $A \in \mathbb{R}^{n \times nL}$ find the signal matrix X such that $B = AX$ where $X = PS$ for some orthogonal $2L$-block diagonal matrix P and k-sparse matrix S.

In this setting the size of the measurement matrix A is $n \times nL$, where n is the number of measurements and L is the compression ratio. The length of the signals is $m = nL$, and the size of the basis P is $nL \times nL$. Since P is $2L$-block diagonal, the size of its blocks is $\frac{n}{2} \times \frac{n}{2}$. Therefore, n must be even.

The following definition will be used to analyze uniqueness in this settings.

Definition 2. *Denote $A = [A_1, ..., A_L]$, such that $A_i \in \mathbb{R}^{n \times n}$ for any $1 \leq i \leq L$. A is called* inter-block diagonal *if there are two indices $i \neq j$ for which the product:*

$$A_i^T A_j = \begin{bmatrix} R_1 & R_2 \\ R_3 & R_4 \end{bmatrix},$$

satisfies: $rank(R_1) = rank(R_4)$, $rank(R_2) = rank(R_3) = \frac{n}{2} - rank(R_1)$. In particular if the product $A_i^T A_j$ is 2-block diagonal then A is inter-block diagonal.

We are ready to state our uniqueness results, for which we provide a sketch of the proof. For the full proof see [6].

Theorem 1. *Under the richness conditions on S, if $A \in \mathbb{R}^{n \times nL}$ is a union of L orthogonal bases, which is not inter-block diagonal, and $\sigma(AP) = n + 1$, then the solution to Problem 5 is unique.*

Proof sketch: Denote the desired solution of Problem 5 by $X = PS$, where the given measurements are $B = AX$, and let

$$A = [A_1, ..., A_L] \ , \ P = \begin{bmatrix} P^1 & & \\ & \ddots & \\ & & P^{2L} \end{bmatrix},$$

where A_i for $i = 1, .., L$ and P^j for $j = 1, ..., 2L$ are all orthogonal matrices.

The DL uniqueness conditions are satisfied, however this time we can ignore only the ambiguity in the scaling but not the ambiguity in the permutation. That is since permutations of P can cancel the block diagonal structure. Therefore, applying DL on B provides $\tilde{D} = APQ$ and $\tilde{S} = Q^t S$ for some unknown permutation matrix Q (a column or row permutation of the identity matrix).

In order to find $X = PS$ we need to extract $\tilde{P} = PQ$ out of $\tilde{D}$ so that $\tilde{P}\tilde{S} = PQQ^T S = X$. Since the number of possible signed permutations is finite, in theory we can go over all permutations Q_D and find one such that $\hat{D} = \tilde{D}Q_D = A\hat{P}$ where $\hat{P}$ is orthogonal $2L$-block diagonal. That is:

$$\hat{D} = [\hat{D}_1, ..., \hat{D}_L] = \left[A_1 \begin{pmatrix} \hat{P}^1 & \\ & \hat{P}^2 \end{pmatrix}, \ldots, A_L \begin{pmatrix} \hat{P}^{2L-1} & \\ & \hat{P}^{2L} \end{pmatrix}\right].$$

Since A_i are orthogonal for all $i = 1, ..., L$, we can recover the blocks of $\hat{P}$ by

$$\begin{bmatrix} \hat{P}^{2i-1} & \\ & \hat{P}^{2i} \end{bmatrix} = A_i^T \hat{D}_i.$$

It can be shown that if A is not inter-block diagonal, P is $2L$-block diagonal and $\sigma(AP) = n + 1$, then the equality $\hat{D} = A\hat{P} = APQQ_D$ implies $\hat{P} = PQQ_D$ so that $\tilde{P} = PQ = \hat{P}Q_D^T$. Therefore, $X = \tilde{P}\tilde{S}$ is unique.

One way to guarantee that A satisfies the conditions of Theorem 1 with probability 1 is to generate it randomly from an i.i.d Gaussian distribution and perform Gram-Schmidt on each block in order to make it orthogonal; See [6].

4 Conclusions

We presented the BCS problem which aims to solve CS problems without the knowledge of the sparsity basis of the signals. We proved that the BCS problem is ill possed in general, and suggested three possible constraints that can be added in order to guarantee the uniqueness of the BCS solution. For each constraint we proved conditions under which the solution is unique, and suggested a way to guarantee these conditions with probability 1. Under each constraint a simple algorithm can be performed in order to recover the signal; See [6].

References

1. Zayyani, H., Babaie-Zadeh, M., Jutten, C.: Bayesian cramer-rao bound for non-blind and blind compressed sensing. Submitted to IEEE Signal. Process. Letters (2009)
2. Bruckstein, A.M., Donoho, D.L., Elad, M.: From sparse solutions of systems of equations to sparse modeling of signals and images. SIAM Review 51(1), 34–81 (2009)
3. Donoho, D.L.: Compressed sensing. IEEE Trans. Info. Theory 52(4), 1289–1306 (2006)
4. Donoho, D.L., Elad, M.: Maximal sparsity representation via l_1 minimization. Proc. Nat. Acad. Sci. 100, 2197–2202 (2003)
5. Aharon, M., Elad, M., Bruckstein, M.: On the uniqueness of overcomplete dictionaries, and practical way to retrieve them. Linear Algebra and Its Applications 416(1), 48–67 (2006)
6. Gleichman, S., Eldar, Y.C.: Blind compressed sending. CCIT Report 759, EE Dept., Technion - Israel Institute of Technology (February 2010)
7. Gribonval, R., Schnass, K.: Dictionary identification from few training samples. In: Proc. 16th EUSIPCO 2008 (August 2008)
8. Kreutz-Delgado, K., Murray, J.F., Rao, B.D., Engan, K., Lee, T.W., Senowski, T.J.: Dictionary learning algorithms for sparse representation. Neural Computation 15(2), 349–396 (2003)
9. Mallat, S.: A wavelet tour of signal processing. Academic Press, London (1999)
10. Mishali, M., Eldar, Y.C.: Sparse sourse seperation from orthogonal mixtures. In: ICASSP, April 2009, pp. 3145–3148 (2009)
11. Candes, E., Romberg, J., Tao, T.: Robust uncertainty principles: Exact signal reconstruction from highly incomplete frequency information. IEEE Trans. Info. Theory 52, 1289–1306 (2006)
12. Rubinstein, R., Zibulevsky, M., Elad, M.: Double sparsity: Learning sparse dictionaries for sparse signal approximation. IEEE Trans. on Signal Processing (to appear)

Blind Extraction of the Sparsest Component

Everton Z. Nadalin[1,4], André K. Takahata[3,4], Leonardo T. Duarte[2,4], Ricardo Suyama[5], and Romis Attux[1,4]

[1] Department of Computer Engineering and Industrial Automation
[2] Department of Microwave and Optics
[3] Department of Communication
[4] Lab. of Signal Processing for Communications (DSPCom)
School of Electrical and Computer Engineering / P.O. Box 6101
University of Campinas – UNICAMP, CEP 13083-970, Campinas, SP, Brazil
[5] Engineering, Modeling and Applied Social Sciences Center (CECS), UFABC, Brazil
{nadalin,attux}@dca.fee.unicamp.br, kazuo@decom.fee.unicamp.br,
ltduarte@dmo.fee.unicamp.br, ricardo.suyama@ufabc.edu.br

Abstract. In this work, we present a discussion concerning some fundamental aspects of sparse component analysis (SCA), a methodology that has been increasingly employed to solve some challenging signal processing problems. In particular, we present some insights into the use of ℓ_1 norm as a quantifier of sparseness and its application as a cost function to solve the blind source separation (BSS) problem. We also provide results on experiments in which source extraction was successfully made when we performed a search for sparse components in the mixtures of sparse signals. Finally, we make an analysis of the behavior of this approach on scenarios in which the source signals are not sparse.

1 Introduction

Studies related to Blind Source Separation (BSS) problems have been increasingly present in the signal processing literature since the first efforts in the area. The BSS approach based on Independent Component Analysis (ICA) is certainly consolidated as a fundamental unsupervised method, and this is a consequence of works like [1], which showed *inter alia* that a contrast function based on the quantification of independence between separated signals has the potential of leading to effective source recovery.

ICA, however, is not the only possible approach to solve the BSS problem, and there are instances in which it is not even a viable choice: this is the case, for example, in scenarios characterized by the presence of more sources than mixtures.

Interestingly, in such scenarios, the idea of sparseness is particularly attractive. Exploring the fact that the sources have a certain degree of sparseness in some domain and that, therefore, not all sources are active at the same time, it is possible to identify the mixing model or even to separate the underlying sources.

When the concept of sparseness is applied to BSS, it typically related to the notion of Sparse Component Analysis (SCA)[2]. In this paper, we present a different

V. Vigneron et al. (Eds.): LVA/ICA 2010, LNCS 6365, pp. 394–401, 2010.

approach exploring the sparseness of the sources, which, in some sense, is closer to the idea underlying ICA: to perform blind separation aiming to recover signals that exhibit the same characteristics of the sources. In the case of ICA, this characteristic is related to their mutual independence. In our case, we assume that the sources are sparse, and thus, try to recover signals as sparse as possible.

We also show that, if we apply the commonly employed ℓ_1 norm to the search for sparse components in a mixture, we have the potential of separating the sources. Afterwards, we discuss under what conditions this norm is an effective contrast function for BSS.

2 Blind Source Separation and Sparse Component Analysis

The goal in the source separation problem is to recover a set of signals (sources) based on the observed signals (mixtures) and a minimum amount of information regarding some particular characteristic of the sources. Mathematically, the observed signals, represented by an N-dimensional vector $\mathbf{x}(k)$, can be expressed as

$$\mathbf{x}(k) = \mathbf{A}\mathbf{s}(k) \tag{1}$$

where $\mathbf{s}(k)$ represents an M-dimensional vector containing the sources and $\mathbf{A}$ denotes the mixing matrix.

In the case in which there are more sensors than sources ($M \leq N$), perfect source recovery is possible given that $\mathbf{A}$ can be inverted. If this is the case, the separation process is based on the task of finding a separation matrix $\mathbf{W}$ such that

$$\mathbf{y}(k) = \mathbf{W}\mathbf{x}(k) \tag{2}$$

represent the source signals, up to permutation and scale ambiguities [1].

The model just described is the most studied in BSS, and, under the hypothesis that the sources are mutually independent, the problem has been successfully solved using, as previously stated, ICA [1]. However, this model is based on hypotheses that are not valid in some practical situations, such as the one that assumes that the mixture matrix $\mathbf{A}$ is considered invertible, which, for instance, is not valid in cases of underdetermined problems. Under this situation, it is not so straightforward to deal with the separating matrix $\mathbf{W}$ because the mixture matrix is not invertible.

Interestingly, a number of works show that, in some underdetermined contexts, it is possible to assume that the sources are sparse [2,3,4]. In these situations, the fact that not all sources are active at the same moment can bring enough information to solve the problem. When not all sources are active at the same time, the underdetermined problem becomes locally determined, thus being possible to identify the mixture matrix $\mathbf{A}$, and, in some situations, to build estimates of the sources.

The first works using SCA to solve the BSS problem [3] assumed the absence of overlap between the sources in the mixture, i.e. disjoint orthogonality [4]. In this case, it is possible to perform source separation and mixing matrix identification even when there are more sources than mixtures. Afterwards, some works extended these methods to non-disjoint sources [2]. These approaches, however, face a relevant difficulty whenever noise is present in the mixtures.

3 Toward a Contrast Function Based on Sparseness

Despite the fact that all the above-mentioned approaches take into account the sparseness of the sources, they do not directly try to estimate sparse signals using a cost function based on a measure of sparseness. Hence, we raise the following question: assuming that the sources are sparse signals, under what conditions adjusting the separating matrix according to a criterion based on sparseness leads to source separation? In other words, when is it possible to separate sparse sources by retrieving sparse signals? Note that such an approach is analogous to the ICA paradigm, being the difference that ICA criteria are based on the recovery of statistical independence.

To answer that question, we consider the extraction of a single source as follows

$$y(k) = \mathbf{w}^T \mathbf{x}(k) \tag{3}$$

where $\mathbf{w}^T$ is a row vector that will be adjusted according to a sparseness criterion and, without loss of generality, $\mathbf{x}(k)$ is a zero-mean process. A first attempt here is to consider the ℓ_1 norm, a common way to measure sparseness. Thus, for a given signal $g(k)$, the "sparseness measure" Ψ_g is defined as

$$\Psi_g = \sum_k |g(k)| \tag{4}$$

In order to gain some insight into (4), let us express it as function of the ℓ_1 norm of the sources. This can be done as follows

$$\Psi_y = \sum_k |y(k)| = \sum_k \left|\mathbf{w}^T \mathbf{x}(k)\right| = \sum_k \left|\mathbf{w}^T \mathbf{A}\mathbf{s}(k)\right| = \sum_k \left|\mathbf{h}^T \mathbf{s}(k)\right| \tag{5}$$

where $\mathbf{h}^T = \mathbf{w}^T\mathbf{A}$ represents the global mapping between the sources and the estimated signal. By using the triangle inequality in (5), we first obtain

$$\Psi_y = \sum_k \left|\sum_j h_j s_j(k)\right| \leq \sum_k \sum_j \left|h_j s_j(k)\right| \tag{6}$$

where equality in (6) holds if at most one source has a non-zero value for each instant k. When this condition is observed, the sources are said to possess disjoint orthogonality [3]. Under such assumption we have

$$\Psi_y = \sum_k \sum_j \left|h_j s_j(k)\right| = \sum_j \left|h_j\right| \sum_k \left|s_j(k)\right| = \sum_j \left|h_j\right| \Psi_{s_j} \tag{7}$$

Henceforth, in order to deal with the fact that the ℓ_1 norm is not invariant with respect to scale, we impose $\|\mathbf{h}\|_2 = 1$, which corresponds to assuming that the source and the recovered signals have the same power. Note that, under such hypothesis, the ℓ_1 norm of $y(k)$ becomes the ℓ_1 norm of the projections of the sources onto $\mathbf{h}$.

Let us now consider that the i-th source is the sparsest one, i.e., $\Psi_{si} \geq \Psi_{sj}$ for all j. In such a situation, we have

$$\Psi_y = \sum_j |h_j| \Psi_{s_j} \geq \Psi_{s_i} \sum_j |h_j| = \Psi_{s_i} \|\mathbf{h}\|_1 \geq \Psi_{s_i} \|\mathbf{h}\|_2 = \Psi_{s_i} \tag{8}$$

where the first inequality comes again from the triangle inequality while the second one is a result of the equivalence of norms [5]. Note that Equation (8) points out that Ψ_{si} is a lower bound for Ψ_y. Moreover, since the equality in (8) is attained for $\mathbf{h} = \mathbf{e}_i$, where $\mathbf{e}_j = 0$ for $i \neq j$ and $\mathbf{e}_i = 1$, we conclude that the minimum of Ψ_y is given by Ψ_{si} and, in this situation, $y(k) = s_i(k)$.

In order to illustrate the result given by (8), we consider a scenario with two mixtures and two sources, being mixed by an orthonormal matrix **A**. The separating vector in this case, **w**, has unit length and, therefore, can be represented by a polar angle Θ_w. In this case, the optimal $\mathbf{h}=\mathbf{e}_i$ is attained when **w** is orthogonal to all the columns of **A** but the one associated with the sparsest signal.

In Fig. 1, we show the values of Ψ_y - Ψ_{si} versus Θ_w. In this example, the direction of the column of **A** associated with the sparsest source is $-\pi/6$, while the direction of the second column is given by $\pi/3$. Note that the difference Ψ_y - Ψ_{si} is null when the separating vector has a direction given by $\Theta_w = \pi/3$, and, thus, is indeed orthogonal to the least sparse source.

If we look again to Fig. 1, we will notice that there is another local minimum in the function, and that it corresponds to the solution **w** that extracts the other source. This fact indicates that all minima of Ψ_y are related to the columns of the mixing matrix. An analysis for the 2-dimensional case is provided in the following.

3.1 Analysis of Minima in 2-Dimensional Cases

Let us now discuss in more detail the case of m sources and 2 mixtures. In our analysis, we maintain the assumption of disjoint orthogonal sources. In such a scenario, it can be shown that

$$\Psi_y = \sum_k \left| \sum_j \mathbf{w}^T \mathbf{a}_j s_j(k) \right| = \sum_k \sum_j \left| \mathbf{w}^T \mathbf{a}_j s_j(k) \right| = \|\mathbf{w}\|_2 \sum_j \|\mathbf{a}_j\|_2 \left| \cos(\theta_w - \theta_j) \right| \Psi_{s_j} \tag{9}$$

where $\mathbf{a}_j$ denotes the j-th column of **A,** Θ_j the angle of $\mathbf{a}_j$, and Θ_w the angle of **w**. We observe that, for each $\mathbf{a}_j$, there is a vector **w** with angle $\Theta_w = \Theta^*_j$ that is orthogonal to $\mathbf{a}_j$, that is, $\cos(\Theta^*_j - \Theta_j) = 0$. It is important to notice that such **w** vector can be used to eliminate the signal associated to $\mathbf{a}_j$ from the mixture of signals. Another interesting fact here is that Ψ_y is non differentiable only at these angles. Moreover, it can be shown that for all $\Theta_w \neq \Theta^*_j$, one has $\partial^2\Psi_y/\partial\Theta_w^2<0$, which means that there is no minimum for all $\Theta_w \neq \Theta^*_j$. Therefore, we can conclude that all local minima of Ψ_y correspond to vectors **w** with $\Theta_w = \Theta^*_j$, for some j. In other words, **w** is orthogonal to one of the columns of **A** at these points. Consequently, one can estimate $\mathbf{a}_j$ vectors by finding all minima of Ψ_y. Moreover, in the case of 2 sources and 2 mixtures, this fact allows us to perform the separation perfectly.

3.2 Beyond Disjoint Orthogonality

In practice, we cannot guarantee the disjoint orthogonality of the source signals. However, we present in this section a set of simulations indicating that even if there is some overlap between source signals, it may still be possible to extract the sparsest source. In these numerical experiments, the mixtures undergo a whitening pre-processing step so $\mathbf{h} = \mathbf{w}^T\mathbf{TA}$, where $\mathbf{T}$ is a whitening matrix. Moreover, we impose $\|\mathbf{w}\|_2=1$. These two requirements are related to the assumption that $\|\mathbf{h}\|_2=1$. Of course, imposing $\|\mathbf{h}\|_2=1$ is impossible since $\mathbf{A}$ is not known in advance. However, whitening the data and imposing $\|\mathbf{w}\|_2=1$ leads to a constant $\|\mathbf{h}\|_2$, for which the derivation done previously is also valid.

In our numerical experiments, we considered sources that were sparse but not orthogonally disjoint. In a first moment, we tested different scenarios in which the number of sparse sources varied from 2 to 15, and the matrix $\mathbf{A}$ was randomly generated. The number of mixtures was the same as the number of sources. In order to conduct this procedure, we assumed that the properties of Ψ_y could be approximated to the one presented in Section 3.1 so that we would be able to separate a source from the mixture by finding a $\mathbf{w}$ that locally minimizes Ψ_y. In order to perform the optimization process we applied a bio-inspired algorithm called opt-aiNet [6], which is known to provide good heuristic solutions in problems with multimodal objective functions.

In Fig. 2 (a), we show the mean value of the signal-to-interference ratio (SIR) for 35 different simulations associated with the sources obtained from the optimization of Ψ_y via the opt-aiNet. The results showed that it was possible to extract a source from the mixture in all cases. In other words, the extraction process was successful in reaching the neighborhood of a minimum that led to the recovery of a sparse source.

We also conducted some simulations in a case with more sources than mixtures. In Fig. 2 (b), we plot the function Ψ_y in an example where there are 2 mixtures and 15 sources with Θ_j equally spaced. It is interesting to note that the 15 minima present in this figure are associated with the angles Θ_w that are orthogonal to each column of $\mathbf{A}$, similarly to the results in Section 3.1, derived for the disjoint orthogonal case. With the angles Θ_w at hand, it becomes possible to estimate the mixing matrix $\mathbf{A}$, in an analogous manner as done by SCA techniques in sub-parametrized cases.

3.3 Beyond Sparseness

So far, we showed that the ℓ_1 norm can be used as a cost function to separate sparse signals which have disjoint orthogonality. Furthermore, in practical cases with sparse sources that are non-disjoint, the same results can be used to perform source extraction and mixing matrix estimation. In contrast to the sparse case, we show now the behavior of the ℓ_1 norm at a system with non-sparse sources.

Given that, let us consider a system with two mixtures and two sources, which, in our case, were generated by a uniform distribution. Fig. 3 (a) depicts the sample space of the sources, which, in this case, is a square centered in the origin.

First we notice that, in equation (5), $\Psi_y \rightarrow n.\mathrm{E}\{|\mathbf{h}^\mathrm{T}\mathbf{s}|\}$ for a large n. Without loss of generality, we consider an orthonormal mixing matrix. Using Fig. 3(a) it is possible to conclude that E{|$\mathbf{h}^T\mathbf{s}$|} can be evaluated by Equation (10), for $0 \le \Theta \le \pi/4$, in which A is the area of the sample space and the integration is carried out over the shaded area.

$$E\{|\mathbf{h}^T\mathbf{s}|\} = \frac{2}{A}\iint x_1\cos\theta + x_2\sin\theta dA = \frac{2}{3}\sin\theta\tan\theta + \frac{1}{2}\cos\theta(1-\tan^2\theta) \quad (10)$$

Fig. 3 (b) shows the result of equation (10), where the maximum value is found at Θ = 0 (the direction of one of the sources) and the minimum value when $\Theta = \pi/4$. We than notice that in this case it is necessary to look for the maximum value of the contrast function Ψ_y to separate the sources. Intuitively, this means that we are now minimizing the sparseness of the resulting signal.

By comparing these results, we notice that, in both cases, separation is achieved at points where Ψ_y attains its extreme values. In the sparse signal case, separation is achieved at the minimum of Ψ_y, and in the uniformly-distributed signal case, it is achieved at the maximum.

To understand this fact, we will now study now some intermediate cases by using signals generated with the generalized Gaussian distribution described by

$$p(x) = \frac{\beta}{2\alpha\Gamma(1/\beta)} e^{-(|x-\mu|/\alpha)^\beta} \quad (11)$$

When $\beta = 1$, $p(x)$ describes a Laplacian distribution, which is considered sparse [7], and the uniformly-distributed case is achieved for $\beta \to \infty$.

Fig. 4 (a) shows the ℓ_1 norm of unit power vectors generated using (11) for $1<\beta<4$, $\alpha = 1$ and $\mu = 0$. As expected, as the signal becomes less sparse, the ℓ_1 norm increases, and the smallest value, as expected, is achieved when $\beta = 1$.

In order to verify the effects of the ℓ_1 norm in source separation, one scenario with 2 sources and 2 mixtures was analyzed. The mixture matrix **A** is orthonormal, and the sources were generated using (11) with the same parameters used before. Fig. 4 (b) shows the SIR values for the cases of minimizing and maximizing the ℓ_1 norm. We observed that the extraction was performed by minimizing the ℓ_1 norm for $\beta < 1.9$ and maximizing the norm when $\beta > 2.1$. Between these two values, the separation is not performed correctly, and the worst case, when both approaches have the same results, occurs for $\beta = 2$, when (11) represents a Gaussian distribution.

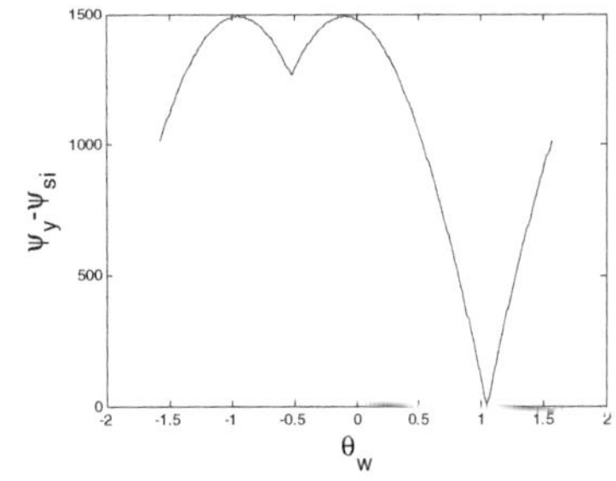

Fig. 1. Value of Ψ_y - Ψ_{si} versus Θ_w. When **w** is orthogonal to the least sparse source, the extraction is done.

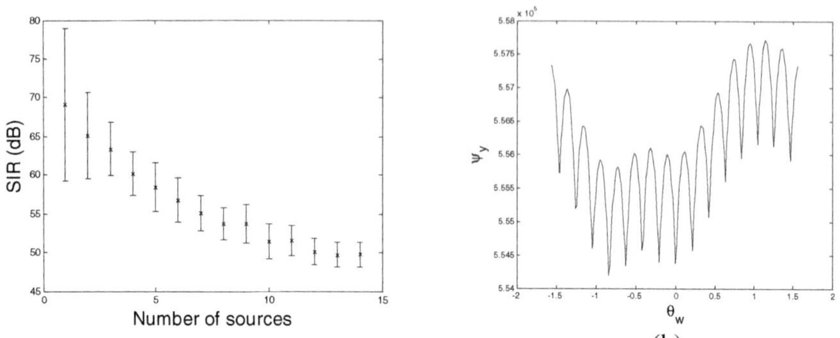

(a) (b)

Fig. 2. (a) SIR value of the extracted source in a determined case, with mixtures varying between 2 and 15. **(b)** Value of Ψ_y versus Θ_w.

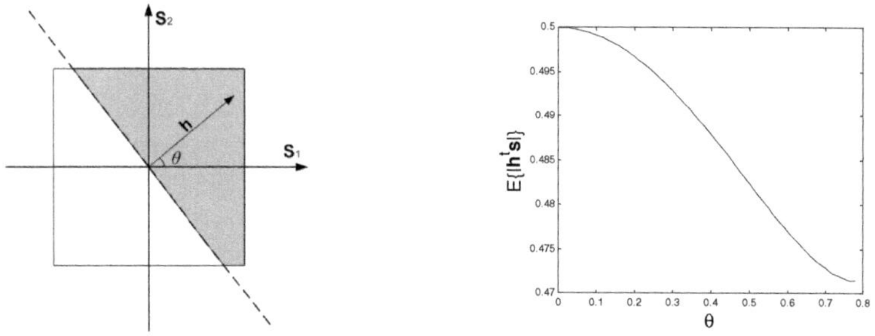

(a) (b)

Fig. 3. (a) The square centered in the origin describes the sample space of the random variable **s** and the shaded area depicts the region where the double integral in (10) is calculated. **(b)** E{|$\mathbf{h}^T\mathbf{s}$|} versus Θ.

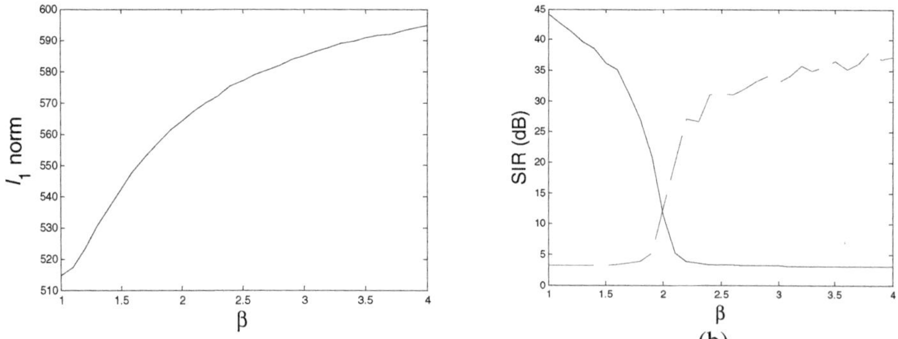

(a) (b)

Fig. 4. (a) Value of ℓ_1 norm of unit power vectors generated using (11) for 1<β<4. **(b)** SIR of extracted sources versus β in a scenario with two sources and two mixtures.

4 Conclusions

The fact that all source signals present some degree of sparseness creates new possibilities to solve complex in BSS problems, such as the undetermined BSS problem. The approach is called Sparse Component Analysis (SCA), but it was verified that, in practice, most works up to this date do not directly search for sparse components in the mixture, but solely rely on the assumption of sparseness of the sources. We showed that it is possible to separate the sources in the determined case by applying an optimization process that uses a measure of sparseness, such as the ℓ_1 norm, as a cost function. We also showed that, for underdetermined cases, the ℓ_1 norm can be also be used to estimate the mixture matrix. In further works, we intend to use other kinds of measures of sparseness, and to construct a method of source separation based on the search for sparse components in the underdetermined case.

Acknowledgments. The authors would like to thank CAPES, CNPq and FAPESP for the financial support, and Renato R. Lopes and João M. T. Romano for their comments and suggestions.

References

1. Comon, P.: Independent component analysis, A new concept? Signal Processing 36, 287–314 (1994)
2. Georgiev, P., Theis, F., Cichocki, A.: Sparse component analysis and blind source separation of underdetermined mixtures. IEEE Transactions on Neural Networks 16(4), 992–996 (2005)
3. Bofill, P., Zibulevsky, M.: Blind separation of more sources than mixtures using sparsity of their short-time Fourier transform. In: Proceedings of the ICA 2000, pp. 87–92 (2000)
4. Rickard, S.: Sparse sources are separated sources. In: Proceedings of the 16th Annual European Signal Processing Conference, Florence, Italy (2006)
5. Boyd, S., Vandenberghe, L.: Convex Optimization. Cambridge University Press, New York (2004)
6. de Castro, L.N., Timmis, J.: An artificial immune network for multimodal function optimization. In: Proc. of IEEE International Conference on Evolutionary Computation, Honolulu, USA, pp. 699–704 (2002)
7. Fevotte, C., Godsill, S.J.: A Bayesian Approach for Blind Separation of Sparse Sources. IEEE Trans. on Audio, Speech and Language Processing 14(6), 2174–2188 (2006)

Blind Extraction of Intermittent Sources

Bertrand Rivet[1], Leonardo T. Duarte[2], and Christian Jutten[1,⋆]

[1] GIPSA-Lab , CNRS UMR-5216, University of Grenoble
Domaine Universitaire, BP 46, 38402 Saint Martin d'Hères cedex, France
[2] DSPCom-Lab, DMO-FEEC, University of Campinas (Unicamp)
Caixa Postal 6101, CEP 13083-970 Campinas, Brazil

Abstract. In this work, we tackle the problem of blind extraction of intermittent sources. Our approach is based on the generalized eigenvector decomposition of covariance matrices and extends previous works in two aspects: by developing a more precise technique to detect inactive periods and by building a more general yet more precise strategy to estimate the vectors that lead to the separation of the intermittent sources. Simulations are carried out to illustrate the effectiveness of our proposal.

Keywords: Blind source separation, intermittent sources, second-order methods, generalized eigenvector decomposition.

1 Introduction

Blind source separation (BSS) concerns the retrieval of a set of signals (sources) by considering only mixed versions of these original sources. When a linear model is assumed, the mixtures $\mathbf{x}(t) = [x_1(t), \cdots, x_N(t)]^T$ are linked to the unknown sources $\mathbf{s}(t) = [s_1(t), \cdots, s_M(t)]^T$ by

$$\mathbf{x}(t) = A\mathbf{s}(t), \tag{1}$$

where $A \in \mathbb{R}^{N\times M}$ is the unknown mixing matrix. When $N \geq M$, independent component analysis (ICA) [9,3,2] methods can be employed to perform source separation in (1). In short, ICA, which works under the assumption that the sources are statistically mutually independent, looks for a separating matrix B that makes the retrieved sources $\mathbf{y}(t) = B\mathbf{x}(t)$ as independent as possible. Alternatively, in second-order methods, BSS is accomplished by exploiting the time structure of the sources (coloration [1] or non-stationarity [12]). In these approaches, as well as in ICA, neither the scaling of the sources nor their original order can be identified.

More recent studies in BSS have been suggesting that it is possible to obtain a better performance or even tackle underdetermined cases by considering prior information that are not present in the basic ICA and second-order methods framework. For instance, one can make use of the fact that the sources can be represented, possibly in a transformed domain, by sparse signals [8].

⋆ Christian Jutten is also with Institut Universitaire de France.

V. Vigneron et al. (Eds.): LVA/ICA 2010, LNCS 6365, pp. 402–409, 2010.

In this paper, we address the extraction of intermittent sources. This assumption, which is closely related to non-stationarity and to sparsity, is based on the observation that the sources are inactive in some windows, possible in a transformed domain —such an hypothesis holds, for instance, when representing speech signals in the temporal or time-frequency domains [13,4], or chemical signals in the frequency domain [5,6]. Source separation of intermittent sources can be carried out by means of a generalized eigenvalue decomposition (GEVD) of two covariance matrices. A first attempt in this direction could be found in [11]. However, this method does not explain how to choose practically these two covariance matrices. To overcome this difficulty one may use a joint diagonalization of several covariance matrices [12]. Nevertheless, there is an overhead in this later method as it separates all the sources and not only the intermittent sources.

More recent studies [13,6] proposed methods based on the GEVD that are specially tailored for the extraction of intermittent sources. The method introduced in the present work extends [13,6] in two aspects: 1) by improving the detection of inactivity periods and 2) by exploiting in a better way these inactivity periods (*i.e.* without the need of a deflation procedure as in [6] or the restriction that, to separate a given source, say s_i, there should be a period when only s_i is inactive [13]). This article is organized as follows. Section 2 presents the proposed approach to exploit the inactivity periods of intermittent sources. The proposed algorithm is introduced in Section 3. Numerical experiments and results are given in Section 4 before conclusion and perspectives in Section 5.

2 Basics

In this section, the principles underlying the proposed method to extract intermittent sources are presented. It is based on a second-order framework. Let consider the linear instantaneous mixing model (1) with as many observations as sources ($M = N$). Also, let us represent the covariance matrix of mixtures $\mathbf{x}(t)$ at sample t by

$$R_{\mathbf{x}}(t) \triangleq E[\mathbf{x}(t)\mathbf{x}(t)^T] = \sum_{i=1}^{N} \sigma_i^2(t)\mathbf{a}_i\mathbf{a}_i^T, \tag{2}$$

where $\sigma_i^2(t) = E[s_i(t)^2]$ is the power of the i-th source at sample t and $\mathbf{a}_i$ is the i-th column of mixing matrix $A = [\mathbf{a}_1, \cdots, \mathbf{a}_N]$.

The proposed method is based on the assumption that there exist some samples where at least one source is inactive: *i.e.* for $t = \tau$, $\exists n \;/\; s_n(\tau) = 0$. Let suppose in this section that all the sources are stationary excepted N_1 sources, say $s_1(t), \cdots, s_{N_1}(t)$ without loose of generality. Moreover, we assume that the first N_1 sources are simultaneously inactive at sample τ: *i.e.* for $t = \tau$, $1 \leq i \leq N_1$, $s_i(\tau) = 0$. Therefore the covariance matrix of observations $\mathbf{x}(t)$ at sample τ can be written as

$$R_{\mathbf{x}}(\tau) = \sum_{i=N_1+1}^{N} \sigma_i^2\mathbf{a}_i\mathbf{a}_i^T, \tag{3}$$

where for the $N - N_1$ stationary sources $\sigma_i^2 = \sigma_i^2(\tau) = \sigma_i^2(t)$. The proposed method is based on the generalized eigenvalue decomposition of the couple $(R_{\mathbf{x}}(\tau), R_{\mathbf{x}}(t))$. It is easy to show that $(R_{\mathbf{x}}(\tau), R_{\mathbf{x}}(t))$ admits only two distinct generalized eigenvalues: 0 degenerated N_1 times whose eigensubspace $\mathcal{E}_0$ is orthogonal to the space spanned by $\{\mathbf{a}_{N_1+1}, \cdots, \mathbf{a}_N\}$, and 1 degenerated $N - N_1$ times whose eigensubspace $\mathcal{E}_1$ is complementary to $\mathcal{E}_0$ in $\mathbb{R}^N$. As a consequence, the projection of the observations $\mathbf{x}(t)$ onto $\mathcal{E}_0$ can be used to cancel the contribution of the sources $s_i(t)$, $N_1 + 1 \leq i \leq N$. This means that all separation vectors $\mathbf{b}_i$, $1 \leq i \leq N_1$, lie in $\mathcal{E}_0$, where $\mathbf{b}_i$ is the i-th column of the separation matrix B. In other words, the space spanned by $\{\mathbf{b}_1, \cdots, \mathbf{b}_{N_1}\}$ is $\mathcal{E}_0$. It is worth noting that when only one source is inactive (let say $s_i(t)$), then $\mathcal{E}_0$ is unidimensional and the corresponding generalized eigenvector is aligned with $\mathbf{b}_i$.

This method allows us firstly to detect how many sources are vanishing by testing the generalized eigenvalues and then to extract the space spanned by the corresponding sources by projecting the observations onto the generalized eigenvectors associated with the generalized eigenvalues equal to zero.

3 Algorithm to Extract Intermittent Sources

The previous section discussed how to extract the space spanned by intermittent sources but not how to extract these sources. Moreover, the inactivity periods are unknown. Given that, we propose, in a first part of this section, a strategy to detect inactivity periods that extends the one proposed in [13]. Then, we show how to estimate the extraction vectors $\mathbf{b}_i$ based on the subspaces observed during the inactive periods.

3.1 Estimation of Inactivity Periods

In order to detect block samples where at least one source is inactive, we proposed to compute for different samples τ the generalized eigenvalue decomposition of couples $\{(R_{\mathbf{x}}(\tau), R_{\mathbf{x}})\}_\tau$, where $R_{\mathbf{x}}$ is the covariance matrix of observations $\mathbf{x}(t)$ estimated with all samples, and $R_{\mathbf{x}}(\tau)$ is the covariance matrix of the observations estimated on windowed samples around τ. The generalized eigendecomposition of $(R_{\mathbf{x}}(\tau), R_{\mathbf{x}})$ provides

$$R_{\mathbf{x}}(\tau)V(\tau) = R_{\mathbf{x}}V(\tau)\Lambda(\tau), \tag{4}$$

where $\Lambda(\tau)$ is a diagonal matrix whose diagonal entries $\lambda_1(\tau) \leq \cdots \leq \lambda_N(\tau)$ are the generalized eigenvalues and $V(\tau)$ is an orthonormal matrix whose columns $\mathbf{v}_i(\tau)$ are the generalized eigenvectors. Let us define g_k such that

$$g_k(\tau) = \prod_{i=1}^{k} f_0(\lambda_i(\tau)) \prod_{j=k+1}^{N} f_1(\lambda_j(\tau)), \tag{5}$$

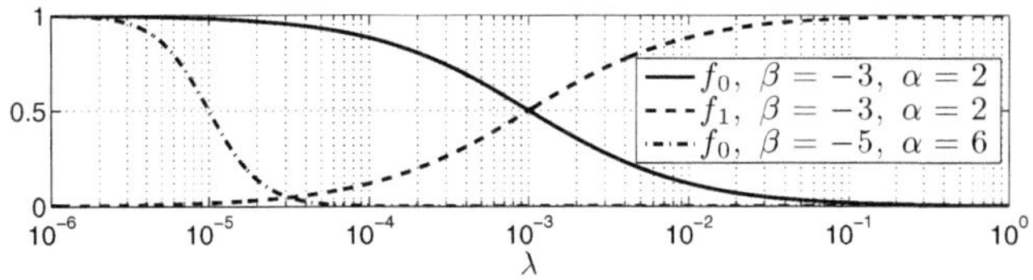

Fig. 1. Detection of inactive sources

where

$$f_0(\lambda) = 1 - \frac{1}{1+\exp(-\alpha(\log(\lambda) - \beta))} \quad \text{and} \quad f_1(\lambda) = \frac{1}{1+\exp(-\alpha(\log(\lambda) - \beta))}.$$

Note that these functions are sigmoids (Fig. 1): if $\log(\lambda) - \beta$ is large compared to $1/\alpha$ then $f_0(\lambda)$ (resp. $f_1(\lambda)$) is about 0 (resp. 1). Accordingly,

$$\hat{k}(\tau) = \arg_k g_k(\tau) > .5$$

is an estimation of the number of inactive sources at sample τ (*i.e.* less powerful than 10^β times their average power), and the subspace $\mathcal{E}_0(\tau, \hat{k}(\tau))$ spanned by $\{\mathbf{v}_1(\tau), \cdots, \mathbf{v}_{\hat{k}(\tau)}(\tau)\}$ is also spanned by $\hat{k}(\tau)$ of the separation vectors $\mathbf{b}_i$. Finally, the identification of inactivity periods $\Theta = \{\tau \mid g_k(\tau) > .5\}$ also provides couples of weight and subspace[1] $\{(g_{\hat{k}(\tau)}(\tau), \mathcal{E}_0(\tau, \hat{k}(\tau))\}$. We shall discuss later how to make use of these weights to improve the estimation of the extraction vectors $\mathbf{b}_i$.

3.2 Estimation of Extraction Vectors

Once the inactivity periods Θ are detected by the previous step, the set of subspaces $\{\mathcal{E}_0(\tau, \hat{k}(\tau))\}_{\tau\in\Theta}$ is used to estimate the separation vectors $\mathbf{b}_i$. It is worth noting that when $\hat{k}(\tau) = 1$ then the corresponding $\mathbf{v}_1(\tau)$ is directly align with one of the separation vectors $\mathbf{b}_i$. However, when $\hat{k}(\tau) \neq 1$, then k separation vectors $\mathbf{b}_i$ lie in the k-dimensional subspace $\mathcal{E}_0(\tau, \hat{k}(\tau))$ but are not necessary align with the eigenvectors $\mathbf{v}_i(\tau)$. To overcome this difficulty, our proposal considers the method for finding the intersection of subspaces described in [7] (see also Appendix). It is interesting to observe that, if $\dim(\mathcal{E}_0(\tau, k) \cap \mathcal{E}_0(\tau', k')) = 1$, then the support vector $\mathbf{u}_{\tau,\tau'}$ lying in the intersection between these two subspaces must be aligned with one of the separation vector $\mathbf{b}_i$. Accordingly searching all the intersections between $\mathcal{E}_0(\tau, \hat{k}(\tau))$ and $\mathcal{E}_0(\tau', \hat{k}(\tau'))$, with $(\tau, \tau') \in \Theta^2$, such that $\dim(\mathcal{E}_0(\tau, \hat{k}(\tau)) \cap \mathcal{E}_0(\tau', \hat{k}(\tau'))) = 1$ provides a set of support vectors

$$\mathcal{U} = \Big\{\mathbf{u}_{\tau,\tau'} \mid \dim\Big(\mathcal{E}_0(\tau, \hat{k}(\tau)) \bigcap \mathcal{E}_0(\tau', \hat{k}(\tau'))\Big) = 1\Big\}. \tag{6}$$

[1] In this article, by sake of simplicity we do not make the difference between a subspace and its matrix representation.

Note, however, that more support vectors than separation vectors might be identified. Therefore, in an ideal situation, some separation vectors are repeated. Of course, when there is some noise in the observed data or when the detection of inactivity periods is not completely perfect, which is usually the case, it is expected that these repeated vectors be actually concentrated around the optimum direction. Therefore, our method is completed by a clustering stage whose goal is exactly to estimate the separation vectors $\mathbf{b}_i$ used in the extraction of the intermittent sources.

In this study, we consider kernel-PCA [10] for performing the clustering. The chosen kernel is

$$\psi(\mathbf{u}_{\tau_1,\tau_2}, \mathbf{u}_{\tau_3,\tau_4}) = \begin{cases} \dfrac{|\mathbf{u}_{\tau_1,\tau_2}^T \mathbf{u}_{\tau_3,\tau_4}| - \cos(\theta_0)}{1 - \cos(\theta_0)}, & \text{if } |\mathbf{u}_{\tau_1,\tau_2}^T \mathbf{u}_{\tau_3,\tau_4}| \geq \cos(\theta_0) \\ 0, & \text{else} \end{cases} \tag{7}$$

where $\mathbf{u}_{\tau_i,\tau_j} \in \mathcal{U}$. θ_0 is an *a priori* chosen angle which defines the minimum angle between two separation vectors. In practice, the accuracy of support vector $\mathbf{u}_{\tau_i,\tau_j}$ depends on the values of the corresponding eigenvalues $\lambda_l(\tau_i)$ and $\lambda_l(\tau_j)$. It can be shown from performance analysis that the accuracy of $\mathcal{E}_0(\tau_i, \hat{k}(\tau_i))$ increases as the related eigenvalues $\lambda_1(\tau_i), \cdots, \lambda_{\hat{k}(\tau_i)}(\tau_i)$ decrease to zero. In view of the observation of the last paragraph, we propose the following weighted version of the kernel (7)

$$\psi'(\mathbf{u}_{\tau_1,\tau_2}, \mathbf{u}_{\tau_3,\tau_4}) = w_{\tau_1,\tau_2}\, w_{\tau_3,\tau_4}\, \psi(\mathbf{u}_{\tau_1,\tau_2}, \mathbf{u}_{\tau_3,\tau_4}), \tag{8}$$

where w_{τ_i,τ_j} is a measure of the inactivity accuracy defined by

$$w_{\tau_i,\tau_j} = \sqrt{g_{k_i}(\tau_i) g_{k_j}(\tau_j)}. \tag{9}$$

Note here that, by proceeding this way, the support vectors $\mathbf{u}_{\tau_i,\tau_j}$ that are associated with lower eigenvalues through $g_{\hat{k}(\tau)}$ are somehow more important in the clustering step. The kernel PCA consists in performing an eigenvalue decomposition of matrix $\Psi' \in \mathbb{R}^{\mathrm{card}(\mathcal{U}) \times \mathrm{card}(\mathcal{U})}$ whose entries are $\psi'(\mathbf{u}_{\tau_1,\tau_2}, \mathbf{u}_{\tau_3,\tau_4})$:

$$\Psi' = \Phi \Delta \Phi^T, \tag{10}$$

where Δ is a diagonal matrix of eigenvalues and Φ is an orthonormal matrix whose columns are eigenvectors of Ψ'. Let $W = [\phi_1, \cdots, \phi_n]$ be the matrix of the n eigenvectors related to the n largest eigenvalues. The extraction matrix $B \in \mathbb{R}^{n \times N}$ is then estimated by

$$B = W^T \Psi' U^T, \tag{11}$$

where U is the matrix obtained by the concatenation of all support vectors in $\mathcal{U}$. Note that the exact number n of intermittent sources has not to be known *a priori* since it can be estimated by checking the eigenvalues $\Delta_{i,i}$ (10). The intermittent sources are finally estimated thanks to

$$\hat{\mathbf{s}}(t) = B\mathbf{x}(t), \tag{12}$$

for all samples t, including those when the intermittent sources are active. The final algorithm is summarized in Algo. 1.

Algorithm 1. Proposed algorithm

{Estimation of the inactivity periods}
  Compute covariance matrix $R_{\mathbf{x}}$ from all samples
  for each τ **do**
    Compute covariance matrix $R_{\mathbf{x}}(\tau)$ from sliding windows centered on τ
    Compute generalized eigenvalue decomposition (4) and for $1 \leq k \leq N$, $g_k(\tau)$ (5)
    Estimate the number $\hat{k}(\tau)$ of inactive sources thanks to $\hat{k}(\tau) = \arg_k g_k(\tau) > .5$
    $\Rightarrow \left(g_{\hat{k}(\tau)}(\tau), \mathcal{E}_0(\tau, \hat{k}(\tau)\right)$
  end for
  Estimate the set of inactivity periods by $\Theta = \{\tau \mid \hat{k}(\tau) \geq 1\}$

{Estimation of the subspaces intersections}
  for each $1 \leq i, j \leq \text{card}(\Theta)$ **do**
    Dimension of the intersection: $d(i,j) = \dim(\mathcal{E}_0(\tau_i, \hat{k}(\tau_i)) \cap \mathcal{E}_0(\tau_j, \hat{k}(\tau_j)))$
  end for
  Estimate the set of support vectors $\mathcal{U}$ (6) such that $d(i,j) = 1$

{Estimation of extraction vectors}
  Compute weighted kernel PCA of $\{(\mathbf{u}_{i,j}, w_{i,j})\}$ (7) to (11)
  Estimate intermittent sources by $\hat{\mathbf{s}}(t) = B\mathbf{x}(t)$ (12)

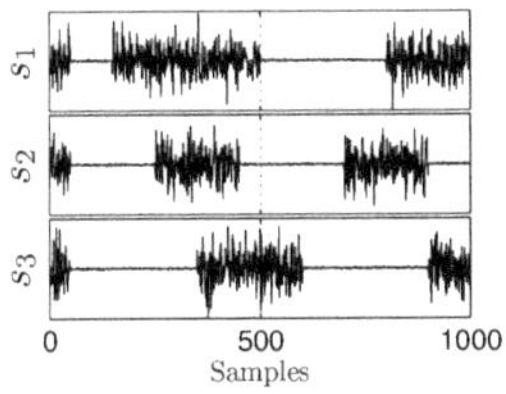

(a) Actual sources

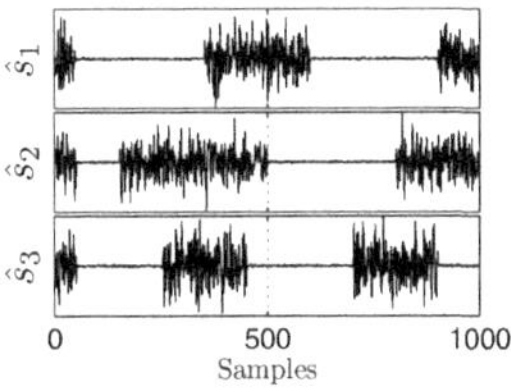

(b) Estimated sources

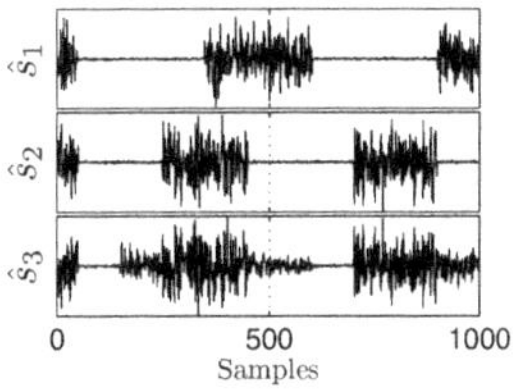

(c) Estimated sources [13]

Fig. 2. Illustration of the proposed methodology

4 Numerical Experiments

The first experiment illustrates the influence of the new methodology to estimate extraction vectors (Fig. 2). Three artificial intermittent sources (Fig. 2(a)) are linearly mixed. As one can see, the three sources are chosen so that the first one is never the only inactive source. In this case, our previous method [13] (Fig. 2(c)) has failed to extract the first actual source: the third estimated source is still a mixture of several actual sources. On the contrary, the proposed methodology (Fig. 2(b)) has succeeded to extract all the three sources without deflation.

The second experiment compares the extraction of 5 speech signals from 70 mixtures of audio signals by the proposed algorithm and a more classical estimation of the separation matrix by joint-diagonalization of covariance matrices based on non stationarity [12] (refered as SONS). In this experiment, the results

Table 1. Comparison of speech sources extraction

	Proposed method	SONS [12]
PI_{dB} (13)	-34	-41
Time [s]	1.1	847

are averaged over 50 randomly chosen configurations: the 5 speech signals, the 65 audio (musical) signals and the mixing matrix are randomly chosen. All the signals are sampled at 16kHz. The time sliding windows have a 40ms length with an overlap of 50%. To evaluate the estimation of the extraction vectors $\mathbf{b}_i$, we have used the performance index defined by

$$PI_{dB} = 10 \log \left(\frac{1}{\text{card}(\mathcal{S})} \sum_{i \in \mathcal{S}} \left(\sum_j \left| \frac{C_{i,j}}{\max_k |C_{i,k}|} \right|^2 - 1 \right) \right), \text{ with } C = B\,A, \quad (13)$$

where $\mathcal{S}$ denotes the set of speech sources. So the smaller the performance index is, the better the extraction is. As one can see (Tab 1), the proposed method is slightly less performant than the SONS method while keeping quite good performance. Indeed, it only exploits a part of the signal while the SONS used the overall signals. However, it is worth noting that the proposed method has a less computational coast.

5 Conclusions and Perspectives

In this paper, we introduced an algorithm to extract intermittent sources from linear mixtures. It is based on second order statistics: the detection of inactivity periods allows to estimate the separation matrix which is used to extract the intermittent sources when they are active. Simulations in different configurations pointed out that our proposal is efficient and presents a low computational cost. Even if in this study the purpose was to extract speech signals, the proposed algorithm can be used in a more general context. Future works include the derivation of an automatic strategy for adjusting parameters. Moreover, a more robust clustering algorithm will also be studied. Finally, this proposed method could be used as an initialization of joint diagonalization of covariance matrices to only extract intermittent sources.

References

1. Belouchrani, A., Abed-Meraim, K., Cardoso, J.-F.: A blind source separation technique using second-order statistic. IEEE Transactions on Signal Processing 45(2), 434–444 (1997)
2. Cardoso, J.-F.: Blind signal separation: statistical principles. Proceedings of the IEEE 86(10), 2009–2025 (1998)

3. Comon, P.: Independent component analysis, a new concept? Signal Processing 36(3), 287–314 (1994)
4. Deville, Y., Puigt, M.: Temporal and time-frequency correlation-based blind source separation methods. Part I: Determined and underdetermined linear instantaneous mixtures. Signal Processing 87(3), 374–407 (2007)
5. Duarte, L.T., Jutten, C., Moussaoui, S.: A bayesian nonlinear source separation method for smart ion-selective electrode arrays. IEEE Sensors Journal 9(12), 1763–1771 (2009)
6. Duarte, L.T., Rivet, B., Jutten, C.: Blind extraction of smooth signals based on a second-order frequency identification algorithm. IEEE Signal Processing Letters 17(1), 79–82 (2010)
7. Golub, G.H., Van Loan, C.F.: Matrix Computation, 3rd edn. Johns Hopkins University Press, Baltimore (1996)
8. Gribonval, R., Lesage, S.: A survey of Sparse Component Analysis for Blind Source Separation: principles, perspectives, and new challenges, Bruges, pp. 323–330 (April 2006)
9. Jutten, C., Hérault, J.: Blind separation of sources. Part I: An adaptive algorithm based on a neuromimetic architecture. Signal Processing 24(1), 1–10 (1991)
10. Müller, K.-R., Mika, S., Rätsch, G., Tsuda, K., Schölkopf, B.: An introduction to kernel-based learning algorithms. IEEE Transactions on Neural Networks 12(2), 181–201 (2001)
11. Parra, L., Sajda, P.: Blind source separation via generalized eigenvalue decomposition. Journal of Machine Learning Research 4, 1261–1269 (2003)
12. Pham, D.-T., Cardoso, J.-F.: Blind separation of instantaneous mixtures of nonstationary sources. IEEE Transactions on Signal Processing 49(9), 1837–1848 (2001)
13. Rivet, B.: Blind non-stationnary sources separation by sparsity in a linear instantaneous mixture. In: Adali, T., Jutten, C., Romano, J.M.T., Barros, A.K. (eds.) ICA 2009. LNCS, vol. 5441, pp. 314–321. Springer, Heidelberg (2009)

Appendix: Intersection of Subspaces

This appendix summarizes some useful considerations about intersection of subspaces [7]. Let $\mathcal{S}_1$ and $\mathcal{S}_2$ be two subspaces in $\mathbb{R}^m$. The principal angles θ_k between subspaces $\mathcal{S}_1$ and $\mathcal{S}_2$ are defined by $\cos(\theta_k) = \max_{\mathbf{u}\in\mathcal{S}_1} \max_{\mathbf{v}\in\mathcal{S}_2} \mathbf{u}^T\mathbf{v} = \mathbf{u}_k^T\mathbf{v}_k$,subject to $\|\mathbf{u}\|_2 = \|\mathbf{v}\|_2 = 1$, $\mathbf{u}^T\mathbf{u}_i = 0$, and $\mathbf{v}^T\mathbf{v}_i = 0$, $\forall i = \{1, \cdots, k-1\}$. Let $Q_1 \in \mathbf{R}^{m\times p}$ and $Q_2 \in \mathbf{R}^{m\times q}$ be two orthonormal basis of $\mathcal{S}_1$ and $\mathcal{S}_2$, respectively. Assume that $p \geq q$. The angles between subspaces can be efficiently obtained by the singular value decomposition of $Q_1^T Q_2$:

$$Y^T\big(Q_1^T Q_2\big)Z = \mathrm{diag}(\lambda_1, \cdots, \lambda_q).$$

The principal vectors and angles are then obtained by $[\mathbf{u}_1, \cdots, \mathbf{u}_p] = Q_1 Y$, $[\mathbf{v}_1, \cdots, \mathbf{v}_q] = Q_2 Z$ and $\cos(\theta_k) = \lambda_k$, $k = \{1, \cdots, q\}$, respectively. As a consequence, if there exist some principal angles such that $\cos(\theta_k) = 1$, then they define the intersection between subspaces $\mathcal{S}_1$ and $\mathcal{S}_2$. The dimension of the intersection is thus defined by the number of principal angles such that $\cos(\theta_k) = 1$. Also, the intersection is spanned by $\{\mathbf{v}_k\}_k$ or by $\{\mathbf{u}_k\}_k$.

Dictionary Learning for Sparse Representations: A Pareto Curve Root Finding Approach

Mehrdad Yaghoobi* and Mike E. Davies

Institute for Digital Communications (IDCom),
the University of Edinburgh, EH9 3JL, UK
{m.yaghoobi-vaighan,mike.davies}@ed.ac.uk
http://www.see.ed.ac.uk/research/IDCOM

Abstract. A new dictionary learning method for exact sparse representation is presented in this paper. As the dictionary learning methods often iteratively update the sparse coefficients and dictionary, when the approximation error is small or zero, algorithm convergence will be slow or non-existent. The proposed framework can be used in such a setting by gradually increasing the fidelity of the approximation. This technique has previously been used for the convex sparse representations. It has been extended here to the non-convex dictionary learning problem by allowing the dictionary be modified.

Keywords: Sparse Inverse Problems, Dictionary Learning for Sparse Representations, Pareto Curves, Gradient Projection Method.

1 Introduction

Inverse problems are the subject of different areas in science and engineering. We name computational tomography, seismology and radar as some examples. The objective is to recover the parameters which give the observed data by applying the forward operator. As the problem is often ill-posed, we need to assume a model for the parameters to resolve the recoverability ambiguity. The sparsity model, in which we assume few parameters are non-zero, can successfully model a variety of observed natural data. The forward operator is often assumed to be linear and finite dimensional. Hence the forward operator can be presented as a *fat* matrix, which is called a dictionary and each column is called an atom[1]. The inverse problem with this setting is an NP-hard problem, but many practical algorithms have been proposed to solve it approximately or exactly in some occasions[1, 2].

When the forward operator in this setting is not given, we can use the domain knowledge to select a *good* model. The term *good* means that such sparse

* This work is supported by EU FP7, FET-Open grant number 225913. MED acknowledges support of his position from the Scottish Funding Council and their support of the Joint Research Institute with the Heriot-Watt University as a component part of the Edinburgh Research Partnership.

V. Vigneron et al. (Eds.): LVA/ICA 2010, LNCS 6365, pp. 410–417, 2010.

parameters (coefficients) are possible to find for the observed data. The domain knowledge can be incorporated as a parametric model for the dictionary [3] or the dictionary can be adapted using a sample based dictionary learning method [4–8]. In the sample based dictionary learning, the algorithm starts with an initial guess for the dictionary and gradually changes it to provide sparser representations or less model mismatch. The learning process has thus two distinct steps, solving a sparse inverse problem with the current dictionary, then updating the dictionary to reduce model mismatch with fixed coefficients[1]. When the dictionary is normal, *i.e.* a dictionary with unit norm atoms, it is recoverable if the observed data uniquely represents the dictionary up to some column permutations and atom sign flips. The uniqueness condition has been explored in [9, 10]. Aharon *et al.* intuitively suggested the K-SVD dictionary learning algorithm as a candidate for the dictionary recovery. This method has been introduced for dictionary learning for sparse approximation, where there exists some model mismatch, *i.e.* approximation error. Although good results have been reported in [7, 8], the relation between exact recovery and K-SVD has not been shown. Recently a new framework for dictionary recovery has been introduced in [11, 12], which is based on dictionary learning for ℓ_1 exact sparse representations. Gribonval *et al.*[12] shows that the generative dictionary is a local minimum of the proposed (non-convex) optimization problem with high probability, when the coefficients and the dictionary follow some distributions. A difficulty with the proposed recovery framework is that most of the dictionary learning methods are not able to work in an exact sparse representation setting. An ℓ_1 exact dictionary learning [13] has been proposed recently. Unfortunately the stability and the practical performance of this method have not been explored. Here we present a new dictionary learning method which can be used for a high fidelity sparse representation. It is based on generalizing the Pareto Curve root finding technique for ℓ_1 sparse representation[14], to a dictionary learning framework. The new framework can also be used in dictionary learning for fixed levels of fidelity sparse *approximations*.

2 Sparse Representation Using Pareto Curve Root Finding

Let $\mathbf{y} \in \mathbb{R}^m$, $\mathbf{x} \in \mathbb{R}^n$ and $f : \mathbb{R}^n \mapsto \mathbb{R}^m$ be respectively the observed data, the coefficient vector and the forward operator. As the forward operator lies in the finite dimensional space here, we represent it using a matrix $\mathbf{D} \in \mathbb{R}^{m \times n}$. The inverse problem can now be formulated as finding $\mathbf{x}$ such that $\mathbf{y} = \mathbf{Dx}$. As this problem often is ill-posed/underdetermined, *e.g* $m < n$, the solution is not unique. By assuming a sparsity model for the coefficients, the sparse inverse problem can be solved by finding the sparsest solution in $\{\mathbf{x}|\mathbf{y} = \mathbf{Dx}\}$. The convex envelope of the sparsity function, constrained to an ℓ_∞ ball, is the ℓ_1

[1] The K-SVD [7] is slightly different, as it also allows the coefficients to adapt in the dictionary update steps.

norm, which is now the most popular regularization factor for sparse coding [2]. The sparse inverse problem can now be found by minimizing the following optimization problem,

$$\min_{\mathbf{x}} \|\mathbf{x}\|_1 \quad s.t \quad \mathbf{y} = \mathbf{D}\mathbf{x}. \tag{1}$$

This optimization problem, which is called Basis Pursuit (BP), is convex and can be solved using the Linear Programing (LP) method[2]. Although LP is one the most powerful methods for *exactly* solving (1), we practically need to solve it up to few significant figures. The gradient based methods converge fast with this precision. The main issue with solving (1) is that the objective is not differentiable and the gradient projection method [15], which is an efficient method to solve convexly constrained continuously differentiable objectives, can not be applied. For the moment let $\mathbf{y} \approx \mathbf{D}\mathbf{x}$ and distance in the observation space be Euclidean, $d(\mathbf{y}, \tilde{\mathbf{y}}) = \|\mathbf{y} - \tilde{\mathbf{y}}\|_2$. A sparse approximation can be found using the following optimization problem, which is called LASSO [16],

$$\min_{\mathbf{x}} \|\mathbf{y} - \mathbf{D}\mathbf{x}\|_2^2 \quad s.t. \quad \|\mathbf{x}\|_1 \leq \tau, \tag{2}$$

where $\tau \in \mathbb{R}^+$ is the radius of the ℓ_1 ball. Let the solution of (2), for each τ, be called $\phi(\tau)$, *i.e.* $\phi(\tau) := \|\mathbf{y} - \mathbf{D}\mathbf{x}^*\|_2^2$, where $\mathbf{x}^*$ is the minimizer of (2). $\phi(\tau)$ is a non-increasing function and when $\mathbf{D}$ is full rank, the set $\mathcal{A} = \{\tau | \phi(\tau) = 0\}$ is non-empty. The *Pareto Curve* is generated by plotting $\phi(\tau)$ for $\tau \in \mathbb{R}^+$. It actually presents the optimal trade-off between the sparsity, here the ℓ_1 norm of the coefficients, and the ℓ_2 norm of the approximation error. Let τ^* be the smallest $\tau \in \mathcal{A}$. It is straight forward to show that any solution of (2) with τ^* is also a solution of (1). The good news is that (2) can efficiently be solved using the gradient projection method, for any given τ. We thus only need to find τ^*. Van den berg *et al.* showed in [14] that $\phi(\tau)$ is convex and differentiable, where $\phi(\tau) \neq 0$, and used a Newton's root finding method to iteratively update $\tau^{[n]}$, which is τ at the nth iteration, such that $\lim_{n\rightarrow\infty} \tau^{[n]} = \tau^*$. The Newton's method is guaranteed to converge with this setting, *i.e.* convex and differentiable function.

3 Dictionary Learning for ℓ_1 Exact Sparse Representation

In sample based dictionary learning, a set of training samples $\mathcal{Y} = \{\mathbf{y}_l\}_{l\in\mathcal{L}}$, where $|\mathcal{L}| = L$, is given. When L is large and the sparse signals are rich enough to uniquely define the dictionary, up to some permutations in columns of $\mathbf{D}$ and atom sign flips, Gribonval *et al.* [12] suggest to solve the following problem to recover the generative dictionary,

$$\min_{\mathbf{X},\mathbf{D}} \|\mathbf{X}\|_1 \quad s.t. \quad \mathbf{D}\mathbf{X} = \mathbf{Y}, \mathbf{D} \in \mathcal{D}, \tag{3}$$

where $\mathbf{X}$, $\mathbf{Y} = [\mathbf{y}_l]_{l\in\mathcal{L}}$ are respectively coefficient and observation matrices, $\mathcal{D}$ is the dictionary admissible set and $\|\cdot\|_1 = \sum_{i,j} |\{\cdot\}_{i,j}|$ is an element-wise one norm in the matrix space. An admissible set $\mathcal{D}$, which constrains the amplitude, has been used to resolve the scale ambiguity, which prevents the dictionary and the coefficients to scale by $\alpha > 1$ and $1/\alpha$ respectively, to reduce the sparsity penalty $\|\cdot\|_1$, while preserving the admissibility $\mathbf{DX} = \mathbf{Y}$. A common choice for $\mathbf{D}$ is a fixed column ℓ_2 norm or a bounded column ℓ_2 norm to make a convex admissible set. A similar argument which was presented earlier, about the non-differentiability of the objective in (1), can be presented here. Let $\psi(\mathbf{X}, \mathbf{D}) = \|\mathbf{Y} - \mathbf{DX}\|_F^2$, where $\|\cdot\|_F$ is the Frobenius norm. A new formulation for the dictionary learning can be presented as,

$$\min_{\mathbf{X},\mathbf{D}} \psi(\mathbf{X}, \mathbf{D}) \quad s.t. \quad \mathbf{X} \in \mathcal{B}_1^\tau, \quad \mathbf{D} \in \mathcal{D}, \tag{4}$$

where $\mathcal{B}_1^\tau := \{\mathbf{X} \mid \|\mathbf{X}\|_1 \leq \tau\}$ is the ℓ_1 ball with radius τ and $\mathcal{D}$ is a convex admissible set. Let $\phi(\tau)$ similarly be the optimum value of (4) for each τ. Although $\phi(\tau)$ is non-increasing and $\mathcal{A}$ is not empty, it might now be non-convex.

The gradient projection can be used again, as the constraints are convex sets, to solve (4) for a fixed τ. The uniform convergence of this method can be shown if $\nabla\psi$ is locally Lipschitz continuous[15]. As $\mathcal{D}$ and $\mathcal{B}_1^\tau$, for a finite τ, are compact, it is straightforward to shows locally Lipschitz continuity of $\nabla\psi$. However in practice this results in a small gradient step, which slow down the convergence of the algorithm. Instead we use a block relaxation technique and keep fixed $\mathbf{X}$ or $\mathbf{D}$ while gradient projecting the other parameter. It allows us to choose a larger gradient step size at each step. The down side of this technique is that we only can show the *convergence of* $(\mathbf{X}, \mathbf{D})$ *to a set of accumulation points*, see for example [8, Appendices A and B]. The gradient steps in directions $\frac{\partial}{\partial \mathbf{X}}\psi$ and $\frac{\partial}{\partial \mathbf{D}}\psi$ are respectively smaller than $(\sigma_{max}(\mathbf{D}))^{-2}$ and $(\sigma_{max}(\mathbf{X}))^{-2}$, where σ_{max} operator finds the largest singular value.

Another difficulty with solving (4), with a fixed τ, is that the (global) minimum might not be found using a gradient projection method. This raises a big issue in the convergence proof of the overall algorithm, where the achievable *local* minimum of $\psi(\mathbf{X}, \mathbf{D})$ might increase in the next minimization step, after increasing τ. To resolve this issue, we initiate the gradient projection algorithm by the (local) minimum found using the previous τ. The algorithm is now guaranteed to reduce the objective after each gradient projection step[15]. As the objective is lower bounded, the stability of the algorithm is guaranteed[2].

When $\phi(\tau)$ is convex, as it has been shown to be in the ℓ_1 sparse representation, Newton's method finds the root. This is not true in the dictionary learning as $\phi(\tau)$ might not be convex. We can use the non-increasing feature of ϕ, by updating $(\mathbf{X}, \mathbf{D})$ as explained earlier, and find the root by applying a line search method. Although such an update scheme for τ may not be as efficient as using Newton's method, *i.e.* more updates needed to find τ^*, we practically found that the proposed line search method in Algorithm 1 converges fast. Note that when τ gets large enough, $\phi(\tau) \to 0$ and the gradient of objective in (4), with respect to each parameter, tends to zero. This is enough to show the convergence of the algorithm to some local minima $\mathbf{X}^*$, $\mathbf{D}^*$.

[2] The stability in a Lyapunov sense, which provides boundedness of the solutions (for $\mathbf{X}$, which might become infinitely large in general).

Algorithm 1. Pareto Curve root finding based Dictionary Learning (PCDL)

initialization: $0 < \delta \ll 1$, $\mathbf{D}_\tau = \mathcal{P}_\mathcal{D}\left([d_{i,j} = \mathcal{N}(1,0)]_{i,j}\right)$, $\mathbf{X}_\tau = \mathbf{D}_\tau^\dagger \mathbf{Y}$, $\tau = \|\mathbf{X}_\tau\|_1/N,K$, $\mu = 1.5$
while $\left(\|\mathbf{D}_\tau \mathbf{X}_\tau - \mathbf{Y}\|_F^2 - \epsilon\right)^2 > .01\epsilon$ **do**
  $\mathbf{X}^{[0]} = \mathbf{X}_\tau$, $\mathbf{D}^{[0]} = \mathbf{D}_\tau$
  for $n = 1$ **to** K **do**
    $\mathbf{A} = \mathbf{X}^{[n-1]} - \frac{2}{\sigma_{max}^2\left(\mathbf{D}^{[n-1]}\right)+\delta} \mathbf{D}^{[n-1]^T}\left(\mathbf{D}^{[n-1]}\mathbf{X}^{[n-1]} - \mathbf{Y}\right)$
    $\mathbf{X}^{[n]} = \mathcal{P}_{\mathcal{B}_1^\tau}(\mathbf{A})$
    $\mathbf{B} = \mathbf{D}^{[n-1]} - \frac{2}{\sigma_{max}^2\left(\mathbf{X}^{[n]}\right)+\delta}\left(\mathbf{D}^{[n-1]}\mathbf{X}^{[n]} - \mathbf{Y}\right)\mathbf{X}^{[n]}$
    $\mathbf{D}^{[n]} = \mathcal{P}_\mathcal{D}(\mathbf{B})$
  end for
  if $\|\mathbf{D}_\tau \mathbf{X}_\tau - \mathbf{Y}\|_F^2 < \epsilon$ **then**
    $\tau = \tau/\mu$
    $\mu = \mu^{1/3}$
  else
    $\mathbf{X}_\tau = \mathbf{X}^{[K]}$, $\mathbf{D}_\tau = \mathbf{D}^{[K]}$
  end if
  $\tau = \mu\tau$
end while
output: $\mathbf{D}_\tau$

Algorithm 1 presents a pseudocode for the Pareto Curve root finding based Dictionary Learning (PCDL) method. It initiates with a random dictionary after projecting onto the admissible set $\mathcal{D}$, $\mathcal{P}_\mathcal{D}(\cdot)$. The initial $\mathbf{X}_\tau$ was selected to be the minimum ℓ_2 norm inverse solution. The algorithm starts with a τ, which is a division of the ℓ_1 norm of the current solution, *i.e.* least square solution. The "for" loop includes K iterations of gradient-projection steps. $\mathcal{P}_{\mathcal{B}_1^\tau}$ is the projection onto the ℓ_1 ball, with radius τ. The "if" part is the line search for updating τ. For a given precision ϵ, if the approximation error is less than this precision, the algorithm steps back and choose a smaller scale factor μ. Otherwise it updates $(\mathbf{X}_\tau, \mathbf{D}_\tau)$. The algorithm stops when $\left(\|\mathbf{D}_\tau \mathbf{X}_\tau - \mathbf{Y}\|_F^2 - \epsilon\right)^2 \leq .01\epsilon$, which practically seems to be an acceptable accuracy.

4 Simulations

We demonstrate the dictionary recovery with a toy example in the first experiment. A normalized (normally distributed) random dictionary $\mathbf{D} \in \mathbb{R}^{20\times 40}$ and a set of $L = 1280$ training samples are generated, where the sparsity is changing from 3 to 7 in different experiments. The locations and magnitude of non-zero coefficients are respectively selected uniformly random and bounded in $[.2, 1]$ with a random sign pattern. By following the definition of the atom recovery from [7, 8], we call an atom is recovered if the Euclidean distance of the atom and one of the recovered atoms is not greater than $0.1\sqrt{2}$. The simulations were repeated 5 times, with new initial random dictionaries, for $K = 500$. The average atom

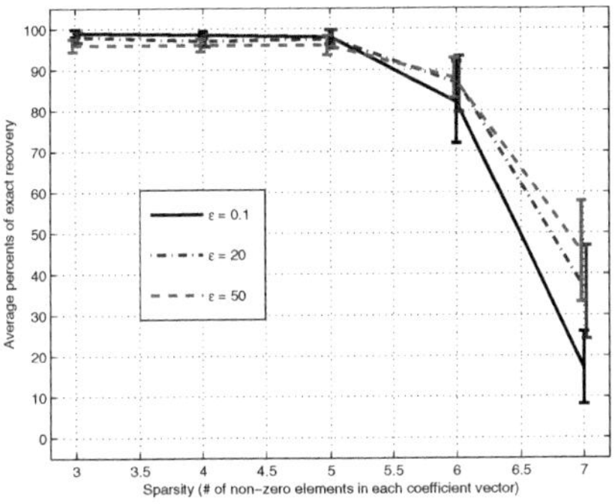

Fig. 1. Exact dictionary recovery with different approximation errors ϵ

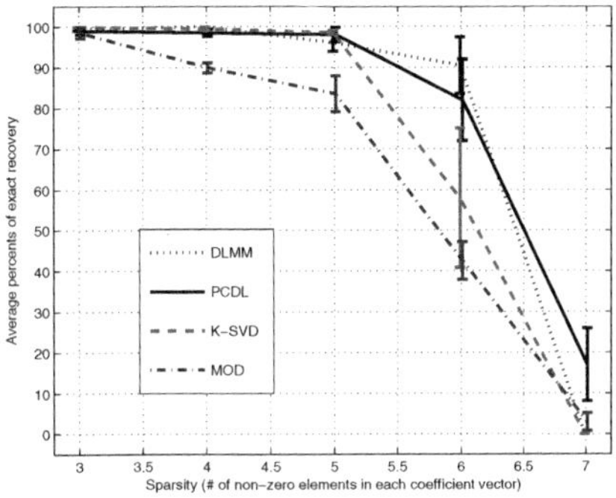

Fig. 2. Exact dictionary recovery using PCDL, MMDL, K-SVD and MOD methods

recovery percentages for three different $\epsilon = 0.1, 20$ and 50, which are roughly corresponding to $42, 19$ and $11dB$ signal to noise ratios, are plotted in Figure 1. The variance of the success is also shown by error bars. It shows that although sparse approximation with high fidelity improves the success rate for highly sparse coefficients, the noisy sparse approximation provides better recovery with less sparse coefficients. The success of dictionary recovery using sparse *approximations*, *i.e.* large ϵ, in a less sparse setting may be caused by the fact that small coefficients, which may actually be zero in the generated sparse coefficients, are now assumed to be noise. The algorithm thus learn a dictionary based on the coefficients which we are more confident to be non-zero. Figure 2 compares the proposed method, $\epsilon = 0.1$, with some other dictionary learning methods. The λ parameter, which is the Lagrange multiplier in the sparse approximation[2], is 0.4 in K-SVD[7], MOD[6] and DLMM[8]. An ℓ_1 sparse approximation with an extra de-biasing step, which is simply an orthogonal projection of the observed data onto the linear span of subdictionary indexed by non-zero coefficients, has been used in the sparse approximation steps of these methods. It demonstrates that PCDL performs almost the same as the best current methods for very sparse and it also shows superior performance for less sparse data. It deserves to mention that the computational complexity of PCDL is often higher than the standard fixed-sparsity dictionary learning methods, as it needs to iteratively reduce τ and learn a dictionary within new ℓ_1 ball. The total computational cost of the algorithm directly depends on the number of iterations in the outer loop, *i.e.* "while" loop in Algorithm 1. In practice we found that after 10 to 20 updates of τ, the algorithm converges to a solution in this example. We also observed that the inner loop, which includes K iterations of coefficients and dictionary updates, is faster than MMDL, for small λ's.

In the second experiment we applied PCDL to the dictionary learning for audio data, which has been shown to have some sparse structures. We chose a 256 by 512 dictionary and randomly select 16384 audio samples, of length 256, from a long

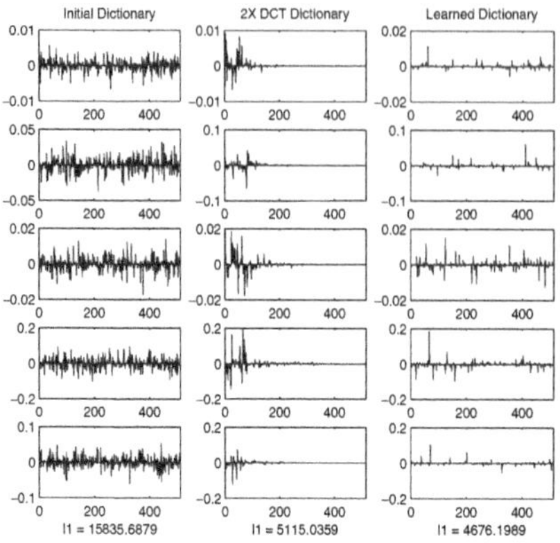

Fig. 3. Sparse representations of five different training data, using three different dictionaries

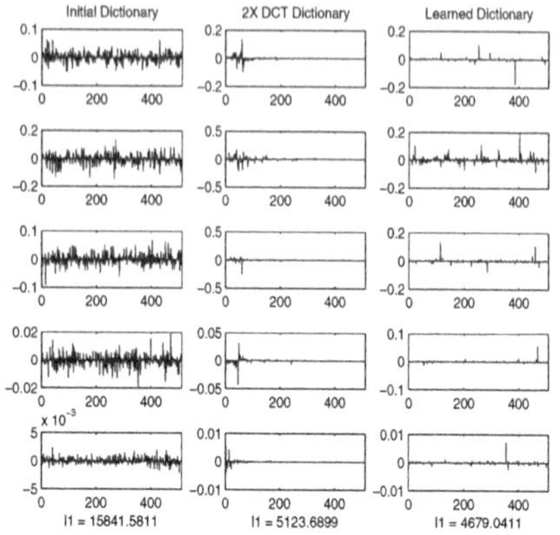

Fig. 4. Sparse representations of five different observed data, out of training set, using three different dictionaries

audio record from BBC radio 3, which plays classical music. The sparse representations of 5 random samples from training and some independent data are respectively shown in Figure 3 and 4. The initial random, a two times overcomplete DCT (oversampled frequency) and the learned dictionaries have been used to find the sparse codes in the left to right columns respectively. The total ℓ_1 of 16384 sparse codes are mentioned in the bottom line. The minimum ℓ_1 sparse representation, using the learned dictionary, provides good sparse codes, even though it may not be the optimum dictionary. The consistency of the learning is also demonstrated, as the learned dictionary works well for independently selected audio blocks. It is also demonstrated that the learned dictionary is superior to the two times overcomplete DCT in providing less ℓ_1, with the same ϵ.

5 Conclusions

We have introduced a new dictionary learning framework for sparse representation. It is based on Pareto Curve root finding which has previously been used for sparse representation. The new algorithm is guaranteed to be stable and we can also show the convergence to a set of fixed points. As the new framework needs to update τ using a line search method, a more efficient method may provide faster convergence using fewer updates of τ. We chose the current technique for updating τ as it provides a uniform reduction of a lower bounded objective.

The proposed algorithm can also be used in a dictionary learning for sparse approximation framework, by using an extra parameter ϵ, which measures the deviation from the exact representation subspace, *i.e.* $\mathbf{DX} = \mathbf{Y}$. This is particularly useful, when ϵ is small, as current dictionary learning methods often converge very slowly when using a small sparsity penalty, *i.e.* small λ.

References

1. Mallat, S., Zhang, Z.: Matching Pursuits with time frequency dictionaries. IEEE Trans. on Signal Processing 41(12), 3397–3415 (1993)
2. Chen, S.S., Donoho, D.L., Saunders, M.A.: Atomic Decomposition by Basis Pursuit. SIAM Journal on Scientific Computing 20(1), 33–61 (1998)
3. Yaghoobi, M., Daudet, L., Davies, M.: Parametric Dictionary Design for Sparse Coding. IEEE Trans. on Signal Processing 57(12), 4800–4810 (2009)
4. Olshausen, B.A., Field, D.J.: Sparse coding with an overcomplete basis set: a strategy employed by V1? Vision Research 37(23), 3311–3325 (1997)
5. Lewicki, M.S., Sejnowski, T.J.: Learning Overcomplete Representations. Neural Comp. 12(2), 337–365 (2000)
6. Engan, K., Aase, S.O., Hakon-Husoy, J.: Method of optimal directions for frame design. In: International Conference on Acoustics, Speech, and Signal Processing (ICASSP), pp. 2443–2446 (1999)
7. Aharon, M., Elad, E., Bruckstein, A.M.: K-SVD: an algorithm for designing of over complete dictionaries for sparse representation. IEEE Trans. on Signal Processing 54(11), 4311–4322 (2006)
8. Yaghoobi, M., Blumensath, T., Davies, M.: Dictionary Learning for Sparse Approximations with the Majorization Method. IEEE Trans. on Signal Processing 57(6), 2178–2191 (2009)
9. Georgiev, P., Theis, F., Cichocki, A.: Sparse component analysis and blind source separation of underdetermined mixtures. IEEE Trans. on Neural Networks 16(4), 992–996 (2005)
10. Aharon, M., Elad, M., Bruckstein, A.M.: On the uniqueness of overcomplete dictionaries and a practical way to retrieve them. Journal of Linear Algebra and Applications 416, 48–67 (2006)
11. Gribonval, R., Schnass, K.: Some Recovery Conditions for Basis Learning by L1-Minimization. In: International Symposium on Communications, Control and Signal Processing, ISCCSP (2008)
12. Gribonval, R., Schnass, K.: Dictionary Identification: Sparse Matrix-Factorisation via ℓ_1 Minimisation. IEEE Trans. on Information Theory 56(7), 3523–3539 (2010)
13. Plumbley, M.D.: Dictionary Learning for l1-Exact Sparse Coding. In: International Conference on Independent Component Analysis and Signal Separation (ICA), pp. 406–413 (2007)
14. Van den Berg, E., Friedlander, M.P.: Probing the Pareto frontier for basis pursuit solutions. SIAM Journal on Scientific Computing 31(2), 890–912 (2008)
15. Goldstein, A.A.: Convex programming in Hilbert space. Bulletin of the American Mathematical Society 70(5), 709–710 (1964)
16. Tibshirani, R.: Regression shrinkage and selection via the Lasso. Journal of Royal Statistical Society Series B 58, 267–288 (1996)

SMALLbox - An Evaluation Framework for Sparse Representations and Dictionary Learning Algorithms

Ivan Damnjanovic, Matthew E.P. Davies, and Mark D. Plumbley

Queen Mary University of London, Centre for Digital Music,
Mile End Road, London, E1 4NS, United Kingdom
{ivan.damnjanovic,matthew.davies,mark.plumbley}@elec.qmul.ac.uk

Abstract. SMALLbox is a new foundational framework for processing signals, using adaptive sparse structured representations. The main aim of SMALLbox is to become a test ground for exploration of new provably good methods to obtain inherently data-driven sparse models, able to cope with large-scale and complicated data. The toolbox provides an easy way to evaluate these methods against state-of-the art alternatives in a variety of standard signal processing problems. This is achieved trough a unifying interface that enables a seamless connection between the three types of modules: problems, dictionary learning algorithms and sparse solvers. In addition, it provides interoperability between existing state-of-the-art toolboxes. As an open source MATLAB toolbox, it can be also seen as a tool for reproducible research in the sparse representations research community.

Keywords: Sparse representations, Dictionary learning, Evaluation framework, MATLAB toolbox.

1 Introduction

Sparse representations has become a very active research area in recent years and many toolboxes implementing a variety of greedy or other types of sparse algorithms have become freely available in the community [1-4]. As the number of algorithms has grown, there has become a need for a proper testing and benchmarking environment. This need was partially addressed with the SPARCO framework [5], which provides a large collection of imaging, signal processing, compressed sensing, and geophysics sparse reconstruction problems. It also includes a large library of operators that can be used to create new test problems. However, using SPARCO with other sparse representations toolboxes, such as SparseLab [1] is non-trivial because of inconsistencies in the APIs of the toolboxes.

Many algorithms exist that aim to solve the sparse representation dictionary learning problem [6-7, 10]. However, no comprehensive means of testing and benchmarking these algorithms exists, in contrast to the sparse representation problem when the dictionary is known. The main driving force for this work is the lack of a toolbox similar to SPARCO for dictionary learning problems. Recognising the need of the community for such a toolbox, we set out to design a MATLAB toolbox with three main aims:

V. Vigneron et al. (Eds.): LVA/ICA 2010, LNCS 6365, pp. 418–425, 2010.

- to enable an easy way of comparing dictionary learning algorithms,
- to provide a unifying API that will enable interoperability and re-use of already available toolboxes for sparse representation and dictionary learning,
- to aid the reproducible research effort in the sparse signal representations and dictionary learning [12].

In the section 2 of this paper, we give a short overview of sparse representations and dictionary learning. In section 3, the SMALLbox toolbox design approach is presented, with implementation details in section 4, followed by usage examples in section 5 and conclusions in section 6.

2 Sparse Representations and Dictionary Learning

One of the main requirements in many signal processing applications is to represent the signal in a transformed domain where it can be expressed as a linear combination of a small number of coefficients. In many research areas such as compressed sensing, image de-noising and source separation, these sparse structured signal representations are sought-after signal models. Depending on the application, we seek either an exact solution for a noise-free models or an approximate sparse reconstruction of the signal in the presence of noise:

$$\mathbf{b} = \mathbf{A}\mathbf{x} \tag{1}$$

$$\mathbf{b} = \mathbf{A}\mathbf{x} + \mathbf{n} \tag{2}$$

where $\mathbf{b} \in R^m$ is the signal of interest, $\mathbf{A} \in R^{m \times n}$ is transformation matrix (or dictionary), $\mathbf{x} \in R^n$ is the sparse coefficients vector and $\mathbf{n} \in R^m$ is noise vector. When $m<n$ the problem is *undetermined* and there is no unique solution. Evidently, additional constraint needs to be imposed to the signal model to find the solution of interest. In this sense, probably the most studied and well-understood constraint is to assume a Gaussian distribution on the coefficients and to minimise the l_2 norm of the vector $\mathbf{x}$. However, in applications such as compression for example, it is more appropriate to impose the sparsity assumption on the coefficients, i.e. to minimise the l_0 norm of the vector $\mathbf{x}$. Since l_0 norm minimisation is known to be an NP-hard problem, one can try finding an approximate solution using greedy algorithms such as Matching Pursuit (MP) [8] or Orthogonal Matching Pursuit (OMP) [9]. Alternatively, we can relax the sparsity assumption by imposing a Laplace distribution on the coefficients and minimise the l_1 norm, which can be solved using different convex optimisation methods.

In the SPARCO toolbox, the problems to be solved are given through a consistent interface represented in the form of a problem structure that contains a measurement vector $\mathbf{b}$, operator $\mathbf{A}$ and all other components of the test problem. Operator $\mathbf{A}$ is given in the following form:

$$\mathbf{A} = \mathbf{M}\mathbf{B} \tag{3}$$

The measurement operator **M** describes how the signal was sampled and operator **B** represents a basis with which the signal can be sparsely represented [5]. It is assumed that the basis that can give a sparse solution is known in advance. The success of the sparse representation heavily depends on the choice of the basis and the transform dictionary **A** and how well the dictionary reflects the structure to be found in the signal. Learning the matrix **A** from the data itself is a key to finding a sparse representation of the new classes of data. Dictionary learning for a sparse representation can be formulated as a problem of the following type:

$$\min_{\mathbf{A},\mathbf{X}} \left\| \mathbf{Y} - \mathbf{A}\mathbf{X} \right\|_F^2 \quad \text{subject to} \quad \forall i \quad \left\| \mathbf{x_i} \right\|_0^0 \leq s \tag{4}$$

where **Y** is a matrix with vectors of training data and $\mathbf{x_i}$ are sparse representations of the training vectors. We want to choose a transform matrix **A** that will minimise the residual, given that the training data representation vectors $\mathbf{x_i}$ are sparse with a maximum of s non-zero coefficients.

Reflecting high activity in the research area, many dictionary learning algorithms are available, but currently no evaluation framework exists for testing them.

3 Design Approach to SMALLbox

The SMALLbox framework is designed to fulfil two main goals: (1) to provide a set of test problems that permit formative evaluation of the techniques and algorithms to be developed elsewhere, and (2) to be a framework within which to build demonstrator applications. The design of the SMALLbox toolbox was constructed to allow easy portability of existing algorithms and new algorithms to be developed, taking into account the experiences in using toolboxes such as SPARCO [5] and SparseLab [1]. A graphical overview of the design of SMALLbox is shown in Fig 1.

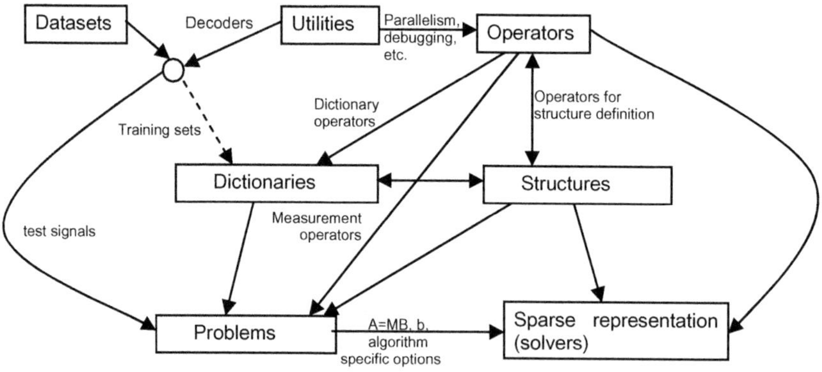

Fig. 1. Design of the Evaluation Framework

The main interoperability of the design is given through the *Problems* part that can be defined either as sparse representations or dictionary learning. In generating a problem, some *utilities* can be used to decode a *dataset* and prepare a *test signal* or a *training set* for dictionary learning. The *dictionaries* can be either defined or learned using dictionary learning algorithms. In the former case, they can be given as implicit dictionaries as a combination of the given *operators* and *structures* or explicit in the form of dictionary matix. In the latter case, they are leaned from the training data. Once the *dictionary* is set in the *problem*, the *problem* is ready to be solved by one of the *sparse representation* algorithms.

SMALLbox is designed to enable an easy exchange of information and comparison of different modules developed through a unified API data structure. The structure is made to fulfil two main goals. The first goal is to separate typical sparse signal processing problems into three meaningful units:

a) problem specification (preparing data for learning the structures, representation and reconstruction),
b) dictionary learning (using a prepared training set to learn the natural structures in the data) and
c) sparse representation (representing the signal with a pre-specified or learned dictionary).

The second goal is a seamless connection between the three types of modules and ease of communication of data between the problem, dictionary learning and sparse representation parts of the structure.

4 SMALLbox Implementation

The evaluation framework is implemented as a Matlab toolbox called SMALLbox. To enable easy comparison with the existing state-of-the-art algorithms, during the installation procedure SMALLbox checks the Matlab path for existence of the following freely available toolboxes and will automatically download and install them, as required:

- SPARCO (v.1.2) - set of sparse representation problems[5]
- SparseLab (v.2.1) - set of sparse solvers [1]
- Sparsify (v.0.4) - set of greedy and hard thresholding algorithms [2]
- SPGL1 (v.1.7) - large-scale sparse reconstruction solver [3]
- GPSR (v.6.0) - Gradient projection for sparse reconstruction [4]
- KSVD-box (v.13) and OMP-box (v.10) - dictionary learning [6]
- KSVDS-box (v.11) and OMPS-box (v.1) - sparse dictionary learning [7].[1]
- SPAMS - Online dictionary learning [10] [2]

The SMALLbox provides a "glue" structure to allow algorithms from those toolboxes to be used with a common API. The structure consists of three main sub-structures: ***Problem*** structure, ***DL*** (dictionary learning) structures and ***solver*** structures. Since the ***Problem*** structure is design to be backward compatible with the SPARCO problem structure [5], it can be filled with SPARCO generateProblem or one of the dictionary

---

[1] The list of 3rd party toolboxes included in SMALLbox version 1.0.

[2] An API for SPAMS is included, but due to licensing issues this toolbox needs to be installed by the user.

learning problems provided in SMALLbox. If the problem is dictionary learning, one or more ***DL*** structures can be specified, so [6-7] or any other dictionary learning technique can be compared with specified set of parameters. Finally, to sparsely represent the signal in a dictionary (either defined in the ***Problem*** structure or learned in the previous step), one or more ***solver*** structures can be used to specify any solver from [1-4] or any of the solvers provided in SMALLbox.

4.1 Generating Problems (*Problem* Structure)

The ***Problem*** structure defines all necessary aspects of a problem to be solved. To be compatible with the SPARCO, it needs to have five fields defined prior to any sparse representation of the data:

- *A* – a matrix or operator representing dictionary in which signal is sparse
- *b* – a vector or matrix representing signal or signals to be represented
- *reconstruct* – a function handle to reconstruct the signal from coefficients
- *signalSize* – the dimension of the signal
- *sizeA* – if matrix **A** is given as an operator the size of the dictionary needs to be defined in advance.

Other fields that further describe the problem, which are useful for either reconstruction of the signal or representation of the results, might be generated by the SPARCO generateProblem function or the SMALLbox problem functions. The new problems implemented in the SMALLbox version 1.0 are: Image De-noising [6], Automatic Music Transcription [11] and Image Representation using another image as a dictionary.

In the case of a dictionary learning problem, fields *A* and *reconstruct* are not defined while generating the problem, but after the dictionary is learned and prior to the sparse representation. In this case, field *b* needs to be given in matrix form to represent the training data and another field *p* defining the number of dictionary elements to be learned needs to be specified.

4.2 Dictionary Learning (*DL* Structure)

The structure for dictionary learning - ***DL*** is a structure that defines dictionary learning algorithm to be used. It is initialised with a utility function SMALL_init_DL, which will define five mandatory fields:

- *toolbox* - a field used to discriminate the API
- *name* - the name of dictionary learning function from the particular toolbox
- *param* - a field containing parameters for the particular DL technique and in the form given by the toolbox API
- *D* - a field where the learned dictionary will be stored
- *time* - a field to store learning time.

After *toolbox*, *name* and *param* fields are set, the function SMALL_learn is called with ***Problem*** and ***DL*** structures as inputs. According to the *DL.toolbox* field, the function calls the *DL.name* algorithm with its API and outputs learned dictionary *D* and *time* spent. The *DL.param* field contains parameters such as dictionary size, the number of iterations, the error goal or similar depending on the particular algorithm used. To compare a new dictionary learning algorithm against existing ones, the algorithm needs to be in the MATLAB path and introduced to SMALLbox by defining two parameters *<Toolbox ID>* and *<Preferred API>* in the SMALL_learn function, where examples and a simple explanation are provided. Once the new dictionary is learned, field *A* of the *Problem* structure is defined to be equal to *DL.D* and also the reconstruction function is instructed to use this particular dictionary. In this way, a SPARCO compatible Problem structure is defined and ready for use.

4.3 Sparse Representation (*Solver* Structure)

Similar to dictionary learning every instance of the sparse representation needs to be initialised with the SMALL_init_solver function. It will define mandatory fields of the ***solver*** structure:

- *toolbox* - a field with toolbox name (e.g. sparselab)
- *name* - the name of solver from the particular toolbox (e.g. SolveOMP)
- *param* - the parameters in the form given by the toolbox API
- *solution* - the output representation
- *reconstructed* - the signal reconstructed from solution
- *time* - the time spent for sparse representation.

With the input parameters of the *solver* structure set, the SMALL_solve function is called with ***Problem*** and ***solver*** structure as inputs. The function calls *solver.name* algorithm with API specified by *sover.toolbox* and outputs *solution*, *reconstructed* and *time* fields. To introduce a new sparse representation algorithm it needs to be in the MATLAB path and *<Toolbox ID>* and *<Preferred API>* need to be defined for the algorithm in the SMALL_solve function. Three solvers that can find a sparse representation of the whole training set matrix in one go are included in SMALLbox (SMALL_MP, SMALL_chol and SMALL_cgp).

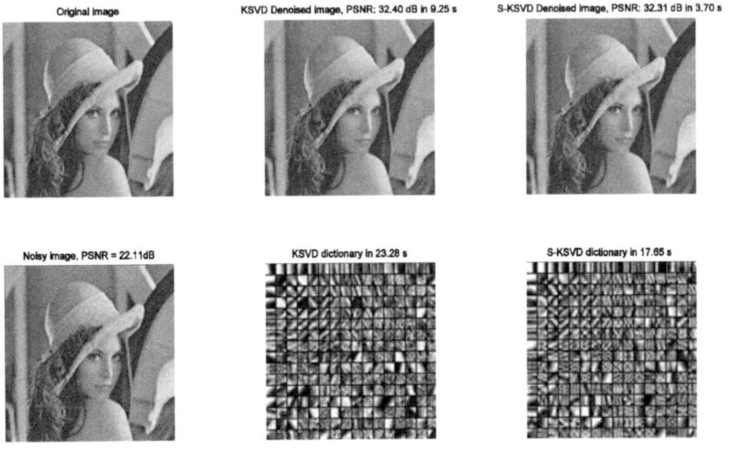

Fig. 2. SMALLbox example results - KSVD [6] versus S-KSVD [7] in image de-noising

5 Examples

As a part of SMALLbox, a variety of examples on how included dictionary learning and sparse representation techniques can be used and compared on SPARCO and SMALLbox problems. As an example, small_solver_test.m will generate SPARCO problem 6 (sparse representation of **b** – a piecewise cubic polynomial signal, in **B** – a Daubechies basis with **M** – a Gaussian ensemble measurement matrix), test four solvers on the problem (SMALL_cgp, SMALL_chol, Solve_OMP from SparseLab and greed_pcgp from Sparsify), show computational time and plot solutions and reconstructed signals against the original.

Two examples of dictionary learning for image de-noising are presented in Figs 2 and 3. In the first example, we compared the KSVD algorithm [6] with S-KSVD [7]. The main idea presented in [7] is that if an implicit dictionary (in this case an overcomplete DCT) is used as base dictionary on which the sparse dictionary is learned, much better computational time can be achieved while still keeping adaptability and performance characteristics of explicit dictionaries. The example and results in Figure 2 support this claim. De-noising in the S-KSVD is almost 3 times faster while the PSNR is only 0.09 dB lower.

The example in Figure 3 presents a comparison of online dictionary learning [10] and KSVD [6] and shows how SMALLbox can be used to easy change the parameters of the problem (in this case the training size). It supports the claim from [10] that, in contrast to iterative algorithms [6], online dictionary learning does not depend on the size of the training set.

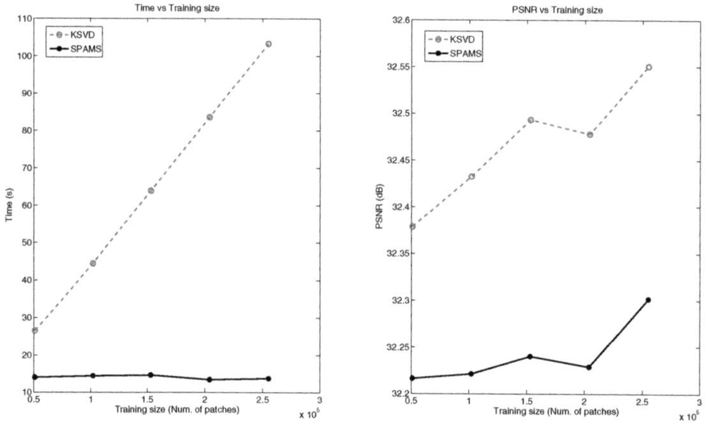

Fig. 3. SMALLbox example results - KSVD [6] versus SPAMS [10] with variable training size

6 Conclusions

We have introduced SMALLbox - an Evaluation Framework that enables easy prototyping, testing and benchmarking of sparse representation and dictionary learning algorithms. This is achieved through a set of test problems and an easy evaluation against state-of-the-art algorithms. As a part of the EU FET SMALL project, more problems, solvers and dictionary learning techniques that are developed will be included in SMALLbox as the project proceeds.

For instructions how to download the SMALLbox and reproduce the figures in this paper, please visit: http://small-project.eu/.

Acknowledgments. This research is supported by EU FET-Open ProjectFP7-ICT-225913 "SMALL", and ESPRC Platform Grant EP/045235/1.

The authors would also like to thank to all researchers working on the SMALL project especially Miki Elad and Pierre Vandergheynst for fruitful discussion and help in developing the SMALLbox.

References

1. Donoho, D., Stodden, V., Tsaig, Y.: Sparselab (2007), `http://sparselab.stanford.edu/`
2. Blumensath, T., Davies, M.E.: Gradient pursuits. IEEE Transactions on Signal Processing 56(6), 2370–2382 (2008)
3. Berg, E.v., Friedlander, M.P.: Probing the Pareto frontier for basis pursuit solutions. SIAM Journal on Scientific Computing 31(2), 890–912 (2008)
4. Figueiredo, M.A.T., Nowak, R.D., Wright, S.J.: Gradient projection for sparse reconstruction: Application to compressed sensing and other inverse problems. Journal of Selected Topics in Signal Processing: Special Issue on Convex Optimization for Signal Processing (December 2007)
5. Berg, E.v., Friedlander, M. P., Hennenfent, G., Herrmann, F., Saab, R., Yılmaz, O.: SPARCO: A testing framework for sparse reconstruction. ACM Trans. on Mathematical Software 35(4), 1–16 (2009)
6. Aharon, M., Elad, M., Bruckstein, A.M.: The K-SVD: An algorithm for designing of overcomplete dictionaries for sparse representation. IEEE Transactions on Signal Processing 54(11), 4311–4322 (2006)
7. Rubinstein, R., Zibulevsky, M., Elad, M.: Double Sparsity: Learning Sparse Dictionaries for Sparse Signal Approximation. IEEE Transactions on Signal Processing 58(3), 1553–1564 (2010)
8. Mallat, S.G., Zhang, Z.: Matching Pursuits with Time-Frequency Dictionaries. IEEE Transactions on Signal Processing, 3397–3415 (December 1993)
9. Pati, Y.C., Rezaiifar, R., Krishnaprasad, P.S.: Orthogonal matching pursuit: Recursive function approximation with applications to wavelet decomposition. In: Singh, A. (ed.) Proc. 27th Asilomar Conference on Signals, Systems and Computers. IEEE Computer Society Press, Los Alamitos (1993)
10. Mairal, J., Bach, F., Ponce, J., Sapiro, G.: Online Learning for Matrix Factorization and Sparse Coding. Journal of Machine Learning Research 11, 19–60 (2010)
11. Bertin, N., Badeau, R., Richard, G.: Blind signal decompositions for automatic transcription of polyphonic music: NMF and K-SVD on the benchmark. In: Proc. Of International Conference on Acoustics, Speech, and Signal Processing, ICASSP 2007, Honolulu, Hawaii, USA, April 15-20, vol. I, pp. 65–68 (2007)
12. Vandewalle, P., Kovacevic, J., Vetterli, M.: Reproducible Research in Signal Processing - What, why, and how. IEEE Signal Processing Magazine 26(3), 37–47 (2009)

Fast Block-Sparse Decomposition Based on SL0

Sina Hamidi Ghalehjegh[1], Massoud Babaie-Zadeh[1,⋆], and Christian Jutten[2]

[1] Department of Electrical Engineering, Sharif University of Technology, Tehran, Iran
[2] GIPSA-Lab, Grenoble, and Institut Universitaire de France, France
shghamidi@gmail.com, mbzadeh@yahoo.com, Christian.Jutten@inpg.fr

Abstract. In this paper we present a new algorithm based on Smoothed ℓ^0 (SL0), called Block SL0 (BSL0), for Under-determined Systems of Linear Equations (USLE) in which the nonzero elements of the unknown vector occur in clusters. Contrary to the previous algorithms such as Block Orthogonal Matching Pursuit (BOMP) and mixed ℓ^2/ℓ^1 norm, our approach provides a fast algorithm, while providing the same (or better) accuracy. Moreover, we will see experimentally that BSL0 has better performance than SL0, BOMP and mixed ℓ^2/ℓ^1 norm when the number of nonzero elements of the source vector approaches the upper bound of uniqueness theorem.

Keywords: Under-determined Systems of Linear Equations, Block-Sparsity, Sparse Decomposition, SL0 Algorithm.

1 Introduction

Sparse solutions of USLE have recently attracted a lot of attentions because of their potential applications in many different areas. They are used, for example, in compressed sensing [1,2], under-determined Sparse Component Analysis (SCA) and source separation [3,4] and atomic decomposition on overcomplete dictionaries [5,6]. Generally, an USLE has infinitely many solutions but it is shown in [7] that under some conditions the sparsest solution of the system is unique.

More concretely, let $\mathbf{s}$ be an $m \times 1$ unknown vector that is observed through an $n \times m$, $n < m$, measurement matrix $\mathbf{A}$ according to $\mathbf{x} = \mathbf{A}\mathbf{s}$. Since this system is an USLE, there are infinitely many possible solutions of $\mathbf{s}$ that satisfy it for a given $\mathbf{x}$ and $\mathbf{A}$. Therefore, we must add some extra conditions to insure the uniqueness of $\mathbf{s}$. Let Spark($\mathbf{A}$) be the smallest number r such that there exists a set of r columns in $\mathbf{A}$ which are linearly dependent [7]. Moreover, define the ℓ^0 norm of $\mathbf{s}$ as its number of nonzero elements. $\mathbf{s}$ is called k-sparse if its ℓ^0 norm is k. It is shown in [7] that if k is less than $r/2$, then sparsest solution of the system is unique.

⋆ This work has been partially funded by Iran Telecom Research Center (ITRC), and also by center for International Research and Collaboration (ISMO) and French embassy in Tehran in the framework of a GundiShapour collaboration program.

V. Vigneron et al. (Eds.): LVA/ICA 2010, LNCS 6365, pp. 426–433, 2010.

The above uniqueness theorem of USLE has led to development of many different recovery algorithms [1,5,8,9]. Most of the algorithms find the sparsest solution by optimizing a cost function that measures the distance of $\mathbf{s}$ from being sparse. Therefore, the more precise our cost function involves the sparsity concept, the closer the recovered vector will be to the exact solution. Two successful studied algorithms are Basis Pursuit (BP) [5,9] and SL0 [8,10] which are based on minimization of ℓ^1 and smoothed-ℓ^0 norm, respectively.

In this paper, we consider recovery of an unknown solution whose nonzero elements occur in clusters. Such a signal is called block-sparse [11,12]. For this purpose, we modify the SL0 idea to use this block structure in the recovery algorithm. We will see in Sect. 4 that taking this structure into account can yield better reconstruction results.

Block-sparsity occurs when dealing with multi-band signals [13] and equalization of sparse communication channels [14]. Furthermore, it was shown in [13] that block-sparsity model can be used to treat the problem of sampling signals that lie in a union of subspaces. A block-sparse signal can be stated as follow:

$$\mathbf{s} = [\underbrace{s_{11}, \cdots, s_{1d}}_{\mathbf{s}_1}, \underbrace{s_{21}, \cdots, s_{2d}}_{\mathbf{s}_2}, \cdots, \underbrace{s_{N1}, \cdots, s_{Nd}}_{\mathbf{s}_N}]^T \quad , \tag{1}$$

where $\mathbf{s}_i$, $i = 1, \cdots, N$ is called the ith block of $\mathbf{s}$ and d is the block size. A signal of dimension m which consists of N blocks of size $d = m/N$ is k sparse if at most k blocks of the signal out of N are nonzero. It is clear that if $d = 1$, block-sparsity reduces to conventional sparsity.

Using ℓ^1 relaxation [5,9] for reconstructing $\mathbf{s}$ does not exploit the fact that the signal is block-sparse, i.e., that the nonzero entries occur in consecutive positions. Therefore, different techniques were suggested in recent years. Stojnic *et. al.* in [15] modify the ℓ^1 norm cost function and call it mixed ℓ^2/ℓ^1 norm, to exploit the block-sparsity. They suggest the following optimization problem for the recovery of $\mathbf{s}$:

$$\underset{\mathbf{s}}{\operatorname{argmin}} \, \|\mathbf{s}_1\|_2 + \|\mathbf{s}_2\|_2 + \cdots + \|\mathbf{s}_N\|_2 \quad \text{s.t.} \quad \mathbf{x} = \mathbf{A}\mathbf{s} \ . \tag{2}$$

In [11] it is shown that under some conditions on measurement matrix, (2) is guaranteed to recover any block-sparse signal, irrespectively of the locations of the nonzero blocks. Furthermore, recovery will be robust in the presence of noise and modeling error. [11] uses semi-definite programming to find the sparsest solution of (2). However, it is still very slow and becomes worse as the dimension increases.

Another approach presented in [16], called Block Orthogonal Matching Pursuit (BOMP), modifies standard Orthogonal Matching Pursuit (OMP) algorithm [17] to use the block structure. BOMP is fast, but is a greedy algorithm and does not provide good estimation of the sources.

Contrary to previous approaches, the method we present in this paper tries to directly work with the block version of ℓ^0 norm and is based on the idea of smoothed ℓ^0 (SL0) [8]. We will see experimentally that the proposed algorithm outperforms both mixed ℓ^2/ℓ^1 norm and BOMP methods.

The paper is organized as follows. The next section introduces the basic principles of our approach. The final algorithm is then stated in Sect. 3. Finally, Sect. 4 provides some experimental results of our algorithm and its comparison with conventional SL0, mixed ℓ^2/ℓ^1 norm and BOMP algorithms.

2 Basic Principles of Our Approach

As our proposed method is based on SL0, we first have a brief review on this algorithm. Let $\mathbf{s}$ be an unknown $m \times 1$ vector and related to $n \times 1$ measurement vector $\mathbf{x}$, $n < m$, through the measurement matrix $\mathbf{A}$ by:

$$\mathbf{x} = \mathbf{A}\mathbf{s} \ . \tag{3}$$

A recovery algorithm tries to find the sparsest solution of (3). The reconstruction can be then stated as the following optimization problem:

$$\underset{\mathbf{s}}{\operatorname{argmin}} \quad \|\mathbf{s}\|_0 \quad \text{s.t.} \quad \mathbf{x} = \mathbf{A}\mathbf{s} \ . \tag{4}$$

To solve (4) one may search for a solution with minimal ℓ^0 norm. This exhaustive search will have complexity of $O(\binom{m}{k}n^3)$ [15] and hence it is an intractable problem as the dimension increases. Recently, [8] has used a smoothed version of ℓ^0 norm, called SL0, which is experimentally shown that is a very fast algorithm and has a very good performance.

The main idea of SL0 is to approximate ℓ^0 *discontinuous* function by a suitable *continuous* one. For example, consider the following one-variable function:

$$f_\sigma(s) \triangleq e^{-s^2/2\sigma^2} \ . \tag{5}$$

It is clear that:

$$\lim_{\sigma \to 0} f_\sigma(s) = \begin{cases} 1 & s = 0 \\ 0 & s \neq 0 \end{cases} \ . \tag{6}$$

Then, by defining $F_\sigma(\mathbf{s}) \triangleq \sum_{i=1}^{m} f_\sigma(s_i)$ we can see that:

$$\|\mathbf{s}\|_0 \approx m - F_\sigma(\mathbf{s}) \ , \tag{7}$$

for small values of σ. Consequently, [8] finds the solution of (4) by maximizing $F_\sigma(\mathbf{s})$ (subject to $\mathbf{x} = \mathbf{A}\mathbf{s}$) for a very small value of σ. The value of σ determines how smooth the function F_σ is: the larger value of σ, the smoother F_σ (but worse approximation of ℓ^0 norm) and vice versa. For small values of σ, F_σ is highly non-smooth and contains a lot of local maxima, and hence its maximization is not easy. On the other hand, for larger values of σ, F_σ is smoother and contains less local maxima. Consequently, to avoid trapping in local maxima, the authors of [8] propose to use a 'decreasing' sequence for σ: for maximizing F_σ for each value of σ (using e.g. gradient algorithms), the initial value of the maximization algorithm is the maximizer of F_σ for the previous (larger) value of σ. More details about this concept and also theoretical analysis of the algorithm is stated in [8,10]. Moreover, [8] suggests that the best initial value of $\mathbf{s}$ for the algorithm is the minimum ℓ^2 norm solution of $\mathbf{x} = \mathbf{A}\mathbf{s}$, i.e. the solution given by the pseudo-inverse of $\mathbf{A}$.

We now modify the above idea for reconstruction of *block-sparse* signals from their measurements. Let $\mathbf{s}$ be as (1) and $I(x)$ be as follow:

$$I(x) \triangleq \begin{cases} 1 & x \neq 0 \\ 0 & x = 0 \end{cases} .$$

By denoting

$$\|\mathbf{s}\|_{2,0} \triangleq \sum_{i=1}^{N} I(\|\mathbf{s}_i\|_2) , \tag{8}$$

a vector $\mathbf{s}$ is block k-sparse if $\|\mathbf{s}\|_{2,0} \leq k$. Therefore, the optimization problem will be:

$$\operatorname*{argmin}_{\mathbf{s}} \quad \|\mathbf{s}\|_{2,0} \quad \text{s.t.} \quad \mathbf{x} = \mathbf{A}\mathbf{s} . \tag{9}$$

By defining $b_i \triangleq \|\mathbf{s}_i\|_2$ we can rewrite (8) as follows:

$$\|\mathbf{s}\|_{2,0} = \sum_{i=1}^{N} I(b_i) \triangleq \|\mathbf{b}\|_0 , \tag{10}$$

where $\mathbf{b} \triangleq [b_1, b_2, \cdots, b_N]^T$. Using (7) it is clear that $\|\mathbf{b}\|_0 \approx N - F_\sigma(\mathbf{b})$ where $F_\sigma(\mathbf{b}) = \sum_{i=1}^{N} f_\sigma(b_i)$ and $f_\sigma(b_i) \triangleq e^{-b_i^2/2\sigma^2}$. Therefore, we can write:

$$\|\mathbf{s}\|_{2,0} \approx N - \sum_{i=1}^{N} e^{-\frac{\sum_{j=1}^{d} s_{ij}^2}{2\sigma^2}} \triangleq N - H_\sigma(\mathbf{s}) , \tag{11}$$

and the final optimization problem will be:

$$\operatorname*{argmax}_{\mathbf{s}} \quad H_\sigma(\mathbf{s}) \quad \text{s.t.} \quad \mathbf{x} = \mathbf{A}\mathbf{s} . \tag{12}$$

To solve the optimization problem (12) we use steepest ascent method. So, we need to compute gradient vector of $H_\sigma(\mathbf{s})$. Because of the block nature of $\mathbf{s}$, it is better to express the gradient vector in a block notation. So, we denote ∇H_σ as follow:

$$\nabla H_\sigma = [\underbrace{h_{11}, \cdots, h_{1d}}_{1^{st}\text{ block}}, \underbrace{h_{21}, \cdots, h_{2d}}_{2^{nd}\text{ block}}, \cdots, \underbrace{h_{N1}, \cdots, h_{Nd}}_{N^{th}\text{ block}}]^T . \tag{13}$$

Therefore we can write:

$$h_{uv} = \frac{\partial H_\sigma(\mathbf{s})}{\partial s_{uv}} = -\frac{s_{uv}}{\sigma^2} e^{-\frac{\sum_{j=1}^{d} s_{uj}^2}{2\sigma^2}} . \tag{14}$$

3 The Final Algorithms

The final algorithm is given in Fig. 1. Like SL0, we choose minimum ℓ^2 norm solution of $\mathbf{x} = \mathbf{A}\mathbf{s}$, obtained by pseudo-inverse of $\mathbf{A}$, as the initial value of the algorithm. Also, as stated in [8], to avoid trapping in local maxima we use a 'decreasing' sequence for σ. Also, steepest ascent consists of iterations of the form $\mathbf{s} \leftarrow \mathbf{s} + \mu_j \nabla H_\sigma$. As said in [8], the step-size parameters μ_j should be decreasing, i.e., for smaller values of σ, smaller values of μ_j should be applied. This is because for smaller values of σ the function H_σ is more 'fluctuating', and hence smaller step-size should be used for its maximization. It is shown in [8] that μ_j should be proportional to σ^2. Therefore, we choose $\mu_j = \mu\sigma^2$.

- Initialization
  1. Let $\hat{\mathbf{s}}_0$ be equal to the minimum ℓ^2 norm solution of $\mathbf{A}\mathbf{s} = \mathbf{x}$ obtained by pseudo-inverse of $\mathbf{A}$.
  2. Choose a suitable decreasing sequence for $\sigma : [\sigma_1, \cdots, \sigma_J]$.
- For $j = 1, \cdots, J$:
  1. Let $\sigma = \sigma_j$
  2. Maximize (approximately) the function F_σ on the feasible set $\{\mathbf{s}|\mathbf{A}\mathbf{s} = \mathbf{s}\}$ using L iterations of steepest ascent algorithm (followed by projection onto the feasible set):
     - Initialization: $\mathbf{s} = \hat{\mathbf{s}}_{j-1}$
     - For $l = 1 \cdots L$ (loop L times):
       * Let ∇H_σ as (13).
       * Let $\mathbf{s} \leftarrow \mathbf{s} + (\mu\sigma^2)\nabla H_\sigma$ (where μ is a small positive constant).
       * Project $\mathbf{s}$ back onto the feasible set:

$$\mathbf{s} \leftarrow \mathbf{s} - \mathbf{A}^T(\mathbf{A}\mathbf{A}^T)^{-1}(\mathbf{A}\mathbf{s} - \mathbf{x}) \ .$$

  3. Set $\hat{\mathbf{s}}_j = \mathbf{s}$
- Final answer is $\hat{\mathbf{s}} = \hat{\mathbf{s}}_J$

Fig. 1. The final BSL0 algorithm

4 Experimental Results

In this section, we discuss the performance of our proposed algorithm and compare it with conventional SL0, BOMP and mixed ℓ^2/ℓ^1 norm methods. In all of the experiments, block-sparse sources are artificially produced. A block k-sparse signal is created as follow: First, the locations of k nonzero blocks are randomly chosen. Each element in chosen blocks is 'active' and the rest elements of $\mathbf{s}$ are 'inactive'. Active elements have normal distribution of $\mathcal{N}(0, 1)$ and inactive elements have been set zero.

Each column of the mixing matrix is randomly generated using a normal distribution with zero mean and variance of 1 and then is normalized to unity. Then the mixtures are generated using $\mathbf{x} = \mathbf{A}\mathbf{s} + \mathbf{n}$ where $\mathbf{n}$ is an additive white

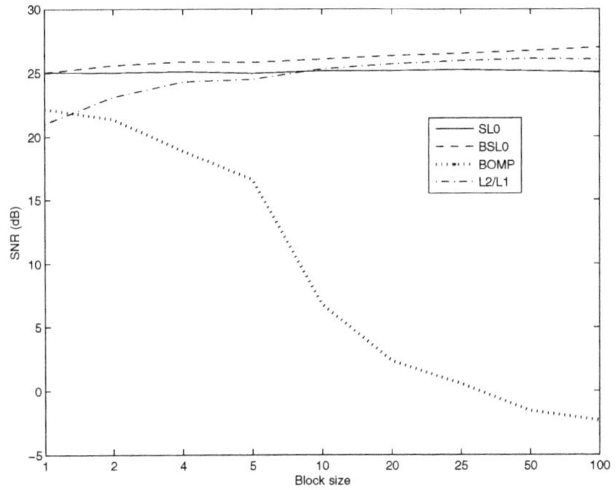

Fig. 2. Performance of BSL0, conventional SL0, BOMP and mixed ℓ^2/ℓ^1 norm as function of block size (d) for $k \times d = 100$

Gaussian noise with covariance matrix $\sigma_n \mathbf{I}_n$ (where $\mathbf{I}_n$ stands for the $n \times n$ identity matrix).

To evaluate the estimation quality, Signal-to-Noise Ratio (SNR) is used. SNR (in dB) is defined as $20 \log(\|\mathbf{s}\|/\|\mathbf{s} - \hat{\mathbf{s}}\|)$, where $\mathbf{s}$ and $\hat{\mathbf{s}}$ denote the actual source and its estimation, respectively.

Experiment 1. In this experiment, we study the computational cost of the presented method, and compare it with conventional SL0, mixed ℓ^2/ℓ^1 norm and BOMP methods. The values used for this experiment are $m = 1000, n = 400, \sigma_n^2 = 0.01$ and $k \times d = 100$. We change k and d during the simulation such that their product stays constant (and equal to 100). Therefore, the possible values for (k, d) will be (100,1), (50,2), (25,4), (20,5), (10,10), (5,20), (4,25), (2,50) and (1,100). For example (k, d) =(25,4) means that $\mathbf{s}$ is a 1000×1 vector with 100 nonzero elements occurring in 25 blocks which contain 4 nonzero elements each.

We use the CPU time as a measure of complexity. Although it is not an exact measure, it gives a rough estimation of the complexity. Our simulations are performed in MATLAB 7.6 environment using an Intel Core 2 Duo 2GHz processor with 3GB of memory, and under Microsoft Windows XP operating system.

The experiment was then repeated 100 times (with the same parameters, but for different randomly generated sources and mixing matrices) and the values of SNR (in dB) and time (in seconds) obtained over these simulations were averaged. Figure 2 shows the result. The averaged CPU time (for $d = 100$) for SL0, BSL0 , ℓ^2/ℓ^1 and BOMP are 0.104, 0.111, 14.623 and 0.070 second, respectively. It is clear that BSL0 has better performance and as we increase the block size, its performance becomes better. This is because of BSL0 cost function in which we directly work with the block version of ℓ^0 norm. Also, as BSL0 is based on the fast SL0 algorithm [8], it is faster than ℓ^2/ℓ^1.

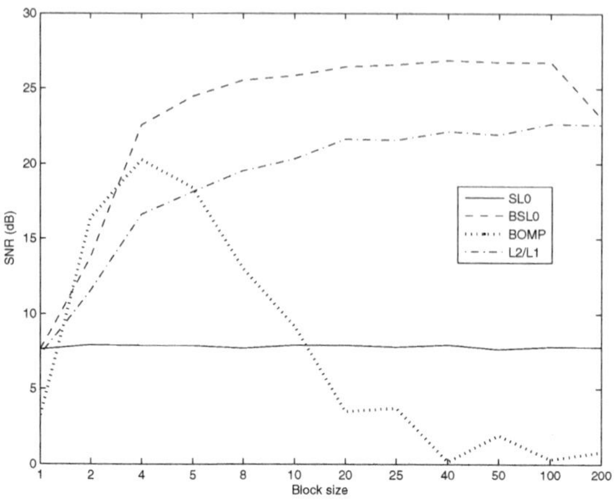

Fig. 3. Performance of BSL0, conventional SL0, BOMP and mixed ℓ^2/ℓ^1 norm as function of block size (d) for $k \times d = 200$

Experiment 2. As said in uniqueness theorem [7], the sparsest solution is unique when $\|\mathbf{s}\|_0 = (k \times d) < n/2 = 200$. So, we set $k \times d = 200$ and chose (k, d) pair from (200,1), (100,2), (50,4), (40,5), (25,8), (20,10), (10,20), (8,25), (5,40), (2,100) and (1,200) to examine its effect on the performance of the algorithms. The other parameters of this experiment are the same as Experiment 1. We can see from Fig. 3 that the conventional SL0 cannot recover the sparsest solution when $\mathbf{s}$ has 200 nonzero elements, but for $d > 4$, BSL0 has a performance of 25 dB or even better. The averaged CPU time (for $d = 200$) for SL0, BSL0, ℓ^2/ℓ^1 and BOMP are 0.091, 0.096, 11.879 and 0.035 second, respectively.

5 Conclusion

In this paper we studied the efficient recovery of block sparse signals using an under-determined system of equations generated from random Gaussian matrices. The motivation for considering block sparse signals is that in many applications the nonzero elements of sparse vectors tend to cluster in blocks. We showed experimentally that BSL0 is highly faster than ℓ^2/ℓ^1, while producing even better estimation accuracy. We also saw experimentally that BSL0 has better performance than SL0 when the number of nonzero elements of the source vector approaches the upper bound of uniqueness theorem.

References

1. Donoho, D.L.: Compressed sensing. IEEE Trans. Info. Theory 52(4), 1289–1306 (2006)
2. Baraniuk, R.G.: Compressive sensing. IEEE Signal Processing Magazine 24(4), 118–124 (2007)

3. Gribonval, R., Lesage, S.: A survey of sparse component analysis for blind source separation: principles, perspectives, ans new challenges. In: ESANN 2006, pp. 323–330 (April 2006)
4. Li, Y., Cichocki, A., Amari, S.: Sparse component analysis for blind source separation with less sensors than sources. In: ICA 2003, pp. 89–94 (2003)
5. Chen, S.S., Donoho, D.L., Saunders, M.A.: Atomic decomposition by basis pursuit. SIAM Journal on Scientific Computing 20(1), 33–61 (1999)
6. Donoho, D.L., Elad, M., Temlyakov, V.: Stable recovery of sparse overcomplete representations in the presence of noise. IEEE Trans. Info. Theory 52(1), 6–18 (2006)
7. Donoho, D.L., Elad, M.: Optimally sparse representation in general (nonorthogonal) dictionaries via l_1 minimization. Proc. Natl. Acad. Sci. 100(5), 2197–2202 (2003)
8. Mohimani, H., Babaie-Zadeh, M., Jutten, C.: A fast approach for overcomplete sparse decomposition based on smoothed ℓ^0 norm. IEEE Trans. Signal Processing 57, 289–301 (2009)
9. Candès, E.J., Romberg, J.K., Tao, T.: Robust uncertainty principles: Exact signal representation from highly incomplete frequency information. IEEE Trans. Info. Theory 52(2), 489–509 (2006)
10. Mohimani, H., Babaie-Zadeh, M., Gorodnitsky, I., Jutten, C.: Sparse recovery using smoothed ℓ^0 (sl0): Convergence analysis. arXiv:cs.IT/1001.5073
11. Eldar, Y.C., Mishali, M.: Robust recovery of signals from a structured union of subspaces. IEEE Trans. Info. Theory 55, 5302–5316 (2009)
12. Eldar, Y.C., Kuppinger, P., Bölcskei, H.: Compressed sensing of block-sparse signals: Uncertainty relations and efficient recovery. arXiv:cs.IT/0906.3173
13. Mishali, M., Eldar, Y.C.: Blind multi-band signal reconstruction: Compressed sensing for analog signals. IEEE Trans. Signal Processing 57(3), 993–1009 (2009)
14. Cotter, S., Rao, B.: Sparse channel estimation via matching pursuit with application to equalization. IEEE Trans. on Comm. (March 2002)
15. Stojnic, M., Parversh, F., Hassibi, B.: On the reconstruction of block-sparse signals with an optional number of measurements. IEEE Trans. Sig. Proc. (2009)
16. Majumdar, A., Ward, R.: Fast group sparse classification. In: PACRIM, pp. 11–16 (2009)
17. Mallat, S., Zhand, Z.: Matching pursuits with time-frequency dictionaries. IEEE Trans. on Signal proc. 41(12), 3397–3415 (1993)

Second-Order Source Separation Based on Prior Knowledge Realized in a Graph Model

Florian Blöchl[1,3], Andreas Kowarsch[1,3], and Fabian J. Theis[1,2]

[1] Institute for Bioinformatics and Systems Biology,
Helmholtz Zentrum München, 85764 Neuherberg, Germany
[2] Institute for Mathematical Sciences, TU München, 85747 Garching, Germany
[3] Equal contributors
{florian.bloechl,andreas.kowarsch,fabian.theis}@helmholtz-muenchen.de

Abstract. Matrix factorization techniques provide efficient tools for the detailed analysis of large-scale biological and biomedical data. While underlying algorithms usually work fully blindly, we propose to incorporate prior knowledge encoded in a graph model. This graph introduces a partial ordering in data without intrinsic (e.g. temporal or spatial) structure, which allows the definition of a *graph-autocorrelation* function. Using this framework as constraint to the matrix factorization task we develop a second-order source separation algorithm called graph-decorrelation algorithm (GraDe). We demonstrate its applicability and robustness by analyzing microarray data from a stem cell differentiation experiment.

1 Introduction

Autocorrelations allow us to identify repeating patterns, such as the presence of periodic signals. In the case of multivariate signals, the cross-correlation with itself after a time shift is called *auto-cross-correlation*, or simply *autocorrelation*. Here, off-diagonal elements detect shifted correlations between different sample dimensions. Various source separation techniques based on intrinsic autocorrelation structure have shown a fast and robust performance [1–4].

However, there exist many examples where data samples do not follow a natural order, such as temporal or spatial ordering, which is necessary for defining a generic kind of autocorrelation function. This holds true especially for large-scale biological data like gene expression levels from microarrays and lipidomic or metabolomic profiles from mass spectrometry. Here, samples are clearly not i.i.d. since they are the read-outs of different states of a complex dynamical system. For instance, genes obey dynamics along a transcription factor network.

We therefore generalize the concept of autocorrelation by using prior knowledge of these interactions to shift samples along a pre-defined underlying graph. Thereby prior information may be of different granularity: edges can be binary (e.g. interactions) or weighted (interaction strength, reaction rates). Based on this concept of graph-autocorrelation we formulate a second-order source separation algorithm called GraDe. We apply Grade to microarray data from a stem cell differentiation experiment. In contrast to other factorization techniques, it finds a structured and detailed separation of known biological processes.

V. Vigneron et al. (Eds.): LVA/ICA 2010, LNCS 6365, pp. 434–441, 2010.

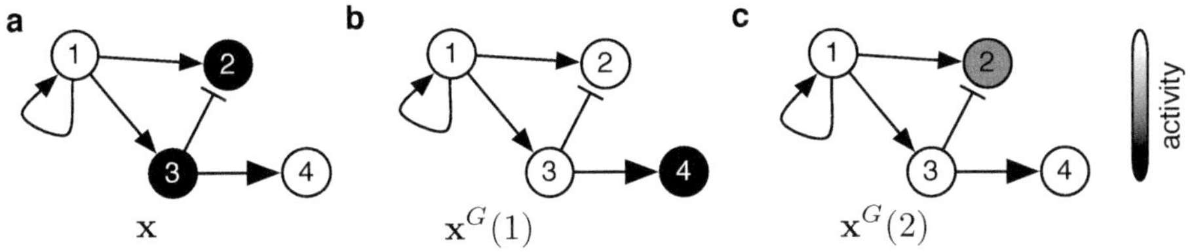

Fig. 1. Illustration of G-shifts with an interaction graph G. Here, arrows symbolize activating (+1), stopped edges inhibitory character (−1) of the interaction. Starting with an initial node activity $\mathbf{x}$ depicted in **a**, after one and two positive G-shifts we find the activity patterns $\mathbf{x}^G(1)$ and $\mathbf{x}^G(2)$ in (**b**) and (**c**), respectively.

2 Graph–Autocorrelation

We encode prior knowledge in a directed, weighted graph $G := (\mathcal{V}, \mathcal{E}, w)$, defined on vertices $\mathcal{V} = \{1, \ldots, l\}$ corresponding to l samples of our data set under consideration. Edges are weighted with weights $w : \mathcal{E} \to \mathbb{R}$ that may be negative. G may also contain self-loops. All weights are collected in a *weight matrix* $\mathbf{W} \in \mathbb{R}^{l \times l}$, where $\mathbf{W}_{ij}$ specifies the weight of edge $i \to j$. Any vertex $i \in \mathcal{V}$ has a set of *successors* $S(i) := \{j | (i, j) \in \mathcal{E}\}$ and *predecessors* $P(i) := \{j | (j, i) \in \mathcal{E}\}$.

We want to shift samples along the partial order implied by the graph G. Note that besides the graph topology also the linkage of different predecessors of a node is crucial. In the following, we choose a linear superposition of inputs which is an appropriate approximation in many applications like e.g. gene expression [5] as analyzed in Section 4. With this choice we do not require detailed information on individual linkages and ensure robustness to noise, i.e. false edges. We point out that any other logic better adapted to certain data can be used instead.

For $\mathbf{x} \in \mathbb{R}^l$, we define the *G-shift* $\mathbf{x}^G$ of $\mathbf{x}$ as the vector with components

$$x_i^G := \sum_{j \in P(i)} \mathbf{W}_{ji} \, x_j \, . \tag{1}$$

Recursively, we define any positive shift $\mathbf{x}^G(\tau)$. This concept is illustrated in Figure 1. For negative shifts we replace predecessors by successors. This flipping of edge directions formally corresponds to a transposition of $\mathbf{W}$. If we assign zero weights to non-existing edges, the above sum can be extended to all vertices. Gathering observations (as rows) into a multivariate data matrix $\mathbf{X}$, we obtain the simple, convenient formulation of the G-shifted data set

$$\mathbf{X}^G(\tau) = \begin{cases} \mathbf{X}\mathbf{W}^{\tau} & \tau \geq 0 \\ \mathbf{X}(\mathbf{W}^{\top})^{\tau} & \tau < 0 \end{cases} . \tag{2}$$

After mean removal, each row of $\mathbf{X}$ is centered. Then—in analogy to the unbiased estimator for cross-correlations—we define the *graph-(cross)-autocorrelation*

$$\mathbf{C}_{\mathbf{X}}^G(\tau) := \frac{1}{l-1} \mathbf{X}^G(\tau) \mathbf{X}^{\top} = \frac{1}{l-1} (\mathbf{X}\mathbf{W}^{\tau}\mathbf{X}^{\top}) \, . \tag{3}$$

The last step is valid for positive shifts with our linkage (2). Note that this definition includes the standard non-graph autocorrelation by shifting along the line graph $1 \to 2 \to \ldots \to l-1 \to l$.

The graph-autocorrelation in (3) is not symmetric, except if the graph G is symmetric. We therefore introduce the *symmetrized graph-correlation*

$$\overline{\mathbf{C}}_{\mathbf{X}}^{G}(\tau) = \frac{1}{2}\left(\mathbf{C}_{\mathbf{X}}^{G}(\tau) + \mathbf{C}_{\mathbf{X}}^{G}(\tau)^{\top}\right) . \tag{4}$$

Since we chose the logic from (2), $\mathbf{C}_{\mathbf{X}}^{G}(\tau)^{\top} = \mathbf{C}_{\mathbf{X}}^{G}(-\tau)$ and therefore we focus on the mean correlation shifting both in positive and negative direction.

3 Source Separation by Graph-Decorrelation

Now, let us consider the general linear source separation problem

$$\mathbf{X} = \mathbf{A}\mathbf{S} + \mathbf{n} \tag{5}$$

for a data set $\mathbf{X} \in \mathbb{R}^{m\times l}$ of m observations and l samples. Here, the mixing matrix $\mathbf{A} \in \mathbb{R}^{m\times n}$ $(m \geq n)$ has full rank. The sources $\mathbf{S} \in \mathbb{R}^{n\times l}$ are uncorrelated, zero-mean stationary processes with nonsingular covariance matrix. The noise term $\mathbf{n} \in \mathbb{R}^{m\times l}$ is a stationary, white zero-mean process with variance σ^2. Without additional restrictions, this factorization problem is, of course, underdetermined.

Here, we assume that the sources have *vanishing G-cross-correlation* with respect to a given graph G and all shifts τ, i.e. $\overline{\mathbf{C}}_{\mathbf{S}}^{G}(\tau)$ is diagonal. We observe

$$\overline{\mathbf{C}}_{\mathbf{X}}^{G}(\tau) = \begin{cases} \mathbf{A}\overline{\mathbf{C}}_{\mathbf{S}}^{G}(\tau)\mathbf{A}^{\top} + \sigma^2\mathbf{I}, & \tau = 0 \\ \mathbf{A}\overline{\mathbf{C}}_{\mathbf{S}}^{G}(\tau)\mathbf{A}^{\top} & \tau \neq 0 \end{cases} . \tag{6}$$

3.1 Identifiability

Clearly, a full identification of $\mathbf{A}$ and $\mathbf{S}$ is impossible, because Equation (5) defines them only up to scaling and permutation of columns: Multiplication of a source by a constant can be compensated by dividing the corresponding row of $\mathbf{A}$. We take advantage of this scaling indeterminacy by requiring that the sources have unit variance, i.e. $\overline{\mathbf{C}}_{\mathbf{S}}^{G}(0) = \mathbf{I}$. We assume white unperturbed data $\tilde{\mathbf{X}} := \mathbf{A}\mathbf{S}$ (possibly after whitening transformation). Then, we see that $\mathbf{A}\mathbf{A}^{\top} = \mathbf{I}$, i.e. $\mathbf{A}$ is orthogonal. Hence, for $\tau > 0$ the factorization (6) represents the eigenvalue decomposition of the symmetric matrix $\overline{\mathbf{C}}_{\mathbf{X}}^{G}(\tau)$. If additionally $\overline{\mathbf{C}}_{\mathbf{S}}^{G}(\tau)$ has pairwise different eigenvalues, the spectral theorem ensures that $\mathbf{A}$ – and hence $\mathbf{S}$ – is uniquely determined by $\mathbf{X}$ up to permutation. However, care has to be taken, because we cannot expect $\overline{\mathbf{C}}_{\mathbf{X}}^{G}(\tau)$ to be of full rank. Obviously, we require more samples than observations $(l \gg m)$, so $\text{rank}(\mathbf{X}) = m$. If G contains an adequate amount of information, $\text{rank}(\mathbf{W})$ is of order l. Then $\text{rank}(\overline{\mathbf{C}}_{\mathbf{X}}^{G}(\tau))$ is essentially determined by (the upper bound) m. Hence, with sufficient prior knowledge we can extract as many sources as observations are available.

3.2 Graph–Decorrelation: The GraDe Algorithm

Equation (6) gives an indication of how to solve the source separation task in our setting. First, we whiten the no-noise term $\tilde{\mathbf{X}} = \mathbf{AS}$ of the observed mixtures. The whitening matrix can be estimated from $\mathbf{X}$ by diagonalizing the symmetric matrix $\overline{\mathbf{C}}^G_{\tilde{\mathbf{X}}}(0) = \overline{\mathbf{C}}^G_{\mathbf{X}}(0) - \sigma^2\mathbf{I}$, provided that the noise variance σ^2 is known or reasonably estimated. If we have more sources than signals, dimension reduction is possible in this step. Insignificant eigenvalues then allow estimation of σ^2 [2]. Now, we estimate the sources by diagonalization of the single, symmetric autocorrelation matrix $\overline{\mathbf{C}}^G_{\mathbf{X}}(\tau)$. This procedure generalizes AMUSE [1], which employs standard autocorrelation, to the extended definition of graph-correlation.

The performance of AMUSE is relatively sensitive to additive noise. Moreover, the estimation by a finite amount of samples may lead to a badly estimated autocorrelation matrix [3]. To alleviate these problems, algorithms like SOBI [2] or TDSEP [4] extend this approach by *joint diagonalization* of *multiple* autocorrelation matrices, calculated with a set of different time-delays. In a similar manner we jointly diagonalize multiple G-autocorrelation matrices obtained from G-shifts with different lags. This approximative joint diagonalization can be achieved by a variety of methods. We use the Jacobi-type algorithm proposed in [6], since we later compare GraDe's performance to the classic SOBI algorithm.

Altogether, we subsume this procedure in the graph-decorrelation algorithm (GraDe). At `http://cmb.helmholtz-muenchen.de/grade` an implementation is freely available. When shifting along the line graph, GraDe with a single lag reduces to AMUSE, and GraDe with multiple shifts corresponds to SOBI.

3.3 Evaluation on Artificial Data

In order to evaluate GraDe, we generated random mixtures of artificial G-decorrelated signals. A common way to create standard-autocorrelated signals are *moving average* (MA) models [7]: For a white noise process ϵ and real coefficients $\theta_1 \dots \theta_q$, a MA(q) model $\mathbf{x}$ is defined by $\mathbf{x}_t = \epsilon_t + \theta_1\epsilon_{t-1} + \cdots + \theta_q\epsilon_{t-q}$. In our notation, we interpret this MA signal $\mathbf{x}$ as a weighted sum of G-shifted versions of ϵ, shifted q times along the line graph G. Therefore, for an arbitrary graph G we define a q-th order *G–MA(q) model* as

$$\mathbf{x} = \epsilon + \theta_1\epsilon^G(1) + \theta_2\epsilon^G(2) + \ldots + \theta_q\epsilon^G(q) \tag{7}$$

Any G-MA(q) process is equivalent to a G-MA(1) process with a modified graph.

In a first simulation, we used directed Erdös-Rényi random graphs [8] with mean connectivity 17.5 and random weights in $(-1, 1)$ to generate $m = 2$ G–decorrelated G-MA(1) signals with $l = 5000$ samples. Data were normalized to unit variance and mixed with a random mixing matrix. We added Gaussian uncorrelated noise of variable strength σ and applied GraDe (without noise estimation) with one and 30 shifts, respectively. Reconstruction quality was estimated using the Amari-index that quantifies the deviation between the correct and the estimated mixing matrix [9]. From Figure 2a we see that for G-MA(1) processes

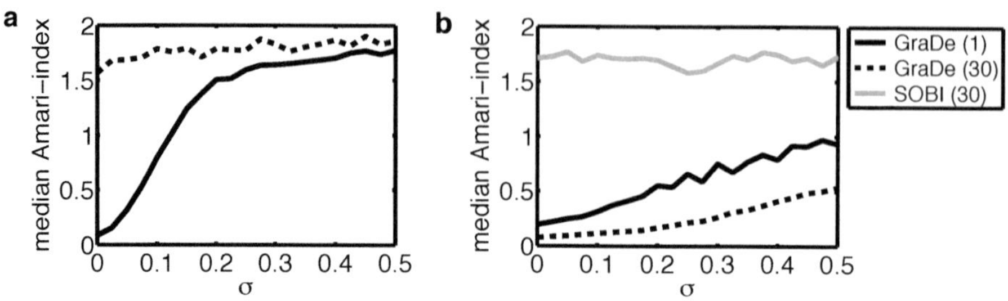

Fig. 2. Performance on artificial data: mixtures of (**a**) two G-MA(1) processes with random graphs G, (**b**) mixtures of two G-MA(20) processes with signed line graphs. The plots show the dependence of median Amari–indices on the noise level σ over 1000 runs. We compare GraDe with one and 30 shifts, in (**b**) in addition SOBI with 30 shifts.

GraDe with a single-shift performs well in the low-noise setting, in contrast to multiple shifts. This is a consequence of the complex short-distance, but vanishing long-distance autocorrelation structures. When performing multiple shifts, each lag is weighted equally, which deteriorates the algorithm's performance.

Accordingly, as shown in Figure 2b, GraDe with multiple shifts outperformed single–shift GraDe when applied to higher order G–MA processes. We generated G–MA(20) processes of sample size $l = 1500$ with a signed line graph, where the edges had weights ± 1 with equal probability. The unsigned line graph used by SOBI was not sufficient to reconstruct these signals in a proper way, whereas GraDe with the true graph separated them even using a single shift only. However, similar to the behavior in the standard-autocorrelation case, here multiple shifts dramatically enhance GraDe's robustness against additive noise.

4 A Microarray Experiment on Stem Cell Differentiation

The regulation of gene expression is essential to proper cell functioning. The first step of gene expression is mRNA transcription. Here, a copy of a gene from the DNA to messenger RNA (mRNA) is made, encoding a chemical "blueprint" for a protein product. Microarrays are the state-of-the-art technology for the genome-wide measurement of these transcript levels. They are known to be quite noisy, and the still high costs keep the number of replicates small. This makes gene expression analysis a particular challenge for machine learning. Matrix factorization techniques are currently explored as unsupervised approaches to such data [10]. Here, the extracted gene expression sources (GES) can be interpreted as distinct biological processes, which are active on a level quantified in the mixing matrix. Applying GraDe, we require that biological processes that can be explained by the underlying network are not split up between different GES.

In the following, we interpret a microarray experiment investigating the crucial role of the transcription factor STAT5 during hematopoietic stem cell differentiation. STAT5 is strongly activated in 30% of patients with acute myelogenous

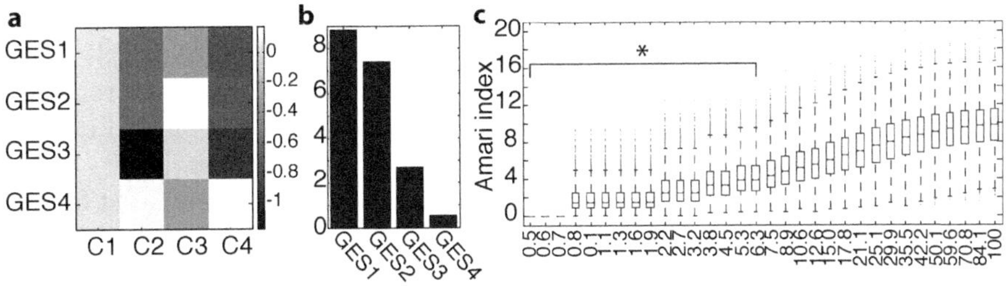

Fig. 3. GraDe result and robustness analysis: The heatmap (**a**) shows the mixing matrix (centered to C1). Conditions C1–C4 correspond to stimulated/unstimulated GMPs (C1-C2) and STAT5–ko cells (C3-C4). (**b**): Eigenvalues of the four GES. (**c**) Amari-indices for the randomized networks against the fraction of randomized edges (in %). The $*$ indicates significant Amari-indices.

leukemia (AML) [11]. The cytokine GM-CSF activates STAT5 and controls the differentiation of progenitor cells (GMPs) into granulocytes and macrophages, two types of white blood cells. AML cells are mostly from the granulocyte lineage, hence it is important to elucidate the role of STAT5 upon GM-CSF stimulation.

In [11], both normal and STAT5-knockout GMPs were stimulated with GM-CSF or left unstimulated. RNA from these four samples was measured with microarrays. Data are available at GEO under accession number GSE14698. We used 1601 differentially expressed genes for further analysis (t-test on stimulated vs. unstimulated gene expression significant with a p-value< 0.05).

To interconnect these genes we used the known gene–gene interactions that are collected in the database TRANSPATH [12]. Applying GraDe, we preferred a single shift, as multiple shifts may lead to an accumulation of errors in the interaction network. We obtained four GES (Figure 3b). We selected genes in the GES that were expressed above the threshold ± 2 and mapped these sets onto biological processes by performing Gene-Ontology (GO) enrichment analysis.

4.1 GraDe Separates GM-CSF and STAT5 Dependent Processes

The sources extracted by GraDe separate GM-CSF and STAT5 dependent biological processes. Figure 3a shows that the contribution of GES 1 differs between stimulated and unstimulated GMPs, but also between wild-type and STAT5–ko. Enriched GO–Terms correspond to responses triggered by external stimuli, activation or regulation of signal transduction as well as responses to STAT5 such as the MAPK or JNK signaling cascades. This is line with previous work [13]. In addition, we found processes linked to cell differentiation, the primary response to GM-CSF stimulation [11]. GES 2 separates stimulated and unstimulated cells, independently of STAT5 condition. Consequently, we identifed biological processes linked to immune response. GM-CSF is an important hematopoietic growth factor for enhancing immune responses and is known to recruit and activate antigen-presenting cells [14]. It also has profound effects on the functional activities of various circulating leukocytes, which are involved in defense processes [13].

Genes in GES 3 can be linked to telomere organization and maintenance. Telomerase activity is reported to be nearly 3-fold higher in GM-CSF stimulated cells [15]. Telomere length is a critical factor in determining the replicative potential of mitotic cells. The accelerated telomere shortening due to excessive replication may hint at hematopoietic stem cell premature aging.

GES 4 has a different contribution in stimulated and unstimulated STAT5–knockouts. Similarly to GES 2, we found several processes linked to immune responses. Additionally, GES 4 contains genes associated with the activation of macrophages or myeloid leukocytes. GM-CSF is involved in the generation of granulocytes and macrophages, responsible for non-specific defense processes.

Applying PCA or ICA [7] to the data, we found only in two sources significantly enriched GO–Terms. Note that we cannot employ SOBI since a gene's position on the microarray chip is completely arbitrary. PCA extracted a source linked to immune response and aggregates various biological processes in a second one. We obtained a similar result performing ICA with JADE [16], which also separates the immune response from other biological processes. Thus, GraDe finds a much more structured, detailed response than fully blind approaches.

4.2 Robustness to Graph Errors

The quality of the employed regulatory network is not perfect. It may contain false interactions and is far from complete. To analyze the robustness of GraDe, we randomized between 0.5 and 100% of the edges in the TRANSPATH graph. In each step we reshuffled 10.000 times the corresponding percentage of edges using degree-preserving rewiring [17]. Applying GraDe with these graphs we obtained new factorizations. We used the Amari index to quantify changes and determined a p-value to detect significantly low changes. This p-value was calculated by comparing the 95% quantile of Amari-indices for each randomization step with Amari-indices for normally distributed random separating matrices. We obtained significantly low ($p < 0.05$) Amari-indices for up to 6% of rewired edges (Figure 3c). Thus, the underlying graph has obviously a strong influence on, but we showed that GraDe is robust against a reasonable amount of graph errors.

5 Conclusion

We have introduced the new measure of graph-autocorrelation. This measure allows us to formulate GraDe, a second-order matrix factorization algorithm that incorporates prior knowledge into the source separation task. In both, simulations with artificial data and a stem cell differentiation microarray experiment, we demonstrated applicability as well as robustness of the proposed approach. In future work we will investigate whether GraDe can be used for model selection when given different alternative underlying graphs of small-scale models.

Acknowledgements. We thank Rodolphe Sepulchre and Dominik Wittmann for stimulating remarks. This work was supported by the Federal Ministry of Education and Research (BMBF) and its MedSys initiative (projects 'LungSys' and 'SysMBo') and the Helmholtz Alliance on Systems Biology (project 'CoReNe').

References

1. Tong, L., Liu, R.W., Soon, V., Huang, Y.F.: Indeterminacy and identifiability of blind identification. IEEE Transactions on Circuits and Systems 38, 499–509 (1991)
2. Belouchrani, A., Abed-Meraim, K., Cardoso, J.F., Moulines, E.: A blind source separation technique using second-order statistics. IEEE Transactions on Signal Processing 45(2), 434–444 (1997)
3. Theis, F., Meyer-Bäse, A., Lang, E.: Second-order blind source separation based on multi-dimensional autocovariances. In: Puntonet, C.G., Prieto, A.G. (eds.) ICA 2004. LNCS, vol. 3195, pp. 726–733. Springer, Heidelberg (2004)
4. Ziehe, A., Mueller, K.R.: TDSEP – an efficient algorithm for blind separation using time structure. In: Niklasson, L., Bodén, M., Ziemke, T. (eds.) Proc. of ICANN 1998, Skövde, Sweden, pp. 675–680. Springer, Berlin (1998)
5. Snoussi, E.H., Thomas, R.: Logical identification of all steady states: the concept of feedback loop characteristic states. Bulletin of Math. Biology 55(5), 973–991 (1993)
6. Cardoso, J., Souloumiac, A.: Jacobi angles for simultaneous diagonalization. SIAM J. Mat. Anal. Appl. 17(1), 161–164 (1995)
7. Hyvärinen, A., Karhunen, J., Oja, E.: Independent component analysis. John Wiley & Sons, Chichester (2001)
8. Erdős, P., Rényi, A.: On random graphs. Publicationes Mathematicae 6, 290–297 (1959)
9. Cichocki, A., Amari, S.: Adaptive Blind Signal and Image Processing: Learning Algorithms and Applications. Wiley, New York (2002)
10. Schachtner, R., Lutter, D., Knollmüller, P., Tomé, A., Theis, F., Schmitz, G., Stetter, M., Vilda, P.G., Lang, E.: Knowledge-based gene expression classification via matrix factorization. Bioinformatics 24(15), 1688–1697 (2008)
11. Kimura, A., Rieger, M.A., Simone, J.M., Chen, W., Wickre, M.C., Zhu, B.M., Hoppe, P.S., O'Shea, J.J., Schroeder, T., Hennighausen, L.: The transcription factors STAT5A/B regulate GM-CSF-mediated granulopoiesis. Blood 114(21), 4721–4728 (2009)
12. Krull, M., Pistor, S., Voss, N., Kel, A., Reuter, I., Kronenberg, D., Michael, H., Schwarzer, K., Potapov, A., Choi, C., Kel-Margoulis, O., Wingender, E.: TRANSPATH: an information resource for storing and visualizing signaling pathways and their pathological aberrations. Nucleic acids research 34, D546–D551(2006)
13. Shi, Y., Liu, C.H., Roberts, A.I., Das, J., Xu, G., Ren, G., Zhang, Y., Zhang, L., Yuan, Z.R., Tan, H.S.W., Das, G., Devadas, S.: Granulocyte-macrophage colony-stimulating factor (GM-CSF) and T-cell responses: what we do and don't know. Cell research 16(2), 126–133 (2006)
14. Krakowski, M., Abdelmalik, R., Mocnik, L., Krahl, T., Sarvetnick, N.: Granulocyte macrophage-colony stimulating factor (GM-CSF) recruits immune cells to the pancreas and delays STZ-induced diabetes. The Journal of pathology 196(1), 103–112 (2002)
15. Flanary, B.E., Streit, W.J.: Progressive telomere shortening occurs in cultured rat microglia, but not astrocytes. Glia 45(1), 75–88 (2004)
16. Cardoso, J.F.: High-order contrasts for independent component analysis. Neural Comput. 11(1), 157–192 (1999)
17. Wong, P., Althammer, S., Hildebrand, A., Kirschner, A., Pagel, P., Geissler, B., Smialowski, P., Blöchl, F., Oesterheld, M., Frishman, D.: An evolutionary and structural characterization of mammalian protein complex organization. BMC Genomics 9(1), 629 (2008)

Noise Adjusted PCA for Finding the Subspace of Evoked Dependent Signals from MEG Data

Florian Kohl[1,2], Gerd Wübbeler[2], Dorothea Kolossa[1], Clemens Elster[2], Markus Bär[2], and Reinhold Orglmeister[1]

[1] Technische Universität Berlin, Straße des 17. Juni 135, 10623 Berlin, Germany
[2] Physikalisch-Technische Bundesanstalt (PTB), Abbestraße 2-12, 10587 Berlin, Germany
{florian.kohl,reinhold.orglmeister}@tu-berlin.de,
d.kolossa@ee.tu-berlin.de,
{gerd.wuebbeler,clemens.elster,markus.baer}@ptb.de

Abstract. Evoked signals that underlie multi-channel magnetoencephalography (MEG) data can be dependent. It follows that ICA can fail to separate the evoked dependent signals. As a first step towards separation, we adress the problem of finding a subspace of possibly mixed evoked signals that are separated from the non-evoked signals. Specifically, a vector basis of the evoked subspace and the associated mixed signals are of interest.

It was conjectured that ICA followed by clustering is suitable for this subspace analysis. As an alternative, we propose the use of noise adjusted PCA (NAPCA). This method uses two covariance matrices obtained from pre- and post-stimulation data in order to find a subspace basis. Subsequently, the associated signals are obtained by linear projection onto the estimated basis. Synthetic and recorded data are analyzed and the performance of NAPCA and the ICA approach is compared.

Our results suggest that ICA followed by clustering is a valid approach. Nevertheless, NAPCA outperforms the ICA approach for synthetic and for real MEG data from a study with simultaneous visual and auditory stimulation. Hence, NAPCA should be considered as a viable alternative for the analysis of evoked MEG data.

1 Introduction

Multi-channel magnetoencephalography (MEG) is a non-invasive technique to record magnetic fields produced by neuronal signals in the human brain. Often, stimulus experiments are conducted to investigate and to map brain functioning. For this, the human subject is exposed to one (or several) external stimuli, such as tones or pictures. A stimulus is commonly repeated for many times (>100) and a time interval including one stimulus is called a trial.

The ultimate goal is to extract the locations and time dynamics of the evoked neuronal activities for each trial. Single-trial analysis is challenging as the MEG recordings appear noisy. Indeed, the recordings consist of superimposed fields of many brain and non-brain signals. Independent component analysis (ICA) methods [1] are frequently used in order to decompose the signal mixture.

V. Vigneron et al. (Eds.): LVA/ICA 2010, LNCS 6365, pp. 442–449, 2010.

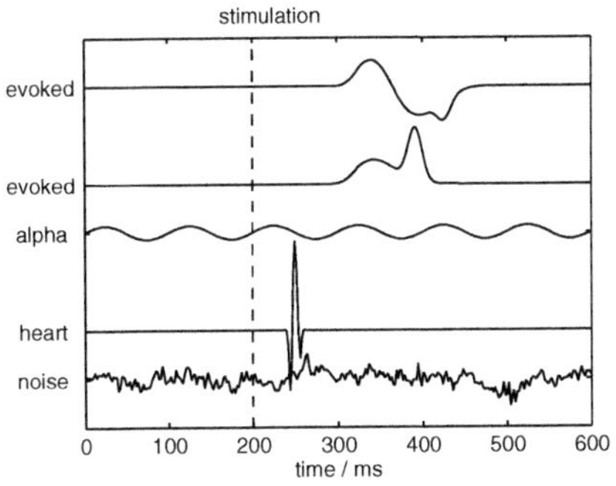

Fig. 1. One trial of synthetic evoked MEG signals. The first two rows depict typical evoked signals, while the remaining rows depict background activities, which reflect possible non-evoked signals. NAPCA uses the estimation of two covariance matrices by considering pre- and post-stimulation data, i.e. data with and without contribution of the evoked signals.

Here, MEG data from simultaneous auditory and visual stimulation experiments are considered. As more than one neuronal signal is evoked at the same time, the signals are assumed *not* to be independent and ICA can fail in separation [2]. As a first step towards separation of dependent signals, we aim at finding the subspace of the evoked signals, i.e. a vector basis and the associated possibly mixed signals of the subspace. The recovered signal subspace is of interest for further MEG data analysis, e.g. for source localisation or as preprocessing step for other separation techniques.

Two methods are investigated. The first is ICA followed by clustering. Based on the conjecture of Cardoso [3] that ICA separates all independent signals from the group of dependent signals, ICA is run on the MEG data with underlying dependent signals. Subsequently, the signal to noise ratio (SNR) of the estimated ICA signals is used to cluster the ICA components into the subspaces of evoked and non-evoked signals.

The second method is noise adjusted principal component analysis (NAPCA) [4,5]. This method was introduced for image processing [4] and has been used as a spatial filter for evoked MEG data [5]. Here, we advocate the use of NAPCA for the analysis of evoked MEG data.

Note that both the basis and the associated signals are of interest. The basis is optimal if it contains linearly independent vectors that equal linear combinations of the evoked vectors. The associated signals are optimal if these are corresponding linear combinations of the evoked signals. Hence, two measures of performance are considered: Nolte's performance index for the quality of the estimated basis [6] and Lemm's SNR meausure [7] for the quality of the recovered signals.

Synthetic and real experiments are assessed. For synthetic data, NAPCA outperforms ICA followed by clustering. For real data, NAPCA recovers clearer dipolar field distributions (the basis) and higher SNRs of the signals, which illustrates its potential.

The paper is organized as follows. Section 2 presents performance measures and theory. Section 3 is on methods assessment with a virtual experiment. Section 4 investigates real evoked MEG data. We conclude in Section 5.

2 Theory

Let $\mathbf{X} \in \mathbb{R}^{m \times L}$ denote the recorded MEG data, which is modeled as

$$\mathbf{X} = \mathbf{AS} + \mathbf{E}, \tag{1}$$

where $\mathbf{A} \in \mathbb{R}^{m \times n}$ is the mixing matrix, $\mathbf{S} \in \mathbb{R}^{n \times L}$ is the signal matrix and $\mathbf{E} \in \mathbb{R}^{m \times L}$ is the sensor noise matrix with m sensors, n signals and L available samples. All matrices are assumed to have full rank. Let the i'th column of the mixing matrix be $\mathbf{A^i}$ and the i'th row of the signal matrix be $\mathbf{S_i}$. Let further Ω be the set of signal indices and EV the subset of evoked signal indices with $EV \subset \Omega$. The data may be given by two parts, the evoked signal part

$$\mathbf{X^S} = \sum_{\mathbf{i} \in \mathbf{EV}} \mathbf{A^i S_i} \tag{2}$$

and the noise part

$$\mathbf{X^N} = \sum_{\mathbf{i} \notin \mathbf{EV}} \mathbf{A^i S_i} + \mathbf{E} \tag{3}$$

which summarize all unwanted contributions to the MEG data. This may now be expressed as

$$\mathbf{X} = \mathbf{X^S} + \mathbf{X^N}. \tag{4}$$

The task of subspace analysis is to find a basis, i.e. any set of linear independent vectors that are linear combinations of the evoked mixing vectors $\mathbf{A^i}$, and to find the corresponding linear combinations of the evoked signals $\mathbf{S_i}$, where $i \in EV$.

2.1 Performance Measures

Nolte's Performance Index. Let $\mathbf{B}$ be the matrix that contains the true basis vectors of the signal subspace and let $\hat{\mathbf{B}}$ be its estimate. Let further $P_{\mathbf{B}} = \mathbf{B}(\mathbf{B^T B})^{-1}\mathbf{B^T}$ and $P_{\hat{\mathbf{B}}} = \hat{\mathbf{B}}(\hat{\mathbf{B}}^T \hat{\mathbf{B}})^{-1} \hat{\mathbf{B}}^T$ be projector matrices. Then Nolte's subspace performance index is given by the M_e'th largest eigenvalue λ_{M_e} of the matrix

$$P_{\mathbf{B}} P_{\hat{\mathbf{B}}} P_{\mathbf{B}}, \tag{5}$$

where M_e is the dimension of the evoked signal subspace. The performance index λ_{M_e} is between 0 and 1 and equals 1 iff the estimated basis spans the true subspace. This index was proposed for the Signal Separation Evaluation Campaign (SiSEC) 2010 by Nolte (cf. also [6]).

Signal to Noise Ratio. Let $\hat{\mathbf{S}}_i$ be i'th estimated signal and let the operator $\overline{[\cdot]}$ denote averaging over trials. Let further $\tilde{\mathbf{S}}_i$ denote i'th estimated signal with subtracted trial average $\overline{\hat{\mathbf{S}}}_i$ from each trial. Following [7], we define the signal to noise ratio (SNR) of an evoked signal as

$$SNR(\hat{\mathbf{S}}_i) = \frac{var(\overline{\hat{\mathbf{S}}_i})}{\overline{var(\tilde{\mathbf{S}}_i)}}. \tag{6}$$

2.2 ICA Followed by Clustering

ICA followed by clustering assumes that ICA can separate the group of dependent signals from the independent signals [3]. If this conjecture is valid, the evoked signal subspace is obtained by running ICA on the MEG data. Subsequently, a suitable choice of M_e number of estimated ICA signals give an estimate to the subspace signals. We chose SNR-based (cf. Equation 6) clustering. The corresponding ICA estimated mixing vectors give an estimate of the basis of the subspace. In this work FastICA [1] has been used for all experiments.

2.3 Noise Adjusted Principal Component Analysis

According to Equation 4, the MEG data can be split into a noise and a signal part. NAPCA assumes that the noise part is stationary, that the signal part is zero at known data periods, and that the noise and signal part are uncorrelated. This setting is depicted by Figure 1. Let $\mathbf{C_X}$ and $\mathbf{C_{X^N}}$ be the covariance matrix of the data and noise part, respectively. NAPCA then consists of two linear transformations. The first transformation whitens the noise part by left-multiplication of Equation 4 with a whitening matrix $\mathbf{V}$ obtained from the covariance matrix $\mathbf{C_{X^N}}$ of the noise part [1]. The transformed data is given by

$$\mathbf{V^T X} = \mathbf{V^T X^S} + \mathbf{V^T X^N}, \tag{7}$$

where the transformed covariance matrix is $\mathbf{V^T C_X V}$. The transformed noise part $\mathbf{V^T X^N}$ is white, thus $\mathbf{V^T C_{X^N} V} = \mathbf{I}$. The second transformation consists of an ordinary PCA, given by

$$\mathbf{E^T V^T C_X V E} = \mathbf{\Lambda}, \tag{8}$$

where $\mathbf{E}$ is the matrix of orthogonal eigenvectors and $\mathbf{\Lambda}$ the matrix of the corresponding eigenvalues, sorted by decreasing order.

The NAPCA estimated subspace basis is given by

$$\hat{\mathbf{B}} = (\mathbf{V^T})^{-1}\tilde{\mathbf{E}}, \tag{9}$$

where $\tilde{\mathbf{E}}$ is the matrix that contains the first M_e eigenvectors.The corresponding signals are estimated by linearly projecting the data onto the subspace basis without inverting the first NAPCA transformation, which may be expressed as

$$\hat{\mathbf{S}} = \tilde{\mathbf{E}}^T \mathbf{V^T X}. \tag{10}$$

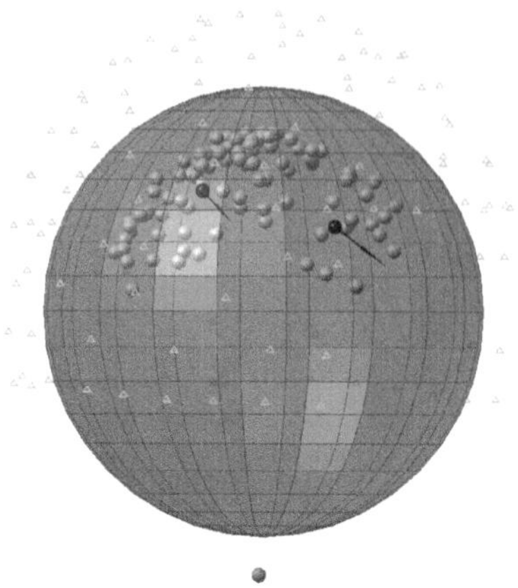

Fig. 2. Sphere model of the human head. A 2-dimensional evoked subspace is given by two equivalent current dipoles (ECDs) (black) with dependent signals. 100 background ECDs (gray) model alpha, eye, heart and other non-evoked independent signals. Sensor locations (diamonds) correspond to the PTB 125-channel MEG system.

3 Virtual Experiment

A virtual experiment was set up with a 2-dimensional evoked subspace. Figure 2 depicts the utilized sphere model. Equivalent current dipoles (ECD) model neuronal currents. Small black spheres depict one realization of evoked ECDs and gray spheres the fixed locations of the noise ECDs modeling heart, alpha, eyes, and brain background signals. 100 such ECDs with corresponding unwanted signals are simulated with different energies. Magnetic fields are calculated at MEG sensor coordinates (diamonds). The superposition of all field patterns gives the synthetic MEG data. The mean SNR of the simulated MEG data is dependent on the realization of the experiment and ranges between -10 dB and -15 dB. Sensor noise with 30 dB simulated data to (sensor) noise ratio was added.

100 realizations are evaluated, each having the same evoked signal forms and signal power, while the location of the evoked signals in the sphere is chosen at random. For each realization, 100 trials with 400 samples are considered and ICA followed by SNR-based clustering and NAPCA are run. As reference, the performance of PCA followed by SNR-based clustering is given, too. For each method, the number of components to be estimated is fixed at two. The data dimensions have been reduced to 40 by PCA, which preserves $> 98\%$ of the total variance.

3.1 Results

Figure 3 (a) and (b) depict Nolte's performance index and the SNR of the recovered signals. PCA is observed to fail. ICA yields significantly better results. The median performance of finding the subspace basis is 0.94 and the SNR is improved compared to the PCA results. However, NAPCA clearly outperforms the ICA approach. The median

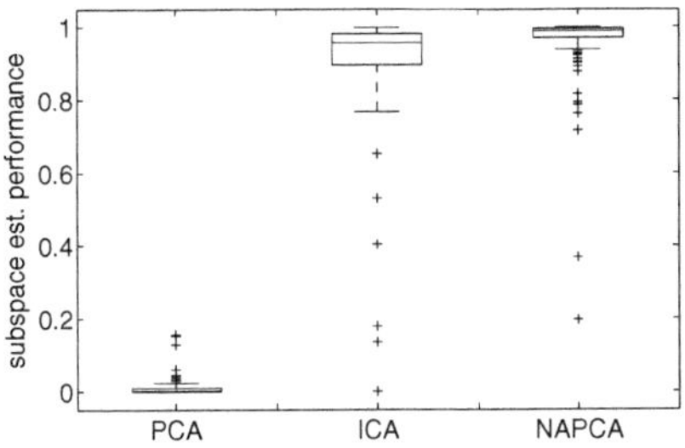

(a)

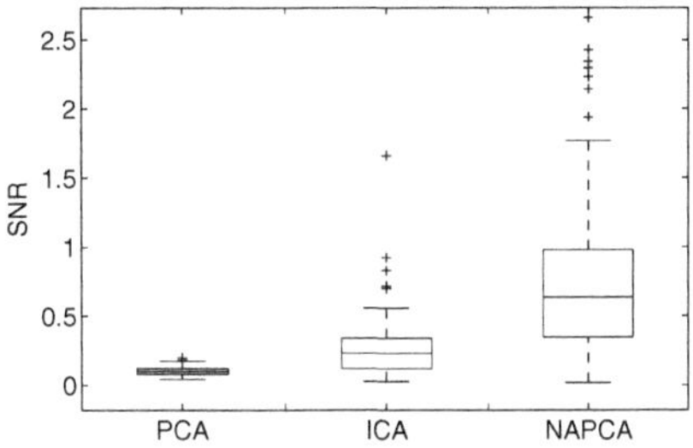

(b)

Fig. 3. Results from synthetic experiments. (a) Subspace recovery performance using Nolte's performance index. (b) SNR of the recovered signals.

performance of finding the subspace basis is 0.99. Furthermore, the SNR has improved significantly, the median SNR of NAPCA is by a factor three larger than the median SNR of the ICA approach.

4 Recorded Data Experiments

Evoked MEG data with simultaneous auditory and visual stimulation have been recorded using a full head 125 channel MEG system. Presentation® (Neurobehavioral Systems, Inc.) was used with the following settings. Tones of 1 KHz and pictures of human faces were presented simultaneously. Stimulation was optimized to evoke signals mainly in the left cerebral hemisphere. The duration of stimulation was 300 ms for each single-trial of 2 seconds duration. The sampling frequency was 500 Hz. 300 trials were recorded and subsequently cut from the continuous recordings. The concatenation of single-trials with 100 ms pre-stimulation duration and 300 ms post-stimulation duration give the MEG data utilized. Here, we show the results of ICA followed by SNR-based clustering and NAPCA for data from two human subjects ('data 1', 'data 2'). The data dimensions have been reduced to 40 by PCA. The subspace was assumed to be of dimension two and, hence, two components were estimated.

4.1 Results

Figure 4 depicts the results of ICA followed by SNR based clustering. Likewise, Figure 5 depicts the NAPCA results. The colorcoded field distributions represent the estimated basis of the subspace. The single-trial plot with trial-average plot illustrates the quality of the associated recovered signal. From single stimulation experiments, it is known that auditory stimulation yields a dipolar field pattern over the auditory cortex which is located near the ears. Likewise, visual stimulation produces a dipolar pattern over the visual cortex located in the occipital region, i.e. in the back of the head. Visual inspection of the field distributions show that all field maps have plausible patterns for the first estimated component. However, for the second estimated component NAPCA seems to be more focal and plausible. For instance, in the second data set, one clearly

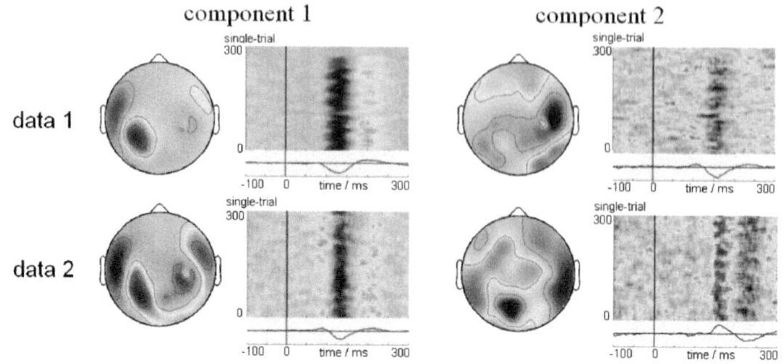

Fig. 4. Field maps, single-trial and trial-average plots for the ICA followed by SNR-based clustering approach. Each component is normalized to unit variance and each field map is normalized to a maximal field value equal to one.

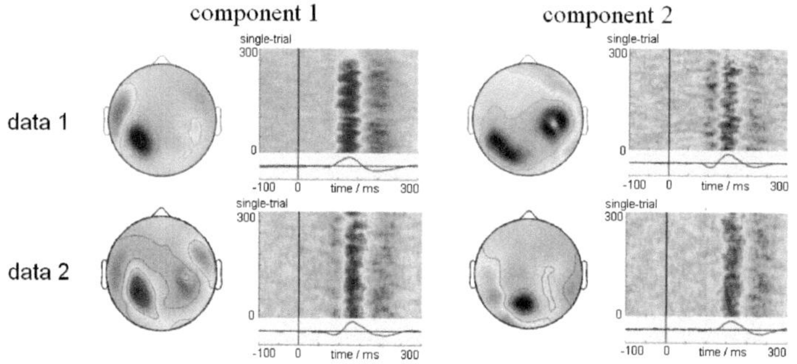

Fig. 5. Field maps, single-trial and trial-average plots for the NAPCA approach. Each component is normalized to unit variance and each field map is normalized to a maximal field value equal to one.

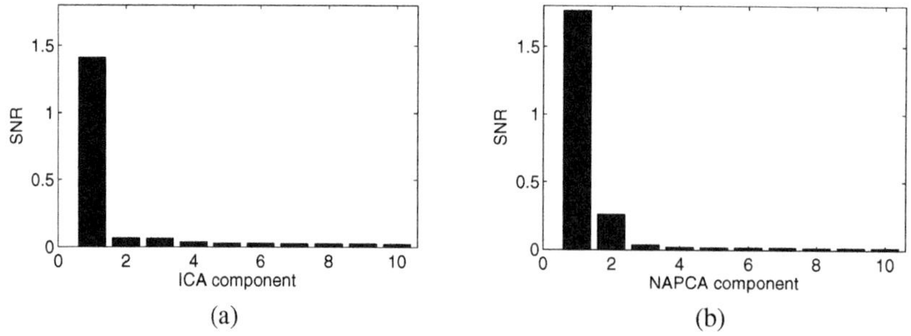

Fig. 6. Signal to noise ratios of subspace components estimated by NAPCA and ICA using the recordings 'data 1'. NAPCA shows two componets with higher SNR values. ICA only has one higher SNR value. The figure suggest that NAPCA outperforms the ICA approach.

observes a dipolar pattern in the back of the head. The single-trial plots suggest that NAPCA performs better than ICA followed by clustering. Figure 6 shows the SNR (cf. Eq. 6) for NAPCA and ICA components for the first data set. We assumed a 2-dimensional subspace. In fact, the NAPCA analysis, Figure 6 (b), supports this assumption as it has two prominent SNR values. In contrast, from the ICA analysis, Figure 6 (a), the correct dimension of the subspace could not have been deduced. Only one prominent SNR value exists. The values for the first two components are 1.8 and 0.25 for NAPCA and 1.2 and 0.06 for ICA. Hence, visual inspection and the SNR analysis suggest that NAPCA yields better results.

5 Conclusions

Evoked MEG data with more than one source of stimulation can lead to dependent signals. Hence, ICA can fail in separating these signals. Here, the problem of finding the subspace of signals was discussed. Specifically, the estimation of vector basis in the measurement space and the quality of the associated signals were assessed. Recently, ICA followed by clustering was proposed for finding the subspaces. We propose NAPCA as an alternative subspace method. NAPCA was shown to yield better results for both the synthetic and recorded MEG data sets. We conclude that NAPCA should be considered for the analysis of evoked MEG data.

Acknowledgment

Financial support of the Deutsche Forschungsgemeinschaft DFG (Grant OR 99/4-1) is gratefully acknowledged. We would like to thank T. Sander from PTB for his assistance in recording the evoked MEG data.

References

1. Hyvärinen, A., Karhunen, J., Oja, E.: Independent component analysis. Wiley, Chichester (2001)
2. Kohl, F., Wübbeler, G., Kolossa, D., Elster, C., Bär, M., Orglmeister, R.: Non-independent BSS: A model for evoked MEG signals with controllable dependencies. In: Adali, T., Jutten, C., Romano, J.M.T., Barros, A.K. (eds.) ICA 2009. LNCS, vol. 5441, pp. 443–450. Springer, Heidelberg (2009)
3. Cardoso, J.F.: Multidimensional independent component analysis. In: Proc. ICASSP, vol. 4, pp. 1941–1944 (1998)
4. Lee, J.B., Woodyatt, A.S., Berman, M.: Enhancement of high spectral resolution remote sensing data by a noise-adjusted principal component transform. IEEE Trans. Geosci. Remote Sensing 28, 295–304 (1990)
5. Wübbeler, G., Link, A., Burghoff, M., Trahms, L., Elster, C.: Latency analysis of single auditory evoked M 100 responses by spatio-temporal filtering. Phys. Med. Biol. 52, 4383–4392 (2007)
6. Nolte, G., Curio, G.: The effect of artifact rejection by signal-space projection on source localization accuracy. IEEE Trans. Biomed. Eng. 46, 400–408 (1999)
7. Lemm, S., Curio, S., Hluschchuk, Y., Müller, K.R.: Enhancing the signal-to-noise ratio of ICA-based extracted ERPs. IEEE Trans. Biomed. Eng. 53, 601–607 (2006)

Binary Sparse Coding

Marc Henniges[1], Gervasio Puertas[1], Jörg Bornschein[1], Julian Eggert[2], and Jörg Lücke[1]

[1] FIAS, Goethe-Universität Frankfurt am Main, Germany
[2] Honda Research Institute Europe, Offenbach am Main, Germany

Abstract. We study a sparse coding learning algorithm that allows for a simultaneous learning of the data sparseness and the basis functions. The algorithm is derived based on a generative model with binary latent variables instead of continuous-valued latents as used in classical sparse coding. We apply a novel approach to perform maximum likelihood parameter estimation that allows for an efficient estimation of all model parameters. The approach is a new form of variational EM that uses truncated sums instead of factored approximations to the intractable posterior distributions. In contrast to almost all previous versions of sparse coding, the resulting learning algorithm allows for an estimation of the optimal degree of sparseness along with an estimation of the optimal basis functions. We can thus monitor the time-course of the data sparseness during the learning of basis functions. In numerical experiments on artificial data we show that the algorithm reliably extracts the true underlying basis functions along with noise level and data sparseness. In applications to natural images we obtain Gabor-like basis functions along with a sparseness estimate. If large numbers of latent variables are used, the obtained basis functions take on properties of simple cell receptive fields that classical sparse coding or ICA-approaches do not reproduce.

1 Introduction

The mammalian brain encodes sensory stimuli by distributed activities across neural populations. Different neurons or different populations of neurons are hereby found to code for different aspects of a presented stimulus. Such distributed or factorial codes can (A) reliably encode large numbers of stimuli using relatively few computational elements and (B) they can potentially make use of the representation of individual components for further processing. In Machine Learning, factorial codes are closely associated with what is often called multiple-causes models. That is, they are related to probabilistic generative models which assume a data point to be generated by a combination of different hidden causes or *hidden variables*. Two very influencial models, that can be regarded as such multiple-causes models, are independent component analysis (ICA) [1] and sparse coding (SC) [2]. Indeed, since it was first suggested [2] sparse coding has become the standard model to explain the response properties of cortical simple cells. In its generative formulation, SC optimizes the parameters of a generative model with a sparse prior $p(\boldsymbol{s} \mid \lambda)$ and a Gaussian noise

V. Vigneron et al. (Eds.): LVA/ICA 2010, LNCS 6365, pp. 450–457, 2010.

model $p(\boldsymbol{y} \,|\, \boldsymbol{s}, W, \sigma)$. In the last two decades, many different variants of the basic SC model have been introduced and discussed in the literature. These variants focused on different ways to train the parameters of the model (e.g., MAP estimates [2], sampling approaches [3] and many more). However, almost none of the approaches studied in the past estimated the data sparseness. This highlights that learning the sparseness seems a much more challenging task than learning the basis functions although usually just a single sparseness parameter has to be estimated. The learning algorithm studied in this paper will be shown to successfully estimate the sparseness. In applications to natural images we, furthermore, show that the algorithm reproduces simple cell properties that have only recently been observed [4].

2 Sparse Coding with Binary Hidden Variables

Consider a set of N independent data points $\{\boldsymbol{y}^{(n)}\}_{n=1,\ldots,N}$ where $\boldsymbol{y}^{(n)} \in \mathbb{R}^D$ (D is the number of observed variables). For these data the studied learning algorithm seeks parameters $\Theta = (W, \sigma, \pi)$ that maximize the data likelihood $\mathcal{L} = \prod_{n=1}^{N} p(\boldsymbol{y}^{(n)} \,|\, \Theta)$ under the generative model:

$$p(\boldsymbol{s}|\pi) \,=\, \prod_{h=1}^{H} \pi^{s_h} \big(1-\pi\big)^{1-s_h}, \qquad p(\boldsymbol{y} \,|\, \boldsymbol{s}, W, \sigma) \,=\, \mathcal{N}(\boldsymbol{y};\, W\boldsymbol{s}, \sigma^2 \mathbb{1}), \quad (1)$$

where $W \in \mathbb{R}^{D \times H}$ and H denotes the number of hidden variables s_h. For small values of π the latent variables are sparsely active. The basis functions $\boldsymbol{W}_h = (W_{1h}, \ldots, W_{Dh})^T$ combine linearly and (given the latents) each observed variable y_d is independently and identically drawn from a Gaussian distribution with variance σ^2. The only difference to the generative model of classical sparse coding is thus the choice of binary latent variables (distributed according to a Bernoulli distribution) instead of latents with continuous values.

To optimize the parameters Θ, we use a variational EM approach (see, e.g., [5]). That is, instead of maximizing the likelihood directly we maximize the free-energy:

$$\mathcal{F}(q,\Theta) = \sum_{n=1}^{N} \Big[\sum_{\boldsymbol{s}} q^{(n)}(\boldsymbol{s}\,;\Theta^{\mathrm{old}}) \Big[\log\big(p(\boldsymbol{y}^{(n)} \,|\, \boldsymbol{s}, W, \sigma)\big) + \log\big(p(\boldsymbol{s} \,|\, \pi)\big) \Big] \Big] + H(q), \quad (2)$$

where $q^{(n)}(\boldsymbol{s}\,;\Theta^{\mathrm{old}})$ is an approximation to the exact posterior and $H(q)$ denotes the Shannon entropy. In the variational EM scheme $\mathcal{F}(q,\Theta)$ is maximized alternately with respect to q in the E-step (while Θ is kept fixed) and with respect to Θ in the M-step (while q is kept fixed). Parameter update rules (M-step equations) are obtained by setting the derivatives of (2) w.r.t. the different parameters to zero. The obtained update rules contain expectation values such as $\langle \boldsymbol{s} \rangle_{q^{(n)}}$ and $\langle \boldsymbol{s}\boldsymbol{s}^T \rangle_{q^{(n)}}$ which are intractable for large H if $q^{(n)}$ is chosen to be the exact posterior ($q^{(n)}(\boldsymbol{s}\,;\Theta^{\mathrm{old}}) = p(\boldsymbol{s} \,|\, \boldsymbol{y}^{(n)}, \Theta^{\mathrm{old}})$). To derive an efficient learning algorithm, our approach approximates the intractable expectation values by truncating the sums over the hidden space of $\boldsymbol{s}$:

$$\langle g(\boldsymbol{s})\rangle_{q^{(n)}} = \frac{\sum_{\boldsymbol{s}} p(\boldsymbol{s}, \boldsymbol{y}^{(n)} \,|\, \Theta^{\mathrm{old}})\; g(\boldsymbol{s})}{\sum_{\tilde{\boldsymbol{s}}} p(\,\tilde{\boldsymbol{s}}\,, \boldsymbol{y}^{(n)} \,|\, \Theta^{\mathrm{old}})} \approx \frac{\sum_{\boldsymbol{s}\in\mathcal{K}_n} p(\boldsymbol{s}, \boldsymbol{y}^{(n)} \,|\, \Theta^{\mathrm{old}})\; g(\boldsymbol{s})}{\sum_{\tilde{\boldsymbol{s}}\in\mathcal{K}_n} p(\,\tilde{\boldsymbol{s}}\,, \boldsymbol{y}^{(n)} \,|\, \Theta^{\mathrm{old}})}, \tag{3}$$

where $g(\boldsymbol{s})$ is a function of $\boldsymbol{s}$ (and potentially the parameters), and where $\mathcal{K}_n$ is a small subset of the hidden space. Eqn. 3 represents a good approximation if the set $\mathcal{K}_n$ contains most of the posterior probability mass. The approach will be referred to as *Expectation Truncation* and can be derived as a novel form of a variational EM approach (compare [6]). For other generative models similar truncation approaches have successfully been used [7, 8]. For the learning algorithm, $\mathcal{K}_n$ in (3) is chosen to contain hidden states $\boldsymbol{s}$ with at most γ active causes $\sum_h s_h \leq \gamma$. Furthermore, we only consider the combinatorics of $H' \geq \gamma$ hidden variables that are likely to have contributed to generating a given data point $\boldsymbol{y}^{(n)}$. More formally we define:

$$\mathcal{K}_n = \{\boldsymbol{s} \mid \big(\textstyle\sum_j s_j \leq \gamma \text{ and } \forall i \notin I : s_i = 0\big) \text{ or } \textstyle\sum_j s_j \leq 1\}, \tag{4}$$

where the index set I contains those latent indices h with the H' largest values of a *selection function* $\mathcal{S}_h(\boldsymbol{y}^{(n)})$. This function is given by:

$$\mathcal{S}_h(\boldsymbol{y}^{(n)}) = \tfrac{\boldsymbol{W}_h^{\mathrm{T}}}{||\boldsymbol{W}_h||}\,\boldsymbol{y}^{(n)}, \qquad \text{with } ||\boldsymbol{W}_h|| = \sqrt{\textstyle\sum_{d=1}^{D}(W_{dh})^2}\,. \tag{5}$$

A large value of $\mathcal{S}_h(\boldsymbol{y}^{(n)})$ signals a high likelihood that $\boldsymbol{y}^{(n)}$ contains the basis function $\boldsymbol{W}_h$ as a component. The last term in (4) assures that all states $\boldsymbol{s}$ with just one non-zero entry are also evaluated. In numerical experiments on ground-truth data we can verify that for most data points the approach (3) with (4) and (5) indeed approximates the true expectation values with high accuracy. By applying this approximation, exact EM (which scales exponentially with H) is altered to an algorithm which scales polynomial with H' (approximately $\mathcal{O}\left(H'^{\gamma}\right)$) and linear with H. Note, however, that in general larger H also require larger amounts of data points.

With the tractable approximations for the expectation values $\langle g(\boldsymbol{s})\rangle_{q^{(n)}}$ computed with (3) to (5) the update equations for W and σ are given by:

$$W^{\mathrm{new}} = \left(\sum_{n\in\mathcal{M}} \boldsymbol{y}^{(n)} \langle \boldsymbol{s} \rangle_{q_n}^{T}\right)\left(\sum_{n'\in\mathcal{M}} \left\langle \boldsymbol{s}\,\boldsymbol{s}^{T}\right\rangle_{q_{n'}}\right)^{-1} \tag{6}$$

$$\sigma^{\mathrm{new}} = \sqrt{\frac{1}{|\mathcal{M}|\,D} \sum_{n\in\mathcal{M}} \left\langle \left|\left|\boldsymbol{y}^{(n)} - W\,\boldsymbol{s}\right|\right|^2 \right\rangle_{q_n}} \tag{7}$$

Note that we do not sum over all data points $\boldsymbol{y}^{(n)}$ but only over those in a subset $\mathcal{M}$ (note that $|\mathcal{M}|$ is the number of elements in $\mathcal{M}$). The subset contains those data points for which (3) finally represents a good approximation. It is defined to contain the N^{cut} data points with highest values $\sum_{\tilde{\boldsymbol{s}} \in \mathcal{K}_n} p(\tilde{\boldsymbol{s}}, \boldsymbol{y}^{(n)} \,|\, \Theta^{\text{old}})$, i.e., with the highest values for the denominator in (3). N^{cut} is hereby the expected number of data points that have been generated by states with less or equal γ non-zero entries: $N^{\text{cut}} = N \sum_{\boldsymbol{s},\, |\boldsymbol{s}| \leq \gamma} p(\boldsymbol{s} \,|\, \pi) = N \sum_{\gamma'=0}^{\gamma} \binom{H}{\gamma'} \pi^{\gamma'} (1-\pi)^{H-\gamma'}$.

Update equations (6) and (7) were obtained by setting the derivatives of Eqn. 2 (w.r.t. W and σ) to zero. Similarly, we can derive the update equation for π. However, as the approximation only considers states $\boldsymbol{s}$ with a maximum of γ non-zero entries, the update has to correct for an underestimation of π. If such a correction is taken into account, we obtain the update rule:

$$\pi^{\text{new}} = \frac{A(\pi)\,\pi}{B(\pi)} \frac{1}{|\mathcal{M}|} \sum_{n \in \mathcal{M}} \langle |\boldsymbol{s}| \rangle_{q_n} \quad \text{with } |\boldsymbol{s}| = \sum_{h=1}^{H} s_h \text{ and} \tag{8}$$

$$A(\pi) = \sum_{\gamma'=0}^{\gamma} \binom{H}{\gamma'} \pi^{\gamma'} (1-\pi)^{H-\gamma'} \quad \text{and} \quad B(\pi) = \sum_{\gamma'=0}^{\gamma} \gamma' \binom{H}{\gamma'} \pi^{\gamma'} (1-\pi)^{H-\gamma'}.$$

Note that if we allow all possible states (i.e., $\gamma = H$), the correction factor $\frac{A(\pi)\,\pi}{B(\pi)}$ in (8) is equal to one over H and the set $\mathcal{M}$ becomes equal to the set of all data points (because $N^{\text{cut}} = N$). Equation (8) then falls back to the exact EM update rule that can canonically be derived by setting the derivative of (2) w.r.t. π to zero (using the exact posterior). Also the update equations (6) and (7) fall back to their canonical form for $\gamma = H$. By choosing a γ between one and H we can thus choose the accuracy of the used approximation. The higher the value of γ the more accurate is the approximation but the larger are also the computational costs. For intermediate values of γ we can obtain very good approximations with small computational costs.

3 Numerical Experiments

The update equations (6), (7), and (8) together with approximation (3) define a learning algorithm that optimizes the full set of parameters of the generative model (1). In order to numerically evaluate the algorithm we ran several tests on artificial and natural data.

Linear bars test. We applied the algorithm to artificial bars data as shown in Fig. 1A. To generate this data we created $H = 10$ basis functions $\boldsymbol{W}_h$ in the form of horizontal and vertical bars. Each bar occupied 5 pixels on a $D = 5 \times 5$ grid. Bars were chosen to be either positive (i.e. $W_h^d \in \{0.0, 10.0\}$) or negative ($W_h^d \in \{0.0, -10.0\}$). Half of the basis functions was randomly assigned the negative values and the other half the positive values. Data points were generated by linearly superimposing these basis functions (compare, e.g., [9] for a similar task) with a sparseness value of $\pi H = 2.0$ (i.e., two active causes per image on average). To this data we added iid Gaussian noise (mean $= 0.0$, std $= 2.0$). After each trial we tested whether each basis function was uniquely represented

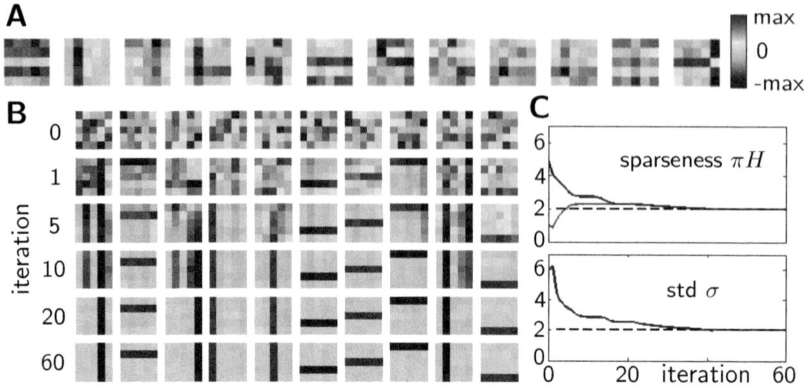

Fig. 1. Linear bars test with $H = 10$, $D = 5 \times 5$, and $N = 500$. **A** 12 examples for data points. **B** Basis functions for iterations given on the left. **C** Sparseness and standard deviation plotted over the iterations. Data for same experiment as in B in blue. Data for a run with initial sparseness value of 1.0 in red. Ground-truth indicated by dashed horizontal line.

by a single bar in oder to compute the success-rate, i.e. the reliability of the algorithm.

The approximation parameters were set to $\gamma = 3$ and $H' = 5$. We started with 20 iterations in which we set $|\mathcal{M}| = N$, then linearly decreased the amount of used data points in the next 20 iterations to $|\mathcal{M}| = N^{\text{cut}}$ where we kept it constant during the last 20 iterations, thus using a total of 60 iterations. The parameters W were initialized by drawing randomly from a Gaussian distribution with zero mean and a standard deviation of 2.0 (compare [6]). Sparseness was initialized at $\pi H = 5.0$, thus assuming that five of the causes contributed to an image on average. The standard deviation was initialized by calculating the sum over all squared data points which led to a value of $\sigma \approx 6.0$. After each iteration we added iid Gaussian parameter noise to the learned basis functions (mean = 0.0, std = 0.05).

We ran the algorithm with the above parameters 1000 times, each time using a newly generated set of $N = 1000$ data points. In 978 of these trials we recovered all bars ($\approx 98\%$ reliability) and obtained a mean value of $\pi H = 2.0$ (± 0.01 std) for the sparseness and $\sigma = 2.0 \pm 0.06$ for the data noise. Reliabilities increased when more data points were used (e.g., $\approx 99\%$ for $N = 4000$) and decreased for lower initial values of πH (e.g., $\approx 96\%$ and $\approx 84\%$ for $\pi H = 3$ and $\pi H = 1$, $N = 2000$, respectively). Figures 1B and 1C show the typical development of the parameters W, πH, and σ over the 60 iterations.

Natural image patches. In order to perform the experiment on natural images, we sampled $N = 200\,000$ patches of $D = 26 \times 26$ pixels from the van Hateren image database [10] (while constraining random selection to patches of images without man-made structures). As a form of preprocessing, we used

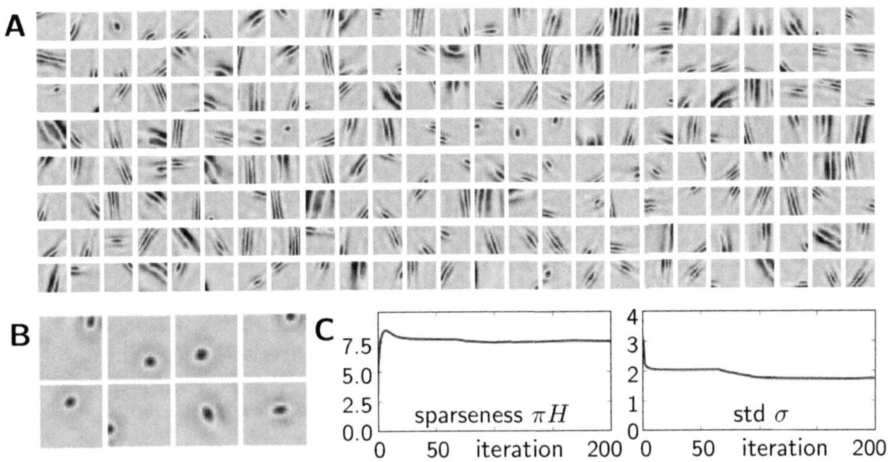

Fig. 2. Numerical experiment on image patches. **A** 200 basis functions randomly selected out of the $H = 700$ used. **B** The most globular of the $H = 700$ fields. **C** Time-courses of sparseness (πH) and data noise (in terms of standard deviation σ).

a Difference of Gaussians (DoG) technique[1]. According to the previous experiment, the initial condition for each basis function was set to the average over the preprocessed input patches plus small Gaussian white noise. The initial noise parameter σ was set following equation 7 by using all data points ($|\mathcal{M}| = N$). Finally, the initial sparseness value was taken to be $\pi H = 1$. The approximation parameters for the algorithm were set to $\gamma = 8$ and $H' = 10$. This choice reflects the relatively high average number of components that we expect for the relatively large patch size used (experiments with different γ and H values have all suggested an average of approximately 6 to 10 components per patch). The number of data points used was $|\mathcal{M}| = N$ during the first 66 iterations, decreased to $|\mathcal{M}| = N^{\text{cut}}$ from iteration 66 to iteration 100 and kept at this value for the remaining 100 iterations. Fig. 2 shows the learned parameters for a run of the algorithm with $H = 700$ hidden variables. Fig. 2A shows a random selection of 200 of the 700 obtained basis functions. In Fig. 2B the most globular of the 700 basis functions are displayed. The monitored time-course of the data sparseness ($\pi\, H$) and the time-course of the data noise (σ) are displayed in Fig. 2C. As can be observed, we obtain Gabor-like basis functions with different orientations, frequencies, and phase as well as globular basis functions with no or very little orientation preferences (compare [4]). Along with the basis functions we obtain an estimate for the noise and, more importantly, for the data sparseness of $\pi\, H = 7.49$ active causes per 26×26 patch. Runs of the algorithm with H

[1] Filter parameters were chosen as in [11]; before the brightest 2% of the pixels were clamped to the maximal value of the remaining 98% (influence of light-reflections were reduced in this way).

smaller than 700 (e.g. $H = 200$) resulted in similar basis functions. However, for smaller H, basis functions had the tendency to be spatially more constrained.

4 Discussion

We have studied a learning algorithm for sparse coding with binary hidden variables. In contrast to almost all the numerous SC and ICA variants, it is capable of learning the full set of parameters. To the knowledge of the authors there are only two previous SC versions that also estimate the sparseness: the algorithm in [3] which assumes a mixture of Gaussians prior, and the algorithm in [12] assuming a Student-t prior. To estimate the sparseness the approach in [3] uses sampling while the approach in [12] screens through discrete values of different sparseness levels to estimate the optimal one by comparison. In contrast, we use an update equation derived from a deterministic approximation (Expectation Truncation; [6]) which represents a novel form of variational EM. Another difference between the approaches [3] and [12] and our approach is the assumption of continuous-valued latents in those, and of binary latents in our case. Binary latents have frequently been used in the past ([13–15] and many more). The approach most similar to ours is hereby [15] which assumes the same underlying generative model. However, in none of these approaches the data sparseness is learned. The presented approach is thus the first algorithm that infers the appearance probabilities and data noise in a linear bars test (but compare [7] which learns the sparseness for non-linear superpositions with a different method). Also in applications to image patches, our approach estimates the sparseness in parallel to learning the basis functions and data noise (Fig. 2). The basis functions hereby take the form of Gabor-like wavelets and of globular functions with no or little orientation tuning. Interestingly, simple cell receptive fields that correspond to such globular functions were observed in *in vivo* recordings in [4]. Modelling approaches have only very recently reproduced such fields [11, 16, 17]. The system in [11, 16] is a neuro-dynamic approach that models cortical microcircuits. The model described in [17] is, notably, a SC version whose hidden variables have a more binary character than those of classical SC: they use latents with continuous values but explicitly exclude values in an interval around zero (while allowing zero itself). If applied to image patches, globular basis functions are obtained in [17] alongside Gabor-like basis functions. In that study the sparseness parameter had to be chosen by hand, however. The algorithm in [15] uses binary latents but, although applied to image patches, globular fields were not reported. This might be due to a relatively small number of hidden units used there. Also in [15] the sparseness level had to be set by hand.

Parameter optimization as studied in this paper can in future work be applied to SC models with continuous latents (e.g., with Laplacian or Student-t prior). Based on such models, the difference between binary and continuous latents can be investigated further. The observation that globular basis functions are obtained with the here presented algorithm might be taken as an indication that the assumption of binary or more binary latents at least facilitates the emergence of localized and circular symmetric basis functions. The observation that such globular functions also describe the response properties of many cortical simple cells [4] might have interesting implications for theories on neural coding.

Acknowledgement. We gratefully acknowledge funding by the German Federal Ministry of Education and Research (BMBF) in the project 01GQ0840 (BFNT Frankfurt), by the Honda Research Institute Europe GmbH, and by the German Research Foundation (DFG) in the project LU 1196/4-1. Furthermore, we gratefully acknowledge support by the Frankfurt Center for Scientific Computing.

References

1. Comon, P.: Independent Component Analysis, a new concept? Signal Process 36(3), 287–314 (1994)
2. Olshausen, B., Field, D.: Emergence of simple-cell receptive field properties by learning a sparse code for natural images. Nature 381, 607–609 (1996)
3. Olshausen, B., Millman, K.: Learning sparse codes with a mixture-of-Gaussians prior. In: Proc NIPS, vol. 12, pp. 841–847 (2000)
4. Ringach, D.L.: Spatial Structure and Symmetry of Simple-Cell Receptive Fields in Macaque Primary Visual Cortex. J. Neurophysiol. 88, 455–463 (2002)
5. Neal, R., Hinton, G.: A view of the EM algorithm that justifies incremental, sparse, and other variants. In: Learning in Graphical Models, pp. 355–369 (1998)
6. Lücke, J., Eggert, J.: Expectation Truncation And the Benefits of Preselection in Training Generative Models. J. Mach Learn Res., revision in review (2010)
7. Lücke, J., Sahani, M.: Maximal causes for non-linear component extraction. J. Mach. Learn. Res. 9, 1227–1267 (2008)
8. Lücke, J., Turner, R., Sahani, M., Henniges, M.: Occlusive Components Analysis. In: Proc NIPS, vol. 22, pp. 1069–1077 (2009)
9. Hoyer, P.: Non-negative sparse coding. In: Neural Networks for Signal Processing XII: Proceedings of the IEEE Workshop, pp. 557–565 (2002)
10. Hateren, J., Schaaf, A.: Independent Component Filters of Natural Images Compared with Simple Cells in Primary Visual Cortex. Proc. Biol. Sci. 265(1394), 359–366 (1998)
11. Lücke, J.: Receptive Field Self-Organization in a Model of the Fine Structure in V1 Cortical Columns. Neural Computation (2009)
12. Berkes, P., Turner, R., Sahani, M.: On sparsity and overcompleteness in image models. In: Proc NIPS, vol. 20 (2008)
13. Hinton, G., Ghahramani, Z.: Generative models for discovering sparse distributed representations. Phil. Trans. R Soc. B 352(1358), 1177 (1997)
14. Harpur, G., Prager, R.: Development of low entropy coding in a recurrent network. Network-Comp. Neural. 7, 277–284 (1996)
15. Haft, M., Hofman, R., Tresp, V.: Generative binary codes. Pattern Anal. Appl. 6(4), 269–284 (2004)
16. Lücke, J., Sahani, M.: Generalized softmax networks for non-linear component extraction. In: de Sá, J.M., Alexandre, L.A., Duch, W., Mandic, D.P. (eds.) ICANN 2007. LNCS, vol. 4668, pp. 657–667. Springer, Heidelberg (2007)
17. Rehn, M., Sommer, F.T.: A network that uses few active neurones to code visual input predicts the diverse shapes of cortical receptive fields. J. Comput. Neurosci. 22(2), 135–146 (2007)

A Multichannel Spatial Compressed Sensing Approach for Direction of Arrival Estimation

Aris Gretsistas and Mark D. Plumbley*

Queen Mary University of London
Centre for Digital Music
Mile End Road, E1 4NS, London, UK
aris.gretistas@elec.qmul.ac.uk

Abstract. In this work, we present a direction-of-arrival (DOA) estimation method for narrowband sources impinging from the far-field on a uniform linear array (ULA) of sensors, based on the multichannel compressed sensing (CS) framework. We discretize the angular space uniformly into a grid of possible locations, which is much larger than the number of sensors, and assume that only a few of them will correspond to the active sources. As long as the DOAs of the sources are located at a few locations on the angular grid, they will share a common spatial support. To exploit this joint sparsity, we take several time snapshots and formulate a multichannel spatial compressed sensing (SM-CS) problem. Simultaneous Orthogonal Matching Pursuit (SOMP) is used for the reconstruction and the estimation of the angular power spectrum. The performance of the proposed method is compared against standard spectral-based approaches and other sparsity based methods.

Keywords: Array signal processing, direction of arrival estimation, sparse representations, compressed sensing, joint sparsity.

1 Introduction

One of the main objectives in array processing is to estimate the spatial energy spectrum and therefore determine the number and location of the sources of energy. This source localization problem using sensor arrays, usually referred to as direction-of-arrival (DOA) estimation, arises in many applications such as radar, communications, seismology and sonar.

Among the various methods that has been proposed for the DOA estimation problem, conventional (or Bartlett) beamforming provides a spatial spectral estimate, which suffers from its inability to resolve closed spaced sources [1]. These resolution limitations of conventional beamforming can be overcome using Capon's method (MVDR) or the MUSIC algorithm. Both techniques can achieve superresolution by focusing on a small number of "search directions"

* This research is supported by ESPRC Leadership Fellowship EP/G007144/1 and Platform Grant EP/045235/1, and EU FET-Open Project FP7-ICT-225913 "SMALL".

V. Vigneron et al. (Eds.): LVA/ICA 2010, LNCS 6365, pp. 458–465, 2010.

where the sources are present. Therefore, this implies some underlying sparsity in the spatial domain.

The emerging field of sparse representations has given renewed interest to the problem of source localization. The concept of spatial sparsity for DOA estimation was first introduced in [2], where it was shown that the source localization problem can be cast as a sparse representations recovery problem in a redundant dictionary using the ℓ_1-SVD method. More recently, in a similar manner, spatial sparsity was linked to the theoretical results of the compressed sensing (CS) framework, utilizing a spatial CS approach for DOA estimation [3]. That work demonstrated that a simple implementation of the CS-based approach can achieve high angular resolution.

In this paper, we attempt to further exploit the CS framework by formulating the problem of DOA estimation of far-field narrowband sources as a multichannel spatial compressed sensing (MS-CS) or distributed compressed sensing reconstruction problem using the joint sparsity model of common sparse support [4]. To achieve this, we uniformly discretize the space into a grid of possible angles of arrival, assuming that only a few of them will correspond to real source locations. A uniform linear array is used to capture the observed time snapshots of the impinging far-field sources.

Assuming that the DOAs of the sources will share a common spatial support at some locations on the grid, we observe several time snapshots and attempt to solve a mixed norm minimization problem. The simultaneous Orthogonal Matching Pursuit (SOMP) algorithm [5] is used for the reconstruction of the common sparse support and yields an estimate of the angular signal energy spectrum, in which the high energy spikes correspond to the sources angles of arrival.

We will show that the proposed method can achieve superresolution and under certain conditions outperform the single-channel spatial CS (SS-CS) approach, while being less vulnerable to highly correlated sources or limited data when compared to MUSIC algorithm.

2 Distributed Compressed Sensing with Common Sparse Support

Consider L linear measurement sensors that each of them obtains M samples or observations of the discrete-time signal of interest $\mathbf{x} = [x_1, \ldots, x_N]^T$, where $M < N$. The observations or measurements vector can be written as

$$\mathbf{Y} = \mathbf{\Phi X} \tag{1}$$

where $\mathbf{\Phi}$ is a $M \times N$ matrix representing the measurement process, $\mathbf{Y}$ is the $M \times L$ measurements and $\mathbf{X}$ an and $N \times L$ matrix.

The main assumption is that the discrete-time signal $\mathbf{x}$ is sparse in some domain $\mathbb{R}^N$ and can be decomposed and accurately represented in the square basis $\mathbf{\Psi} \in \mathbb{R}^{N \times N}$. The resulting system of linear equations

$$\mathbf{Y} = \mathbf{\Phi X} = \mathbf{\Phi \Psi S} = \mathbf{AS} \tag{2}$$

is underdetermined as the dimensionality of the sparse $N \times L$ matrix $\mathbf{S}$ is larger than the dimensionality of the observed signal space. In the case that $L = 1$, the problem is reduced to single channel standard compressed sensing problem.

However, considering the common sparse support distributed compressed sensing model introduced in [4], sparsity is enforced across the columns of the sparse matrix $\mathbf{S}$. Therefore, we need to solve the mixed $\ell_{1,2}$ norm minimization problem defined as:

$$\min \|\mathbf{S}\|_{1,2} = \sum_{j=0}^{N} \|\mathbf{S}^j\|_2 \quad \text{such that} \quad \mathbf{Y} = \mathbf{AS} \tag{3}$$

The above optimization problem is convex and can be solved using interior-point methods. Alternatively, greedy approaches as simultaneous Orthogonal Matching Pursuit (SOMP) [5], which is the joint sparsity version of Orthogonal Matching Pursuit (OMP) [6] can provide faster convergence for an approximate solution.

The theoretical recovery results of compressed sensing have been generalized in the multiple sensor case [7]. Therefore, if matrix $\mathbf{A}$ obeys the *restricted isometry property* (RIP)

$$(1 - \delta_k)\|\mathbf{s}\|_2^2 \leq \|\mathbf{As}\|_2^2 \leq (1 + \delta_k)\|\mathbf{s}\|_2^2 \tag{4}$$

where δ_k is a small constant and if $\delta_{2k} < \sqrt{2} - 1$ then the the solution $\overline{\mathbf{S}}$ obeys

$$\|\mathbf{S} - \overline{\mathbf{S}}\|_F \leq C_0 k^{-1/2} \|\mathbf{S} - \hat{\mathbf{S}}^{(k)}\|_{1,2} \tag{5}$$

for some constant C_0, where $\mathbf{S}^{(k)}$ is the matrix $\mathbf{S}$ with all but the largest k components set to 0 and $\|.\|_F$ is the frobenius matrix norm. The recovery will be exact as long as $|\text{supp}\mathbf{X}| \leq k$.

Put another way, if the measurement matrix $\mathbf{\Phi}$ and the representation matrix $\mathbf{\Psi}$ are sufficiently incoherent and the number of measurements $M < N$ is sufficiently large, the original signal can be recovered with high probability. It is well known that random Gaussian or Bernoulli matrices obey the RIP condition with overwhelming probability as long as the number of measurements for each sensor is $M \geq Ck \log(N/k)$ [8]. In the case of partial Fourier matrices the bound on measurements will be $M \geq Ck \log^4(N)$ [9].

However, one can notice that these bounds are the same for the single channel compressed sensing problem. The above results based on RIP follow the worst case analysis scenario and therefore predict no performance gain resulting in theoretical equivalence between single and multi-channel sparsity. In practice, the worst case scenario is not always the case and empirical results have shown that as the number of channel $L \to \infty$, $k + 1$ measurements per channel will be sufficient for the exact reconstruction. As stated in [10], average case analysis can give more insight into joint sparsity and its superiority to the standard single channel case. Average case analysis provides the theoretical framework for joint sparsity and proves that the probability of recovery failure gets increasingly smaller as the number of channels increases [7].

3 Source Localization via Multichannel Compressed Sensing

Consider k narrowband sources with known central frequency f propagating plane waves from the far-field which impinge on an uniform linear array (ULA) of M sensors with inter-sensor spacing d. The sources arrive on the array from the unknown angles $\theta_1, \theta_2, \ldots, \theta_k$ that we wish to estimate. The linear array response to the impinging plane waves can be expressed as:

$$\boldsymbol{a}(\theta_n) = [1, e^{-j\omega_c \tau_2(\theta_n)}, \ldots, e^{-j\omega \tau_M(\theta_n)}] \tag{6}$$

or after the substitution of $\omega = 2\pi f$ and $\tau_p = pd\cos(\theta_n)/c$, where d is the sensor spacing which is chosen at half the wavelength $d = \lambda/2$, $\lambda = c/f$ is the wavelength and c is the speed of the propagation, we end up with the formula:

$$\boldsymbol{a}(\theta_n) = [1, e^{-j\pi\cos(\theta_n)}, \ldots, e^{-j\pi\cos(\theta_n)(M-1)}] \tag{7}$$

Therefore, the array output can be expressed as:

$$\mathbf{y}(t) = \mathbf{A}(\theta)\mathbf{s}(t) + \mathbf{n}(t) \tag{8}$$

whre $\boldsymbol{s}(t)$ is a $k \times 1$ vector containing the k sources, $\boldsymbol{y}(t)$ is a $M \times 1$ vector of the measurements of the array sensors, $\boldsymbol{n}(t)$ is the noise vector and $\mathbf{A}(\theta)$ is a $M \times k$ matrix containing the array responses for the k plane waves

$$\mathbf{A}(\theta) = [\boldsymbol{a}(\theta_1), \ldots, \boldsymbol{a}(\theta_k)]. \tag{9}$$

The above problem, presented in equation (8), is an overdetermined or evendetermined system of linear equations as long as $M \geq k$ and can be easily solved by inverting the matrix when the DOAs of the sources are known. However, the DOAs are unknown and need to be estimated. Following a spatial sparsity approach as in [2], we can uniformly discretize the bearing space into $N >> k$ possible angles of arrival and construct a redundant matrix of N atoms corresponding to the array responses of the respective angles of arrival

$$\mathbf{\Phi} = [\boldsymbol{a}(\theta_1), \ldots, \boldsymbol{a}(\theta_N)]. \tag{10}$$

This discretization provides the sparsity requirement of the compressed sensing framework. However as has already been stated above, in order for the reconstruction to be exact the chosen sensing matrix must obey the RIP condition and contain elements that are not correlated. According to the recent spatial compressed sensing approach, the source localization problem can be formulated as:

$$\mathbf{y}(t) = \mathbf{\Phi}\mathbf{s}(t) + \mathbf{n}(t) \tag{11}$$

where $\mathbf{\Phi}$ can be viewed as the sensing system of the compressed sensing problem. Considering that the expansion matrix $\mathbf{\Psi}$ is the Dirac matrix, the resulting pair of matrices is highly incoherent [3]. In fact, it has been proven that the pair of

Dirac basis and Fourier basis exhibits *maximal incoherence* [8] and therefore the compressed sensing framework can be applied to the above problem for the one time sample case and can be solved by *Basis Pursuit* (BP) ℓ_1 minimization [11].

However, for the multiple time samples case we intend to follow a MS-CS approach under the common sparse support model rather than solving the BP problem for each time sample separately. In that case the problem can be formulated:

$$\mathbf{Y} = \mathbf{\Phi S} + \mathbf{N}. \tag{12}$$

We are interested in enforcing spatial sparsity and not temporal, so we need to solve the mixed norm minimization problem for noisy data:

$$\min \|\mathbf{S}\|_{1,2} \quad \text{such that} \quad \|\mathbf{\Phi S} - \mathbf{Y}\|_F \leq \epsilon. \tag{13}$$

The mixed $\ell_{1,2}$ norm minimization can be solved using linear programming methods e.g. interior-point methods. Our approach will focus on the fast convergence of joint sparsity greedy approaches and more specifically the SOMP algorithm will be used. The proposed method assumes that the additive noise at the sensors of the array is white gaussian noise and the standard deviation σ is known. Therefore, the stopping criteria for the SOMP algorithm can be adjusted according to the noise level, very much like in [12].

It has already been stated that the compressed sensing worst case analysis shows theoretical equivalence for the single and multi-channel case. However, the worst case can occur when there is no additional information on the support from the multiple channels. This implies that each of them will contain the same information, which typically is not the case in the source localization problem, as the received signals' time samples are expected to vary. Consequently, the proposed multi-channel spatial sparsity model exploits the joint support and is expected to require fewer sensors than the $M \geq CK \log^4(N)$ required for the standard compressed sensing approach.

4 Experimental Results

In this section we present experimental results of the proposed method of joint spatial sparsity compressed sensing. The performance of our method is compared against the array processing spectral based approaches of conventional beamforming and MUSIC and the spatial sparsity compressed sensing method. We consider five narrowband far-field sources located at angles 30°, 45° , 50°, 100° and 130° respectively. The plane waves impinge on an ULA of $M = 40$ sensors equally spaced at the half wavelength $d = \lambda/2$. The sensors acquire $L = 200$ time snapshots of the superimposed signals and the signal-to-noise ratio (SNR) is 10dB. The additive noise is gaussian and MS-CS method assumes the knowledge of its standard deviation for the adjustment of the stopping criteria of SOMP. Empirically, we have found that a good choice is 0.8σ. The angular resolution is 1° and subsequently considering omnidirectional sensors the number of potential locations is $N = 180$.

Figure 1 illustrates the estimated angular power spectrum of all tested methods. Conventional beamforming exhibits the worst performance in terms of resolution. MUSIC and both SS-CS and MS-CS methods are able to estimate the DOAs. However, the proposed method gives the sharper peaks and showcases the best performance.

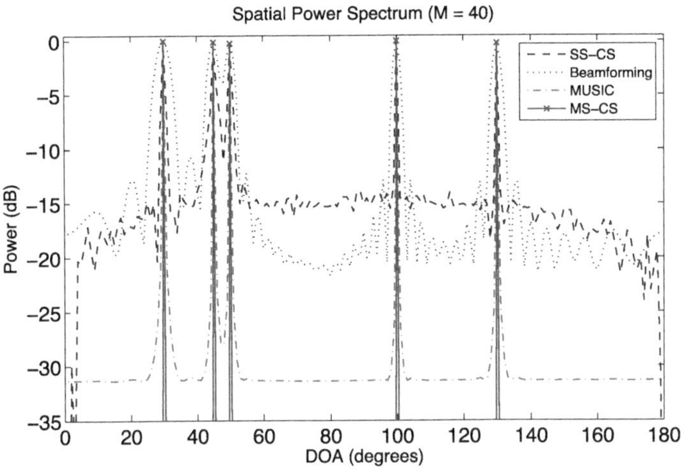

Fig. 1. Comparison of the proposed method (MS-CS) with the single channel CS (SS-CS) approach, conventional beamforming and MUSIC algorithm. The number of sensors is $M = 40$ and the SNR is 10dB.

Next we reduce the number of sensors to $M = 30$ and repeat the same experiment. Figure 2 shows that for the two closed spaced sources the performance of the SS-CS method decreases significantly. However, the performance of MS-CS method and MUSIC algorithm remain intact.

The elapsed time of each method is shown in Table 1. Clearly, the spectral based methods are faster than the sparsity based approaches. However, the proposed method improves on the convergence speed of the single channel case by exploiting the joint spatial sparsity.

Table 1. Elapsed time for the spatial spectrum estimation

No of sensors	Beamformer	MUSIC	SS-CS	MS-CS
30	0.0041s	0.0054s	1.1040s	0.0401s
40	0.0047s	0.0068s	1.5831s	0.0705s

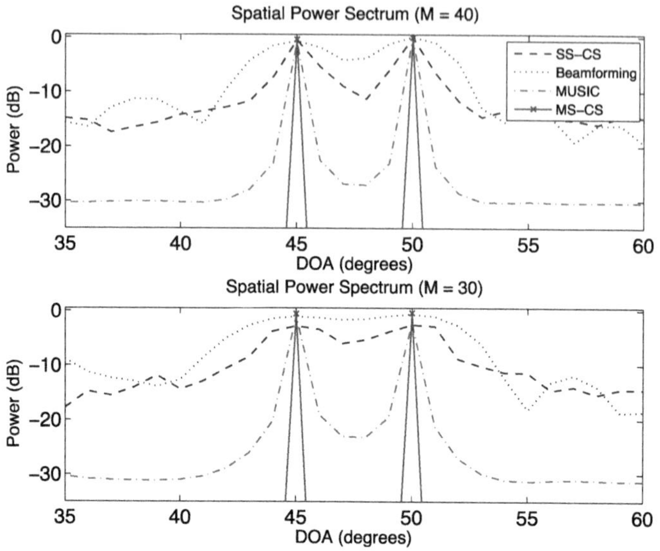

Fig. 2. Comparison of the proposed method (MS-CS) with the single channel CS (SS-CS) approach, conventional beamforming and MUSIC algorithm. The number of sensors is decreased from $M = 40$ to $M = 30$. The SNR is 10dB.

5 Conclusions

We have proposed a multi-channel spatial compressed sensing technique that approaches the problem of source localization as a multiple vector sparse recovery problem with common sparse support. The proposed method is based on the fact that the unknown DOAs share the same sparse support, as plane waves impinge on the array from fixed locations. Simultaneous Orthogonal Matching Pursuit is used to solve the mixed norm minimization problem and assumes that the standard deviation of the additive noise is known. We have shown that the exploitation of this joint sparsity results in high resolution performance. Simulations show that our MS-CS method outperforms the standard compressed sensing method and other spectral based approaches such as beamforming or MUSIC algorithm.

Interestingly, although the worst case theoretical analysis of compressed sensing results in equivalence between the single and multi-channel case, simulations show that joint sparsity provides performance gain. This is in agreement with the average case analysis, which typically happens to be the case in the problem of sparse directional of arrival estimation, and provides the theoretical explanation on the decrease of performance of the single channel method when the number of sensors decreased.

Our future work will investigate the applicability of the proposed method in the case of a room acoustics environment and focus on the development of the MS-CS version in the near-field scenario for wideband sources.

References

1. Krim, H., Viberg, M.: Two decades of array signal processing research: the parametric approach. Signal Processing Magazine, IEEE 13(4), 67–94 (1996)
2. Malioutov, D., Cetin, M., Willsky, A.S.: A sparse signal reconstruction perspective for source localization with sensor arrays. IEEE Transac- tions on Signal Processing 53(8), 3010–3022 (2005)
3. Bilik, I.: Spatial compressive sensing approach for eld directionality estimation. In: Radar Conference, 2009 IEEE, May 2009, pp. 1–5 (2009)
4. Duarte, M.F., Sarvotham, S., Baron, D., Wakin, M.B., Baraniuk, R.G.: Dis- tributed compressed sensing of jointly sparse signals. In: Conference Record of the Thirty-Ninth Asilomar Conference on, Signals, Systems and Computers, November 1, pp. 1537–1541 (2005)
5. Tropp, J.A., Gilbert, A.C., Strauss, M.J.: Simultaneous sparse approximation via greedy pursuit, in Acoustics, Speech, and Signal Processing, 2005. In: Proceedings (ICASSP 2005). IEEE International Conference on, March 2005, vol. 5, pp. v/721 –v/724 (2005)
6. Mallat, S.G., Zhang, Z.: Matching pursuits with time-frequency dictionaries. IEEE Transactions on Signal Processing 41(12), 3397–3415 (1993)
7. Eldar, Y.C., Rauhut, H.: Average case analysis of multichannel sparse recovery using convex relaxation. IEEE Transactions on Information Theory 56(1), 505–519 (2010)
8. Candes, E.J., Wakin, M.B.: An introduction to compressive sampling, Signal Processing Magazine. IEEE 25(2), 21–30 (2008)
9. Candes, E.J., Tao, T.: Near-optimal signal recovery from random projections: Universal encoding strategies? IEEE Transactions on Information Theory 52(12), 5406–5425 (2006)
10. Gribonval, R., Rauhut, H., Schnass, K., Vandergheynst, P.: Atoms of all channels, unite! Average case analysis of multi-channel sparse recovery using greedy algorithms. Journal of Fourier analysis and Applications 14(5), 655–687 (2008)
11. Chen, S.S., Donoho, D.L., Saunders, M.A.: Atomic decomposition by basis pursuit. SIAM review 43(1), 129–159 (2001)
12. Elad, M., Aharon, M.: Image denoising via learned dictionaries and sparse representation. In: IEEE Computer Society Conference on, Computer Vision and Pattern Recognition, vol. 1, pp. 895–900 (June 2006)

Robust Second-Order Source Separation Identifies Experimental Responses in Biomedical Imaging

Fabian J. Theis[1], Nikola S. Müller[2], Claudia Plant[3], and Christian Böhm[4]

[1] IBIS, Helmholtz Zentrum Munich, Germany
[2] Max Planck Institute for Biochemistry, Martinsried, Germany
[3] Florida State University, USA
[4] University of Munich, Germany

Abstract. Multidimensional biomedical imaging requires robust statistical analyses. Corresponding experiments such as EEG or FRAP commonly result in multiple time series. These data are classically characterized by recording response patterns to any kind of stimulation mixed with any degree of noise levels. Here, we want to detect the underlying signal sources such as these experimental responses in an unbiased fashion, and therefore extend and employ a source separation technique based on temporal autodecorrelation. Our extension first centers the data using a multivariate median, and then separates the sources based on approximate joint diagonalization of multiple sign autocovariance matrices.

We consider the following *blind source separation* (BSS) problem: Let $\mathbf{x}(t)$ be an (observed) stationary m-dimensional stochastical process (with not necessarily discrete time t) and $\mathbf{A}$ a full rank matrix such that

$$\mathbf{x}(t) = \mathbf{A}\mathbf{s}(t) + \mathbf{n}(t). \tag{1}$$

A common assumption in so-called second-order BSS is to require the n-dimensional source signals $\mathbf{s}(t)$ to have diagonal autocovariances $\mathbf{R}_\tau(\mathbf{s})$ for all τ. The additive noise $\mathbf{n}(t)$ is modelled by a stationary, temporally and spatially white zero-mean process with variance σ^2. $\mathbf{x}(t)$ is observed, and the goal is to recover $\mathbf{A}$ and $\mathbf{s}(t)$. Having found $\mathbf{A}$, $\mathbf{s}(t)$ can be estimated by $\mathbf{A}^\dagger\mathbf{x}(t)$, where † denotes the pseudo-inverse of $\mathbf{A}$, equalling the inverse if $m = n$. So the BSS task reduces to the estimation of the mixing matrix $\mathbf{A}$.

1 Robust Second-Order Blind Source Separation

By centering the processes, we can assume that $\mathbf{x}(t)$ and hence $\mathbf{s}(t)$ have zero mean. The autocovariances then have the following structure

$$\mathbf{R}_\tau(\mathbf{x}) = \begin{cases} \mathbf{A}\mathbf{R}_0(\mathbf{s})\mathbf{A}^\top + \sigma^2\mathbf{I} & \tau = 0 \\ \mathbf{A}\mathbf{R}_\tau(\mathbf{s})\mathbf{A}^\top & \tau \neq 0 \end{cases} \tag{2}$$

V. Vigneron et al. (Eds.): LVA/ICA 2010, LNCS 6365, pp. 466–473, 2010.

so they are affine equivariant for $\tau \neq 0$. This functional is denoted as scatter matrix [1, 2]. In the following, we will extend the autocovariance to a robust version, which only is orthogonally equivariant; this results in some additional effort in the prewhitening phase of the method.

1.1 Robust Centering Using a Spatial Median

The (one-dimensional) median of a random variable $\mathbf{X}$ with finite first moment is defined as the interval of points where its distribution function equals 1/2. Our first goal is to replace centering by mean removal by a robust version based on the multivariate median. Instead of defining a marginal median on each component, we will use a spatial version for the centering proceduce instead. We remark that other notions for multivariate or spatial medians have been proposed, such as Oja's simplex median, half-space median, simplicial depth median.

Given a n-dimensional random vector $\mathbf{X}$ with distribution $P_{\mathbf{X}}$, a global minimum of

$$f(\mathbf{y}) := E(d(\mathbf{X}, \mathbf{y})) = \int d(\mathbf{x}, \mathbf{y})\, dP_{\mathbf{X}}(\mathbf{x}) \tag{3}$$

is called an L_1-*median* of $\mathbf{X}$, in the following often for simplicity denoted as median of $\mathbf{X}$. Here d may be any metric, but for now denotes the Euclidean distance $d(\mathbf{x}, \mathbf{y}) := \|\mathbf{x} - \mathbf{y}\|$ of two vectors $\mathbf{x}, \mathbf{y} \in \mathbb{R}^n$.

This also called *continuous median* [3] is estimated by the finite-sample median typically studied in the literature. Given a set $\mathcal{X}$ of points $\mathbf{x}_1, \ldots, \mathbf{x}_T \in \mathbb{R}^n$, equivalently, a global minimum of $f(\mathbf{y}) := \sum_{t=1}^{T} d(\mathbf{x}_t, \mathbf{y})$ is said to be an L_1-*median* of $\mathcal{X}$. In industrial applications, this is known as the optimal location problem initially proposed by [4].

Lemma 1. *The spatial median of a finite point set $\mathcal{X}$ always exists. It is unique if the points are not collinear.*

Proof. The function $f(\mathbf{y})$ is continuous in $\mathbf{y}$, and clearly $\lim_{|\mathbf{y}| \to \infty} f(\mathbf{y}) = \infty$. So $f(\mathbf{y})$ takes its infimum, hence a global minimum exists. For fixed $\mathbf{x}_t$, $\mathbf{y} \mapsto d(\mathbf{x}_t, \mathbf{y})$ is convex, so also $f(\mathbf{y})$ as sum of convex functions is convex. Assume we have two minima $\mathbf{y}$ and $\mathbf{z}$ of f. f is convex, so it takes the same value along the line segment joining $\mathbf{y}$ and $\mathbf{z}$. Now, $\lambda \mapsto d(\mathbf{x}_t, \lambda\mathbf{y} + (1 - \lambda)\mathbf{z})$ is strictly convex, then so is $\lambda \mapsto f(\lambda\mathbf{y} + (1 - \lambda)\mathbf{z})$. But a strictly convex function has a unique minimum, so $\mathbf{y} = \mathbf{z}$.

The proof has been adopted from [5]. It can also easily be generalized to the asymptotic, random vector case, see [5]. The calculation of the L_1-median is a nonlinear optimization problem; a popular method is given by the Weiszfeld algorithm [6], which we use in modified form according to [7].

1.2 SAM: Time-Delayed Sign Covariance Matrices

In [8], Visuri et al. introduced a set of robust estimators of covariance matrices. Many extensions have since then been studied, and have already found applications in BSS in the context of independent component analysis [1, 2, 9]. Here we will use these robust covariance matrices to robustly estimate autocovariances.

For this assume a given stationary n-dimensional time series $\mathbf{x}(t)$, centered according to a chosen method e.g. via removal of spatial median. Then we define the *spatial sign autocovariance matrix of* $\mathbf{x}(t)$ *with lag* τ as

$$\mathrm{SAM}_\tau(\mathbf{x}) := \mathbf{E}\left(\frac{\mathbf{s}(t+\tau)\,\mathbf{s}(t)^\top}{\|\mathbf{s}(t+\tau)\|\,\|\mathbf{s}(t)\|}\right).$$

In contrast to the normal autocovariances, $\mathrm{SAM}_\tau(\mathbf{x})$ is robust against outliers, since sample scale is removed by definition. As we will see in the simulations, this estimator is robust against perturbation of a sample (influence function), in contrast to standard covariance. SAM_τ is related to scatter matrices, but only orthogonally invariant i.e.

$$\mathrm{SAM}_\tau(\mathbf{Ax}) = \mathbf{A}\,\mathrm{SAM}_\tau(\mathbf{x})\mathbf{A}^\top \tag{4}$$

for orthogonal $\mathbf{A}$. Other robust shape estimators such as Spearman rank covariance and Kendall's tau together with spatial extensions are defined in [8, 1] and can be extended to autocovariance estimators in a similar fashion.

1.3 BSS Using SAM

According to equation (4), SAM_τ is orthogonally equivariant. This gives an indication of how to perform BSS i.e. how to recover $\mathbf{A}$ from $\mathbf{x}(t)$. The common first step consists of whitening the no-noise term $\tilde{\mathbf{x}}(t) := \mathbf{As}(t)$ of the observed mixtures $\mathbf{x}(t)$ using an invertible matrix $\mathbf{V}$ such that $\mathbf{V}\tilde{\mathbf{x}}(t)$ has unit covariance. For this we can do eigenvalue decompositon of any fully equivariant robust covariance estimate $\mathbf{C}(\mathbf{x})$ as proposed in [2] e.g. Dümbgen's shape matrix or Tyler's estimate. Instead we decided to use the original fully equivariant shape estimator proposed in [8].

So by construction $\tilde{\mathbf{x}}(t)$ fulfills $\mathbf{I} = \mathbf{C}(\tilde{\mathbf{x}}(t))$. Due to the scaling indeterminacy of BSS, we can moreover assume that this also holds for $\mathbf{s}(t)$, so due to the above $\mathbf{I} = \mathbf{AC}(\tilde{\mathbf{x}}(t))\mathbf{A}^\top = \mathbf{AA}^\top$ and we can assume orthogonal $\mathbf{A}$. Now define the *symmetrized SAM* of $\mathbf{x}(t)$ as $\overline{\mathrm{SAM}}_\tau(\mathbf{x}) := \frac{1}{2}\left(\mathrm{SAM}_\tau(\mathbf{x}) + (\mathrm{SAM}_\tau(\mathbf{x}))^\top\right)$. Equation (4) shows that also the symmetrized autocovariance $\mathbf{x}(t)$ is orthogonally equivariant, and we get

$$\overline{\mathrm{SAM}}_\tau(\mathbf{x}) = \mathbf{A}\overline{\mathrm{SAM}}_\tau(\mathbf{s})\mathbf{A}^\top \tag{5}$$

for $\tau \neq 0$. Our model assumption is to have (robustly) autodecorrelated sources, so $\overline{\mathrm{SAM}}_\tau(\mathbf{s})$ is diagonal. Hence equation (5) is an eigenvalue decomposition of the symmetric matrix $\overline{\mathrm{SAM}}_\tau(\mathbf{x})$. If we furthermore assume that $\overline{\mathrm{SAM}}_\tau(\mathbf{x})$ or equivalently $\overline{\mathrm{SAM}}_\tau(\mathbf{s})$ has n different eigenvalues, then the above decomposition i.e. $\mathbf{A}$ is uniquely determined by $\overline{\mathrm{SAM}}_\tau(\mathbf{x})$ except for orthogonal transformation in each eigenspace and permutation. The robust extension of the AMUSE algorithm [10] is given by diagonalization of $\overline{\mathrm{SAM}}_\tau(\mathbf{x})$ after whitening — the diagonalizer yields the desired separating matrix.

However, AMUSE decisively depends on the choice of τ — if an eigenspace of $\overline{\mathrm{SAM}}_\tau(\mathbf{x})$ is higher dimensional, the algorithm fails. Moreover the autocovariance

Algorithm 1: Sign autocovariance SOBI algorithm (SAM-SOBI)

Input: $d \times T$ sample matrix $\mathbf{X}$ of a multivariate time series, number of autocovariances K
Output: separation matrix $\mathbf{W}$

pre-whiten data
calculate eigenvalue decomposition $\mathbf{E}_0 \boldsymbol{\Lambda}_0 \mathbf{E}_0^\top$ of a robust covariance estimate
$\mathbf{V} \leftarrow \boldsymbol{\Lambda}_0^{-1/2} \mathbf{E}_0^\top$
$\mathbf{Y} \leftarrow \mathbf{VX}$

estimate symmetrized sign autocovariance matrices
for $t \leftarrow 1 \ldots T$ **do**
  $\mathbf{Z}(t) \leftarrow \mathbf{Y}(t)/\|\mathbf{Y}(t)\|$
end
for $\tau \leftarrow 1 \ldots K$ **do**
  $\mathrm{SAM}_\tau \leftarrow (T-\tau)^{-1}\mathbf{Z}(:, 1 : T-\tau+1)\mathbf{Z}(:, \tau : T)^\top$
  $\overline{\mathrm{SAM}}_\tau \leftarrow (\mathrm{SAM}_\tau + \mathrm{SAM}_\tau^\top)/2$
end

joint diagonalization
$\mathbf{U} \leftarrow$ approximate joint diagonalizer of $\{\overline{\mathrm{SAM}}_\tau\}$ such that $\mathbf{U}\,\overline{\mathrm{SAM}}_\tau \mathbf{U}^\top$ is approximately diagonal
$\mathbf{W} \leftarrow \mathbf{UV}$

matrices are only estimated by a finite number of samples, and due to possibly colored noise, the autocovariance at τ could be badly estimated. Hence we extend the SOBI algorithm [11] to the robust situation by taking a whole set of sign autocovariance matrices $\mathrm{SAM}_i := \overline{\mathrm{SAM}}_i(\mathbf{x})$, $i = 1, \ldots, K$, of $\mathbf{x}(t)$ with varying time lags. Approximate joint diagonalization [12] yields the desired separation matrix. We call the resulting algorithm *SAM-SOBI*. It is summarized in the algorithm table 1.

2 Simulations

We systematically evaluate the effect of robust autocovariance estimation and the influence of the number of (sign) autocovariance matrices on simulated data. We generated $(n = 3)$-dimensional correlated data by filtering a decorrelated multivariate i.i.d Gaussian signal with a uniform random finite-impulse-response filter of length $w = 10$. We set the 0-th filter coefficient to 1, which implies that the signal follows an MA-model. Our data set consists of $N = 1000$ time points. We compared the four methods Amuse, SAM-Amuse, SOBI and SAM-SOBI in the case of additive noisy mixtures $\mathbf{X}(t) = \mathbf{AS}(t) + \mathbf{n}(t)$ and low SNR of 29dB. Figure 1(a) shows the performance of mixing matrix recovery measured by the Amari error gathered over 1000 runs. For visualizing separation quality, we also plot the Amari error of random matrices. Clearly, both Amuse methods are only slightly better than chance, in contrast to the SOBI methods, which make use of the additional information of the $K = 10$ used autocovariance matrices. SOBI

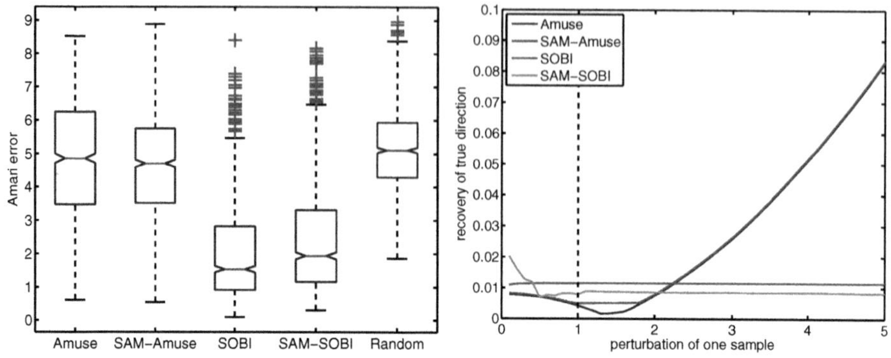

(a) recovery of data with additive noise (b) influence function

Fig. 1. Simulations. In (a) we compare algorithm performances on $n = 3$-dimensional random filtered Gaussian data with some additive noise (SNR=29dB). In (b), we determine the influence function of a sample; for this we compare algorithm performance on $n = 2$ sinusiodals, when a single sample is increased in scale.

slightly outperforms the robust SOBI, which is to be expected since no outliers are present. So in the case of mixtures with additive white noise without outliers, multiple autocovariance matrices are needed, but robust estimation thereof does not help.

The situation changes if we add outliers to the data. We determine the so-called *influence function*, which is related to the breaking point but nicely visualizes performance with respect to outlier strength. An empirical influence function at a given observation i is defined as the function $f(x)$ that gives the estimators output when replacing the i-th sample by x. In figure 1(b), we plot the empirical influence function on a process of two sinusoidal signals in $[-1, 1]$ with 1000 time points. We perturb a single sample with factors from $[0, 5]$. The two non-robust estimation algorithms quickly decay in terms of quality when increasing the perturbation of the single sample. This is due to the fact that mean and covariance are sensitive to large values. In contrast the SAM versions almost ignore the single sample perturbation, with SAM-SOBI outperforming the Amuse version as expected.

3 Biomedical Applications

EEG Data. We validated SAM-SOBI on two real datasets to demonstrate the strength of SAM-SOBI. First data of electroencephalography (EEG) in response to a painful laser stimuli were analysed. Time series of all 64 electrodes are basically comprised of three phases: a reference phase of one second before the pain stimuli (named type0) and two neuronal response phases with "type1' a phase of high neuronal activity (150-500ms) followed by a neuronal response named "type2" (500ms-1.5seconds) [13]. For EEG data analysis ICA is one of

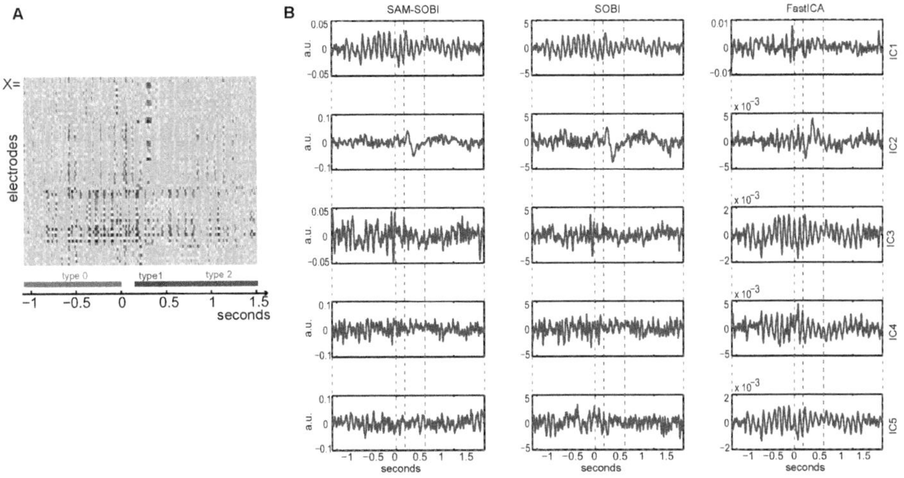

Fig. 2. EEG Data. A. Pseudocoloring of the input EEG data time series. Three time frames were defined as reference neuronal activity (type0) and two latencies of neuronal responses (type1/2). **B.** ICA results of three algorithms resulting in five strongest components (IC1-5). Time frames of type0-2 are indicated in green, red and blue.

the common techniques. To validate SAM-SOBI with respect to SOBI and fastICA we performed ICA segregating time (Fig. 3BA). The three algorithms were applied with standard parameter to estimate five independent components (IC) (Fig. 3B). This dimensionality reduction enabled us to analyse only the strongest components.

To quantify the demixed response patterns, we focused on the variance chance of both neuronal responses (Tab. 1). The strongest IC (IC1) showed a decrease in variance mainly in type2 response, thus a reduced neuronal activity after pain perception in type1. IC2 then detected the typical pain perception pattern in type1. The strongest amplitude (here measured by variance) was detected by SAM-SOBI. FastICA showed the less pronounced pain perception. We observed that SAM-SOBI was able to capture pain perception in just one component with only little noise, or, influences of other underlying components. SOBI components showed a stronger influence of the natural EEG noise level, compared to SAM-SOBI. FastICA was less effective not only in detecting the typical pain perception (in IC2) but also in demixing of the signals into independent components. To conclude, SAM-SOBI showed a better demixing as well as a higher ability to capture the real components of EEG data then SOBI and fastICA.

FRAP Data. To demonstrate that SAM-SOBI is a all-round ICA method, we collected microscopy data of flourescence recovery after photobleaching (FRAP) experiments. Yeast cells expressing Cdc42-GFP flourescent protein accumulate Cdc42 in the bud tip during cell division [14]. Data was generated by measuring the Cdc42-flourescence of this cap whereby the cap was bleached (t=0). After 25 seconds the flourescence of the cap recovered (Fig. 3A). Again, we applied the

Table 1. Variances of type1 and type2 time frames of IC1-5 of EEG data

	IC1		IC2		IC3		IC4		IC5	
Algorithm	type.1	type.2	type.1	type.2	type.1	type.2	type.1	type.2	type.1	type.2
SAM-SOBI	.37	**.21**	**.67**	.18	.16	.26	.18	.28	.27	.36
SOBI	.37	**.19**	**.62**	.20	.22	.27	.17	.31	.25	.26
fastICA	.32	**.19**	**.67**	.18	.35	.22	.29	.26	.31	.22

ICA algorithms to maintain underlying five components (Fig. 3B) after correcting for intensity loss over time. All three ICA algorithms were able to detect the FRAP-typical flourescence recovery in the strongest IC (IC1). Closer inspection revealed that fastICA was not able to really demix the FRAP data since the typical recovery is present in four components. We further elucidated the performance of IC1. The typical FRAP analysis includes fitting of a single exponential to the recovery curve ($t \leq 0$, Fig. 3C). We inverted and rescaled IC1 to (0,1) with respect to the basal value ($t = 0$) and pre-bleach values ($t < 0$). For noise-robust de-mixing algorithms we expect a precise fit. Thus, we benchmarked the algorithms with the root mean squared error (RMSE) indicating the goodness of the fit (n=3). SAM-SOBI performed better than fastICA and SOBI. Again, we can conclude that SAM-SOBI does not only effectively determine independent components but describe data in a noise-robust manner.

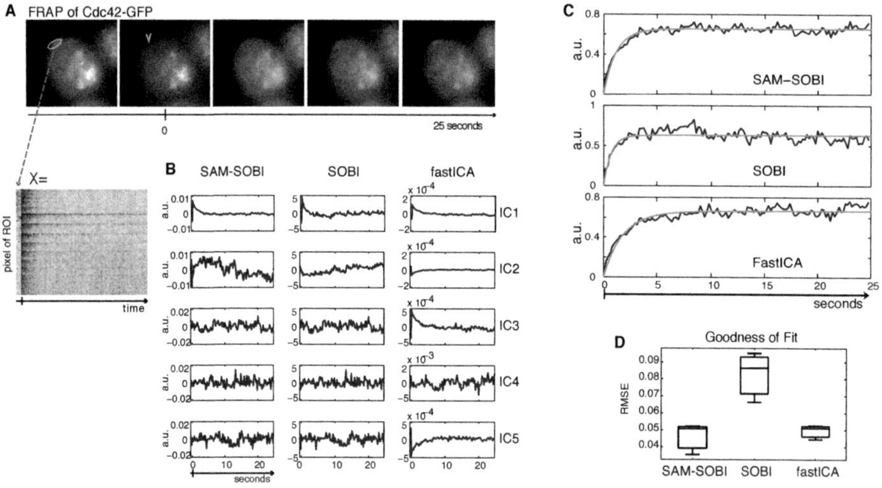

Fig. 3. FRAP Data. A. FRAP experiment as a timelaps movie over 30 seconds. The bud tip of the cell was linearized and used as a times series for the ICA algorithms. **B.** ICA results of three algorithms resulting in five strongest components (IC1-5). **C.** IC1 starting at timepoint 0 rescaled (blue line) and fitted with an single exponential (green). **D.** Distribution of goodness of fits of three different FRAP experiments shown by root mean squared error.

4 Conclusions

We have presented a robust method for second-order blind source separation, extending the classical SOBI algorithm by using sign autocovariance matrices. The robust estimators outperform the standard methods on data with outliers, but in particular also on two real, novel biomedical applications. It is important to note that there exist many more, statistically more sophisticated techniques for robust covariance estimation, which are currently being imported into the ICA field, see e.g. [2]. The present contribution shows that this can easily be done also for second-order BSS methods.

Acknowledgements. The authors thank Enrico Schulz, Laura Tiemann, Tibor Schuster, Joachim Gross and Markus Ploner for providing the EEG data as well as Tina Freisinger and Roland Wedlich-Söldner (MPI for Biochemistry) for generating and providing the FRAP data. FJT gratefully acknowledges partial support by the Helmholtz Alliance on Systems Biology (project 'CoReNe').

References

1. Oja, H., Sirkiä, S., Eriksson, J.: Scatter matrices and independent component analysis. Austrian Journal of Statistics 35(2), 175–189 (2006)
2. Nordhausen, K., Oja, H., Ollila, E.: Robust independent component analysis based on two scatter matrices. Austrian Journal of Statistics 37(1), 91–100 (2008)
3. Fekete, S., Mitchell, J., Weinbrecht, K.: On the continuous weber and k-median problems. In: Proc. sixteenth SoCG, pp. 70–79 (2000)
4. Weber, A.: Über den Standort der Industrien. Tübingen (1909)
5. Dudley, R.: Department of mathematics, MIT, course 18.465 (2005)
6. Weiszfeld, E.: Sur le point par lequel la somme des distances de n points donnés est minimum. Tohoku Mathematics Journal 43, 355–386 (1937)
7. Vardi, Y., Zhang, C.H.: The multivariate L_1-median and associated data depth. Proc. Nat. Acad. Sci. USA 97(4), 1423–1426 (2000)
8. Visuri, S., Koivunen, V., Oja, H.: Sign and rank covariance matrices and rank covariance matrices. Journal of Statistical Planning and Inference 91(2), 557–575 (2000)
9. Kirshner, S., Poczos, B.: ICA and ISA using schweizer-wolff measure of dependence. In: Proc. ICML 2008, vol. 307 (2008)
10. Tong, L., Liu, R.W., Soon, V., Huang, Y.F.: Indeterminacy and identifiability of blind identification. IEEE Transactions on Circuits and Systems 38, 499–509 (1991)
11. Belouchrani, A., Meraim, K.A., Cardoso, J.F., Moulines, E.: A blind source separation technique based on second order statistics. IEEE Transactions on Signal Processing 45(2), 434–444 (1997)
12. Cardoso, J., Souloumiac, A.: Blind beamforming for non gaussian signals. IEE Proceedings - F 140(6), 362–370 (1993)
13. Iannetti, G.D., Zambreanu, L., Cruccu, G., Tracey, I.: Operculoinsular cortex encodes pain intensity at the earliest stages of cortical processing as indicated by amplitude of laser-evoked potentials in humans. Neuroscience 131, 199–208 (2005)
14. Wedlich-Söldner, R., Wai, S.C., Schmidt, T., Li, R.: Robust cell polarity is a dynamic state established by coupling transport and GTPase signaling. J. Cell Biol. 166, 889–900 (2004)

Decomposition of EEG Signals for Multichannel Neural Activity Analysis in Animal Experiments

Vincent Vigneron[1,⋆], Hsin Chen[2], Yen-Tai Chen[3], Hsin-Yi Lai[3], and You-Yin Chen[3]

[1] IBISC CNRS FRE 3190, Université d'Evry, France
vincent.vigneron@ibisc.univ-evry.fr
[2] Dept. of Electrical Engineering, National Tsing Hwa University, Taiwan
hchen@ee.nthu.edu.tw
[3] Dept. of Electrical Engineering, National Chiao-Tung University, Taiwan
kenchen@cn.nctu.edu.tw

Abstract. We describe in this paper some advanced protocols for the discrimination and classification of neuronal spike waveforms within multichannel electrophysiological recordings. Sparse decomposition was used to serarate the linearly independent signals underlying sensory information in cortical spike firing patterns. We introduce some modifications in the the IDE algorithm to take into account prior knowledge on the spike waveforms. We have investigated motor cortex responses recorded during movement in freely moving rats to provide evidence for the relationship between these patterns and special behavioral task.

Keywords: Sparse decomposition, classification, semi-supervised learning, Atomic Decomposition, IDE akgorithm.

1 Introduction

The analysis of continuous and multichannel neuronal signals is complex, due to the large amount of information received from every electrode. Neural spike recording have shown that the primary motor cortex (M1) encodes information about movement direction [2] limb velocity, force [3] and individual muscle activity [11]. To further investigate the relationship between cortical spike firing patterns and special behavioral task, we have investigated motor cortex responses recorded during movement in freely moving rats: Fig. 1 shows the experimental setup for neural activities recording and the video captures related animal behavioral task simultaneously.

The study, approved by the Institutional Animal Care and Use Committee at the National Chiao Tung University, was conducted according to the standards established in the Guide for the Care and Use of Laboratory Animals. Four male Wistar rats weighing 250-300 g (BioLASCO Taiwan Corp., Ltd.) were individually housed on a 12 h light/dark cycle, with access to food and water *ad libitum*. Dataset was collected from the motor cortex of awake animal performing a simple reward task. In this task, male rats (BioLACO Taiwan Co.,Ltd) were trained to press a lever to initiate a trial in return

⋆ This project was supported in part by fundings from the Hubert Curien program of the Foreign French Minister and from the Taiwan NSC. The neural activity recordings were kindly provided by the Neuroengineering lab. of the National Chiao-Tung University.

V. Vigneron et al. (Eds.): LVA/ICA 2010, LNCS 6365, pp. 474–481, 2010.

Fig. 1. The experimental setup (top). Light-color (red virtual ring) was belted up the right forelimb to be recognized the trajectory by video tracking system. The sequence images were captured the rat performing the lever press tasks in return for a reward of water drinking (bottom).

for a water reward. The animals were water restricted 8-hour/day during training and recording session but food were always provided to the animal ad lib every day.

The animals were anesthetized with pentobarbital (50 mg/kg i.p.) and placed on a standard stereotaxic apparatus (Model 9000, David Kopf, USA). The dura was retracted carefully before the electrode array was implanted. The pairs of 8 microwire electrode arrays (no.15140/13848, 50m in diameter; California Fine Wire Co., USA) are implanted into the layer V of the primary motor cortex (M1). The area related to forelimb movement is located anterior 2-4 mm and lateral 2-4 mm to Bregma. After implantation, the exposed brain should be sealed with dental acrylic and a recovery time of a week is needed.

During the recording sessions, the animal was free to move within the behavior task box (30 cm$\times$30 cm$\times$ 60 cm), where rats only pressed the lever via the on right forelimb, and then they received 1-ml water reward as shown in Fig. 1. A Multi-Channel Acquisition Processor (MAP, Plexon Inc., USA) was used to record neural signals. The recorded neural signals were transmitted from the headstage to an amplifier, through a band-pass filter (spike preamp filter: 450-5 kHz; gain: 15,000-20,000), and sampled at 40 kHz per channel. Simultaneously, the animals'behavior was recorded by the video tracking system (CinePlex, Plexon Inc., USA) and examined to ensure that it was consistent for all trials included in a given analysis [13].

Neural activity was collected from 400-700ms before to 200-300 ms after lever release for each trail. Data was recorded from 33 channels, action potentials (spikes) crossing manually set thresholds were detected, sorting and the firing rate for each neuron was computed in 33 ms time binsSince the signals are collected with 10 nanometers invasive probes, the noise effects are limited.

The main processing steps of EEG signal decomposition are (i) spike detection, (ii) spike feature extraction, and finally (iii) spike clustering. They are discussed in the next sections.

2 Blind Source Separation for Spikes Detection

Blind source separation (BSS) has been widely used for processing EEG signals. The most simple but widely used case is the instantaneous case, where the source signals arrive at the sensors at the same time; here, the signals have narrow bandwidths and the sampling frequency is normally low. The BSS model in this case can be easily formulated as

$$\mathbf{x}(n) = A\mathbf{s}(n) + \mathbf{v}(n) \tag{1}$$

where $m \times 1$ $\mathbf{s}(n)$, $n_e \times 1$ $\mathbf{x}(n)$, and $n_e \times 1$ $\mathbf{v}(n)$ denote respectively the vectors of source signals, observed signals, and noise at discrete time n and n_e denotes the number of electrodes. A is the mixing matrix of size $n_e \times m$. Although the main assumptions about the source signals, such as uncorrelatedness or independency of such signals, have not yet been verified, the empirical results illustrate the effectiveness of such methods. EEG signals are noisy and nonstationary signals, which are normally affected by one or more types of internal artefacts. In the case of brain signals, the independency or uncorrelatedness conditions for the sources may not be satisfied. A topographic ICA method proposed by Hyvarinen *et al.* [7] incorporates the dependency among the nearby sources in not only grouping the independent components related to nearby sources but also separating the sources originating from different regions of the brain. When the sources are sparse, i.e. at each time instant, the number of non-zero values are less or equal to the number of mixtures. The columns of the mixing matrix may be calculated individually, which makes the solution to the underdetermined case possible. The problem can be stated as a clustering problem since the lines in the scatter plot can be separated based on their directionalities by means of clustering. The same idea has been followed more comprehensively by Li *et al* [8]. In their method, however, the separation has been performed in two different stages. First, the unknown mixing matrix is estimated using the k-means clustering method. Then, the source matrix is estimated using a standard linear programming algorithm. To maximize the sparsity of the extracted signals in the output of the separator. The columns of the mixing matrix A assign each observed data point to only one source based on some measure of proximity to those columns. Therefore the mixing system can be presented as

$$x_i(n) = \sum_{j=1}^{M} a_{ji} s_j(n), = 1, \ldots, N \tag{2}$$

where in an ideal case $a_{ji} = 0$ for $i \neq j$. For finding the sparse solution of (2), one may search for a solution for which the ℓ^0 norm of s, i.e. the number of non-zero components of s, is minimized. This is written as:

$$\text{minimize} \sum_{i=1}^{m} |s_i|^0 \quad \text{subject to} \quad \mathbf{As} = \mathbf{x} \tag{3}$$

Direct solution of this problem needs a combinatorial search and is NP-hard. The ℓ_0 norm is also sensitive to noise. Many different algorithms have been proposed in recent years for finding the sparse solution of (3) Some examples are Basis Pursuit (BP) [4], Smoothed ℓ^0 (SL0) [9,10], and FOCUSS [6]. Many of these algorithms, replace the ℓ^0 norm in (3) by another function of $\mathbf{s}$, and solve the problem:

$$\text{minimize } f(\mathbf{s}) \quad \text{subject to} \quad \mathbf{As} = \mathbf{x} \tag{4}$$

Minimization of the ℓ_1-norm is one of the most performant methods for estimation of the sources. The ℓ_1 -norm minimization can be manifested as

$$\min ||s(n)||_1 \text{ subject to } \mathbf{As} = \mathbf{x} \tag{5}$$

A detailed discussion of signal recovery using ℓ_1-norm minimization is presented by Takigawa et al. [12]. Minimal ℓ_1 norm solution may be found by Linear Programming (LP). Fast algorithms for Linear programing are Simplex and Interior Point method. LP provides in practice very good practical results and is supported by Dohono theorem [5] according to "for 'most' 'large' underdetermined systems of linear equations, the minimal ℓ_1 norm solution is also the sparsest solution". Therefore clustering the observation points in the space-time-frequency domain can be effectively used for separation of spikes. The main drawback is that LP is very time-consuming.

IDE (Iterative Detection Estimation), proposed by [1] is an effective tool in detection and classification of sparse sources in a multidimensional space. The main idea is twofold:

- Step 1 (Detection): Starting with a previous estimate (or initial guess) of the source vector, $\hat{\mathbf{s}}$ first (roughly) detect which sources are 'active' and which are 'non-active'
- Step 2 (Estimation): Knowing active sources, obtain a new estimate of $\mathbf{s}$ by finding a solution of $x = As$ whose active indices better coincide with those predicted by the detection step.

This two step iteration continues until no further improvement is possible (i.e. until a given level of accuracy is reached). The term 'active' is used to refer to sources having 'considerable values'. The test for activity is carried for one source at time. The detection step results from a binary hypothesis testing, with a mixture of Gaussian source model to model sparsity. Further details and discussion on the solution can be found in [8]. The outcome is highly valuable for detection of spikes corresponding in practice to left and right finger movements in the context of brain-computer interfacing.

3 Spike Sorting and Clustering: Estimating Sources and Classifying Them at the Same Time

Clustering refers to a self-organizing process which arranges feature vectors into clusters so that the members of a cluster are closer to each other than to the centers of other clusters. Each new feature vector is either assigned to the cluster being closest in the feature space or appointed as a representative of a new cluster. A clustering algorithm may be designed to group feature vectors based on prior knowledge of the number of clusters. However, such knowledge is not available in the EEG application, so the number of clusters is only determined once the clustering process has been fully completed. We propose in this section to combine the IDE functionality and statistical learning theory to designed a semi-supervised IDE algorithm. Conversely to the classical IDE, the approach proposed does rely on labeled data which could be well adapted in some complex situations. The main idea of our approach is to compare the supervised information given by the learning data with an unsupervised modeling based on the Gaussian mixture model. With such an approach, if some learning data have wrong labels, the comparison of the supervised information with an unsupervised modeling of the data allows to detect the inconsistent labels. It is possible afterward to build a supervised classifier by giving a low confidence to the learning observations with inconsistent labels. The main advantages of the proposed approach compared to previous works are the explicit modeling of more than two classes and the flexibility of the method due to the use of a global mixture model. Let us consider a mixture model in which two different structures coexist: an unsupervised structure of K clusters (represented by the random discrete variable S) and a supervised structure, provided by the learning data – labeled spikes -, of k classes (represented by the random discrete variable C). As in the standard mixture model, we assume that the data $(x_1, \ldots, x_n)$ are independent realizations of a random vector $X \in \mathbb{R}^p$ with density function:

$$p(x) = \sum_{j=1}^{K} p(S = j)p(x|S = j) \tag{6}$$

where $p(S = j)$ is the prior probability of the jth cluster and $p(x|S = j)$ is the corresponding conditional density. Let us now introduce the supervised information carried by the learning data. Since $\sum_{j=1}^{k} p(C = i|S = j) = 1$ for all $j = 1, \ldots, K$, we can plug this quantity in Eq. (6) to obtain

$$p(x) = \sum_{i=1}^{k} \sum_{j=1}^{K} p(C = i|S = j)p(S = j)p(x|S = j) \tag{7}$$

where $p(C = i|S = j)$ can be interpreted as the probability that the jth cluster belongs to the ith class and thus measures the consistency between classes and clusters. Using the classical notations of parametric mixture models and introducing the notation $r_{ij} = P(C = i|S = j)$, we can reformulate

$$p(x) = \sum_{i=1}^{k} \sum_{j=1}^{K} r_{ij} \pi_j p(x|S = j) \tag{8}$$

where $\pi_j = P(S = j)$. Therefore, Eq. (8) exhibits both the "modeling" part of our approach, based on the mixture model, and the "supervision" part through the parameters r_ij . Since the modeling introduced in this section is based on the mixture model, we can use any conditional density to model each cluster.

Estimation procedure. Due to the nature of the model proposed, the estimation procedure is made of two steps corresponding, respectively, to the unsupervised and to the supervised part of the comparison.The first step consists in estimating the parameters of the mixture model in an unsupervised way leading to the clustering probabilities $P(S = j|X = x)$. In the second step, the parameters r_{ij} linking the mixture model with the information carried by the labels of the learning data are estimated by maximization of the likelihood. The classification step mainly consists in calculating the posterior probability $P(C = i|X = x)$ for each class $i = 1, \dots, k$. In the case of the model described in this section, this posterior probability can be expressed as follows using Bayes' rule. We finally obtain

$$P(C = i|X = x) = \sum_{j=1}^{K} r_{ij} P(S = j|X = x). \tag{9}$$

Therefore, the classification step of the algorithm requires the estimation of the probabilities r_{ij} as well as the unsupervised classification probabilities $P(S = j|X = x)$. In the first step of the estimation procedure, the labels of the data are discarded to form K homogeneous groups. Therefore, this step consists in estimating the parameters of the chosen mixture model. In the case of the usual Gaussian model (step 1 of IDE algorithm), the classical procedure for estimating the proportions π_j, the means μ_j and the variance matrices Σ_j , for $j = 1, \dots, K$ is the expectation-maximization algorithm proposed by Dempster et al. In this second step of the procedure, the labels of the data are introduced to estimate the $k \times K$ matrix of parameters $R = \{r_i j\}$ and we use the parameters learned in the previous step as the mixture parameters. The parameter r_{ij} modify the unsupervised model for taking account of the label information. The log-likelihood associated to our model can be expressed as

$$\ell(R) = \sum_{i=1}^{k} \sum_{j=1}^{K} \log p(X = x, C = i) \propto \sum_{i=1}^{k} \sum_{j=1}^{K} \log \left(\sum_{j=1}^{K} r_{ij} p(S = j|X = x) \right). \tag{10}$$

Consequently, we end up with a constrained optimization problem

$$\text{maximise} \sum_{i=1}^{k} \sum_{j=1}^{K} \log p(X = x, C = i) \propto \sum_{i=1}^{k} \sum_{j=1}^{K} \log \left(\sum_{j=1}^{K} r_{ij} p(S = j|X = x) \right) \tag{11}$$

with respect to $r_{ij} \in [0, 1]$ and $\sum_{i=1}^{k} r_{ij} = 1, \forall j = 1, \dots, K$.

4 Experimental Results

In this section, we present experimental results on artificial and real datasets in situations illustrating the problem of supervised classification under uncertainty (see Fig. 2). The plots show the estimated source vectors – spikes $\hat{\mathbf{s}}$, obtained at each iteration. The data used in Fig. 2 are for a single (typical) realization of the experiment on one channel. Ensembles of neurons were simultaneously recorded in this during long periods of spontaneous behaviour. Waveforms provided from the neural activity were then processed and classified. Typically, most information correlated across neurons in the ensemble were concentrated in a small number of signals. The experimental studies show that the algorithm is as efficient as fully supervised techniques when the label noise is low and very robust to label noise.

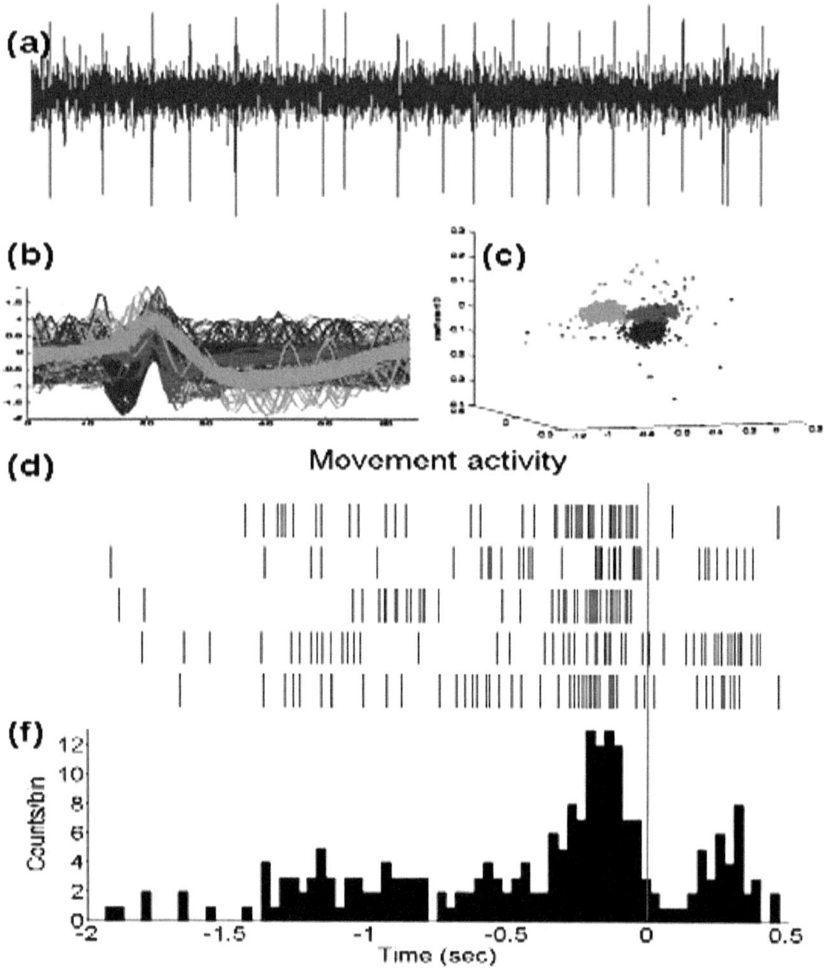

Fig. 2. The raw data of neural recording and an example of spike sorting. (a) Real raw data recording in M1 area. (b) Three individual spikes are detected from the raw data from (a). Visualized result for spike sorting by using principal component analysis. All spike timestamps were displayed for one trail a single neural recording (d) and the firing activity histograms (f) around the time of behavioral task. The bin is 33 ms. The red line denotes that the animal presses the lever.

5 Conclusion

The semi-supervised classification algorithm proposed here is designed to performing classification in the presence of noisy labelled data. Ensembles of neurons were simultaneously recorded in this during long periods of spontaneous behaviour. Waveforms provided from the neural activity were then processed and classified. Typically, most information correlated across neurons in the ensemble were concentrated in a small number of signals.

References

1. Amin, A.A., Babaie-Zadeh, M., Jutten, C.A.: A fast method for sparse component analysis-based on iterative detection-estimation. In: Proceedings of Maxent (2006)
2. Amirikian, B., Georgopoulus, A.P.: Motor Cortex: Coding and Decoding of Directional Operations. In: The Handbook of Brain Theory and Neural Networks, pp. 690–695. MIT Press, Cambridge (2003)
3. Bodreau, M., Smith, A.M.: Activity in rostal motor cortex in response to predicatel force-pulse pertubations in precision grip task. J. Neurophysiol. 86, 1079–1085 (2005)
4. Donoho, D.L.: For most large underdetermined systems of linear equations the minimal ℓ_1-norm solution is also the sparsest solution. Technical report (2004)
5. Donoho, D.L., Huo, X.: Uncertainty Principles and Ideal Atomic Decomposition. IEEE Trans. Inform. Theory 47(7), 2845–2862 (2001)
6. Gorodnitsky, I.F., Rao, B.D.: Sparse signal reconstruction from limited data using FOCUSS, a re-weighted minimum norm algorithm. IEEE Transactions on Signal Processing 45(3), 600–616 (1997)
7. Hyvarinen, A., Hoyer, P.O., Inkl, M.: Topographic independent component analysis. Neural Computation 13, 1527–1558 (2001)
8. Li, Y., Cichocki, A., Amari, S.: Sparse component analysis for blind source separation with less sensors than sources. In: ICA 2003, pp. 89–94 (2003)
9. Mohimani, G.H., Babaie-Zadeh, M., Jutten, C.: Fast sparse representation based on smoothed ℓ_0 norm. In: Davies, M.E., James, C.J., Abdallah, S.A., Plumbley, M.D. (eds.) ICA 2007. LNCS, vol. 4666, pp. 389–396. Springer, Heidelberg (2007)
10. Mohimani, H., Babaie-Zadeh, M., Jutten, C.: A fast approach for overcomplete sparse decomposition based on smoothed l0 norm. Accepted in IEEE Trans. on Signal Processing
11. Morrow, M.M., Miller, L.E.: Prediction of muscle activity by populations of sequentially recorded primary motor cortex neurons. J. Neurophysiol. 89, 1079–1085 (2003)
12. Takigawa, I., Kudo, M., Nakamura, A., Toyama, J.: On the minimum ℓ_1-norm signal recovery in underdetermined source separation. In: Puntonet, C.G., Prieto, A.G. (eds.) ICA 2004. LNCS, vol. 3195, pp. 193–200. Springer, Heidelberg (2004)
13. Van Staveren, G.W., Buitenweg, J.R., Heida, T., Ruitten, W.L.C.: Wave shape classification of spontaneaous neural activity in cortical cultures on micro-electrode arrays. In: Proceedings of the Second joint EMBS/BMES Conference, Houston, TX, USA, October 23-26 (2002)

Using Non-Negative Matrix Factorization for Removing Show-Through

Farnood Merrikh-Bayat[1,*], Massoud Babaie-Zadeh[1], and Christian Jutten[2]

[1] Department of Electrical Engineering, Sharif University of Technology, Azadi Avenue, Tehran, Iran

[2] GIPSA-lab, Depart. of Images and Signal, Grenoble, France and Institut Universitaire de France

f_merrikhbayat@ee.sharif.edu, mbzadeh@sharif.edu, christian.jutten@gipsa-lab.grenoble-inp.fr

Abstract. Scanning process usually degrades digital documents due to the contents of the backside of the scanned manuscript. This is often because of the show-through effect, *i.e.* the backside image that interferes with the main front side picture mainly due to the intrinsic transparency of the paper used for printing or writing.

In this paper, we first use one of Non-negative Matrix Factorization (NMF) methods for canceling show-through phenomenon. Then, non-linearity of show-through effect is included by changing the cost function used in this method. Simulation results show that this proposed algorithm can remove show-through effectively.

1 Introduction

In this paper, we are going to consider one of the most common degradations, called show-through usually appearing in ancient documents which are written or printed on both sides of the page. Show-through is an undesired appearance of a printed image or text of the reverse side of the paper which can significantly degrade the readability of the document.

Several approaches for show-through reduction have been already investigated. Some authors have used only one side of the document and they have tried to distinguish show-through from foreground image by using various features in document and presented show-through removal techniques [1]. Although these methods certainly perform better than simple thresholding, there is no way to unambiguously differentiate foreground from show-through without comparing both sides of the document. In [2], authors process both sides of the document simultaneously, in order to identify regions that are mainly show-though, and replace them by an estimate of the background. Most of these works have the drawback that they can only be used with texts or handwritings. Interests

* This work has be partially supported by center for International Research and Collaboration (ISMO) and French embassy in Tehran in the framework of a GundiShapour collaboration program.

V. Vigneron et al. (Eds.): LVA/ICA 2010, LNCS 6365, pp. 482–489, 2010.

in applying Blind Source Separation (BSS) algorithms for solving this problem have been increased noticeably nowadays. In [3,4], authors modeled the show-through effect as a superimposition of back and front sides printed images. Then, by assuming that the scanned front side image (corrupted by backside) and the scanned back side image (corrupted by front side) are linear mixtures of the *independent* front and back sides images, they used BSS techniques for estimating the pure sources. Tonazzini *et al.* in [5] represent an effective method for removing show-through in color images by using only one side of the paper. One of the most disadvantages of these methods in spite of their good results is that the results are not perfect especially in regions where the images of the front and back sides of the paper have overlaps and the front side's image is near to black. In such regions, the recovered front image is whiter than other sections where there is no overlap. The main reason of having poor results in these areas is that show-through is in fact a nonlinear phenomenon as illustrated in [6,7]. Sharma in [8] considered a nonlinear model for this phenomenon and proposed to compensate this effect by using adaptive filters. Almeida suggested in [9] MISEP method based on Multi-Layer Perceptron networks for separating real-word nonlinear image mixtures. The main drawback of using MISEP or other universal nonlinear networks for BSS is the separability issue: the ICA does not lead necessarily to BSS using such networks. In addition to BSS techniques, wavelet-based method has also been recently proposed for separation of nonlinear show-through and bleed-through image mixtures [10].

In this paper, we first consider linear mixing model for show-through and use one of non-negative matrix factorization methods for separating the pure sources. Then, by performing simple simulations, we show that using linear model for show-through is not correct. Therefore, non-linearity of show-through is included in our method by changing the cost function used in selected NMF algorithm.

The paper is organized as follows. In Section 2, we assume linear model for show-through and use NMF for its removal. Section 3 describes a procedure of modifying the selected NMF algorithm for taking into account non-linearity of show-through. Finally, a few experimental results with real printed or manuscript documents are presented in Section 4, before conclusions and perspectives in Section 5.

2 Modeling Show-through as a Linear Phenomenon and Using NMF for Separating the Sources

In this section, assuming that show-through is a linear phenomenon, we will try to apply NMF to this particular application as another method for separating the sources and decreasing the degradation caused by show-through.

Let vectors $\mathbf{s}_1$ and $\mathbf{s}_2$ be the front side and backside pure images written or printed on sides of the paper respectively, and vectors $\mathbf{x}_1$ and $\mathbf{x}_2$ be the scanned images of sides of the paper. Note that these vectors are obtained by concatenating rows of the corresponding image matrixes. Then, assuming the show-through is linear, we have

$$\mathbf{X} = \begin{bmatrix} \mathbf{x}_1 \\ \mathbf{x}_2 \end{bmatrix} = \begin{bmatrix} a_{11} \; a_{12} \\ a_{21} \; a_{22} \end{bmatrix} \times \begin{bmatrix} \mathbf{s}_1 \\ \mathbf{s}_2 \end{bmatrix} = \mathbf{AS}. \tag{1}$$

If the size of images obtained through the scanning process is $m \times n$, the size of vectors $\mathbf{s}_1$, $\mathbf{s}_2$, $\mathbf{x}_1$ and $\mathbf{x}_2$ will be $1 \times mn$. Here, the objective is to find the original sources, *i.e.* $\mathbf{s}_1$ and $\mathbf{s}_2$, and it should be noted that in (1), both of the mixing matrix, $\mathbf{A}$, and pure sources, $\mathbf{S}$, are unknown.

Instead of somewhat complex BSS techniques, here we will show how NMF can be used for separating the sources in linear mixtures [11].

The main idea behind NMF methods is to factorize a non-negative matrix $\mathbf{X}$ as a product of two other non-negative matrices, $\mathbf{A}$ and $\mathbf{S}$ [11]. Sometimes, these non-negativity constraints are only placed on matrices $\mathbf{S}$ and $\mathbf{X}$ as we will do in this paper, too. Note that these non-negativity constraints are essential in our case because we are trying to extract the original sources which are non-negative pure images. A conventional approach to find $\mathbf{A}$ and $\mathbf{S}$ is by minimizing the difference between $\mathbf{X}$ and $\mathbf{AS}$ [12]

$$\min_{\mathbf{S},\mathbf{A}} J_1(\mathbf{S},\mathbf{A}) = \|\mathbf{X} - \mathbf{AS}\|_F^2 \qquad \text{subject to:} \quad s_{ij} \text{ and } a_{kl} > 0, \ \forall i, j, k, l, \tag{2}$$

where in this cost function $\|\cdot\|_F$ denotes the Frobenius norm and s_{ij} and a_{ij} are respectively the element of matrixes $\mathbf{S}$ and $\mathbf{A}$ in the ith row and the jth column. The gradient descent techniques has been previously proposed for minimization of this cost function [12] and we restate it here for convenience as follows

$$\mathbf{A}^{k+1} = \mathbf{A}^k - \mu \nabla_{\mathbf{A}} J_1(\mathbf{S}^k, \mathbf{A}^k) \quad \text{and} \quad \mathbf{S}^{k+1} = \mathbf{S}^k - \mu \nabla_{\mathbf{S}} J_1(\mathbf{S}^k, \mathbf{A}^k), \tag{3}$$

where k and μ are the index of iteration and fixed step size, respectively, and gradients of J_1 with respect to $\mathbf{S}$ and $\mathbf{A}$ are

$$\nabla_{\mathbf{S}} J_1(\mathbf{S},\mathbf{A}) = \mathbf{A}^T(\mathbf{AS} - \mathbf{X}) \quad \text{and} \quad \nabla_{\mathbf{A}} J_1(\mathbf{S},\mathbf{A}) = (\mathbf{AS} - \mathbf{X})\mathbf{S}^T, \tag{4}$$

where T stands for transposition. Due to the subtraction operation used in each iteration, the non-negative property of $\mathbf{S}$ and $\mathbf{A}$ cannot be guaranteed, and a projection has to be introduced to project any negative elements back to non-negative regions. This has be done in each iteration in this method by

$$\mathbf{A}^{k+1} = \max(0, \mathbf{A}^{k+1}) \quad \text{and} \quad \mathbf{S}^{k+1} = \max(0, \mathbf{S}^{k+1}), \tag{5}$$

Therefore, this algorithm is essentially a simple Projected Gradient (PG) method [12]. Hence, we can summarize the application of the projected gradient based NMF algorithm of [12] for removing show-through as:

- Initialize matrices $\mathbf{A}^0$ and $\mathbf{S}^0$ by positive random values;
- Consider a suitable value for step size, *i.e.* μ;
- For $k = 0, 1, \ldots$ until the convergence of the algorithm:

- Calculate the gradients of cost function J_1 presented in (4);
- Update matrices $\mathbf{A}$ and $\mathbf{S}$ using equation (3);
- Replace the negative elements of $\mathbf{A}$ and $\mathbf{S}$ by zero.

Figures 1(c) and 1(d) show the result of applying this NMF method to Figs. 1(a) and 1(b) which belong to a manuscript degraded by show-through effect. The results illustrate the capability of NMF algorithms in removing show-through and improving the readability of the document. However, a new degradation is still visible in the results of this simulation. This degradation is associated with linearly modeling of show-through and for better visibility of its effect on the outputs, an interesting region of one of the output images is magnified which is shown in Fig. 1(e). Here, it can be seen that in areas where two images printed on sides of the paper have overlap with each other (shown by arrows), recovered pixels are whiter than what they should be. This is because of the fact that show-through is a nonlinear phenomenon as we have shown in our previous paper experimentally either [6]. Using linear model for show-through means that during the scanning process, all of the pixels of the backside image will be added with their corresponding pixels in the front side image by the same factor. Whereas in reality, this factor is different for each pixel and depends on the grayscale of the front side pixels; as the front side image becomes darker, backside image is added to the front image with lower factor.

In the next section, we will change the cost function used in the NMF method to compensate the degradation caused by linearly modeling of show-through.

3 Compensating the Degradation Caused by Using Linear Model for Show-through

Figure 1(e) demonstrated the problem of using linear model for show-through. We propose here that this degradation can be compensated by adding some regularizing terms to the cost function of (2). These added terms should have some properties such as:

- Where there is no overlap between the printed images on sides of the paper, the results obtained through the minimization of non-regularized cost function, J_1, are fine. So minimization of these added terms should have no effect on output results;
- Where there is overlap between the printed images on sides of the paper, the results obtained through the minimization of non-regularized cost function, J_1, are whiter than what they should be. Therefore, minimization of these added terms should cause the recovered images to become darker in these areas (see such areas that are shown by arrows in Fig. 1(e)).

Here, we will show that the following suggested regularized cost function has the above mentioned properties (its reason will be explained after the calculation of the gradients of this new cost fuction)

$$J_2(\mathbf{S}, \mathbf{A}) = \|\mathbf{X} - \mathbf{AS}\|_F^2 + w_1 \left(\mathbf{s}_1 \mathbf{v}^T\right) + w_2 \left(\mathbf{s}_2 \mathbf{v}^T\right) \qquad \text{subject to:} \quad s_{ij} > 0, \ \forall i, j, \tag{6}$$

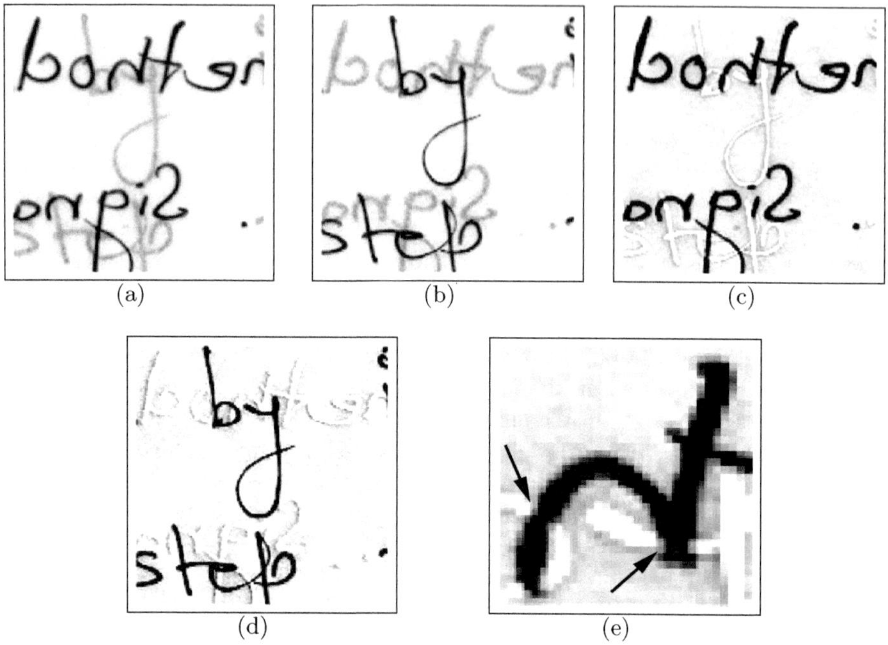

Fig. 1. Results of NMF method while assuming linear model for show-through. (a) and (b) are recto and verso images having show-through (obtained from *http://www.site.uottawa.ca/~edubois/documents*). (c) and (d) are recovered recto and verso images. Figure (e) shows that using linear model for show-through has some problems.

where $w_1 > 0$ and $w_2 > 0$ are regularization weight parameters and $\mathbf{v}$ is a spatially varying weight vector whose ith element is defined as

$$v(i) = exp(-\frac{(x_1^2(i) + x_2^2(i))}{2\sigma^2}) \qquad \text{for} \qquad i = 1, 2, \ldots, nm, \tag{7}$$

where σ is a constant that determines the variance of the above 2-dimensional Gaussian function and $x_1(i)$ and $x_2(i)$ are the ith element of vectors $\mathbf{x}_1$ and $\mathbf{x}_2$ respectively. It is clear that $v(i)$ will have high value only if both of $x_1(i)$ and $x_2(i)$ have small values.

Again, $J_2(\mathbf{S}, \mathbf{A})$ can be minimized by using a PG approach. Gradient of the above cost function, *i.e.* $J_2(\mathbf{S}, \mathbf{A})$, with respect to $\mathbf{A}$ is the same as (4). Gradient of this cost function with respect to $\mathbf{S}$ is

$$\nabla_{\mathbf{S}} J_2(\mathbf{S}, \mathbf{A}) = \mathbf{A}^T(\mathbf{AS} - \mathbf{X}) + \begin{bmatrix} w_1\mathbf{v} \\ w_2\mathbf{v} \end{bmatrix}. \tag{8}$$

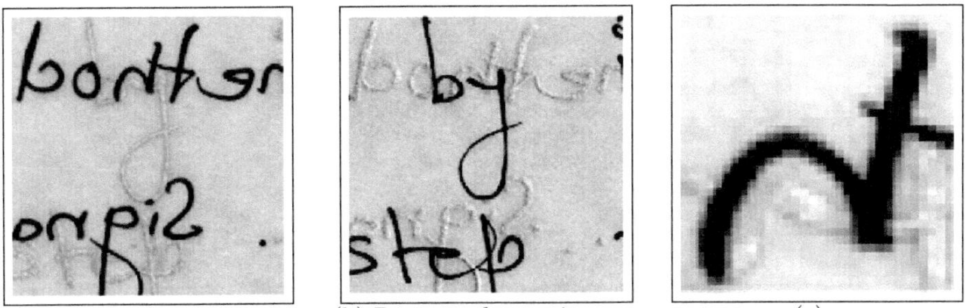

(a) Recovered recto image. (b) Recovered verso image. (c)

Fig. 2. Results obtained by using NMF method while assuming non-linear model for show-through. Figure (c) confirms that using non-linear mixing model for show-through has improved the output results.

Therefore, updating matrices **A** and **S** in each iteration before projection can be written as

$$\begin{aligned} \mathbf{A}^{k+1} &= \mathbf{A}^{k} - \mu \nabla_{\mathbf{A}} J_2(\mathbf{S}^{k}, \mathbf{A}^{k}) \\ \mathbf{S}^{k+1} = \mathbf{S}^{k} - \mu \nabla_{\mathbf{S}} J_2(\mathbf{S}^{k}, \mathbf{A}^{k}) &= \mathbf{S}^{k} - \mu \left[(\mathbf{A}^{k})^T \left(\mathbf{A}^{k}\mathbf{S}^{k} - \mathbf{X} \right) \right] - \mu \begin{bmatrix} w_1 \mathbf{v} \\ w_2 \mathbf{v} \end{bmatrix} . \end{aligned} \tag{9}$$

The reason that using cost function of (6) instead of (2) will compensate the degradations cause by linearly modeling show-through becomes clear from this equation and the definition of vector **v**. In equation (9), in those pixels of scanned images where front and backside images are nearly black (*i.e.* $x_1(i)$ and $x_2(i)$ have small values), the third term in right-hand side of this equation, *i.e.* $\left[v(i) \quad v(i)\right]^T$, will have a large value and therefore will decrease more $s_1(i)$ and $s_2(i)$ in these kind of pixels (will cause them to become more black). However, if at least one of the front or backside scanned images be nearly white (have high pixel value), $e^{-\alpha(x_1^2(i)+x_2^2(i))}$ and consequently $\left[v(i) \quad v(i)\right]^T$ will be small and therefore the third term will have no effect on $s_1(i)$ and $s_2(i)$. This means that in those pixels where the writings of two sides of the paper overlap with each other, recovered images will be darker compared to those images recovered by using (3). But in those pixels where the writings of two sides of the paper do not overlap with each other, recovered images will have no difference from those obtained in previous section.

Figures 2(a) and 2(b) again show the result of applying the algorithm proposed in previous section to Figs. 1(a) and 1(b) but this time by using equation (9) instead of equation (3) (note that in this case, the other steps such as initialization and projection are remained unchanged). Figure 2(c) shows the same part previously shown in Fig. 1(e) for having a better comparison. As this image indicates, degradation caused by using linear mixing model for show-through (shown in Fig. 1(e) by arrows) has been removed considerably.

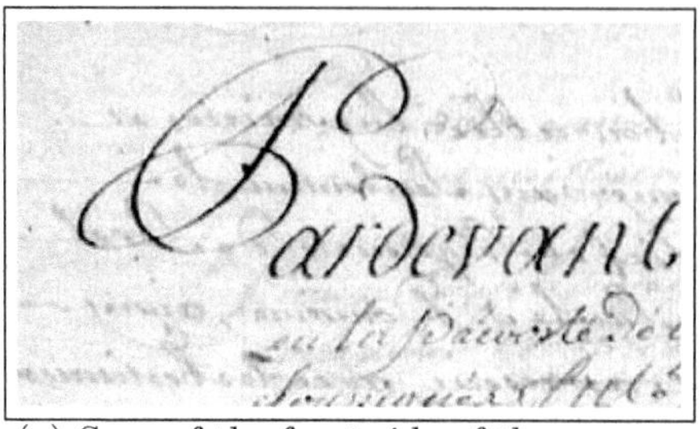

(a) Scan of the front side of the paper.

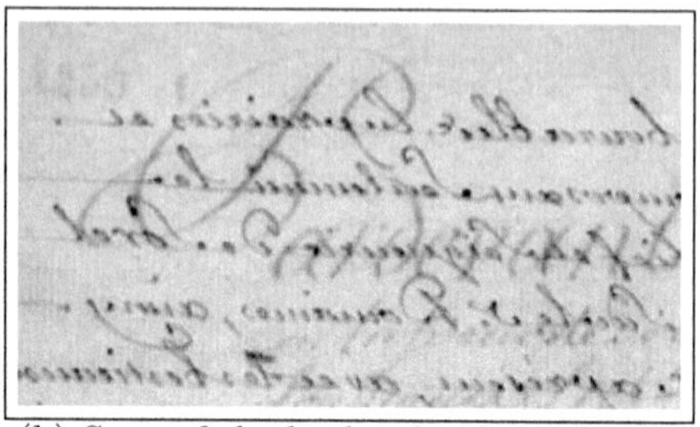

(b) Scan of the back side of the paper.

(c) recovered front side image.

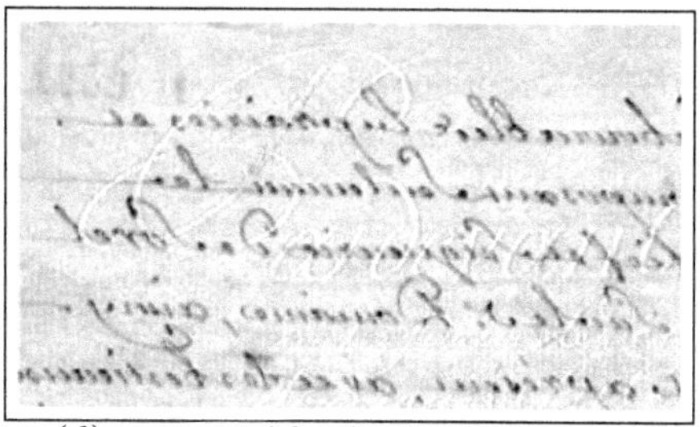

(d) recovered backside side image.

Fig. 3. Show-through cancelation on actual scanned images. (a) and (b) show the scanned input images applied to our method (obtained from *http://www.site.uottawa.ca/~edubois/documents*). Output images are shown in (c) and (d) indicate that this scheme can remove show-through almost completely.

4 Experimental Results

In this section, we will demonstrate the advantages of the proposed algorithm by performing an experiment on real scanned images. In the following simulation, σ and μ are set to 22 and 10^{-6} respectively. A suitable value for w_1 and w_2 is 6×10^5 which is obtained by trial and error and this value is used in all of the simulations performed in this paper. In the following experiments, we ran our proposed algorithm in MATLAB 7.1 on a Windows Vista PC with Intel 2.10 GHz Core 2 Duo CPU and 3 GB RAM. The results of the simulation that we performed by using the presented method are shown in Figs. 3(c) and 3(d). The original registered images (Figs. 3(a) and 3(b)) are ancient document which are degraded by show-through severely. Improvement in document readability is obvious in Figs. 3(c) and 3(d). The size of the input images was 150×570. For these particular input images, the algorithm converged after about 55 seconds.

Like BSS techniques, in NMF methods, separation is achieved up to some indeterminacies: The order of the sources and their amplitudes remain unknown. Scaling indeterminacy of sources is a drawback of our proposed method because regularization parameters w_1 and w_2 depend on the amplitudes of sources $\mathbf{s}_1$ and $\mathbf{s}_2$ which are unknown. Fortunately after performing some experiments, we recognized that in our case, amplitudes of the recovered sources converge to approximately the same values for different input images. We are not sure but maybe, this is because of the added regularizing terms or because the input images are bounded between 0 and 255. As a result, it seems that fix values can be used for w_1 and w_2.

5 Conclusion

In this paper, we used gradient based non-negative matrix factorization algorithm for removing show-through. By performing an experiment, we showed the disadvantage of using linear mixing model for show-through and therefore, we changed it into the nonlinear mixing model simply by modifying the cost function used in NMF method. Finally, we justified the effectiveness of this new method by performing an experiment on actual scanned images.

References

1. Nishida, H., Suzuki, T.: A Multi-Scale Approach to Restoring Scanned Colour Documents with Show-Through Effects. In: Proc. Seventh International Conference on Document Analysis and Recognition, vol. 1, pp. 584–588 (2003)
2. Wang, Q., Tan, C.L.: Matching of Double-sided Document Images to Remove Interference. In: IEEE CVPR 2001 (December 2001)
3. Tonazzini, A., Bianco, G., Salerno, E.: Registration and enhancement of double-sided degraded manuscripts acquired in multispectral modality. In: 10th International Conference on Document Analysis and Recognition, Spain, July 2009, pp. 546–550 (2009)
4. Tonazzini, A., Salerno, E., Bedini, L.: Fast Correction of Bleed-through Distortion in Grayscale Documents by a Blind Source Separation Technique. Int. Journal of Document Analysis IJDAR 10(1), 17–25 (2007)
5. Tonazzini, A., Salerno, E., Mochi, M., Bedini, L.: Bleed-Through Removal from Degraded Documents Using a Color Decorrelation Method. In: Marinai, S., Dengel, A.R. (eds.) DAS 2004. LNCS, vol. 3163, pp. 229–240. Springer, Heidelberg (2004)
6. Merrikh-Bayat, F., Babaie-Zadeh, M., Jutten, C.: A Nonlinear Blind Source Separation Solution for Removing the Show-through Effect in the Scanned Documents. In: 16th European Signal Processing Conference (EUSIPCO 2008), Lausanne, Switzerland (August 2008)
7. Almeida, M.S.C., Almeida, L.B.: Separating nonlinear image mixtures using a physical model trained with ICA. In: IEEE International Workshop on Machine Learning for Signal Processing, Maynooth, Ireland (2006)
8. Sharma, G.: Cancellation of Show-through in Duplex Scanning. In: Proc. IEEE Int. Conf. Image Processing, vol.2, pp. 609–612 (September 2000)
9. Almeida, L.B.: Separating a Real-life Nonlinear Image Mixture. Journal of Macine Learning Research 6, 1199–1232 (2005)
10. Almeida, M.S.C., Almeida, L.B.: Wavelet-based separation of nonlinear show-through and bleed-through image mixtures. Journal of Neurocomputing 72, 57–70 (2008)
11. Chu, M., Diele, F., Plemmons, R., Ragni, S.: Optimality, Computation and Interpretation of Non-negative Matrix Factorizations. Wake Forest University (2004)
12. Lin, C.J.: Projected Gradient Methods for Non-negative Matrix Factorization. Neural Computation 19, 2756–2779 (2007)

Nonlinear Band Expansion and 3D Nonnegative Tensor Factorization for Blind Decomposition of Magnetic Resonance Image of the Brain

Ivica Kopriva[1] and Andrzej Cichocki[2,3]

[1] Rudjer Bošković Institute, Bijenička cesta 54, P.O. Box 180, 10002, Zagreb, Croatia
[2] Laboratory for Advanced Brain Signal Processing
Brain Science Institute, RIKEN, Saitama, 351-0198, Japan
[3] Warsaw University of Technology and Systems Research Institute, PAN, Poland
ikopriva@irb.hr, cia@brain.riken.jp

Abstract. α- and β-divergence based nonnegative tensor factorization (NTF) is combined with nonlinear band expansion (NBE) for blind decomposition of the magnetic resonance image (MRI) of the brain. Concentrations and 3D tensor of spatial distributions of brain substances are identified from the Tucker3 model of the 3D MRI tensor. NBE enables to account for the presence of more brain substances than number of bands and, more important, to improve conditioning of the expanded matrix of concentrations of brain substances. Unlike matrix factorization methods NTF preserves local spatial structure in the MRI. Unlike ICA-, NTF-based factorization is insensitive to statistical dependence among spatial distributions of brain substances. Efficiency of the NBE-NTF algorithm is demonstrated over NBE-ICA and NTF-only algorithms on blind decomposition of the realistically simulated MRI of the brain.

Keywords: Nonnegative tensor factorization, nonlinear band expansion, magnetic resonance image, multi-spectral image.

1 Introduction

Blind or unsupervised multi-spectral and hyper-spectral image (MSI&HSI) decomposition attracts increased attention due to its capability to discriminate materials present in the MSI/HSI without knowing their spectral profiles [1]. For the purpose of blind decomposition, MSI is commonly represented in a form of the linear mixture model (LMM), whereupon vectorized version of the MSI is modeled as a product of the basis matrix, also known as spectral reflectance matrix, and matrix of vectorized spatial distributions of the materials present in the MSI, [1,2]. Magnetic resonance image (MRI) samples are images acquired by pulse sequences specified by three MR tissue parameters, spin-lattice (T_1) and spin-spin (T_2) relaxation times, and proton density (PD), [3,4]. Although MRI has different physical interpretation than MSI, it can also be represented using LMM, [3], and an analogy can be made with the three spectral MSI, such as red-green-blue (RGB) image, [5-7]. Here, basis matrix is interpreted as matrix of concentrations of the brain substances and source matrix represents vectorized spatial distributions of the brain substance present in the MRI. Each of the three

V. Vigneron et al. (Eds.): LVA/ICA 2010, LNCS 6365, pp. 490–497, 2010.

modalities of the MRI is designed for enhancing contrast of specific substance: PD for gray matter (GM), T_1 for white matter (WM) and T_2 for cerebral fluid (CF). Yet, each modality image still contains visible remnants of other substances, [3,4]. This appears to be especially true for PD image. Therefore, there is a need for additional (blind) post-processing (decomposition) of the acquired MRI.

Thanks to equivalence of the representations, advanced algorithms developed for the MSI data analysis can in principle be used for MRI analysis as well. In this paper we are particularly interested in blind decomposition methods due to their capabilities to estimate concentrations and spatial distributions of the brain substances having at disposal acquired MRI only. A standard tool for the solution of the related blind source separation (BSS) problem is independent component analysis (ICA), [8], which is based upon assumption that spatial distributions of the brain substances are non-Gaussian and statistically independent. However, as shown in [9,10], statistical independence assumption is not fulfilled for the hyper-spectral and multi-spectral data due to the fact that sum of the materials present in the pixel footprint must be constant. Thus, materials must necessarily be statistically related. Violation of statistical independence assumption is greater when dimensionality (number of spectral bands) of MSI is low, [6], which is also the case with 3-band MRI. Therefore, an algorithm for blind decomposition of the MRI the performance of which does not depend on statistical relations among brain substances is of great interest.

One avenue of research is related to algorithms that rely on sparseness between spatial distributions of the brain substances. The sparseness assumption implies that only small number of substances occupies each pixel footprint. Sparseness-based methodology has been applied in [5] for the purpose of blind decomposition of low-dimensional MSI and it is known as sparse component analysis (SCA), [11]. In this paper we rely on tensor representation of the MRI. As opposed to matrix representation, employed by ICA, SCA and/or nonnegative matrix factorization (NMF) algorithms, tensor representation preserves local spatial structure in the MRI. The main contribution of this manuscript is new method for blind decomposition of the 3-band MRI of the brain that combines nonlinear band expansion (NBE) and nonnegative tensor factorization (NTF). NBE has been used previously for blind MRI decomposition in combination with ICA in [3], and for blind MSI decomposition in combination with ICA and dependent component analysis (DCA) in [6]. NTF has been also used recently in [7] for blind MSI decomposition. We are unaware of previous combined application of NBE and NTF for blind MRI or MSI decomposition. NBE-NTF decomposition brings two improvements in relation to matrix factorization decomposition schemes: (*i*) NBE enables to account for the presence of more brain substances than number of bands, [3], and, more important, to improve conditioning of the expanded matrix of concentrations of brain substances. This is important for contrast enhancement between substances with similar concentration profiles. This has been used in [6] for robust blind decomposition of the low-dimensional low-intensity multi-spectral fluorescent image of the basal cell carcinoma; (*ii*) unlike ICA-, NTF-based factorization is insensitive to statistical dependence among spatial distributions of brain substances.

2 Theory and Algorithm

2.1 Magnetic Resonance Image and Linear Mixture Model

In analogy with MSI, MRI is represented in a form of matrix-based LMM [1-4]:

$$\mathbf{X}_{(3)} \approx \mathbf{AS} \tag{1}$$

where $\mathbf{X}_{(3)} \in R_{0+}^{3\times I_1 I_2}$ represents 3-mode unfolded version, [12], of the original MRI tensor $\underline{\mathbf{X}} \in R_{0+}^{I_1\times I_2\times 3}$ consisted of three (T1, T2 and PD weighted) images with the size of $I_1 \times I_2$ pixels. Here, R_{0+} is a real manifold with nonnegative elements. Columns of $\mathbf{A} \in R_{0+}^{3\times J}$ represent concentrations of the J brain substances present in the MRI, while rows of $\mathbf{S} \in R_{0+}^{J\times I_1 I_2}$ represent spatial distributions of these substances. Thus, BSS methods can be applied to factorize MRI $\mathbf{X}_{(3)}$ for enhancing contrast of the brain substances, [1,3], in the same spirit as it has been done in MSI analysis, [1,2,5,6]. However, matrix factorization problem implied by LMM (1) has infinitely many solutions unless additional constraints are imposed on model variables **A** or **S** in (1). NMF algorithms mostly impose sparseness constraints on $\{\mathbf{s}_j\}_{j=1}^{J}$, while ICA algorithms impose statistical independence constraints on $\{\mathbf{s}_j\}_{j=1}^{J}$. Sparseness constraints imply that at each pixel location (i_1,i_2) only few object exists. In medical imaging applications where pixel footprint is small it is justified to assume that only one object occupies each pixel footprint. Following described interpretation of the LMM (1), it is easy to verify that concentration similarity of the sources $\mathbf{s}_m$ and $\mathbf{s}_n$ affects the condition number of the mixing matrix **A**. This is because the corresponding column vectors $\mathbf{a}_m$ and $\mathbf{a}_n$ become close to collinear. As shown in [6], in the context of MSI analysis, in addition to deteriorate conditioning of the mixing matrix concentration similarity between the sources makes them also statistically dependent. Hence, fundamental requirement imposed by the ICA algorithms fails when sources have similar concentration profiles and this occurs increasingly more often when number of bands is small. To relax constraint-based requirements we adopt 3D tensor representation of the MRI, whereupon MRI tensor $\underline{\mathbf{X}}$ is represented in a form of Tucker3 model [12-14]:

$$\underline{\mathbf{X}} \approx \underline{\mathbf{G}} \times_1 \mathbf{A}^{(1)} \times_2 \mathbf{A}^{(2)} \times_3 \mathbf{A}^{(3)} \tag{2}$$

Here, $\underline{\mathbf{X}} \in R_{0+}^{I_1\times I_2\times 3}$, $\underline{\mathbf{G}} \in R_{0+}^{J\times J\times J}$ is core tensor, $\left\{\mathbf{A}^{(n)} \in R_{0+}^{I_n\times J}\right\}_{n=1}^{3}$ are array factors and $\times_n$ denotes n-mode product of a tensor with a matrix $\mathbf{A}^{(n)}$. 3-mode unfolded version of tensor $\underline{\mathbf{X}}$ in (2) is:

$$\mathbf{X}_{(3)} \approx \mathbf{A}^{(3)}\mathbf{G}_{(3)}\left[\mathbf{A}^{(2)} \otimes \mathbf{A}^{(1)}\right]^{\mathrm{T}} \tag{3}$$

where '⊗' denotes Kronecker's product and $\mathbf{G}_{(3)}$ represent 3-mode unfolded core tensor $\underline{\mathbf{G}}$. In direct comparison between (1) and (3) we arrive at:

$$\mathbf{A} \approx \mathbf{A}^{(3)}$$
$$\underline{\mathbf{S}} \approx \underline{\mathbf{G}} \times_1 \mathbf{A}^{(1)} \times_2 \mathbf{A}^{(2)} = \underline{\mathbf{X}} \times_3 \left(\mathbf{A}^{(3)}\right)^{\dagger} \tag{4}$$

where $\underline{\mathbf{S}} \in R_{0+}^{I_1 \times I_2 \times J}$ represents tensor of spatial distributions of J brain substances contained in MRI and '†' denotes the Moore-Penrose pseudo-inverse. Thus, NTF-based blind decomposition of MRI yields both matrix of concentrations and tensor of spatial distributions of the brain substances present in the MRI.

2.2 Nonlinear Band Expansion

Although the GM, WM and CF brain substances are of the main interest in MRI analysis, possible presence of other substances such as muscle, skin, fat, etc. as signal sources can violate the fundamental condition required by majority the BSS methods: the number of sources J must be less than or equal to the number of measurements. In the case of MRI number of measurements equals number of bands that is 3. Thus, number of sources most likely exceeds number of bands. To overcome this limitation, NBE has been used in combination with ICA to perform blind decomposition of MRI, [3]. However, in regard to the NBE there are two important points still missed in [3]: (*i*) in addition to increasing number of measurements NBE also improves conditioning of the matrix of the concentrations of brain substances: **A**. This has been discussed in [6] in the context of low-dimensional MSI decomposition and is important for enhancing contrast between substances with similar concentration profiles; (*ii*) after NBE, sources in expanded mixture model still remain statistically dependent. This deteriorates accuracy of the ICA-based decomposition. To partially neutralize this problem, DCA has been used in [6]. Here, we propose combined use of NBE and NTF to perform blind MRI decomposition for enhancing contrast of WM, GM and CF brain substances. Let $\left\{\mathbf{X}_{i_3} \in R_{0+}^{I_1 \times I_2}\right\}_{i_3=1}^{3}$ be set of original band (T_1, T_2 and PD) MR images. Through NBE we introduce new sets of images, [3,6]: set of auto-correlated band images $\left\{\mathbf{X}_{i_3}^2\right\}_{i_3=1}^{3}$, and set of cross-correlated band images $\left\{\mathbf{X}_{i_3}\mathbf{X}_{j_3}\right\}_{i_3=1, j_3=i_3+1}^{2,3}$. The 3-mode unfolded version of the NBE MRI is then obtained as:

$$\overline{\mathbf{X}}_{(3)} \approx \overline{\mathbf{A}}\overline{\mathbf{S}} \tag{5}$$

where $\overline{\mathbf{X}}_{(3)} \in R_{0+}^{9 \times I_1 I_2}$, $\overline{\mathbf{A}} \in R_{0+}^{9 \times \bar{J}}$ and $\overline{\mathbf{S}} \in R_{0+}^{\bar{J} \times I_1 I_2}$. Thus, original 3-band MRI is transformed into 9-band image containing: 3 original band images, 3 auto-correlated band images and 3 cross-correlated band images. We now assume that only one brain substance occupies each pixel footprint. It is then straightforward to show that extended source matrix $\overline{\mathbf{S}}$ in addition to original source images $\left\{\mathbf{s}_j\right\}_{j=1}^{J}$ contains also their squares $\left\{\mathbf{s}_j^2\right\}_{j=1}^{J}$. Thus, $\bar{J} = 2J$. Now, the BSS requirement $J \leq 3$ transforms into $J \leq 4.5$. It is however even more important that $\overline{\mathbf{A}}$ in (5) is better conditioned than **A** in (1), [6]. This enhances contrast between brain substances with similar concentrations across T_1, T_2 and PD bands. Due to its insensitivity to statistical dependence among the sources and preservation of the local spatial structure of the MRI, we apply NTF-based decomposition on NBE MRI. Therefore, NBE MRI is represented in a tensor format using Tucker3 model:

$$\underline{\overline{\mathbf{X}}} \approx \underline{\overline{\mathbf{G}}} \times_1 \overline{\mathbf{A}}^{(1)} \times_2 \overline{\mathbf{A}}^{(2)} \times_3 \overline{\mathbf{A}}^{(3)} \tag{6}$$

where $\underline{\overline{\mathbf{X}}} \in R_{0+}^{I_1 \times I_2 \times 9}$, $\underline{\overline{\mathbf{G}}} \in R_{0+}^{2J \times 2J \times 2J}$ is core tensor and $\left\{\overline{\mathbf{A}}^{(n)} \in R_{0+}^{I_n \times 2J}\right\}_{n=1}^{3}$ are factors. 3-mode unfolded version of tensor $\underline{\overline{\mathbf{X}}}$ in (6) is:

$$\overline{\mathbf{X}}_{(3)} \approx \overline{\mathbf{A}}^{(3)} \overline{\mathbf{G}}_{(3)} \left[\overline{\mathbf{A}}^{(2)} \otimes \overline{\mathbf{A}}^{(1)}\right]^{\mathrm{T}} \tag{7}$$

where $\overline{\mathbf{G}}_{(3)}$ represent 3-mode unfolded core tensor $\underline{\overline{\mathbf{G}}}$. In direct comparison between (7) and (5) we obtain:

$$\begin{aligned} &\overline{\mathbf{A}} \approx \overline{\mathbf{A}}^{(3)} \\ &\underline{\overline{\mathbf{S}}} \approx \underline{\overline{\mathbf{G}}} \times_1 \overline{\mathbf{A}}^{(1)} \times_2 \overline{\mathbf{A}}^{(2)} = \underline{\overline{\mathbf{X}}} \times_3 \left(\overline{\mathbf{A}}^{(3)}\right)^{\dagger} \end{aligned} \tag{8}$$

where $\underline{\overline{\mathbf{S}}} \in R_{0+}^{I_1 \times I_2 \times 2J}$. Improved conditioning of $\overline{\mathbf{A}}^{(3)}$ is of great importance for increasing accuracy of the NBE-NTF-based decomposition because its pseudo-inverse is involved in calculation of the expanded tensor of spatial distributions of brain substances $\underline{\overline{\mathbf{S}}}$.

3 Experiment

MRI used for comparative performance analysis was obtained from the MRI simulator of McGill University, [15,16]. Three MR images of the brain corresponding to modalities PD, T_1 and T_2 are obtained under the following specifications: protocol=International Consortium of Brain Mapping (ICBM), phantom name=normal, slice thickness=5 mm, noise=0%, INU=0%. Generated volume contained 36 MR image slices in Z direction with a size of each slice of 217×181 pixels. We have used MRI slice number 13 to demonstrate NBE-NTF-, NBE-ICA- and NTF-based blind MRI decompositions. Implementations of Tucker3 α- and β-NTF algorithms, [12,13], were based on MATLAB Tensor Toolbox provided at [17]. For ICA-based decompositions enhanced version (EFICA), [18], of the FastICA algorithm with *tanh* nonlinearity has been used. Efficiency of NBE-NTF decomposition of the MRI image against NBE-ICA and NTF decompositions is demonstrated on the synthetic brain images as in [3]. Figure 1 shows three simulated MR images of the brain with PD,

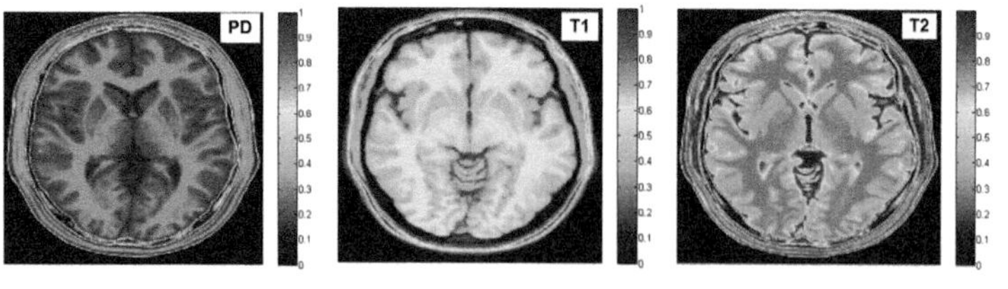

Fig. 1. (color online). Realistically simulated MRI of the brain from left to right: PD (GM), T_1 (WM) and T_2 (CF).

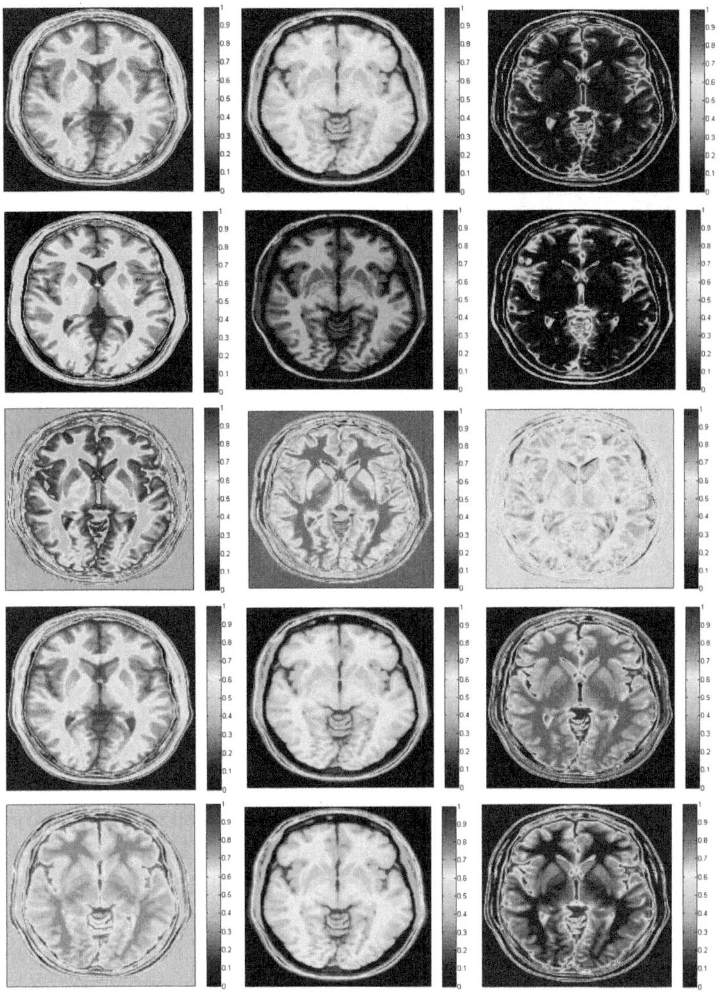

Fig. 2. (color online). GM, WM and CF brain substance images obtained by, from top to bottom: NBE and α-NTF blind decomposition with α=2; NBE and β-NTF blind decomposition with β=1; NBE and EFICA algorithm with *tanh* nonlinearity; α-NTF algorithm with α=2; β-NTF algorithm with β=1. All NTF-based results are obtained after 5000 iterations.

T_1 and T_2 modalities. Comparative performance analysis is focused on the spatial distributions of GM, WM and CF because they are of actual interest in MRI. Thus, Figure 2 shows spatial distributions of GM, WM and CF brain substances extracted respectively by: NBE and Tucker3 α-NTF algorithm; NBE and Tucker3 β-NTF algorithm; NBE and EFICA algorithm; Tucker3 α-NTF algorithm, and Tucker3 β-NTF algorithm. Combined NBE and Tucker3 NTF takes approximately 5 minutes in MATLAB environment on PC running under the Windows XP operating system using Intel Core 2 Quad Processor Q6600 operating with clock speed of 2.4 GHz and 16GB of RAM installed. Images in Figures 1 and 2 are rescaled to interval [0,1] and

Fig. 3. From left to right: edges of the GM spatial map extracted by Canny's algorithm with a threshold set to 0.2 from original PD image, NBE and β-NTF segmented image and β-NTF only segmented image

shown in pseudo-color scale. Probability of substance presence 0 is shown in dark blue color and probability of substance presence 1 is shown in dark red color. Except rescaling to [0,1] interval no additional post-processing has been done on decomposition results shown in Figure 2. Binary maps of decomposed images could be easily obtained through intensity-based segmentation of decomposed images. α- and β-NBE-NTF algorithms extracted all three substances correctly, while NBE-ICA and α- and β-NTF algorithms failed to do so. NBE-ICA failed due to the violation of the statistical independence assumption. Worse performance of α- and β-NTF algorithms is caused by poorer conditioning of the matrix of concentrations that was caused by similarity of concentrations of brain substances. This is demonstrated in Figure 3 where edge maps extracted by Canny's algorithm with a threshold set to 0.2 are shown for GM substance obtained from original PD image, NBE and β-NTF obtained spatial map and β-NTF only obtained spatial map. Due to the poor contrast boundary edges were completely missed for original PD image, were better detected from β-NTF segmented image and even more better from NBE and β-NTF segmented image.

4 Conclusion

Following analogy with MSI, MRI of the brain is represented by LMM using Tucker3 model of the 3D MRI tensor. Matrix of the concentrations of brain substances is identified as mode-3 array factor of the 3D MRI tensor, while 3D tensor of spatial distributions of the brain substances is obtained through 3-mode tensor product between MRI tensor and pseudo-inverse of the mode-3 array factor. Tensor-based representation of the MRI preserves local spatial structure of the images and yields factorization insensitive to statistical dependence among the spatial distributions of the brain substances present in the MRI. To further improve accuracy of MRI decomposition, the MRI tensor is expanded through nonlinear band expansion yielding new MRI tensor with 9-bands (as opposed to original 3-band MRI). In relation to original MRI tensor, the band-expanded tensor is characterized with mode-3 array factor that is better conditioned which is important for enhancing contrast between brain substances with similar concentrations across T_1, T_2 and PD bands.

Acknowledgment. Part of this work was supported in part through grant 098-0982903-2558 funded by the Ministry of Science, Education and Sport, Republic of Croatia.

References

1. Chang, C.I. (ed.): Hyperspectral Data Exploitation: Theory and Applications, New York. John Wiley, Chichester (2007)
2. Du, Q., Kopriva, I., Szu, H.: Independent-component analysis for hyperspectral remote sensing imagery classification. Opt. Eng. 45, 1–13 (2006)
3. Ouyang, Y.C., Chen, H.M., Chai, J.W., Chen, C.C.C., Poon, S.K., Yang, C.W., Lee, S.K., Chang, C.I.: Band Expansion-Based Over-Complete Independent Component Analysis for Multispectral Processing of Magnetic Resonance Image. IEEE Trans. Biomed. Eng. 55, 1666–1677 (2008)
4. Nakai, T., Muraki, S., Bagarinao, E., Miki, Y., Takehara, Y., Matsuo, K., Kato, C., Sakahara, H., Isoda, H.: Application of independent component analysis to magnetic resonance imaging for enhancing the contrast of gray and white matter. Neuroimage 21, 251–260 (2004)
5. Kopriva, I., Cichocki, A.: Blind decomposition of low-dimensional multi-spectral image by sparse component analysis. J. Chemometrics 23, 590–597 (2009)
6. Kopriva, I., Peršin, A.: Unsupervised decomposition of low-intensity low-dimensional multi-spectral fluorescent images for tumour demarcation. Med. Image Analysis 13, 507–518 (2009)
7. Kopriva, I., Cichocki, A.: Blind Multi-spectral Image Decomposition by 3D Nonnegative Tensor Factorization. Opt. Lett. 34, 2210–2212 (2009)
8. Cichocki, A., Amari, S.: Adaptive Blind Signal and Image Processing. John Wiley, Chichester (2002)
9. Nascimento, J.M.P., Dias, J.M.B.: Does Independent Component Analysis Play a Role in Unmixing Hyperspectral Data? IEEE Trans. Geosci. Remote Sens. 43, 175–187 (2005)
10. Nascimento, J.M.P., Dias, J.M.B.: Vertex Component Analysis: A Fast Algorithm to Unmix Hyperspectral Data. IEEE Trans. Geosci. Remote Sens. 43, 898–910 (2005)
11. Li, Y., Cichocki, A., Amari, S.: Analysis of Sparse Representation and Blind Source Separation. Neural Comput. 16, 1193–1234 (2004)
12. Cichocki, A., Zdunek, R., Phan, A.H., Amari, S.I.: Nonnegative Matrix and Tensor Factorizations - Applications to Exploratory Multi-way Data Analysis and Blind Source Separation. John Wiley, Chichester (2009)
13. Cichocki, A., Phan, A.H.: Fast Local Algorithms for Large Scale Nonnegative Matrix and Tensor Factorizations. IEICE Trans. Fundamentals E92-A(3), 708–721 (2009)
14. Tucker, L.R.: Some mathematical notes on three-mode factor analysis. Psychometrika 31, 279–311 (1966)
15. Kwan, R.K.S., Evans, A.C., Pike, G.B.: MRI Simulation-Based Evaluation of Image-Processing and Classification Methods. IEEE Trans. Med. Imag. 18, 1085–1097 (1999)
16. `http://www.bic.mni.mcgill.ca/brainweb/`
17. `http://csmr.ca.sandia.gov/~tkolda/TensorToolbox`
18. Koldovský, Z., Tichavský, P., Oja, E.: Efficient Variant of Algorithm for FastICA for Independent Component Analysis Attaining the Cramér-Rao Lower Bound. IEEE Trans. Neural Net. 17, 1265–1277 (2006)

Informed Source Separation Using Latent Components*

Antoine Liutkus, Roland Badeau, and Gaël Richard

Institut Telecom, Telecom ParisTech, CNRS LTCI

Abstract. We address the issue of source separation in a particular *informed* configuration where both the sources and the mixtures are assumed to be known during a so-called *encoding* stage. This knowledge enables the computation of a side information which ought to be small enough to be watermarked in the mixtures. At the *decoding* stage, the sources are no longer assumed to be known, only the mixtures and the side information are processed to perform source separation.

The proposed method models the sources jointly using latent variables in a framework close to multichannel nonnegative matrix factorization and models the mixing process as linear filtering. Separation at the decoding stage is done using generalized Wiener filtering of the mixtures. An experimental setup shows that the method gives very satisfying results with mixtures composed of many sources. A study of its performance with respect to the number of latent variables is presented.

1 Introduction

This study concerns a special case of source separation, called *informed source separation* (ISS), that was introduced by PARVAIX in [7]. ISS can be understood as an encoding/decoding framework in which both the sources and the mixtures are available at the encoder's side, but only the mixtures are available at the decoder's side, as well as some *side information* that may have been created by the encoder and transmitted along with the mixtures to assist the separation process. ISS thus aims at making source separation robust by providing adequate prior knowledge to the separation algorithms, and allows applications such as *active listening* that consists in being able to mute tracks as in classical Karaoke applications or to add separate effects to them.

The main advantage of ISS is that it permits to reliably recover the separated tracks from the mixtures with only a very small amount of side information. The method we propose here allows to control the quantity of information that is sent to the decoder. As highlighted by PARVAIX in [7], if the side information is sufficiently small, it can be directly embedded in the mixture signals by watermarking, allowing active listening of recordings on conventional stereophonic audio CD.

* This work is supported by the French National Research Agency (ANR) as a part of the DReaM project (ANR-09-CORD-006-03) and partly supported by the Quaero Program, funded by OSEO.

V. Vigneron et al. (Eds.): LVA/ICA 2010, LNCS 6365, pp. 498–505, 2010.

We propose a new method for informed source separation that is based on jointly modeling the sources at the encoder side using latent additive independent Gaussian variables in a framework that is very similar to Non-negative Matrix Factorization (NMF). Then, the mixing process is modeled via linear filters that are estimated using techniques inspired from the automatic mixing literature [1]. The *side information* considered in our proposed system thus consists of the spectrum and activation coefficients of each latent component as well as the mixing parameters. The number of necessary bits to store this information is mainly controlled by the number of latent variables considered. At the decoder side, the separation process involves generalized Wiener filtering [2] and allows reaching excellent performance provided that enough latent variables have been chosen.

This paper is organized as follows. In section 2 we detail the model used for representing the source signals as well as the mixing process at the encoder side. In section 3 we describe the estimation method and outline the separation technique induced by the model at the decoder side. Finally, we give some experimental results along with a study of the influence of the number of latent variables on the separation quality in section 4.

2 Model

2.1 Introduction

We consider a set of M source signals $(s_{m,t})_{m=1\cdots M,t=1\cdots L}$ and a set of K mixture signals $(x_{k,t})_{k=1\cdots K,t=1\cdots L}$ of same lengths L. We define $S_{m,\omega n} = [\mathcal{F}(s_{m,.})]_{\omega,n}$ and $X_{k,\omega n} = [\mathcal{F}(x_{k,.})]_{\omega,n}$ as the complex-valued Short Time Fourier Transforms (STFT) of the signals $s_{m,.}$ and $x_{k,.}$ for frequency bin $\omega \in [1 : N_\omega]$ and frame index $n \in [1 : N_n]$.

2.2 Source Signals

Complex Gaussian Model. Following the formalism introduced by BENAROYA in [2], the signals of interest are locally modeled as independent wide-sense stationary centered random variables and can thus be characterized by their covariance. In the spectral domain, this can be expressed by writing that the STFT $Y_{\omega n}$ of some signal y_t for frame index n and frequency bin ω obeys[1]:

$$Y_{\omega n} \sim \mathcal{N}_c(0, \sigma^2_{Y_{\omega n}})$$

where $\sigma^2_{Y_{\omega n}}$ is the power spectral density of y_t for the frame n at frequency ω. We further assume that $\{Y_{\omega n}\}_{\omega,n}$ are independent, which stands asymptotically for all (ω, n).

[1] $\mathcal{N}_c$ is the *proper Gaussian complex distribution* and is defined on the plane by its probability density function $f(z) = \frac{1}{\pi\sigma^2}\exp\left(-\frac{|z|^2}{\sigma^2}\right)$.

Mixture of Latent Components. The study presented here considers R underlying latent independent centered random Gaussian variables $c_{r,t}$ called the *latent components*, each of which has a power spectral density $W_{\omega r}$ and is only modulated in time by some frame-dependent activation coefficients H_{rn}. In the frequency domain, the STFT $C_{r,\omega n}$ of each latent component $c_{r,t}$ for frame n and frequency ω is thus modeled as:

$$C_{r,\omega n} \sim \mathcal{N}_c(0, W_{\omega r} H_{rn}) \tag{1}$$

Each source signal $s_{m,t}$ is then simply modeled as a weighted sum of these R latent components:

$$s_{m,t} = \sum_{r=1}^{R} \sqrt{Q_{mr}} c_{r,t} \tag{2}$$

where the non-negative coefficient $\sqrt{Q_{mr}}$ is the contribution of the latent component r to source m. As can be seen, all the sources are modeled as a sum of the *same* underlying components. Combining (1) and (2), we have, for frequency bin ω and frame index n:

$$S_{m,\omega n} \sim \mathcal{N}_c(0, \sum_{r=1}^{R} Q_{mr} W_{\omega r} H_{rn}) \tag{3}$$

We can readily see that the model (3) is equivalent to the multichannel NMF model presented in [6]. Indeed, the source signals $s_{m,t}$ are modeled as linear instantaneous mixtures of R latent components. An interesting feature of our model is to allow one single number of latent components for all the source signals.

2.3 Mixing Process

Following [1], we will model each mixture signal as a sum of filtered versions of the sources

$$x_{k,t} = \sum_{m=1}^{M} \sum_{\tau=0}^{P} a_{km,\tau} s_{m,t-\tau} = \sum_{m=1}^{M} s_{km,t} \tag{4}$$

where P is the order of the mixing filters, $(a_{km,\tau})_{\tau=0..P}$ is the impulse response of the filter from source m to mixture k and $s_{km,t}$ is called the contribution of source m to mixture k for time t. We will only consider causal and Finite Impulse Response (FIR) filters here. This model can be approximated in the spectral domain as $X_{k,\omega n} = \sum_{m=1}^{M} A_{km,\omega} S_{m,\omega n}$, where $A_{km,\omega}$ is the frequency response of the filter $(a_{km,\tau})_{\tau=0..P}$ at the frequency corresponding to bin ω.

2.4 Unmixing Process

During the estimation process, we aim at recovering the original sources $s_{m,t}$ given their contributions $s_{km,t}$ in the mixtures. To that purpose, we follow a *beamforming* approach that consists in estimating $s_{m,t}$ as the sum of filtered versions of $s_{km,t}$: $\hat{s}_{m,t} = \sum_{k=1}^{K} \sum_{\tau=0}^{P_u} u_{mk,\tau} s_{km,t-\tau}$, where u_{mk} is the FIR *unmixing filter* of length P_u from mixture k to source m. If U_{mk} is the frequency response of u_{mk}, this can be approximated in the spectral domain as:

$$\hat{S}_{m,\omega n} = \sum_{k=1}^{K} U_{mk,\omega} S_{km,\omega n}. \tag{5}$$

2.5 Set of Parameters

The total set Θ of parameters is $\Theta = \{W, H, Q, \mathbf{A}, \mathbf{U}\}$, where $\mathbf{A}$ and $\mathbf{U}$ are respectively composed of all the $M \times K$ impulse responses of the *mixing* and *unmixing* filters from the M sources to the K mixtures and *vice versa*. The total number of parameters is then:

$$\#\Theta = \underbrace{N_\omega \times R}_{\text{for } W} + \underbrace{N_n \times R}_{\text{for } H} + \underbrace{M \times R}_{\text{for } Q} + \underbrace{M \times K \times (P + P_u)}_{\text{for } \mathbf{A} \text{ and } \mathbf{U}} \tag{6}$$

As can be seen from (6), using fixed parameters for the STFT, $\#\Theta$ is controlled by the number R of latent components and the orders P and P_u of the mixing and unmixing filters.

3 Parameters Estimation

3.1 Multichannel NMF for Source Signals

Multiplicative Update Rules for the Parameters. For only one source, the model (3) is equivalent to the NMF approach that was popularized by LEE and SEUNG in [5] when using a particular measure called the Itakura-Saito divergence, which is a special case of β-divergence for $\beta = 0$ (see [4] on this point).

Algorithms in the aforementioned papers can be generalized to the case of M sources and the corresponding update rules for the parameters are summarized in Algorithm 1 for any β-divergence[2].

[2] Notations :

- . denotes element-wise product
- $\frac{A}{B}$ denotes element-wise division
- $M_{m.}$ is the mth row of matrix M
- $[A.^{\alpha}]_{mn} = [A]_{mn}^{\alpha}$
- $\mathbf{S}_m = |S_m|^{.2}$ is the power spectrum of source m
- diag(D) is a column vector containing the diagonal elements of D if D is a matrix or is the matrix whose diagonal elements are composed of the elements of D if D is a vector.
- $\hat{\mathbf{S}}_m = W\text{diag}(Q_{m.})H$ is the estimated power spectrum of source m with current model parameters.

Algorithm 1. Update rules for the parameters of the source model (3) for one iteration

- Q update:
  $Q_{m.} \leftarrow \text{diag}\left(\text{diag}(Q_{m.}).\frac{W^T\left(\hat{S}_m^{.\beta-2}.\mathbf{S}_m.(WH)\right)H^T}{W^T\hat{S}_m^{.\beta-1}H^T}\right)$
- W update:
  $W \leftarrow W.\frac{\sum_{m=1}^M\left(\hat{S}_m^{.\beta-2}.\mathbf{S}_m\right)(\text{diag}(Q_{m.})H)^T}{\sum_{m=1}^M \hat{S}_m^{.\beta-1}(\text{diag}(Q_{m.})H)^T}$
- H update:
  $H \leftarrow H.\frac{\sum_{m=1}^M(W\text{diag}(Q_{m.}))^T\left(\hat{S}_m^{.\beta-2}.\mathbf{S}_m\right)}{\sum_{m=1}^M(W\text{diag}(Q_{m.}))^T\hat{S}_m^{.\beta-1}}$
- Normalization of W and Q and scaling of H accordingly.

As pointed out by BERTIN in [3], better results can be obtained if optimization is first performed with convex cost functions such as the Kullback-Leibler generalized divergence ($\beta = 1$) and then with the Itakura-Saito distance ($\beta = 0$), which is not convex. Such a tempering approach was hence used in this study and indeed proved to show better performance.

3.2 Estimation of the Mixing Filters

The problem of estimating the mixing filters of different sources in a mixture has already been addressed in so-called *automatic mixing* studies such as [1]. The main idea of these techniques is to choose the mixing filters so as to minimize the mean squared error $\frac{1}{L}\sum_t \left|x_{k,t} - \sum_{m=1}^M (a_{km} * s_m)(t)\right|^2$ for all k. This is done using standard least-squares methods.

3.3 Source Separation at the Decoder

Sources Contributions in Mixtures. When the parameters of the model have been estimated, we no longer suppose that the source signals $s_{m,t}$ are available. We then focus here on the decoder side, where only the mixtures $x_{k,t}$ and the parameters Θ are available. Considering the mixing model given in 2.3 and the source model (3), we have $S_{km,\omega n} \sim \mathcal{N}_c\left(0, |A_{km,\omega}|^2 \sum_{r=1}^R Q_{mr}W_{\omega r}H_{rn}\right)$. If we define $\sigma^2_{km,\omega n} = |A_{km,\omega}|^2 \sum_{r=1}^R Q_{mr}W_{\omega r}H_{rn}$, we then have $X_{k,\omega n} \sim \mathcal{N}_c\left(0, \sum_{m=1}^M \sigma^2_{km,\omega n}\right)$ and the minimum mean square error (MMSE) estimate of $S_{km,\omega n}$ is thus given by (see also [2,6,4]):

$$\hat{S}_{km,\omega n} = \frac{\sigma^2_{km,\omega n}}{\sum_{m'=1}^M \sigma^2_{km',\omega n}} X_{k,\omega n} \tag{7}$$

Sources Estimates through Beamforming. Given all the $\hat{s}_{km,t}$, our objective is now to estimate $s_{m,t}$. Using (5) and (7), we readily see that the estimate

$\hat{S}_{m,\omega n}$ of source m for time-frequency bin (ω, n) given Θ and the mixtures is given by:

$$\hat{S}_{m,\omega n} = \sum_{k=1}^{K} U_{mk,\omega} \hat{S}_{km,\omega n} \tag{8}$$

This computation at the decoder side does not require much computational resource as its complexity is $\mathcal{O}(R \times M \times K)$.

The decoder requires the the unmixing filters u_{mk} in order to compute (8). They are included in the parameters set Θ by the encoder, which also computes $\hat{s}_{mk,t}$ following (8) and then chooses u_{mk} so as to minimize the squared error $\frac{1}{L}\sum_t |s_{m,t} - \hat{s}_{m,t}|^2$ for all m.

4 Evaluation

4.1 Corpus and Metrics

Corpus. Experiments were done with the internal Source Separation Corpus gathered for the Quaero program [3], from which 9 different excerpts were chosen of various musical styles along with their constitutive separated tracks. The corpus includes excerpts constituted of 5 to 11 separated tracks, which are of many kinds, including acoustic instruments such as piano, guitar, male and female singers, distorted sounds/voices, digital effects, etc.

All mixing was done in stereo on real Digital Audio Workstations. It includes equalizing and panning. All sampling rates were set to 44.1kHz and signals are approximately 30s long.

Metrics. Objective criteria to evaluate the quality of the separation were used as defined in the *bsseval* toolbox [8] and include the Source to Distortion Ratio (SDR), the Source to Interference Ratio (SIR) and the Source to Artifacts Ratio (SAR). All values are in dB. In order to assess the quality of separation, we have compared the results given by the proposed method to results given by the idealized (oracle) time-frequency mask expressed as follows:

$$\hat{S}_{km,\omega n} = \frac{\|S_{km,\omega n}\|^2}{\sum_{m'=1}^{M} \|S_{km',\omega n}\|^2} X_{k,\omega n}$$

For each excerpt, statistics are averaged over all its constitutive sources in order to give a general overview of the performance of the method. Complete evaluation along with sample signals can be downloaded from our website.

Models Parameters. All the STFT were computed for frames of 70ms, with 30% overlap. The order of the mixing and unmixing filters a_{km} and u_{mk} were all set to $P = 150$ and the number of iterations for Algorithm 1 was set to 60, the first 30 iterations used $\beta = 1$ and the last 30 iterations used $\beta = 0$. As $\#\Theta$ is mainly controlled by the number R of latent components, we have studied the performance of the method with respect to R.

[3] www.quaero.org

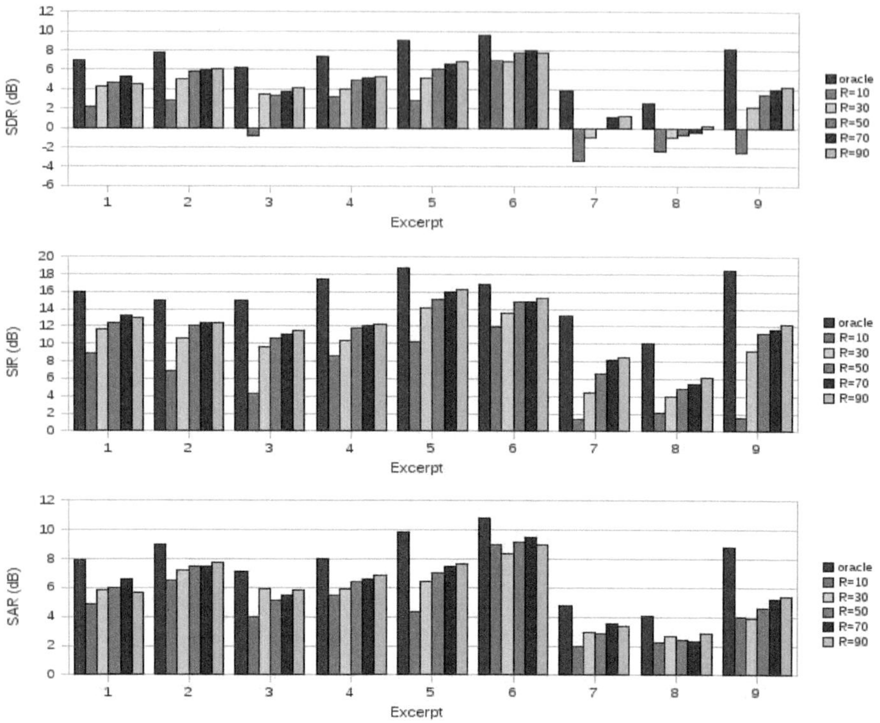

Fig. 1. Average SDR/SIR/SAR scores (in dB) for the estimation of the individual sources. Results are averages over the sources for each excerpt of the corpus.

4.2 Sources Estimates

The results for estimating the sources from the mixtures are given in Figure 1.

4.3 Discussion

Several remarks can be made when considering the results given in Figure 1. First, it is perceptually very hard to notice any difference between the original signals and the sources recovered using the oracle method.

Secondly, the quality of the separation is directly controlled by the number R of latent components. As R increases, performance gets closer to the oracle method. There is thus a trade-off between the quality of the separation and the weight of the models parameters. For the results given here, $\#\Theta$ ranges from 1% for $R = 10$ to 7.5% for $R = 90$ of the number of samples in the mixtures.

Finally, even if damaged for small R, the sources are very well isolated one from another, as confirmed by the very high SIR scores.

5 Conclusion

Informed source separation consists in providing valuable prior knowledge to a source separation algorithm. This study considers the case where this knowledge has been computed at an *encoding* stage where both the mixtures and the original sources are known. It then jointly models the source signals through additive latent variables and models the mixing and unmixing processes as linear filters. At the *decoding* stage, separation is performed using generalized Wiener filtering of the mixtures signals.

The total weight of the parameters is extremely small compared to that of the mixtures, typically less than 5 percents. Even though this information is neither quantized nor compressed, this aready allows hiding it directly in the mixture signals through watermarking.

The proposed method allows reaching excellent performance and managed to successfully separate up to 11 sources in stereophonic mixtures during our experiments. The quality of the separation is directly related to the number of latent components used for modeling the sources and can be reliably known by the encoder.

References

1. Barchiesi, D., Reiss, J.: Automatic target mixing using least-squares optimization of gains and equalization settings. In: Proc. of the 12th Conf. on Digital Audio Effects (DAFx 2009), Como, Italy, September 2009, pp. 7–14 (2009)
2. Benaroya, L., Bimbot, F., Gribonval, R.: Audio source separation with a single sensor. IEEE Trans. on Audio, Speech and Language Processing 14(1), 191–199 (2006)
3. Bertin, N., Févotte, C., Badeau, R.: A tempering approach for Itakura-Saito non-negative matrix factorization. With application to music transcription. In: Proc. IEEE Intl. Conf. Acoust. Speech Signal Processing (ICASSP 2009), Washington, DC, USA, April 2009, pp. 1545–1548 (2009)
4. Févotte, C., Bertin, N., Durrieu, J.-L.: Nonnegative matrix factorization with the Itakura-Saito divergence with application to music analysis. Neural Computation 21(3), 793–830 (2009)
5. Lee, D.D., Seung, H.S.: Algorithms for non-negative matrix factorization. Advances in Neural Information Processing Systems 13, 556–562 (2001)
6. Ozerov, A., Févotte, C.: Multichannel nonnegative matrix factorization in convolutive mixtures for audio source separation. IEEE Trans. on Audio, Speech and Language Processing 18(3), 550–563 (2010)
7. Parvaix, M., Girin, L., Brossier, J.-M.: A watermarking-based method for informed source separation of audio signals with a single sensor. IEEE Transactions on Audio, Speech and Language Processing (2010) (to be published)
8. Vincent, E., Févotte, C., Gribonval, R.: Performance measurement in blind audio source separation. IEEE Trans. on Audio, Speech and Language Processing 14(4), 1462–1469 (2006)

Non-stationary t-Distribution Prior for Image Source Separation from Blurred Observations

Koray Kayabol* and Ercan E. Kuruoglu

ISTI, CNR, via G. Moruzzi 1, 56124, Pisa, Italy
{koray.kayabol,ercan.kuruoglu}@isti.cnr.it

Abstract. We propose a non-stationary spatial image model for the solution of the image separation problem from blurred observations. Our model is defined on first order image differentials. We model the image differentials using t-distribution with space varying scale parameters. This prior image model has been used in the Bayesian formulation and the image source are estimated using a Langevin sampling method. We have tested the proposed model on astrophysical image mixtures and obtained better results regarding stationary model for the maps which have high intensity changes.

1 Introduction

In blind signal separation, one of the three general assumptions is sufficient for separability of the signals namely, non-Gaussianity, non-whiteness and non-stationarity [1]. In this paper, we propose an image prior in Bayesian framework that includes the three separability conditions mentioned above. If we consider the image sources, the non-Gaussianity and non-whiteness have been already studied in the literature by using data fit model [2], [3] and Bayesian methods [4], [5]. Most of the previous studies for image source separation have assumed that the source signals are stationary. In [7], a neural network approach is used for non-stationary blind source separation. The method proposed in [6] is based on the non-stationarity in Fourier domain. For image separation part, non-stationary image separation using particle filter has been proposed in [8] and 1-dimensional non-stationary parametric image model has been used.

In this paper, we propose a Bayesian approach for the non-stationary image source separation problem which takes the non-stationarity into consideration using multi-variate t-distribution with space varying scale parameter. The prior densities are constituted by modeling the image differentials in different directions as Multivariate Student's t-distributions [9]. A non-stationary t-distribution model has been used in [10] for Bayesian image restoration. We exploit the non-stationary image model in [10] to extend the method in [9]. The non-stationary t-distribution model has the edge-preserving property and its parameters can be more easily calculated than the edge-preserving Markov

* Koray Kayabol is supported by the International Centre for Theoretical Physics, Trieste, Italy, Training and Research in Italian Laboratories program.

V. Vigneron et al. (Eds.): LVA/ICA 2010, LNCS 6365, pp. 506–513, 2010.

Random Field (MRF) prior. We use the joint posterior density of the complete variable set to obtain the joint estimate of all the variables. For estimation of the sources, we use an efficient Markov Chain Monte Carlo (MCMC) sampling method in which we resort to the Langevin stochastic equation [9]. To estimate the scale and degree of freedom (dof) parameters of t-distribution, we use the ML estimation via Expectation-Maximization (EM) algorithm as in [11]. We also find the non-stationarity parameter using ML estimation.

We test our model on the separation of the convolutional mixtures of astrophysical maps [12], which have non-stationarity in their nature. We gain a perceptible improvement in the sense of Peak Signal-to-Inference Ratio (PSIR).

In Section 2, the problem definition is given in a Bayesian framework. The simulation results are presented in Section 4 and interpreted in Section 5.

2 Problem Definition in the Bayesian Framework

We assume that the observed images, $y_k, k \in \{1, 2, \ldots, K\}$, are linear combinations of L source images. Since we consider the astrophysical map separation, in our case $K > L$. Let the kth observed image be denoted by $y_{k,n}$, where $n \in \{1, 2, \ldots, N\}$ represents the lexicographically ordered pixel index. Taking into account the effect of the point spread function $\mathbf{h}_k$ of the aperture, and by denoting $\mathbf{s}_l$ and $\mathbf{y}_k$ as $N \times 1$ vector representations of source and observation images, respectively, the observation model can be written as

$$\mathbf{y}_k = \mathbf{h}_k * \sum_{l=1}^{L} a_{k,l}\mathbf{s}_l + \mathbf{n}_k \tag{1}$$

where the asterisk denotes convolution. The observation model is not an instantaneous linear mixing, since $\mathbf{h}_k$ changes for each channel. The vector $\mathbf{n}_k$ represent an iid zero-mean noise with $\Sigma = \sigma_k^2\mathbf{I}_N$ covariance matrix where $\mathbf{I}_N$ is an identity matrix. Although the noise is not homogeneous in the astrophysical maps, we assume that the noise variance is homogeneous within each sky patch. The the point spread functions and the noise variances of the channels are known.

2.1 Non-stationary Source Model

In this paper, we extend the stationary t-distribution image model previously proposed in [9] using the non-stationary model proposed in [10]. For this purpose, we write an auto-regressive source model using the first order neighbors of the pixel in the direction d:

$$\mathbf{s}_l = \alpha_{l,d}\mathbf{G}_d\mathbf{s}_l + \mathbf{t}_{l,d} \tag{2}$$

where the maximum number of first order neighbors is 8 but we use only 4 neighbors, $d \in \{1, \ldots, 4\}$, in the vertical and horizontal directions. Matrix $\mathbf{G}_d$ is a linear one-pixel shift operator, $\alpha_{l,d}$ is the regression coefficient and the regression error $\mathbf{t}_{l,d}$ is an independently but not identically t-distributed zero-mean vector with degree of freedom parameter $\beta_{l,d}$ and scale parameter matrix $\Delta_{l,d}$. The matrix $\Delta_{l,d}$ is an $N \times N$ diagonal matrix with element $\delta_{l,d,n}$. The multivariate probability density function of an image modelled by a t-distribution can be written as

$$p(\mathbf{t}_{l,d}|\alpha_{l,d},\beta_{l,d},\Delta_{l,d}) = \frac{\Gamma((N+\beta_{l,d})/2)}{\Gamma(\beta_{l,d}/2)(\pi\beta_{l,d}\delta_{l,d})^{N/2}} \times \left[1 + \frac{(\mathbf{s}_l - \alpha_{l,d}\mathbf{G}_d\mathbf{s}_l)^T \Delta_{l,d}^{-1}(\mathbf{s}_l - \alpha_{l,d}\mathbf{G}_d\mathbf{s}_l)}{\beta_{l,d}}\right]^{-(N+\beta_{l,d})/2} \quad (3)$$

where $\Gamma(.)$ is the Gamma function. The t-distribution can be also written in implicit form using a Gaussian and a Gamma density [11]:

$$p(\mathbf{t}_{l,d}|\alpha_{l,d},\beta_{l,d},\Delta_{l,d}) = \int \mathcal{N}\left(\mathbf{t}_{l,d}|0, \frac{\Delta_{l,d}}{\xi_{l,d}}\right) \mathcal{G}\left(\xi_{l,d}|\frac{\beta_{l,d}}{2}, \frac{\beta_{l,d}}{2}\right) d\xi_{l,d}. \quad (4)$$

We use the representation in (4) for calculation of the parameters using EM method. We can write the density of $\mathbf{s}_l$ by using the image differentials in different directions, by assuming the directional independence, as $p(\mathbf{s}_l|\alpha_{l,d},\beta_{l,d},\Delta_{l,d}) = \prod_{d=1}^{4} p(\mathbf{t}_{l,d}|\alpha_{l,d},\beta_{l,d},\Delta_{l,d})$.

We assume uniform priors for $\alpha_{l,d}$ and $\beta_{l,d}$. For space varying scale parameter $\delta_{l,d,i}$ of t-distribution, we use an inverse-Gamma prior with shape parameter $\nu_{l,d} - 1$ and scale parameter $(\nu_{l,d} - 1)\bar{\delta}_{l,d}$ such that

$$p(\delta_{l,d,i}|\nu_{l,d},\bar{\delta}_{l,d}) = \frac{((\nu_{l,d}-1)\bar{\delta}_{l,d})^{\nu_{l,d}-1}}{\Gamma(\nu_{l,d}-1)\delta_{l,d,i}^{\nu_{l,d}}} e^{-(\nu_{l,d}-1)\frac{\bar{\delta}_{l,d}}{\delta_{l,d,i}}} \quad (5)$$

where $\bar{\delta}_{l,d}$ is the global scale variable over all the pixels and $\nu_{l,d}$ is the parameter which control the homogeneity of the scale parameter. Its role is explained in Section 3.3. We assume a Laplace prior for $\nu_{l,d}$ such as $\lambda_{l,d}\exp\{-\lambda_{l,d}(\nu_{l,d}-1)\}$ for $\nu_{l,d} > 1$.

2.2 Posteriors

To define the BSS problem in the Bayesian framework, the joint posterior density of all of the unknowns is written as $p(\mathbf{s}_{1:L},\mathbf{A},\Theta|\mathbf{y}_{1:K}) \propto p(\mathbf{y}_{1:K}|\mathbf{s}_{1:L},\mathbf{A})p(\mathbf{s}_{1:L},\mathbf{A},\Theta)$ where $\Theta = \{\alpha_{1:L,1:4},\beta_{1:L,1:4},\delta_{1:L,1:4,1:N},\lambda_{1:L,1:4},\nu_{1:L,1:4},\bar{\delta}_{1:L,1:4}\}$, $p(\mathbf{y}_{1:K}|\mathbf{s}_{1:L},\mathbf{A})$ is the likelihood and $p(\mathbf{s}_{1:L},\mathbf{A},\Theta)$ is the joint prior density of unknowns. The joint prior can be factorized as $p(\mathbf{s}_{1:L}|\alpha_{1:L,1:4},\beta_{1:L,1:4},\delta_{1:L,1:4,1:N})\, p(\nu_{1:L,1:4}|\lambda_{1:L,1:4})\, p(\delta_{1:L,1:4,1:N}|\nu_{1:L,1:4},\bar{\delta}_{1:L,1:4})$ by assuming the uniform priors for $\mathbf{A}$, α, λ, β and $\bar{\delta}$'s. Furthermore, since the sources are assumed to be independent, the joint probability density of the sources is also factorized as $p(\mathbf{s}_{1:L}|\Theta) = \prod_{l=1}^{L} p(\mathbf{s}_l|\Theta)$.

To estimate all the unknowns, we write their conditional posteriors as

$$p(a_{k,l}|\mathbf{y}_{1:K},\mathbf{s}_{1:L},\mathbf{A}_{-a_{k,l}},\Theta) \propto p(\mathbf{y}_{1:K}|\mathbf{s}_{1:L},\mathbf{A})$$

$$
\begin{aligned}
p(\alpha_{l,d}|\mathbf{y}_{1:K}, \mathbf{s}_{1:L}, \mathbf{A}, \Theta_{-\alpha_{l,d}}) &\propto p(\mathbf{t}_{l,d}|\Theta) \\
p(\beta_{l,d}|\mathbf{y}_{1:K}, \mathbf{s}_{1:L}, \mathbf{A}, \Theta_{-\beta_{l,d}}) &\propto p(\mathbf{t}_{l,d}|\Theta) \\
p(\delta_{l,d,n}|\mathbf{y}_{1:K}, \mathbf{s}_{1:L}, \mathbf{A}, \Theta_{-\delta_{l,d,n}}) &\propto p(t_{l,d,n}|\Theta)p(\delta_{l,d,n}|\nu_{l,d}, \bar{\delta}_{l,d}) \\
p(\bar{\delta}_{l,d}|\mathbf{y}_{1:K}, \mathbf{s}_{1:L}, \mathbf{A}, \Theta_{-\bar{\delta}_{l,d}}) &\propto p(\delta_{l,d,1:N}|\nu_{l,d}, \bar{\delta}_{l,d}) \\
p(\nu_{l,d}|\mathbf{y}_{1:K}, \mathbf{s}_{1:L}, \mathbf{A}, \Theta_{-\nu_{l,d}}) &\propto p(\delta_{l,d,1:N}|\nu_{l,d}, \bar{\delta}_{l,d})p(\nu_{l,d}|\lambda_{l,d}) \\
p(\mathbf{s}_l|\mathbf{y}_{1:K}, \mathbf{s}_{(1:L)-l}, \mathbf{A}, \Theta) &\propto p(\mathbf{y}_{1:K}|\mathbf{s}_{1:L}, \mathbf{A})p(\mathbf{s}_l|\Theta)
\end{aligned} \tag{6}
$$

where $t_{l,d,n}$ is an element of the vector the $\mathbf{t}_{l,d}$ and "–variable" expressions in the subscripts denote the removal of that variable from the variable set.

The Maximum Likelihood (ML) estimations of the parameters $\alpha_{l,d}$, $\beta_{l,d}$, $\delta_{l,d,n}$, $\bar{\delta}_{l,d}$ and $\nu_{l,d}$ are obtained using an EM method [9]. We tune the parameter $\lambda_{l,d}$ experimentally. To estimate the source images, we use Langevin sampler, whose details are given in Section 3.3.

3 Estimation of Sources and Parameters

In this section, we give the estimation of the mixing matrix, source maps and their parameters.

3.1 Mixing Matrix

We assume that the prior of $\mathbf{A}$ is uniform between 0 and ∞. From the conditional density of $a_{k,l}$ in (6), it is estimated in each iteration as

$$
a_{k,l} = \frac{1}{\mathbf{s}_l^T \mathbf{H}_k^T \mathbf{H}_k \mathbf{s}_l} \mathbf{s}_l^T \mathbf{H}_k^T \left(\mathbf{y}_k - \mathbf{H}_k \sum_{i=1, i \neq l}^{L} a_{k,i} \mathbf{s}_i \right) u(a_{k,l}) \tag{7}
$$

where $u(a_{k,l})$ is the unit step function.

3.2 Astrophysical Map Estimation

In the estimation of the source, we use an efficient Monte Carlo simulation whereby the Langevin stochastic equation, which exploits the gradient information of the energy function to produce a new proposal. The Langevin equation used in this study is written as

$$
\mathbf{s}_l^{k+1} = \mathbf{s}_l^k - \frac{1}{2}\mathbf{D}\mathbf{g}(\mathbf{s}_l^k) + \mathbf{D}^{\frac{1}{2}}\mathbf{w}_l \tag{8}
$$

where the diagonal matrix $\mathbf{D}^{\frac{1}{2}}$ contains the discrete time steps $\tau_{l,n}$, so that, for the ith pixel, the diffusion coefficient is $\mathbf{D}_{n,n} = \tau_{l,n}^2$. Matrix $\mathbf{D}$ is referred to here as diffusion matrix. We determine it by taking the inverse of the diagonal of the Hessian matrix of the energy $E(\mathbf{s}_l) = -\log p(\mathbf{s}_l|\mathbf{y}_{1:K}, \mathbf{s}_{(1:L)-l}, \mathbf{A}, \Theta)$.

Since the random variables for the image pixel intensities are produced in parallel by (8), this procedure is faster than the random walk adopted in [15]. The derivation details of the equation can be found in [9]. The samples produced using (8) are applied to a Metropolis-Hastings scheme pixel-by-pixel as in [9]. The acceptance probability of any proposed sample is defined as $\min\{\varphi(s_{l,n}^{k+1}, s_{l,n}^k), 1\}$, where $\varphi(s_{l,n}^{k+1}, s_{l,n}^k) \propto e^{-\Delta E(s_{l,n}^{k+1})} q(s_{l,n}^k | s_{l,n}^{k+1}) / q(s_{l,n}^{k+1} | s_{l,n}^k)$, $\Delta E(s_{l,n}^{k+1}) = E(s_{l,n}^{k+1}, s_{(1:L)-l,n}^k) - E(s_{1:L,n}^k)$. The proposal density is $q(s_{l,n}^{k+1} | s_{l,n}^k) = \mathcal{N}\left(s_{l,n}^{k+1} | s_{l,n}^k + \tau_{l,n}^2 g(s_{1:L,n}^k), \tau_{l,n}^2\right)$.

3.3 Parameter Estimation

From (4), the ML estimation of the parameters are found using EM algorithm as

$$\alpha_{l,d} = \frac{\mathbf{s}_l^T \mathbf{G}_d^T \mathbf{s}_l}{\mathbf{s}_l^T \mathbf{G}_d^T \mathbf{G}_d \mathbf{s}_l} \tag{9}$$

$$\delta_{l,d,n} = \frac{1}{0.5 + \nu_{l,d}} \left(\langle \xi_{l,d} \rangle \frac{\mathbf{t}_{l,d,n}^2}{2} + (\nu_{l,d} - 1)\bar{\delta}_{l,d} \right) \tag{10}$$

$$\bar{\delta}_{l,d} = \frac{N(\nu_{l,d} - 1)}{\mathrm{tr}(\Delta_{l,d}^{-1})\nu_{l,d}} \tag{11}$$

From (10), we can see the role of $\nu_{l,d}$, when $\nu_{l,d}$ equals to 1, $\delta_{l,d,n}$ is strictly space varying. If $\nu_{l,d}$ goes to infinity, $\delta_{l,d,n}$ becomes homogenous over space.

The maximization with respect to $\beta_{l,d}$ and $\nu_{l,d}$ do not have simple solutions. It can be solved by setting their first derivatives to zero:

$$-\psi_1(\beta_{l,d}/2) + \log \beta_{l,d} + \langle \log \xi_{l,d} \rangle - \langle \xi_{l,d} \rangle + 1 + \frac{1}{\beta_{l,d}} = 0 \tag{12}$$

$$-N\psi_1(\nu_{l,d} - 1) + N\log(\nu_{l,d} - 1) - \sum_{n=1}^{N} \log \delta_{l,d,n} - \sum_{n=1}^{N} \frac{\bar{\delta}_{l,d}}{\delta_{l,d,n}} + N - \lambda_{l,d}(\nu_{l,d} - 1) = 0 \tag{13}$$

where $\psi_1(.)$ is the first derivative of $\log \Gamma(.)$ and it is called digamma function. We set the parameter $\lambda_{l,d} = 5$ after some experiments.

4 Simulation Results

This section presents some astrophysical image separation results of the proposed non-stationary model along with the stationary model and non Bayesian Least Square (LS) method. The method with stationary priors is called ALS (Adaptive Langevin Sampler) and the proposed method is denoted as ALSnonS. In the first competitor method, we apply the pre-estimated separation matrix to find LS solution. As a pre-estimation method, one can use Independent Component

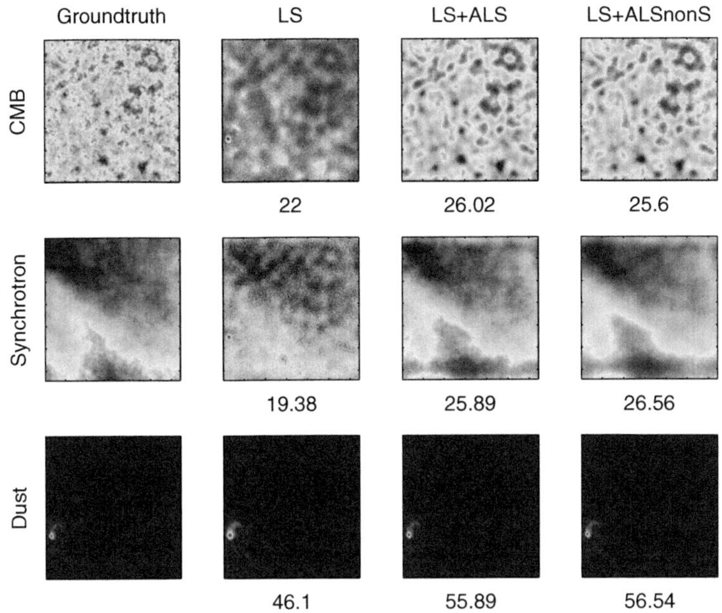

Fig. 1. Reconstructed astrophysical maps from blurred and noisy observations by the LS, LS+ALS and LS+ALSnonS methods. The numbers under the reconstructed maps are the PSIR values in dB. The map is located at 0° longitude and 20° latitude and sized 128×128.

Analysis (ICA) [2], Spectral Matching ICA (SMICA) [6] or Fourier Domain Correlated Component Analysis (FDCCA) [13]. For de-blurring, we use the Wiener filter with known psf and noise covariance [14]. The LS solution methods includes the FDCCA, ICA, SMICA and the methods which estimate the mixing matrix. We call the competitor methods LS and LS+ALS, respectively. We use the solution of LS as initial value for our methods, so we call the proposed method as LS+ALSnonS. We have tested our algorithm on a sky patch that is located at (0,20) galactic coordinates. The observation images are generated by using a 9×3 mixing matrix simulating nine images at frequencies 30, 44, 70, 100, 143, 217, 353, 545, and 857 GHz. The size of the channel maps is 128×128.

Fig. 1 shows the ground truth astrophysical source images and the estimated ones with LS, LS+ALS and LS+ALSnonS. The mixed sources are CMB, synchrotron and dust maps. The Peak Signal-to-Inference Ratio PSIR $= 20\log\{\sqrt{N} \max(\mathbf{s}_l^*)/||\mathbf{s}_l^* - \hat{\mathbf{s}}_l||\}$ is used as a numerical performance indicator. $\mathbf{s}_l^*$ and $\hat{\mathbf{s}}_l$ are the ground-truth and the estimated images respectively.The PSIR values of the estimated maps are denoted under each result in Fig. 1.

The PSIR results show that the non-stationary model provides significant improvement for synchrotron and dust maps, but the results for CMB map is not better than those obtained with stationary model. For this sky patch, we can say the CMB is almost stationary. We can localize the non-stationary regions as

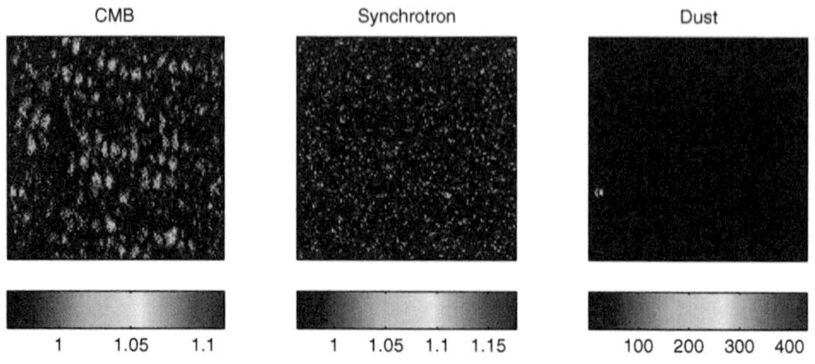

Fig. 2. Normalized space varying scale parameter maps for CMB, synchrotron and dust for direction $d = 1$. $\delta_{l,1,1:N}/\bar{\delta}_{l,1}$ where $l = \{1, 2, 3\}$

shown in Fig. 2 that shows the normalized space varying scale parameter maps $\delta_{l,1,1:N}/\bar{\delta}_{l,1}$, where $l = \{1, 2, 3\}$, for CMB, synchrotron and dust for direction $d = 1$. From Fig. 2, it is seen that the normalized scale parameter map of dust map has very high values which are 400 times greater than the stationary scale parameter. The intensities of the normalized scale parameter map of CMB and synchrotron are distributed around 1.

5 Conclusion

In this study, we have proposed statistical models for separation of non-stationary source images, and have shown that taking into account the non-stationarity of the sources results in quite higher performance as compared to the stationary model. The advantage of the proposed model comes out by applying the model to the more non-stationary sources, i.e. point sources. A by-product of the algorithm is the non-stationarity map of the images. By using these maps, It is also possible to search the non-stationarity over images. An application of the non-stationarity can be found in the analysis of CMB radiation.

Acknowledgment

The authors would like to thank Diego Herranz for his valuable discussions and providing the simulated maps. The simulated maps are courtesy of the Planck working group on diffuse component separation (WG2.1) [16].

References

1. Cardoso, J.-F.: The three easy routes to independent component analysis; Contrasts and Geometry. In: Int. Conf. on Indepen. Comp. Anal. ICA 2001, San Diego (2001)
2. Hyvarinen, A., Oja, E.: Fast fixed-point algorithm for independent component analysis. Neural Computation 9(7), 1483–1492 (1997)

3. Belouchrani, A., Abed-Meraim, K., Cardoso, J.-F., Moulines, E.: A Blind source separation technique using second-order statistics. IEEE Trans. Signal Process. 45(2), 434–444 (1997)
4. Snoussi, H., Patanchon, G., Macias-Pérez, J., Mohammad-Djafari, A., Delabrouille, J.: Bayesian blind component separation for cosmic microwave background observations. In: MaxEnt Workshops, pp. 125–140 (2001)
5. Tonazzini, A., Bedini, L., Kuruoglu, E.E., Salerno, E.: Blind separation of auto-correlated images from noisy mixtures using MRF models. In: Int. Sym. on Indepen. Comp. Anal and Blind Sig. ICA 2003 (September 2003)
6. Cardoso, J.-F., Snoussi, H., Delabrouille, J., Patanchon, G.: Blind separation of noisy Gaussian stationary sources: Application to cosmic microwave background imaging. In: European. Conf. on Signal Processing, EUSIPCO 2002, pp. 561–564 (2002)
7. Matsuoka, K., Ohya, M., Kawamoto, M.: A neural net for blind separation of nonstationary signals. Neural Networks 8(3), 411–419 (1995)
8. Costagli, M., Kuruoglu, E.E.: Image separation using particle filters. Digital Signal Process. 17, 935–946 (2007)
9. Kayabol, K., Kuruoglu, E.E., Sanz, J.L., Sankur, B., Salerno, E., Herranz, D.: Adaptive Langevin sampler for separation of t-distribution modelled astrophysical Maps. IEEE Trans. Image Process. 19(9) (2010)
10. Chantas, G.K., Galatsanos, N.P., Nikas, A.C.: Bayesian restoration using a new nonstationary edge-preserving image prior. IEEE Trans. Image Process. 15(10), 2987–2997 (2006)
11. Liu, C., Rubin, D.B.: ML estimation of the t distribution using EM and its extensions, ECM and ECME. Statistica Sinica 5, 19–39 (1995)
12. Kayabol, K., Kuruoglu, E.E., Sanz, J.L., Sankur, B., Salerno, E., Herranz, D.: Astrophysical map reconstruction from convolutional mixtures. In: Astronomical Data Analysis. Conf., ADA 2010, Monastir, Tunisia (2010)
13. Bedini, L., Salerno, E.: Extracting astrophysical sources from channel-dependent convolutional mixtures by correlated component analysis in the frequency domain. In: Apolloni, B., Howlett, R.J., Jain, L. (eds.) KES 2007, Part III. LNCS (LNAI), vol. 4694, pp. 9–16. Springer, Heidelberg (2007)
14. Bonaldi, A., Ricciardi, S., Leach, S., Stivoli, F., Baccigalupi, C., De Zotti, G.: WMAP 3yr data with the CCA: anomalous emission and impact of component separation on the CMB power spectrum. Mon. Not. R. Astron. Soc. 382(4), 1791–1803 (2007)
15. Kayabol, K., Kuruoglu, E.E., Sankur, B.: Bayesian separation of images modelled with MRFs using MCMC. IEEE Trans. Image Process. 18(5), 982–994 (2009)
16. Planck Science Team.: PLANCK: The scientific programme. European Space Agency, ESA (2005), `http://www.esa.int/SPECIALS/Planck/index.html`

Automatic Rank Determination in Projective Nonnegative Matrix Factorization

Zhirong Yang, Zhanxing Zhu, and Erkki Oja

Department of Information and Computer Science*
Aalto University School of Science and Technology
P.O.Box 15400, FI-00076, Aalto, Finland
{zhirong.yang,zhanxing.zhu,erkki.oja}@tkk.fi

Abstract. Projective Nonnegative Matrix Factorization (PNMF) has demonstrated advantages in both sparse feature extraction and clustering. However, PNMF requires users to specify the column rank of the approximative projection matrix, the value of which is unknown beforehand. In this paper, we propose a method called ARDPNMF to automatically determine the column rank in PNMF. Our method is based on automatic relevance determination (ARD) with Jeffrey's prior. After deriving the multiplicative update rule using the expectation-maximization technique for ARDPNMF, we test it on various synthetic and real-world datasets for feature extraction and clustering applications to show the effectiveness of our algorithm. For FERET faces and the Swimmer dataset, interpretable number of features are obtained correctly via our algorithm. Several UCI datasets for clustering are also tested, in which we find that ARDPNMF can estimate the number of clusters quite accurately with low deviation and good cluster purity.

1 Introduction

Since its introduction by Lee and Seung [1] as a new machine learning method, Nonnegative Matrix Factorization (NMF) has been applied successfully in many applications, including signal processing, text clustering and gene expression studies, etc. (see [2] for a survey). Recently much progress for NMF has been reported both in theory and practice. Also there are several variants to extend original NMF, (e.g. [3–5]). *Projective Nonnegative Matrix Factorization* (PNMF), introduced in [6–8], approximates a data matrix by its nonnegative subspace projection. Compared with NMF, the PNMF has a number of benefits such as better generalization, a sparser factorizing matrix without ambiguity, and close relation to principal component analysis, which are advantageous in both feature extraction and clustering [8].

However, a remaining difficult problem is how to determine the dimensionality of the approximating subspace in PNMF in practical applications. In most cases,

* Supported by the Academy of Finland in the project *Finnish Centre of Excellence in Adaptive Informatics Research*.

V. Vigneron et al. (Eds.): LVA/ICA 2010, LNCS 6365, pp. 514–521, 2010.

one has to guess a suitable component number, e.g. the number of features needed to encode facial images. Such trial-and-error procedures can be tedious in practice. In this work, we propose a variant of PNMF called ARDPNMF that can automatically determine dimensionality of factorizing matrix. Our method is based on the *automatic relevance determination* (ARD) [9] technique which has been used in Bayesian PCA [10] and adaptive sparse supervised learning [11]. The proposed algorithm is free of user-specified parameters. Such property is especially desired for exploratory analysis of the data structure. Empirical results on several synthetic and real-world datasets demonstrate that our method can effectively discover the number of features or clusters.

This paper is organized as follows. In Section 2, we summarize the essence of PNMF and model selection in NMF. Then, we derive our algorithm ARDPNMF in Section 3. In Section 4, the experimental results of the proposed algorithm on a variety of synthetic and real datasets for feature extraction and clustering are presented. Section 5 concludes the paper.

2 Related Work

2.1 Projective Nonnegative Matrix Factorization

Given a nonnegative input matrix $\mathbf{X} \in \mathbb{R}_+^{m\times n}$, Projective Nonnegative Matrix Factorization (PNMF) seeks a nonnegative matrix $\mathbf{W} \in \mathbb{R}_+^{m\times r}$ such that

$$\mathbf{X} \approx \mathbf{W}\mathbf{W}^T\mathbf{X}. \tag{1}$$

Compared with the NMF approximation scheme, $\mathbf{X} \approx \mathbf{WH}$, PNMF replaces $\mathbf{H}$ matrix with $\mathbf{W}^T\mathbf{X}$. As a result, PNMF has a number of advantages over NMF [8], including high sparseness in the factorizing matrix $\mathbf{W}$, closer equivalence to clustering, easy nonlinear extension to a kernel version, and fast approximation of newly coming samples without heavy re-computation. The name "projective" comes from the fact that $\mathbf{W}\mathbf{W}^T$ is very close to a projection matrix because the $\mathbf{W}$ learned by PNMF is highly orthogonal. It can be made fully orthogonal by post-processing.

PNMF based on the Euclidean distance solves the following optimization problem:

$$\underset{\mathbf{W}\geq 0}{\text{minimize}}\, J_F(\mathbf{W}) = \frac{1}{2}\sum_{ij}\left[X_{ij} - \left(\mathbf{W}\mathbf{W}^T\mathbf{X}\right)_{ij}\right]^2. \tag{2}$$

Previously, Yuan and Oja [6] presented a multiplicative algorithm that iteratively applies the following update rule for the above minimization:

$$W'_{ik} \leftarrow W_{ik}\frac{A_{ik}}{B_{ik}} \tag{3}$$

$$\mathbf{W}^{\text{new}} \leftarrow \mathbf{W}'/\|\mathbf{W}'\|, \tag{4}$$

where $\mathbf{A} = 2\mathbf{X}\mathbf{X}^T\mathbf{W}$, $\mathbf{B} = \mathbf{W}\mathbf{W}^T\mathbf{X}\mathbf{X}^T\mathbf{W} + \mathbf{X}\mathbf{X}^T\mathbf{W}\mathbf{W}^T\mathbf{W}$, and $\|\mathbf{W}'\|$ calculates the square root of the maximal eigenvalue of $\mathbf{W}'^T\mathbf{W}'$.

2.2 Model Selection in NMF

In NMF, Tan and Févotte [12] addressed the model selection problem based on automatic relevance determination. First, a prior is added on the columns and rows of matrix **W** and **H**. A Bayesian NMF model with the prior is then built. After maximizing the posterior, they obtain a multiplicative update rule to do both factorization and determination of component number simultaneously. The limitation of this method is that the prior distribution still depends on the hyper-parameters. For real-world applications, the hyper-parameters must be chosen suitably in advance to obtain reasonable results. In this sense, this method is not totally automatic for determining the component number.

In the following section, we overcome this problem and apply the ARD method to PNMF by selecting Jeffrey's prior [13] to get rid of hyper-parameters. Then our algorithm is totally automatic without any user-specified parameters.

3 ARDPNMF

Firstly, we construct a generative model for PNMF based on the Euclidean distance, where the likelihood function is a normal distribution.

$$p(X_{ij}|\mathbf{W}) = \mathcal{N}\left(X_{ij}|\left(\mathbf{W}\mathbf{W}^T\mathbf{X}\right)_{ij}, \mathbf{I}\right) \tag{5}$$

Following the approach of Bayesian PCA [10], we give a normal prior on the kth column of $\mathbf{W}$ with variance γ_k. Due to the nonnegativity in PNMF, we treat the distribution of each column of $\mathbf{W}$ as half-normal distribution.

$$p(W_{ik}|\gamma_k) = \mathcal{HN}(W_{ik}|0,\gamma_k) = \frac{\sqrt{2}}{\sqrt{\pi\gamma_k}} \exp\left(-\frac{W_{ik}^2}{2\gamma_k}\right) \tag{6}$$

for $W_{ik} \leq 0$, and zero otherwise.

Similar to [13], we impose a non-informative Jeffreys' hyper prior on the variances $\boldsymbol{\gamma}$ to control the sparseness of $\mathbf{W}$:

$$p(\gamma_k) \propto \frac{1}{\gamma_k} \tag{7}$$

We choose this prior because it expresses ignorance with respect to scale and the resulting model is parameter-free, which plays a significant role in determining the component number automatically.

The posterior of $\mathbf{W}$ for the above model is given by

$$p(\mathbf{W}|\mathbf{X},\boldsymbol{\gamma}) \propto p(\mathbf{X}|\mathbf{W})p(\mathbf{W}|\boldsymbol{\gamma}) \tag{8}$$

Because $\boldsymbol{\gamma}$ is unobserved, we apply the *Expectation-Maximization* (EM) algorithm by regarding $\boldsymbol{\gamma}$ as a hidden variable.

E-step. Given the current parameter estimates and observed data, E-step computes the expectation of the complete log-posterior, which is known as Q-function:

$$Q(\mathbf{W}|\mathbf{W}^{(t)}) = \int \log p(\mathbf{W}|\mathbf{X},\boldsymbol{\gamma})p(\boldsymbol{\gamma}|\mathbf{W}^{(t)},\mathbf{X})d\boldsymbol{\gamma} \tag{9}$$

Thanks to the property of Jeffrey's prior, we have a concise form of Q-function following the derivation in [13]:

$$Q(\mathbf{W}|\mathbf{W}^{(t)}) = -J_F(\mathbf{W}) - \frac{1}{2}\mathrm{Tr}(\mathbf{W}\mathbf{V}^{(t)}\mathbf{W}^T), \tag{10}$$

where $J_F(\mathbf{W})$ is the original objective function in PNMF (see Equation (2)), $\mathbf{V}^{(t)}$ is a diagonal matrix with $V_{ii}^{(t)} = \left\|\mathbf{w}_i^{(t)}\right\|^{-2}$, and $\left\|\mathbf{w}_i^{(t)}\right\|$ is the L_2-norm of the ith column of matrix $\mathbf{W}^{(t)}$. Note that we ignore the constants independent of $\mathbf{W}$ to present a simplified version of the Q-function.
M-step. This step maximizes the Q-function w.r.t parameters.

$$\mathbf{W}^{(t+1)} = \arg\max_{\mathbf{W}} Q(\mathbf{W}|\mathbf{W}^{(t)}), \tag{11}$$

which is equivalent to minimizing its negative form

$$Q_{\mathrm{ard}}(\mathbf{W}|\mathbf{W}^{(t)}) = -Q(\mathbf{W}|\mathbf{W}^{(t)}) = J_F(\mathbf{W}) + \frac{1}{2}\mathrm{Tr}(\mathbf{W}\mathbf{V}^{(t)}\mathbf{W}^T). \tag{12}$$

The derivative of $Q_{\mathrm{ard}}(\mathbf{W}|\mathbf{W}^{(t)})$ with respect to $\mathbf{W}$ is

$$\frac{\partial Q_{\mathrm{ARD}}(\mathbf{W}|\mathbf{W}^{(t)})}{\partial \mathbf{W}_{ik}} = - A_{ik} + B_{ik} + \left(\mathbf{W}\mathbf{V}^{(t)}\right)_{ik}. \tag{13}$$

For $\mathbf{A}, \mathbf{B}$, see eq. (3).

A commonly used principle that forms multiplicative update rule in NMF is

$$W'_{ik} \leftarrow W_{ik}^{(t)} \frac{\nabla_{ik}^-}{\nabla_{ik}^+}, \tag{14}$$

where ∇^- and ∇^+ denote the negative and positive parts of the derivative [1]. Applying this principle to the gradient given in Equation (13), we obtain the multiplicative update rule for ARDPNMF:

$$W'_{ik} \leftarrow W_{ik}^{(t)} \frac{A_{ik}^{(t)}}{B_{ik}^{(t)} + \left(\mathbf{W}^{(t)}\mathbf{V}^{(t)}\right)_{ik}}. \tag{15}$$

The ARDPNMF algorithm is summarized in Algorithm 1. After the algorithm converges, we apply a simple thresholding to keep the $\mathbf{W}$ columns whose norm is larger than a small constant ϵ. In practice such thresholding is insensitive to ϵ because the ARD prior forces these norms towards two extremes, as demonstrated in Section 4.1.

4 Experimental Results

We have implemented the ARDPNMF algorithm and tested it on various synthetic and real-world datasets to find out the effectiveness of our algorithm. The focus is on feature extraction and clustering.

Algorithm 1. ARDPNMF based on Euclidean distance

Usage: $\mathbf{W} \leftarrow \text{ARDPNMF}(\mathbf{X}, r)$, where $r < m$ is a large initial component number.
Initialize $\mathbf{W}^{(0)}$, $t \leftarrow 0$.
repeat
$\quad \mathbf{V}^{(t)} \leftarrow \text{diag}(\|\mathbf{w}_1^{(t)}\|^{-2}, \ldots, \|\mathbf{w}_r^{(t)}\|^{-2})$
$\quad W'_{ik} \leftarrow W_{ik}^{(t)} \dfrac{A_{ik}^{(t)}}{B_{ik}^{(t)} + (\mathbf{W}^{(t)}\mathbf{V}^{(t)})_{ik}}$
$\quad \mathbf{W}^{(t+1)} \leftarrow \mathbf{W}'/\|\mathbf{W}'\|$
$\quad t \leftarrow t + 1$
until convergent conditions are satisfied
Check the diagonal elements in matrix $\mathbf{V}$, and keep the columns of $\mathbf{W}$ with large L_2-norms as the effective components.

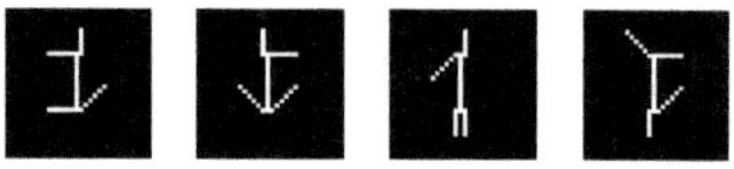

Fig. 1. Some sample images of Swimmer dataset

4.1 Swimmer Dataset

Swimmer dataset [14] consists of 256 images, each of which depicts a figure with one static part (torso)and four moving parts (limbs) with size 32×32. Each moving part has four different positions. Four of the 256 images are displayed in Figure 1. The task here is to extract the 16 limb positions and 1 torso position. Firstly, we vectorized each image matrix and treated it as one column of input matrix $\mathbf{X}$. The initial component number was set to $r = 36$. Each column of $\mathbf{W}$ learned by ARDPNMF has the same dimensionality as the input column vectors and thus can be displayed as base images in Figure 2. We found that our algorithm can correctly extract all the 17 desired features. The L_2-norms of all the columns of $\mathbf{W}$ are shown in Figure 3. We can easily see that the L_2-norms of ineffective basis images are equal to zero or very close to zero. The three values between 0 and 1 correspond to three duplicates of the torsos.

4.2 FERET Faces Dataset

The FERET face dataset [15] for feature extraction consists of the inner part of 2409 faces with size of 32×32. We normalized the images via dividing the pixel values by their maximal value 255. In ARDPNMF, the initial component number was chosen as $r = 64$. Figure 4 shows the resulting base images, which demonstrates high sparseness in the factorizing matrix $\mathbf{W}$ and captures nearly all facial parts.

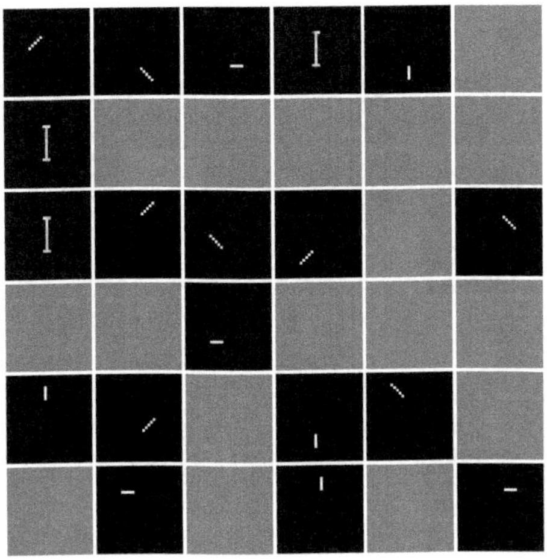

Fig. 2. 36 basis images of Swimmer dataset. The gray cells correspond to columns whose L_2-norms are zero or very close to zero.

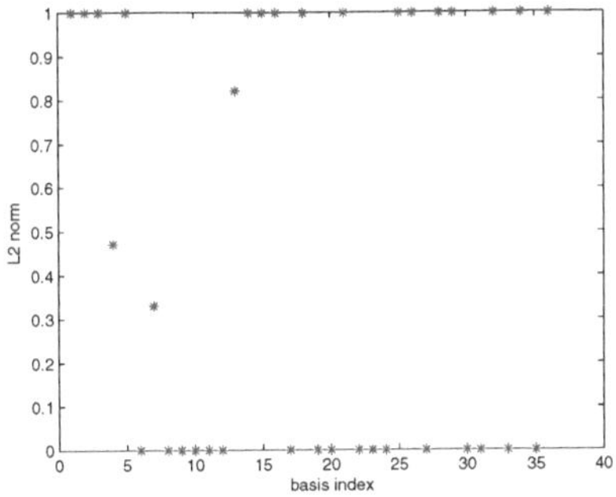

Fig. 3. L_2-norm of 36 basis images in Swimmer dataset

4.3 Clustering for UCI Datasets

Clustering is another important application of PNMF. We construct the input matrix $\mathbf{X}$ by treating each sample vector as a row. Then the index of the maximal value in a row of $\mathbf{W}$ indicates the cluster membership of the corresponding sample [8]. We have adopted a widely-used measurement called *purity* [8] for quantitatively analyzing clustering results, which is defined as follows:

$$\text{purity} = \frac{1}{n} \sum_{k=1}^{r'} \max_{1 \leq l \leq q} n_k^l, \tag{16}$$

where q is the true number of classes, r' is the effective number of components (clusters), n_k^l is the number of samples in the cluster k that belongs to original

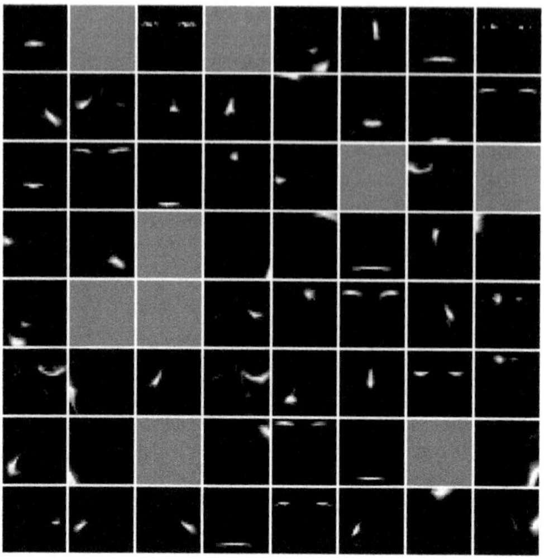

Fig. 4. 64 basis images of FERET dataset. 55 of them are effective basis. Remaining gray ones' L_2-norms are zero or close to zero.

Table 1. Clustering Performance

Datasets	iris	ecoli	glass	wine	parkinsons
Number of classes	3	5	6	3	2
Estimated cluster number	4.34 ± 0.71	2.74 ± 0.60	3.34 ± 0.61	3 ± 0.40	4.37 ± 0.58
Purity	0.95 ± 0.01	0.68 ± 0.06	0.67 ± 0.05	0.9 ± 0.09	0.77 ± 0.02

class l, and n is the total number of samples. Larger purity value indicates better clustering results, and value 1 indicates total agreement with the ground truth.

We chose several commonly used datasets in the UCI repository[1] as experimental data. In each dataset, ARDPNMF was run 100 times with different random seeds for $\mathbf{W}$ initialization, and we set the initial cluster number r as 36. Table 1 shows the mean and standard deviation of the number of clusters and purities, as well as the numbers of ground truth classes. ARDPNMF can automatically estimate the cluster number which is not far from the true class number, with small deviations. Furthermore, our method can achieve reasonably good clustering performance especially when the estimated r value is close to the ground truth.

5 Conclusion

In this paper, using Bayesian construction and EM algorithm, we have presented the ARDPNMF algorithm which can automatically determine the rank of the projection matrix in PNMF. By using Jeffreys' prior as the model prior, we

[1] `http://www.ics.uci.edu/~mlearn/MLRepository.html`

have made our algorithm totally free of human tuning in finding algorithm parameters. Through experiments on various synthetic and real-world datasets for feature extraction and clustering, ARDPNMF demonstrates its effectiveness in model selection for PNMF. Moreover, our algorithm is readily extended to other dissimilarity measures, such as the α or β divergences [2]. Our method however could be sensitive to the initialization of the factorizing matrix in some cases, which we should improve in the future for a more robust estimate of the rank.

References

1. Lee, D.D., Seung, H.S.: Learning the parts of objects by non-negative matrix factorization. Nature 401, 788–791 (1999)
2. Cichocki, A., Zdunek, R., Phan, A.H., Amari, S.: Nonnegative Matrix and Tensor Factorizations: Applications to Exploratory Multi-way Data Analysis. John Wiley, Chichester (2009)
3. Dhillon, I.S., Sra, S.: Generalized nonnegative matrix approximations with bregman divergences. Advances in Neural Information Processing Systems 18, 283–290 (2006)
4. Choi, S.: Algorithms for orthogonal nonnegative matrix factorization. In: Proceedings of IEEE International Joint Conference on Neural Networks, pp. 1828–1832 (2008)
5. Ding, C., Li, T., Jordan, M.I.: Convex and semi-nonnegative matrix factorizations. IEEE Transactions on Pattern Analysis and Machine Intelligence 32(1), 45–55 (2010)
6. Yuan, Z., Oja, E.: Projective nonnegative matrix factorization for image compression and feature extraction. In: Kalviainen, H., Parkkinen, J., Kaarna, A. (eds.) SCIA 2005. LNCS, vol. 3540, pp. 333–342. Springer, Heidelberg (2005)
7. Yang, Z., Yuan, Z., Laaksonen, J.: Projective non-negative matrix factorization with applications to facial image processing. International Journal on Pattern Recognition and Artificial Intelligence 21(8), 1353–1362 (2007)
8. Yang, Z., Oja, E.: Linear and nonlinear projective nonnegative matrix factorization. IEEE Transaction on Neural Networks 21(5), 734–749 (2010)
9. Mackay, D.J.C.: Probable networks and plausible predictions – a review of practical bayesian methods for supervised neural networks. Network: Computation in Neural Systems 6(3), 469–505 (1995)
10. Bishop, C.M.: Bayesian pca. Advances in Neural Information Processing Systems, 382–388 (1999)
11. Tipping, M.E.: Sparse bayesian learning and the relevance vector machine. Journal of Machine Learning Research 1, 211–244 (2001)
12. Tan, V.Y.F., Févotte, C.: Automatic relevance determination in nonnegative matrix factorization. In: Proceedings of 2009 Workshop on Signal Processing with Adaptive Sparse Structured Representations, SPARS 2009 (2009)
13. Figueiredo, M.A.: Adaptive sparseness for supervised learning. IEEE Transactions on Pattern Analysis and Machine Intelligence 25(9), 1150–1159 (2003)
14. Donoho, D., Stodden, V.: When does non-negative matrix factorization give a correct decomposition into parts? Advances in Neural Information Processing Systems 16, 1141–1148 (2003)
15. Phillips, P.J., Moon, H., Rizvi, S.A., Rauss, P.J.: The FERET evaluation methodology for face recognition algorithms. IEEE Trans. Pattern Analysis and Machine Intelligence 22, 1090–1104 (2000)

Non-negative Independent Component Analysis Algorithm Based on 2D Givens Rotations and a Newton Optimization

Wendyam Serge Boris Ouedraogo[1,2,3,⋆], Antoine Souloumiac[1], and Christian Jutten[3,⋆⋆]

[1] CEA, LIST, Laboratoire d'Outils pour l'Analyse de Données, Gif-sur-Yvette, F-91191, France
[2] Unité Signaux et Systèmes, National School of Engineers of Tunis, BP 37, 1002 Tunis, Tunisia
[3] GIPSA-lab, UMR 5216 CNRS, University of Grenoble, 961 rue de la Houille Blanche BP 46 F-38402 Grenoble Cedex, France
{wendyam-serge-boris.ouedraogo}@cea.fr

Abstract. In this paper, we consider the Independent Component Analysis problem when the hidden sources are non-negative (Non-negative ICA). This problem is formulated as a non-linear cost function optimization over the special orthogonal matrix group $SO(n)$. Using Givens rotations and Newton optimization, we developed an effective axis pair rotation method for Non-negative ICA. The performance of the proposed method is compared to those designed by Plumbley and simulations on synthetic data show the efficiency of the proposed algorithm.

Keywords: Non-negative ICA, Givens rotations, Newton optimization, Complexity calculation.

1 Introduction

We consider the batch mode of ICA. Lets $S = \begin{bmatrix} s_1 \; s_2 \; \cdots \; s_n \end{bmatrix}^T$ be the n hidden sources observed through a mixing matrix $A = [a_{ij}]$, $1 \leq i \leq m$ and $1 \leq j \leq n$. The noiseless model of ICA can be written:

$$X = AS \tag{1}$$

where $X = \begin{bmatrix} x_1 \; x_2 \; \cdots \; x_m \end{bmatrix}^T$. We consider a square system where $m = n$.

The task of ICA is to find A and S given X.

In "classical" ICA the sources are required to be independent and non-Gaussian. Under these conditions, many algorithms based on maximization of the source

⋆ This study is conducted in the context of co-tutelle Ph-D program between CEA List, GIPSA Lab of University of Grenoble (France) and Unité Signaux et Systèmes of National School of Engineers of Tunis (Tunisia).

⋆⋆ Christian JUTTEN is also with Institut Universitaire de France.

V. Vigneron et al. (Eds.): LVA/ICA 2010, LNCS 6365, pp. 522–529, 2010.

non-Gaussianity [5][4] or independence [3] have been developed for estimating the hidden sources up to the permutation and scaling indeterminations.

Subsequently, using a priori knowledge on the sources, some constraints such as sparsity have been added in ICA to favour particular types of solutions [17].

In many real world applications such as biomedical imaging, music or spectrum analysis, the sources are known to be non-negative. This a priori must be taken into account when estimating the sources.

Several authors have proposed methods for solving equation (1) under non-negativity constraint on S and/or A. The most used approach is Non-negative Matrix Factorization (NMF) [14][6][7][13] where the estimated sources and mixing matrix are all constrainted to be non-negative. However, the non-negativity alone is not sufficient to guarantee the uniqueness of the solution [8][9][10]. So depending on the application some constraints such as sparseness and/or smoothness also been incorporated in NMF to improve the parts based representation and reduce the range of admissible solutions [18].

For estimating the sources and/or mixing matrix under non-negativity constraint, another approach uses the a priori knowledges of the variables distribution to design a Bayesian method [19][20]. This approach needs however to make a "good" choice of A and S prior density and can be time-consuming.

Slightly relaxing the non-negativity constraint, Plumbley introduced Non-negative Independent Component Analysis [1][2] for solving (1) under non-negativity constraint on S, A being positive or of mixed sign. This approach requires the sources to be *non-negative* ($Pr(s_i < 0) = 0, \forall\ 1 \leq i \leq n$), *independent* ($Pr(\prod s_i) = \prod Pr(s_i)$), and *well grounded* ($\forall\ \delta > 0,\ Pr(s_i < \delta) > 0,\ \forall\ 1 \leq i \leq n$).

In this paper, we use Givens rotations and Newton optimization to develop an efficient axis pair rotation method for non-negative ICA.

The rest of the paper is organized as follows. Section 2 recapitulates Non-negative ICA problem and its formulation as a non-linear cost function optimization. In section 3, we describe the proposed axis pair rotation method. The computational complexity is evaluated in section 4 and we compare it with the geodesic search method and the axis pair rotation method designed by Plumbley. Section 5 discusses the simulation results and finally section 6 presents the conclusions.

2 Non-negative Independent Component Analysis

Under the independence and well grounded assumptions, the non-negative hidden sources S can be estimated by whitening the observations X and rotating the whitened data to fit them on the positive orthant.

In fact, let Z be the whitened observations, $Z = VX = VAS$ where V is a whitening matrix. Assuming that the sources are unit variance or transformed to be so, the covariance matrix of Z is given by $C_Z = I_n = (VA)(VA)^T$, then VA is an orthonormal matrix.

Let $Y = WZ$ and W be a rotation matrix ($W^TW = WW^T = I_n$ and det $W = 1$).

$$Y = WVAS = US \text{ where } U = WVA$$

Plumbley showed [1] that U is a permutation matrix if and only if Y is positive (i.e. each element of Y is positive) with probability 1.

It is then sufficient to find a rotation matrix W so that the components of $Y = WZ$ are positive.

We consider the following negativeness measure criterion defined in [2]:

$$J(W) = \frac{1}{2} \left\| Z - W^T Y^+ \right\|_F^2$$

where $Y^+ = \left[y_1^+ y_2^+ \cdots y_n^+ \right]^T$, $y_i^+ = \max(0, y_i)$ and $\|\|_F$ is the Frobenius norm.

$$J(W) = \frac{1}{2} \left\| Z - W^T Y^+ \right\|_F^2 = \frac{1}{2} \left\| W^T Y - W^T Y^+ \right\|_F^2 = \frac{1}{2} \left\| Y - Y^+ \right\|_F^2 = \frac{1}{2} \left\| Y^- \right\|_F^2$$

where $Y^- = \left[y_1^- y_2^- \cdots y_n^- \right]^T$ and $y_i^- = \min(0, y_i)$.

One can prove that $J(W) = 0 \Leftrightarrow Y^- = 0 \Leftrightarrow Y$ is positive with probability 1.

In pratical algorithm, the task of Non-negative ICA is to find a rotation matrix W that minimizes the criterion J. This is equivalent to solve the optimization problem (2) on the group of rotation matrices $SO(n)$:

$$W^* = \arg \min_{W \in SO(n)} J(W) \tag{2}$$

Several methods such as non-negative PCA [11][12], axis pair method [2] or geodesic search [15][16] have been proposed for solving (2). In the next section we propose an efficient axis pair rotation method for solving optimization (2).

3 Givens Parametrization and Newton Optimization for Non-negative ICA

When the sources are independent and well grounded, the task of solving the Non-negative Independent Component Analysis problem reduces to finding the rotation matrix W^* which minimizes the criterion J (resolving equation (2)).

Noting that any general n-dimensional rotation can be written as a product of *Givens rotations* $G(i_l, j_l, \theta_l)$ where

$$G(i_l, j_l, \theta_l) = \begin{pmatrix} 1 \ldots & 0 & \ldots & 0 & \ldots 0 \\ \vdots \ \ddots & \vdots & & \vdots & \vdots \\ 0 \ldots & \cos(\theta_l) & \cdots & \sin(\theta_l) & \ldots 0 \\ \vdots & \vdots & \ddots & \vdots & \vdots \\ 0 \ldots & -\sin(\theta_l) & \cdots & \cos(\theta_l) & \ldots 0 \\ \vdots & \vdots & & \vdots & \ddots \ \vdots \\ 0 \ldots & 0 & \ldots & 0 & \ldots 1 \end{pmatrix} \begin{matrix} \\ \\ i_l \\ \vdots \\ j_l \\ \\ \\ \end{matrix} \tag{3}$$

the task of computing the optimal rotation W^* is iteratively performed by several sweep of the $\frac{n(n-1)}{2}$ rotations, each rotation $G(i, j, \theta_{i,j}^k)$ decreasing the criterion for the axis pair (i,j), $1 \le i < j \le n$ at sweep k. The whole rotation W^* is performed by multiplying the individual one. Note that $G(i,j,\theta^k_{i,j})$ do not commute and the product is written from rigth to left.

$$W^* = \prod_k \prod_{i=1}^{n-1} \prod_{j=i+1}^{n} G(i,j,\theta^k_{i,j})$$

3.1 Computing the Rotation $G(i,j,\theta^k_{i,j})$

For fixed (i,j), the optimal rotation $G(i,j,\theta^k_{i,j})$ is determined by the angle $\theta^k_{i,j}$. For updating Y, when multiplying by $G(i,j,\theta^k_{i,j})$, the components of Y remain unchanged excepted for rows i and j, the computing of the optimal angle $\theta^k_{i,j}$ is done on the reduced 2D data given by (4):

$$Y^k_{i,j} = \begin{bmatrix} (Y^k_{i,j})_1 \\ (Y^k_{i,j})_2 \end{bmatrix} = \begin{pmatrix} \cos(\theta^k_{i,j}) & \sin(\theta^k_{i,j}) \\ -\sin(\theta^k_{i,j}) & \cos(\theta^k_{i,j}) \end{pmatrix} \begin{pmatrix} Y(i,.) \\ Y(j,.) \end{pmatrix} \tag{4}$$

To simplify notation, we replace $\theta^k_{i,j}$, $(Y^k_{i,j})_1$, $(Y^k_{i,j})_2$ respectively by θ, Y_1 and Y_2.
The criterion on reduced 2D data is given by:

$$J(\theta) = \frac{1}{2}\left\|(Y^k_{i,j})^-\right\|^2_F = \frac{1}{2}\sum_l \left(Y^2_{1l}\mathbb{1}_{Y_{1l}<0} + Y^2_{jl}\mathbb{1}_{Y_{jl}<0}\right) \tag{5}$$

where $\mathbb{1}_{Y_x<0} = \begin{cases} 1 & \text{if } Y_x < 0 \\ 0 & \text{otherwise} \end{cases}$

Differentiating (5) with respect to θ and noticing that

$$Y_{1l} = Y(i,l)\cos(\theta) + Y(j,l)\sin(\theta) \Longrightarrow \frac{dY_{1l}}{d\theta} = -Y(i,l)\sin(\theta) + Y(j,l)\ cos(\theta) = Y_{2l}$$

$$Y_{2l} = -Y(i,l)\sin(\theta) + Y(j,l)\cos(\theta) \Longrightarrow \frac{dY_{2l}}{d\theta} = -Y(i,l)\cos(\theta) - Y(j,l)\sin(\theta) = -Y_{1l}$$

we get

$$\frac{dJ}{d\theta} = \sum_l Y_{1l}Y_{2l}\left[\mathbb{1}_{Y_{1l}<0}\mathbb{1}_{Y_{2l}>0} - \mathbb{1}_{Y_{1l}>0}\mathbb{1}_{Y_{2l}<0}\right] \tag{6}$$

and

$$\frac{d^2J}{d\theta^2} = \sum_l \left(Y^2_{2l} - Y^2_{1l}\right)\left[\mathbb{1}_{Y_{1l}<0}\mathbb{1}_{Y_{2l}>0} - \mathbb{1}_{Y_{1l}>0}\mathbb{1}_{Y_{2l}<0}\right] \tag{7}$$

A Newton method is used for computing θ leading to:

$$\theta = -\frac{dJ}{d\theta} / \frac{d^2J}{d\theta^2} \tag{8}$$

Note that $\frac{d^2J}{d\theta^2} = 0$ if all the samples are in the positive and/or the negative quadrant or all the samples are on the first and/or the second bisector. In this case it is not necessary to perform the rotation.

3.2 Proposed Algorithm

The Givens based parametrization/Newton optimization method is described in the following algorithm:

```
Start whith whitening the data
   Z=VX
Initialization
   W = I_{n}, Y = Z, k = 1
Begin
   Repeat
      For i=1 to n-1
         For j=i+1 to n
            compute dJ=dJ/dtheta as in (6)
            compute d2J=d^{2}J/dtheta^{2} as in (7)
            If (d2J==0)
               continue
            Else
              theta_min=-dJ/d2J, form G(i,j,theta_min) as in (3)
              W = G(i,j,theta_min) W, Y = G(i,j,theta_min) Y
            End
         End
      End
   k=k+1
   Until J(W) is less than a tolerance
End
```

4 Computational Complexity

The proposed algorithm complexity is evaluated by counting the number of floating point operations (flops). One flop corresponds to one multiplication followed by one addition. We suppose that we have p samples ($p >> n$), the whole computational complexity is computed by adding the individual complixity term and keeping only the term having p. For axis pair rotation methods, when updating W and Y by rotating axis (i, j), it is only necessary to update rows i and j.

Proposed Axis Pair Rotation method complexity for one sweep
1. $1_{Y_1<0} \rightarrow O(p)$
2. $1_{Y_2<0} \rightarrow O(p)$
3. $\frac{dJ}{d\theta} \rightarrow O(4p)$
4. $\frac{d^2J}{d\theta^2} \rightarrow O(4p)$
5. $\theta = -\frac{dJ}{d\theta} / \frac{d^2J}{d\theta^2} \rightarrow O(1)$
6. updating $W \rightarrow O(4n)$
7. updating $Y \rightarrow O(4p)$

The proposed algorithm has a complexity of $O(\frac{n(n-1)}{2}14p) = O(7n(n-1)p)$

Geodesic search method [15] complexity for one iteration

1. $Y^- = min(0, Y) \to O(np)$
2. $\delta = \frac{1}{2}\left\|Y^-Y^T - Y(Y^-)^T\right\|_F \to O(n^2p)$
3. $H = \left[Y^-Y^T - Y(Y^-)^T\right]/\delta \to O(n^2)$
4. $\frac{dJ}{dt} = -2\delta \to O(1)$
5. $\frac{d^2J}{dt^2} = \|KoHY\|_F + < Y^-, H^2Y > \to O(4n^2p)$
6. $t = -\arctan\left(\frac{dJ}{dt}/\frac{d^2J}{dt^2}\right) \to O(1)$
7. $B = tH \to O(n^2)$
8. $R = \exp(B) \to O(n^3)$
9. $W = RW \to O(n^3)$
10. $Y = RY \to O(n^2p)$

The geodesic search method has a complexity of $O((6n^2 + n)p)$.

Plumbley's Axis Pair Rotation method [2] complexity for one iteration

1. $Y^+ = max(0, Y) \longrightarrow O(np)$
2. $Y^- = Y - Y^+ \longrightarrow O(np)$
3. $G = Y^+(Y^-)^T - Y^-(Y^+)^T \longrightarrow O(n^2p)$
4. find (i, j) for which $|G_{ij}|$ is maximun $\longrightarrow O(n^2)$
5. using reduced 2D data, compute $J_{i,j} \longrightarrow O(2p)$
6. $\frac{dJ_{i,j}}{d\theta} = G_{ij} \longrightarrow O(1)$
7. $\theta = -J/\frac{dJ}{d\theta} \to O(1)$
8. $W = G(i, j, \theta)W \longrightarrow O(4n)$
9. $Y = G(i, j, \theta)Y \longrightarrow O(4p)$

Plumbley's axis pair rotation method has a complexity of $O((n^2 + 2n + 6)p)$.

Comparing to the Plumbley's axis pair rotation method, the proposed method seems to be more complex for one sweep, however before concluding one must take into account the total number of iterations necessary for each method to converge. Comparing to geodesic search, the proposed method have similar computational complexity and becomes better for low scale problem ($n \leq 8$).

5 Simulations and Discussions

The proposed algorithm is evaluated on synthetic data. We use $n = 8$ synthetic sources of $p = 1000$ samples as illustrated on Fig. 1. The source are mixed with Gaussian random matrix A giving the observations X. To measure the separation performance, we use two performance measures, the reconstruction error and separation error defined respectively by $E_{rec} = \frac{1}{np}\|Y^-\|_F^2$ and $E_{sep} = \frac{1}{n(n-1)}\left[\sum_i\left(\sum_j \frac{|(WVA)_{ij}|}{\max_k|(WVA)_{ik}|} - 1\right) + \sum_j\left(\sum_i \frac{|(WVA)_{ij}|}{\max_k|(WVA)_{kj}|} - 1\right)\right]$ Fig. 2. shows the source estimated by proposed axis pair rotation method, we can see that the source are successfully recovered. Fig.3. shows the reconstruction error versus iteration number. The proposed axis pair rotation method takes about 20

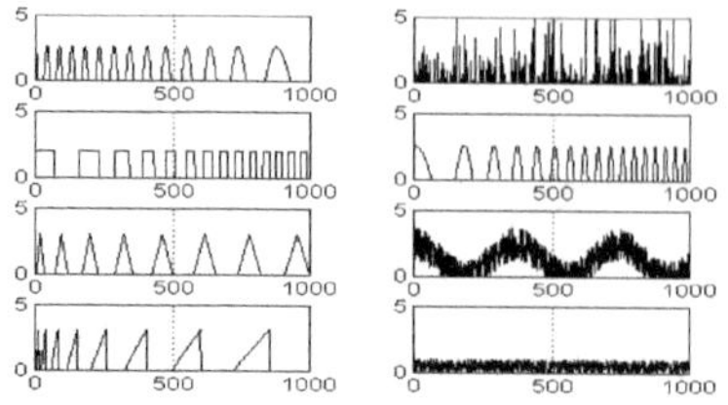

Fig. 1. Original sources

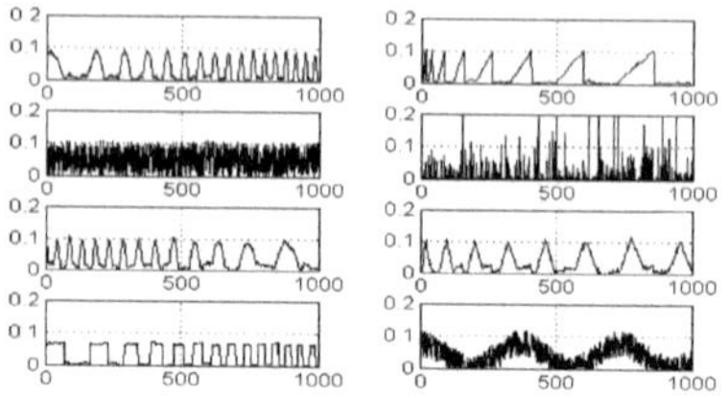

Fig. 2. Estimated sources using proposed axis pair rotation method

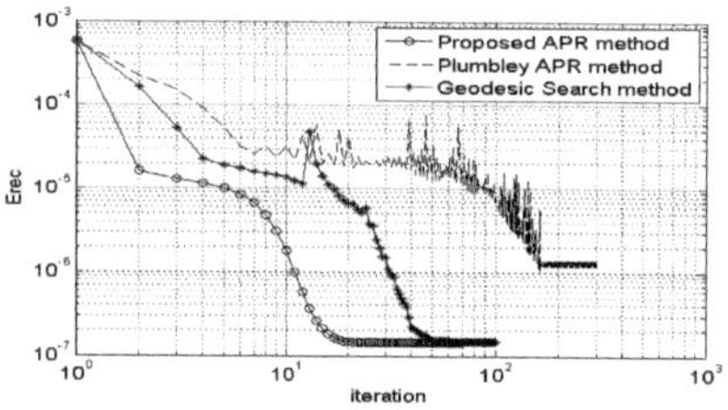

Fig. 3. Reconstruction error

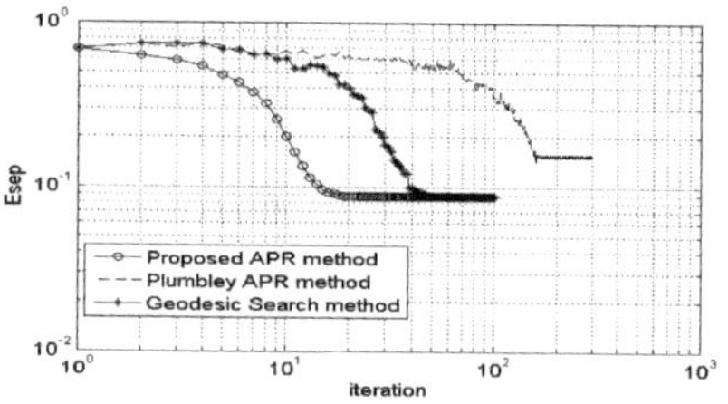

Fig. 4. Separation error

sweeps to reach the reconstruction error less than 2×10^{-7} while the geodesic search takes 60 iterations. The two methods seem to have same performance when converging. Plumbley's axis pair rotation method takes about 200 iterations to converge and is worser than proposed method and geodesic search one. The same result is observed on separation error (Fig. 4.).

6 Conclusions

We have considered the problem of Independent Component Analysis when the hidden sources are non-negative. Assuming that the sources are independent and well grounded, this problem is formulated as a non-linear cost function optimization over the rotation matrices group.

Using Givens 2D rotations and Newton optimization, we developed an efficient axis pair rotation method for non negative ICA. Calculating the computational complexity, we show that the proposed method is less complex than the geodesic search method especially for low scale problem. Simulations on synthetic data show that the proposed method is more efficient that axis pair method proposed by Plumbley.

As a future issue, it is interesting to investigate how to reduce the independence constraint to uncorrelation and how to relax the well grounded constraint for taking into account a great number of real applications. Considering noisy mixtures instead of (1) and more general case ($m \neq n$) can also be interesting.

References

1. Plumbley, M.: Conditions for Nonnegative Independent Component Analysis. IEEE Signal Processing Letters 9, 177–180 (2002)
2. Plumbley, M.: Algorithms for Nonnegative Independent Component Analysis. IEEE Transactions on Neural Networks 14, 534–543 (2003)
3. Comon, P.: Independent component analysis. A new concept? Signal Processing 36, 287–314 (1994)
4. Cardoso, J.F., Souloumiac, A.: Blind Beamforming for Non Gaussian Signal. IEE Proceedings-F 140, 362–370 (1993)
5. Hyvärinen, A.: Fast and Robust Fixed-Point Algorithms for Independent Component Analysis. IEEE Transactions on Neural Networks 10, 626–634 (1999)
6. Lee, D.D., Seung, H.S.: Algorithms for Non-negative Matrix Factorization. In: 14th Annual Neural Information Processing Systems Conference. NIPS, Denver (2000)
7. Lin, C.J.: Projected Gradient Methods for Non-negative Matrix Factorization. Neural Computation 19, 2756–2779 (2007)
8. Laurberg, H.: Uniqueness of Non-negative Matrix Factorization. In: IEEE Statistical Signal Processing Workshop, pp. 49–53 (2007)
9. Theis, F.J., Stadlthanner, K., Tanaka, T.: First Results on Uniqueness of Sparse Non-negative Matrix Factorization. In: EUSIPCO 2005 (2005)
10. Donoho, D., Stodden, V.: When Does Non-Negative Matrix Factorization Give a Correct Decomposition into Parts? Advances in Neural Information Processing Systems 16, 1141–1148 (2003)
11. Plumbley, M., Oja, E.: A "Nonnegative PCA" Algorithm for Independent Component Analysis. IEEE Transactions on Neural Networks 15, 66–76 (2004)
12. Li, Y., Zheng, H.: Improvement for Nonnegative PCA Algorithm for Independent Component Analysis. In: Neural Networks and Brain, ICNN&B, International Conference on, 2000–2002 (2005)
13. Cichocki, A., Zdunek, R., Phan, A.H., Amari, S.: Nonnegative Matrix and Tensor Factorizations Applications to Exploratory Multi-way Data Analysis and Blind Source Separation. John Wiley & Sons, Ltd., Chichester (2009)
14. Paatero, P.: Least Squares Formulation of robust non-negative factor analysis. Chemometrics and Intelligent Laboratory Systems 37, 23–35 (1997)
15. Plumbley, M.: Optimization using Fourier Expansion over a Geodesic for Non-Negative ICA. In: Puntonet, C.G., Prieto, A.G. (eds.) ICA 2004. LNCS, vol. 3195, pp. 44–56. Springer, Heidelberg (2004)
16. Plumbley, M.: Geometrical methods for non-negative ICA: Manifolds, Lie groups and toral subalgebras. Neurocomputing 67, 161–197 (2005)
17. Bronstein, A.M., Bronstein, M.M., Zibulevsky, M., Zeevi, Y.Y.: Sparse ICA for Blind Separation of Transmitted and Reflected Images. Wiley Periodicals, 84–91 (2005)
18. Hoyer, P.O.: Non-negative Matrix Factorization with Sparseness Constraints. Journal of Machine Learning Research 5, 1457–1469 (2004)
19. Moussaoui, S., Brie, D., Mohammad-Djafari, A., Carteret, C.: Separation of Non-Negative Mixture of Non-Negative Sources Using a Bayesian Approach and MCMC Sampling. IEEE Transactions on Signal Processing 54, 4133–4145 (2006)
20. Duarte, L.T., Jutten, C., Moussaoui, S.: A Bayesian Nonlinear Source Separation Method for Smart Ion-Selective Electrode Arrays. IEEE Sensors Journal 9, 1763–1771 (2009)

A New Geometrical BSS Approach for Non Negative Sources

Cosmin Lazar[1], Danielle Nuzillard[2], and Ann Nowé[1]

COMO Lab, Department of Computer Science, VUB, Pleinlaan 2,
B-1050 Brussels, Belgium,
CReSTIC, URCA, Moulin de la Housse, BP 1039, 51089 Reims cedex 2, France
{vlazar,ann.nowe}@vub.ac.be,
{danielle.nuzillard}@univ-reims.fr

Abstract. A new blind source separation method for non-negative sources based on geometrical evidences of the linear mixing model is presented. We show that the proposed method is able to find the mixing matrix as well as the original sources from an observation matrix under the assumption that for every source there is at least one instance where the underlined source is active and all the others are not. One major advantage of our proposal is that the number of sources is found automatically as being the number of extreme data in a set of points. Under the assumption mentioned above, our approach outperforms two well known implementations for NNMF BSS (ALS and multiplicative update algorithms).

Keywords: Blind source separation, convex hull, probability density function, extreme data.

1 Introduction

The idea of using geometry for blind source separation has been firstly introduced by Puntonet *et al.* in [1,2,3] and then by [4,5]. These methods have become quite popular due to their visual explanation making them easier to understand and implement. The first geometrical approach for BSS consists in finding the slopes of the parallelepiped enclosing all instances of the observation matrix. The method is investigated in the case of two sources and two observations and assumes that the sources have bounded *pdf*. In [6] a new geometrical algorithm is proposed based on the fact that a random vector is independent if and only if the Hessian of its logarithmic density (resp. characteristic function) is diagonal everywhere.

There are several works addressed to the BSS of non negative sources. Some of them are represented by non-negative matrix factorization algorithms (NNMF). Alternative least square (ALS) [7] and multiplicative updates [8,9] algorithms are only two examples of this class of NNMF approaches while Hoyer's implementation [10] incorporates sparseness constraints.

Another direction is nonnegative Independent Component Analysis (nICA) [11] which derives from more general BSS problems where signals can be negative. The non-negativity of sources for BSS has also been explored in [12] and [13]

V. Vigneron et al. (Eds.): LVA/ICA 2010, LNCS 6365, pp. 530–537, 2010.

based on Second Order Blind Identification (SOBI) [14] following by Alternative least square (ALS) iterations.

Most methods for NNMF approximate the mixing matrix as a product of two non negative matrices [8,9] by randomly initialisation, one serving as the estimate of the sources while the other represents the mixing matrix. The main drawback of these methods is that the solution is not unique, which may result in indeterminacy of the sources and mixing matrix. This literature survey is far from complete but from our knowledge there are few works [2,4,5,3] exploring the BSS problem of non-negative sources from a geometrical point of view. The novelty of our approach consists in the fact that we are looking for extreme points in the space of observations. Under certain week assumptions and without imposing any constraint on the *pdf* of sources the proposed method provides a unique solution for the separation of non-negative sources from non-negative observations regardless the number of sources or observations. Moreover, the number of sources is determined automatically as being the number of extreme points in the observation matrix projected on the unit hyperplane.

Our paper is organized as follows: in the next section we will show some geometrical insights with respect to the linear mixing model of non-negative sources on which our method is based. We will describe then our current proposal and finally the method is tested on an image separation application. Results are compared with those obtained by using the non-negative matrix factorization mentioned above.

2 Linear Mixing Model - Geometrical Insights

We deal with the linear mixing model from a geometrical point of view. For the sake of simplicity and visualisation reasons we limit our examples to a maximum number of 3 observations, where the number of sources equals either 2 or 3. We consider first the case of 2 sources with uniform distribution, non-negatives and bounded in the range [0 1], figure 1 (a). Now, every instance i in the source matrix $S = \begin{pmatrix} s_1(i) \\ s_2(i) \end{pmatrix}$ is represented by a point in the space defined by (s_1, s_2) and all points in this space are bounded by the cone defined by the segments $[(0,0)\ (0,1)]$ and $[(0,0)\ (1,0)]$. In the underlined space, the angle between these two segments has the maximal admitted value and all pairs which can be written as $(0, s_2^i), (s_1^j, 0)$ have this property. Note that, all pairs of segments defined by the origin and points inside the maximal cone have an angle smaller than the maximum value. The points lying on the edges of the maximal cone are called *extreme data*.

Obviously, the effect of the linear mixing model is a geometrical transformation of the original cloud of points, following the directions of the mixing vectors. This effect is shown in figure 1 (b) and (c). After transformation, all points will lie inside the maximal cone resulted after the mixing process and since we deal with a linear transformation, the extreme data will keep the *extreme data property*. Following we will show that under certain hypotheses, this extreme points

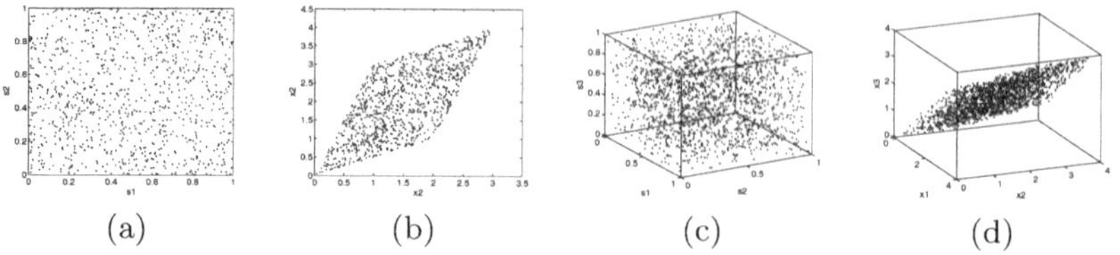

Fig. 1. Illustration of the linear mixing model on uniform sources : a) 2 uniform sources, b) 2 linear mixtures of 2 uniform sources, c) 3 uniform sources and d) 3 linear mixtures of 3 uniform sources

have the same direction as the mixing vectors. Figure 1 shows the geometry of the linear mixing model for 2 and 3 sources with uniform *pdf*.

In the following we express in mathematical terms the geometrical facts observed above. Let $X^{2\times n} = A^{2\times 2}S^{2\times n}$ be a non-negative observations matrix. We suppose that:

1. the sources s_1 and s_2 are extracted from a uniform *pdf* and $s_1 \geq 0$ and $s_2 \geq 0$,
2. $\exists$ at least one instance i such that $(s_1^i \neq 0)\&(s_2^i = 0)$ (the first source is active and the second one is not active),
3. $\exists$ at least one instance j such that $(s_1^j = 0)\&(s_2^j \neq 0)$ (the first source is not active and the second one is active).

If conditions 1, 2 and 3 are satisfied, we can state the following:

Proposition 1. *The extreme points in the observation matrix X have the same orientation as the mixing vectors (the columns of the mixing matrix A). Hence, one can retrieve the mixing matrix by finding the extreme points in the observation matrix. We remind that in a 2D observation space, a pair of points (X_k, X_l), $X_{k,l} \in X$ is said to be a set of extreme points if :*

$$\alpha(X_k, X_l) = \max \arccos \frac{X_k^T X_l}{\|X_k\|\|X_l\|}$$

Demonstration 1. *As we stated above, the linear mixing model consists in a geometrical transformation of the original cloud of data. Without loss of generality, one can project the original data or the sources either on the plane either on unity sphere. Obviously, the points resulting after the mixture of the normalized sources S_{norm} with a mixing matrix A will have the same orientation as the points resulted after the mixture of S with A as it is shown in figure 2 for 2D case and figure 3 for 3D case. The identification of mixing matrix columns is up to a scale factor and a permutation.*

Let $A^{(2\times 2)}$ be a non-negative mixing matrix $A = \begin{pmatrix} a_{11} & a_{12} \\ a_{21} & a_{22} \end{pmatrix}$, let $S^{(2\times n)}$ be the source matrix $S = \begin{pmatrix} s_1^1 \dots 0 \dots s_1^i \dots s_1^n \\ s_2^1 \dots s_2^j \dots 0 \dots s_2^n \end{pmatrix}$, and the observation matrix $X =$

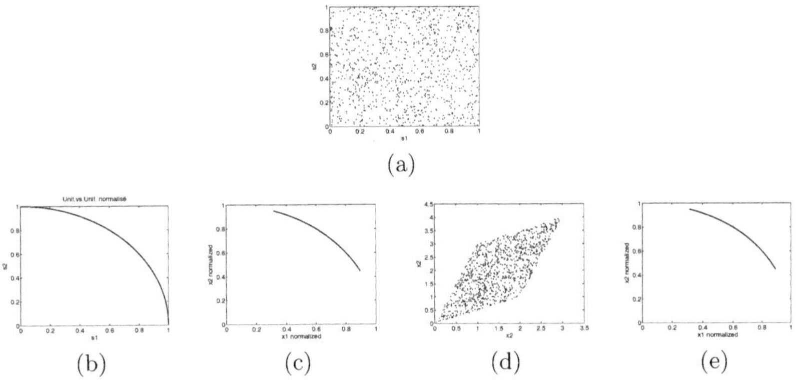

Fig. 2. First line : a) 2 sources with uniform *pdf*, b) 2 sources with uniform *pdf* projected on the unity disk, c) mixture of 2 sources with uniform *pdf* projected on the unity sphere, d) mixture of 2 sources with uniform *pdf*, e) projection on the unity disk of a mixture of 2 sources with uniform *pdf*

AS. In a detailed form :

$$X = \begin{pmatrix} x_1^1 \dots a_{12}s_2^j \dots a_{11}s_1^i \dots x_1^n \\ x_2^1 \dots a_{22}s_2^j \dots a_{21}s_1^i \dots x_2^n \end{pmatrix} \tag{1}$$

Now, the normalized source matrix (projected on the unity plane) is given by :

$$S_{norm} = \begin{pmatrix} s_{1norm}^1 \dots 0 \dots 1 \dots s_{1norm}^n \\ s_{2norm}^1 \dots 1 \dots 0 \dots s_{2norm}^n \end{pmatrix} \tag{2}$$

and the correspondent observation matrix X_{norm} is given by :

$$X_{norm} = \begin{pmatrix} x_{1norm}^1 \dots a_{12} \dots a_{11} \dots x_{1norm}^n \\ x_{2norm}^1 \dots a_{22} \dots a_{21} \dots x_{2norm}^n \end{pmatrix} \tag{3}$$

In the matrix X, equation 1, i-th and j-th columns represent the column-vectors of the mixing matrix, multiplied by a scale factor. In the matrix X_{norm}, equation 3, i-th and j-th columns are the column-vectors of the mixing matrix. Thus the mixing vectors can be determined as extreme points of the matrix X_{norm}. These points represent the pair of points (X_{inorm}, X_{jnorm}) for which the angle between them defined by $\arccos(X_{i\,norm}^T X_{j\,norm})$ is maximal among all the others pairs of points in the set X_{norm}. By substituting $X_{i\,norm}$ and $X_{j\,norm}$ by $\frac{X_i}{\|X_i\|}$ and by $\frac{X_j}{\|X_j\|}$, the proposition is demonstrated.

Now we are going to generalize the proposition 1 for p sources: let $X \in \mathbb{R}^{p \times n}$ be a observation matrix representing the linear mixing between a source matrix $S \in \mathbb{R}^{p \times n}$ and a non-negative mixing matrix $A^{(p \times p)}$. Supposing that:

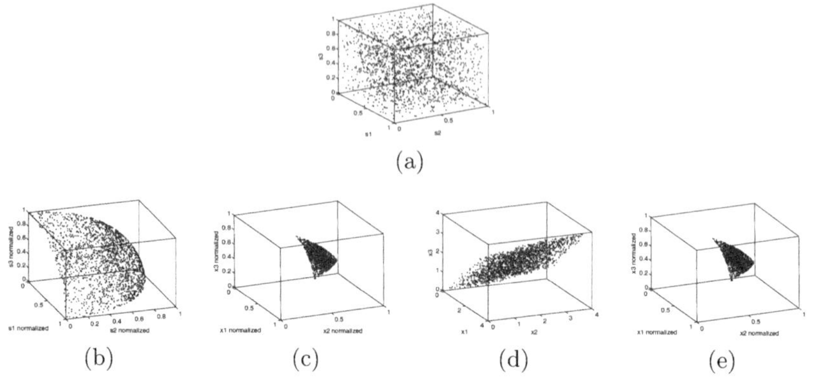

Fig. 3. First line : a) 3 sources with uniform *pdf*, b) 3 sources with uniform *pdf* projected on the unity disk, c) mixture of 3 sources with uniform *pdf* projected on the unity sphere, d) mixture of 3 sources with uniform *pdf*, e) projection on the unity disk of a mixture of 3 sources with uniform *pdf*

1. the line vectors s_i of the matrix S are drawn from any distribution, positively defined $s_i \geq 0$,
2. for each source i there exists an instant k where it is active and all the others are inactive $(s_i^k \neq 0) \& (s_j^k = 0)$, $j = 1 : p$, $j \neq i$.

If these conditions are satisfied then one can state the following proposition:

Proposition 2. *The extreme points of the observation matrix X represent the columns vectors of the mixing matrix A multiplied by a scale factor. They can be found as being the vertices of the convex hull which encloses the data points from the normalized observation matrix X_{norm}.*

3 The Algorithm

From the geometrical insights of the linear mixing model, we resume our approach to the search of the extreme data from a given set of points. Here we propose a solution to the underlined problem. First all points are projected on the unity hyperplane. Then we estimate the convex hull of the set of points in a multidimensional space. A convex hull is defined by its vertices and faces. Since all points are projected on the unity hyperplane, the vertices of the convex hull will be the extreme data of the original set of points. We will call our approach **Ex**treme **P**oint **B**lind **S**ource **S**eparation) (ExPBSS). For the estimation of the convex hull we have used the *quickhull algorithm* [15].

Given an observation matrix X obtained as a result of a linear mixture of sources and if the assumptions 1, 2 and 3 hold, our approach can be resume in 2 steps :

Algorithm 1. The pseudocode of the algorithme ExPBSS (**Ex**treme **P**oint **B**lind **S**ource **S**eparation)

1: Projection of the observations X_i on the unity hyperplane :

$$X_i = \frac{X_i}{\|X_i\|_{L_1}} \tag{4}$$

2: Estimate the convex hull of the new set of points using quick hull algorithm. The vertices of the convex hull will contain the extreme data of the original set of points and the origin which has been inserted in the previous step.

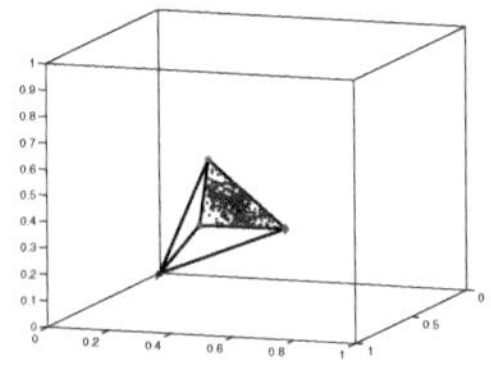

Fig. 4. Observation matrix X : the extreme points of the mixing matrix are actually the columns vectors of the mixing matrix.

4 Evaluation and Comparison

We evaluate our method for blind source separation of linear mixtures of grey scale images. We have used four images as sources which have been mixed with a random mixing matrix $A^{5\times 4}$ resulting a mixture matrix X with 5 observations. Source images as well as the observations resulted after the linear mixing process are shown in figure 5 while the recovered sources are shown in figure 6. Results are compared with those obtained by using two NMF algorithms - alternative least square (ALS) [7] and multiplicative updates using euclidean distance [8].

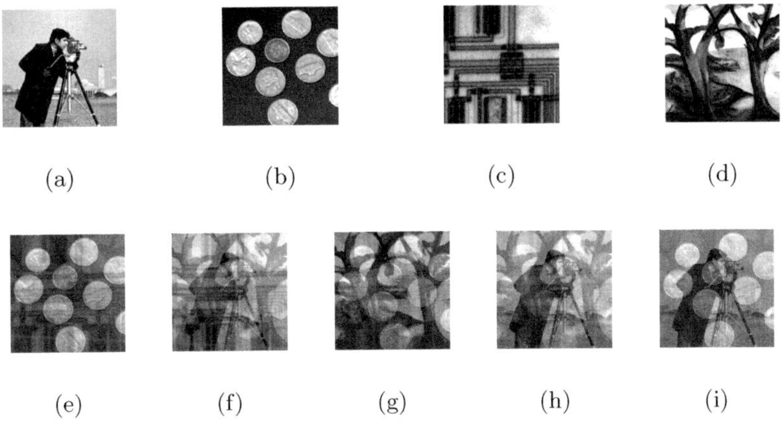

Fig. 5. First row: source images. Second row: observation images.

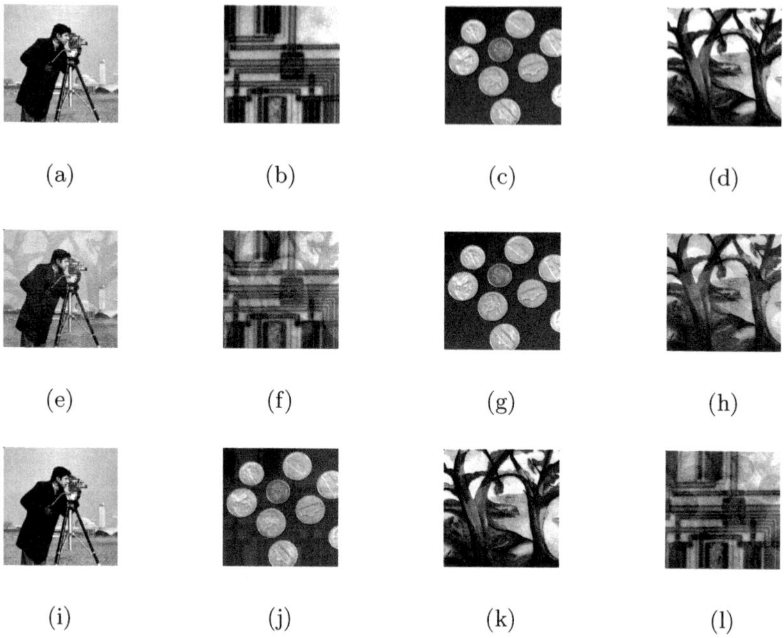

Fig. 6. First line - recovered images using ExPBSS; second line - recovered images using NMF (alternative least square algorithm); third line - recovered images using NMF (multiplicative updates algorithm). Both NMF algorithms have been run over 1000 iterations.

5 Conclusion

A new algorithm for blind source separation of non negative sources is described. The idea comes from geometrical insights of the linear mixing model. It has been shown that under certain quite week assumptions (*e.g.* for every source there exists at least one instance where it is active while all others are inactive), the mixing vectors have the same orientation as the extreme data in the normalized observation matrix. No assumption on the statistical properties of sources has been made. We propose to find the extremities of a set of points by finding the vertices of the convex hull which encloses the underlined distribution of points. Our algorithm provides a unique solution under the assumptions mentioned above. The number of sources is not previously known. It derives automatically as being the number of extreme data in the normalized observation matrix. We tested our algorithm in an application for image unmixing and results are very encouraging. The current approach outperforms two well known implementations for NNMF BSS (ALS and NMF with multiplicative update rules). Currently we are testing it to estimate its robustness to noise.

Acknowledgements

We would like to thank to Stijn Maganck - post-doctoral researcher at COMO Lab VUB and Jonatan Taminau - PhD student at COMO Lab VUB for useful discussions.

References

1. Puntonet, C.G., Pristo, A., Jutten, C., Rodriguez Alvarez, M., Ortega, J.: Separation of sources: A geometry - based procedure for reconstruction of n-valued signals. Signal Processing 46(3), 267–284 (1995)
2. Puntonet, C., Mansour, A., Jutten, C.: A geometrical Algorithm for Blind Separation of Sources. In: Gretsi 1995, Juan-les-Pins, September 1995, pp. 273–276 (1995)
3. Puntonet, C., Prieto, A.: An adaptive geometrical procedure for blind separation of sources. Neural Processing Letters 2(5), 23–27 (1995)
4. Babaie-Zadeh, M., Mansour, A., Jutten, C., Marvasti, F.: A geometrical approach for separating several speech signals. In: Puntonet, C.G., Prieto, A.G. (eds.) ICA 2004. LNCS, vol. 3195, pp. 798–806. Springer, Heidelberg (2004)
5. Babaie-Zadeh, M., Jutten, C., Mansour, A.: Sparse ICA via cluster-wise PCA. Neurocomputing (69), 1458–1466 (2006)
6. Theis, F.J.: A new concept for separability problems in blind source separation. Neural Comput. 16(9), 1827–1850 (2004)
7. Paatero, P.: Least-squares formulation of robust nonnegative factor analysis. Chemometrics Intelligent Lab. Syst. 37, 23–35 (1997)
8. Lee, D., Seung, H.: Learning the parts of objects by non-negative matrix factorization. Nature 401, 788–791 (1999)
9. Lee, D.D., Seung, H.S.: Algorithms for non-negative matrix factorization. Advances in Neural Information Processing 13 (2001)
10. Hoyer, P.O.: Non-negative Matrix Factorization with Sparseness Constraints. Journal of Machine Learning Research 5, 1457–1469 (2004)
11. Plumbley, M.D.: Algorithms for non-negative independent component analysis. IEEE Trans. Neural Netw. 14(3), 534–543 (2003)
12. Nuzillard, D.: Separation of non-orthogonal data, ICA 2000, HUT Helsinki, pp. 321–326 (June 2000)
13. Bonnet, N., Nuzillard, D.: Independent component analysis: A new possibility for analysing series of electron energy loss spectra. Ultramicroscopy 102(4), 327–337 (2005)
14. Belouchrani, A., Meraim, K., Cardoso, J.-F., Moulines, E.: A blind source separation technique using second order statistics. IEEE Trans. Signal Process. 45(2), 434–444 (1997)
15. Bradford Barber, C., Dobkin, D.P., Huhdanpaa, H.: The Quickhull Algorithm for Convex Hulls. ACM Transactions on mathematical software 22(4), 469–483 (1996)

Dependent Component Analysis for Cosmology: A Case Study

Ercan E. Kuruoglu

Instituto di Scienze e Tecnologie dell'Informazione (ISTI), "A. Faedo",
Italian National Council of Research (CNR), via G. Moruzzi 1, 56124, Pisa, Italy
ercan.kuruoglu@isti.cnr.it

Abstract. In this paper, we discuss various dependent component analysis approaches available in the literature and study their performances on the problem of separation of dependent cosmological sources from multichannel microwave radiation maps of the sky. Realisticaly simulated cosmological radiation maps are utilised in the simulations which demonstrate the superior performance obtained by tree-dependent component analysis and correlated component analysis methods when compared to classical ICA.

Keywords: Dependent component analysis, cosmic microwave background radiation, cosmological source separation.

1 Introduction

Independent component analysis has found various applications in diverse areas ranging from audio processing to fMRI and from astrophysical image processing to telecommunications. The central assumption of ICA is independence. In almost all published work, the approach is to form a reconstruction of the sources by enforcing the independence of the sources either by maximizing non-Gaussianity or minimizing the mutual information between sources.

While the independence assumption provides us with means of developing mathematical means to solve the source separation problem, its validity is highly questionable. In audio signals, the component signals from different instrument that make up the mixture have a certain degree of correlation with other instrument signals [1]. In the case of fMRI, many activation patterns in brain signals show significant correlation and are difficult to isolate independent components. Recent work on fMRI aim at clustering dependent components together [2]. Similarly, in gene expression data, it is difficult to isolate components which are independent from each other. Several genes seem to be functioning in combination with each other and a simple ICA approach does not lead to meaningful components [3].

Yet another area where the components that make up the mixing show dependence among themselves are astronomy images. The images made by satellite missions at microwave ranges over various frequency channels contain cosmological components which are generally assumed to be independent. However, some

V. Vigneron et al. (Eds.): LVA/ICA 2010, LNCS 6365, pp. 538–545, 2010.

of these cosmological components have related generation mechanisms and therefore are statistically dependent and therefore, an ICA approach cannot provide a successful model for isolation of each source. In this paper, we will consider this problem and test a number of dependent component analysis algorithms available in the literature.

In the next section, background information is provided on the cosmological source separation problem. In Section 3, we present some of the existing dependent component analysis techniques in the literature, namely multivariate ICA, topographic ICA, correlated component analysis and tree-dependent component analysis. In Section 4, a simulation study is provided comparing the performances of the methods discussed in the previous section.

2 Cosmological Component Separation

2.1 CMB

The most important source among those that make up the celestial microwave radiation is the cosmic microwave radiation and is the principal objective of the WMAP and Planck satellite missions. It is widely accepted that CMB is Gaussian [4]. It is also widely expected to be stationary, and that CMB anisotropies can be represented as the multiplication of a spatial template with a nonlinear function of the frequency.

2.2 Galactic Components

There are three main galactic radiations: synchrotron, galactic dust and free-free emission.

Synchrotron radiation is generated by the electrons accelerating along magnetic fields. Although synchrotron radiation originates in the galaxy, it extends also to outside the galactic plane and is less concentrated in the galactic plane when compared to other galactic foreground components.

Galactic dust is made up of small particles which range from the order of nanometers to micrometers. Their radiation is dominant especially in very high frequency channels.

Free-free emission or "Bremsstrahlung" emission is caused by the collision of free electrons with heavy ions in the ionised medium. Electrons loose energy in these collisions and emit photons.

2.3 Extragalactic Sources

The main extragalactic sources are the Sunyaev-Zeldovich effect and the point sources. The Sunyaev-Zeldovich (SZ) effect is generated with the inverse Compton scattering of photons from CMB on electrons. They can be observed in the presence of a cluster of galaxies although they can also be caused by any large body with hot ionised gas. The SZ effect is of major cosmological importance since it can help determine the value of Hubble's constant.

Point sources are caused by distant stars or galaxies which appear as localised, impulsive bursts of radiation. Unlike the sources discussed above, they are not diffuse sources: it is not possible to consider templates that scale in different frequencies and each channel needs to be considered separately. Due to these properties, the general approach in the literature is to detect and remove them from radiation maps before starting the component separation task.

For more detailed information on cosmological sources, the reader is referred to [6].

2.4 Observation Model

Assuming that the effects of the telescope beam on the angular resolution at different measurement channels have been equalized, the observed signal at M different frequencies can be modelled as

$$\mathbf{x}(\xi,\eta) = \mathbf{A}\mathbf{s}(\xi,\eta) + \mathbf{e}(\xi,\eta) \tag{1}$$

where $\mathbf{x} = \{x_i, i = 1, ..., M\}$ is the M-vector of the observations, i being a channel index, $\mathbf{A}$ is an $M \times N$ mixing matrix, $s = s_c, c = 1, ..., N$ is the N-vector of the individual source processes, ξ and η are angular coordinates on the celestial sphere and $e = e_i, i = 1, ..., M$ is the M-vector of instrumental noise.

3 Dependent Source Separation Methods

A common assumption among works in the literature is the independency of the cosmological sources. Although it is well known that CMB is independent from the rest of the sources, the galactic sources demonstrate significant statistical dependence among themselves. Recently a small number of researchers have started addressing this problem [9]. However, other dependent component analysis methods exist in the literature and we discuss them next.

3.1 Multidimensional ICA

Multidimensional ICA (MICA) [7], whihc is one of the first independent subspace analysis methods, transforms the concept of separation of independent sources to the separation of the observations into independent subspaces by the reformulation:

$$\mathbf{x} = \mathbf{A}\mathbf{s} = \sum_{p=1}^{m} \mathbf{x}_p \tag{2}$$

where $\mathbf{x}_p = s_p\mathbf{a_p}$, s_p are the individual sources and $\mathbf{a_p}$ are the corresponding columns of the mixing matrix $\mathbf{A}$. All components $\mathbf{x}_p$ belong to the space generated by the corresponding column of $\mathbf{A}$, i.e. $x_p \in Span\,(a_p)$. Therefore, each $\mathbf{x_p}$ is the orthogonal projection of $\mathbf{x}$ onto the space defined by $\mathbf{a_p}$ and the projection operator is $\mathbf{\Pi}_p = \frac{a_p a_p^t}{\|a_p\|^2}$.

Although, this is a simple reparametrization of ICA, it provides an additional possibility, that sources which originally belong to the same space, and hence not independent, can be accommodated by the vector (multivariate) extension of the formulation in (2) allowing s_p to be a vector that contains the dependent sources belonging to the same subspace . In that case, the projection matrix to recover the components becomes: $\boldsymbol{\Pi}_p = \mathbf{A_p}\left(\mathbf{A_p^t A_p}\right)^{-1}\mathbf{A_p^t}$ where $\mathbf{A_p}$ is such that $\mathbf{x_p} = \mathbf{A_p s_p}$. The subspace components can be calculated from:

$$x_p = \tilde{\Pi}_p x = \Pi_p \left(\sum_{q=1}^{m} \Pi_q\right)^{-1} x. \tag{3}$$

One important drawback of this method is that one can recover components coming from independent subspaces although we cannot uniquely determine the dependent components that compose independent subspace component in (2).

3.2 Topographic ICA

Topographic ICA (TICA) [8] can be considered as an extension of independent subspace analysis. TICA aims at providing an auxiliary variable to model the dependence between sources. Each source is expressed as the multiplication of a variance (scale parameter) and an auxiliary variable, i.e. $s_i = z_i\sigma_i$. The variances which are random variables themselves are utilized as a tool to model the interaction or a topographic relation between sources as in:

$$\sigma_i = \phi\left(\sum_{k}^{n} h(i,k)u_k\right) \tag{4}$$

where $h(i,k)$ defines the topographic relation, ϕ is some nonlinear function and u_k independent components used to generate the variances.

The topography is desired to be such that components s_i and s_j that are far from each other are almost independent. That is s_i and s_j have almost non-overlapping neighbourhoods. If on the contrary s_i and s_j have significant dependence then they should be near according to the chosen topography, that is $h(i,j)$ has a significant non-zero value. Clearly based on the choice of h, ϕ and u, a wide range of nonlinearities can be modelled. A simple choice is: $h(i,j) = 1, |i-j| < K$ and zero elsewhere. However, the likelihood function obtained based on this model is very complicated and is very difficult to maximize. Hyvarinen et al. therefore utilize approximations to it and learn the model utilizing a gradient descent search algorithm.

TICA assumes that the sources are uncorrelated, however have higher order dependencies. Therefore, second order dependent (correlated) components are not handled.

3.3 Correlated Component Analysis

Bedini et al. [9,5] proposed a second order statistics based method, namely Correlated Component Analysis (CorCA) that models also the correlation between galactic components. The algorithm utilizes a method based on second order statistics as in SOBI; however, it does not assume the covariance between different sources to be zero.

The technique exploits the fact that the parametrization with spectral indices reduces the number of unknowns in the mixing matrix, which compensates for the increase in the number of parameters in the covariance matrix due to the correlations between some of the sources. To the best of our knowledge, this is the only work in the literature that explicitly aimed at separating dependent cosmological components. The drawback of the method is that it models on second order dependence (covariance) between sources, while we do not have any indications of linear dependence between galactic sources.

3.4 Tree-Dependent Component Analysis

While TICA utilizes a topography map for modelling the dependence and diversity between sources, Tree-Dependent Component Analysis (TDCA), utilizes tree structures for modelling the conditional probabilities between different sources [10]. In this sense, TDCA can be considered as an extension of classical ICA where one looks for a linear transform for converting the observations to independent sources to where one looks for a tree structure that best fits the graphical model described by the conditional densities.

Let's consider a tree structure defined by $T(\mathcal{V}, \mathcal{E})$, where $\mathcal{V}$ is the set of nodes and $\mathcal{E} \subset \mathcal{V} \times \mathcal{V}$ is the set of branches. We can associate every node $v \in \mathcal{V}$ with a random variable s_v . The topology of the graph characterizes the existing relations between various random variables. Every branch in the graph indicates that the nodes connected by the branch are dependent random variables. To model a probability density on a tree, we define the probability density functions on the branches and the nodes as $p_{uv}(x_u, x_v)$ and $p_u(x_u)$ for $(u,v) \in \mathcal{E}$ and $u \in \mathcal{V}$. The joint probability density over this tree is defined as ([10]):

$$p(x) = \prod_{(u,v)\in\mathcal{E}} \frac{p_{uv}(x_u, x_v)}{p_u(x_u)\, p_v(x_v)} \prod_{u\in\mathcal{V}} p_u(x_u) \tag{5}$$

where D is a normalisation constant. Fixing a tree T and varying the potentials we obtain a family of densities which factorise on T.

In TDCA, the aim is to model the variable $\mathbf{x}$ using the model $\mathbf{x} = \mathbf{A}\mathbf{s}$, where $\mathbf{A}$ is an invertible mixing matrix and $\mathbf{s}$ factorizes in a tree T. Let $\mathbf{W} = \mathbf{A}^{-1}$, and $D^{W;T}$ denote the set of all distributions that factorise on T. If $x \in \mathbb{R}^m$ has the probability density p_x, the minimum of the Kullback-Leibler divergence between p_x and a density $q_x \in \mathcal{D}^{W,T}$ is equal to the T-mutual information of $\mathbf{s} = \mathbf{W}\ \mathbf{x}$, that is [10]:

$$J(x, W, T) = \min_{q\in D^{W,T}} D(p \parallel q) = I^T(s) = I(s_1, \ldots, s_m) - \sum_{(u,v)\in\mathcal{E}} I(s_u, s_v) \tag{6}$$

This expression provides us with a cost function $J(x, W, T)$, which can be minimised with respect to $\mathbf{W}$ and T to learn the model parameters, i.e. the tree structure, the conditional probabilities between sources and the mixing matrix.

4 Simulation Study

In this section, we present results on realistic simulations of astrophysical data. In these simulations, we compare the performances of FastICA, multidimensional ICA, Topographic ICA and correlated component analysis and tree-dependent component analysis. We use a realistic mixing matrix as reported in [11]. The sources we utilise are CMB, free-free emission, galactic dust and synchrotron which are realistic simulations of those expected to be obtained by the Planck mission. It is well known that CMB is independent from other sources while, the galactic components have dependence among them.

The observation channels we consider are 30, 44, 70, 100 GHz in accordance with 4 of 9 Planck satellite mission channels. The patches we use in this simulation are from outside the galactic plane (about 20 degrees off galactic axis).

We first compare the Amari divergences for various source separation methods. The Amari divergence is for a matrix $\mathbf{A}$ given by

$$\mathcal{D}(B) = \sum_{i=1}^{n} \left(\frac{\sum_{j=1}^{n} |B_{ij}|}{\max_j |B_{ij}|} - 1 \right) + \sum_{j=1}^{n} \left(\frac{\sum_{i=1}^{n} |A_{ij}|}{\max_i |B_{ij}|} - 1 \right) \tag{7}$$

where $\mathbf{B} = WA$ and $W = \hat{A}^{-1}$. Amari divergence is a measure of the distance between the estimated matrix and the original matrix. It also takes care of the scale and permutation ambiguities in the estimate of the mixing matrix. As is clear from Table 1, FastICA shows the worst performance in estimating the mixing matrix while CorCA has significantly better performance from the rest. MICA provides only moderate improvement over ICA, while TICA and TDCA perform significantly better. One should keep in mind when evaluating the superior performance of CorCA that, the technique makes use of astrophysical prior knowledge to reparametrise the mixing matrix which significantly reduces the number of unknowns.

Next, we give the source maps obtained by each technique in Figure 1. The superior performance of TDCA and CorCA is clear. Other techniques have difficulty in recovering one or more galactic sources while they all seem to be performing well for CMB. A possible explanation is that the patch used in the

Table 1. Amari divergences for of mixing matrix estimates obtained by FastICA, multivariate ICA, topographic ICA, tree-dependent component analysis, correlated component analysis

	Amari divergence
FastICA	6.7795
MICA	6.4167
TICA	4.7032
TDCA	3.4024
CorCA	2.0632

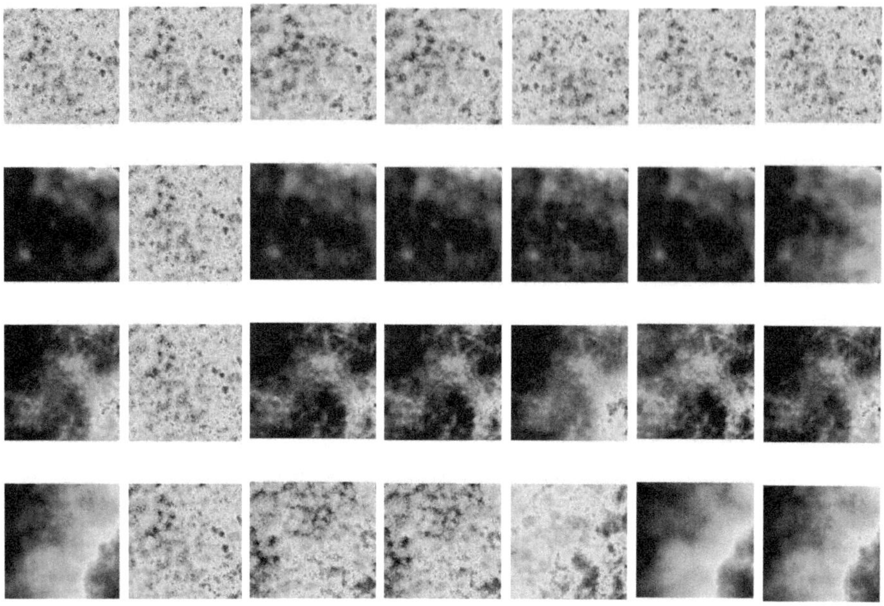

Fig. 1. 1st column: original sources (CMB, free-free emission, dust and synchrotron); 2nd column: observations, 3rd column: estimates by FastICA, 4th column: estimates by multidimensional ICA, 5th column: estimates by topographic ICA, 6th column: estimates by tree-dependent CA, 7th column: estimates by correlated CA.

simulation comes from outside the galactic plane where the galactic sources are weak and CMB is dominant. On a patch from the galactic plane we expect improvement in galactic source estimates and deterioration in CMB estimates.

Finally, the mutual information estimated between the sources by TDCA are given in Table 2. As expected, TDCA finds very small dependence (not zero due to finite sample size) between CMB and other sources, while significant dependence between synchrotron and galactic dust. Non-zero dependences are also detected between free-free emission and dust and synchrotron.

Table 2. Mutual information estimated by TDCA between sources on a patch outside the galactic plane

Mutual Information	Dust	Free-free	Synchrotron	CMB
Dust	-	0.0931	0.2864	0.0174
Free-free	0.0931	-	0.2514	0.0391
Synchrotron	0.2864	0.2514	-	0.0403
CMB	0.0174	0.0391	0.0403	-

5 Conclusions

We have studied various dependent component separation methods in the test case of cosmological source separation where galactic components possess dependence among them. The improvement of separation performance over classical ICA is very significant. Among DCA methods, MICA showed relatively less improvement while TDCA and CorCA have proved to be the most successful. The reason behind the success of TDCA is capturing successfully the dependence structure while behind the success of CorCA, it is the significant reduction of number of unknowns by utilising prior astrophysical information. Incorporating this prior information in TDCA can lead to better results. The modest performance of TICA can be improved by a more elaborated choice of the neighbourhood function. In this work, we have ignored the antenna noise or the beam affects. In future work, these practical issue also need to be considered.

References

1. Casey, M.A., Westner, A.: Separation of Mixed Audio Sources By Independent Subspace Analysis. In: Int. Computer Music Conference. U. Michigan Press, Ann Arbor (2000)
2. Savia, E., Klami, A., Kaski, S.: Fast dependent components for fMRI analysis. In: ICASSP 2009. IEEE Press, New York (2009)
3. Kim, J.K., Choi, S.: Tree-Dependent Components of Gene Expression Data for Clustering. In: Kollias, S.D., Stafylopatis, A., Duch, W., Oja, E. (eds.) ICANN 2006. LNCS, vol. 4132, pp. 837–846. Springer, Heidelberg (2006)
4. Hu, W., Dodelson, S.: Cosmic microwave background anisotropies. Annual Reviews Astronomy Astrophysics 40, 171–216 (2002)
5. Bonaldi, A., Bedini, L., Salerno, E., Baccigalupi, C., De Zotti, G.: Estimating the spectral indices of correlated astrophysical foregrounds by a second-order statistical approach. Monthly Notices of the Royal Astronomical Society 373, 271–279 (2006)
6. ESA, Planck surveyor homepage, `http://www.rssd.esa.int/index.php?project=PLANCK`
7. Cardoso, J.F.: Multidimensional independent component analysis. In: Proc. ICASSP 1998, Seattle, WA (1998)
8. Hyvarinen, A., Hoyer, P.O., Inki, M.: Topographic Independent Component Analysis. Neural Computation 13(7), 1527–1558 (2001)
9. Bedini, L., Herranz, D., Salerno, E., Baccigalupi, C., Kuruoglu, E.E., Tonazzini, A.: Separation of correlated astrophysical sources using multiple-lag data covariance matrices. EURASIP J. Applied Signal Processing (15), 2400–2412 (2005)
10. Bach, F., Jordan, M.: Beyond independent components: trees and clusters. Journal of Machine Learning Research 4(7-8), 1205–1233 (2004)
11. Baccigalupi, C., et al.: Neural networks and the separation of cosmic microwave background and astrophysical signals in sky maps. Monthly Notices of the Royal Astronomical Society 318, 769–780 (2000)

A Time-Frequency Technique for Blind Separation and Localization of Pure Delayed Sources

Dimitri Nion, Bart Vandewoestyne, Siegfried Vanaverbeke,
Koen Van Den Abeele, Herbert De Gersem, and Lieven De Lathauwer*

K. U. Leuven Campus Kortrijk, Group Science, Engineering and Technology,
Etienne Sabbelaan 53, 8500 Kortrijk, Belgium
{Dimitri.Nion,Bart.Vandewoestyne,Siegfried.Vanaverbeke,Koen.VanDenAbeele,
Herbert.DeGersem,Lieven.DeLathauwer}@kuleuven-kortrijk.be

Abstract. In this paper we address the problem of overdetermined blind separation and localization of several sources, given that an unknown scaled and delayed version of each source contributes to each sensor recording. The separation is performed in the time-frequency domain via an Alternating Least Squares (ALS) algorithm coupled with a Vandermonde structure enforcing strategy across the frequency mode. The latter allows to update the delays and scaling factors of each source with respect to all sensors, up to the ambiguities inherent to the mixing model. After convergence, a reference sensor can be chosen to remove these ambiguities and the Time Difference of Arrival (TDOA) estimates can be exploited to localize the sources individually.

1 Introduction

Assume that N unknown sources $s_n(t)$, $n = 1, \ldots, N$, are simultaneously and isotropically broadcasting in an anechoic propagation environment. The noise-free signal $r_m(t)$ received by the mth sensor, $m = 1, \ldots, M$, is

$$r_m(t) = \sum_{n=1}^{N} a_{mn} s_n(t - \tau_{mn}) = \sum_{n=1}^{N} (a_{mn} \delta(t - \tau_{mn})) \star s_n(t), \tag{1}$$

where $a_{mn} \in \mathbb{R}$ and $\tau_{mn} \in \mathbb{R}^+$ are the attenuation factor and the propagation delay (in seconds), respectively, between the nth source and the mth sensor, $\delta(t)$ is the Dirac impulse function and $\star$ is the linear convolution operator. The separation, time delays estimation and localization of several source signals propagating in an open-space acoustics environment is an important problem in signal

* Research supported by: (1) Research Council K.U.Leuven: GOA-Ambiorics, GOA-MaNet, CoE EF/05/006 Optimization in Engineering (OPTEC), CIF1, STRT1/08/023, (2) F.W.O.: (a) projects G.0321.06 and G.0427.10N, (b) Research Communities ICCoS, ANMMM and MLDM, (3) the Belgian Federal Science Policy Office: IUAP P6/04 (DYSCO, "Dynamical systems, control and optimization", 2007–2011), (4) EU: ERNSI.

V. Vigneron et al. (Eds.): LVA/ICA 2010, LNCS 6365, pp. 546–554, 2010.

processing, finding applications in seismics, biomedicine, sonar, radar and communications. For existing methods to handle this problem, we refer to [4,8,1,5]. In this paper, we will propose a new separation technique that belongs to the class of time-frequency algorithms, as the extended AC-DC algorithm of [8]. However, contrarily to the latter method, our approach is not limited to the case of two sources. Moreover, we do not use the second-order statistics of the observed signals. Instead, we exploit the algebraic structure of the data, by embedding a Vandermonde structure enforcing strategy within an ALS updating scheme.

Notation. The pseudo-inverse of a matrix $\mathbf{Y}$ is denoted by $\mathbf{Y}^{\dagger}$, its transpose by $\mathbf{Y}^T$ and its Frobenius norm by $\|\mathbf{Y}\|$. The diagonal matrix $\text{diag}(\mathbf{y})$ holds the entries of $\mathbf{y}$ on its diagonal and $\text{diag}(\mathbf{Y})$ is the vector consisting of the diagonal entries of the square matrix $\mathbf{Y}$. We will also use a Matlab-type notation for matrix sub-blocks, i.e., $[\mathbf{A}]_{l:m,:}$ represents the matrix built after selection of $m-l+1$ rows of $\mathbf{A}$, from the lth to the mth, and all columns of $\mathbf{A}$. The mode-2 product of a third-order tensor $\mathcal{Y} \in \mathbb{C}^{L\times M\times N}$ by a matrix $\mathbf{B} \in \mathbb{C}^{J\times M}$, denoted $\mathcal{Y} \bullet_2 \mathbf{B}$, is an $(L \times J \times N)$ tensor defined, for all index values, by $(\mathcal{Y} \bullet_2 \mathbf{B})_{ljn} = \sum_{m=1}^{M} y_{lmn} b_{jm}$.

2 Time-Frequency Formulation

Let F_s denote the sampling frequency and the $N_s \times 1$ vectors $\mathbf{r}_m$ and $\mathbf{s}_n$ denote the discrete-time versions of $r_m(t)$ and $s_n(t)$, respectively. Consider a partition of these vectors into P (possibly overlapping) frames of F samples each and compute the Discrete Fourier Transform (DFT) of each frame, to get a collection of PF time-frequency samples $\underline{r}_m(p, f)$, $p = 1, \ldots, P$, $f = 1, \ldots, F$, for each sensor. From the Fourier transform shift-theorem, the time-frequency discrete version of (1) can be written as

$$\underline{r}_m(p, f) \simeq \sum_{n=1}^{N} a_{mn} \omega^{(f-1)D_{mn}} \underline{s}_n(p, f), \quad f = 1, \ldots, F, \tag{2}$$

where $\omega = \exp(-2j\pi/F)$ and $D_{mn} = F_s \tau_{mn}$ is the Time Of Arrival (TOA), in number of samples, between source n and sensor m. Note that the approximation (2) is exact only for periodic signals $s_n(t)$, or equivalently, if the time-convolution is circular. This approximation is satisfactory if F is significantly larger than the maximum delay [6]. To limit the circularity effect, a spectral smoothing approach is commonly used. In practice, we will compute the DFT of consecutive overlapping windowed frames (a Hanning window will be used). The time-frequency model (2) can be written as

$$\underline{\mathrm{R}}(f) = \mathbf{H}(f) \cdot \underline{\mathrm{S}}(f), \quad f = 1, \ldots, F, \tag{3}$$

where $[\underline{\mathrm{R}}(f)]_{m,p} \stackrel{\text{def}}{=} \underline{r}_m(p, f)$, $[\underline{\mathrm{S}}(f)]_{n,p} \stackrel{\text{def}}{=} \underline{s}_n(p, f)$, $[\mathbf{H}(f)]_{m,n} \stackrel{\text{def}}{=} a_{mn}\omega^{(f-1)D_{mn}}$. Let the third-order tensor $\mathcal{H} \in \mathbb{C}^{M\times N\times F}$ be defined as $[\mathcal{H}]_{m,n,f} \stackrel{\text{def}}{=} [\mathbf{H}(f)]_{m,n}$. For any sensor-source pair (m,n), the vector

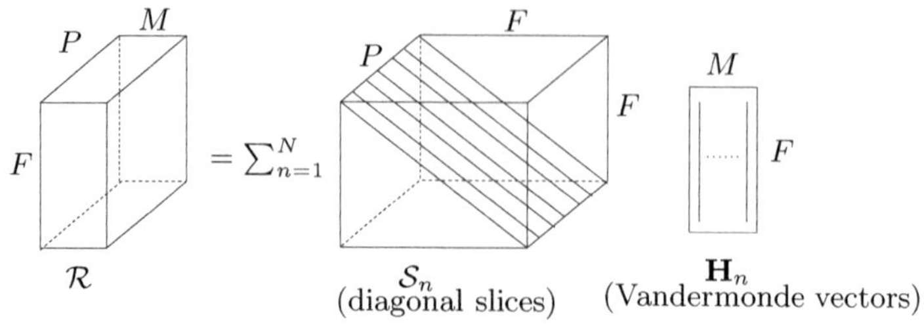

Fig. 1. Tensor view of the problem

$$\mathbf{h}_{mn} \stackrel{\text{def}}{=} [\mathcal{H}]_{m,n,1:F} = [a_{mn}, a_{mn}\omega^{D_{mn}}, \ldots, a_{mn}\omega^{(F-1)D_{mn}}]^T, \tag{4}$$

is a Vandermonde vector. This specific structure will be enforced on its estimate $\hat{\mathbf{h}}_{mn}$ at each step of the iterative algorithm proposed in Section 4. In the following, we will work under the following assumptions:

(A1) $P \geq N$ and $M \geq N$, i.e., we work in the overdetermined case,

(A2) $\mathbf{H}(f)$ and $\underline{\mathrm{S}}(f)$ are rank-N, for $f = 1, \ldots, F$, which is generically satisfied in practice.

3 Model Ambiguities

Let $\mathcal{R} \in \mathbb{C}^{F \times M \times P}$ and $\mathcal{S}_n \in \mathbb{C}^{F \times F \times P}$ denote the third-order tensors defined by $[\mathcal{R}]_{f,m,p} \stackrel{\text{def}}{=} \underline{\mathrm{r}}_m(p, f)$ and $[\mathcal{S}_n]_{:,:,p} \stackrel{\text{def}}{=} \mathrm{diag}([\underline{\mathrm{s}}_n(p,1), \underline{\mathrm{s}}_n(p,2), \ldots, \underline{\mathrm{s}}_n(p,F)])$, respectively. Let $\mathbf{H}_n \in \mathbb{C}^{F \times M}$ be the channel Vandermonde matrix associated to the nth source, such that $[\mathbf{H}_n]_{:,m} = \mathbf{h}_{mn}$, i.e., $\mathbf{H}_n$ is a slice of $\mathcal{H}$ obtained by fixing the source index to n. It follows that the time-frequency mixing models (2) and (3) can be written in tensor format (see Fig. 1) as

$$\mathcal{R} = \sum_{n=1}^{N} \mathcal{S}_n \bullet_2 \mathbf{H}_n^T. \tag{5}$$

Even in case of perfect separation, i.e., when the contribution $\mathcal{H}_n \stackrel{\text{def}}{=} \mathcal{S}_n \bullet_2 \mathbf{H}_n^T$ of each source to the observed tensor $\mathcal{R}$ is perfectly estimated, it is clear that, for any arbitrary non-singular matrices $\mathbf{Z}_n \in \mathbb{C}^{F \times F}$, $n = 1, \ldots, N$, Eq. (5) is equivalent to

$$\mathcal{R} = \sum_{n=1}^{N} (\mathcal{S}_n \bullet_2 \mathbf{Z}_n^{-1}) \bullet_2 (\mathbf{H}_n^T \cdot \mathbf{Z}_n). \tag{6}$$

However, for the tensor $(\mathcal{S}_n \bullet_2 \mathbf{Z}_n^{-1})$ to have the same structure as $\mathcal{S}_n$, i.e., for the P slices of $(\mathcal{S}_n \bullet_2 \mathbf{Z}_n^{-1})$ to be diagonal matrices, the matrix $\mathbf{Z}_n$ has to be diagonal. Moreover, for the Vandermonde structure of $\mathbf{H}_n$ to be preserved, the vector $\mathbf{u}_n \stackrel{\text{def}}{=} \mathrm{diag}(\mathbf{Z}_n)$ has to be a Vandermonde vector. In other words, if the respective structures of $\mathcal{S}_n$ and $\mathbf{H}_n$ are enforced on their respective estimates in the computational strategy, the remaining ambiguity on $\hat{\mathbf{H}}_n$ in case of perfect separation is

$$\hat{\mathbf{H}}_n = \mathrm{diag}([\alpha_n, \alpha_n \omega^{\phi_n}, \ldots, \alpha_n \omega^{(F-1)\phi_n}])\mathbf{H}_n, \tag{7}$$

with unknown arbitrary scaling factor α_n and phase factor ϕ_n[1]. This shows that, for a given source n, the coefficients a_{mn} and D_{mn} w.r.t. all sensors can only be recovered up to these ambiguities:

$$\hat{\mathbf{h}}_{mn} = [\tilde{a}_{mn}, \tilde{a}_{mn}\omega^{\tilde{D}_{mn}}, \ldots, \tilde{a}_{mn}\omega^{(F-1)\tilde{D}_{mn}}], \tag{8}$$

where $\tilde{a}_{mn} \stackrel{\text{def}}{=} a_{mn}\alpha_n$ and $\tilde{D}_{mn} \stackrel{\text{def}}{=} D_{mn} + \phi_n$. Since the ambiguities $\{\alpha_n, \phi_n\}$ only depend on the source index, this suggests that they can be removed by choosing a reference sensor. Therefore, given the estimates $\tilde{a}_{mn}$ and $\tilde{D}_{mn}$, and a reference sensor, say $\underline{\mathrm{M}}$ (not necessarily the same for each source), one can compute the relative attenuation factor $\underline{a}_{mn} \stackrel{\text{def}}{=} \frac{\tilde{a}_{mn}}{\tilde{a}_{\underline{\mathrm{M}}n}} = \frac{a_{mn}}{a_{\underline{\mathrm{M}}n}}$ and the relative Time Difference Of Arrival (TDOA) $\underline{D}_{mn} \stackrel{\text{def}}{=} \tilde{D}_{mn} - \tilde{D}_{\underline{\mathrm{M}}n} = D_{mn} - D_{\underline{\mathrm{M}}n}$. As illustrated in Section 5, estimation of the relative TDOAs w.r.t. a reference sensor is sufficient to localize the sources.

4 ALS Algorithm with Vandermonde Structure Enforcing

Estimation of $\mathbf{H}(f)$ and $\underline{\mathrm{S}}(f)$, $f = 1, \ldots, F$, can be achieved by solving the following optimization problem

$$\min_{\{\mathbf{H}(f), \underline{\mathrm{S}}(f)\}_{f=1}^{F}} \gamma$$
$$\text{s.t. } \mathbf{h}_{mn} \text{ defined in (4) is a Vandermonde vector}, \forall m, \forall n,$$

where $\gamma \stackrel{\text{def}}{=} \sum_{f=1}^{F} \|\underline{\mathrm{R}}(f) - \mathbf{H}(f) \cdot \underline{\mathrm{S}}(f)\|^2$. In this section, we propose an algorithm that consists of three steps at each iteration. In the first step, given the previous estimates $\hat{\underline{\mathrm{S}}}(f)$, $f = 1, \ldots, F$, the matrices $\hat{\mathbf{H}}(f)$ are updated in the least squares sense:

$$\hat{\mathbf{H}}^{(LS)}(f) = \underline{\mathrm{R}}(f) \cdot \hat{\underline{\mathrm{S}}}(f)^{\dagger}, \; f = 1, \ldots, F. \tag{9}$$

In the second step, the purpose is to enforce the Vandermonde structure on the MN vectors $\hat{\mathbf{h}}_{mn}^{(LS)} \stackrel{\text{def}}{=} [\hat{\mathcal{H}}^{(LS)}]_{m,n,1:F}$, for $m = 1, \ldots, M$, $n = 1, \ldots, N$, where $[\hat{\mathcal{H}}^{(LS)}]_{m,n,f} \stackrel{\text{def}}{=} [\hat{\mathbf{H}}^{(LS)}(f)]_{m,n}$. Several algorithms have been proposed in the literature for the latter task (see, e.g., [3] and references therein). In practice, we will use the popular periodogram-based technique proposed in [7]. This consists of the computation of the FFT of the zero-padded sequence $\hat{\mathbf{h}}_{mn}^{(LS)}$. For each sensor-source pair (m, n), $\tilde{D}_{mn}$ is then updated as the index l for which the modulus of the FFT takes its maximum value, whereas $\tilde{a}_{mn}$ is updated as the

[1] Of course, the source components are estimated in an arbitrary order since one can arbitrarily permute the N terms of the sum in (5).

Algorithm 1. ALS algorithm with Vandermonde structure enforcing.

STEP 1: Time-frequency computation

Build $\underline{\mathrm{R}}(f) \in \mathbb{C}^{M \times P}$, $f = 1, \ldots, F$ from FFT of P overlapping windowed frames of recorded signals. (Typical parameters: $F = 2048$, Hanning window, 50% overlap).

STEP 2: Blind separation

—— Initialization ———-

stop=0, $k = 1$, K_{max} (e.g., $K_{max} = 200$) and ϵ (e.g., $\epsilon = 10^{-6}$). Randomly generate unitary matrices $\hat{\underline{\mathrm{S}}}(f) \in \mathbb{C}^{N \times P}$, $f = 1, \ldots, F$. Possibly try several random starting points.

—– Start alternating updates ———

while stop=0

  $k = k + 1$

  (2.a). $\hat{\mathbf{H}}^{(LS)}(f) = \underline{\mathrm{R}}(f) \cdot \hat{\underline{\mathrm{S}}}(f)^{\dagger}, \ f = 1, \ldots, F.$

  (2.b). $\{\hat{\tilde{D}}_{mn}, \hat{\tilde{a}}_{mn}\} \leftarrow \text{periodogram}(\hat{\mathbf{h}}_{mn}^{(LS)}), \ m = 1, \ldots, M, \ n = 1, \ldots, N.$

    $\hat{\mathbf{h}}_{mn}^{(VDM)} \leftarrow [\hat{\tilde{a}}_{mn}, \hat{\tilde{a}}_{mn} w^{\hat{\tilde{D}}_{mn}}, \ldots, \hat{\tilde{a}}_{mn} w^{(F-1)\hat{\tilde{D}}_{mn}}], \ m = 1, \ldots, M, \ n = 1, \ldots, N.$

  (2.c). $\hat{\underline{\mathrm{S}}}(f) = \hat{\mathbf{H}}^{(VDM)}(f)^{\dagger} \cdot \underline{\mathrm{R}}(f), \ f = 1, \ldots, F.$

  if ($k = K_{max}$) or ($|\gamma^{(k)} - \gamma^{(k-1)}| \leq \epsilon$)

    stop=1;

  end

end

If several starting points are used, keep the estimates associated to the smallest final value of γ.

Choose reference sensor $\underline{\mathrm{M}}$ and remove ambiguities: $\hat{\underline{\mathrm{a}}}_{mn} = \frac{\hat{\tilde{a}}_{mn}}{\hat{\tilde{a}}_{\underline{\mathrm{M}}n}}$ and $\hat{\underline{\mathrm{D}}}_{mn} = \hat{\tilde{D}}_{mn} - \hat{\tilde{D}}_{\underline{\mathrm{M}}n}$.

value taken by the real part of the FFT at index l. Each vector $\hat{\mathbf{h}}_{mn}^{(LS)}$ is then substituted by the Vandermonde vector $\hat{\mathbf{h}}_{mn}^{(VDM)}$, built from the estimate of $\tilde{D}_{mn}$ and $\tilde{a}_{mn}$ as in (8). The matrices $\hat{\mathbf{H}}^{(LS)}(f)$, $f = 1, \ldots, F$, are substituted by $\hat{\mathbf{H}}^{(VDM)}(f)$ accordingly. In the last step, the matrices $\hat{\underline{\mathrm{S}}}(f)$ are updated in the least squares sense as

$$\hat{\underline{\mathrm{S}}}(f) = \hat{\mathbf{H}}^{(VDM)}(f)^{\dagger} \cdot \underline{\mathrm{R}}(f), \ \ f = 1, \ldots, F. \tag{10}$$

The scaling and phase ambiguities α_n and ϕ_n are removed after convergence, as explained in Section 3. Note that convergence of the resulting algorithm is not guaranteed to be monotonic. Although the least squares updates of $\hat{\mathbf{H}}(f)$ and $\hat{\underline{\mathrm{S}}}(f)$ can only decrease or maintain the current value of γ, this is not guaranteed for the Vandermonde structure enforcing step. However, we observed through numerical experiments that our algorithm converges monotonically in many practical situations. A summary of the proposed technique is given in Algorithm 1.

5 Source Localization

The purpose of this stage is to localize the N sources from the TDOA estimates $\hat{\underline{\mathrm{D}}}_{mn}$, in number of samples, w.r.t. the reference sensor $\underline{\mathrm{M}}$. Let $\mathbf{u}_n = [x_n, y_n]^T$ denote the unknown vector of Cartesian coordinates of the nth source in a bi-dimensional propagation medium[2] and $\tilde{\mathbf{u}}_m = [\tilde{x}_m, \tilde{y}_m]^T$ the vector of known coordinates of the mth sensor. Choose the reference sensor $\underline{\mathrm{M}}$ as the origin of

[2] For simplicity, the localization task is formulated for a 2D medium. It can easily be generalized to the 3D case.

the new system of coordinates, in which the nth source and mth sensor have coordinates $\mathbf{z}_n \stackrel{\text{def}}{=} \mathbf{u}_n - \tilde{\mathbf{u}}_{\underline{\text{M}}}$ and $\tilde{\mathbf{z}}_m \stackrel{\text{def}}{=} \tilde{\mathbf{u}}_m - \tilde{\mathbf{u}}_{\underline{\text{M}}}$, respectively. Let us compute the relative range difference estimates $\hat{\underline{\text{d}}}_{mn} \stackrel{\text{def}}{=} \hat{\underline{\text{D}}}_{mn} v / F_s$, where v denotes the wave velocity in the propagation medium. In case of perfect estimation, $\hat{\underline{\text{d}}}_{mn}$ satisfies

$$\hat{\underline{\text{d}}}_{mn} + \|\mathbf{z}_n\| = \|\tilde{\mathbf{z}}_m - \mathbf{z}_n\|. \tag{11}$$

Squaring both sides of (11) yields

$$\hat{\underline{\text{d}}}_{mn}\|\mathbf{z}_n\| + \tilde{\mathbf{z}}_m^T \mathbf{z}_n = \frac{1}{2}\left(\|\tilde{\mathbf{z}}_m\|^2 - \hat{\underline{\text{d}}}_{mn}^2\right). \tag{12}$$

Considering all sensors except the reference sensor, $m \in \{1, 2, \ldots, M\} - \{\underline{\text{M}}\}$, (12) is equivalent to

$$\begin{bmatrix} \tilde{z}_1(1) & \tilde{z}_1(2) & \hat{\underline{\text{d}}}_{1n} \\ \vdots & \vdots & \vdots \\ \tilde{z}_M(1) & \tilde{z}_M(2) & \hat{\underline{\text{d}}}_{Mn} \end{bmatrix} \begin{bmatrix} z_n(1) \\ z_n(2) \\ \|\mathbf{z}_n\| \end{bmatrix} = \frac{1}{2} \begin{bmatrix} \|\tilde{\mathbf{z}}_1\|^2 - \hat{\underline{\text{d}}}_{1n}^2 \\ \vdots \\ \|\tilde{\mathbf{z}}_M\|^2 - \hat{\underline{\text{d}}}_{Mn}^2 \end{bmatrix},$$

which is compactly written as

$$\mathbf{Z}_n \theta_n = \mathbf{p}_n, \tag{13}$$

where $\mathbf{Z}_n \in \mathbb{R}^{(M-1)\times 3}$ and $\mathbf{p}_n \in \mathbb{R}^{(M-1)\times 1}$ are known. For each source index n, (13) can be solved in the least squares sense. Assuming that $\mathbf{Z}_n$ is of rank three, we get

$$\hat{\theta}_n = \mathbf{Z}_n^{\dagger} \mathbf{p}_n, \tag{14}$$

where $\|\mathbf{z}_n\|$ is treated as a variable independent from $z_n(1)$ and $z_n(2)$. A better option is to solve (13) as a constrained minimization problem

$$\min_{\theta_n} \psi$$
$$\text{s.t. } \theta_n(3) = \sqrt{\theta_n(1)^2 + \theta_n(2)^2}, \tag{15}$$

where

$$\psi \stackrel{\text{def}}{=} \|\mathbf{Z}_n \theta_n - \mathbf{p}_n\|^2. \tag{16}$$

We refer to [2,9] and references therein for solutions to this problem. In practice, we will use the Quadratically Constrained Least Squares (QCLS) method of [9]. The localization procedure is repeated independently for each source. Finally, the coordinates of the nth source in the initial Cartesian system are obtained by $\hat{\mathbf{u}}_n = \hat{\mathbf{z}}_n + \tilde{\mathbf{u}}_{\underline{\text{M}}}$, where $\hat{\mathbf{z}}_n = [\hat{\theta}_n(1), \hat{\theta}_n(2)]^T$. Note that the accuracy of the coordinate estimates relies on the accuracy of the TDOA estimates. In practice, it may happen that several TDOA estimates are significantly more accurate than others. This suggests that a subset of $\tilde{M} \geq 3$ reliable estimates among $M-1$ should be used in the localization process. In order to find the $\tilde{M}$ most

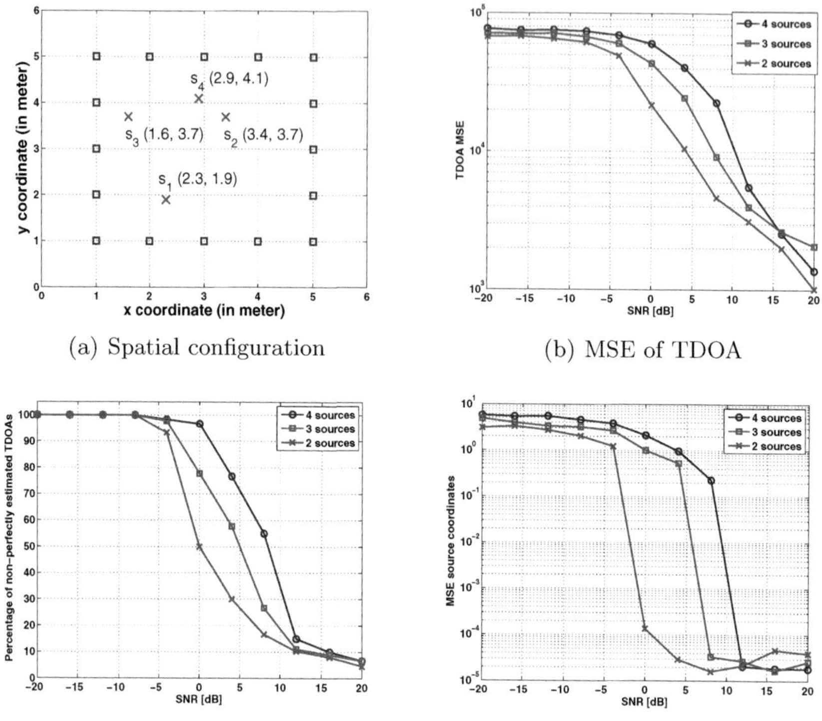

(a) Spatial configuration (b) MSE of TDOA

(c) % of non-perfectly estimated TDOAs (d) MSE of source coordinates

Fig. 2. Spatial configuration and results of Monte-Carlo experiments

reliable estimates, one can select $\tilde{M}$ rows of $\mathbf{Z}_n$, then solve (15) with the resulting reduced-size matrix, and repeat the procedure for all possible combinations of $\tilde{M}$ rows chosen among $M-1$. The final estimate of θ_n is the one associated to the smallest value of ψ.

6 Numerical Experiments

Let the noise-free signal $\mathbf{r}_m$ at receiver m be corrupted by Additive White Gaussian Noise (AWGN), $\tilde{\mathbf{r}}_m = \mathbf{r}_m + \sigma_m \mathbf{v}_m$, where the noise vector $\mathbf{v}_m$ is generated from a zero-mean unit-variance Gaussian distribution and σ_m is computed at each receiver to ensure a chosen Signal to Noise Ratio (SNR), $\text{SNR}_m = 10\,\log_{10}(\|\mathbf{r}_m\|^2/\|\sigma_m \mathbf{v}_m\|^2)$ [dB]. SNR_m is fixed here to the same value for all receivers and is further denoted SNR. We simulate the 2D propagation environment of Fig. 2(a), with $N = 4$ sources and $M = 16$ sensors. The sources consist of 10000 samples of different speech signals, with a sampling frequency $F_s = 16$ kHz. The wave speed is $v = 340$ m.s^{-1}. In this section, we illustrate the performance of our algorithm via Monte-Carlo simulations consisting of 100 independent trials for each value of the SNR. The noise vector $\mathbf{v}_m$ and the scaling factors a_{mn} for $m = 1, \ldots, M$, $n = 1, \ldots, N$, are randomly generated for each trial (a_{mn} is drawn between -3 and 3 with a uniform distribution). All experiments are conducted with the reference sensor located at $(1, 1)$. Fig. 2(b) illustrates the evolution of the Mean Square Error (MSE) ζ of the TDOA estimates, $\zeta = \frac{1}{N(M-1)} \sum_{n=1}^{N} \sum_{m=1}^{M-1} (\hat{\underline{\mathrm{D}}}_{mn} - \underline{\mathrm{D}}_{mn})^2$ for $N = \{2, 3, 4\}$. Fig. 2(c) illustrates the evolution of the percentage of non-perfectly estimated TDOAs ($\hat{\underline{\mathrm{D}}}_{mn} - \underline{\mathrm{D}}_{mn} \neq 0$), computed over the $N(M-1)$ estimates. It can be observed that, for SNR=20 dB, more than 90% of the TDOAs are perfectly estimated, even with $N = 4$ sources. Fig. 2(d) illustrates the evolution of the MSE ρ of the source coordinates, $\rho = \frac{1}{2N} \sum_{n=1}^{N} (x_n - \hat{x}_n)^2 + (y_n - \hat{y}_n)^2$, the latter being computed from the $\tilde{M} = 6$ most reliable sensors. It can be observed that, above a SNR threshold, the value of which depends on the number of sources, the localization of all sources is almost perfect. For instance, with $N = 2$, $\rho \simeq 0$ for SNR ≥ 0 dB, whereas the associated percentage of non-perfectly estimated TDOAs on Fig. 2(c) is 50% for this SNR value. This shows the benefit of searching for the most reliable TDOAs to be used in the localization process.

7 Conclusion

In this paper, we have proposed a novel time-frequency technique to deal with the problem of blind separation and localization of pure delayed sources. The core idea of the separation task is to interleave a Vandermonde structure enforcing strategy on the channel updates across the frequency mode with alternating least squares updates of the source and channel matrices. The localization task relies on the selection of the most reliable subset of TDOA estimates. Monte-Carlo experiments with two, three, and four sources have been conducted to corroborate our findings.

References

1. Chabriel, G., Barrère, J.: An Instantaneous Formulation of Mixtures for Blind Separation of Propagating Waves. IEEE Trans. Sig. Proc. 54(1), 49–58 (2006)
2. Cheung, K.W., So, H.C., Ma, W.K., Chan, Y.T.: A Constrained Least Squares Approach to Mobile Positioning: Algorithms and Optimality. EURASIP J. on Applied Sig. Proc. 2006(ID 20858), 1–23 (2006)
3. Clarkson, I.V.L.: Frequency estimation, phase unwrapping and the nearest lattice point problem. In: ICASSP 1999, pp. 1609–1612 (1999)
4. Emile, B., Comon, P.: Estimation of time delays between unknown colored signals. Signal Proc. 69(1), 93–100 (1998)
5. Omlor, L., Giese, M.: Blind source separation for over-determined delayed mixtures. In: Advances in Neural Information Processing Systems, vol. 19, pp. 1049–1056. MIT Press, Cambridge (2007)
6. Parra, L., Spence, C.: Convolutive blind separation of non-stationary sources. IEEE Trans. on Speech and Audio Processing 8(3), 320–327 (2000)

7. Rife, D.C., Boorstyn, R.R.: Single-tone parameter estimation from discrete-time observations. IEEE Trans. Inform. Theory IT 20(5), 591–598 (1974)
8. Yeredor, A.: Blind Source Separation with Pure Delays Mixture. In: ICA 2001 (2001)
9. Zhou, Y., Lamont, L.: Constrained least squares approach for TDOA localization: a global optimum solution. In: ICASSP 2008. pp. 2577–2580 (2008)

Joint Eigenvalue Decomposition Using Polar Matrix Factorization

Xavier Luciani[1,2] and Laurent Albera[1,2]

[1] Inserm, UMR 642, Rennes, F-35000, France
[2] Université de Rennes 1, LTSI, Rennes, F-35000, France
http://perso.univ-rennes1.fr/laurent.albera/

Abstract. In this paper we propose a new algorithm for the joint eigenvalue decomposition of a set of real non-defective matrices. Our approach resorts to a Jacobi-like procedure based on polar matrix decomposition. We introduce a new criterion in this context for the optimization of the hyperbolic matrices, giving birth to an original algorithm called JDTM. This algorithm is described in detail and a comparison study with reference algorithms is performed. Comparison results show that our approach provides quicker and more accurate results in all the considered situations.

Keywords: Joint diagonalization by similarity, joint eigenvalue decomposition, Jacobi method, polar matrix decomposition.

1 Introduction

In this study, we investigate the problem of Joint EigenValue Decomposition (JEVD) of a set of real matrices, which is encountered in different contexts such as 2-D DOA estimation [1], joint angle-delay estimation [2], multidimensional harmonic retrieval [3], Independent Component Analysis (ICA) [4] or Multi-way analysis [5]. JEVD consists in finding an eigenvector matrix $\boldsymbol{A}$ from a set of non-defective matrices $\boldsymbol{M}^{(k)}$ verifying:

$$\forall k = 1 \cdots K, \quad \boldsymbol{M}^{(k)} = \boldsymbol{A}\boldsymbol{D}^{(k)}\boldsymbol{A}^{-1}, \tag{1}$$

where the K diagonal matrices $\boldsymbol{D}^{(k)}$ are unknown.

This problem should not be confused with the classical problem of Joint Diagonalization by Congruence (JDC), for which $\boldsymbol{A}^{-1}$ is replaced by $\boldsymbol{A}^{\mathsf{T}}$, except when $\boldsymbol{A}$ is an orthogonal or unitary matrix [6]. JDC is the core of many ICA algorithms [7,8]. A large majority of these algorithms resorts to a suitable factorization of $\boldsymbol{A}$, performed by means of a Jacobi-like procedure. Such an approach has been naturally adapted to JEVD, although few papers have addressed this problem. Two main kinds of Jacobi-like algorithms have been developed in this context, based on different matrix factorizations. Originally, several authors had recourse to the QR factorization of $\boldsymbol{A}$ in order to compute the different sets of eigenvalues [3, 9]. Arguing that these QR-algorithms suffer from convergence

V. Vigneron et al. (Eds.): LVA/ICA 2010, LNCS 6365, pp. 555–562, 2010.

problems, Fu and Gao proposed an effective sh-rt algorithm [10] based on the polar decomposition. Indeed the polar decomposition has been used favourably for eigenvalue decomposition purpose since a long time [11, 12, 13] and also for JDC [14]. In a recent paper, Iferroudjene *et al.* introduced an alternative version of the sh-rt algorithm called JUST [15]. Our turn, we propose in this work an improvement of the sh-rt technique with significant numerical results.

2 Notations

In the following, scalars, vectors and matrices are denoted by lower case (a), lower case boldface ($\boldsymbol{a}$) and upper case boldface ($\boldsymbol{A}$) letters, respectively. The i-th entry of vector $\boldsymbol{a}$ is denoted by a_i while A_{ij} is the (i, j)-th component of matrix $\boldsymbol{A}$. Diagonal matrices are denoted by $\boldsymbol{D}$, Givens and hyperbolic rotation matrices are denoted by $\boldsymbol{G}$ and $\boldsymbol{H}$, respectively. For instance $\boldsymbol{G}(\theta_{ij})$ and $\boldsymbol{H}(\phi_{ij})$ are equal to the identity matrix, at the exception of the following components:

$$G(\theta_{ij})_{ii} = \cos(\theta_{ij});\ \ G(\theta_{ij})_{ij} = \sin(\theta_{ij}).\ \ H(\phi_{ij})_{ii} = \cosh(\phi_{ij});\ H(\phi_{ij})_{ij} = \sinh(\phi_{ij}).$$
$$G(\theta_{ij})_{ji} = -\sin(\theta_{ij});\ G(\theta_{ij})_{jj} = \cos(\theta_{ij}).\ \ H(\phi_{ij})_{ji} = \sinh(\phi_{ij});\ H(\phi_{ij})_{jj} = \cosh(\phi_{ij}).$$

3 A Novel Algorithm for JEVD

3.1 A Jacobi-Like Computation of the Polar Matrix Decomposition

In this subsection, all matrices are square matrices of dimensions N. Polar matrix decomposition states that any non-singular real matrix can be factorized into the product of an orthogonal matrix $\boldsymbol{Q}$ and a positive symmetric matrix $\boldsymbol{S}$. It is well known that $\boldsymbol{Q}$ can be decomposed into a product of Givens rotation matrices and a unitary diagonal matrix. In the same way, it has been shown that $\boldsymbol{S}$ can be decomposed into a product of hyperbolic rotation matrices and diagonal matrices [14]. Since (1) admits a scaling indeterminacy, the eigenvector matrix can only be estimated up to a diagonal scaling matrix. Therefore, taking into account that diagonal, hyperbolic and Givens matrices commute, one can reasonably assume that a solution matrix $\boldsymbol{A}$ can be found as a product of Givens and hyperbolic rotation matrices:

$$\boldsymbol{A} = \prod_{i=1}^{N-1} \prod_{j=i+1}^{N} \boldsymbol{G}(\theta_{ij})\boldsymbol{H}(\phi_{ij}). \tag{2}$$

The main point is that any Givens or hyperbolic matrix is defined by only one parameter (angle). Therefore we have now to find a set of $M = N(N-1)/2$ couple of parameters $\{\theta_{ij}, \phi_{ij}\}_{1\leq i<j\leq N}$ in order to get (1). Jacobi-like procedures achieve this problem by optimizing a suitable criterion with respect to each parameter, one by one. Consequently, inserting (2) into (1) we get:

$$\forall k = 1 \cdots K,\quad \boldsymbol{D}^{(k)} = \left(\prod_{m=1}^{M} \boldsymbol{H}(\phi_m)^{-1}\boldsymbol{G}(\theta_m)^{\mathsf{T}}\right) \boldsymbol{M}^{(k)} \left(\prod_{m=1}^{M} \boldsymbol{G}(\theta_m)\boldsymbol{H}(\phi_m)\right), \tag{3}$$

where each index m $(m = 1 \cdots M)$ corresponds to a couple (i, j) $(1 \leq i < j \leq N)$. Then, the algorithm scheme is simple. It consists in iteratively diagonalizing the $\boldsymbol{M}^{(k)}$ matrices by a successive optimization with respect to $\boldsymbol{G}(\theta_m)$ and $\boldsymbol{H}(\phi_m)$:

$$\forall k = 1 \cdots K, \quad \boldsymbol{M}^{(k,0)} = \boldsymbol{M}^{(k)}, \tag{4}$$

$$\forall k = 1 \cdots K, \quad \forall m = 1 \cdots M, \quad \boldsymbol{N}^{(k,m)} = \boldsymbol{G}(\theta_m)^{\intercal} \boldsymbol{M}^{(k,m-1)} \boldsymbol{G}(\theta_m), \tag{5}$$

$$\forall k = 1 \cdots K, \quad \forall m = 1 \cdots M, \quad \boldsymbol{M}^{(k,m)} = \boldsymbol{H}(\phi_m)^{-1} \boldsymbol{N}^{(k,m)} \boldsymbol{H}(\phi_m). \tag{6}$$

Thereby, at each stage m, the optimal corresponding Givens and hyperbolic matrices are successively computed in order to get K diagonal matrices $\boldsymbol{M}^{(k,M)}$ at the end of the process.

3.2 Optimization Step

A natural criterion to compute the optimal m-th Givens angle θ_m is thus to minimize the sum of the euclidean norms of the off-diagonal terms of the K matrices $\boldsymbol{N}^{(k,m)}$:

$$\zeta_G(\theta_m) = \sum_{k}^{K} \sum_{\substack{p,q \\ p \neq q}}^{N,N} \left(N_{pq}^{(k,m)} \right)^2 . \tag{7}$$

This criterion is the generalization of the original Jacobi criterion to the JD context. Since Givens matrices are orthogonal, the same definition of $\boldsymbol{N}^{(k,m)}$ holds in both the JDC and JEVD cases and thus the same optimization algorithms can be used. For instance, our proposed algorithm resorts to the same approach than the JAD algorithm described in [16] whereas the sh-rt and JUST algorithms use an other minimization scheme.

Once the optimal Givens matrix $\boldsymbol{G}(\theta_m)$ is computed, different criteria can be used for the optimal computation of $\boldsymbol{H}(\phi_m)$. This is the main difference between the three JEVD algorithms. The JUST algorithm resorts to criterion (7) by replacing $\boldsymbol{N}^{(k,m)}$ by $\boldsymbol{M}^{(k,m)}$, whereas the sh-rt method aims at minimizing the Frobenius norm of $\boldsymbol{M}^{(k,m)}$.

Instead of minimizing all the (off-diagonal) entries, we propose to target two particular off-diagonal entries of $\boldsymbol{M}^{(k,m)}$: if m corresponds to the $(i, j)_{i<j}$ couple, we simply aim at computing the optimal $M_{ij}^{(k,m)}$ and $M_{ji}^{(k,m)}$ components by using a "targeting" hyperbolic matrix. It is noteworthy that the transformation (6) affects the i-th and j-th rows and the i-th and j-th columns of $\boldsymbol{M}^{k,m}$ but only the (i, j) and the (j, i) components are twice affected by the hypebolic matrix and its inverse. Hence our choice to focus on the latter. Therefore, we use the following alternative criterion ζ_H^{JDTM} for the computation of the hyperbolic matrix:

$$\zeta_H^{JDTM}(\phi_m) = \sum_{k}^{K} \left(M_{ij}^{(k,m)} \right)^2 + \left(M_{ji}^{(k,m)} \right)^2 , \tag{8}$$

giving birth to our Joint Diagonalization algorithm based on Targeting hyperbolic Matrices (JDTM). A similar approach has been used with success in a different context [14]. Note that in the case of Givens matrices we showed that the optimizations of criteria (7) and (8) were mathematically equivalent.

Now, let us look at the components of $\boldsymbol{M}^{k,m}$. As previously mentioned, we only consider the (i,j)-th and (j,i)-th components which are given by:

$$M_{ij}^{(k,m)} = \left(N_{ii}^{(k,m)} - N_{jj}^{(k,m)}\right) \frac{\sinh(2\phi_m)}{2} + N_{ij}^{(k,m)} \cosh(\phi_m)^2 - N_{ji}^{(k,m)} \sinh(\phi_m)^2, \tag{9}$$

$$M_{ji}^{(k,m)} = \left(N_{jj}^{k,m} - N_{ii}^{(k,m)}\right) \frac{\sinh(2\phi_m)}{2} - N_{ij}^{(k,m)} \sinh(\phi_m)^2 + N_{ji}^{(k,m)} \cosh(\phi_m)^2. \tag{10}$$

Furthermore we can write that:

$$\left(M_{ij}^{(k,m)}\right)^2 + \left(M_{ji}^{(k,m)}\right)^2 = \frac{\left(M_{ij}^{(k,m)} + M_{ji}^{(k,m)}\right)^2}{2} + \frac{\left(M_{ij}^{(k,m)} - M_{ji}^{(k,m)}\right)^2}{2} \tag{11}$$

where the first term of the right-hand side is constant. Indeed, we derive from (9) and (10) the following equality:

$$\frac{\left(M_{ij}^{(k,m)} + M_{ji}^{(k,m)}\right)^2}{2} = \frac{\left(N_{ij}^{(k,m)} + N_{ji}^{(k,m)}\right)^2}{2} \tag{12}$$

where the right-hand side clearly does not depend on ϕ_m. Thereby minimizing ζ_H^{JTDM} is equivalent to minimize the λ function defined by:

$$\lambda(\phi_m) = \sum_k^K \left(M_{ij}^{(k,m)} - M_{ji}^{(k,m)}\right)^2. \tag{13}$$

We denote by $\boldsymbol{y}^{(m)}$ the column vector of $\mathbb{R}^K$ defined by $y_k^{(m)} = M_{ij}^{(k,m)} - M_{ji}^{(k,m)}$, so that $\lambda(\phi_m) = \boldsymbol{y}^{(m)\mathsf{T}}\boldsymbol{y}^{(m)}$. From (9) and (10) we obtain:

$$\boldsymbol{y}^{(m)} = \boldsymbol{W}^{(m)}\boldsymbol{x}(\phi_m), \tag{14}$$

with:

$$\boldsymbol{W}^{(m)} = \begin{bmatrix} N_{ii}^{1,m} - N_{jj}^{1,m} & N_{ij}^{1,m} - N_{ji}^{1,m} \\ \vdots & \vdots \\ N_{ii}^{K,m} - N_{jj}^{K,m} & N_{ij}^{K,m} - N_{ji}^{K,m} \end{bmatrix}; \boldsymbol{x}(\phi_m) = \begin{bmatrix} \sinh(2\phi_m) \\ \cosh(2\phi_m) \end{bmatrix}.$$

Now defining the diagonal (2×2) matrix $\boldsymbol{J}$ such that $J_{11} = -J_{22} = -1$ and observing that $\boldsymbol{x}(\phi_m)^{\mathsf{T}}\boldsymbol{J}\boldsymbol{x}(\phi_m) = 1$, we have thus to minimize the quantity $\boldsymbol{x}(\phi_m)^{\mathsf{T}}\boldsymbol{W}^{(m)\mathsf{T}}\boldsymbol{W}^{(m)}\boldsymbol{x}(\phi_m)$ under the constraint that $\boldsymbol{x}(\phi_m)^{\mathsf{T}}\boldsymbol{J}\boldsymbol{x}(\phi_m) = 1$. This can be done using the Lagrange multipliers strategy. Thereby, we have to minimize the L function given by:

$$L(\boldsymbol{x}(\phi_m), \mu(\phi_m)) = \boldsymbol{x}(\phi_m)^{\mathsf{T}}\boldsymbol{W}^{(m)\mathsf{T}}\boldsymbol{W}^{(m)}\boldsymbol{x}(\phi_m) - \mu(\phi_m)\boldsymbol{x}(\phi_m)^{\mathsf{T}}\boldsymbol{J}\boldsymbol{x}(\phi_m). \tag{15}$$

This leads to:

$$\boldsymbol{JW}^{(m)\top}\boldsymbol{W}^{(m)}\boldsymbol{x}(\phi_m) = \mu(\phi_m)\boldsymbol{x}(\phi_m). \tag{16}$$

Thus,$\mu(\phi_m)$ and $\boldsymbol{x}(\phi_m)$ are associated eigenvalue and eigenvector of matrix $\boldsymbol{JW}^{(m)\top}\boldsymbol{W}^{(m)}$ and it is easily shown that $\lambda(\phi_m) = \mu(\phi_m)$. $\boldsymbol{JW}^{(m)\top}\boldsymbol{W}^{(m)}$ is diagonalizable by construction and it can be shown that it has two non-null eigenvalues of opposite sign (the proof is not given here due to lack of space). Since the Gram matrix $\boldsymbol{W}^{(m)\top}\boldsymbol{W}^{(m)}$ is a positive semi-definite matrix, $\boldsymbol{x}(\phi_m)^\top\boldsymbol{W}^{(m)\top}\boldsymbol{W}^{(m)}\boldsymbol{x}(\phi_m)$ is positive and hence $\lambda(\phi_m)$. As a consequence $\boldsymbol{x}(\phi_m)$ is the eigenvector associated to the positive eigenvalue of $\boldsymbol{JW}^{(m)\top}\boldsymbol{W}^{(m)}$. Finally we have:

$$\phi(m) = \frac{1}{2}\text{atanh}\left(\frac{x(\phi_m)_1}{x(\phi_m)_2}\right). \tag{17}$$

We have just shown how the Givens and hyperbolic matrices are computed by the JDTM algorithm for the M couples (i,j). Actually, since the previous procedure follows an alternate optimization scheme, it has to be repeated several times before convergence. Each repetition is called a "sweep". In other words, if N_s is the number of sweeps, $\boldsymbol{A}$ is actually estimated by:

$$\widehat{\boldsymbol{A}} = \prod_{n_s}^{N_s}\prod_{i=1}^{N-1}\prod_{j=i+1}^{N}\boldsymbol{G}(\theta_{ij}^{n_s})\boldsymbol{H}(\phi_{ij}^{n_s}). \tag{18}$$

This iterative approach is common to all Jacobi-like algorithms. Finally, the K matrices, $\widehat{\boldsymbol{D}}^{(k)} = \widehat{\boldsymbol{A}}^{-1}\boldsymbol{M}^{(k)}\widehat{\boldsymbol{A}}$ are approximately diagonalized. One can measure the diagonalization achievement by using the following criterion:

$$r_D = \sum_{k}^{K}\sum_{\substack{p,q\\p\neq q}}^{N,N}\left(\widehat{\boldsymbol{D}}_{pq}^{(k)}\right)^2. \tag{19}$$

4 Simulations

The proposed algorithm has been validated and compared to the sh-rt and JUST algorithms by varying values of i) the Signal-to-Noise Ratio (SNR), ii) the matrix dimension N and iii) the number K of matrices to be diagonalized. Three kinds of simulation were conducted in order to quantify the relative effects of these quantities. The matrix set to be diagonalized is buit according to the following model:

$$\forall k = 1\cdots K,\ \boldsymbol{M}^{(k)} = \frac{\boldsymbol{X}^{(k)}}{\|\boldsymbol{X}^{(k)}\|_F} + \sigma\frac{\boldsymbol{E}^{(k)}}{\|\boldsymbol{E}^{(k)}\|_F}\ \text{with}\ \boldsymbol{X}^{(k)} = \boldsymbol{A}\boldsymbol{D}^{(k)}\boldsymbol{A}^{-1}. \tag{20}$$

$\boldsymbol{A}$, $\boldsymbol{D}^{(k)}$ and $\boldsymbol{E}^{(k)}$ entries are drawn randomly according to a standard normal distribution. $\boldsymbol{E}^{(k)}$ simulates a Gaussian additive noise. We define the SNR as $-20\log_{10}(\sigma)$. Hence, σ is chosen in order to obtained the desired value of SNR.

At the end of each sweep the euclidean norm of the off-diagonal components of $\boldsymbol{M}^{(k,M)}$ is computed and compared to the value computed at the previous sweep. Algorithms are stopped when the relative deviation between two successive values is smaller than 10^{-7}. N_s is then the number of sweeps needed to reach the convergence. We define r_A as the relative squared error between the true eigenvector matrix and its estimate after having removed the scaling and permutation indeterminacies. Therefore, algorithm results are judged according to three criteria: r_D, N_s and r_A. Note that the three algorithms have comparable numerical complexities, thereby N_s is a pertinent criterion to judge the convergence speed.

Each situation is repeated 100 times with a new draw of the $\boldsymbol{A}$, $\boldsymbol{D}^{(k)}$ and $\boldsymbol{E}^{(k)}$ matrices at each time. We present here the median values of r_D and r_A along with the mean values of N_s obtained from each algorithm.

4.1 Influence of the SNR

This first scenario varies the SNR from 10 dB to 100 dB whereas K and N are fixed to 50 and 10, respectively. Results are reported in figure 1(a). JDTM and sh-rt provide very close results in terms of diagonalization achievement whatever the SNR value, while the JUST algorithm is not as good in case of high SNR. At the exception of the 10 dB case (for which no algorithm converges) the JDTM algorithm requires at most half less sweeps than the other algorithms to reach the convergence. Furthermore, JDTM consistently provides a better estimation of the eigenvector matrix. Notably, at 20 and 30 dB one can note that at least 50% of the eigenvector matrices are well estimated by the JDTM algorithm, whereas this is not the case with the other algorithms.

4.2 Influence of the Matrix Size

Now we vary the matrix dimension N from 5 to 60 whereas the size of the matrix set and the SNR are fixed to 50 and 80 dB respectively. Results are reported in figure 1(b). JDTM and sh-rt outclass the JUST algorithm. Both provides very satisfying results, even in the case of large matrices. Diagonalization achievement of both algorithms are quite similar but the distance between the two curves increases with the matrix size in favour of the JDTM algorithm. In addition, the latter needs between 5 to 8 sweeps to converge against 10 to 17 for sh-rt and this gap also increases with the matrix size. In the same way, the comparison of the r_A median values highlights the efficiency of the JDTM algorithm which clearly improves sh-rt results, whatever the considered matrix size.

4.3 Influence of the Size of the Matrix Set to Be Diagonalized

Finally we vary the number of matrices from 5 to 40 whereas the matrix size and the SNR are fixed to 20 and 80 dB, respectively. Results are reported in figure 1(c). These are similar to those observed previously. The JDTM algorithm

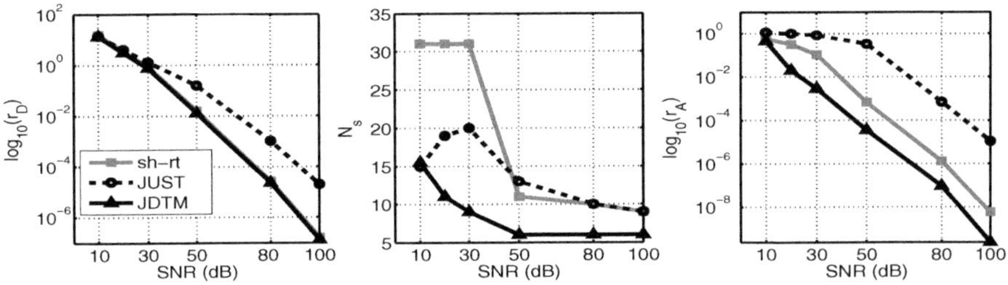

(a) according to the SNR

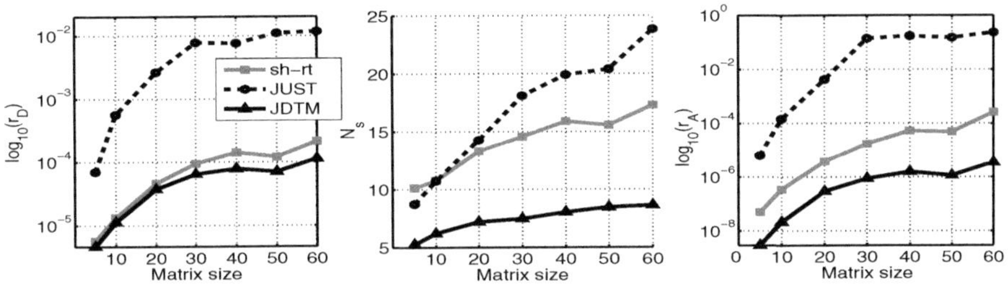

(b) according to the matrix size

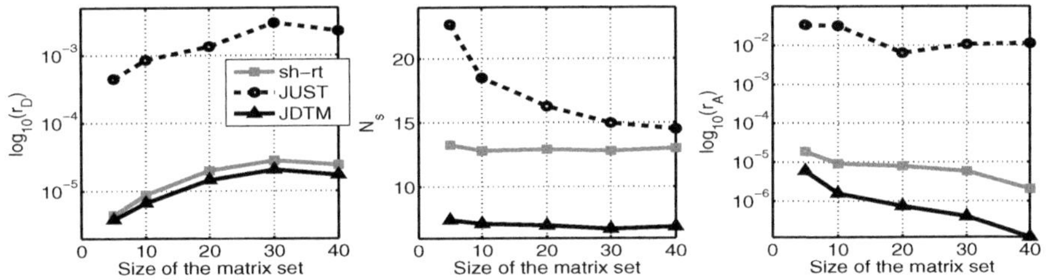

(c) according to the size of the number of matrices to be diagonalized

Fig. 1. Evolution of the three comparison criteria: r_D (median values), Ns (mean values) and r_A (median values)

consistently surpasses the other algorithms. Notably, it only requires around 7 sweeps to converge against 13 for sh-rt, whatever the size of the matrix set. In addition, regarding the eigenvector matrix estimation, the gap observed between the two curves in favour of the JDTM algorithm gradually increases with the size of the matrix set.

5 Conclusion

In spite of its simplicity we observed that the proposed algorithm invariably outperforms the reference algorithms in all the considered situations and according to both classical criteria, which are the diagonalization achievement and the number of sweeps. In addition it has been shown that the JDTM algorithm also provides a better estimation of the eigenvector matrix. From the presented results, it notably appears that the JDTM algorithm is a fast algorithm with a good estimation precision of the eigenvector matrix, particularly in the most difficult cases. Finally this study also highlights the sensitivity of Jacobi-like algorithms to the choice of the optimization criterion.

References

1. van der Veen, A.J., Ober, P.B., Deprettere, E.F.: Azimuth and elevation computation in high resolution DOA estimation. IEEE Trans. Signal Proc. 40, 1828–1832 (1992)
2. Lemma, A.N., van der Veen, A.J., Deprettere, E.F.: Analysis of joint angle-frequency estimation using ESPRIT. IEEE Trans. Signal Proc. 51, 1264–1283 (2003)
3. Haardt, M., Nossek, J.A.: Simultaneous Schur decomposition of several nonsymmetric matrices to achieve automatic pairing in multidimensional harmonic retrieveal problems. IEEE Trans. Signal Proc. 46, 161–169 (1998)
4. Albera, L., Ferréol, A., Chevalier, P., Comon, P.: ICAR, a tool for blind source separation using fourth order statistics only. IEEE Trans. Signal Proc. 53(10-1), 3633–3643 (2005)
5. Roemer, F., Haardt, M.: A closed-form solution for multilinear PARAFAC decompositions. In: IEEE SAM 2008, pp. 487–491 (2008)
6. Bunse-Gerstner, A., Byers, R., Mehrmann, V.: Numerical Methods for Simultaneous Diagonalization. SIAM J. Matrix Anal. Applicat. 14 (4), 927–949
7. Yeredor, A.: Non-Orthogonal Joint Diagonalization in the Least-Squares Sense with Application in Blind Source Separation. IEEE Trans. Signal Proc. 50(7), 1545–1553 (2002)
8. Karfoul, A., Albera, L., Birot, G.: Blind underdetermined mixture identification by joint canonical decomposition of HO cumulants. IEEE Trans. Signal Proc. 58(2), 638–649 (2010)
9. Strobach, P.: Bi-iteration multiple invariance subspace tracking and adaptive ESPRIT. IEEE Trans. Signal Proc. 48, 442–456 (2000)
10. Fu, T., Gao, X.: Simultaneous Diagonalization with Similarity Transformation for Non-defective Matrices. In: IEEE ICASSP 2006, pp. 1137–1140 (2006)
11. Goldstine, H.H., Horwitz, L.P.: A procedure for the diagonalization of normal matrices. J. ACM 6(2), 176–195 (1959)
12. Eberlein, P.J.: A Jacobi-like method for the automatic computation of eigenvalues and eigenvectors of an arbitrary matrix. Journal of the Society for Industrial and Applied Mathematics 10(1), 74–88 (1962)
13. Ruhe, A.: On the quadratic convergence of a generalization of the Jacobi method to arbitrary matrices. BIT Numerical Mathematics 8, 210–231 (1968)
14. Souloumiac, A.: Nonorthogonal joint Diagonalization by Combining Givens and Hyperbolic Rotations. IEEE Trans. Signal Proc. 57(6), 2222–2231 (2009)
15. Iferroudjene, R., Abed Meraim, K., Belouchrani, A.: A New Jacobi-like Method for Joint Diagonalization of Arbitrary non-defective Matrices. Applied Mathematics and Computation 211, 363–373 (2009)
16. Cardoso, J.-F., Souloumiac, A.: Jacobi Angles for Simultaneous Diagonalization. SIAM Journal on Matrix Analysis and Applications 17(1), 161–164 (1996)

Joint SVD and Its Application to Factorization Method

Gen Hori[1,2]

[1] Faculty of Business Administration, Asia University, Tokyo 180-8629, Japan
[2] Lab. for Advanced Brain Signal Processing, BSI, RIKEN, Saitama 351-0198, Japan
hori@brain.riken.jp

Abstract. This paper introduces an application of joint SVD to the factorization method which is a standard method in computer vision for the estimation of the camera motion and the object shape from an image stream taken by a camera that moves around the object. For computer vision systems with several cameras installed for the same direction, we implement a new algorithm for estimation of camera motion matrix that utilizes the measurement matrices from all the cameras based on joint SVD.

1 Introduction

The factorization method proposed by Tomasi and Kanade[1] is a method for the estimation of the camera motion and the object shape from an image stream taken by a camera that moves around the object. While almost previous methods for the same purpose are weak for noise and give unstable solution, the factorization method has a large advantage in that it is robust under noisy situation and gives a stable solution because it is based on well-established SVD (singular value decomposition).

On the other hand, joint SVD is a problem of finding a pair of unitary matrices which simultaneously diagonalizes several (possibly non-square) matrices. Joint SVD problem appears as a natural extension of joint diagonalization problem which has applications in signal separation methods and structured eigenvalue problems. Pesquet-Popescu et al.[2] introduced a Jacobi-like algorithm for joint SVD problem and proposed to apply joint SVD to obtaining separable representation of images. Hori[4] proposed a pair of matrix gradient flows to solve joint SVD problem.

This paper proposes to apply joint SVD to the estimation of the camera motion matrix for computer vision systems with several cameras installed for the same direction. The proposed method utilizes the measurement matrices from all the cameras based on joint SVD. A preliminary experiment using a computer-generated measurement matrix shows that the proposed method improves the estimation of the camera motion matrix comparing to the factorization method based on the standard SVD.

V. Vigneron et al. (Eds.): LVA/ICA 2010, LNCS 6365, pp. 563–570, 2010.

2 Factorization Method

This section summarizes the factorization method introduced by Tomasi and Kanade[1] which recovers the camera motion and the object shape from the image stream taken by the camera that moves around the object. For simplicity, we restrict our attention to the orthographic projection model (in which the three dimensional space is orthogonally projected to the image plane) throughout the study.

Suppose that we have an image stream which consists of F images taken by the camera that moves around the object. From the first image of the stream, we find P feature points on the object and track them throughout the F images to obtain the coordinates

$$\{(x_{fp}, y_{fp}) \mid f = 1, \ldots, F, p = 1, \ldots, P\}.$$

From the coordinates, we define a $2F \times P$ matrix

$$\tilde{A} = \begin{pmatrix} x_{11} & \cdots & x_{1P} \\ \vdots & & \vdots \\ x_{F1} & \cdots & x_{FP} \\ y_{11} & \cdots & y_{1P} \\ \vdots & & \vdots \\ y_{F1} & \cdots & y_{FP} \end{pmatrix}.$$

Next, we calculate the row-wise averages of the matrix $\tilde{A}$,

$$\bar{x}_f = \frac{1}{P}\Sigma_{p=1}^{P} x_{fp}, \quad \bar{y}_f = \frac{1}{P}\Sigma_{p=1}^{P} y_{fp},$$

and subtract them from the corresponding rows of $\tilde{A}$ to define a new matrix A

$$A = \tilde{A} - \begin{pmatrix} \bar{x}_1 \\ \vdots \\ \bar{x}_F \\ \bar{y}_1 \\ \vdots \\ \bar{y}_F \end{pmatrix} (1 \cdots 1)$$

whose rows are zero-mean. The matrix A is called "measurement matrix" and consists of the relative coordinates of the feature points to the centroid,

$$(x_{fp} - \bar{x}_f, y_{fp} - \bar{y}_f).$$

The measurement matrix A is factorized into two matrices M and S that correspond to the camera motion and the object shape respectively. As described in

Fig.1, we denote the coordinate of the camera and two unit vectors that define the camera direction for the f-th image by

$$t_f = (t_{fx}, t_{fy}, t_{fz})^T, \quad i_f = (i_{fx}, i_{fy}, i_{fz})^T, \quad j_f = (j_{fx}, j_{fy}, j_{fz})^T$$

where ^{T} denotes the transpose and the coordinate of the p-th feature point in the three dimensional space by

$$s_p = (s_{px}, s_{py}, s_{pz})^T,$$

from which we define a $2F \times 3$ matrix M that corresponds to the camera motion

$$M = \begin{pmatrix} {i_1}^T \\ \vdots \\ {i_F}^T \\ {j_1}^T \\ \vdots \\ {j_F}^T \end{pmatrix} = \begin{pmatrix} i_{1x} & i_{1y} & i_{1z} \\ \vdots & \vdots & \vdots \\ i_{Fx} & i_{Fy} & i_{Fz} \\ j_{1x} & j_{1y} & j_{1z} \\ \vdots & \vdots & \vdots \\ j_{Fx} & j_{Fy} & j_{Fz} \end{pmatrix},$$

and a $3 \times P$ matrix S that corresponds to the object shape

$$S = \begin{pmatrix} s_1 & \cdots & \cdots & s_P \end{pmatrix} = \begin{pmatrix} s_{1x} & \cdots & \cdots & s_{Px} \\ s_{1y} & \cdots & \cdots & s_{Py} \\ s_{1z} & \cdots & \cdots & s_{Pz} \end{pmatrix}.$$

Here we set the origin of the three dimensional space to the centroid of the feature points so that the rows of S are zero-mean,

$$\sum_{p=1}^{P} s_p = 0.$$

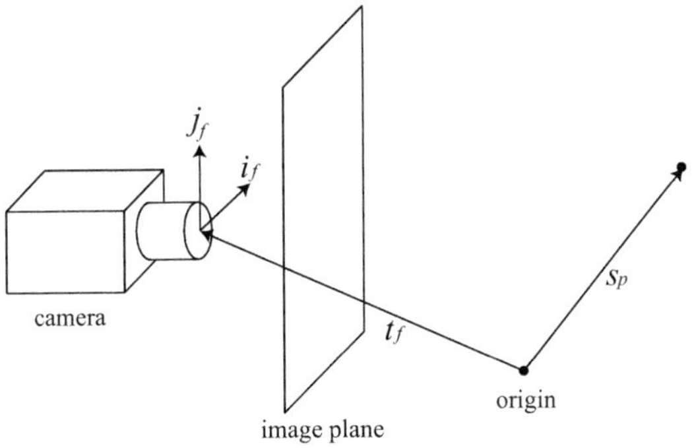

Fig. 1. Coordinate system

Because the coordinate of the p-th feature point on the f-th image is given as

$$x_{fp} = {i_f}^T(s_p - t_f), \quad y_{fp} = {j_f}^T(s_p - t_f),$$

and the rows of A and S are zero-mean, A is factorized as

$$A = MS \tag{1}$$

not depending on t_f. This leads to the fact that the rank of the measurement matrix A is 3 when it is not influenced by any noise. In practice of the factorization method, we obtain the decomposition (1) using the standard SVD (singular value decomposition) as follows. The first step is to decompose the $2F \times P$ measurement matrix A by SVD as

$$A = \tilde{U}\tilde{\Sigma}\tilde{V}^T$$

where $\tilde{U}$ is a $2F \times P$ matrix, $\tilde{\Sigma}$ a $P \times P$ diagonal matrix and $\tilde{V}$ a $P \times P$ matrix. Even when the measurement matrix A is influence by some noise, it is almost rank-3, that is, the diagonal elements of $\tilde{\Sigma}$ other than the dominant three are negligible so that we can truncate $\tilde{U}$ to a $2F \times 3$ matrix U, $\tilde{\Sigma}$ to a 3×3 matrix Σ and $\tilde{V}$ to a $P \times 3$ matrix V to have

$$A = U\Sigma V^T.$$

The second step is to find an invertible matrix T that gives the solution as

$$\begin{aligned} M &= UT \\ S &= T^{-1}\Sigma V^T \end{aligned}$$

using the fact that the rows of the matrix M need to meet the following conditions

$${i_f}^T i_f = 1, \quad {j_f}^T j_f = 1, \quad {i_f}^T j_f = 0 \qquad (f = 1, \ldots, F). \tag{2}$$

It is shown that the conditions (2) give a unique solution for the matrix T therefore the unique M and S up to the ambiguity related to the choice of the coordinate system.

3 Joint SVD

Given a set of K real or complex $m \times n$ matrices

$$\{A_1, A_2, \ldots, A_K\},$$

joint SVD or simultaneous SVD is a problem of finding orthogonal or unitary matrices U and V which make

$$\{U^*A_1V, U^*A_2V, \ldots, U^*A_KV\}$$

as diagonal as possible simultaneously. Joint SVD problem appears as a natural extension of joint diagonalization problem which has applications in signal separation methods and structured eigenvalue problems. We say that a joint SVD problem is *left-exact* (*right-exact*) when all the given matrices share a common set of left (right) singular vectors. We say that the problem is *exact* when it is *left-exact* and *right-exact* and the left and right singular vectors have one-to-one correspondence over the given matrices. From a purely mathematical viewpoint, a joint SVD problem can be solved exactly only if the problem is *exact*, where the SVD of an arbitrary single matrix A_k gives the solution to the problem. From a practical viewpoint, however, the problem is rarely *exact* and we have to find optimal solution with respect to some diagonality criterion. One of the typical situations in practical applications is that additive noise on observed matrices makes the problem *non-exact* although the problem is theoretically *exact*.

Hori[4] proposed to use the gradient ascent equations of the potential function

$$\varphi(U,V) = \sum_{k=1}^{K} \phi_k(U^* A_k V) \tag{3}$$

to solve a joint SVD problem where each $\phi_k(A)$ is a diagonality criterion supposed to take its maximum when A is an extended diagonal matrix. The general gradient ascent equations are derived as

$$\begin{aligned} \frac{dU}{dt} &= \frac{1}{2} U \sum_{k=1}^{K} ((U^* A_k V)(\frac{d\phi_k}{dA}(U^* A_k V))^* \\ &\qquad -(\frac{d\phi_k}{dA}(U^* A_k V))(U^* A_k V)^*), \end{aligned} \tag{4}$$

$$\begin{aligned} \frac{dV}{dt} &= \frac{1}{2} V \sum_{k=1}^{K} ((U^* A_k V)^*(\frac{d\phi_k}{dA}(U^* A_k V)) \\ &\qquad -(\frac{d\phi_k}{dA}(U^* A_k V))^*(U^* A_k V)). \end{aligned} \tag{5}$$

For example, we define the following diagonality criterion

$$\phi_k(A) = \sum_{1 \le j \le n} |a_{jj}|^2, \qquad k = 1, \ldots, K,$$

which takes its maximum when A is an extended diagonal matrix, and substitute it in (4) and (5) to obtain a pair of gradient flows for joint SVD,

$$\begin{aligned} \dot{U} &= U \sum_{k=1}^{K} ((U^* A_k V)\mathrm{diag}(U^* A_k V)^* \\ &\qquad -\mathrm{diag}(U^* A_k V)(U^* A_k V)^*), \end{aligned} \tag{6}$$

$$\begin{aligned} \dot{V} &= V \sum_{k=1}^{K} ((U^* A_k V)^*\mathrm{diag}(U^* A_k V) \\ &\qquad -\mathrm{diag}(U^* A_k V)^*(U^* A_k V)). \end{aligned} \tag{7}$$

There are other possible approaches to the joint SVD problem. For example, consider two joint diagonalization problems of Hermitian matrices,

$$\{A_1{A_1}^*, A_2{A_2}^*, \ldots, A_K{A_K}^*\}$$

and

$$\{{A_1}^*A_1, {A_2}^*A_2, \ldots, {A_K}^*A_K\},$$

and suppose that U and V are the solutions to the respective problems obtained by some standard joint diagonalization procedure, then

$$\{U^*A_1V, U^*A_2V, \ldots, U^*A_KV\}$$

is expected to be the joint SVD of the given matrices. Hori[3] compares the approach to joint SVD via joint diagonalization and the direct approach to join SVD to conclude that the direct approach gives more accurate solution.

4 Factorization Method for Several Cameras

In certain kinds of computer vision systems including two-eyed robot vision systems, several cameras are installed for the almost same direction. This section considers the factorization method for such systems in which K cameras are installed. Fig.2 describes the case of $K = 2$. We consider the case where the relative positions of the cameras are invariant throughout the image stream and all the cameras are set to the same direction, that is,

$$\begin{matrix} {i_f}^{(1)} = {i_f}^{(2)} = \cdots = {i_f}^{(K)} \\ {j_f}^{(1)} = {j_f}^{(2)} = \cdots = {j_f}^{(K)} \end{matrix} \quad (f = 1, \ldots, F).$$

In this case, the camera motion matrix M should be common to all the cameras. However each camera gives each measurement matrix A_k which is subject to noise and the camera-wise SVD of the measurement matrix

$$A_k = U_k \Sigma_k {V_k}^T \ (k = 1, \ldots, K)$$

possibly results in different U_k therefore different camera motion matrix M_k. For this situation, we propose to consider the real joint SVD of $A_1, \ldots, A_K$ which makes

$$U^T A_k V \ (k = 1, \ldots, K)$$

as diagonal as possible simultaneously. This gives a common U and therefore a common camera motion matrix M and the resulting camera motion matrix is expected to be more accurate than the one obtained from the SVD of a single measurement matrix of some representative camera because the proposed method utilizes the information from all the cameras.

Fig.3 shows the result of a preliminary experiment with a computer-generated 250×125 measurement matrix $A = MS$ where the camera motion matrix M is generated by changing the yaw, pitch and roll angles from $-10°$ to $10°$ and the

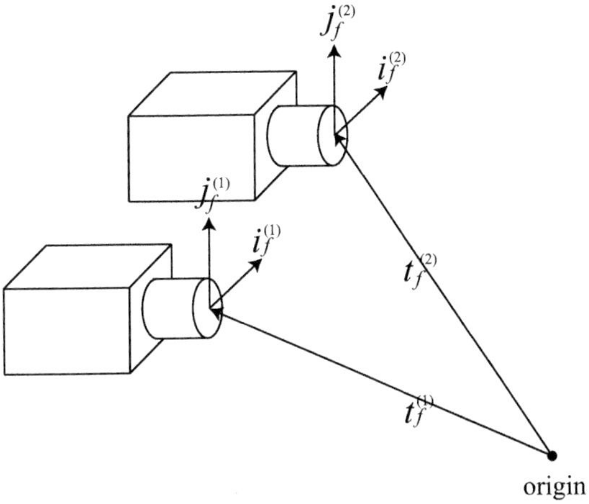

Fig. 2. Parallel cameras

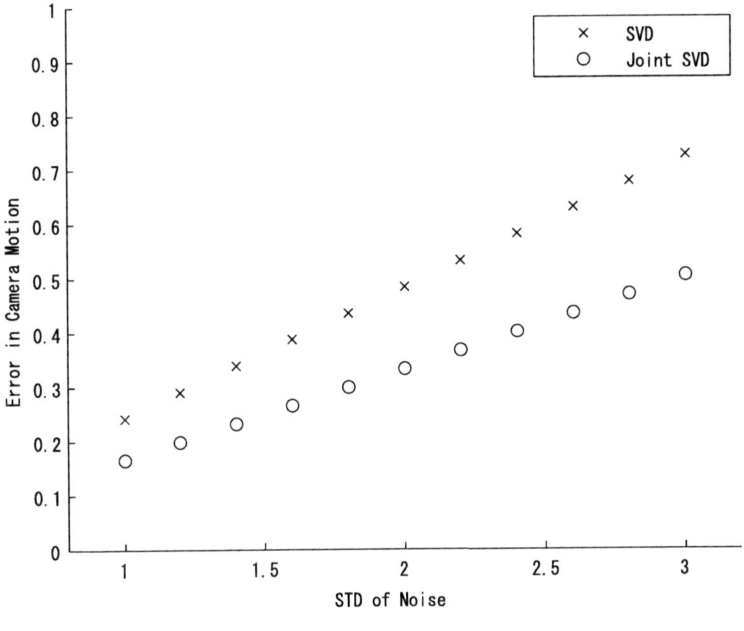

Fig. 3. Preliminary experiment

object shape matrix S is of 125 equally-spaced lattice points in the three dimensional space. We suppose that the whole object is projected onto a quarter of the image plane of 1000×1000 pixels and change the standard deviation of the Gaussian noise added to the elements of the measurement matrix A from 1 pixel to 3 pixels. The camera motion matrix is estimated using the factorization method based on the standard SVD of a single noisy measurement matrix and joint SVD of two noisy measurement matrices. The difference between the true camera motion matrix and the estimated one is measured using the matrix Frobenius norm and averaged over 10 trials. From the result, we see that the proposed method based on the joint SVD of two measurement matrices improves the estimation of the camera motion matrix comparing to the factorization method based on the SVD of a single measurement matrix.

5 Concluding Remarks

We have proposed to apply joint SVD to the estimation of the camera motion matrix in the computer vision systems with several cameras installed for the same direction. A preliminary experiment using a computer-generated measurement matrix has shown that the proposed method improves the estimation of the camera motion matrix comparing to the factorization method based on the standard SVD. Our future study includes the extension of the proposed method to more precise projection models such as the scaled orthographic projection model and the paraperspective projection model as well as the experiments using real image stream data taken by parallel cameras.

References

1. Tomasi, C., Kanade, T.: Shape and motion from image streams under orthography - A factorization method. Int. J. Comput. Vision 9-2, 137–154 (1992) Technical Report CMU-CS-91-172, CMU (1991)
2. Pesquet-Popescu, B., Pesquet, J.-C., Petroplu, A.P.: Joint singular value decomposition - a new tool for separable representation of images. In: Proc. 2001 Intl. Conf. on Image Processing, vol. 2, pp. 569–572 (2001)
3. Hori, G.: Comparison of Two Main Approaches to Joint SVD. In: Proc. Intl. Workshop Independent Component Analysis Blind Signal Separation, pp. 42–49 (2009)
4. Hori, G.: A general framework for SVD flows and joint SVD flows. In: Proc. IEEE Intl. Conf. Acoustics Speech Signal Processing, vol. 2, pp. 693–696 (2003)
5. Hori, G.: A new approach to joint diagonalization. In: Proc. Intl. Workshop Independent Component Analysis Blind Signal Separation, pp. 151–155 (2000)
6. Hori, G., Tanaka, T.: Finding initial values for time-varying joint diagonalization. In: Proc. IEEE Intl. Conf. Acoustics and Speech Signal Processing, pp. 2018–2021 (2010)
7. Cardoso, J.-F., Souloumiac, A.: Blind beamforming for non Gaussian signals. IEEE Proc.-F 140, 362–370 (1993)
8. Yeredor, A.: Non-orthogonal joint diagonalization in the least-squares sense with application in blind source separation. IEEE Trans. Signal Processing 50(7), 1545–1553 (2002)
9. Pham, D.T.: Joint approximate diagonalization of positive definite Hermitian matrices. SIAM J. Matrix Anal. Appl. 22, 1136–1152 (2001)
10. Ziehe, A., Laskov, P., Nolte, G., Müller, K.-R.: A fast algorithm for joint diagonalization with non-orthogonal transformations and its application to blind source separation. J. Machine Learning Research 5, 777–800 (2004)

Double Sparsity: Towards Blind Estimation of Multiple Channels

Prasad Sudhakar[1], Simon Arberet[2], and Rémi Gribonval[1]

[1] METISS Team, Centre de recherche INRIA Rennes - Bretagne Atlantique
Rennes CEDEX 35042, France
{firstname.lastname}@inria.fr
[2] Institute of Electrical Engineering
École Polytechnique Fédérale de Lausanne (EPFL), CH-1015 Lausanne, Switzerland
{firstname.lastname}@epfl.ch

Abstract. We propose a framework for blind multiple filter estimation from convolutive mixtures, exploiting the time-domain sparsity of the mixing filters and the disjointness of the sources in the time-frequency domain. The proposed framework includes two steps: (a) a clustering step, to determine the frequencies where each source is active alone; (b) a filter estimation step, to recover the filter associated to each source from the corresponding incomplete frequency information. We show how to solve the filter estimation step (b) using convex programming, and we explore numerically the factors that drive its performance. Step (a) remains challenging, and we discuss possible strategies that will be studied in future work.

Keywords: Blind filter estimation, sparsity, convex optimisation.

1 Introduction

Source separation systems have several applications such as speech processing, music transcription, biomedical signal processing, etc. In a general setting, we consider M mixtures $x_i(t), i = 1 \dots M$ of N source signals $s_j(t), j = 1 \dots N$, given by the convolutive model

$$x_i(t) = \sum_{j=1}^{N} (a_{ij} \star s_j)(t) + v_i(t) \tag{1}$$

where $a_{ij}(t)$ is a filter of length L which models the impulse response between the jth source and the ith microphone, and $v_i(t)$ is the noise at the ith microphone. For brevity, we denote the sources, filters, noise and mixtures by s_j, a_{ij}, v_i and x_i respectively, by dropping the time index.

Blind source separation (BSS) systems attempt to estimate the sources given only the mixtures. This is often done in two stages: the mixing filters are estimated first, and subsequently they are used for source estimation. In case of instantaneous and anechoic mixtures, the filters are simply scalars or time-delayed

V. Vigneron et al. (Eds.): LVA/ICA 2010, LNCS 6365, pp. 571–578, 2010.

scalars and several methods [1] (and references within) have been proposed to estimate the mixing parameters. Many methods such as DUET [2] rely on the sparsity and disjointness of the sources in a transform domain to estimate the parameters and the sources.

The problem gets more complicated with convolutive mixtures. Frequency domain techniques transform convolutive mixtures problem into multiple complex-valued instantaneous mixtures problem, under the narrowband approximation. But, this approach suffers from the ambiguities of arbitrary scaling and permutations of the sources and solving them is a challenging problem in itself [3].

On the other hand, when there is only one source, the problem of blindly estimating filters from the filtered versions of the source is a well studied problem. In addition, if the filters are sparse then the filter estimation problem can be cast as a standard sparse vector recovery problem [4]. Subsequently, sparse recovery algorithms can be used to solve it.

In a nutshell, there are techniques exploiting source sparsity to estimate instantaneous and anechoic mixing parameters and there are techniques to estimate sparse filters blindly in a single source setting. The grand goal of our effort is to combine these two strands of work and propose a blind mixing filter estimation framework which exploits the time-frequency domain source sparsity and time domain filter sparsity simultaneously. The proposed framework involves a source activity estimation step (clustering) and a filter estimation step.

Ideally, in such a framework the clustering step has to be performed blindly using the mixtures, and the filter recovery process depends on this clustering step. However, as the contribution in this paper, we focus on the formulation and experimental validation of the filter recovery step by solving the clustering step with strong side information. This serves as the first step in realising a completely blind system.

2 Sparse Filter Estimation

Before we present our contributions, let us first describe some existing work on blind estimation of sparse filters in single and multiple sources settings.

Case of a single source: Let us start with the simplest case of estimating filters when there is only one source s and two outputs x_1 and x_2. This is the single-input-two-output (SITO) case and we have: $x_i = a_i \star s + v_i, i = 1, 2$. Consider a single frame of s of length T and the length of the filters be L, then the length of x_i will be $T + L - 1$. In the absence of noise, we have the following cross-relation (CR) [5].

$$x_2 \star a_1 = x_1 \star a_2 \tag{2}$$

For convenience, let us associate the signal a_i to the column vector $\mathbf{a}_i = [a_i(t)]_{t=1}^{L}$ and likewise s to $\mathbf{s}$ and x_i to $\mathbf{x}_i$.

The convolution $x_i \star a_j$ is associated to the multiplication between the Toeplitz matrix[1]

$$\mathcal{T}[\mathbf{x}_i] = \begin{bmatrix} x_i(0) & 0 & \cdots & 0 \\ x_i(1) & x_i(0) & \cdots & 0 \\ \vdots & & \ddots & \vdots \\ x_i(T+L-3) & \cdots & \cdots & x_i(0) \end{bmatrix}, \tag{3}$$

and the vector $\mathbf{a}_j$. By using the shorthand $\mathcal{B}[\mathbf{x}_1, \mathbf{x}_2] = \big[\mathcal{T}[\mathbf{x}_2], -\mathcal{T}[\mathbf{x}_1]\big]$, we can write the CR (2) as

$$\mathcal{B}[\mathbf{x}_1, \mathbf{x}_2] \cdot \mathbf{a} = \mathbf{0}, \text{ where } \quad \mathbf{a} = \begin{bmatrix} \mathbf{a}_1 \\ \mathbf{a}_2 \end{bmatrix}. \tag{4}$$

This relation has inspired several methods (named CR methods) to estimate the filters blindly from the observations [5]. These methods generally do not assume anything about the nature of the filters, however in scenarios such as underwater/wide band wireless communications, the filters that model the channels are often sparse in the time domain.

With the additional sparsity assumption, the SITO filter estimation problem can be formulated as the following ℓ^1 minimisation problem [4], with $\mathbf{B} := \mathcal{B}[\mathbf{x}_1, \mathbf{x}_2]$.

$$\text{minimize} \ \ \|\mathbf{a}\|_1 \quad \text{subject to} \quad \|\mathbf{B} \cdot \mathbf{a}\|_2 \leq \epsilon \ \text{ and } \ \|\mathbf{a}\|_2 = 1. \tag{5}$$

The normalisation $\|\mathbf{a}\|_2 = 1$ mentioned in [4] is to avoid the trivial zero-vector solution. However, this makes the problem non-convex, and there remains a shift ambiguity of the solution. As an alternative, we use the constraint $a_1(t_0) = 1$, where t_0 is an arbitrarily chosen time index. The resulting problem is convex:

$$\text{minimize} \ \ \|\mathbf{a}\|_1 \quad \text{subject to} \quad \|\mathbf{B} \cdot \mathbf{a}\|_2 \leq \epsilon \ \text{ and } \ \mathbf{a}_1(t_0) = 1 \tag{6}$$

It can be solved using any standard convex optimisation algorithm, and we chose to use the `CVX` software package [8].

Case of multiple sources: When dealing with multiple sources, the CR formulation (2) cannot be directly used without further assumptions. Aïssa-El Bey *et al.* [6] have extended the above described SITO approach to N sources, by assuming that it is possible to identify time segments where only one source contributes to the mixtures. Then a SITO problem can be formulated locally at such segments and solved to obtain the filters for that particular source.

Let us describe this with an illustration. Fig. 1(a) is the time-domain plot of two sources s_1 and s_2. Fig. 1(b) shows the plot of their mixtures, x_1 and x_2, obtained by convolving the sources with the filters $a_{ij}, i = 1, 2$. These mixtures do not satisfy the CR (4) over the entire time frame, but the sources are such that at the time interval I_j, only source j is active. If we extract the mixtures at

[1] Calligraphic letters will denote operators that map a vector to a matrix, e.g. $\mathcal{T}[\mathbf{x}_i]$.

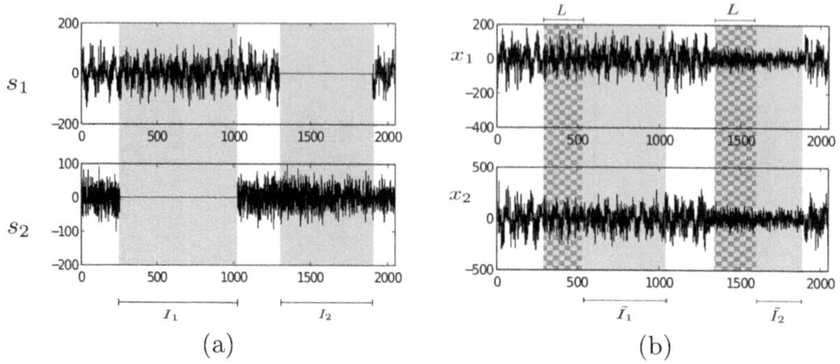

Fig. 1. (a) Sources with intervals where only one source is active. (b) Mixtures from the sources.

the appropriate time segments $\tilde{I}_j$, then we obtain the segments of the mixtures $y_i^{(j)} = \{x_i(t)\}_{t\in\tilde{I}_j}$ that depend on a single source j: $y_i^{(j)} = a_{ij} \star \tilde{s}_j$, where $\tilde{s}_j$ is the restriction of the source to a certain time interval. The vectors $\mathbf{y}_i^{(j)}$ corresponding to $y_i^{(j)}$ now satisfy the CR $\mathcal{B}[\mathbf{y}_1^{(j)}, \mathbf{y}_2^{(j)}] \cdot \mathbf{a}^{(j)} = \mathbf{0}$ and so this can be used to estimate the filters for source j by solving the optimisation problem (6) with $\mathbf{B} = \mathcal{B}[\mathbf{y}_1^{(j)}, \mathbf{y}_2^{(j)}]$. The authors of [6] have explicitly presented a technique to identify the intervals $\tilde{I}_j$ and have experimentally demonstrated the results of this approach.

3 Proposed Framework

In general, the sources may overlap in the time domain, and the approach described in the previous section might not be suitable for the filter estimation task, even if the filters are sparse. Instead of disjoint time supports, it is a common assumption in BSS to consider sources with almost disjoint time-frequency (T-F) representations, in the short-time Fourier transform (STFT) domain.

T-F domain SITO: Let us start with the single source setting again. Let $\hat{\mathbf{x}}_i$ be the STFT of the vector $\mathbf{x}_i$, and $\hat{\mathbf{a}}_i$ be the Fourier transform of $\mathbf{a}_i$ (appropriately zero padded) such that $\hat{\mathbf{x}}_i = [\hat{\mathbf{x}}_i(\tau, f)]_{\tau,f}$ and $\hat{\mathbf{a}}_i = [\hat{\mathbf{a}}_i(f)]_f$. Let us consider STFT frames $1 \leq \tau \leq N_T$. In each frame, the cross relation (2) becomes

$$\hat{\mathbf{x}}_2(\tau, f) \cdot \hat{\mathbf{a}}_1(f) \simeq \hat{\mathbf{x}}_1(\tau, f) \cdot \hat{\mathbf{a}}_2(f), \quad \forall f. \tag{7}$$

Defining $\mathcal{D}[\hat{\mathbf{x}}_i](\tau) = \mathrm{diag}\left([\hat{\mathbf{x}}_i(\tau, f)]_f\right)$, the CR in the STFT domain will be

$$\widehat{\mathcal{B}}[\hat{\mathbf{x}}_1, \hat{\mathbf{x}}_2] \cdot \mathbf{a} \simeq \mathbf{0} \text{ with } \widehat{\mathcal{B}}[\hat{\mathbf{x}}_1, \hat{\mathbf{x}}_2] = \begin{bmatrix} \mathcal{D}[\hat{\mathbf{x}}_2](1), & -\mathcal{D}[\hat{\mathbf{x}}_1](1) \\ \vdots & \vdots \\ \mathcal{D}[\hat{\mathbf{x}}_2](N_T), & -\mathcal{D}[\hat{\mathbf{x}}_1](N_T) \end{bmatrix} \begin{bmatrix} \mathbf{F}^* & \mathbf{0} \\ \mathbf{0} & \mathbf{F}^* \end{bmatrix}, \tag{8}$$

where $\mathbf{F}^*$ is the Fourier matrix of appropriate size. Using this CR, the optimisation problem (6) with $\mathbf{B} := \widehat{\mathcal{B}}[\hat{\mathbf{x}}_1, \hat{\mathbf{x}}_2]$ can be solved to obtain the filters.

Multiple sources: In the case of N sources, the above formulation can be generalised. In the time domain, the intervals I_j enabled us to formulate the CR for each source. Likewise, if we can identify a set of T-F points Ω_j for each source j, where only one source is active, then these sets can be used to formulate the CR and estimate the filters as described previously. For each source j, one can build a matrix $\widehat{\mathcal{B}}_{\Omega_j}[\hat{\mathbf{x}}_1, \hat{\mathbf{x}}_2]$ that contains the rows of $\widehat{\mathcal{B}}[\hat{\mathbf{x}}_1, \hat{\mathbf{x}}_2]$ indexed by the T-F points $(\tau, f) \in \Omega_j$ and form the cross relation $\widehat{\mathcal{B}}_{\Omega_j}[\hat{\mathbf{x}}_1, \hat{\mathbf{x}}_2] \cdot \mathbf{a}^{(j)} = 0$. Then for each source j, we can estimate the filter $\mathbf{a}^{(j)}$ by solving (6) with $\mathbf{B} := \widehat{\mathcal{B}}_{\Omega_j}[\hat{\mathbf{x}}_1, \hat{\mathbf{x}}_2]$.

Proposed framework: The mixing filter estimation process using time-frequency domain cross relation can be summarized in the following steps.

---

1. Compute the time-frequency representations $\hat{\mathbf{x}}_i$, $i = 1, 2$.
2. For each source j,
   P1 { (a) Identify the set Ω_j.
   P2 { (b) Build the matrix $\mathbf{B} := \widehat{\mathcal{B}}_{\Omega_j}[\hat{\mathbf{x}}_1, \hat{\mathbf{x}}_2]$.
   (c) Solve (6) to obtain the estimated filter $\tilde{\mathbf{a}}^{(j)}$.

In the rest of this paper, we focus on the evaluation of the performance of the proposed framework *when the solution to* **P1** *is known.* Further work will be devoted to addressing **P1** and **P2** simultaneously, and preliminary ideas will be discussed in the conclusion.

4 Experimental Verification

The framework presented in the previous section was experimentally verified in the multiple source setting for $N = 2$ and $N = 3$. The paragraphs below present the details of data generation, the experimental protocol and the results.

4.1 Data Generation

Filters: The length of the individual filters $\mathbf{a}_i$ was set to $L = 256$, and their sparsity was set to $\|\mathbf{a}_i\|_0 = k/2$, for various values of k. So the unknown vector $\mathbf{a}$ was of length $2 * L = 512$ and its sparsity was k. The $k/2$ support indices on each channel were chosen uniformly at random in the set $(\frac{L}{4}, \frac{3L}{4})$. The filter coefficients were generated i.i.d. Gaussian with zero mean, unit variance and sorted to have decreasing magnitudes along the time axis within the support. Every filter $a_{1j}(t)$ was finally normalised and shifted to have $a_{1j}(L/2) = 1$.

Sources: For each source j, we generated N_T independent time frames $s_j^\tau(t)$, $1 \leq \tau \leq N_T$. Each frame $s_j^\tau(t)$ of length $T = 3 * L$ was a sum of N_F sinusoids:

$$s_j^\tau(t) = \sum_{w=1}^{N_F} A_{jw}^\tau \sin(2\pi f_{jw}^\tau t + \phi_{jw}^\tau), \tag{9}$$

where the frequencies f_{jw}^τ were chosen uniformly at random in $[0, 1/2]$. The amplitudes A_{jw}^τ were generated i.i.d. Gaussian with zero mean, unit variance, and the phases ϕ_{jw}^τ were chosen uniformly at random in $[0, 2\pi]$.

Performance: By nature, the estimated solution $\tilde{a}^{(j)}(t)$ suffers from shift and scaling ambiguity (to satisfy the normalisation constraint). The following definition of the output signal-to-noise ratio (SNR), which accounts the scaling and shift ambiguity, was used as a recovery performance measure.

$$\mathrm{SNR}_{out} = 10 \log_{10} \left(\frac{\sum_j \sum_t \|a^{(j)}(t)\|_2^2}{\min_{t',\mu} \sum_j \sum_t \left(\|a^{(j)}(t) - \mu \cdot \tilde{a}^{(j)}(t - t')\|_2^2 \right)} \right). \tag{10}$$

4.2 Experimental Protocol

We performed experiments in the ideal single source setting for various values of filter sparsity k and number of sinusoids per frame N_F.

We determined the number of frames required to recover the filters with an output SNR (defined above) of 20dB. This number depends on the sparsity k and the number of sinusoids per frame N_F, and let us denote this by $\#_{20}(k, N_F)$. In the experiments for multiple source setting, we arbitrarily chose $N_T(k, N_F) = \#_{20}(k, N_F) \times 2$.

For every combination of k and N_F, 20 independent trials were done and the performance was averaged. In each trial, the sources and filters were generated as described previously, the mixtures x_i were obtained according to (1) with $v_i = 0$ and the vectors $\hat{\mathbf{x}}_i$ were formed. For each source j, **P1** and **P2** were solved as described below.

P1: Obtaining Ω_j using side information: The set Ω_j was constructed using as side information the frequencies $\{f_{jw}^\tau\}_{w=1}^{N_F}$ that are used to generate the source in (9). Let $\hat{\mathbf{s}}_j^\tau$ be the time-frequency domain vector of length $F = T + L - 1$ obtained by the appropriate zero-padding and transformation of the frame τ of the source s_j^τ. We defined, using $\theta = 10$dB,

$$(\tau, f) \in \Omega_j \iff f \in \{f_{jw}^\tau\}_{w=1}^{N_F} \text{ and } 20 \cdot \log_{10} \left(\frac{|\hat{\mathbf{s}}_j^\tau(f)|}{|\hat{\mathbf{s}}_{j'}^\tau(f)|} \right) \geq \theta, \ \forall j' \neq j. \tag{11}$$

P2: Filter estimation by convex optimization: For each source j, the matrix $\widehat{\mathcal{B}}_{\Omega_j}[\hat{\mathbf{x}}_1, \hat{\mathbf{x}}_2]$ was built using the set Ω_j. Then, the resulting convex optimisation problem (6) was solved to obtain the filter estimates $\tilde{\mathbf{a}}^{(j)}$. The value of

ϵ used in (6) actually tells us about the amount of imperfection that we would like to allow in the CR. Setting ϵ is a challenging task. In these experiments, we relied upon an oracle setting of ϵ. For each source j, we used $\epsilon_j = \|\mathbf{B} \cdot \mathbf{a}^{(j)}\|_2$, with $\mathbf{a}^{(j)}$ the true filter.

Results: Tables 1 and 2 show the average output SNR for various (k, N_F) for 2 and 3 sources respectively. In both cases, the anechoic filters are recovered with very high SNR, and the output SNR is at least 10dB when $N_F \leq 3$ and $k \leq 10$. For a given sparsity k, the output SNR drops as the number of sinusoids per frames N_F increases.We experimented with higher values of N_F and we found that the performance continues to degrade. This is because the sources tend to interfere more as N_F increases, thereby violating the CR badly. Indeed, even though we generated sums of sinusoids, their Fourier transform has peaks at the associated frequencies that can have a large main lobe and secondary lobes, leading to interferences. This could be compensated by setting a higher threshold θ to compute the set Ω_j, at the price of a smaller number of "visible" frequencies per time frame, which in turn could be compensated by increasing the number of observed time frames $N_T(k, N_F)$.

5 Discussion and Future Work

We have described the existing work on the time-domain CR method to estimate sparse filters from convolutive mixtures. The method exploits the time domain disjointness of the sources, which is rather a restrictive scenario. By making a more realistic assumption that the sources are disjoint in the time-frequency domain, we have extended the CR formulation to the time-frequency domain and proposed a framework to estimate sparse filters. The framework contains a time-frequency clustering step followed by a filter estimation step. In a setting where the clustering is performed using strong side information, we have presented the results of experimental evaluation of the filter estimation step.

In the future, we would like to understand how to set ϵ with less or no prior information. As mentioned earlier, the recovery performance could be improved by using a higher threshold θ and a larger number of time frames $N_T(k, N_F)$. Also, the run-time of the algorithm for large problem sizes is an issue and we would like to explore alternative fast algorithms.

Table 1. Average output SNR for $N = 2$

	$k = 2$ (Anechoic)	$k = 4$	$k = 6$	$k = 8$	$k = 10$	$k = 12$	$k = 14$	$k = 16$
$N_F = 1$	60.76	29.07	19.24	18.96	18.68	10.95	13.39	13.84
$N_F = 2$	47.98	23.85	18.15	15.51	15.00	9.08	9.51	10.37
$N_F = 3$	46.38	24.99	15.77	17.35	13.98	9.59	12.15	9.79
$N_F = 4$	44.57	20.18	14.84	16.88	14.74	9.18	8.22	9.01
$N_F = 5$	42.30	21.75	11.87	15.37	13.82	10.30	8.21	7.52

Table 2. Average output SNR for $N = 3$

	$k = 2$ (Anechoic)	$k = 4$	$k = 6$	$k = 8$	$k = 10$	$k = 12$	$k = 14$	$k = 16$
$N_F = 1$	52.54	26.34	20.74	20.64	17.79	13.41	11.76	13.05
$N_F = 2$	51.49	23.85	16.89	16.99	15.30	9.95	10.49	9.28
$N_F = 3$	44.91	22.74	13.36	13.62	15.04	10.87	10.51	8.60
$N_F = 4$	44.00	21.34	13.73	13.61	10.98	10.05	9.38	8.91
$N_F = 5$	39.68	23.50	13.52	13.41	11.21	8.97	9.04	7.60

We will also consider solving the clustering and filter estimation problems simultaneously. Given an estimate of the filters, we would like to formulate and solve a convex problem to accomplish the clustering task. We intend to estimate the clusters and filters by solving the corresponding convex problems alternatively.

Acknowledgements

This work was supported in part by Agence Nationale de la Recherche (ANR), project ECHANGE (ANR-08-EMER-006) and by the EU FET-Open project FP7-ICT-225913-SMALL.

References

1. Arberet, S., Gribonval, R., Bimbot, F.: A Robust Method to Count and Locate Audio Sources in a Multichannel Underdetermined Mixture. IEEE Trans. on Signal Processing 58(1), 121–133 (2010)
2. Jourjine, A., Rickard, S., Yilmaz, O.: Blind separation of disjoint orthogonal signals: Demixing n sources from 2 mixtures. In: ICASSP 2000, vol. 5, pp. 2985–2988 (2000)
3. Pedersen, M.S., Larsen, J., Kjems, U., Parra, L.C.: A survey of convolutive blind source separation methods. In: Springer Handbook of Speech Processing, Springer, New York (2007)
4. Aïssa-El-Bey, A., Abed-Meraim, K.: Blind SIMO channel identification using a sparsity criterion. In: SPAWC 2008, pp. 271–275 (2008)
5. Liu, H., Xu, G., Tong, L.: A deterministic approach to blind identification of multichannel FIR systems. In: ICASSP 1994, vol. 4, pp. 581–584 (1994)
6. Aïssa-El-Bey, A., Abed-Meraim, K., Grenier, Y.: Blind Separation of Underdetermined Convolutive Mixtures Using Their Time–Frequency Representation. IEEE TASLP 15(5), 1540–1550 (2007)
7. Candès, E.J., Romberg, J., Tao, T.: Robust uncertainty principles: Exact signal reconstruction from highly incomplete frequency information. IEEE Trans. on Info. Theory 52(2), 489–509 (2006)
8. Grant, M., Boyd, S.: CVX: Matlab software for disciplined convex programming (web page and software) (June 2009), http://stanford.edu/~boyd/cvx

Adaptive and Non-adaptive ISI Sparse Channel Estimation Based on SL0 and Its Application in ML Sequence-by-Sequence Equalization

Rad Niazadeh[1], Sina Hamidi Ghalehjegh[1],
Massoud Babaie-Zadeh[1,⋆], and Christian Jutten[2]

[1] Department of Electrical Engineering, Sharif University of Technology, Tehran, Iran
[2] GIPSA-Lab, Grenoble, and Institut Universitaire de France, France
radniazadeh@gmail.com, shghamidi@gmail.com, mbzadeh@yahoo.com,
Christian.Jutten@inpg.fr

Abstract. In this paper, we firstly propose an adaptive method based on the idea of Least Mean Square (LMS) algorithm and the concept of smoothed l_0 (SL0) norm presented in [1] for estimation of sparse Inter Symbol Interface (ISI) channels which will appear in wireless and acoustic underwater transmissions. Afterwards, a new non-adaptive fast channel estimation method based on SL0 sparse signal representation is proposed. ISI channel estimation will have a direct effect on the performance of the ISI equalizer at the receiver. So, in this paper we investigate this effect in the case of optimal Maximum Likelihood Sequence-by-sequence Equalizer (MLSE) [2]. In order to implement this equalizer, we propose a new method called pre-filtered Parallel Viterbi Algorithm (or pre-filtered PVA) for *general ISI sparse channels* which has much less complexity than ordinary Viterbi Algorithm (VA) and also with no considerable loss of optimality, which we have examined by doing some experiments. Indeed, Simulation results clearly show that the proposed concatenated estimation-equalization methods have much better performance than the usual equalization methods such as Linear Mean Square Equalization (LMSE) for ISI sparse channels, while preserving simplicity at the receiver with the use of PVA.

1 Introduction

An sparse channel is a channel whose impulse response has only a few significant coefficients. More precisely, in the framework of digital communication there are special scenarios in which one can model the overall channel as a Finite Impulse Response (FIR) sparse filter which will produce interference with previous samples. Such channels may be encountered, for example, in wireless multipath fading channels, acoustic underwater channels, etc. In such scenarios, the operation of estimating the channel and equalizing the produced ISI is an important

⋆ This work has been partially funded by Iran Telecom Research Center (ITRC), and also by center for International Research and Collaboration (ISMO) and French embassy in Tehran in the framework of a GundiShapour collaboration program.

V. Vigneron et al. (Eds.): LVA/ICA 2010, LNCS 6365, pp. 579–587, 2010.

task of the receiver. Considering the problem of estimating the channel, many efforts have been done to design batch estimation algorithms [3,4,5]. Most of these algorithms try to exploit the sparsity of the channel or detecting the locations of non-zero taps of the Channel Impulse Response(CIR). In addition to these, adaptive algorithms are also presented in the literature which are based on iterative estimating of the sparse filter [6]. At the other side, the problem of equalizing the output of sparse channel is in the point of interest. Specially optimal Viterbi equalization is desired [7].

In this work, we try to investigate these two problems (which are so related to each other) in the case of ISI sparse channels with a novel approach. According to the best knowledge of the authors, all of the works around the problems of estimating the channel and equalization have considered ISI channel as 'FIR causal filter' with no constraint on the taps. The first contribution of our work is to solve these two problems in the case of using a matched filter structure at the receiver. In this situation, the resulting sparse FIR channel can be assumed to be minimum phase [2] (the reason of this assumption is described in Sect. 2). This assumption can make our problem much easier to solve (as will be shown in Sect. 3 of this work). So, by the use of this assumption we develop our algorithms in order to find the solutions to both of the estimation and equalization problems. Followed by these, we experimentally examine the efficiency of the proposed algorithms in a concatenation of estimation and equalization.

In the first section of this paper, motivated by Smoothed l_0-norm (SL0) algorithm [1] which is known as a fast sparse representation technique, we propose two new approaches for sparse channel estimation problem. Firstly, by modifying LMS algorithm introduced by Widrow and Hoff [8], we propose an adaptive algorithm similar to the algorithms proposed in [6] (which are named as Zero-Attracting LMS (ZA-LMS) and Re-weighted Zero-Attracting LMS (RZA-LMS) in [6]) called SL0-LMS. We show that by exploiting the sparsity information and using the concept of Smoothed l_0-norm we can improve the filtering performance. The second proposed algorithm is a non-adaptive algorithm which uses SL0 in a direct manner to estimate channel coefficients by finding the sparsest solution[1] of an under-determined system of linear equations [1]. Although this algorithm is non-adaptive, it uses the speed of SL0 while preserving enough accuracy for the act of equalization. It is important to mention that none of the proposed algorithms in this section depends on the assumption of modelling the channel as a minimum phase filter and can be used in the cases when we do not use the matched filter structure at the receiver.

In the next section, efficient equalization based on Maximum Likelihood Sequence-by-sequence Equalization (MLSE) will be described. In conventional ISI sparse channels, implementation of MLSE using ordinary VA is almost impossible (because of the fact that computational complexity of the receiver grows exponentially with the channel memory [2]). However, in some special cases known as zero-pad ISI channels, Parallel Viterbi Algorithm (PVA) is used instead of ordinary VA [9] which has much less complexity. In the case of *general*

[1] Solution with minimum l_0-norm, i.e. minimum number of non-zero coefficients.

ISI sparse channels, at our best knowledge, no such solution is yet presented in the literature. So, we propose a new method for using PVA in *general* ISI sparse channels based on the modelling of the ISI channel with a minimum-phase FIR filter. Our idea is based on the usage of a pre-filter at the receiver. In fact, this filter will re-shape the channel structure to a zero-pad channel which is presented in [9] and so, applying the PVA method will be possible afterwards. Although this method is not optimal, as we will see in experimental results, we do *not* have considerable amount of loss in optimality in most practical cases.

In the last section, simulation results will verify the efficiency of the presented algorithms in both fields of estimation of the channel and equalization of the produced ISI. It is important to note that the overall performance of the receiver system (as a concatenation of channel estimation and ISI equalization) depends on both of these parts which will be evaluated in these simulations and it is considered as the overall measure of efficiency of the proposed algorithms for both estimation and equalization tasks.

2 ISI Sparse Channel Estimation

In this section, we would like to estimate the ISI sparse channel coefficients while transmitting digital bits through this channel using Binary Phase Shift Keying (BPSK) modulation. Including the BPSK modulator and de-modulator with the channel and assuming that we are using matched filter structure at the receiver, the total ISI channel can be modelled as an FIR minimum-phase filter $\mathbf{f} = [f_0, \cdots f_{M-1}]^T$, which has only a few non-zero coefficients, followed by an additive white Gaussian noise $(v(n))$ which is independent of the input [2]. To show that this assumption is reasonable, we note that by sampling at the output of the matched filter with a suitable period, it is convenient to model the overall channel as a symmetric non-causal FIR filter with taps equal to $\{x(n) : -M \leq n \leq M\}$ and an additive coloured noise $z(n)$ with spectral density of $X(z) = \mathscr{Z}\{x(n)\}$ in which $\mathscr{Z}\{.\}$ denotes the Z-transform operator [2]. This model of describing an ISI channel is called the "X" model in the area of communication [2]. Additionally, we force the sparsity constraint on the taps of $x(n)$ which is a valid constraint in the scenarios that were mentioned in the previous section. Although this model is very common in the literature, we did not use this model in our work. Instead, we use a simpler model called "F" model [2] which can be made by decomposing the Z-transform of channel impulse response into a minimum phase filter and its conjugate reciprocal $(X(z) = F(z)F^*(\frac{1}{z^*}))$[2]. So by concatenating a whitening filter $G(z) = \frac{1}{F^*(\frac{1}{z^*})}$ to the end of sampler, the equivalent discrete channel would be a minimum phase FIR filter with taps equal to $\{f(n) : 0 \leq n \leq M\}$ and $F(z) = \mathscr{Z}\{f(n)\} = G(z)X(z)$.

According to the relation between $x(n)$ and $f(n)$ in the frequency domain, we have $x(n) = f(n) * f^*(-n)$ in which $*$ denotes the convolution operation. So according to this relation, when $x(n)$ is very sparse it is very unlikely that

[2] The possibility of this decomposition is due to the symmetry of $x(n)$.

$f(n)$ would not be sparse. Although we have currently no mathematical proof for this, it can heuristically seen from the convolution sum in which it is very unlikely that the convolution of two non sparse signal produces an sparse signal. It can also be experimentally verified. According to all of these, we use the above mentioned minimum phase sparse FIR model for the ISI channel which will make our problem much easier to solve.

Now, let $d(n)$ be the last observed sample of the noisy output signal of the channel and let $\mathbf{u}(n)$ be a vector that contains the last M samples of the input signal of the ISI sparse channel, that is:

$$\mathbf{u}(n) = [u(n), u(n-1), \cdots, u(n-M+1)]^T \ . \tag{1}$$

Consequently, we have the following input-output relation for the channel:

$$d(n) = \mathbf{f}^T \mathbf{u}(n) + v(n) \ . \tag{2}$$

In order to estimate the channel, we use a semi-random training sequence (which is known to the receiver) that is generated by producing a random block of 0, 1 bits with length equal to M and transmitting it periodically (note that the BPSK modulator will map these bits to ± 1 symbols).

2.1 Adaptive Sparse Channel Estimation

Review of ZA-LMS and RZA-LMS. In the state of transmitting the training sequence, the standard LMS, which is based on iteratively minimizing the cost function $J(\mathbf{w}(n)) = E\{e^2(n)\}$, adaptively estimates $\mathbf{f}$ by the following recursion:

$$\mathbf{w}(n+1) = \mathbf{w}(n) + \mu e(n)\mathbf{u}(n) \ . \tag{3}$$

where $\mathbf{w}(n)$ is the estimated adaptive filter at the nth iteration, μ is the step size parameter and $e(n) = d(n) - \mathbf{w}(n)^T \mathbf{u}(n)$ [8]. Then, in ZA-LMS, the cost function is modified by adding a penalty term based on l_1 norm to enforce some sparsity on $\mathbf{w}(n)$ which is an estimation of the sparse vector $\mathbf{f}$:

$$J_1(\mathbf{w}(n)) = \frac{1}{2} E\{e^2(n)\} + \gamma \|\mathbf{w}(n)\|_1 \ . \tag{4}$$

Using steepest descent, the channel coefficients update equation will then be [6]:

$$\mathbf{w}(n+1) = \mathbf{w}(n) + \mu e(n)\mathbf{u}(n) - \mu\gamma \operatorname{sgn}(\mathbf{w}(n)) \ . \tag{5}$$

In RZA-LMS the new cost function is defined as [6]:

$$J_2(\mathbf{w}(n)) = \frac{1}{2} E\{e^2(n)\} + \gamma \sum_{i=1}^{M} \log(1 + \varepsilon |w_i|) \ . \tag{6}$$

The log-sum term has been used because it behaves more similarly to the l_0 norm than $\|\mathbf{w}(n)\|_1$. So, the update equation will be

$$w_i(n+1) = w_i(n) + \mu e(n) u_i(n) - \mu\gamma\varepsilon \frac{\operatorname{sgn}(w_i(n))}{1 + \varepsilon |w_i(n)|} \ . \tag{7}$$

The New Smoothed l_0-LMS Algorithm (SL0-LMS). Inspired from the idea of the SL0 algorithm [1], we propose replacing the above mentioned cost function by:

$$J_3(\mathbf{w}(n)) = \frac{1}{2}E\{e^2(n)\} + \gamma\|\mathbf{w}(n)\|_0 \ , \tag{8}$$

in which, $\|\mathbf{w}\|_0$ is replaced by its smooth approximation as [1] in order to exploit the sparsity nature of the estimated channel, i.e. we use this approximation:

$$\|\mathbf{w}\|_0 \approx M - \sum_{i=1}^{M} e^{-w_i^2/2\sigma^2} \ . \tag{9}$$

So, the update equation will be

$$w_i(n+1) = w_i(n) + \mu e(n)u_i(n) - \frac{\rho\gamma}{\sigma^2}w_i(n)e^{-w_i^2(n)/2\sigma^2} \ . \tag{10}$$

As shown in [1], (9) tends to equality when $\sigma \to 0$. Consequently, we expect the SL0-LMS have a better performance than RZA-LMS and the experimental results prove this. Unfortunately, this algorithm requires adjusting the parameter σ, which involves nested loops and thus increases the computational complexity.

2.2 Non-Adaptive SL0 Based Sparse Channel Estimation

In this section, we propose a non-adaptive channel estimation algorithm which will estimate the channel coefficients *after* observing $m \ll M$ successive samples of the output signal of the channel. The relation of theses observation to the input sequence can be described in the matrix form as follows :

$$\mathbf{d}_{m\times 1} = \mathbf{A}_{m\times M}.\mathbf{f}_{M\times 1} + \mathbf{v} \ . \tag{11}$$

in which, $\mathbf{v}$ is a vector including samples of the Gaussian noise, $\mathbf{d}$ is the vector of observations, $\mathbf{f}$ is the vector of channel taps and $\mathbf{A}$ is a Toeplitz random matrix known to the receiver whose rows are circular shift of $\mathbf{q} = [q_0, \cdots, q_{M-1}]^T$ and $\mathbf{q}$ is a M-length random block of ± 1s (see Sect. 2). Consequently, if we start observing the channel output after l transmissions, then $\mathbf{A}$ can be expressed as the following matrix:

$$\mathbf{A} = \begin{bmatrix} q_l & q_{l-1} & q_{l-2} & \cdots & q_0 & q_{M-1} & \cdots & q_{l+1} \\ q_{l+1} & q_l & q_{l-1} & \cdots & q_1 & q_0 & \cdots & q_{l+2} \\ \vdots & \vdots & \vdots & \vdots & \vdots & \vdots & \ddots & \vdots \\ q_{l-1} & q_{l-2} & q_{l-3} & \cdots & q_{M-1} & q_{M-2} & \cdots & q_l \end{bmatrix} . \tag{12}$$

Now, (11) is a noisy under-determined system of linear equations which has to be solved under the sparsity constraints. Now, we want to find the sparsest solution of (11) while having a constraint on square error, i.e. we are trying to solve the following optimization problem:

$$\text{argmin } \|\tilde{f}\|_0 \quad s.t \quad \|\mathbf{A}\tilde{\mathbf{f}} - \mathbf{d}\|_2^2 \leq \epsilon \ . \tag{13}$$

in which we assume that in (11), $\|\mathbf{v}\|_2^2 \leq \epsilon$. There are many methods for finding the sparsest solution of a noisy under-determined system of linear equations such as Basis Pursuit De Noising (BPDN) [10], Least-Absolute Shrinkage and Selection Operator (LASSO) [11] and Robust SL0 [1,12]. By the use of simulations, we have seen that there is not much difference in the accuracy of Robust SL0, LASSO and BPDN in our application and so, we use Robust-SL0 according to its speed in comparison to the other two algorithms. So, in our experimental results we just mentioned the results of Robust-SL0 algorithm.

3 Pre-Filtered PVA Equalization in General ISI Sparse Channels

After estimating channel coefficients using one of the methods of the previous section, in this part, efficient equalization in *general* ISI sparse channels will be investigated (in which the channel has a large amount of memory while having a few number of significant coefficients). In these channels, implementation of the optimum ISI equalizer (MLSE) by the use of Viterbi algorithm is almost impossible (because computational complexity of the receiver grows exponentially with channel memory [2]). PVA which uses parallel trellises can be used in a special case of such channels known as *the zero-pad channel* [9] in which channels have equally spaced coefficients. Assume that $\mathbf{w}$ is a zero-pad channel with the length $M = K.L + 1$, then it has a structure as follows:

$$\mathbf{w} = [\underbrace{w_0 \; 0 \; 0 \; \ldots}_{K \text{ coefficients}} \underbrace{w_K \; 0 \; 0 \; \ldots}_{K \text{ coefficients}} \underbrace{w_{2K} \; 0 \; 0 \; \ldots}_{K \text{ coefficients}} \ldots \; 0 \; 0 \; w_{L.K}] \; . \tag{14}$$

As in [9], it is evident that received symbols $\{u_0, u_K, u_{2K} \ldots\}$ will have interference between themselves while having no interference on any other symbols and they will generate output symbols $\{d_0, d_K, d_{2K} \ldots\}$. Similarly, symbols $\{u_1, u_{K+1}, u_{2K+1} \ldots\}$ have ISI between themselves and have no interference with any other symbols and so on. So, we can equalize this channel using K parallel trellises each using Viterbi algorithm for a channel with coefficients as:

$$\mathbf{w}' = [w_0 \quad w_K \quad w_{2K} \quad \ldots \quad w_{L.K}]^T \; . \tag{15}$$

and the input of the jth trellis are symbols $\{d_{i.k+j} : \forall i \in \mathrm{N}\}$. So, this PVA structure will have less overall complexity than a single trellis in these channels. In fact, the complexity of a single trellis is of $\mathrm{O}(2^M)$ and the complexity of PVA is of $\mathrm{O}(K.2^L)$ which is much less than the single trellis case (for example, in our experiment $K = L = 64$ and hence, implementation of normal Viterbi algorithm which requires 2^{64} states is impossible. But having 8 trellises each with $2^8 = 64$ states is practical and possible). In the case of general ISI sparse channels, reducing the complexity of Viterbi algorithm is much more sophisticated and according to the authors' best knowledge, no exact solution has been presented in the literature [9]. To find a solution for the case of general ISI sparse channel, firstly we use the matched filter structure at the receiver and so, as was mentioned before, we can use the "F" model of the channel in which the channel impulse response can be assumed to be an FIR minimum-phase sparse filter. In this way, we can change the channel to be similar to the *comb* shape channel described above. Precisely, our idea is based on a pre-filtered equalization. This pre-filter will *re-shape* the channel impulse response to a zero-pad channel which has coefficients similar to the original channel, but they are moved so that the coefficients of the resulting channel will be equally spaced. In order to build such a filter, we use an IIR filter structure which has the estimated channel as its denominator and re-shaped channel as its numerator. In other words we use the following filter:

$$W_{\mathrm{pre}}(z) = \frac{\tilde{F}(z)}{F(z)} \ . \tag{16}$$

in which $F(z)$ and $\tilde{F}(z)$ are the Z-transforms of the channel impulse response and its reshaped version, respectively. According to the fact that the channel impulse response is minimum-phase, this IIR filter will be stable and causal. So, PVA can be implemented afterwards using a few number of trellises which will cause a great amount of computational complexity reduction, but the equalization structure will be the dependent to the CIR which is somehow impractical.

4 Experimental Results

In this section, firstly we compare the performance of channel estimators mentioned in Sect. 2.1 which are SL0-LMS (Eq. 10), ZA-LMS (Eq. 5), RZA-LMS (Eq. 7) and standard LMS. In this experiment, the input signal is an equiprobable random vector of ± 1 with length 5000 and additive noise is white Gaussian random sequence of length 5000 and variance 10^{-3}. The channel has 256 taps where only 28 of them are non zero (which are selected random). The result of various estimation methods with equal step-sizes is shown in Fig. 1. It is clear that for this long sparse channel, the SL0-LMS is drastically better than

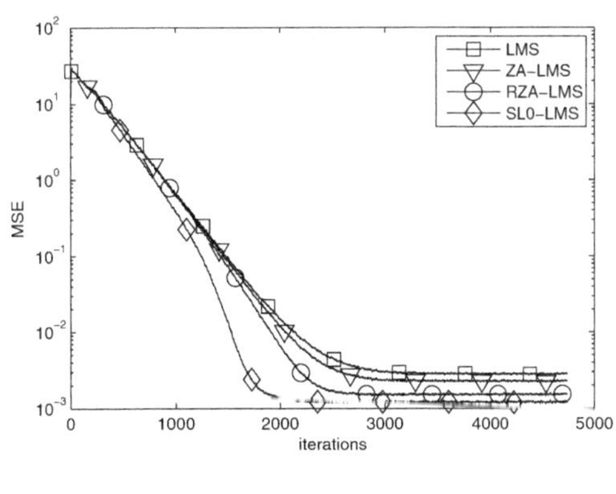

Fig. 1. Tracking and steady-state behaviors of estimation methods

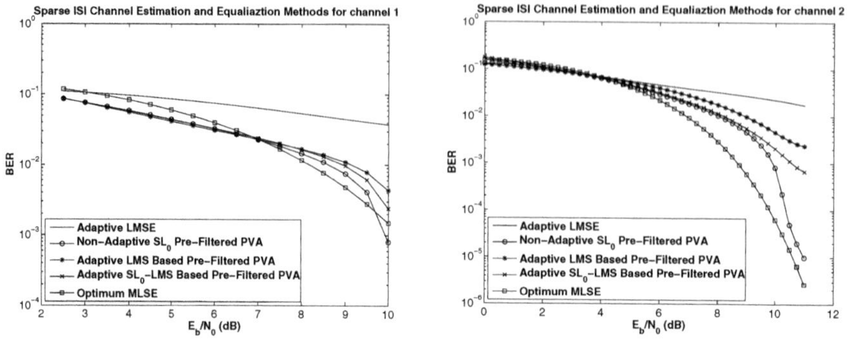

(a) experiment 1. (b) experiment 2.

Fig. 2. Comparison of the BER-SNR curves for our estimation/equalization methods

the other algorithms, both in steady-state and convergence rate behaviour while needs a pre-adjustment of parameter σ.

After that, concatenation of the proposed channel estimation methods with PVA will be applied in two experiments in order to test the efficiency of the proposed channel estimation algorithms and pre-filtered PVA. In these experiments, our equalization method will be compared to adaptive LMS Equalizer (LMSE) and approximate Bit Error Rate (BER) bounds for MLSE introduced in [2]. Channel taps for these two experiments are chosen at random, but with the constraint that the channel is minimum-phase according to our model. So, we generate random channels and select two of those which are minimum phase. In the case of LMSE, we have used an adaptive filter that it has the same number of taps comparing to the actual channel. Additionally the one for experiment 1 is so close to the comb shape, while the one for experiment 2 is not. The resulting BER vs Signal to Noise Ratio (SNR) per Bit is shown in Fig. 2(a) and Fig. 2(b). It is important to note that the derivation of the optimum MLSE's BER curves is impossible during the simulation (according to its complexity), and so we use approximate bounds for "Optimum MLSE" curves in these figures instead of using the exact curves. Advantages of our estimation methods and the proposed pre-filtered PVA equalizer could be seen in these results.

5 Conclusion

In this paper, we showed that with the use of proposed estimation/equalization methods, we benefit the speed of adaptive filtering, the optimality of ML equalizer and the complexity reduction of PVA. According to that, we have no significant loss of performance at the receiver while having much reduction in complexity. In fact, pre-filtering may increase the noise power, but as we have shown experimentally, the performance of our method is not much less than MLSE bound and so is appropriate in the sense of error performance.

References

1. Mohimani, H., Babaie-Zadeh, M., Jutten, C.: A fast approach for overcomplete sparse decomposition based on smoothed l_0-norm. IEEE Trans. Signal Processing 57(1), 289–301 (2009)
2. Proakis, J.: Digital Communication, 4th edn. Mc-Graw Hill, New york (2001)
3. Carbonelli, C., Vedantam, S., Mitra, U.: Sparse channel estimation with zero tap detection. IEEE Trans. Communication 6(5), 1743–11754 (2007)
4. Cotter, S., Rao, B.: Sparse channel estimation via matching pursuit with application to equalization. IEEE Trans. Communication 50(3), 374–378 (2002)
5. Li, W., Vedantam, S., Preisig, J.: Estimation of rapidly time-varying sparse channels. IEEE Journal of Oceanic Engineering 32(4), 927–940 (2007)
6. Chen, Y., Gu, Y., Hero III, A.O.: Sparse LMS for system identification. In: ICASSP, pp. 3125–3128 (2009)
7. Mcginty, N.C., Kennedy, R., Hoeher, P.: Parallel trellis viterbi algorithm for sparse channels. IEEE Communication Letters 2(5), 143–145 (1998)
8. Widrow, B., Streams, S.: Adaptive Signal Processing. Prentice-Hall, New Jersey (1985)
9. Mietzner, J., Hoeher, S., Land, I., Hoeher, P.: Trellis-based equalization for sparse ISI channels revisited, ISIT, 229–233 (2005)
10. Chen, S., Donoho, D., Saunders, M.: Atomic decomposition by basis pursuit. SIAM review 43(1), 129–159 (2001)
11. Tibshirani, R.: Regression shrinkage and selection via the lasso. Journal of the Royal Statistical Society. Series B (Methodological) 58(1), 267–288 (1996)
12. Eftekhari, A., Babaie-Zadeh, M., Jutten, C., Moghaddam, H.: Robust-SL0 for stable sparse representation in noisy settings. In: ICASSP 2009, pp. 3433–3436 (2009)

Extraction of Foetal Contribution to ECG Recordings Using Cyclostationarity-Based Source Separation Method

Michel Haritopoulos[1], Cécile Capdessus[1], and Asoke K. Nandi[2]

[1] Institut PRISME
21, rue de Loigny la Bataille, 28000, Chartres
{Michel.Haritopoulos,Cecile.Capdessus}@univ-orleans.fr
http://www.univ-orleans.fr/prisme/

[2] Signal Processing and Communications group, Department of Electrical Engineering and Electronics
The University of Liverpool, Brownlow Hill, Liverpool L69 3GJ, U.K.
A.Nandi@liverpool.ac.uk

Abstract. In this paper we propose a cyclostationary approach to the problem of the foetal electrocardiogram (FECG) extraction from a set of cutaneous potential recordings of an expectant mother. We adopted a semi-blind source separation (BSS) method for which the only necessary prior knowledge is that of the fundamental cyclic frequency of the cyclostationary process to be estimated. Using this technique, the estimated cyclostationary FECG source of interest is found to be free from any interferences with the mother's ECG (MECG) signal. Experimental results and perspectives for future research conclude this paper.

Keywords: Cyclostationarity, Semi-Blind Source Extraction, Foetal Electrocardiogram.

1 Introduction

Electrocardiograms (ECG) are very often used as diagnostic tools for heart monitoring. In the case of pregnant women, the recorded ECGs contain information concerning both mother's and foetus' condition. Early diagnosis in the case of the mother's ECGs (MECG) can be helpful for medical assessment of possible diseases, while in the foetus case (FECG), any prenatal information that can be provided may be of great importance to prevent complications. ECGs acquired with non-invasive techniques do not give direct access to FECGs, because those are hidden by the maternal ECGs of higher amplitude and contaminated by various sources of disturbances (biological or other).

It is shown in the literature ([1], [2], [3], [4]), that this problem can be formulated in blind source separation (BSS) terms. It consists of separating or extracting one or more signals of interest from a set of observations which contain a linear or non-linear mixing of the unknown source signals that one wants to estimate.

V. Vigneron et al. (Eds.): LVA/ICA 2010, LNCS 6365, pp. 588–595, 2010.

In the case of ECG signals, m electrodes are placed at several key locations on the mother's body and the observations carried out by these electrodes are modelled in terms of BSS as a vector $\mathbf{x}$ assumed to be a linear mixture of the unknown source vector $\mathbf{s}$:

$$\mathbf{x}(t) = A\mathbf{s}(t) \tag{1}$$

where $\mathbf{s}(t)$ are the n unknown sources $[s_1(t), s_2(t), ..., s_n(t)]^\dagger \in \mathbb{R}^{n\times 1}$, $\mathbf{x}(t)$ outputs the m linearly mixed observations $[x_1(t), x_2(t), ..., x_m(t)]^\dagger \in \mathbb{R}^{m\times 1}$, A stands for the $m \times n$ mixing matrix and $\dagger$ denotes the transpose operator. Equation (1) is the well known *noiseless linear instantaneous* model of the BSS problem.

Various approaches have been used to solve the FECG extraction problem by using BSS techniques. One can find in the literature second-order statistics (SOS) based BSS methods [5] or adaptive noise cancelling methods compared with higher-order statistics (HOS) based BSS techniques [6]. Sparse representations [7] as well as wavelet based ICA [8] applied to this particular problem have also been reported.

Here we choose to apply an algorithm that allows to extract one specific cyclostationary source whose cyclic frequency is *a priori* known. A preliminary study of the data indicates that the observations exhibit cyclostationarity at the frequency of the foetal heartbeat rate, which is estimated and then used for the extraction.

2 Methods

2.1 Extraction Principle

The extraction method was previously presented in [9]. It aims at extracting a cyclostationary source of known cyclic frequency α_0 from a set of observations. We thus look for a $1 \times m$ extraction vector B such that:

$$z(t) = B\mathbf{x}(t) \tag{2}$$

is an estimate of the cyclostationary source. Hypotheses are as follows:

- The observations are instantaneous and additive mixtures of the sources.
- There are at least as many sensors as sources.
- The sources are supposed to be zero mean and uncorrelated to each other.
- The source to be extracted is cyclostationary at a frequency α_0 that is *a priori* known, i.e. it can either be measured or computed.
- The other sources can be either stationary or cyclostionary, provided that none of them is cyclostationary at the same frequency α_0.

The extraction is performed by minimising over B a criterion $C(B)$ based on second-order statistics of the observations. Let us denote by $R_z(t,0)$ the autocorrelation function of the estimate $z(t)$ at zero time-lag. Then:

$$C(B) = \left| \frac{R_z(0)}{R_z^{\alpha_0}(0)} \right| \tag{3}$$

where $R_z(0)$ and $R_z^{\alpha_0}(0)$ are the coefficients of the Fourier series decomposition of $R_z(t,0)$ at respective frequencies 0 and α_0.

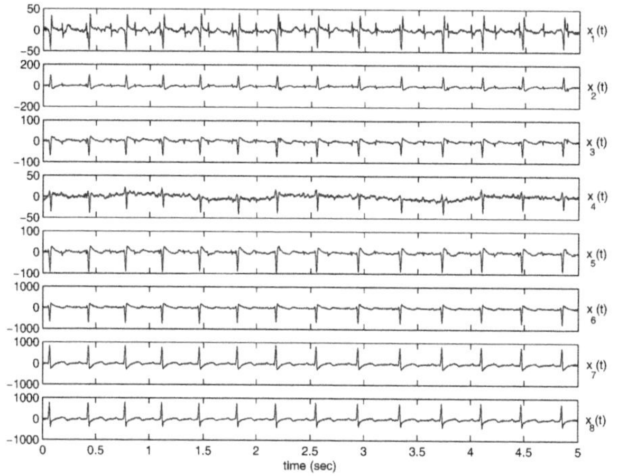

Fig. 1. Eight skin electrodes recordings form a pregnant woman

2.2 Algorithm

Let us denote by $\widehat{R}_{\mathbf{x}}(0) = \langle \mathbf{x}(t)\mathbf{x}^\dagger(t)\rangle_\theta$ and $\widehat{R}_{\mathbf{x}}^{\alpha_0}(0) = \langle \mathbf{x}(t)\mathbf{x}^\dagger(t)e^{-2\pi j\alpha_0 t}\rangle_\theta$ the estimates of the covariance matrix of the observations and their cyclic covariance matrix at frequency α_0, respectively. The $\langle . \rangle_\theta$ operator stands for the temporal averaging over θ seconds. The criterion is estimated from the observations as follows:

$$\widehat{C}(B) = \left| \frac{B\widehat{R}_{\mathbf{x}}(0)B^\dagger}{B\widehat{R}_{\mathbf{x}}^{\alpha_0}(0)B^\dagger} \right| . \tag{4}$$

It has been shown in [9] that minimising eq. (4) over B leads to the extraction of the corresponding cyclostationary component.

3 Methodology and Results

The proposed algorithm is very well adapted to the problem of FECG extraction as shown below. We apply it to a commonly used dataset in relation to FECG extraction research [10].

3.1 Experimental Dataset

The dataset we used is illustrated in Fig. 1. These are eight-channel cutaneous potential recordings from eight skin electrodes placed at different positions on an expectant mother's body. The first five observations denoted $[x_1(t), ..., x_5(t)]$ were recorded from the abdominal area, while the last three $[x_6(t), ..., x_8(t)]$ correspond to electrodes placed at the mother's thoracic region. All signals were sampled at frequency $f_e = 500\ Hz$ during $5\ s$, so that the observations vector $X(t)$ is $N = 2500$ samples long. The foetal heartbeat component can be perceived together with MECG contribution and noise in the $x_{1-5}(t)$ observations subset, while in the $x_{6-8}(t)$, the FECG contributions are barely visible because of the much longer distance between the foetus and the electrodes located on the mother's chest.

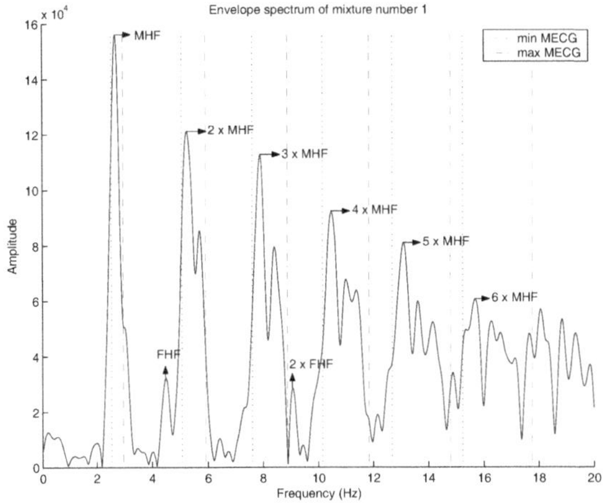

Fig. 2. Mother's (MHF) and foetus' (FHF) heartbeat frequencies, and their harmonics. Theoretical evolution of the MHF within the $k \times [2.538, 2.958]\ Hz$ values range ($k = 1, .., 6$) drawn in dotted and dash-dotted lines (minimum and maximum value, respectively).

We performed envelope spectrum analysis of the observations, i.e. the Fourier transform of the squared signals [11]. All eight channels give substantially the same envelope spectra; but as on the temporal ECG representations, it is at the $x_1(t)$ channel that the foetal heartbeat rate can be distinguished clearly. Fig. 2 is a zoom of the first mixture $x_1(t)$ vector's envelope spectrum; mother's heartbeat frequency (MHF) and its harmonic components $k \times MHF,\ k \in [1, 6]$ are dominating the whole spectrum. The frequency intervals $[min\ MECG, max\ MECG]$ plotted with dotted and dash-dotted lines, respectively, correspond to the $[2.538, 2.958]\ Hz$ frequency range values, within which mother's heartbeats vary. As expected, MHF lies in this interval.

Foetus' heartbeat frequency (FHF) frequency and its second harmonic also appear in this figure. But, the widening of the MHF harmonics at higher frequencies, makes it difficult to point out with certainty which spectral line corresponds to a FHF harmonic. The presence of spectral lines in the envelope spectrum supports the thesis of second order cyclostationarity characterizing the FECG component, with a fundamental cyclic frequency α_0 equal to 4.49 Hz, i.e. the value of the FHF.

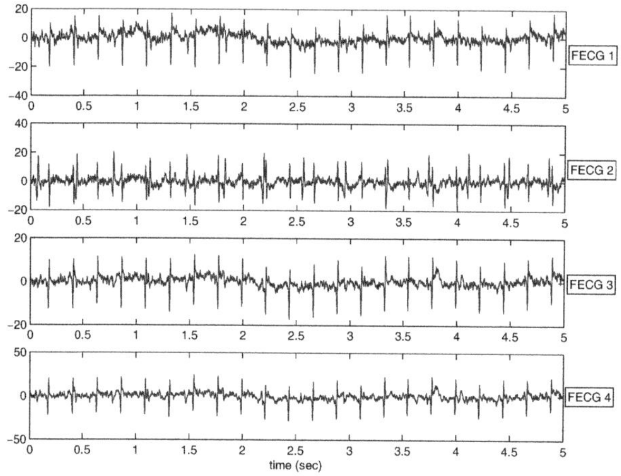

Fig. 3. From the top: four FECG estimations provided by the proposed algorithm with input mixture vectors $X_1(t)$,...,$X_4(t)$, respectively

3.2 Experiments and Results

We carried out experiments by applying our algorithm to different sets of observations. We describe, hereafter, the results obtained by using four different observations vectors. The first mixture vector $X_1(t)$ is composed of observations $[x_1(t), x_2(t), x_3(t)]^\dagger$, the second one $X_2(t) = [x_1(t), x_2(t), x_5(t)]^\dagger$, the third one $X_3(t)$ is $[x_1(t), x_3(t), x_5(t)]^\dagger$ and the last one is made of four observations vectors $X_4(t) = [x_1(t), x_2(t), x_3(t), x_5]^\dagger$ (see Fig. 1).

Note, that, we did not include into our mixtures vectors recorded data from channel number 4, to which FECG contribution is very low. Furthermore, baseline wander is observed in this channel as this is the case in many other ECG datasets. This introduces amplitude variations to all separated FECG contributions after application of a BSS algorithm [7]. One can find in the literature (e.g., [12]) preprocessing methods for baseline wander removal and bandpass filtering of ECG recordings. After running the new cyclostationary criterion based BSS algorithm [9], with fundamental FECG cyclic frequency $\alpha_0 = 4.49$ Hz as estimated from the envelope spectrum of $x_1(t)$, we obtain for each mixture vector $X_i(t)$, $i \in [1, 4]$ an estimate of the foetal electrocardiogram signal. Extraction results are shown in Fig. 3. The extracted FECG that exhibits both very little baseline wandering and the foetus heartbeats, is the last estimate, i.e. the one that used information from 4 channels for minimising criterion of eq. (4).

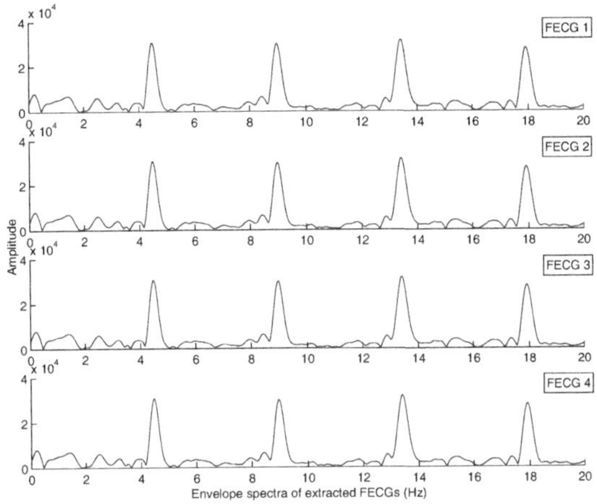

Fig. 4. From top to bottom: zoom to the envelope spectra of the extracted FECGs by using 4 different mixture vectors

Fig. 4 shows the envelope spectra of the extracted FECG signals after application of the proposed estimation criterion algorithm to the mixture vectors $X_i(t)$, $i \in [1, 4]$ (from top to bottom). In order to compare these results with the envelope spectra of Fig. 2, we plotted a zoom for each extracted FECG signature. It is worth pointing out that our algorithm works well for all different combinations of the mixture vector components, as well as for a different observations number m. Indeed, one can see in this figure that, the FHF fundamental frequency and its next three harmonics are not contaminated by any MHF spectral lines. Experiments have been carried out with all eight channels ECG recordings as input mixing vector to our extraction algorithm. A study of the FECG signal spectrum extracted after application of our algorithm to the whole ECG recordings set, confirms the existence of the first MHF harmonic, while no significant MHF components can be found in the spectrum computed from the extracted FECG using $X_4(t)$. This could be justified by assuming that recordings from the mother's thoracic region bring very few new information concerning the foetus' heartbeats.

3.3 Robustness to Foetus's Cyclic Frequency

The only prior knowledge required for our estimation method to work, is that of the foetus's fundamental cyclic frequency value α_0. As explained in section 3.2, we used a value of 4.49 Hz for α_0 as it is estimated from the envelope spectrum of the first ECG channel recordings. Deviations from this value can be detected depending on which channel's envelope spectrum the estimation is based on. Thus, we run some experiments with the previously described $X_4(t)$ mixing vector. At each run, we applied the proposed cyclostationary extraction method to $X_4(t)$ with a different α_0 value and we computed a separation quality criterion for the corresponding estimated FECG signal. This is a periodicity measure (PM $\in [0, 100]$) based on second-order statistical properties and introduced in [12]; it is computed at the mean mother's heartbeat rate without taking into account its variability:

$$PM = \frac{|E\{x(t)x(t+\tau_\mu)\}|}{\sqrt{E\{x(t)^2\}E\{x(t+\tau_\mu)^2\}}} * 100\%, \qquad (5)$$

where τ_μ is the mean mother's heartbeat period. If the extracted FECG is clear of MHF components, then PM should be 0%.

We use the parameter PM to establish two things. First we consider how efficient is the method in removing MECG in the extracted FECG signals. PM values computed from each of the eight observations, namely $x_1(t), ..., x_8(t)$, range from 17% to 29%. On the other hand, when we extract the FECG from the set of four observations $[x_1(t), x_2(t), x_3(t), x_5]$, the corresponding value is 0.3% which indicates essentially no presence in MECG. Next we consider how robust is the extracted FECG in terms of measured PM. We had estimated α_0 to be 4.49 Hz. Now we allow α_0 to vary from 4.4 Hz to 4.6 Hz in steps of 0.01 Hz. In each of these cases PM values are measured and these are all less than 0.5%. This indicates remarkable robustness of extracted FECG in terms of PM values.

4 Discussions and Conclusions

We have introduced in this paper a novel application of our cyclostationarity-based BSS algorithm to the problem of foetus heartbeat signals extraction from a pregnant's woman ECG recordings. We pointed out the main lines of the estimation principle and then successfully applied it to real ECG signals.

To use the proposed method, there is no need to prewhiten (by the principal component analysis, for example) the sensor signals; centering the observations (zero mean) and decorrelation assumption are sufficient. We just have to know/estimate the fundamental cyclic frequency of the cyclostationary process (i.e. the FECG signal, in this study) to extract from the mixtures. It then estimates the source signal of interest. In the case presented in this paper, since the cyclostationary frequency chosen for the extraction was that of the FECG signal, the estimated source fits the heartbeats of the foetus. The indeterminacies inherent to BSS methods, i.e. the sign and the amplitude of the separated source signal, still remain in the proposed method. But post processing of the extracted FECG signal may be used to compute its contribution to each sensor.

Further experiments are under progress in order to be able to compare our algorithm's performance with that of similar work reported in the literature; its robustness to the foetus' cyclic frequency is also studied for different ECG channels and different number of mixtures. Taking into account the presence of baseline wander in some sensors, extracting the mothers's heartbeat signals and the characteristic waveforms associated with each heartbeat are of great interest in order to perform a robust monitoring of the foetus well-being.

References

1. Lathauwer, L.D., Moor, B.C., Vandewalle, J.: Foetal Electrocardiogram Exctraction by Blind Source Subspace Separation. IEEE Trans. Biomed. Eng. 47(5), 567–572 (2000)
2. Zarzoso, Z., Nandi, A.K., Bacharakis, E.: Maternal and Foetal ECG Separation Using Blind Source Separation Methods. IMA Journal of Mathematics Applied in Medicine & Biology 14(3), 207–225 (1997)
3. Bacharakis, E., Nandi, A.K., Zarzoso, V.: Foetal ECG Extraction Using Blind Source Separation Methods. In: Ramponi, G., Sicuranza, G.L., Carrato, S., Marsi, S. (eds.) Edizioni Lint Trieste, Italy, vol. 1, pp. 395–398 (1996)
4. Bacharakis, E., Nandi, A.K., Zarzoso, V.: Foetal ECG Separation from Surface Measurements. In: IMA Conference on Mathematics in Medicine and Biology, Oxford, UK (1996)
5. Jafari, M.G., Wang, W., Chambers, J.A., Hoya, T., Cichocki, A.: Sequential Blind Source Separation Based Exclusively on Second-Order Statistics Developed for a Class of Periodic Signals. IEEE Transaction on Signal Processing 54(3), 1028–1040 (2006)
6. Zarzoso, V., Nandi, A.K.: Noninvasive Foetal Electrocardiogram Extraction: Blind Separation Versus Adaptive Noise Cancellation. IEEE Transactions on Biomedical Engineering 48(1), 12–18 (2001)
7. Blumensath, T., Davies, M.E.: Blind Separation of Maternal and Foetal ECG Recordings using Adaptive Sparse Representations. In: Nandi, A.K., Zhu, X. (eds.) Proceedings of ICA Research Network International Workshop 2006, Liverpool, UK (2006)
8. Azzerboni, D., La Foresta, F., Mammone, N., Morabito, F.C.: A New Approach Based on Wavelet-ICA Algorithms for Foetal Electrocardiogram Extraction. In: Proceedings ESSAN 2005, d-side publi., Bruges, Belgium (2005)
9. Capdessus, C., Nandi, A.K., Bouguerriou, N.: Source Extraction Algorithm Based on Cyclic Properties. In: Proceedings of ICA Research Network International Workshop 2008, Liverpool, UK, pp. 36–39 (2008)
10. (1999, May) DaISy; Database for the Identification of Systems. ESAT/SISTA, K. U. Leuven, Belgium, `http://homes.esat.kuleuven.be/~smc/daisy/`
11. Randall, R.B., Antoni, J., Chobsaard, S.: The Relationship between Spectral Correlation and Envelope Analysis for Cyclostationary Machine Signals. Application to Ball Bearing Diagnostics. Mechanical Systems and System Processing 15(5), 945–962 (2001)
12. Sameni, R., Jutten, C., Shamsollahi, M.B.: A Deflation Procedure for Subspace Decomposition. IEEE Transactions on Signal Processing 58(4), 2363–2374 (2010)

Common SpatioTemporal Pattern Analysis

Ronald Phlypo*, Nisrine Jrad, Bertrand Rivet, and Marco Congedo

Vision and Brain Signal processing (ViBS) Group, GIPSA Lab. Grenoble INP/UMR 5216 CNRS. BP 46, 961, Rue de la Houille Blanche, 38901 Saint Martin d'Hères, France

{firstname.lastname}@gipsa-lab.grenoble-inp.fr

Abstract. In this work we present a method for the estimation of a rank-one pattern living in two heterogeneous spaces, when observed through a mixture in multiple observation sets. Using a well chosen representation for an observed set of second order tensors (matrices), a singular value decomposition of the set structure yields an accurate estimate under some widely acceptable conditions. The method performs a completely algebraic estimation in both heterogeneous spaces without the need for heuristic parameters. Contrary to existing methods, neither independence in one of the spaces, nor joint decorrelation in both of the heterogeneous spaces is required. In addition, because the method is not variance based in the input space, it has the critical advantage of being applicable with low signal-to-noise ratios. This makes this method an excellent candidate ,e.g., for the direct estimation of the spatio-temporal P300 pattern in passive exogenous brain computer interface paradigms. For these applications it is often sufficient to consider quasi-decorrelation in the temporal space only, while we do not want to impose a similar constraint in the spatial domain.

1 Introduction

Passive exogenous brain computer interface (BCI) paradigms, such as the P300 speller, require sound signal processing techniques to identify the typical cerebral responses measured as scalp potential differences. The difficulty of the signal processing is mainly due to the low signal-to-noise ratio of the electroencephalogram (EEG), i.e. the recording containing the temporally and spatially sampled potential field at the scalp. Indeed, the P300 voltage as measured at the scalp electrodes amounts to a few microvolts, while the ongoing spontaneous electroencephalogram is dominated by oscillations extending over several tens of microvolts. Fortunately, signal processing techniques may use the phase-lock between the stimulus onset and the P300 inflection (note that the P300 originates from a positive inflection around 300ms post-stimulus). A straightforward way to estimate the P300 waveform would thus be to mean out the unsynchronised ongoing cerebral activity over a set of observations (trials) aligned to the stimuli onsets.

* This work has been supported through the project grant ANR-09-BLAN-0330 (Gaze & EEG) of the ANR (National Research Agency), France.

V. Vigneron et al. (Eds.): LVA/ICA 2010, LNCS 6365, pp. 596–603, 2010.

However, plain averaging (taking the mean value over the trials) is biased by the appearance of high amplitude artifacts such as eye blinks present in some of the trials (outliers of the distribution over which the mean is taken), and this bias is inversely proportional with the number of trials that are available. One could use robust statistics to estimate the average waveform, or use the more commonly weighted averaging techniques [1]. Weighted averaging techniques optimise the ratio of the estimated signal energy with respect to that of the noise and ongoing background EEG conditional on the availability of the covariance structures of the signal part and the background/noise part. However, care should be taken in the estimation of the weights, since generally the covariance structures of the background EEG (together with that of the noise components) and that of the component of interest can only be roughly estimated, resulting in an estimation of inferior quality with respect to simple averaging [2,3].

It is worthy to note that the above methods make only use of the temporal diversity (the temporal distribution of the samples over a time window), but none does exploit the spatial diversity (the spatial distribution of the samples over the various electrodes). In [4], the careful use of principal component analysis with varimax rotation has been promoted. It has been noted that the variance itself might not be the optimal criterion to obtain a P300 estimate and that the correlation might be more appropriate. In addition, – and confirmed in [5] – the varimax rotation seems to be preferable, since here, compact temporal representations of the components are preferred over smeared out variances. The varimax rotation is also closely related to the independent component analysis (ICA) [6] (applied to event-related potentials in e.g. [7]). Despite the fact that the compactness of the representation uses temporal information, rather than mere spatial decorrelation based on the covariance structure, the methods still either need a proper pre-selection of that trials/electrodes that are not contaminated by artifacts [5] or need a posterior selection of the components [7]. This is because some artifacts, such as blinks, have a similar compact temporal representation.

To avoid being biased by the artifact components in the estimation of the P300 waveform, methods based on spatio-temporal models have been introduced, e.g. [8,9]. Whilst [8] imposes a regular waveform comprising three free parameters (a waveform equivalent to the gamma distribution function with latency, amplitude and form factor as free parameters), [9] derives an iteratively refined waveform obtained through iterations of ICA and a selection procedure reminiscent to an expectation-maximisation. The estimation of the latency in [8] results in an exhaustive search over all possible delays. Moreover, the energy function used in the estimation of the latency needs to be sufficiently smoothened to alleviate spurious minima (a consequence of the noise), which requires the introduction of an extra, heuristically chosen parameter. In [9], the algorithm also requires a heuristically chosen parameter setting, namely a lower threshold on the correlation between the current estimate of the template and the individual estimates. It should be mentioned that both proposed algorithms only have an implicit coupling between the spatial and the temporal subspace and estimate them in an alternating fashion. As a consequence, estimation errors in each of the spaces carry over to the other space.

From the above, we may observe that the attention given in literature to this field of research has increased considerably during the last decades. Despite this interest in the topic, the authors are not aware of any attempt to explicitly estimate the spatio-temporal signal model without either modelling the signal of interest [8] or using a spatial estimation model with posterior calculation of the corresponding temporal waveforms [10,11] or algorithms alternating between spatial and temporal estimates [9,8]. In this contribution we present a method based on a joint spatio-temporal estimation of the P300. Incorporating the spatial distribution and the temporal waveform in a single, non-iterative algorithm without heuristic parameter selection, we aim at a more robust estimator. In addition, since no particular waveform will be imposed, nor a characteristic spatial distribution, the proposed method can be easily adopted for solving similar estimation problems in two coupled, heterogeneous spaces.

2 Methods

2.1 Notational Conventions

Upper case boldface ($\mathbf{A}$) and lower case boldface ($\mathbf{a}$) characters will respectively denote matrices and column vectors. Scalars and constants will be denoted by lower case light face (a) and upper case light face (A) characters, respectively. The i-th column of $\mathbf{A}$ is thus $\mathbf{a}_i$ and the jth entry of $\mathbf{a}$, a_j. An ensemble of matrices $\mathbf{A}^{(k)}$ will be denoted by the calligraphic upper case $\mathcal{A} = \{\mathbf{A}^{(k)} \,|\, k = 1, 2, \ldots, K\}$. The symbol $\otimes$ denotes the tensor product (Kronecker product) and the symbol $(\cdot)^T$ will stand for the matrix transposition operator.

2.2 Model

We suppose to have K samples of a process $\mathbf{X}$ defined on two coupled heterogeneous spaces. To ease the presentation, we will from hereon consider that these spaces are the spatial and the temporal dimensions typical of the P300, where the temporal dimension is taken relative to the stimulus onset. The K samples then simply represent the trials. The process realisations $\mathbf{X}^{(k)} \in \mathbb{R}^{M \times N}$ are matrices, with entries x_{mn} referred to their spatial index m, the arbitrarily chosen electrode index, and their temporal index n relative with respect to the stimulus onset[1]. Suppose the following generative model for $\mathbf{X}$:

$$\mathbf{X} = \begin{cases} \sigma_1 \boldsymbol{\upsilon}\boldsymbol{\nu}^T + \sum\limits_{i=2}^{L} \sigma_i \boldsymbol{\alpha}_{i-1} \boldsymbol{\beta}_{i-1}^T & \text{with probability } p \\ \sum\limits_{i=1}^{L} \sigma_i \boldsymbol{\mu}_i \boldsymbol{\zeta}_i^T & \text{with probability } 1-p \end{cases}, \qquad (1)$$

[1] Although it is common to choose n monotonically increasing with respect to the relative time after onset, this is not necessary for the algorithm to function well. The only pre-requisite is that there is a bijective relation between the relative time after onset and the temporal index.

which says that with probability p we observe the spatio-temporal structure $\mathbf{u}\mathbf{v}^T$ up to some angular noise on $\mathbf{u}$ and $\mathbf{v}$, yielding $\boldsymbol{\upsilon}\boldsymbol{\nu}^T$. Eq. (1) is completely defined by further fixing

$$\begin{array}{ll} \boldsymbol{\upsilon} \sim P_{VMF}(\mathbf{u}, \kappa) & \boldsymbol{\nu} \sim P_{VMF}(\mathbf{v}, \kappa) \\ \boldsymbol{\alpha}_i \sim P_{VMF}([\mathbf{U}^\perp]_i, \kappa_2) & \boldsymbol{\beta}_i \sim P_{VMF}([\mathbf{V}^\perp]_i, \kappa_2) \\ \boldsymbol{\mu}_i \sim P_{VMF}(\mathbf{q}_i, \kappa_2) & \boldsymbol{\zeta}_i \sim P_{VMF}(\mathbf{r}_i, \kappa_2) \end{array} \tag{2}$$

and for σ_i any (non-degenerate) distribution on the (positive) real numbers may be chosen. For our convenience we have chosen the uniform distribution on $(0, 1)$ in the simulations. $\mathbf{U}^\perp \in \mathbb{R}^{M\times(L-1)}$ and $\mathbf{V}^\perp \in \mathbb{R}^{N\times(L-1)}$ are random vectors constraint to form an orthogonal basis for the orthogonal complement of $\mathbf{u}$ and $\mathbf{v}$, respectively; $\mathbf{q}_i$ and $\mathbf{r}_i$ are L vectors from an orthogonal basis for $\mathbb{R}^M$ and $\mathbb{R}^N$, respectively. $P_{VMF}(\boldsymbol{\mu}, \kappa)$ stands for the Von Mises-Fischer probability distribution function with mean $\boldsymbol{\mu} \in \mathbb{S}^{N-1}$ and form parameter κ (in particular, $\kappa = 0$ means a uniform distribution on the hypersphere $\mathbb{S}^{N-1}$ and $\kappa \to +\infty$ means an improper Dirac distribution at $\boldsymbol{\mu}$). Remark that we have only chosen two different values for all κ's (κ and κ_2) to simplify the representation (and the subsequent simulations), however, κ might differ for each of the vectors in the model.

For the specific case of the P300 the above means we suppose that the spatio-temporal pattern – up to some directional noise – is present with probability p in our trials. We will detail this further in the next paragraph.

2.3 Representation of a Matrix

The samples, our trials, simply form a collection of observed matrices $\mathcal{X} = \{\mathbf{X}^{(k)}\}_{k=1}^K$. These matrices can be represented by a multitude of equivalent forms, the most straightforward being

$$\mathbf{X}^{(k)} = \begin{pmatrix} x_{11}^{(k)} & x_{12}^{(k)} & \dots & x_{1N}^{(k)} \\ x_{21}^{(k)} & x_{22}^{(k)} & \dots & x_{2N}^{(k)} \\ \vdots & \vdots & \ddots & \vdots \\ x_{M1}^{(k)} & x_{M2}^{(k)} & \dots & x_{MN}^{(k)} \end{pmatrix} \in \mathbb{R}^{M\times N} .$$

In fact, for matrices, being second order tensors defined on two (possibly heterogeneous) spaces (space and time for the EEG), we have that each element corresponds to a spatial position (the electrode index m) and a relative time (temporal index n). If we take the respective canonical bases $\mathcal{E}^s = \{\mathbf{e}_m^s\}_{m=1}^M$ and $\mathcal{E}^t = \{\mathbf{e}_n^t\}_{n=1}^N$ for each of these spaces, the coefficients of $\mathbf{X}^{(k)}$ with respect to the (tensor product) basis

$$(\mathbf{e}_1^s \otimes \mathbf{e}_1^t, \mathbf{e}_2^s \otimes \mathbf{e}_1^t, \dots, \mathbf{e}_M^s \otimes \mathbf{e}_1^t, \mathbf{e}_1^s \otimes \mathbf{e}_2^t, \mathbf{e}_2^s \otimes \mathbf{e}_2^t, \dots, \mathbf{e}_M^s \otimes \mathbf{e}_N^t)$$

are given as $\left(x_{11}^{(k)}, x_{21}^{(k)}, \dots, x_{M1}^{(k)}, x_{12}^{(k)}, x_{22}^{(k)}, \dots, x_{MN}^{(k)}\right)^T \in \mathbb{R}^{M\cdot N}$.

From the theory of matrix algebra, each matrix can be represented on a maximum of $L = \min(M, N)$ basis vectors (in the product space). A possible reduced representation can be found through the singular value decomposition of $\mathbf{X}^{(k)}$ (where we have dropped the superscripts in what follows to augment readability)

$$\mathbf{X} = \boldsymbol{\Phi}\boldsymbol{\Lambda}\boldsymbol{\Psi}^T$$

from which we obtain that the coordinates of $\mathbf{X}$ in its specific basis

$$(\boldsymbol{\phi}_1 \otimes \boldsymbol{\psi}_1, \boldsymbol{\phi}_2 \otimes \boldsymbol{\psi}_2, \ldots \boldsymbol{\phi}_L \otimes \boldsymbol{\psi}_L, \boldsymbol{\phi}_2 \otimes \boldsymbol{\psi}_1 \ldots \boldsymbol{\phi}_M \otimes \boldsymbol{\psi}_N)$$

are given by $\left(\lambda_1, \lambda_2, \ldots, \lambda_L, \underbrace{0, 0, \ldots 0}_{M \cdot N - L}\right)$. In other words, the matrix $\mathbf{X}^{(k)}$ lies in a L-dimensional subspace of $\mathbb{R}^{M \cdot N}$. This representation will form the basis of the spatio-temporal method that will be developed in the rest of this contribution.

2.4 Some Useful Properties of the Tensor Product Basis

The kronecker product is nowadays a widely used operation in the manipulation of multi-way arrays and tensors and it is known to have some attractive properties (see e.g. [12]), especially for our application.

Property 1 (Transposability of Orthogonality). *If ϕ_m and $\phi_{m'}$ are two orthogonal vectors in one of the two spaces, then the vectors $\phi_m \otimes \psi_n$ and $\phi_{m'} \otimes \psi_n$ are orthogonal vectors in the product space.*

It follows from the distributivity and the associativity of the tensor product that $\langle \phi_m \otimes \psi_n, \phi_{m'} \otimes \psi_{n'} \rangle = \langle \phi_m, \phi_{m'} \rangle \langle \psi_n, \psi_{n'} \rangle$, or, the correlation in the product space is the product of the correlations in the respective spaces. In other words, to have a high correlation in product space, the correlation in both spaces should be high.

Practically, we have that the P300 template should be a spatial as well as a temporal representative and it clearly does not suffice to be a representative in only a single one of these spaces. In addition, for activity other than the evoked potential to be captured in a spatio-temporal representation, it should correlate spatially as well as temporally over the different trials. In other words, background activity may be spatially correlated as long as it is temporally uncorrelated and vice versa.

2.5 Angles between Subspaces

From the representation of $\mathbf{X}^{(k)}$ discussed in section 2.3, we may retain the basis for the subspace on which $\mathbf{X}^{(k)}$ is defined, rather than its coefficients. For two matrices $\mathbf{X}^{(k)}$ and $\mathbf{X}^{(k')}$ we could then try to find the best common representative. Putting $\boldsymbol{\Xi}^{(k)} = \left(\boldsymbol{\xi}_1^{(k)}, \boldsymbol{\xi}_2^{(k)}, \ldots \boldsymbol{\xi}_L^{(k)}\right)$ and $\boldsymbol{\Xi}^{(k')} = \left(\boldsymbol{\xi}_1^{(k')}, \boldsymbol{\xi}_2^{(k')}, \ldots \boldsymbol{\xi}_L^{(k')}\right)$

as some orthonormal subspace bases for $\mathbf{X}^{(k)}$ and $\mathbf{X}^{(k')}$ respectively, we could define the angle between subspaces, analogously as in [13], as

$$\cos(\theta_i) = \max_{\boldsymbol{\omega}_i,\boldsymbol{\rho}_i} \boldsymbol{\omega}_i^T {\Xi^{(k)}}^T \Xi^{(k')} \boldsymbol{\rho}_i, \text{ subject to } \begin{matrix} \langle \boldsymbol{\omega}_i, \boldsymbol{\omega}_{i'} \rangle = 0 \\ \langle \boldsymbol{\rho}_i, \boldsymbol{\rho}_{i'} \rangle = 0 \end{matrix}, \forall i' < i \ .$$

The angle between the subspaces $\Xi^{(k)}$ and $\Xi^{(k')}$ is defined to be θ_L. It can be demonstrated that $0 \leq \cos(\theta_i) \leq 1$ and the index i for which $1 = \cos(\theta_i) > \cos(\theta_i + 1)$ determines the dimension of the common subspace.

However, the angle between subspaces does not easily extend to multiple matrices if we do not restrict ourselves to pairwise comparisons. But we are looking for a common subspace to all observations (i.e. the intersection of all subspaces). Unfrotunately, it suffices that there exists a single observation that does not comprise the subspace of interest to have the zero vector as the resulting intersection. In other words, a single trial not containing a P300 response to the stimulus would compromise the analysis.

2.6 Best Representative Subspace

To avoid the above drawback, we use the following trick:

1. Compose $\mathbf{T} = \left(\Xi^{(1)}, \Xi^{(2)}, \ldots, \Xi^{(k)}, \ldots \Xi^{(K)}\right)$
2. Calculate the major singular value $\lambda_1(\mathbf{T})$ of $\mathbf{T}$ and its corresponding left and right eigenvectors $\boldsymbol{\rho}$ and $\boldsymbol{\omega}$
3. Calculate the best rank-one approximation to the *vectorized matrix* $\boldsymbol{\rho}$
4. Calculate $\tilde{\omega}_k = \sum_{i=(k-1)L+1}^{kL} \omega_i^2$

The best rank one approximation to $\boldsymbol{\rho}$ may be written as $\lambda_1(\boldsymbol{\rho})\, \hat{\mathbf{u}} \otimes \hat{\mathbf{v}}$. Concerning the vector $\tilde{\boldsymbol{\omega}}$, remark that its norm $\|\tilde{\boldsymbol{\omega}}\|_1 = 1$ and that the weights $\tilde{\omega}_\ell$ thus form a partition of 1. In fact, $\lambda_1(\mathbf{T})$ may be seen as an approximation to p in Eq. (1) and $\tilde{\omega}_\ell$ the relative probability that the ℓ-th observation adheres to the upper equation of the model in Eq. (1).

3 Results

In this section we will display the results obtained on synthetic data only, due to a lack of space. All data in the simulations have been generated in accordance to the model in Eq. (1). We have taken $\mathbf{u} \in \mathbb{R}^3$ and $\mathbf{v} \in \mathbb{R}^5$, $K = 25$ and have run 100 Monte Carlo realisations.

In the presentation of the results we have taken the mean over the different values of κ_2 since we have seen that its influence on the end results is less significant than that of p or κ. The results of this study are given in Figure 1(a) in terms of $\sqrt{\rho_{\mathbf{u}}\rho_{\mathbf{v}}} = \sqrt{\langle \hat{\mathbf{u}}, \mathbf{u}\rangle\langle \hat{\mathbf{v}}, \mathbf{v}\rangle}$, i.e. the geometric mean of the correlations, which is simply the square root of the inner product taken in the product space.

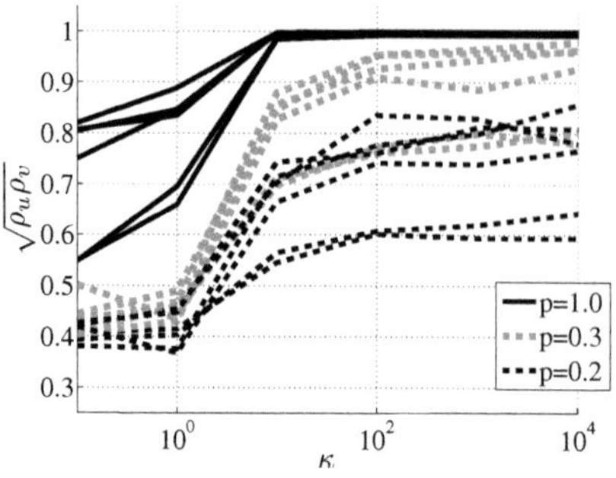

(a)

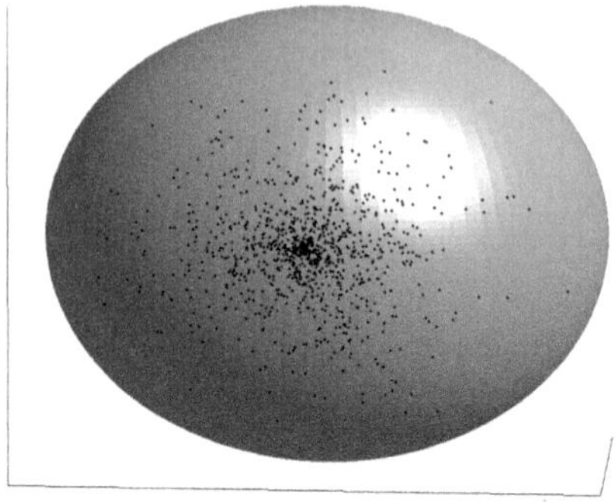

(b)

Fig. 1. (a) The mean performance over 100 Monte Carlo runs of the algorithm when varying κ and p in the model of Eq. (1). $\kappa_2 = 10^{-1}, 10^0 \ldots 10^4$ with lower values in κ_2 resulting in a line closer to the x-axis. (b) 1000 samples from the Von Mises-Fisher distribution on the sphere $\mathbb{S}^2$ for $\kappa = 10$ (the knee in Figure (a)).

4 Discussion

From the Figures 1(a)&1(b), we observe that the performance degrades with an increase in the corruption probability $(1 - p)$. Also, we clearly have a similar behaviour of the correlation as a function of κ. We observed also that for fixed p and κ, augmenting the number of observations results in a better estimation performance (results not shown), which supports the assumption that the used statistics are consistent.

Since in practice the P300 does not correlate spatio-temporally with the background activity and noise, the proposed methodology is promising in the estimation of the P300 and related waveforms. Preliminary results on real data (not shown here) confirm this assumption. The above assumption contrasts with those of the principal component analysis of the observed matrices or that of the best rank-one approximation to the three way array $\mathcal{X}$ composed by stacking matrices $\mathbf{X}^{(k)}$, since the latter algorithms are variance rather than occurrence based. It is in this perspective that our method also resembles (kernel-based) clustering methods. However, in the latter, each of the vectors in $\mathbf{T}$ is attributed a cluster index, while the authors are not aware of any attempt to jointly estimate the best joint rank-1 matrix representation for a *clustering* of the matrices.

Note that the spatio-temporal pattern that results from the method does not need to occur in all observations, even not in one single observation. From Figure 1(a), we observe that acceptable performance is already achieved for $\kappa \gtrsim 10$ and $p > 0.4$. Actually, it suffices that the chosen spatio-temporal pattern is the closest to the majority of the subspaces spanned by the observations. This is reminiscent to the largest singular value $(0 \leq \cos(\theta_1) \leq 1)$ that may be found in the algorithm calculating angles between subspaces [13], an angle whose cosine does not necessarily equal one.

5 Conclusion

The proposed method seems to be promising for estimating a pattern in a product space of two heterogeneous spaces as is the case in the spatio-temporal P300 estimate. The method results in a direct spatio-temporal decomposition, rather than a spatial decomposition with posterior temporal estimation as often witnessed in literature. Moreover, instead of imposing independence, decorrelation, orthogonality or sparsity in one or both of the heterogeneous spaces, the method only relies on the re-occurrence of a spatio-temporal pattern in a subset of the observations, a pattern that may be subjected to noise (partial re-occurence). In addition, the algorithm has an algebraic solution without heuristic parameters to choose.

References

1. Hoke, M., Ross, B., Wickesberg, R., Lütkenhöner, B.: Weighted averaging theory and application to electric response audiometry. Electroencephalography and Clinical Neurophysiology 57(5), 484–489 (1984)
2. Davila, C.E., Mobin, M.S.: Weighted averaging of evoked potentials. IEEE Transactions on Biomedical Engineering 39(4), 338–345 (1992)
3. Lütkenhöner, B., Hoke, M., Pantev, C.: Possibilities and limitations of weighted averaging. Biological Cybernetics 52(6), 40–416 (1985)
4. Chapman, R.M., McCrary, J.W.: EP component identification and measurement by principal component analysis. Brain and Cognition 27, 288–310 (1995)
5. Kayser, J., Tenke, C.E.: Optimizing PCA methodology for ERP component identification and measurement: theoretical rationale and empirical evaluation. Clinical Neurophysiology 114, 2307–2325 (2003)
6. Comon, P.: Independent component analysis, a new concept? Signal Processing 36, 287–314 (1994)
7. Makeig, S., Westerfield, M., Jung, T.-P., Covington, J., Townsend, J., Sejnowski, T.J.: Independent components of the late positive response complex in a visual spatial attention task. Journal of Neuroscience 19, 2665–2680 (1999)
8. Li, R., Keil, A., Principe, J.C.: Single-trial P300 estimation with a spatiotemporal filtering method. Journal of Neuroscience Methods 177, 488–496 (2009)
9. Iyer, D., Zouridakis, G.: Single-trial evoked potential estimation: Comparison between independent component analysis and wavelet denoising. Clinical Neurophysiology 118, 495–504 (2007)
10. Wang, Y., Berg, P., Scherg, M.: Common spatial subspace decomposition applied to analysis of brain responses under multiple task conditions: a simulation study. Clinical Neurophysiology 110, 604–614 (1999)
11. Krusienski, D.J.: A method for visualizing independent spatio-temporal patterns of brain activity. EURASIP Journal on Advances in Signal Processing 2009 (2009)
12. Laub, A.J.: Matrix Analysis for Scientists and Engineers. Society for Industrial and Applied Mathematics (2005)
13. Golub, G.H., Van Loan, C.F.: Matrix Computations, 3rd edn. The Johns Hopkins University Press, Baltimore (1996)

Recovering Spikes from Noisy Neuronal Calcium Signals via Structured Sparse Approximation

Eva L. Dyer[1], Marco F. Duarte[2], Don H. Johnson[1], and Richard G. Baraniuk[1]

[1] Dept. of Electrical and Computer Engineering, Rice University, Houston, TX 77005, USA
[2] Program in Applied and Comp. Math., Princeton University, Princeton, NJ 08544, USA

Abstract. Two-photon calcium imaging is an emerging experimental technique that enables the study of information processing within neural circuits *in vivo*. While the spatial resolution of this technique permits the calcium activity of individual cells within the field of view to be monitored, inferring the precise times at which a neuron emits a spike is challenging because spikes are hidden within noisy observations of the neuron's calcium activity. To tackle this problem, we introduce the use of sparse approximation methods for recovering spikes from the time-varying calcium activity of neurons. We derive sufficient conditions for exact recovery of spikes with respect to (i) the decay rate of the spike-evoked calcium event and (ii) the maximum firing rate of the cell under test. We find—both in theory and in practice—that standard sparse recovery methods are not sufficient to recover spikes from noisy calcium signals when the firing rate of the cell is high, suggesting that in order to guarantee exact recovery of spike times, additional constraints must be incorporated into the recovery procedure. Hence, we introduce an iterative framework for *structured sparse approximation* that is capable of achieving superior performance over standard sparse recovery methods by taking into account knowledge that spikes are non-negative and also separated in time. We demonstrate the utility of our approach on simulated calcium signals in various amounts of additive Gaussian noise and under different degrees of model mismatch.

Keywords: Two-photon calcium imaging, structured sparse approximation, spike recovery, exact support recovery, coherent dictionaries.

1 Introduction

Despite the tight link between chemical and electrical signaling in neural systems, experimental studies of neural information processing have mainly been limited to observations of the electrical activity of single neurons and small neural networks. Recently however, a new experimental technique known as two-photon calcium imaging has enabled the study of the concentration of calcium within the dendritic trees of single neurons as well as the time-varying calcium activity of neural populations *in vivo* [1]. Studies of neuronal calcium activity have already confirmed the existence of functional maps in the visual cortex [2]; however, in order to use this technique to study temporal correlations between distinct neurons, calcium imaging must also be able to uncover the precise times at which each cell emits a spike or action potential (AP). Fortunately, the calcium activity of the cell may be used to infer the times at which the neuron emits

V. Vigneron et al. (Eds.): LVA/ICA 2010, LNCS 6365, pp. 604–611, 2010.

a spike from its cell body due to the fact that immediately after a spike is emitted from the cell, calcium rushes into the cell body and this results in a spike-evoked calcium transient that has roughly exponential decay with a decay time of approximately 0.5–1 second. Due to the long time course of spike-evoked calcium transients and high levels of photon shot noise incurred during the sensing process, the problem of inferring spikes hidden within the calcium activity of a cell is extremely challenging. Furthermore, the question of whether precise spike timing information can be extracted from noisy calcium signals is still open to debate, especially for fast-spiking interneurons.

In this paper, we introduce the use of sparse approximation methods for recovering spike times from noisy observations of the time-varying calcium activity of an individual neuron. By using sparse approximation techniques for spike inference, we are afforded a great deal of theory that enables a principled study of the limits of spike recovery from calcium signals. We use these results to derive sufficient conditions that ensure exact recovery of spike times with respect to (i) the decay rate of the spike-evoked calcium event and (ii) the maximum firing rate of the cell under test. This theoretical study, coupled with our experimental findings, suggests that in order to guarantee exact recovery of spikes from noisy calcium signals, additional constraints must be incorporated into the recovery procedure. Hence, we introduce the use of *structured sparse approximation* for spike recovery and show that, when spikes are separated in time and this knowledge is leveraged in the recovery procedure, spike times can be extracted faithfully in higher levels of noise than when no constraint is imposed. We detail a structured sparse recovery algorithm capable of exploiting this sort of knowledge and study the utility of our approach to structured sparse recovery on simulated calcium signals in various levels of noise.

Related Work. The problem of spike recovery from calcium imaging data was previously studied by Vogelstein et al., who develop sequential Monte Carlo methods [3] and fast methods for non-negative deconvolution [4]. Both of these methods assume that the spikes are distributed according to a Poisson distribution and that the spike-evoked calcium transients have exponential decay. In contrast to both of these approaches, the methods we develop here can be used to perform spike inference for arbitrary calcium waveforms and do not leverage any model of the dynamical behavior of the spikes nor do they assume a particular noise distribution. For this reason, our methods require far fewer parameters than methods that rely on estimates of the distributions of the noise and spiking dynamics. In addition, our work is the first to address the question of whether spike timing information can be reliably extracted from calcium signals with respect to parameters that govern the generation of spike-evoked calcium signals.

Paper Organization. We introduce sparse representations, nonlinear approximation methods for sparse recovery, and metrics required to derive dictionary dependent conditions for exact spike recovery in Section 2. Following this, we show how the problem of spike recovery from neuronal calcium signals can be posed as a sparse approximation problem and derive conditions for exact spike recovery in coherent dictionaries in Section 3. In Section 4, we introduce methods for structured sparse approximation, detail our algorithmic approach to structured sparse signal recovery, and then present our experimental results. We end with conclusions in Section 5.

2 Background

Sparse Representation. In many applications, it is advantageous to transform a signal into a new domain where it admits a compact or *sparse* representation. This idea lies at the heart of transform coding-based compression algorithms such as JPEG, where orthonormal bases are typically employed. When the collection of elements used to represent the signal are redundant, the representation is no longer unique; however, redundant systems often produce even sparser representations than orthonormal systems. To make this precise, we will refer to a finite collection of M unit-norm atoms as a dictionary, $\mathcal{D} = \{\varphi_m\}_{m=1}^{M}$, where $\|\varphi_m\|_2 = 1$, for all m. Upon stacking the atoms in $\mathcal{D}$ into the columns of a dictionary matrix $\mathbf{\Phi} \in \mathbb{R}^{N \times M}$, an exact reconstruction of the input signal $\mathbf{x} \in \mathbb{R}^N$can be obtained by finding a linear combination of the atoms in the dictionary weighted by the coefficient vector $\mathbf{a} \in \mathbb{R}^M$, where $\mathbf{x} = \sum_{m=1}^{M} a(m)\varphi_m = \mathbf{\Phi}\mathbf{a}$. The dictionary is considered complete if it spans $\mathbb{R}^N$ with $M = N$ and overcomplete if it spans $\mathbb{R}^N$ with $M > N$.

When using an overcomplete dictionary, the simplest explanation of the data or the sparsest representation may be found by finding an approximate solution to the following non-convex problem,

$$\min_{\mathbf{a}} \|\mathbf{a}\|_0 \quad \text{subject to} \quad \mathbf{x} = \mathbf{\Phi}\mathbf{a}, \tag{1}$$

where $\|\mathbf{a}\|_0$, denoted as the "ℓ_0-norm" of $\mathbf{a}$, counts the number of non-zeros in $\mathbf{a}$. We call a signal k-sparse if $\|\mathbf{a}\|_0 \leq k$ and refer to the index set containing the indices of atoms with non-zero coefficients as the *support* of the signal or supp($\mathbf{a}$) = Λ. We denote the corresponding sub-dictionary as $\mathbf{\Phi}_\Lambda$, the complement set of atoms as $\mathbf{\Phi}_{\Lambda^c}$, and can express the entire dictionary matrix as, $\mathbf{\Phi} = [\mathbf{\Phi}_\Lambda \; \mathbf{\Phi}_{\Lambda^c}]$. The ℓ_0-norm penalty in (1) is non-convex; however, we can relax (1) by replacing the ℓ_0 term with the ℓ_1-norm $\|\mathbf{a}\|_1 = \sum_{m=1}^{D} |a_m|$, resulting in a method known as basis pursuit (BP) [5]. Alternatively, greedy methods such as orthogonal matching pursuit (OMP) [6] can be employed which select atoms iteratively, subtracting the contribution of each selected atom from the current signal residual.

Exact Recovery Conditions. In order to derive conditions that guarantee that sparse approximation methods like BP and OMP will recover a unique representation of a signal in a particular dictionary, two measures related to the similarity of atoms in the dictionary can be used. For unit-norm atoms, the *maximum coherence* is defined as $\mu \equiv \max_{i \neq j} |\langle \varphi_i, \varphi_j \rangle|$ and the *cumulative coherence* is

$$\mu(k) \equiv \max_{|\Lambda|=k} \max_{m \in \Lambda^c} \sum_{n \in \Lambda} |\langle \varphi_n, \varphi_m \rangle| \leq k\mu. \tag{2}$$

When atoms are sufficiently incoherent, a signal synthesized from a collection of these atoms is guaranteed to be unique and hence recoverable by standard sparse approximation methods like BP and OMP. In [7], Gribonval and Nielsen derive a sufficient condition for exact recovery of a signal drawn from the sub-dictionary $\mathbf{\Phi}_\Lambda$ that depends on what we call the intra-support coherence, μ_Λ, and the inter-support coherence, μ_{Λ^c} of the dictionary.

Proposition 1. (Neumann Exact Recovery Coefficient (ERC) [7]) *Let a signal be supported over the index set Λ. If the Neumann ERC of a dictionary $\mathcal{D} = \{\varphi_m\}$,*

$$\mu_\Lambda + \mu_{\Lambda^c} = \max_{n \in \Lambda} \sum_{m \in \Lambda, m \neq n} |\langle \varphi_m, \varphi_n \rangle| + \max_{n \in \Lambda^c} \sum_{m \in \Lambda} |\langle \varphi_m, \varphi_n \rangle| < 1, \tag{3}$$

holds then OMP will recover the support set Λ exactly.

This result implies a slightly weaker condition stated below in (4) that is also sufficient to guarantee exact recovery of k-sparse signals. In [8], Tropp demonstrates that this condition guarantees that a quantity related to the Neumann ERC is positive, which is sufficient for exact recovery of all k-sparse signals via BP as well.

$$\mu(k) + \mu(k-1) < 1. \tag{4}$$

3 Exact Recovery of Spikes from Neuronal Calcium Signals

Before delving into conditions that ensure exact recovery of spikes from calcium signals, we briefly introduce a model for spike-evoked calcium signals that is a generalization of the linear model introduced in [3].

Signal Model. When a spike is emitted from a neuron, voltage-gated calcium channels across the cell's membrane cause an influx of calcium into the cell. The calcium transients that emerge are marked by a quick rise time and roughly exponential decay on the order of 0.5–1 second. Thus, the detected fluorescence signal $\mathbf{y} \in \mathbb{R}^N$ can be modeled as

$$\mathbf{y} = \sum_{m \in \Lambda} a_m \varphi_m + \eta = \mathbf{\Phi a} + \eta = \mathbf{x} + \eta, \tag{5}$$

where φ_m is a N-dimensional vector of zeros with a spike-evoked waveform placed at the mth sample, $\mathbf{x} \in \mathbb{R}^N$ is a linear combination of calcium waveforms drawn from the dictionary $\mathbf{\Phi}$ weighted by the coefficient vector $\mathbf{a} \in \mathbb{R}^M$, and η is a vector containing additive noise. With this linear model, sparse approximation techniques such as those described in section 2 can be used to recover the index set Λ and corresponding non-zero coefficients $\mathbf{a}_\Lambda \in \mathbb{R}^{|\Lambda|}$ for arbitrary calcium waveforms. We note that because the coefficients in $\mathbf{a}$ correspond to the amplitude of each spike, the coefficients are all non-negative but are not necessarily equal due to slight variations in spike generation process. If we wish to enforce knowledge that coefficients used in the representation must be non-negative, a non-negative version of OMP and BP can easily be obtained. In the case of greedy algorithms like OMP and CoSaMP [9], non-negativity can be imposed by simply selecting atoms with large non-negative coefficients (instead of coefficients that are large in absolute magnitude) and replacing the least-squares step at each iteration with a non-negative least-squares step.

Sufficient Conditions for Exact Spike Recovery. We will now study conditions that allow for exact recovery of spike times from neuronal calcium signals. We say that exact recovery is obtained when the index set Λ used to synthesize a signal can be recovered without any errors. Although one of the advantages of using sparse approximation methods is that they can be used to recover spikes from dictionaries consisting of shifted versions of arbitrary spike-evoked calcium signals, to simplify our subsequent analysis, we will assume that each spike-evoked calcium signal has exponential decay and a finite window of length L over which it is non-vanishing. Hence, each atom $\varphi_m \in \mathbb{R}^N$ can be written as $\varphi_m(n) = \beta^{n-m}$, for all $n = \{m, \ldots, m+(L-1)\}$ and 0 otherwise, where L is chosen so that β^L negligible. Thus, the maximum coherence between any two atoms is simply the normalized inner product between atoms at neighboring points in time, $\mu \approx \beta$, where the approximation is due to the fact that the waveforms are truncated to L samples. Likewise, the cumulative coherence is defined over k calcium waveforms placed at neighboring samples: $\mu(k) \approx \sum_{n=1}^{min(k,L)} \beta^n = \frac{\beta(\beta^k-1)}{\beta-1} = \frac{\mu(\mu^k-1)}{\mu-1} \leq \mu(L)$. By simply plugging this definition into (4), we obtain the following sufficient condition for exact support recovery from shift-invariant dictionaries of spike-evoked calcium transients with exponential decay.

Proposition 2. (Exact Spike Recovery from Calcium Signals) *For any k-sparse signal synthesized from a dictionary with cumulative coherence $\mu(k) = \mu(\mu^k - 1)/\mu - 1$, where μ is the maximum coherence, if $3\mu - \mu^k(\mu - 1) > 1$, then this condition is sufficient to guarantee exact recovery via OMP and BP.*

This implies that to guarantee that two spikes can be recovered faithfully, we need $\mu^2 + 2\mu - 1 < 0$, and for $\mu \in (0,1)$ this implies that $\mu < \sqrt{2} - 1$ or that $\mu < 0.414$. We show the relationship between this condition for exact recovery for different values of k in Figure 1(b).

4 Structured Sparse Approximation for Spike Recovery

In the previous section, we found that in order to guarantee that two spikes that occur at subsequent samples can be reliably separated, $\mu < 0.414$. In practice, we find that this condition is overly pessimistic and that sparse approximation methods are quite robust to high levels of coherence. However, when noise is introduced into observations, standard sparse approximation methods begin to fail when the noise exceeds 20% of the spike amplitude. Hence, to ensure exact recovery of spike times in the levels of noise present in real calcium imaging data, we must go "beyond sparsity" and incorporate additional constraints into our recovery procedure.

Drawing upon inspiration from recent results in structured sparse signal recovery [10], we introduce the use of a flexible iterative framework for structured sparse approximation that can be used to incorporate a number of different model-driven constraints into our recovery procedure. We find that if we leverage the fact that spikes are non-negative, this leads to improved recovery; however, if we also assume that spikes are separated by a minimum number of samples, we can obtain even better recovery performance in noisy settings. Although this type of support constraint may not be applicable in all settings, e.g, when the cell under test is a fast-spiking cell, there are many settings where there exist physical constraints on the spike rate, either due to the refractory period or the maximum firing rate of the cell. When this assumption is satisfied, we find that it is advantageous to leverage this knowledge during recovery.

Algorithm 1. Non-Negative Separated Support Model CoSaMP (SSM-CoSaMP)

Inputs: Dictionary $\mathbf{\Phi}$, observations $\mathbf{y}$, K, Δ, model approximation algorithm $\mathbb{M}(\cdot, K, \Delta)$
Output: K-sparse approximation $\widehat{\mathbf{x}}$ to true signal $\mathbf{x}$
$\widehat{\mathbf{x}}_0 = 0$, $\mathbf{r} = \mathbf{y}$; $\mathbf{b} = 0$; $i = 0$ {initialize}
while halting criterion false **do**
1. $i \leftarrow i + 1$
2. $\mathbf{e} \leftarrow \mathbf{\Phi}^{\dagger}\mathbf{r}$ {form signal residual estimate}
3. $\Omega \leftarrow \text{supp}(\mathbb{M}(\mathbf{e}, 2K, \Delta))$ {prune signal residual estimate according to model}
4. $\Lambda \leftarrow \Omega \cup \text{supp}(\widehat{\mathbf{x}}_{i-1})$ {merge supports}
5. $\mathbf{b}_{\Lambda} \leftarrow \mathbb{N}(\mathbf{y}, \mathbf{\Phi}_{\Lambda})$ {compute non-negative least-squares signal estimate}
6. $\widehat{\mathbf{x}}_i \leftarrow \mathbb{M}(\mathbf{b}, K, \Delta)$ {prune signal estimate according to model}
7. $\mathbf{r} \leftarrow \mathbf{y} - \mathbf{\Phi}\widehat{\mathbf{x}}_i$ {update measurement residual}
end while
return $\widehat{\mathbf{x}} \leftarrow \widehat{\mathbf{x}}_i$

To provide insight into why the introduction of a constraint on the firing rate of the cell leads to improved recovery, we examine the two terms in the Neumann ERC that must be minimized to ensure exact support recovery. For a signal drawn from a collection of atoms that are all separated by at least Δ samples, the intra-support coherence is bounded by $\mu_{\Lambda} \leq \sum_{n=1}^{min(k,L)} \beta^{\Delta n} = \frac{\mu_{\Delta}((\mu_{\Delta})^k - 1)}{\mu_{\Delta} - 1}$, where the maximum coherence $\mu_{\Delta} = \beta^{\Delta}$ (see Fig. 1(a)). Although signals generated under this model exhibit reduced intra-support coherence for all $\Delta > 1$, when we compute the inter-support coherence $\mu_{\Lambda^c} = \max_{n \in \Lambda^c} \sum_{m \in \Lambda} |\langle \varphi_m, \varphi_n \rangle|$, an atom that lies in between any of the spikes in the support set will be selected to maximize this term. This means that even when spikes are separated by Δ samples, if our dictionary is not modified to reflect this constraint then the inter-support coherence remains high. Therefore, when our recovery algorithm is agnostic to the model that generated our observations, we do not observe improved recovery performance. In contrast, when we employ a structured sparse recovery method that leverages knowledge that there is some separation between atoms in the support, we reduce the total coherence of our representation and this leads to better estimates of spike times in noisy settings. To enforce this knowledge within an algorithmic framework, we introduce an iterative method for structured sparse approximation that is detailed in Algorithm 1. We denote the minimizer of the non-negative least-squares problem $\mathbb{N}(\mathbf{s}, \mathbf{\Phi}_{\Lambda}) = \arg\min_{\mathbf{a} \geq 0} \|\mathbf{\Phi a} - \mathbf{s}\|_2^2$ and the model-based pruning algorithm described in [10] as $\mathbb{M}(\mathbf{s}, K, \Delta)$ which finds the closest ℓ_2 approximation to a signal $\mathbf{s}$ that has exactly K coefficients that are at least Δ samples apart from one another.

Results. To generate the data used for our simulations, we generate a sequence of spikes according to a Poisson distribution with rate λ and then pass the spikes through the model pruning algorithm $\mathbb{M}(\cdot, K, \Delta)$ to ensure that all spikes are at least Δ samples apart. We note that we never assume knowledge of the dynamical behavior of the spikes and only use a Poisson distribution to simulate realistic spiking behavior. Afterwards, the spikes are passed through a first-order IIR filter with difference equation, $x(n) = \beta x(n-1)$, and i.i.d. Gaussian noise with zero mean and variance σ^2 is added.

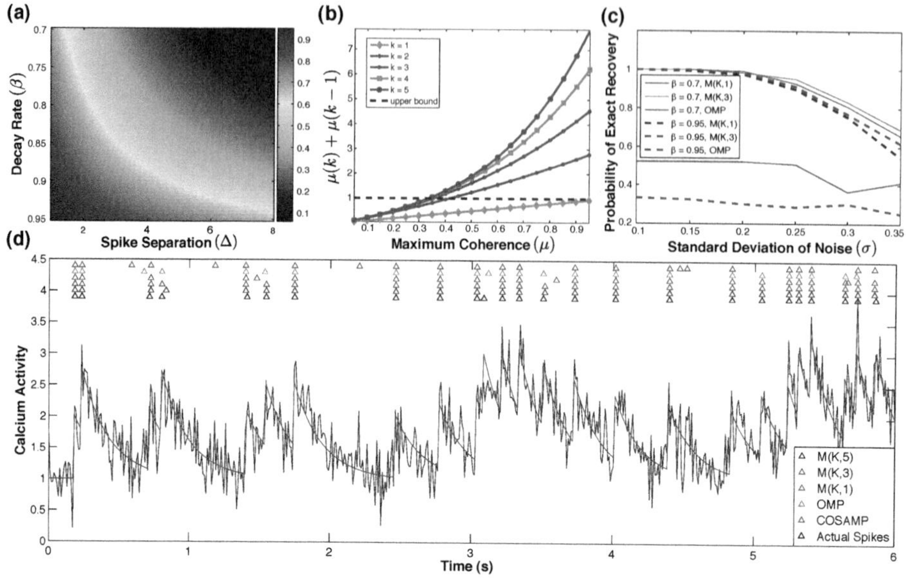

Fig. 1. (a) Maximum coherence μ_Δ for different values of Δ and β; (b) comparison of μ_Δ with $\mu_\Delta(k)+\mu_\Delta(k-1)$ and different sparsity levels (exact recovery is guaranteed when curve lies below one); (c) Probability of exact recovery of a spike train with $\Delta^\star = 3$ and $\beta = \{0.7, 0.95\}$ for different values of σ; (d) Synthetic calcium signals generated with $K = 25$, $\Delta^\star = 5$, $\beta = 0.95$, $\sigma = 0.3$. On top, we show the actual spike train (black) and recovered spikes using Algorithm (1) with $\Delta = \{5, 3, 1\}$ (blue, green, and red) and for OMP and COSAMP (cyan and magenta).

In Fig. 1(d) we show an example of a calcium signal in red, the noisy fluorescence signal in blue, and the corresponding spike trains recovered via Algorithm (1) and compare these results with those obtained via CoSaMP [9] and OMP. For this example, we set $\Delta^\star = 5$ samples (for a sampling rate of 100 Hz, $\Delta = 5 = 50\ ms$) and employ Algorithm (1) for $\Delta = \{1, 3, 5\}$, where $\Delta = 1$ corresponds to the case where no separation constraint is imposed but the non-negativity constraint is in effect. We show improved recovery over conventional methods such as CoSaMP and OMP when a non-negativity constraint is imposed but even further improvement when we assume $\Delta = \{3, 5\}$. In Fig. 1(c) we show the probability of exact recovery for various amounts of additive noise after averaging over 10 trials. We find that near-perfect recovery is obtained until the standard deviation of the noise is equal to 20% of the amplitude of the spikes. In contrast, OMP fails to recover the correct support even in small amounts of noise due to the high levels of coherence in the dictionary.

5 Discussion

In this paper, we have introduced the use of sparse approximation for recovering spikes from calcium imaging data and studied conditions that enable exact recovery of spikes from a superposition of spike-evoked calcium signals. This study marks the first effort of its kind to explore the limits of using calcium imaging data for extracting precise spike timing information from neuronal calcium signals. We show that, although spike times may not be reliably extracted from calcium signals collected from fast-spiking cells, for cells that exhibit slower spiking rates, structured sparse approximation methods may be used to reliably recover spike timing information, even from very noisy data. When the firing rate of the cell is too high to permit exact recovery of spikes, all hope is not lost. If we employ our proposed structured sparse recovery method and let the minimum spacing between spikes (Δ) equal to the window over which we are willing to obtain an estimate of the number of spikes emitted instead of exact spike timing information, then when single spikes are separated by this amount of time, our method will recover these spikes exactly; when short bursts of spikes are emitted, our method will simply produce an estimate of the number of spikes that were emitted. In preliminary studies, we find when the firing rate of the neuron exceeds a rate that enables exact spike recovery and our proposed methods are employed, we obtain more accurate estimates of the firing rate of the cell than when standard histogram-based approaches employed. Our future efforts include: using structured sparse approximation for computing accurate estimates of firing rates, a large-scale study of spike recovery methods for *in vivo* data, and extending our analysis to study exact recovery conditions for structured sparse approximation methods.

Acknowledgments. We thank Dr. Andreas Tolias, James Cotton, Dimitri Yatsenko and Chinmay Hegde for insightful discussions. This work was supported by AFOSR-FA9550-07-1-0301and ONR-N00014-08-1-1112. ED was supported by a training fellowship from NLM-5T15LM007093. MFD was supported by NSF DMS-0439872.

References

1. Göbel, W., Helmchen, F.: In Vivo Calcium Imaging of Neural Network Function. Physiology 22, 358–365 (2007)
2. Ohki, K., Chung, S., Ch'ng, Y.H., Kara, P., Reid, C.: Functional imaging with cellular resolution reveals precise micro-architecture in visual cortex. Nature 433, 597–603 (2005)
3. Vogelstein, J., Watson, B., et al.: Spike inference from calcium imaging using sequential Monte Carlo methods. Biophysical Journal 97(2), 636–655 (2009)
4. Vogelstein, J., Packer, A., et al.: Fast non-negative deconvolution for spike train inference from population calcium imaging. J. Neurophysiology (2010) (in press)
5. Chen, S., Donoho, D., Saunders, M.: Atomic decomposition by basis pursuit. SIAM J. Sci. Comput. 20(1), 33–61 (1999)
6. Davis, G., Mallat, S., Avellaneda, M.: Greedy adaptive approximation. J. Constr. Approx. 13, 57–98 (1997)
7. Gribonval, R., Nielsen, M.: Beyond sparsity: Recovering structured representations by ℓ_1 minimization and greedy algorithms. Adv. in Comp. Math. (28), 23–41 (2008)
8. Tropp, J.A.: Greed is good: Algorithmic results for sparse approximation. IEEE Trans. Info. Theory 50(10), 2231–2242 (2004)
9. Needell, D., Tropp, J.: CoSaMP: Iterative signal recovery from incomplete and inaccurate samples. Appl. Comp. Harmonic Anal. 26, 301–321 (2008)
10. Hegde, C., Duarte, M.F., Cevher, V.: Compressive Sensing Recovery of Spike Trains Using a Structured Sparsity Model. SPARS (2009)

Semi-nonnegative Independent Component Analysis: The (3,4)-SENICA^{exp} Method

Julie Coloigner[1,2], Laurent Albera[1,2], Ahmad Karfoul[5], Amar Kachenoura[1,2], Pierre Comon[3,4], and Lotfi Senhadji[1,2]

[1] Inserm, UMR 642, Rennes, F-35000, France
[2] Université de Rennes 1, LTSI, Rennes, F-35000, France
[3] CNRS, UMR 6070, Sophia Antipolis, F-06903, France
[4] I3S, Université de Nice Sophia Antipolis, F-06903, France
[5] Faculty of Mechanical and Electrical Engineering, University Al-Baath, Homs, Syria

Abstract. To solve the Independent Component Analysis (ICA) problem under the constraint of nonnegative mixture, we propose an iterative algorithm, called (3,4)-SENICA^{exp}. This method profits from some interesting properties enjoyed by third and fourth order statistics in the presence of mixed independent processes, imposing the nonnegativity of the mixture by means of an exponential change of variable. This process allows us to obtain an unconstrained problem, optimized using an ELSALS-like procedure. Our approach is tested on synthetic magnetic resonance spectroscopic imaging data and compared to two existing ICA methods, namely SOBI and CoM2.

1 Introduction

Since the early works of Jutten et al. [1] and the mathematical definition of the concept by Comon [2], Independent Component Analysis (ICA) goes on raising great interest. These problems find their place in numerous applications including telecommunications, speech processing, or biomedical engineering [3]. For instance, ICA methods have been used in Magnetic Resonance Spectroscopic Imaging (MRSI) in order to decompose recorded noisy mixtures of spectra into signals of interest, which are related to some tissue types and allow us to diagnose some tumors [4]. Several ICA techniques are available, depending on the assumptions made. Some methods use Second Order (SO) [5], SO and Fourth Order (FO) [2] or only Higher Order (HO) [6] statistics.

However, in many applications such as MRSI, the source components to be extracted may not be statistically independent enough [7]. Fortunately, in such contexts, recorded data have other properties such as nonnegativity [7]. Consequently, authors opted for Nonnegative Matrix Factorization (NMF) methods to the detriment of ICA in order to extract the source components of interest [7]. NMF techniques consist in factorizing a matrix as a product of two matrices with only nonnegative elements. NMF was introduced by Paatero and Tapper [8] in 1994. But this problem became popular with the work of Lee and Seung [8].

V. Vigneron et al. (Eds.): LVA/ICA 2010, LNCS 6365, pp. 612–619, 2010.

Currently we can define four categories of algorithms: multiplicative rules [8], projected Alternating Least Squares (ALS) methods [8], projected gradient algorithms such as the interior point gradient [9] and techniques based on change of variable [10]. In fact, to our knowledge, only the change of variable $\boldsymbol{A} = \boldsymbol{E} \odot \boldsymbol{E}$ was used [10] where $\odot$ denotes the Hadamard product operator.

Nevertheless, using only nonnegativity may not be sufficient as demonstrated by Moussaoui et al. [11]. Then a first class of methods [12,13] aims at taking into account the mutual independence and the nonnegativity of the sources. However, such techniques do not profit from the potential nonnegativity of the mixing matrix. On the other hand, Moussaoui et al. [11] propose a Bayesian approach, which deals with the ICA problem under the constraint of nonnegative mixtures of nonnegative sources. More precisely, authors incorporate the nonnegativity constraint by considering that both the sources and the mixing coefficients follow a Gamma probability distribution.

But in some applications, only the mixing matrix is nonnegative. For instance, in MRSI, the mixed spectra are not naturally nonnegative. They become nonnegative after a phase shift procedure, which may be complicated technically to execute. Note that such a scheme is absolutely necessary for the above mentionned ICA-NMF methods [11,12,13]. So we propose to relax the constraint of nonnegative sources and to solve the following problem:

Problem 1. Given a real random vector $\boldsymbol{x}$*, find a* $(N \times P)$ *mixing matrix* $\boldsymbol{A}_o$ *and a* P*-dimensional source random vector* $\boldsymbol{s}$ *such that* $\boldsymbol{x} = \boldsymbol{A}_o \boldsymbol{s} + \boldsymbol{\nu}$ *where* $\boldsymbol{A}_o$ *has nonnegative components,* $\boldsymbol{s}$ *has statistically independent components, and* $\boldsymbol{\nu}$ *is an* N*-dimensional Gaussian noise vector, independent of* $\boldsymbol{s}$*.*

The proposed SEmi-Nonnegative ICA algorithm, called (3,4)-SENICA^{exp}, exploits some interesting proprieties enjoyed by Third Order (TO) and FO statistics in the presence of mixed independent processes, imposing the nonnegativity of the mixture by means of an exponential change of variable. This operation allows us to obtain an unconstrained problem, which owns good convergence properties, due to the use of an ELSALS optimization procedure.

2 The (3,4)-SENICA^{exp} Algorithm

The purpose of this section is to show i) how we can combine TO and FO cumulants into one matrix having a special algebraic structure due to the statiscal independence of source components and ii) how to use nonnegativity through this particular structure in order to identify the mixing matrix.

Let's define the entries $\mathcal{C}_{n_1,n_2,n_3,\boldsymbol{x}}$ and $\mathcal{C}_{n_1,n_2,n_3,n_4,\boldsymbol{x}}$ of the TO and FO cumulant arrays, $\boldsymbol{\mathcal{C}}_{3,\boldsymbol{x}}$ and $\boldsymbol{\mathcal{C}}_{4,\boldsymbol{x}}$, respectively, of a zero-mean N-dimensional random vector $\boldsymbol{x}$:

$$\begin{aligned}
\mathcal{C}_{n_1,n_2,n_3,\boldsymbol{x}} &= \mathsf{E}[x_{n_1} x_{n_2} x_{n_3}] \\
\mathcal{C}_{n_1,n_2,n_3,n_4,\boldsymbol{x}} &= \mathsf{E}[x_{n_1} x_{n_2} x_{n_3} x_{n_4}] - \mathsf{E}[x_{n_1} x_{n_2}]\mathsf{E}[x_{n_3} x_{n_4}] \\
&\quad - \mathsf{E}[x_{n_1} x_{n_3}]\mathsf{E}[x_{n_2} x_{n_4}] - \mathsf{E}[x_{n_1} x_{n_4}]\mathsf{E}[x_{n_2} x_{n_3}]
\end{aligned} \tag{1}$$

where $\mathsf{E}[.]$ denotes the mathematical expectation operator. We propose to merge together the entries of both cumulant arrays in the same matrix, $\boldsymbol{T}_{\boldsymbol{x}}^{(3,4)}$, of size $(M \times N^2)$ with $M = N + N^2$. More precisely, the (m_1,m_2)-th entry, $T_{m_1,m_2,\boldsymbol{x}}^{(3,4)}$, of $\boldsymbol{T}_{\boldsymbol{x}}^{(3,4)}$ is given by:

$$T_{m_1,m_2,\boldsymbol{x}}^{(3,4)} = \begin{cases} \mathcal{C}_{n_1,n_2,m_1,\boldsymbol{x}} \text{ for any } m_1 \in \{1,\ldots,N\} \text{ with } m_2 = n_2 + N(n_1-1) \\ \mathcal{C}_{n_1,n_2,n_3,n_4,\boldsymbol{x}} \text{ for any } m_1 \in \{N+1,\ldots,N+N^2\} \\ \quad \text{with } m_1 = n_4 + Nn_3 \text{ and } m_2 = n_2 + N(n_1-1) \end{cases} \tag{2}$$

Under the assumptions made in problem 1 and using the multi-linearity property (3) enjoyed by cumulants, the TO and FO cumulants of problem 1 can be expressed by:

$$\begin{aligned} \mathcal{C}_{n_1,n_2,m_1,\boldsymbol{x}} &= \sum_{p=1}^{P} A_{n_1,p} A_{n_2,p} A_{m_1,p} \mathcal{C}_{p,p,p,\boldsymbol{s}} \\ \mathcal{C}_{n_1,n_2,n_3,n_4,\boldsymbol{x}} &= \sum_{p=1}^{P} A_{n_1,p} A_{n_2,p} A_{n_3,p} A_{n_4,p} \mathcal{C}_{p,p,p,p,\boldsymbol{s}} \end{aligned} \tag{3}$$

By inserting (3) into (2), we obtain the following algebraic structure of the matrix $\boldsymbol{T}_{\boldsymbol{x}}^{(3,4)}$:

$$\boldsymbol{T}_{\boldsymbol{x}}^{(3,4)} = \boldsymbol{C}_o(\boldsymbol{A}_o \oslash \boldsymbol{A}_o)^{\mathsf{T}} \tag{4}$$

with $\boldsymbol{C}_o = [\boldsymbol{C}_{3,\boldsymbol{s}}\boldsymbol{A}_o^{\mathsf{T}}, \boldsymbol{C}_{4,\boldsymbol{s}}(\boldsymbol{A}_o \oslash \boldsymbol{A}_o)^{\mathsf{T}}]^{\mathsf{T}}$ where matrices $\boldsymbol{C}_{3,\boldsymbol{s}} = \mathrm{diag}\,([\mathcal{C}_{1,1,1,\boldsymbol{s}}, \cdots, \mathcal{C}_{P,P,P,\boldsymbol{s}}])$ and $\boldsymbol{C}_{4,\boldsymbol{s}} = \mathrm{diag}\,([\mathcal{C}_{1,1,1,1,\boldsymbol{s}}, \cdots, \mathcal{C}_{P,P,P,P,\boldsymbol{s}}])$ of size $(P \times P)$ are diagonal, and where $\oslash$ denotes the column-wise Kronecker product.

To determine the matrices $\boldsymbol{A}_o$ and $\boldsymbol{C}_o$ from $\boldsymbol{T}_{\boldsymbol{x}}^{(3,4)}$, based on equation (4) we propose to minimize the Frobenius norm of the difference between $\boldsymbol{T}_{\boldsymbol{x}}^{(3,4)}$ and $\boldsymbol{C}(\boldsymbol{A} \oslash \boldsymbol{A})^{\mathsf{T}}$ using the change of variable $\boldsymbol{A} = \exp(\boldsymbol{E})$. The latter allows us to benefit from the nonnegativity of mixture $\boldsymbol{A}_o$ and achieve an unconstrained optimization problem, totally characterized by the following objective function:

$$f(\boldsymbol{E},\boldsymbol{C}) = \|\boldsymbol{T}_{\boldsymbol{x}}^{(3,4)} - \boldsymbol{C}(\exp(\boldsymbol{E}) \oslash \exp(\boldsymbol{E}))^{\mathsf{T}}\|_F^2 \tag{5}$$

where the couple of variables $(\boldsymbol{E},\boldsymbol{C})$ belongs to the open set $\mathbb{R}^{M\times P} \times \mathbb{R}^{N\times P}$ and where $\|.\|_F$ denotes the Frobenius norm. Next, we propose to use an ALS procedure in order to minimise (5) due to its good compromise between simplicity and effectiveness. For the sake of convenience, we chose to alternately minimize the cost function (5) w.r.t. $\boldsymbol{C}$, and each component of $\boldsymbol{E}$, optimizing w.r.t. one variable $E_{m,n}$ while keeping the other ones fixed. In fact, the minimization of $f_{E_{m,n}} : E_{m,n} \mapsto f(\boldsymbol{E},\boldsymbol{C})$, where $E_{m,n}$ is the (m,n)-th component of $\boldsymbol{E}$, is achieved by vanishing its derivative given by:

$$f'_{E_{m,n}}(E_{m,n}) = a_4 \exp(E_{m,n})^4 + a_2 \exp(E_{m,n})^2 + a_1 \exp(E_{m,n}) \tag{6}$$

with:

$$a_1 = -2\sum_{k=1}^{M}\sum_{i\neq m}^{N} T_{k,N(i-1)+m,\boldsymbol{x}}^{(3,4)} C_{k,n} \exp(E_{i,n}) + 2\sum_{k=1}^{M}\sum_{i\neq m}^{N}\sum_{e\neq n}^{P} \exp(E_{i,n}+E_{i,e}+E_{m,e}) C_{k,n} C_{k,e} \tag{7}$$

$$a_2 = 2\sum_{k=1}^{M}\sum_{i\neq m}^{N}\exp(E_{i,n})^2 C_{k,n}^2 - 2\sum_{k=1}^{N} T_{k,N(m-1)+m,\boldsymbol{x}}^{(3,4)} C_{k,n} + 2\sum_{k=1}^{M}\sum_{e\neq n}^{P}\exp(E_{m,e})^2 C_{k,n} C_{k,e} \quad (8)$$

$$a_4 = 2\sum_{k=1}^{M} C_{k,n}^2 \quad (9)$$

It is noteworthy that the $f'_{E_{m,n}}$ is a polynomial in variable $A_{m,n} = \exp(E_{m,n})$ with one zero root. As a result, we can search for the stationnary points of $f_{E_{m,n}}$ by computing the strictly positive roots of a third degree polynomial of the form $h(Z) = Z^3 + pZ + q$, with $p = a_2/a_4$ and $q = a_1/a_4$. Cardano's method can then be used. First, it consists in computing the discriminant, Δ, of Cardano's technique given by $\Delta = q^2 + 4p^3/27$. Depending on the sign of Δ, we may find zero, one or two strictly positive roots. More particularly, if Δ is zero, then there is a unique strictly positive root given by:

$$Z_0^+ = \begin{cases} \frac{3q}{p} \text{ if } p \text{ and } q \text{ have the same sign} \\ \frac{-3q}{2p} \text{ otherwise} \end{cases} \quad (10)$$

If Δ is strictly negative, then there is at least one strictly positive root given by:

$$Z_0^+ = K\cos(u) \text{ where } K = 2\sqrt{\frac{-p}{3}} \text{ and } u = \frac{1}{3}\arccos\left(\frac{-q}{2}\sqrt{\frac{27}{-p^3}}\right) \quad (11)$$

In addition, there may be another strictly positive root equal to $K\cos(u+4\pi/3)$. If need be, among both strictly positive roots, the one which minimizes function $f_{E_{m,n}}$ has to be chosen. Eventually, if Δ is strictly positive, then it exists a unique root in the real field defined by:

$$Z_0 = \sqrt[3]{\frac{-q+\sqrt{\Delta}}{2}} + \sqrt[3]{\frac{-q-\sqrt{\Delta}}{2}} \quad (12)$$

The latter may be negative or strictly positive. We noted by means of simulations that the case $Z_0 < 0$ could happen. Such a case implies that h does not vanish on $[\varepsilon, +\infty[$ for every strictly positive number ε. Now, $f_{E_{m,n}}$ is a continuous function on $[\varepsilon, +\infty[$, diverging to $+\infty$ as $E_{m,n}$ approaches $+\infty$. As a result, $f_{E_{m,n}}$ is a strictly increasing function on $[\varepsilon, +\infty[$ and its minimum on this interval is achieved for $E_{m,n} = \varepsilon$. So, when the case $Z_0 < 0$ happens, we propose to choose a small strictly positive number ε decreasing to zero as a function of the number of iterations. Such a strategy allows us to process mixtures $\boldsymbol{A}_o$ with some zero components in spite of the exponential change of variable. As far as the minimization of $f_{\boldsymbol{C}}$ is concerned, the solution is well-known and given by $\boldsymbol{C} = \boldsymbol{T}_{\boldsymbol{x}}^{(3,4)}((\exp(\boldsymbol{E}) \oslash \exp(\boldsymbol{E}))^{\sharp})^{\mathsf{T}}$ where $^{\sharp}$ denotes the pseudo-inverse operator. On the other hand, the ALS algorithm is sensitive to the initialization and it converges slowly [14]. Consequently, we propose i) to initialize with the absolute value of the mixing matrix estimated by the SOBI technique [5] and ii) to use the Enhanced Line Search (ELS) ALS method [14] in order to exit from potential local minima or cycles. Let's explain how to apply an ELSALS strategy to our problem. Bro [15] and Harshman have proposed an extrapolation given by $\boldsymbol{T}^{(i)}_{\text{new}} = \boldsymbol{T}^{(i)}_{it-2} + \beta_o^{(i)}(\boldsymbol{T}^{(i)}_{it-1} - \boldsymbol{T}^{(i)}_{it-2})$ where $\boldsymbol{T}^{(1)} = \boldsymbol{E}$, $\boldsymbol{T}^{(2)} = \boldsymbol{C}$ and $\beta_o^{(i)}$ is a factor of relaxation associated with $\boldsymbol{T}^{(i)}$ (LSALS), computing by a line search. The LSALS method predicts the parameters many iterations ahead. Then Rajih and Comon proposed to calculate the optimal relaxation factor, by minimizing the following criterion [14]:

$$g(\beta^{(1)}, \beta^{(2)}) = \left\| \boldsymbol{T}_{\boldsymbol{x}}^{(3,4)} - \boldsymbol{G}_{it}^{(2)}(\exp(\boldsymbol{G}_{it}^{(1)}) \oslash \exp(\boldsymbol{G}_{it}^{(1)}))^{\mathsf{T}} \right\|_F^2 \tag{13}$$

at iteration it w.r.t. $\beta^{(1)}$ and $\beta^{(2)}$, where $\boldsymbol{G}_{it}^{(i)} = \boldsymbol{T}^{(i)}_{it-2} + \beta^{(i)}(\boldsymbol{T}^{(i)}_{it-1} - \boldsymbol{T}^{(i)}_{it-2})$, leading to the ELSALS scheme. The latter can escape from a local minimum or cycle in one iteration. Both optimal factors are calculated by an optimization routine.

3 Computer Results

The goal of this section is to illustrate problem 1 through a biomedical application and to assess the behavior of the proposed method, in comparison to two classical ICA methods, namely SOBI [5] and CoM2 [2]. To do so, some experiments are carried out on simulated MRSI data. In such a context, it is assumed that the metabolites do not interact so that the linear mixing model is valid. In addition, the concentrations of the metabolites are positive, ensuring the non-negativity of the mixing matrix. Eight observations (two observation are shown in figure 2), considered as a noisy mixture of one source of interest, namely the N-Acetyl Aspartate (NAA) metabolite, and artifact source such as the Lipid metabolite, are constructed. Each metabolite spectrum (depicted in figure 2) is generated by a lorentzian function where the parameters (location and scale parameters) are fixed to derive realistic NAA and Lipid-like metabolites. Moreover, all reported results are obtained by calculating the median over 100 independent experiments. Finally, the Signal-to-Noise Ratio (SNR) is calculated by fixing the powers of the both sources and varying the power of the noise.

3.1 Performance Criterion

We propose to measure the error between matrix $\boldsymbol{A}_o$ and its estimate $\widehat{\boldsymbol{A}_o}$. The performance criterion must be invariant to scale and permutation indeterminacies inherent to problem 1. Consequently, we choose the measure defined in [14] and given by:

$$D(\boldsymbol{A}_o, \widehat{\boldsymbol{A}_o}) = \min_{\boldsymbol{\Pi}} \Psi(\boldsymbol{A}_o, \widehat{\boldsymbol{A}_o}\boldsymbol{\Pi}) \tag{14}$$

where matrix $\boldsymbol{\Pi}$ belongs to the set of permutations, and where:

$$\Psi(\boldsymbol{M}) = \sum_{p=1}^{P} \| \boldsymbol{m}_p - \frac{\widehat{\boldsymbol{m}_p}^{\mathsf{T}} \boldsymbol{m}_p}{\widehat{\boldsymbol{m}_p}^{\mathsf{T}} \widehat{\boldsymbol{m}_p}} \widehat{\boldsymbol{m}_p} \| \tag{15}$$

with $\boldsymbol{m}_p$ and $\widehat{\boldsymbol{m}}_p$ the pth columns of $\boldsymbol{M}$ and $\widehat{\boldsymbol{M}}$, respectively.

3.2 Convergence of (3,4)-SENICA$^{\exp}$

Four experiments are realized in order to evaluate the behavior of the (3,4)-SENICA$^{\exp}$ algorithm as a function of the number of iterations for four SNR_1 values i.e. SNR_1 = -40, -32.5, -17.5 and -2.5 dB (SNR_1 is related to the first source, namely the NAA metabolite). Figures 1 (a) and (b) display the error $D(\boldsymbol{A}_o, \widehat{\boldsymbol{A}_o})$ and its estimate $\widehat{\boldsymbol{A}}$, and the cost function defined in equation (5), $f(\widehat{\boldsymbol{E}_o}, \widehat{\boldsymbol{C}_o})$, respectively, as a function of the number of iterations. We observe that the behavior of $D(\boldsymbol{A}_o, \widehat{\boldsymbol{A}_o})$ and $f(\widehat{\boldsymbol{E}_o}, \widehat{\boldsymbol{C}_o})$ are globally similar. This expresses the uniqueness of the decomposition (4). Moreover, figure 1 (a) clearly shows that (3,4)-SENICA$^{\exp}$ converges beyond 400 iterations, whatever the used SNR_1. We can also remark that, for SNR_1's -2.5 and -17.5 dB, the values of $D(\boldsymbol{A}_o, \widehat{\boldsymbol{A}_o})$ increase for the first iterations and then decrease to obtain a satisfying solution. This behavior is due to the fact that the matrix used for initialization is close to the optimal solution, in these two cases.

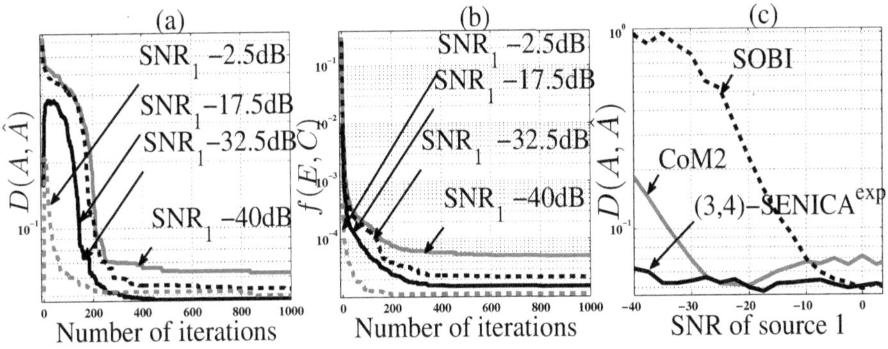

Fig. 1. Performance results: (a) performance criterion 14 at the output of (3,4)-SENICA$^{\exp}$ as a function of number of iterations for four SNR_1 values, (b) cost function (5) as function of number of iterations for four SNR_1 values, and (c) performance criterion 14 at the output of three methods as a function of SNR_1

3.3 Influence of SNR_1 and Comparison to Classical Methods

In this experiment we study the behavior of the (3,4)-SENICA$^{\exp}$, SOBI and CoM2 methods as a function of SNR_1. The results presented by (3,4)-SENICA$^{\exp}$ are obtained for 1000 iterations. Figure 2 shows the median, over 100 realizations, of the estimated sources obtained by the three methods for a SNR_1 equal to -35 dB. Clearly, SOBI fails to separate both sources. Regarding the CoM2 algorithm the separation of the both sources is not perfect as announced in figure 1 (c). Indeed, the NAA metabolite is still present in the second source. As far as

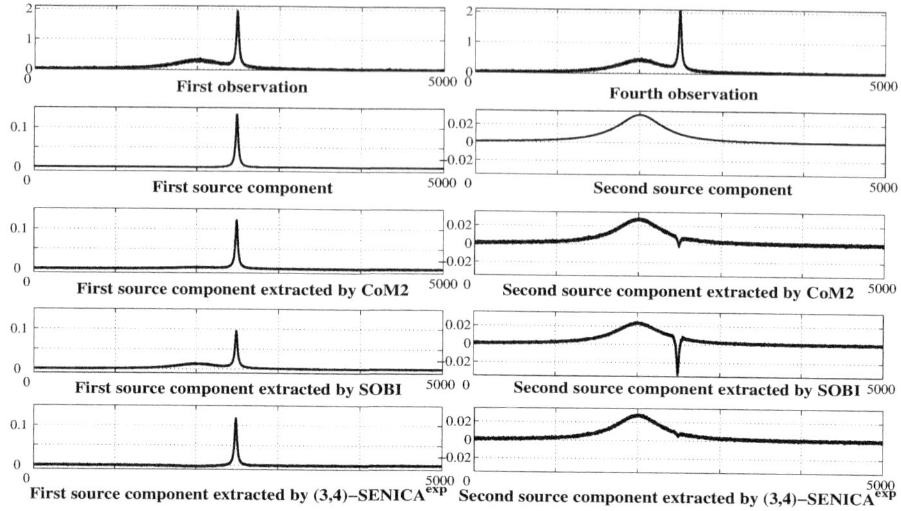

Fig. 2. Median over 100 realizations of the estimated sources obtained by the three considered methods for a SNR_1 equal to -35 dB

(3,4)-SENICA^{exp} is concerned, the separation of both metabolites is quasi-ideal. Figure 1 (c) illustrates, for each methods, the variation of the median of performance criterion over 100 realizations as a function of SNR_1. We observe a good behavior of (3,4)-SENICA^{exp} with a performance stability for the considered SNR_1 range. As previously mentioned, CoM2 is slightly less effective especially for low SNR_1 values. Concerning the SOBI algorithm, Figure 1 (c) displays poor results particularly for low SNR_1's.

4 Conclusion

In this paper, we propose a new method, called (3,4)-SENICA^{exp}, to solve problem 1 taking into account the mutual independence of the sources and the nonnegativity of the mixing matrix. It combines the use of i) the TO and FO statistics showing interesting properties in the presence of mixed independent processes and ii) an exponential change of variable to ensure the nonnegativity of the mixing matrix. Our method is compared in terms of performance to two existing ICA methods for handling simulated MRSI data. The obtained results show that (3,4)-SENICA^{exp} presents a better behavior than classical ICA algorithms, expecially for low SNR's. Forthcoming work will include a theoretical study of convergence and identifiability of (3,4)-SENICA^{exp}, the use of another change of variable in order to express nonnegativity and tests on real MRSI data.

References

1. Herault, J., Jutten, C., Ans, B.: Détection de grandeurs primitives dans un message composite par une architecture de calcul neuromimétique en apprentissage non supervisé. In: GRETSI 1885, Dixième colloque sur le Traitement du Signal et des Images, Nice, France, pp. 1017–1022 (September 1985)
2. Comon, P.: Component Analysis, a new concept ? Signal Processing 36(3), 287–314 (1994)
3. Comon, P., Jutten, C. (eds.): Handbook of Blind Source Separation, Independent Component Analysis and Applications. Academic Press, London (2010) ISBN: 978-0-12-374726-6, hal-00460653
4. Ladroue, C., Howe, F., Griffiths, J., Tate, R.: Independent component analysis for automated decomposition of in vivo magnetic resonance spectra. Magnetic Resonnance in Medicine 50, 697–703 (2003)
5. Belouchrani, A., Abed-Meraim, K., Cardoso, J.F., Moulines, E.: A blind source separation technique using second-order statistics. IEEE Transactions On Signal Processing 45(2), 434–444 (1997)
6. Albera, L., Ferreol, A., Comon, P., Chevalier, P.: Blind Identification of Overcomplete Mixtures of sources (BIOME). Linear Algebra Applications 391C, 3–30 (2004)
7. Sajda, P., Du, S., Brown, T., Stoyanova, R., Shungu, D., Mao, X., Parra, L.: Nonnegative matrix factorization for rapid recovery of constituent spectra in magnetic resonance chemical shift imaging of the brain. IEEE Transactions on Medical Imaging 23(12), 1453–1465 (2004)
8. Berry, M., Browne, M., Langville, A.N., Pauca, P., Plemmons, R.J.: Algorithms and aplications for approximate nonnegative matrix factorization. Computational Statistics and Data Analysis 52(8), 155–173 (2007)
9. Merritt, M., Zhang, Y.: Interior-point gradient method for large-scale totally nonnegative least squares problems. Journal of Optimization Theory and Application 126, 191–202 (2005)
10. Chu, M., Diele, F., Plemmons, R., Ragni, S.: Optimality, computation and interpretation of nonnegative matrix factorizations, `http://www.wfu.edu/~plemmons` (2004)
11. Moussaoui, S., Brie, D., Mohammad-Djafari, A., Carteret, C.: Separation of nonnegative mixture of non-negative sources using a bayesian approach and mcmc sampling. IEEE Trans. Signal Processing 54(11), 4133–4145 (2006)
12. Oja, E., Plumbley, M.: Blind separation of positive sources by globally convergent gradient search. Neural Computation 16(9), 1811–1825 (2004)
13. Zheng, C.H., Huang, D.S., Sun, Z.L., Lyu, M., Lok, T.M.: Nonnegative independent component analysis based on minimizing mutual information technique. Neurocomputing 69(7), 878–883 (2006)
14. Comon, P., Luciani, X., Almeida, A.L.F.D.: Tensor decompositions, alternating least squares and other thales. Journal of Chemometrics 23 (April 2009)
15. Bro, R.: Multi-way Analysis in the Food Industry: Models, Algorithms, and Applications. PhD thesis, University of Amsterdam (1998)

Classifying Healthy Children and Children with Attention Deficit through Features Derived from Sparse and Nonnegative Tensor Factorization Using Event-Related Potential

Fengyu Cong[1], Anh Huy Phan[2], Heikki Lyytinen[3], Tapani Ristaniemi[1], and Andrzej Cichocki[2]

[1] Department of Mathematical Information Technology, University of Jyväskylä, PL 35, (Agora), Jyväskylä, 40014, Finland
[2] Lab for Advanced Brain Signal Processing, Brain Science Institute - RIKEN, Japan
[3] Department of Psychology, University of Jyväskylä, PL 35, 40014, Finland
{fengyu.cong,heikki.lyytinen,tapani.ristaniemi}@jyu.fi, {phan,cia}@brain.riken.jp

Abstract. In this study, we use features extracted by Nonnegative Tensor Factorization (NTF) from event-related potentials (ERPs) to discriminate healthy children and children with attention deficit (AD). The peak amplitude of an ERP has been extensively used to discriminate different groups of subjects for the clinical research. However, such discriminations sometimes fail because the peak amplitude may vary severely with the increased number of subjects and wider range of ages and it can be easily affected by many factors. This study formulates a framework, using NTF to extract features of the evoked brain activities from time-frequency represented ERPs. Through using the estimated features of a negative ERP-mismatch negativity, the correct rate on the recognition between health children and children with AD approaches to about 76%. However, the peak amplitude did not discriminate them. Hence, it is promising to apply NTF for diagnosing clinical children instead of measuring the peak amplitude.

Keywords: Diagnosis, children with attention deficit, clinical, classification, nonnegative tensor factorization, mismatch negativity, event-related potential.

1 Introduction

One goal of the study on event-related potentials (ERPs) [1] is to find good endophenotypes to discriminate the healthy subjects and patients with mental disorders. For example, a negative ERP-mismatch negativity (MMN) [2] has been extensively studied for healthy children and children with disorders [3]. Particularly, it has been found that MMN peak amplitudes of the children with attention deficit (AD) can be significantly smaller than those of the normal children under some specific paradigms [4, 5, 6]. Nevertheless, even using the same

V. Vigneron et al. (Eds.): LVA/ICA 2010, LNCS 6365, pp. 620–628, 2010.

paradigm to elicit MMN, when the number of children increases and the mean ages of children are different, the difference of MMN peak amplitudes between the normal children and children with AD can not be significant [4, 7]. This means that the MMN peak amplitude is not so sufficiently robust a feature to discriminate different groups of children. Consequently, our objective of this study is to extract additional significant features of MMN for the diagnosis of children with AD.

Recently, the nonnegative matrix factorization (NMF) and nonnegative tensor factorization (NTF) have been used in the study of the feature extraction and selection from the time-frequency represented ongoing electroencephalogram (EEG) [8,9,10,11,12]. NMF is a linear transformation model and NTF is a multi-way extension of NMF. They assume that the elements in the linear transformation model are nonnegative and even sparse, and they estimate the basis functions of the dataset in the model [13].

Representations of EEG recordings by the latent variables using multi-linear transformation model allows us to extract components [11, 8]. NTF performs the multi-dimensional decomposition, for example, a three-dimensional array represents three modes: space (channels), time and frequency. In our previous study, we have demonstrated that both hierarchical alternating least squares (HALS) NMF and NTF algorithms [14, 15] are able to extract significant time-frequency represented MMN component from the multichannel time-frequency represented EEG recordings [16]. Moreover, the HALS NTF algorithm provides improved performance in comparison to existing algorithms, especially for large scale problems [17]. To demonstrate the effectiveness of the proposed method, we perform HALS NTF to extract features of MMN, and classify healthy children and children with AD into two discriminative classes.

2 Data Description

In this study, the control group consisted of 11 boys and 10 girls and the mean age of the group was 11 years 6 months (age range 8 years 8 months to 13 years 2 months); AD group consisted of 18 boys and 3 girls and the mean age was 11 years (age range 8 years 4 months to 13 years 5 months).

MMN can be elicited by the oddball paradigm [3, 2]. An uninterrupted sound [4, 7] under the oddball paradigm was used to elicit MMN. Fig.1 illustrates this paradigm (adapted from [4, 7]). There were at least six repetitions of the alternating 100ms tones between deviations. Nine EEG channels (frontal F3, Fz, F4; central C3, Cz, C4; parietal (Pz) and mastoids (M1, M2)) were recorded with Electro-Cap International 20-electrode cap using the standard 10-20 system. The potentials were referred to the tip of nose. The sampling frequency was 200Hz and an analog band pass filter of 0.1-30Hz was used. Recording started 300ms before the onset of a deviant stimulus and lasted 350ms after the onset of a deviant. Thus each trial contained the recordings of 650 ms, i.e., 130 samples. The data was processed offline.

Hereinafter, the data under dev30 was taken for the analysis. In order to remove artifacts, trials exceeding 100 micro volts or with null recordings were rejected. The number of kept trials per subject varied from 300 to 350 trials. After the artifacts rejection, the kept trials were averaged to obtain the averaged trace. Then, the Morlet wavelet transformation (MWT) was performed on the averaged trace to achieve the time-frequency represented MMN [18]. To MWT, the half wavelet length was set to be six for the optimal resolutions of the frequency and time [18]; the frequency range was set from 2 to 8.5Hz, and this was because the optimal frequency band of MMN in our dataset was in this range [19]; 256 frequency bins were calculated within this frequency range.

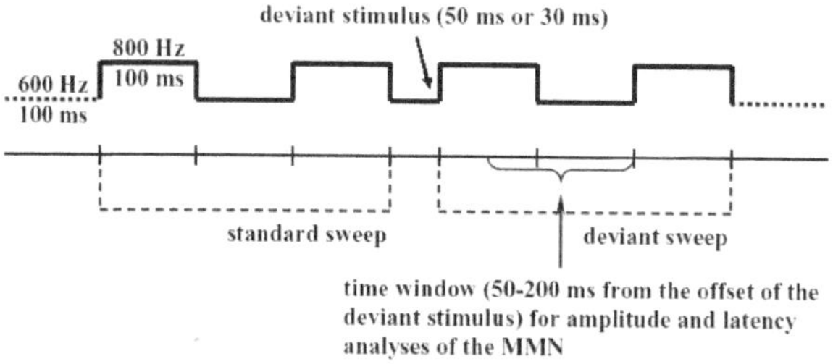

Fig. 1. Illustration for stimulus sequence to elicit MMN through continuous sound

3 Classification of Multiway Dataset

This section briefly introduces the classification method for the dataset described above. The model of tensor factorization will be applied to estimate the common components for all the training samples. Those components can be considered as basis components, and matrices of components are basis factors. The nonnegative tensor factorization (NTF) with the HALS algorithm will be mentioned.

3.1 Nonnegative Tensor Factorization and the HALS Algorithm

The NTF model [20] can be formulated as follows. For a given N-order tensor $\underline{\mathbf{Y}} \in \mathbb{R}^{I_1 \times I_2 \times \cdots \times I_N}$ perform a factorization into a set of N unknown nonnegative matrices: $\mathbf{U}^{(n)} = [\boldsymbol{u}_1^{(n)}, \boldsymbol{u}_2^{(n)}, \ldots, \boldsymbol{u}_N^{(n)}] \in \mathbb{R}^{I_n \times J}, (n = 1, 2, \ldots, N)$, which represent common factors, described as

$$\underline{\mathbf{Y}} \approx \sum_{j=1}^{J} \boldsymbol{u}_j^{(1)} \circ \boldsymbol{u}_j^{(2)} \circ \cdots \circ \boldsymbol{u}_j^{(N)} \tag{1}$$

where $\|\boldsymbol{u}_j^{(n)}\| = 1$, for $n = 1, 2, \ldots, N-1$, and $j = 1, 2, \ldots, J$, and symbol '$\circ$' denotes the outer product of the vectors. The target of NTF is to obtain the suitable $\mathbf{U}^{(n)}$. In the form of tensor products, the NTF model is written as

$$\underline{\mathbf{Y}} \approx \underline{\mathbf{I}} \times_1 \mathbf{U}^{(1)} \times_2 \mathbf{U}^{(2)} \cdots \times_N \mathbf{U}^{(1)} = \hat{\underline{\mathbf{Y}}}, \tag{2}$$

where $\hat{\underline{\mathbf{Y}}}$ is an approximation of the tensor $\underline{\mathbf{Y}}$, and $\underline{\mathbf{I}}$ is an identity tensor [20,21]. Each factor $\mathbf{U}^{(n)}$ explains the data tensor along a corresponding mode. Hence one factor can be considered as features of the data onto the subspace spanned by the others. Most algorithms for NTF are to minimize a squared Euclidean distance as the following cost function [20]

$$D(\underline{\mathbf{Y}}|\hat{\underline{\mathbf{Y}}}) = \frac{1}{2}\|\underline{\mathbf{Y}} - \underline{\mathbf{I}} \times_1 \mathbf{U}^{(1)} \times_2 \mathbf{U}^{(2)} \cdots \times_N \mathbf{U}^{(1)}\|_F^2 . \tag{3}$$

One frequently used optimization procedure for the minimization is the alternating least squares (ALS), i.e., the cost function is alternatively minimized with respect to the sets of parameters and at each time, and one set of parameters is optimized while other parameters keep fixed. With such a global learning rule, the computation and update of the learning are performed on all matrices.

The ALS [22], HALS [14, 15, 17] or multiplicative LS (least squares) algorithms [23,24] can be directly derived from the Frobenius cost function (3). The ALS algorithm has good convergence rate. Unfortunately, the standard ALS algorithm suffers from unstable convergence properties, demands high computational cost due to matrix inverse, often returns suboptimal solutions. Moreover it is quite sensitive with respect to noise, and can be relatively slow in the special case when data are nearly collinear. The multiplicative LS algorithms [23,24,13] have a relative low complexity but they are characterized by rather slow convergence and they sometimes converge to spurious local minima.

In this study, we applied the HALS algorithm whose the simplified version for NMF has been proved to be superior than the multiplicative algorithms by Gillis and Glineur in [25]. The HALS algorithm sequentially updates components $\boldsymbol{u}_j^{(n)}$ by minimizing a set of local squared Euclidean cost functions

$$D^{(j)}(\underline{\mathbf{Y}}^{(j)}|\boldsymbol{u}_j^{(1)} \circ \boldsymbol{u}_j^{(2)} \circ \cdots \circ \boldsymbol{u}_j^{(N)}) = \frac{1}{2}\|\underline{\mathbf{Y}}^{(j)} - \boldsymbol{u}_j^{(1)} \circ \boldsymbol{u}_j^{(2)} \circ \cdots \circ \boldsymbol{u}_j^{(N)}\|_F^2, \tag{4}$$

where the rank-one tensors $\underline{\mathbf{Y}}^{(j)}$ are defined as

$$\underline{\mathbf{Y}}^{(j)} = \sum_{r=1, r\neq j}^{J} \boldsymbol{u}_r^{(1)} \circ \boldsymbol{u}_r^{(2)} \circ \cdots \circ \boldsymbol{u}_r^{(N)} . \tag{5}$$

We calculate the gradient of (4) with respect to vector $\boldsymbol{u}_j^{(n)}$, and set it to zero to obtain a fixed point learning rule for $\boldsymbol{u}_j^{(n)}$ given by

$$\begin{aligned} \boldsymbol{u}_j^{(n)} &\leftarrow \left[\underline{\mathbf{Y}}^{(j)} \bar{\times}_1 \boldsymbol{u}_j^{(1)} \bar{\times}_2 \boldsymbol{u}_j^{(2)} \cdots \bar{\times}_{n-1} \boldsymbol{u}_j^{(n-1)} \bar{\times}_{n+1} \boldsymbol{u}_j^{(n+1)} \cdots \bar{\times}_N \boldsymbol{u}_j^{(N)}\right]_+ \\ &= \left[\underline{\mathbf{Y}}^{(j)} \bar{\times}_{-n} \{\boldsymbol{u}_j\}\right]_+, \qquad n = 1, 2, \cdots, N, \end{aligned} \tag{6}$$

where symbol $\bar{\times}_n$ denotes the n-mode tensor by vector multiplication [21,20]. The factors except the last factor will be normalized to be ℓ_2-unit vectors during the iteration $\boldsymbol{u}_j^{(n)} \leftarrow \boldsymbol{u}_j^{(n)}/\|\boldsymbol{u}_j^{(n)}\|_2$, $n = 1, 2, \ldots, N-1$.

3.2 Decomposition and Feature Extraction

In this study, for the training stage, the data tensor was organized to have five modes including time, frequency, channel, subject and group. The last two modes were merged to reduce the number of basis factors. In fact, a test sample was of only four modes without the group dimension. This means the first three factors corresponding to temporal, spectral and spatial modes would be used as bases of the feature subspace. However, organization of tensor with 5 modes helped us to evidently discriminate features of the two classes as shown in Fig. 2. The last factor consisted of features of the EEG spectral tensors in the subspace spanned by the temporal, spectral and spatial factors.

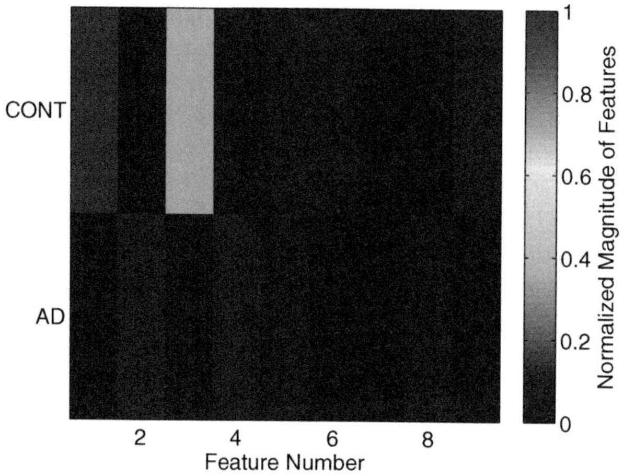

Fig. 2. Features extracted from the 5-D training tensor using the HALS algorithm

Since the leave-one-out policy was used for the training and test of the classifier, 20 subjects' data were used for training, and one subject's data were for the test. Every subject was tested respectively. Generally, the number of components (features) should be chosen to satisfy the uniqueness condition for the PARAFAC [26], that means $\sum_{n=1}^{N} k_{\mathbf{U}^{(n)}} \geq 2J + (N-1)$, where $k_{\mathbf{U}^{(n)}}$, $n = 1, 2, 3, 4$ are defined as the maximum value k such that any k columns are linearly independent. For this study, nine was set for the number of features to be extracted out. Hence, for the training stage, NTF was performed on the matrix with the dimension of 130 samples by 256 frequency bins by 9 channels by 20 subjects-2 groups. Dimensions of $\mathbf{U}^{(1)}_{train}$, $\mathbf{U}^{(2)}_{train}$,$\mathbf{U}^{(3)}_{train}$, $\mathbf{U}^{(4)}_{train}$ were 130 by 9, 256 by 9, 9 by 9, 40 by 9, respectively. For the test stage, NTF was implemented on the matrix with the dimension of 130 samples by 256 frequency bins by 9 channels by 1 subject. The dimensions of $\mathbf{U}^{(1)}_{test}$, $\mathbf{U}^{(2)}_{test}$,$\mathbf{U}^{(3)}_{test}$, $\mathbf{U}^{(4)}_{test}$ and $\mathbf{U}^{(5)}_{test}$ were 130 by 9, 256 by 9, 9 by 9, 1 by 9, respectively. The test features were obtained by projecting the test samples onto the feature subspace of basis factors $\mathbf{U}^{(1)}_{train}$, $\mathbf{U}^{(2)}_{train}$ and $\mathbf{U}^{(3)}_{train}$.

3.3 Classifier

After feature extraction, one tensor has been compressed to be a feature vector of J entries. All the training features then are stacked in one matrix of N samples $\times$ J features. A row of this matrix reflects features of a training sample. Features of the test samples are organized in the same manner. Suitable labels for test samples are identified by using a classifier with set of training features. For this study, the k-nearest neighbor classifier was applied with $k = 3$. In fact, we could use any classifiers from the Weka Machine Learning Toolkit [27], and the resulted performances were not so much different.

4 Experimental Results

Fig.2 demonstrates the features of two groups under the decomposition of all subjects with the data matrix of 130 samples by 256 frequency bins by 9 channels by 21 subjects by 2 groups. The difference between two groups was significant. After the features for the training and test were extracted, they were fed to the classifier. The correct rate of the recognition between control (healthy) children and children with AD was 76.2%. However, with the peak amplitudes of MMN, the difference between two groups was not evident [7].

5 Conclusion and Discussion

In this study, we have applied a NTF algorithm to extract features of the evoked brain activities from the time-frequency represented ERPs and performed preliminary diagnosis of the children with AD. The proposed method significantly outperforms the frequently used method based on the discrimination of peak amplitudes of MMNs. Although the nonnegative matrix decomposition methods have been already used for the classification of ongoing EEG [9,10,11,12], to our knowledge, they have not been studied for the discrimination of ERPs. So, our method is novel, significant, and quite promising in the research to diagnose not only the children with AD but for other mental disorders.

Moreover, this study is based on the average over EEG recordings of single-trials, i.e., the evoked (phase-locked) brain activity of an ERP. This is because most knowledge of ERPs is derived from the averaged EEG recordings over single-trials, i.e., from the evoked responses [1,2,3]. Indeed, during the experiment of an ERP, not only the evoked responses but also the induced (non-phase-locked) responses can be elicited [18]. So, it will be interesting to investigate whether features derived from NTF on the induced responses are also helpful to discriminate the healthy subjects and patients with mental disorders. Furthermore, in the real-time system, the single-trial recordings should be used. Meanwhile, the single-trial recordings contain both the evoked and induced responses. So, which type of responses of an ERP to be chosen and how to extract features of an ERP with our proposed method from single-trial recordings for the diagnosis task can be another challenging and significant topic in the future.

It should be noted that the number of features was set to be identical to the number of channels in this study. This treatment intuitively origins from the fact that the number of sources is often assumed to be the same to the number of channels in the application of independent component analysis on EEG [28]. However, we have not managed to prove that such a treatment is the most reasonable in this study. So, how to design a paradigm to estimate the most appropriate number of features can be a very significant research topic in the application of NTF to extract features from EEG recordings. Optimization of the classification in this study is not discussed due to limited space. However, we have found that the tensor model, feature selection and the type of the classifiers may play key roles. We hope to improve performance by optimizing those important roles in the further study. Moreover, the groups of children were formulated according to the age, and the sex was not balanced in this study. With the limitation of the sexes and ages of the participants, we could not make the balanced groups of the sex with the similar ages in our dataset. In the future, it can be interesting to investigate whether the sex has significant impact on the classification or not.

Acknowledgments. Cong thanks the international mobility grants (Spring-2009) of University of Jyväskylä for sponsoring the research.

References

1. Luck, S.: An Introduction to the Event-Related Potential Technique. MIT Press, Cambridge (2005)
2. Näätänen, R.: Attention and Brain Function. Lawrence Erlbaum, NJ (1992)
3. Duncan, C.C., Barry, R.J., Connolly, J.F., Fischer, C., Michie, P.T., Näätänen, R., Polich, J., Reinvang, I., Van Petten, C.: Event-related potentials in clinical research: Guidelines for eliciting, recording, and quantifying mismatch negativity, p300, and n400. Clin. Neurophysiol. 120(11), 1883–1908 (2009)
4. Huttunen-Scott, T., Kaartinen, J., Tolvanen, A., Lyytinen, H.: Mismatch negativity (mmn) elicited by duration deviations in children with reading disorder, attention deficit or both. Int. J. Psychophysiol. 69(1), 69–77 (2008)
5. Bental, B., Tirosh, E.: The relationship between attention, executive functions and reading domain abilities in attention deficit hyperactivity disorder and reading disorder: a comparative study. Journal of Child Psychology and Psychiatry and Allied Disciplines 48, 455–463 (2007)
6. Purvis, K., Tannock, R.: Phonological processing, not inhibitory control, differentiates adhd and reading disability. Journal of the American Academy of Child and Adolescent Psychiatry 39, 485–494 (2000)
7. Huttunen, T., Halonen, A., Kaartinen, J., Lyytinen, H.: Does mismatch negativity show differences in reading-disabled children compared to normal children and children with attention deficit? Dev. Neuropsychol. 31(3), 453–470 (2007)
8. Cichocki, A., Lee, H., Kim, Y.D., Choi, S.: Non-negative matrix factorization with alpha-divergence. Pattern Recogn. Lett. 29, 1433–1440 (2008)
9. Lee, H., Kim, Y.D., Cichocki, A., Choi, S.: Nonnegative tensor factorization for continuous EEG classification. Int. J. Neural Syst. 17(4), 1–13 (2007)

10. Li, J., Zhang, L.: Regularized tensor discriminant analysis for single trial eeg classification in bci. Pattern Recogn. Lett. 31(7), 619–628 (2010)
11. Lee, H., Cichocki, A., Choi, S.: Kernel nonnegative matrix factorization for spectral eeg feature extraction. Neurocomputing 72, 3182–3190 (2009)
12. Phan, A.H., Cichocki, A.: Tensor decomposition for feature extraction and classification problems (invited paper). IEICE T Fund. Electr. (2010) (accepted)
13. Lee, D., Seung, H.: Learning of the parts of objects by non-negative matrix factorization. Nature 401, 788–791 (1999)
14. Phan, A.H., Cichocki, A.: Multi-way nonnegative tensor factorization using fast hierarchical alternating least squares algorithm (HALS). In: Proc. The 2008 International Symposium on Nonlinear Theory and its Applications, Budapest, Hungary, September 7-10 (2008)
15. Cichocki, A., Phan, A.H.: Fast local algorithms for large scale nonnegative matrix and tensor factorizations. IEICE T Fund. Electr (invited paper) E92A(3), 708–721 (2009)
16. Cong, F., Phan, A.H., Cichocki, A., Lyytinen, H., Ritaniemi, T.: Identical fits of nonnegative matrix/tensor factorization may correspond to different extracted event-related potentials. In: Proc. International Joint Conference on Neural Networks 2010, Barcelonar, Spain, 17-24 (July 2010) (in press)
17. Cichocki, A., Phan, A.H., Caiafa, C.: Flexible HALS algorithms for sparse nonnegative matrix/tensor factorization. In: Proc. 18-th IEEE workshops on Machine Learning for Signal Processing, Cancun, Mexico, October 16-19, pp. 73–78 (2008)
18. Tallon-Baudry, C., Bertrand, O., Delpuech, C., Pernier, J.: Stimulus specificity of phase-locked and non-phase-locked 40 Hz visual responses in human. J. Neurosci. 16(13), 4240–4249 (1996)
19. Kalyakin, I., Gonzalez, N., Joutsensalo, J., Huttunen, T., Kaartinen, J., Lyytinen, H.: Optimal digital filtering versus difference waves on the mismatch negativity in an uninterrupted sound paradigm. Dev. Neuropsychol. 31(3), 429–452 (2007)
20. Cichocki, A., Zdunek, R., Phan, A.H., Amari, S.: Nonnegative Matrix and Tensor Factorizations. Wiley, Chichester (2009)
21. Kolda, T., Bader, B.: Tensor decompositions and applications. Technical Report SAND2007-6702, Sandia National Laboratories, Albuquerque, NM and Livermore, CA (2007)
22. Bro, R.: Multi-way Analysis in the Food Industry - Models, Algorithms, and Applications. PhD thesis, University of Amsterdam, Holland (1998)
23. Kim, Y.D., Choi, S.: Nonnegative Tucker Decomposition. In: Proc. Conf. Computer Vision and Pattern Recognition 2007, Minneapolis, Minnesota, USA, pp. 1–8 (June 18-23, 2007)
24. Mørup, M., Hansen, L., Parnas, J., Arnfred, S.: Decomposing the time-frequency representation of EEG using non-negative matrix and multi-way factorization. Technical report (2006)
25. Gillis, N., Glineur, F.: Nonnegative factorization and the maximum edge biclique problem. Technical Report arXiv:0810.4225. 2008-64 (2008)
26. Stegeman, A.: On uniqueness conditions for Candecomp/Parafac and Indscal with full column rank in one mode. Linear Algebra Appl. 431(1-2), 211–227 (2009)
27. Hall, M., Frank, E., Holmes, G., Pfahringer, B., Reutemann, P., Witten, I.H.: The weka data mining software: An update. SIGKDD Explorations 11(1), 10–18 (2009)
28. Delorme, A., Makeig, S.: EEGLAB: an open source toolbox for analysis of single-trial EEG dynamics. J. Neurosci. Meth. 134, 9–21 (2004)

Appendix: Outer Product

Several special matrix products are important for representation of tensor factorizations and decompositions.

The outer product of the tensors $\underline{\mathbf{Y}} \in \mathbb{R}^{I_1 \times I_2 \times \cdots \times I_N}$ and $\underline{\mathbf{X}} \in \mathbb{R}^{J_1 \times J_2 \times \cdots \times J_M}$ is given by $\underline{\mathbf{Z}} = \mathbf{Y} \circ \mathbf{X} \in \mathbb{R}^{I_1 \times I_2 \times \cdots \times I_N \times J_1 \times J_2 \times \cdots \times J_M}$, where $z_{i_1,i_2,\ldots,i_N,j_1,j_2,\ldots,j_M} = y_{i_1,i_2,\ldots,i_N}\, x_{j_1,j_2,\ldots,j_M}$. Observe that, the tensor $\underline{\mathbf{Z}}$ contains all the possible combinations of pair-wise products between the elements of $\underline{\mathbf{Y}}$ and $\underline{\mathbf{X}}$.

As special cases, the outer product of two vectors $\boldsymbol{a} \in \mathbb{R}^I$ and $\boldsymbol{b} \in \mathbb{R}^J$ yields a rank-one matrix $\mathbf{A} = \boldsymbol{a} \circ \boldsymbol{b} = \boldsymbol{a}\boldsymbol{b}^T \in \mathbb{R}^{I \times J}$ and the outer product of three vectors: $\boldsymbol{a} \in \mathbb{R}^I$, $\boldsymbol{b} \in \mathbb{R}^J$ and $\boldsymbol{c} \in \mathbb{R}^Q$ yields a third-order rank-one tensor: $\underline{\mathbf{Z}} = \boldsymbol{a} \circ \boldsymbol{b} \circ \boldsymbol{c} \in \mathbb{R}^{I \times J \times Q}$, where $z_{ijq} = a_i\, b_j\, c_q$.

Riemannian Geometry Applied to BCI Classification

Alexandre Barachant[1], Stéphane Bonnet[1], Marco Congedo[2], and Christian Jutten[2]

[1] CEA, LETI, DTBS/STD/LE2S, 17 rue des Martyrs, F-38054 Grenoble, France
[2] Team ViBS (Vision and Brain Signal Processing), GIPSA-lab, CNRS, Grenoble Universities. Domaine Univiversitaire, 38402 Saint Martin d'Hères, France

Abstract. In brain-computer interfaces based on motor imagery, covariance matrices are widely used through spatial filters computation and other signal processing methods. Covariance matrices lie in the space of Symmetric Positives-Definite (SPD) matrices and therefore, fall within the Riemannian geometry domain. Using a differential geometry framework, we propose different algorithms in order to classify covariance matrices in their native space.

1 Introduction

A Brain-Computer Interface (BCI) aims at providing an alternative non-muscular communication path and a control system for the individuals with heavy motor disabilities like in Spinal Chord Injury (SCI) or Locked-In Syndrome (LIS) patients. This new interface should perform an automatic decoding of measured brain activity [1].

Non-invasive BCIs use mainly ElectroEncephaloGraphic (EEG) activity recorded from a cap of scalp electrodes [2]. The goal is to detect and classify some specific patterns of EEG activities so as to drive an external effector (e.g. computer mouse, wheelchair, ...) [1]. Different paradigms can be used to activate these brain patterns, either synchronously (evoked potentials) [3] or asynchronously (brain rhythm modulation) after a co-adaptive learning phase. EEG-based features are usually related to the power (or variance) of relevant EEG channels in specific frequency bands (e.g. mu brain oscillation in the frequency band 5-15 Hz) [3]. In this article, we propose a new signal processing framework in BCI applications, which is based on **Riemannian geometry**.

As it will be shown, interesting properties may be derived by considering the space of symmetric positive-definite (SPD) matrices. The main motivation of our work is to make use of the concept of Riemannian distance between SPD matrices in BCI applications. We will introduce basic tools to manipulate EEG data in this Riemannian manifold and illustrate these concepts with a simple and didactic binary classification task. Doing so, EEG data can be manipulated in a convenient way through their spatial covariances, and then detection/classification

V. Vigneron et al. (Eds.): LVA/ICA 2010, LNCS 6365, pp. 629–636, 2010.

can be achieved by measuring, for instance, the Riemannian distance between covariance matrices of signal epochs and and covariance matrices of reference epochs. Indeed, classical methods treats the SPD matrices as if they were naturally lying in the Euclidean space, whereas the natural geometry to be considered is the Riemann geometry.

This approach has already been followed in diffusion tensor imaging to study the statistical properties of a population of geometric objects [4], in image processing to detect pedestrians images [5] or in radar detection and High Resolution Doppler Imagery to achieve a robust statistical estimation of Toeplitz Hermitian positive definite covariance matrices of sensor data time series [6]. First we will present the basic tools of Riemannian geometry in space of SPD matrices, next we will propose some classification and filtering algorithms and finally we will show results on real EEG data.

2 Differential Geometry in Space of SPD Matrices

2.1 Definitions and Properties

This section exposes some **definitions** and **properties** of differential geometry in the Riemannian manifold of SPD matrices. More sophisticated explanations can be found in the reference [7].

A Riemannian manifold is a differentiable manifold in which the tangent space at each point is a finite-dimensional Euclidean space. We denote by $S(n) = \{\mathbf{S} \in M(n), \mathbf{S}^T = \mathbf{S}\}$ the space of all $n \times n$ symmetric matrices in the space of square matrices and denote by $P(n) = \{\mathbf{P} \in S(n), \mathbf{P} > 0\}$ the set of all $n \times n$ symmetric positive-definite (SPD) matrices. The Riemannian distance between two SPD matrices $\mathbf{P}_1$ and $\mathbf{P}_2$ in $P(n)$ is given by [7]:

$$\delta_R(\mathbf{P}_1, \mathbf{P}_2) = \|\mathrm{Log}\left(\mathbf{P}_1^{-1}\mathbf{P}_2\right)\|_F = \left[\sum_{i=1}^{n} \log^2 \lambda_i\right]^{1/2} \tag{1}$$

where the λ_i's are the real strictly positive eigenvalues of $\mathbf{P}_1^{-1}\mathbf{P}_2$ and $\|.\|_F$ is the Frobenius norm of a matrix. In Eq (1),the operator Log (.) is the logarithm of a matrix. Given that SPD matrices are diagonalizable and invertible, $\mathrm{Log}(\mathbf{P})$ can be computed by diagonalization of $\mathbf{P}$: $\mathrm{Log}(\mathbf{P}) = \mathbf{V}\log(\mathbf{D})\mathbf{V}^{-1}$ with $\log(\mathbf{D})$ the logarithm of each element of $\mathbf{D}$ where $\mathbf{D}$ and $\mathbf{V}$ are respectively the diagonal matrix of eigenvalues and the matrix of eigenvectors of $\mathbf{P}$. The geometric mean in the Riemannian sense, i.e. associated with the metric defined in Eq (1), of m given SPD matrices $\mathbf{P}_1, \ldots, \mathbf{P}_m$ is defined as [7]:

$$\mathfrak{G}\left(\mathbf{P}_1, \ldots, \mathbf{P}_m\right) = \underset{\mathbf{P} \in P(n)}{\mathrm{argmin}} \sum_{i=1}^{m} \delta_R^2\left(\mathbf{P}, \mathbf{P}_i\right) \tag{2}$$

There is no closed-form expression for such mean computation, but iterative algorithms can be employed, as demonstrated in section 2.2.

The shortest path between two points in the Riemannian space of SPD matrices is defined by the geodesic $\gamma(t)$ with $t \in [0,1]$:

$$\gamma(t) = \mathbf{P}_1^{1/2}\left(\mathbf{P}_1^{-1/2}\mathbf{P}_2\mathbf{P}_1^{-1/2}\right)^t \mathbf{P}_1^{1/2} \tag{3}$$

Identically to the matrix logarithm, the power of SPD matrices can be computed using a diagonalization: $\mathbf{P}^t = \mathbf{V}\mathbf{D}^t\mathbf{V}^{-1}$.

2.2 Tangent Space

In complete Riemannian space, given a point $\mathbf{P} \in P(n)$, it is possible for every point $\mathbf{P}_i \in P(n)$, to identify a tangent vector $\mathbf{S}_i \in S(n)$ such as $\mathbf{S}_i = \dot{\gamma}(0)$ with $\gamma(t)$ the geodesic between $\mathbf{P}$ and $\mathbf{P}_i$. The Riemannian Log map operator $\text{Log}_\mathbf{P} : P(n) \rightarrow S(n)$ achieves the mapping $\text{Log}_\mathbf{P}(\mathbf{P}_i) = \mathbf{S}_i$.

$T_\mathbf{P}$, the tangent space at $\mathbf{P}$, is the space defined by the whole set of tangent vector $\mathbf{S}_i$, here $S(n)$. In this tangent space, the metric is flat and allows us to use arithmetic mean and other classical tools. The Riemannian Exp map operator $\text{Exp}_\mathbf{P}(\mathbf{S}_i) = \mathbf{P}_i$ allows to go back in the original space of SPD matrices $P(n)$ in a one-to-one mapping. Both operators are crucial in the manipulation of SPD matrices as we will discover. Using the affine-invariant metric [7], we have the expressions:

$$\text{Exp}_\mathbf{P}(\mathbf{S}_i) = \mathbf{P}^{1/2}\text{Exp}\left(\mathbf{P}^{-1/2}\mathbf{S}_i\mathbf{P}^{-1/2}\right)\mathbf{P}^{1/2} \tag{4}$$

$$\text{Log}_\mathbf{P}(\mathbf{P}_i) = \mathbf{P}^{1/2}\text{Log}\left(\mathbf{P}^{-1/2}\mathbf{P}_i\mathbf{P}^{-1/2}\right)\mathbf{P}^{1/2} \tag{5}$$

We can refer to [4] for efficient computation. Figure 1 illustrates these operations.

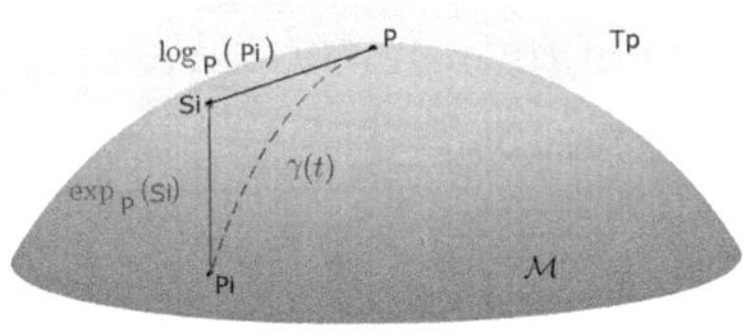

Fig. 1. Tangent space of the manifold $\mathcal{M}$ at point $\mathbf{P}$, $\mathbf{S}_i$ the tangent vector of $\mathbf{P}_i$ and $\gamma(t)$ the geodesic between $\mathbf{P}$ and $\mathbf{P}_i$

The mean of m SPD matrices can be obtained using the concept of tangent space. Using Riemannian Log map, we first project the whole dataset in tangent space. In this Euclidean space, the arithmetic mean is the correct average estimate and can be easily computed. Finally we project the obtained arithmetic mean into SPD space using Riemannian exponential map. After few iterations, we obtain the geometric mean of SPD matrices. Algorithm 1 explains this process.

Algorithm 1. Mean of m SPD matrices

Input: Ω a set of m SPD matrices $\mathbf{P}_i \in P(n)$ and $\epsilon > 0$.
Output: $\mathbf{P}_\Omega$ the estimated mean in $P(n)$.

Initialise $\mathbf{P}_\Omega^{(1)} = \frac{1}{m}\sum_{i=1}^{m}\mathbf{P}_i$
repeat
  $\mathbf{S} = \frac{1}{m}\sum_{i=1}^{m}\mathrm{Log}_{\mathbf{P}_\Omega^{(t)}}(\mathbf{P}_i)$ {Arithmetic mean in tangent space}
  $\mathbf{P}_\Omega^{(t+1)} = \mathrm{Exp}_{\mathbf{P}_\Omega^{(t)}}(\mathbf{S})$
until $\|\mathbf{S}\|_F < \epsilon$

3 Classifying on Riemannian Manifold

We denote by $\mathbf{E} \in R^{n \times t}$ a given EEG recording epoch with n electrodes and t samples. The spatial sample covariance matrix is proportional to: $\mathbf{P} = \mathbf{E}\mathbf{E}^T$ for centered-data matrix is, by construction, lying in $P(n)$. The goal in BCI is to determine what mental task is associated with a segment of data, or in other terms, determine the class $\omega_i \in \{1, 2\}$ of the observation $\mathbf{E}_i$ or equivalently its spatial covariance matrix $\mathbf{P}_i$. In motor imagery BCI paradigm, the mental task would be for instance, the recognition of either left- and right-hand imagery movements [8]. This binary classification problem is usually tackled using supervised learning where the patient undergoes first a training session where the correspondence $\{\mathbf{E}_i, \omega_i\}$ is known by experiment design, so as to produce decision rules for next unknown test sessions [2]. Taking into account covariance matrices as elements of Riemannian space of SPD matrices, spatial information is directly accessible and classification can be performed without preprocessing, in our approach. Based on the concept presented in section 2, a very simple algorithm, given in Algorithm 2, is proposed for illustration purpose. It is merely based on the computation of Riemannian distances to classify a new epoch.

Algorithm 2. Simple classification based on Riemannian distance

Input: Ω a set of m SPD matrices $\mathbf{P}_i \in P(n)$.
Input: $\omega_i \in \{1, 2\}$ the class of $\mathbf{P}_i$.
Input: $\mathbf{P}_x$ a SPD matrix of unknown class.
Output: ω_x the estimated class of test covariance matrix $\mathbf{P}_x$.

$\mathbf{P}_{\Omega_1} = \mathfrak{G}(\mathbf{P}_i)$ with $\{i|\omega_i = 1\}$ {Riemannian mean for class 1}
$\mathbf{P}_{\Omega_2} = \mathfrak{G}(\mathbf{P}_i)$ with $\{i|\omega_i = 2\}$ {Reimannian mean for class 2}
$d = \delta(\mathbf{P}_x, \mathbf{P}_{\Omega_1}) - \delta(\mathbf{P}_x, \mathbf{P}_{\Omega_2})$
if $d \leq 0$ **then**
  $\omega_x = 1$
else
  $\omega_x = 2$
end if
return ω_x

As it can be observed, the whole algorithm 2 relies on the computation of both intra-class SPD means ($\mathbf{P}_{\Omega_1}, \mathbf{P}_{\Omega_2}$) and the shortest Riemannian distance between the test covariance matrix and the two intra-class SPD means. The main limitation of this approach is that it may exist a large part of distance between the two matrices which is not class-related i.e. the class-related information contained in distance can vanish in front of noise.

Therefore, it is preferable to perform some filtering over SPD matrices before applying Algorithm 2. Our approach is inspired by the principal geodesics analysis (PGA) method of Fletcher et. al. [4]. We search first the geodesics that support class-related information and perform filtering along this line in Riemann space to discard irrelevant information. To compute these filters, we propose a supervised algorithm named FGDA for Fisher Geodesic Discriminant Analysis. This algorithm is an extension of Fisher Linear Discriminant Analysis to the tangent space and is given by Algorithm 3.

Algorithm 3. FGDA Filters

Inputs: Ω a set of m SPD matrices $\mathbf{P}_i \in P(n)$, $\omega_i \in \{1, 2\}$ the class of $\mathbf{P}_i$, K number of selected components.

Outputs: $\widetilde{\mathbf{W}}_k, k = 1, \ldots K \in \Re^{n(n+1)/2}$, $\mathbf{P}_\Omega$.

1: $\mathbf{P}_\Omega = \mathfrak{G}(\mathbf{P}_i)$ {Compute Riemannian mean of the whole set}
2: **for** $i = 1$ to m **do**
3: $\quad \mathbf{S}_i = \mathrm{Log}_{\mathbf{P}_\Omega}(\mathbf{P}_i)$ {Apply Riemannian Log map}
4: $\quad \widetilde{\mathbf{S}}_i = vec(\mathbf{S}_i)$ {Keep upper triangular matrix in vector form $n(n+1)/2$}
5: **end for**
6: $\widetilde{\mathbf{W}} = LDA(\widetilde{\mathbf{S}}_i)$ {compute the projection vectors using the Fisher LDA criterion}
7: Select the first K vectors $\widetilde{\mathbf{W}}_k$

In order to project data in tangent space, we compute the Riemannian mean $\mathbf{P}_\Omega$. $\mathbf{P}_\Omega$ is the point where the tangent vectors $\mathbf{S}_i$ will be computed, different points can be used [5], however the use of the Riemannian mean minimizes the approximation caused by the projection in the tangent space of the dataset. After performing data projection into the tangent space, the step 6 of this algorithm compute the different projection vectors using the Fisher LDA criterion [9]. This is a maximisation of the ratio of the *between-class* scatter matrix $\mathbf{\Sigma}_b$ and the *within-class* scatter matrix $\mathbf{\Sigma}_w$ and can be solved easily by eigenvector decomposition of $\mathbf{\Sigma}_w^{-1}\mathbf{\Sigma}_b$ [9]. Interestingly, the exponential map of these filters gives the main geodesics issued from $\mathbf{P}_\Omega$.

Typically, the number K of selected components is low. We could consider that the all class-related information is contained within the first five components. The filtering operation is explained in Algorithm 4. Step 2 computes the variation mode and projection using least-squares estimate. We search $\widetilde{\mathbf{S}}_x$ supported by the K components $\widetilde{\mathbf{W}}$, given by Algorithm 4, which best fit with $\mathbf{S}_x$. We call this operation a filtering operation because there is no dimensional reduction.

Finally, the core algorithm of this work is presented in Algorithm 5. First we compute FGDA filters on training dataset with Algorithm 3 , then we apply

Algorithm 4. Geodesic Filtering

Inputs: $\mathbf{P}_x$, $\mathbf{P}_\Omega$, $\widetilde{\mathbf{W}} = [\widetilde{\mathbf{W}}_1 \dots \widetilde{\mathbf{W}}_K] \in \Re^{n(n+1)/2 \times K}$.
Output: $\widetilde{\mathbf{P}}_x$
1: $\mathbf{S}_x = \text{Log}_{\mathbf{P}_\Omega}(\mathbf{P}_x)$ {Apply Riemannian Log map}
2: $\widetilde{\mathbf{S}}_x = \widetilde{\mathbf{W}}\left(\widetilde{\mathbf{W}}^T\widetilde{\mathbf{W}}\right)^{-1}\widetilde{\mathbf{W}}^T vec(\mathbf{S}_x)$ {Filtering operation}
3: $\widetilde{\mathbf{P}}_x = \text{Exp}_{\mathbf{P}_\Omega}\left(unvec\left(\widetilde{\mathbf{S}}_x\right)\right)$ {Apply Riemannian Exp map}
4: **return** $\widetilde{\mathbf{P}}_x$

those filters on the dataset with Algorithm 4 and finally we use Algorithm 2 to obtain classes of filtered test data. It is also possible to apply a LDA classifier in tangent space without going back to original space. The LDA classifer uses only the first component and gives the same results as our complete method when taking one component.

Algorithm 5. Classification, filtered version

Inputs: Ω a set of m SPD matrices $\mathbf{P}_i \in P(n)$, $\omega_i \in [1:2]$ the class of $\mathbf{P}_i$, K number of kept components.
Input : $\mathbf{P}_x$ the SPD matrix to classify
Output: ω_x the class of $\mathbf{P}_x$
1: Compute FGDA filters (Algorithm 3)
2: Filter all SPD matrices $\mathbf{P}_i, \mathbf{P}_x$ (Algorithm 4)
3: Classify the filtered SPD matrices (Algorithm 2)
4: **return** ω_x

Figure 2 represents these different manipulations over a simulated dataset. 2×2 covariance matrices are generated according to two Wishart distributions, one for each classes. First, data are projected in tangent space using Log. map operator (Fig 2.D). Next, LDA components are computed and applied to the data (two components for Fig 2.E, only one for Fig 2.F). Finally, data are wrapped back to the original space with Exp. map operator (Fig 2.B and Fig 2.C).

4 Classification Results in BCI

4.1 Introduction

In order to evaluate the performances of the proposed methods, we compare them to an implementation of a reference method [8]. This latter method represents the classical signal processing chain in asynchronous BCI. It is composed of frequency filtering, spatial filtering (using CSP approach), Log Variance feature extraction and finally Fisher's LDA classification. Datasets IVa of BCI competition III [10] are used for analysis. Only 9 electrodes are used (F_3, F_Z, F_4, C_3, C_Z, C_4, P_3, P_Z, P_4). This subset of electrodes can represent a typical case of an every-day use

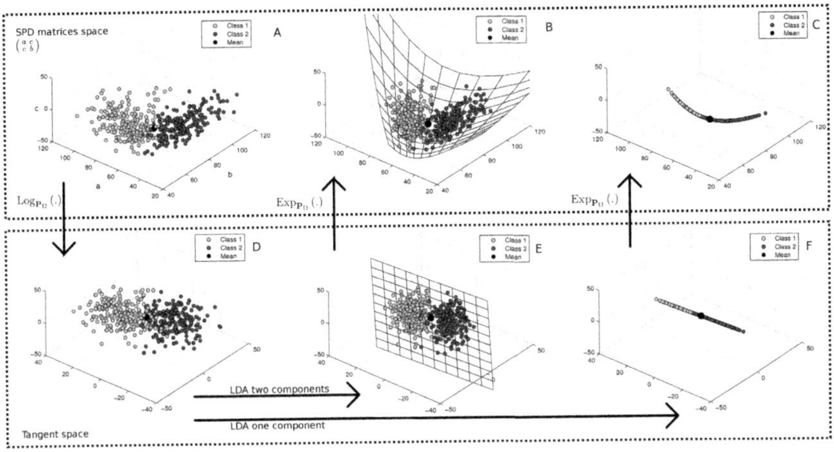

Fig. 2. Manipulations in tangent and original space of 2×2 covariance matrices

BCI. A $(10 - 30$ Hz) band-pass filter has been applied on the original signals for all subjects. This dataset is composed by 5 subjects who performed 280 trials of right-hand or right-foot motor imagery. It was provided by Fraunhofer FIRST, Intelligent Data Analysis Group. We use 10-fold cross-validation to evaluate properly the performance.

4.2 Classification Results

Results are given for both filtered (Algo 5) and unfiltered (Algo 2) version of the proposed classification algorithm based on Riemannian distance. We compare both methods with a reference method as described in Section 4.1. In the results, we kept only 4 FGDA filters for the filtered method and 6 CSP filters for the reference methods. Classification error rates are given in Table 1. The filtered version method (Algo 5) outperforms our implementation of the reference algorithm for almost all the subjects. The unfiltered version (Algo 2) is worst than the reference method but shows good results considering its simplicity. Because a high inter-subject variability, results are not really significant. However, we have benchmark our algorithms on several datasets and we can say that the

Table 1. Classification Error rate. 10-fold Cross-validation.

User	Reference	Algo 2	Algo 5
aa	26	28.9	**22.5**
al	3.2	3.9	**2.8**
av	34.2	39.6	**34.2**
aw	**6.4**	11	7.4
ay	7.4	12.1	**7.1**
Mean	15.5 ± 13.8	19.1 ± 14.7	$\mathbf{14.8} \pm 13.6$

Riemannian approach is effective and shows very good results in difficult cases (i.e. noisy dataset, small amount of data, multi-class).

5 Conclusion

We have presented an useful framework for working in space of SPD matrices, illustrating it with simple methods and giving results that show how promising it is. This approach could be generalized to other signal processing methods in BCI or elsewhere. Methods can be developed to work natively in space of SPD matrices like our Algorithm 2 or in tangent space like Algorithm 3. The Euclidean Tangent Space allows us to use classical methods with the only limitation of the high number of dimension involved. In this work, FGDA is limited by the trend to over-fitting of the LDA algorithm, when the dimension is close to the number of trials. To avoid this effect a regularized LDA or a variable selection approach can be used. Furthermore, it is not always necessary to go back in SPD space, since the tangent space and original space are linked through Log. and Exp. map operators. Finally the main limitation of these methods is the computation time, Riemannian mean requires a large number of diagonalization of $n \times n$ matrices and and it becomes computationally expensive when n is large (> 50).

References

1. Lebedev, M.A., Nicolelis, M.A.L.: Brain-machine interfaces: past, present and future. Trends in Neurosciences (September 2006)
2. van Gerven, M., Farquhar, J., Schaefer, R., Vlek, R., Geuze, J., Nijholt, A., Ramsey, N., Haselager, P., Vuurpijl, L., Gielen, S., Desain, P.: The brain-computer interface cycle. Journal of Neural Engineering (August 2009)
3. Wolpaw, J.R., Birbaumer, N., McFarland, D.J., Pfurtscheller, G., Vaughan, T.M.: Brain-computer interfaces for communication and control. Clinical Neurophysiology (2002)
4. Fletcher, P.T., Joshi, S.: Principal geodesic analysis on symmetric spaces: Statistics of diffusion tensors. Computer Vision and Mathematical Methods in Medical and Biomedical Image Analysis (2004)
5. Tuzel, O., Porikli, F., Meer, P.: Pedestrian detection via classification on riemannian manifolds. Pattern Analysis and Machine Intelligence (2008)
6. Barbaresco, F.: Interactions between symmetric cone and information geometries: Bruhat-Tits and siegel spaces models for high resolution autoregressive doppler imagery. In: Emerging Trends in Visual Computing. Springer, Heidelberg (2009)
7. Moakher, M.: A differential geometric approach to the geometric mean of symmetric Positive-Definite matrices. SIAM J. Matrix Anal. Appl., 26 (2005)
8. Blankertz, B., Tomioka, R., Lemm, S., Kawanabe, M., Muller, K.-R.: Optimizing spatial filters for robust EEG Single-Trial analysis. Signal Processing Magazine. IEEE (2008)
9. Duda, R.O., Hart, P.E., Stork, D.G.: Pattern Classification, ch. 3.8.2, 2nd revised edn., pp. 117–124. J. Wiley & Sons Inc., Chichester (November 2000)
10. BCI competition III, dataset IVa, `http://ida.first.fhg.de/projects/bci/competition_iii`

Separating Reflections from a Single Image Using Spatial Smoothness and Structure Information

Qing Yan[1,2], Ercan E. Kuruoglu[3], Xiaokang Yang[1,2], Yi Xu[1,2], and Koray Kayabol[3]

[1] The Institute of Image Communication and Information Processing
[2] Shanghai Key Lab of Digital Media Processing and Transmission
Shanghai, China
[3] ISTI, CNR, via G.Moruzzi 1, 56124, Pisa, Italy
{yanqing_amy,xkyang,xuyi}@sjtu.edu.cn,
{ercan.kuruoglu,koray.kayabol}@isti.cnr.it

Abstract. We adopt two priors to realize reflection separation from a single image, namely spatial smoothness, which is based on pixels' color dependency, and structure difference, which is got from different source images (transmitted image and reflected image) and different color channels of the same image. By analysing the optical model of reflection, we simplify the mixing matrix further and realize the method for getting spatially varying mixing coefficients. Based on the priors and using Gibbs sampling and appropriate probability density with Bayesian framework, our approach can achieve impressive results for many real world images that corrupted with reflections.

Keywords: Reflection separation, BSS, optical model, Bayesian framework.

1 Introduction

When people take a picture through transparent glass, they will often be disappointed to see that their image is a superposition of a transmitted image and a reflected image. How to remove the reflection is not only an everyday question for photographers, but also a challenging problem for researchers. It calls for the knowledge of optical models, blind source separation (BSS), and image processing.

Approaches based on polarimetric imaging had caught researchers' attention early. Reflection might be removed by incorporating a polarizer into the optical system. However, this happens only when viewing angle equals to the Brewster angle [1], which is hard to achieve.

By analyzing optical model, it is found that the combination of transmitted image and reflected image is a linear mixing process [2]. With this observation, the statistical tool of independent components analysis (ICA) was applied for blindly separating two source images [2]. By the reason that sparseness can significantly improve ICA algorithms, sparse ICA (SPICA) was adopted to handle the problem [3], which ingeniously turned the intractable optical separation into a simple "geometric" separation and yielded very good results. Based on SPICA, a series of approaches were further developed to handle multiple source images with different motions [4]-[6]. However, these approaches assume that there are two or more mixed image observations, and all

V. Vigneron et al. (Eds.): LVA/ICA 2010, LNCS 6365, pp. 637–644, 2010.

pixels share the same mixing coefficients. This is not true in most real-life scenarios. Sarel and Irani proposed an approach to generate spatially varying mixing functions and separated transparent layers through layer information exchange [7]. But this approach still needs at least two input mixed images.

The approaches using two or more mixed images to separate reflections have achieved excellent outcomes. However, it is still quite difficult when only a single image is available. Obtaining two different images from only one image is a seriously ill-posed problem. Levin et al tried to use local features to handle the problem. This approach often fails even when correct decomposition is perceptually obvious [8]. Later, they used sparsity prior and user assisted information to realize such separation [9]. This system is not automatic, and its result is somewhat blurred. Kayabol et al use color channel dependencies to obtain the MAP estimate of separated source images [10]. It restricts reflections are achromatic and achieves compelling result. But this approach still follows the rule of constant mixing coefficients, and the color of its separated result may be not physically valid.

In this paper, we developed Kayabol et al's approach further by incorporating structure information into its original Bayesian model and transforming its constant mixing matrix into physically valid and spatially varying coefficients. In most cases, transmitted image and reflected image are independent and have quite different structures, while different color channels of the same image always share a similar structure. Thus structure information may become an important cue when separating reflection. Moreover, we still suppose reflected image is achromatic as in [10], which happens when grey objects are reflected, like clouds, walls and so on. By using the knowledge of optical model, we can simplify the mixing matrix into a three-element vector. Thus local mixing coefficient is able to obtain easily.

The organization of this paper is as follows: the optical model and the mathematical framework of our approach are proposed in section 2. The probability density formulations in the framework are fully expounded in section 3. In section 4, an experimental comparison is made between our approach and Kayabol's. Finally, conclusion and acknowledgement are given in section 5 and section 6.

2 Problem Formulation

2.1 Optical Model of Transmitted Image and Reflected Image

The views that a person sees are formed by the light that enters person's eyes. When a person looks through a transparent glass (with no light refraction), he/she may find a real object and a virtual object situated on the optical axis. That is a joined effect of transmitted light and reflected light. Thus, we can assume the view seen by observer is a linear interpolation of transmitted image and reflected image. And we can express the observed image I in a single channel as,

$$I = K_t I_{transmit} + (1 - K_t) I_{reflect} \tag{1}$$

where K_t is the glass's transparency, $I_{transmit}$ is the transmitted image and $I_{reflect}$ is the reflected image. Assuming that the RGB channels are independent, we can expand the single channel observation model in (1) to three channel observation model using the matrix-vector notation: $I = A \cdot S + V$. And to every pixel, this equation becomes:

$$\begin{bmatrix} I_r(n) \\ I_g(n) \\ I_b(n) \end{bmatrix} = \begin{bmatrix} A_r(n) & 0 & 0 & 1-A_r(n) \\ 0 & A_g(n) & 0 & 1-A_g(n) \\ 0 & 0 & A_b(n) & 1-A_b(n) \end{bmatrix} \begin{bmatrix} S_r(n) \\ S_g(n) \\ S_b(n) \\ S_m(n) \end{bmatrix} + V \quad (2)$$

where n is pixel index, $S_r(n)$, $S_g(n)$, and $S_b(n)$ are three color channels of transmitted source image, $S_m(n)$ is the achromatic reflected source image, $A_r(n)$, $A_g(n)$, and $A_b(n)$ are glass's transparencies to different color light, and V is a zero-mean Gaussian noise vector for the presence of noise and model imperfection. Hence there are only three variants in every pixel's mixing matrix A .

2.2 Problem Definition in the Bayesian Framework

Both mixing matrix A and source vector S are unknown in (2). We formulate this BSS problem in Bayesian framework, thus the joint posterior density of A and S can be written as:

$$p(A, S \mid I) \propto p(I \mid A, S) p(A, S) \quad (3)$$

Separated sources can be obtained when the maximum-a-posterior (MAP) result of (3) is achieved.

It is obvious that mixing matrix, transmitted source and reflected source are independent. Thus, Gibbs sampling is utilized to obtain the MAP result, which breaks down the multivariate sampling problem into a set of univariate ones as follows:

$$\begin{aligned} &p(A \mid I_{r,g,b}, S_{r,g,b}, S_m) \propto p(I_{r,g,b} \mid A, S_{r,g,b}, S_m) p(A \mid S_{r,g,b}, S_m) \\ &p(S_{r,g,b} \mid I_{r,g,b}, S_m, A) \propto p(I_{r,g,b} \mid A, S_{r,g,b}, S_m) p(S_{r,g,b} \mid S_m, A) \\ &p(S_m \mid I_{r,g,b}, S_{r,g,b}, A) \propto p(I_{r,g,b} \mid A, S_{r,g,b}, S_m) p(S_m \mid S_{r,g,b}, A) \end{aligned} \quad (4)$$

In (4) mixing matrix A is determined by glass, so it has no correlation with S_m and $S_{r,g,b}$ ($p(A \mid S_{r,g,b}, S_m) = p(A)$). Then we can take into consideration of $p(S_{r,g,b} \mid S_m, A)$ and $p(S_m \mid S_{r,g,b}, A)$. Both S_m and $S_{r,g,b}$ should be spatially smooth ($p_{smooth}(S_{r,g,b})$, $p_{smooth}(S_m)$) and their structures should be as different as possible ($p_{structure\_s}(S_{r,g,b}, S_m)$). Moreover, to make the separated transmitted image valid, structure similarity between different color channels ($p_{structure\_c}(S_{rgb})$) is also included into the regularization, as shown in (5).

$$\begin{aligned} &p(A \mid S_{r,g,b}, S_m) = p(A) \\ &p(S_{r,g,b} \mid S_m, A) = p(S_{r,g,b} \mid S_m) \propto p_{smooth}(S_{r,g,b}) \cdot p_{structure\_s}(S_{r,g,b}, S_m) \cdot p_{structure\_c}(S_{rgb}) \\ &p(S_m \mid S_{r,g,b}, A) = p(S_m \mid S_{r,g,b}) \propto p_{smooth}(S_m) \cdot p_{structure\_s}(S_{r,g,b}, S_m) \end{aligned} \quad (5)$$

Table 1. The framework of our Gibbs Sampling algorithm

1. Take initial values of Sr, Sg, Sb, Sm, Ar, Ag, Ab.
2. Repeat for t=1, 2, ...
   For n = 1, 2, . . . , N (N is the total number of pixels of the image)
   For k = 1, 2, 3, 4 (Sr, Sg, Sb, Sm)
   $$S_k^{t+1}(n) = Sample_{S_k(n)}\{p(S_k(n) \mid S_k^t(\backslash n), S^t_{\{r,g,b,m\}\backslash k}, I_{r,g,b}, A^t_{r,g,b})\}$$
   For k=1, 2, 3 (Ar, Ag, Ab)
   $$A_k^{t+1}(n) = Sample_{A_k(n)}\{p(A_k(n) \mid A_k^t(\backslash n), A^t_{\{r,g,b\}\backslash k}, S^t_{\{r,g,b,m\}}, I_{r,g,b})\}$$
3. Continue step 2 until the joint distribution does not change

3 Probability Density Formulation

3.1 Observation and Mixing Matrix Model

To model the presence of noise and model imperfection, there is an observation noise vector in (2). Since the noise is assumed to be iid zero-mean Gaussian at each pixel, the observation model is also Gaussian:

$$p(I_{r,g,b}(n) \mid S_{r,g,b,m}(n), A\ (n)) = \prod_{k\in\{r,g,b\}} \mathcal{N}(I_k(n) \mid \overline{I}_k(n), \sigma_k^2(n)) \tag{6}$$

where $\mathcal{N}(I_k(n) \mid \overline{I}_k(n), \sigma_k^2(n))$ represents a Gaussian density with variance $\sigma_k^2(n)$, which is defined by user , and mean $\overline{I}_k(n)$ ($\overline{I}_k(n) = A_k(n)\cdot S_k(n) + (1-A_k(n))\cdot S_m(n)$).

The value of glass transparency ranges from 0 to 1. With no prior knowledge about glass, we assume uniform distribution in every element of mixing matrix ($p(A)$).

Using the same prior and likelihood in (6), the posterior density of $A_k(n)$ is formed as:

$$p(A_k(n) \mid I_{r,g,b}, S_{r,g,b,m}, A_k(n)) = \mathcal{N}(A_k(n) \mid \mu_k(n), \gamma_k(n)) \quad k = 1,2,3 \tag{7}$$

$$\mu_k(n) = \sum_{l\in\Omega(n)} \frac{1}{S_k(l)S_k(l)} S_k(l)[I_k(l) - (1-A_k(l))\cdot S_m(l)], \quad \gamma_k(n) = \frac{\sigma_k^2}{\sum_{l\in\Omega(n)} S_k(l)S_k(l)} \tag{8}$$

To make A_k spatially smooth, we estimate the mean $\mu_k(n)$ and the variance $\gamma_k(n)$ in a local window ($\Omega(n)$) around the nth pixel in source S_m and the kth channel of input image.

3.2 Smoothness Model

Spatial smoothness is a key feature for inverse problems in image processing. It is used to obtain stable solutions in ill conditioned inverse problem. We adopt Markov Random Field (MRF) model to describe the smoothness probability, and its density is chosen as Gibbs distribution with non-convex energy potential function, which ensures the edge preserving smoothing [10]. The Gibbs distribution is given as:

$$p(S_k(n)) = \frac{1}{Z(\beta_k)} e^{-U(S_k(n))}, \qquad k = c, m \tag{9}$$

where S_c means the vector of $[S_r, S_g, S_b]^T$ from transmitted source, $Z(\beta_k)$ and β_k are the partition function and parameter of MRF respectively, and $U(S_k(n))$ is the energy function of Gibbs distribution given as follows:

$$U(S_k(n)) = \frac{1}{2} \sum_{\{n,q\} \in \mathcal{C}} \beta_k \cdot \rho_k(S_k(n) - S_k^t(q)), \qquad k = c, m \tag{10}$$

where $\rho_k(\cdot)$ is the non-convex potential function and $\mathcal{C}$ is the entire clique set.

In this work we opt to describe pixel differences in terms of iid Cauchy density. Although the pixels are dependent, the pixel differences can often be modelled as an iid process [11].

For reflected source, the clique potential becomes:

$$\rho_m(S_m(n) - S_m^t(q)) = \ln[1 + \frac{(S_m(n) - S_m^t(q))^2}{\delta_m}] \tag{11}$$

where δ_m is the scale parameter of Cauchy distribution.

For transmitted source, we use multivariate Cauchy density to make full use of the dependencies between different color channels. The clique potential can be written as:

$$\rho_c(S_c(n) - S_c^t(q)) = \frac{5}{2} \ln[1 + [S_c(n) - S_c^t(q)]^T \Delta^{-1} [S_c(n) - S_c^t(q)]] \tag{12}$$

where Δ is a 3×3 symmetric matrix defining the correlation between color channels. In this paper both δ_m, β_m, Δ and β_c are assumed homogeneous over the MRF.

3.3 Structure Model

Structure prior is based on the observation that structure varies greatly among different images. We adopt normalized gray-scale correlation (NGC) as the feature to represent structure similarity. It is pointed in [7] that NGC, as a local structure feature, can be little affected by spatially varying deformation, linear gray-scale deformation, and non-linear gray-scale deformation. Thus it can work well during the Gibbs sampling process, when gray-scale is partly changed in the image.

$$NGC(f, g) = \frac{C(f, g)}{\sqrt{V(f) \cdot V(g)}} \tag{13}$$

where f and g are two different images, $C(f, g)$ is the covariance between f and g, $V(f)$ and $V(g)$ are the variance of f and g respectively.

$$C(f, g) = \frac{1}{N} \sum_{j=1}^{N} f_j \cdot g_j - \bar{f} \cdot \bar{g} \qquad V(f) = \frac{1}{N} \sum_{j=1}^{N} f_j^2 - \bar{f}^2 \tag{14}$$

To verify how important structure information is, we collect a dataset of over 100 images from the internet, including nature images, texture images, and some synthetic images. We use $D_{structure} = 1 - NGC$ as the feature of structure distance to measure structure difference in different images and different channels of the same image.

We divide two different images (f and g , which are in the same size) into a series of overlapped blocks. By calculating $D_{structure}$ in every corresponding blocks of f and g , a histogram of block structure distance is generalized. The same method is implemented to different channels of the same image.

Since NGC $\in$ [-1, 1], $D_{structure}$ belongs to [0, 2]. Structure difference reaches its maximum when $D_{structure}$ equals 1, and gets its minimum when $D_{structure}$ equals 0 or 2. Fig.1 shows the example histograms that are obtained by analysing different images (a, b) and different channels of the same image (c, d).

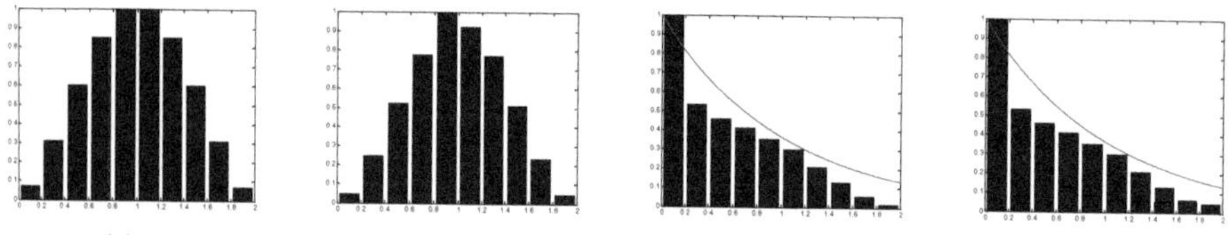

(a) (b) (c) (d)

Fig. 1. Histograms of structure distance in different images (a, b) and different channels (c, d)

By observing Fig.1, we can conclude that structure information varies a lot in different images and varies little in different channels of the same image. So maximizing structure difference between transmitted source and reflected source can be a good cue for separation. And keeping structure similarity between different channels of transmitted source is also a good way to ensure the result's validity.

By observing structure distance distributions, we can generate probability density functions of structure difference in different situations. In this paper, Gaussian model with mean 1 and variance 1 is used to describe structure distance distribution of different sources.

$$p_{structure\_s}(S_k, S_m^t) = \frac{1}{\sqrt{2\pi}} e^{-\frac{(D_{structure}(S_k, S_m^t)-1)^2}{2}} \quad k = r, g, b \qquad (15)$$

And negative exponential distribution (the red line in Fig.4 (c) (d)) can well represent structure distance between different channel images, so we have

$$p_{structure\_c}(S_i, S_j^t) = e^{-D_{structure}(S_i, S_j^t)} \quad i, j \in r, g, b, \; i \neq j \qquad (16)$$

4 Experimental Result

Our approach can achieve impressive separation results to many real world images. In Fig.2 we can see most of reflections are eliminated or weakened, while the tone of transmitted image is kept well. Moreover, spatially varying mixing coefficients can make our approach more flexible to handle different reflections in different regions.

However, we should admit there are some reflections left in the restored image. We believe a stronger modelling of structure and a further study about reflection's attribute in color space is necessary. That has been part of our current research work.

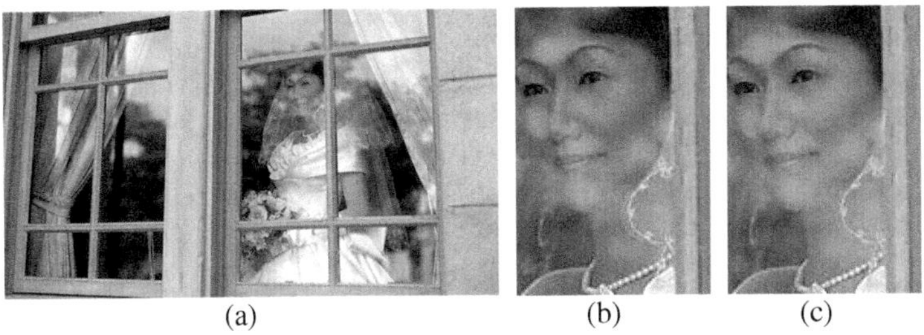

Fig. 2. Separating reflections from a single image. (a) The original image. (b) The input mixed image. (c) The transmitted image got by our approach.

To elaborate the superiority of our approach better, we made a comparison with Kayabol's approach in Fig.3. The image is provided by [10], which is a toy standing behind a transparent CD box with achromatic reflections occurring on the box surface.

With the help of spatial smoothness and color channel dependence, reflections fade in every color channel. Considering structure information, spatial smoothness will be constrained and image will not be blurred very much. Moreover, spatially varying mixing coefficients can make the separation more subtle and preserve the tone of original image well.

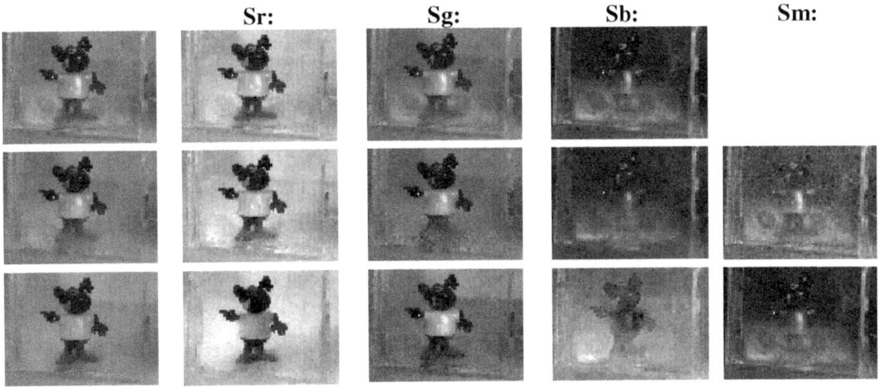

Fig. 3. The separation comparison between our approach and Kayabol's approach. Row 1: the input image and its three color channels. Row 2: the results of our approach (the transmitted image, three channels of transmitted image, and reflected image). Row 3: the corresponding results of Kayabol's approach.

5 Conclusion

We adopt two priors to separate reflections from a single image, namely spatial edge preserving smoothness based on pixels' color dependency, and structure difference in different source images and different color channels of the same image. By analyzing optical model of reflection, we simplify the mixing matrix and realize spatially varying mixing coefficients.

Based on these priors and using Gibbs sampling and appropriate probability density with Bayesian framework, our approach is able to handle real world images that are corrupted with achromatic reflections. Moreover, spatially varying mixing coefficients can help to preserve the tone of original image making results physically valid. Our approach may fail when three color channels of the input mixed image are too similar, since structure information is very hard to obtain in that case. We believe more knowledge about reflection's color attribute will advance this approach.

Acknowledgment

This work was supported in part by Research Fund for the Doctoral Program of Higher Education of China (200802481006, 20090073110022), NSFC (60932006), 973 Program (2010CB731401), and the program for Professor of Special Appointment (Eastern Scholar) at Shanghai Institutions of Higher Learning. The second author gratefully acknowledges the support of the 111 Project (B07022) and CNR's short term mobility program.

References

1. Born, M., Wolf, E.: Principles of Optics. Pergamon, London (1965)
2. Farid, H., Adelson, E.H.: Separating reflections and lighting using independent components analysis. CVPR 1, 262–267 (1999)
3. Bronstein, A.M., Bronstein, M.M., Zibulevsky, M., Zeevi, Y.Y.: Sparse ICA for blind separation of transmitted and reflected images. International Journal of Imaging System and Technology 15(1), 84–91 (2005)
4. Be'ery, E., Yeredor, A.: Blind separation of reflections with relative spatial shifts. In: ICASSP, vol. 5, pp. 625–628 (2006)
5. Gai, K., Shi, Z.W., Zhang, C.S.: Blindly separating mixtures of multiple layers with spatial shifts. In: CVPR, pp. 1–8 (2008)
6. Gai, K., Shi, Z.W., Zhang, C.S.: Blind separation of superimposed images with unknown motions. In: CVPR, pp. 1881–1888 (2009)
7. Sarel, B., Irani, M.: Separating transparent layers through layer information exchange. In: Pajdla, T., Matas, J(G.) (eds.) ECCV 2004. LNCS, vol. 3024, pp. 328–341. Springer, Heidelberg (2004)
8. Levin, A., Zomet, A., Weiss, Y.: Separating reflections from a single image using local features. In: CVPR, pp. 306–313 (2004)
9. Levin, A., Weiss, Y.: User assisted separation of reflections from a single image using a sparsity prior. IEEE Trans. Pattern Analysis and Machine Intelligence 29(9), 1647–1655 (2007)
10. Kayabol, K., Kuruoglu, E.E., Sankur, B.: Image source separation using color channel dependencies. In: Proceedings of the 8th International Conference on Independent Component Analysis and Signal Separation, pp. 499–506 (2009)
11. Kayabol, K., Kuruoglu, E.E., Sankur, B.: Bayesian separation of images modeled with MRFs using MCMC. IEEE Trans. Image Process. 18(5), 982–994 (2009)

ICA over Finite Fields

Harold W. Gutch[1,2], Peter Gruber[3], and Fabian J. Theis[2,4]

[1] Max Planck Institute for Dynamics and Self-Organization, Department of Nonlinear Dynamics, Göttingen, Germany
[2] Technical University of Munich, Germany
[3] University of Regensburg, Germany
[4] Helmholtz Zentrum Neuherberg, Germany
harold.gutch@ds.mpg.de, peter.gruber@biologie.uni-regensburg.de, fabian.theis@helmholtz-muenchen.de

Abstract. Independent Component Analysis is usually performed over the fields of reals or complex numbers and the only other field where some insight has been gained so far is GF(2), the finite field with two elements. We extend this to arbitrary finite fields, proving separability of the model if the sources are non-uniform and non-degenerate and present algorithms performing this task.

Usually, Independent Component Analysis (ICA) is performed over $\mathbb{R}$, the field of reals, and some extensions of this to the complex case have been performed [1,3]. Other than this, the only other case that has been investigated, is the case of GF(2), the field with two elements [7]. However this is not the only case of a finite field, rather generally for any prime power $q = p^n$ there is a finite field with q elements, denoted GF(q). These fields most prominently find application in coding theory, e.g. with low-density parity-check (LDPC) codes, and although here GF(2) finds the most attention, other finite fields also are of interest too, for instance see [5] for a statistical physics based analysis of LDPC codes over finite fields. We present a separation theorem for ICA over arbitrary finite fields, as long as the sources are all non-uniform and have nowhere probability mass 0, generalizing the results in [7], suggest algorithms to efficiently solve the ICA task and show simulations showing the validity of the approach.

1 Finite Fields and Discrete Probability Transformations

1.1 Finite Fields

Fields are the mathematical formalization of the well known operations $+$ and $\cdot$ following the usual conventions.

Definition 1. *A field K is a set of objects (numbers) together with two operations $+$ and $\cdot$ such that*

1. *For every $a, b \in K$, $a + b = b + a$ and $a \cdot b = b \cdot a$ (commutativity)*

V. Vigneron et al. (Eds.): LVA/ICA 2010, LNCS 6365, pp. 645–652, 2010.

2. *For every $a, b, c \in K$, $a + (b + c) = (a + b) + c$ and $a \cdot (b \cdot c) = (a \cdot b) \cdot c$ (associativity)*
3. *For every $a, b, c \in K$, $a \cdot (b + c) = (a \cdot b) + (a \cdot c)$ (distributivity)*
4. *There are elements $0 \in K$ and $1 \in K$ with the property that $0 + a = a$ and $1 \cdot a = a$ for every $a \in K$ (additive. resp. multiplicative neutrals)*
5. *For every $a \in K$ there is some $b \in K$ such that $a + b = 0$; if $a \neq 0$, then there is some $c \in K$ such that $a \cdot c = 1$ (additive resp. multiplicative inverses)*

The usual notations known from the reals apply here too, i.e. writing the additive (resp. multiplicative – if it exists) inverse of some $a \in K$ as $-a$ (resp. a^{-1}), denoting these operations subtraction and division, giving multiplication and division higher precedence than addition and subtraction and silently dropping the $\cdot$ symbol.

The most common fields $\mathbb{Q}, \mathbb{R}$ and $\mathbb{C}$, the fields of rationals, reals and complex numbers, all have an infinite number of elements, but there are also fields with only a finite number of elements, and it can be shown that for a fixed integer q, there is essentially only one finite field, denoted $\mathrm{GF}(q)$. For example, the field with the smallest number of elements is $\mathrm{GF}(2) = \{0, 1\}$ where then $0 + 0 = 1 + 1 = 0$, $0 + 1 = 1$, $0 \cdot 0 = 0 \cdot 1 = 0$ and $1 \cdot 1 = 1$, also denoted the *binary field*, and in this case addition is sometimes also denoted XOR.

It can be shown that if a field contains only a finite number q of elements, then this has to be some prime power: $q = p^n$ with some prime p and some positive integer n. If here $n = 1$, the field is called a *prime field*, and then the operations are simply $+$ and $\cdot$ of the numbers $\{0 \ldots p - 1\}$ *modulo* p. For the general case, $n > 1$ see e.g. [4].

One may define vector spaces over finite fields as usual, and then linear maps between vector spaces can be written as usual by matrices (with entries in $K = \mathrm{GF}(q)$), where as usual $\mathrm{Mat}(m \times n, K)$ is the set of $m \times n$ matrices with entries in K. Again, a matrix is invertible if and only if its determinant (calculated in K) is non-zero, and the set of invertible $n \times n$ matrices is denoted $\mathrm{Gl}(n, K)$. Many known results from linear algebra apply to arbitrary fields, however we note that there is no notion of orthogonality in vector spaces over K.

1.2 Finite Random Variables

A finite random variable X is a random variable whose probability space Ω is finite, i.e. it can take only a finite number of outcomes (states). In this case p_X, the function mapping an outcome ω to its probability is also called *probability mass function (pmf)*. If there is some constant c such that $p_X(t) = c$ for all $t \in \Omega$, then X is called *uniform*. If there is some $t \in \Omega$ such that $p_X(t) = 0$, then X is said to be *degenerate*. An m-dimensional random vector $\mathbf{X}$ over a finite probability space simply is an m-dimensional vector of random variables, each over Ω. If $f : \Omega^m \to \Omega^m$ is some invertible transformation of Ω^m, then the probability mass function of $\mathbf{X}$ transforms accordingly:

$$p_{\mathbf{X}}(\mathbf{x}) = p_{f(\mathbf{X})}(f(\mathbf{x})) .$$

Note that unlike in this equation's continuous counterpart, where instead of a finite probability distribution $p_{\mathbf{X}}$ we have a probability density function, we here do not have to adjust for the density transform by multiplying with $|(f'(\mathbf{x}))^{-1}|$.

If $\mathbf{X}$ is an m-dimensional random vector over a finite probability space Ω, the probability mass functions of the components (marginal random vectors) X_i are defined by the projections $p_{X_i}(t) = \sum_{\mathbf{x}\in\Omega^m, x_i=t} p_{\mathbf{X}}(\mathbf{x})$. Then $\mathbf{X}$ is independent if its probability mass function factorizes into the marginal probability mass functions, $p_{\mathbf{X}}(\mathbf{x}) = \prod_{i=1}^{m} p_{X_i}(x_i)$ for $\mathbf{x} = (x_1, \ldots, x_m)$.

Unlike in the real case, we here have no notion of moments, as e.g. the expression that would denote the mean, $\sum_{k\in K} p_X(k)k$ has no meaning, as here products would be taken over real values and elements of K.

1.3 ICA over Finite Fields

Let $K = \mathrm{GF}(q)$ be a finite field, so $q = p^n$ with some prime p. Assume $\mathbf{S}$ to be an independent m-dimensional random vector over K.

Lemma 1. *Let $\mathbf{v} \in K^m$, $\mathbf{v} \neq 0$. The multiset*[1] *$S := \{\mathbf{v}^\top\mathbf{s}|\mathbf{s} \in K^m\}$ consists of q^{m-1} copies of K.*

The proof of this lemma consists of a straight-forward sum over all such vectors, and with this tool, we can now already prove separability of ICA over $\mathrm{GF}(q)$:

Theorem 1. *Let q be a prime power, $K = \mathrm{GF}(q)$ the finite field with q elements. Assume $\mathbf{S}$ to be an m-dimensional independent random vector over K with probability distribution $p_{\mathbf{S}}$ where the component distributions p_{S_k} are neither uniform ($p_{S_k}(s) = q^{-1}$ for all s) nor anywhere zero. Then if, for some $\mathbf{A} \in \mathrm{Gl}(m, K)$, the probability distribution of $\mathbf{X} := \mathbf{AS}$ again is independent, then $\mathbf{A} = \mathbf{PD}$ for some permutation matrix $\mathbf{P}$ and some diagonal matrix $\mathbf{D}$.*

Proof. Let us assume that this is not the case. Then there is some tuple of $m-1$ indices such that in every row of $\mathbf{A}$ at least one of the elements corresponding to these indices is non-zero. Permuting the columns, if required, we may assume that this is the case for columns $2 \ldots n$, that is there is no row where only the first element is non-zero, or, equivalently, that in every row at least one of elements at positions $2 \ldots n$ is non-zero. We will show that in this case p_{S_1} is constant, so S_1 is uniform. The probability distribution of $\mathbf{X}$ has the form $p_{\mathbf{X}}(\mathbf{As}) = p_{\mathbf{S}}(\mathbf{s})$. Due to independence we have for any $\mathbf{s} \in K^m$:

$$\prod_{i=1}^{m} p_{X_i}(\mathbf{A}_i\mathbf{s}) = p_{\mathbf{X}}(\mathbf{As}) = p_{\mathbf{S}}(\mathbf{s}) = \prod_{i=1}^{m} p_{S_i}(s_i)$$

(note that these are products over reals). By assumption everything here is non-zero, so we may take the logarithm on both sides and, fixing a $t \in K$, we get

[1] Unlike a set, a multiset keeps track of the number of occurances of elements; unlike a sequence however, a multiset is not ordered.

$$\sum_{\substack{\mathbf{s}\in K^m \\ s_1=t}} \sum_{i=1}^{m} \tilde{p}_{X_i}(\mathbf{A}_i\mathbf{s}) = \sum_{\substack{\mathbf{s}\in K^m \\ s_1=t}} \sum_{i=1}^{m} \tilde{p}_{S_i}(s_i) \tag{1}$$

where $\tilde{p} = \log p$. The RHS of equation (1) then is

$$\sum_{i=1}^{m} \sum_{\substack{\mathbf{s}\in K^m \\ s_1=t}} \tilde{p}_{S_i}(s_i) = q^{m-1}\tilde{p}_{S_1}(t) + q^{m-2}\sum_{i=2}^{m}\sum_{s\in K} \tilde{p}_{S_i}(s) .$$

Let us now look at the LHS of equation (1): Given some $1 \le i \le m$, we are summing over all possible values of $\tilde{p}_{X_i}(\mathbf{A}_i\mathbf{s})$ where $\mathbf{s} \in K^m, s_1 = t$, or, equivalently, over all possible values of $\tilde{p}_{X_i}(a_{i1}t + \mathbf{v}^\top\mathbf{s}')$ where $\mathbf{v}^\top = (a_{i2}, \ldots, a_{im})$ and $\mathbf{s}'$ takes all possible values in K^{m-1}. By our assumption, $\mathbf{v}^\top \neq 0$ for all i, so according to Lemma 1 we are summing over q^{m-2} copies of $\sum_{s\in K}\tilde{p}_{X_i}(a_{i1}t + s)$, or, equivalently, over q^{m-2} copies of $\sum_{s\in K}\tilde{p}_{X_i}(s)$. Summing up now over all i, we see that the LHS of equation (1) is equal to

$$q^{m-2}\sum_{i=1}^{m}\sum_{s\in K}\tilde{p}_{X_i}(s) .$$

Collecting both the LHS and the RHS of equation (1), we get

$$\tilde{p}_{S_1}(t) = q^{-1}\sum_{i=1}^{m}\sum_{s\in K}\tilde{p}_{X_i}(s) - q^{-1}\sum_{i=2}^{m}\sum_{s\in K}\tilde{p}_{S_i}(s)$$

which holds for all $t \in K$, so $\tilde{p}_{S_1}$ (and then also p_{S_1}) is constant. Hence S_1 is uniform, contradicting the assumption.

2 Algorithms for ICA over Finite Fields

We first describe algorithmic approaches under the assumption of perfect knowledge of the mixed sources $\mathbf{X}$.

First of all, due to the discreteness of our probability measure, the usual continuity arguments often encountered in ICA algorithms are not applicable here: There is no way to estimate "how good" a candidate demixing matrix $\mathbf{W}$ or an extraction vector is, it is either correct or false. Therefore standard approaches employing that "guess" a demixing matrix or an extraction vector, and then iteratively update it (like e.g. JADE or FastICA) cannot be applied here.

On the other hand $\mathrm{Mat}(m \times m, K)$ is finite, and even more so $\mathrm{Gl}(m, K)$[2], so it would be possible to perform an exhaustive search over all $\mathbf{A} \in \mathrm{Gl}(m, K)$. This however is only possible for very small values of q and m.

In the following we present three alternative algorithms, all of which have a run time far lower than a simple brute force approach.

---

[2] To be precise, there are $q^{\binom{m}{2}}(q^{-m}; q)_m$ such matrices, where $()_j$ is the q-Pochhammer symbol.

2.1 Algorithm A: Entropy Based Extraction/Deflation

The first algorithm essentially is a generalization of the algorithm presented in [7] to the case of an arbitrary finite field. In it, we extract a single source, remove the contribution from this source to the mixtures (deflation [2]), and repeat this process m times after which we have found all m sources.

Entropy Based Extraction. The idea here is that non-trivial linear combinations of independent random variables have a higher entropy than the random variables themselves. We can therefore scan the set of linear combinations of the observations $\mathbf{X}$ searching for the linear combination that has the lowest entropy. In other words, we search for

$$\mathbf{v}_0 = \operatorname{argmin}_{\mathbf{v} \in K^m} H(\mathbf{v}^\top \mathbf{X})$$

where $H(X) = -\sum_{k \in K} p_X(k) \log(p_X(k))$. If $\mathbf{v}_0$ is such a weight vector, then $\mathbf{v}_0 \mathbf{X}$ is equivalent to the source of lowest entropy, so $\mathbf{v}_0$ has to be a row of the inverse of the mixing matrix $\mathbf{A}$.

Entropy Based Deflation. Having recovered a single source, say S_1, we now remove it from all mixtures X_i. This step is similar to the deflation in [7], just that in our more general setting, it is not sufficient to compare $H(X_i)$ and $H(X_i + S_1)$, instead we now have to minimize $H(X_i + kS_1)$ with respect to $k \in K$ and replace X_i then with $X_i + kS_1$.

2.2 Algorithm B: Entropy Based Demixing

Similarly to the entropy based extraction/deflation, we choose two sources, say X_i and X_j and a $k \in K$ such that $H(X_i + kX_j) < H(X_i)$ and then replace X_i with $X_i + kX_j$, so we remove kX_j from X_i, demixing the observations $\mathbf{X}$ step by step. Note that this operation can be represented by the matrix with ones on the diagonal, the entry k at position (i, j) and zeros everywhere else, which clearly is invertible. We repeat this step as long as we can find two indices $i \neq j$ and a scalar $k \neq 0$ such that the inequality above holds. While this algorithm is prone to local minima, it is a lot faster than pure entropy based extraction, as we have a lower search space to scan in every step.

2.3 Algorithm C: Extreme Probability Based Extraction/Deflation

We now present an algorithm that is by many orders of magnitude faster than the previous two entropy based algorithms, as it simply requires ordering of the probabilities in question. However this comes with the cost of decreased accuracy.

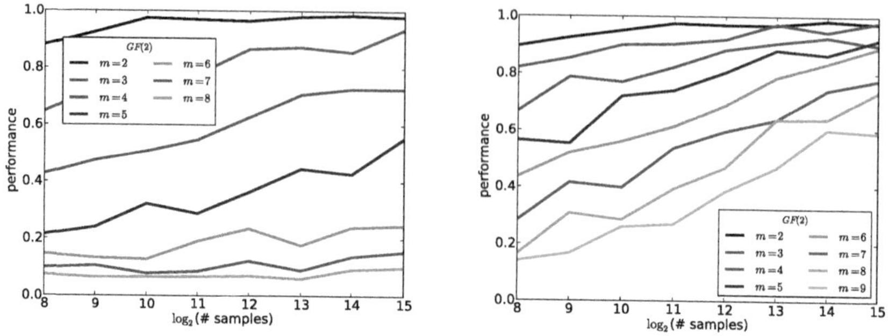

(a) Algorithm A (Entropy based extraction/deflation)

(b) Algorithm B (Entropy based demixing)

Fig. 1. Evaluation of the Algorithms A and B. The x-axis depicts $\log(N)$ for the number N of samples, the y-axis depicts the rate of recovered sources, averaged over 100 runs. For small values of m, even with a moderate number of samples, good reconstruction is possible, outperforming algorithm C.

Extreme Probability Based Extraction. Let us first assume that $p_{\mathbf{S}}(\mathbf{x}) \neq p_{\mathbf{S}}(\mathbf{y})$ for $\mathbf{x} \neq \mathbf{y}$. We sort the points of $\mathbf{V} = K^m$ according to probability in the sources and denote the sorted points $\mathbf{s}_1, \ldots, \mathbf{s}_{q^m}$ (so $i < j \Leftrightarrow p_{\mathbf{S}}(\mathbf{s}_i) < p_{\mathbf{S}}(\mathbf{s}_j)$). As $p_{\mathbf{S}}(\mathbf{s}_1) = p_{S_1}(\mathbf{s}_1)_1 \cdots p_{S_m}(\mathbf{s}_1)_m$ is minimal among all probabilities, for every index k the k-th marginal probability of $p(\mathbf{s}_1)$ also has to be minimal among the possible k-th marginal probabilities, so the coordinates of $\mathbf{s}_1$ correspond to the minimal marginal probabilities. It is then easy to see that the coordinates of $\mathbf{s}_2$ can differ from those of $\mathbf{s}_1$ in only one single component[3] so $\mathbf{s}_2 - \mathbf{s}_1 = \mathbf{e}_k$, the k-th unit vector for some k. Since straight lines are mapped to straight lines by the linear mapping $\mathbf{A}$, the difference of the two points of lowest probability in the observations $\mathbf{X}$ too gives us the direction of the k-th source (in the observations), i.e. a column of the mixing matrix $\mathbf{A}$. The same approach of course also works if we take the points of highest and second highest probability instead.

In the general case where not all probabilities differ, we may still fall back to the computationally more complex entropy based extraction, restricting the search space to the set of directions gained from the maximum/minimum probability extractor.

In fact, closer analysis reveals that the maximum probability based extractor can be seen as an approximation of the entropy based extractor where we approximate the entropy $H(S) = -\sum_{k \in K} p_S(k) \log(p_S(k))$ by $\hat{H}(S) = -p_S(k_0) \log(p_S(k_0))$ where k_0 is chosen among all $k \in K$ such that $\hat{H}(S)$ is maximal. This expression is maximal whenever $p_S(k_0)$ has maximal distance from 0.5, which is the case if we choose maximal or minimal probability in every single coordinate.

[3] E.g., with two sources, if $\mathbf{s}_1$ and $\mathbf{s}_2$ differ in both the first and second coordinate, then $\log p_{S_1}(\mathbf{s}_1)_1 + \log p_{S_2}(\mathbf{s}_1)_2 < \log p_{S_1}(\mathbf{s}_1)_1 + \log p_{S_2}(\mathbf{s}_2)_2 < \log p_{S_1}(\mathbf{s}_2)_1 + \log p_{S_2}(\mathbf{s}_2)_2$.

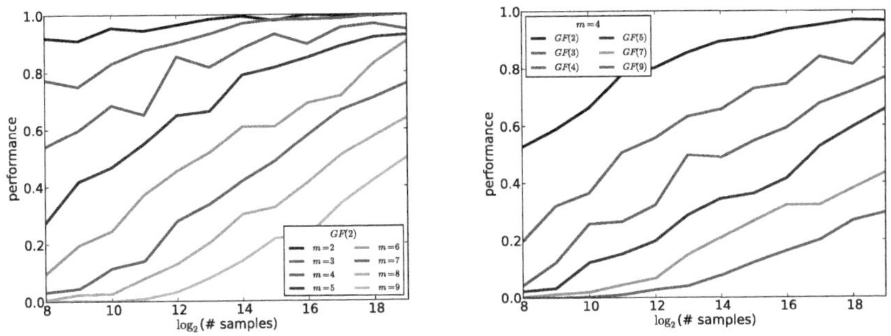

(a) Algorithm C (Probability based extraction/deflation)

(b) Algorithm C (Probability based extraction/deflation)

Fig. 2. Evaluation of Algorithm C for various values of $K = \mathrm{GF}(q)$ and m. The x-axis depicts $\log(N)$ for the number N of samples, the y-axis depicts the rate of recovered sources, averaged over 100 runs. As long as q and m are small, moderate samplesizes are sufficient for good reconstruction at a far lower run time than with Algorithms A and B.

Maximum/Minimum Probability Based Deflation. We now need to remove the recovered source direction $\mathbf{v}$ from the mixtures, i.e. project the mixtures onto a subspace $\mathbf{W}$ complementary to $\mathbf{v}$, i.e. such that $\mathbf{W} \oplus \mathrm{span}(\mathbf{v}) = \mathbf{V}$. Let us first see how we can perform this in terms of the sources $\mathbf{S}$. We so far have recovered a source direction, say $\mathbf{e}_1$ and now wish to remove this component from all sources. We choose s_1^0 where p_{S_1} is minimal. The direction $\mathbf{e}_1$ is independent from the rest, so the probability mass function factorizes, and we can write $p_{\mathbf{S}}(\mathbf{s}) = p_{S_1}(s_1)p_{\mathbf{S}'}(\mathbf{s}')$. For any fixed $(s_2, \ldots, s_m)$, then also $p_{\mathbf{S}}(., s_2, \ldots, s_m)$ is minimal at s_1^0. Therefore collapsing the probability mass of every point $\mathbf{s} \in \mathbf{V} = K^m$ onto the point $\mathbf{s} + l\mathbf{e}_1$ chosen such that $p_{\mathbf{S}}(\mathbf{s} + l\mathbf{e}_1)$ is minimal gives us a plane spanned by the vectors $(\mathbf{e}_2, \ldots, \mathbf{e}_m)$, which is just what we wanted to achieve.

Now, in terms of $\mathbf{X}$ the same is possible: Assume the recovered source direction $\mathbf{v}$. We then iterate over all points $\mathbf{x} \in \mathbf{V} = K^m$ and for every $\mathbf{x}$ we select l such that $p_{\mathbf{X}}(\mathbf{x}+l\mathbf{v})$ is minimal. We then replace $p_{\mathbf{X}}^{new}(\mathbf{x}) = \sum_{l \in K} p_{\mathbf{X}}^{old}(\mathbf{x}+l\mathbf{v})$, leaving us with the remaining directions.

Note that this operation is linear and hence can be expressed by a matrix, so the tedious step of searching for a maximal element along $\mathbf{v}$ only has to be performed for a basis of the complementing subspace.

Similar to before, everything here may be of course be performed just the same by replacing minimal with maximal probabilities.

2.4 Estimating the Probability Mass Function

So far we have assumed perfect (asymptotic) knowledge of the probability mass function. In a usual setting however only a finite number N of i.i.d. samples $\mathbf{x}_n$ are given, and just as in [7] we then get a consistent estimator of $p_{\mathbf{X}}$ by

$$\hat{p}_{\mathbf{X}}(\mathbf{x}) = \frac{1}{N}\sum_{i=1}^{N}\mathcal{I}\{\mathbf{x}_n = \mathbf{x}\}$$

where $\mathcal{I}$ is the indicator function.

2.5 Simulations

We implemented all three algorithms in Sage [6] and present simulations thereof. We independently and uniformly sampled m random variables on GF(q), generated $N = 2^k$ (for values of k up to 19) i.i.d. samples of these and mixed them according to a randomly chosen invertible matrix $\mathbf{A}$. For each value of q, m and N we let the algorithms suggest a recovery matrix $\mathbf{W}$, and we evaluated the quality of $\mathbf{WA}$ by counting the relative number of rows with exactly 1 non-zero entry. Figures 1 and 2 depict the averages over 100 runs.

3 Conclusion

We have transferred the well-known separability theorem of ($\mathbb{R}$-)ICA to the domain of finite fields of arbitrary size, presenting algorithms that solve the task efficiently. Future work may include subspace analysis where not all sources are assumed to be independent, and applications, e.g. in coding.

References

1. Bingham, E., Hyvärinen, A.: A fast fixed-point algorithm for independent component analysis of complex valued signals. International Journal of Neural Systems 10(1), 1–8 (2000)
2. Delfosse, N., Loubaton, P.: Adaptive blind separation of independent sources: a deflation approach. Signal Processing 45(1), 59–83 (1995)
3. Eriksson, J., Koivunen, V.: Complex random vectors and ICA models: identifiability, uniqueness, and separability. IEEE Transactions on Information Theory 52(3), 1017–1029 (2006)
4. Lang, S.: Algebra. Addison-Wesley, Reading (1965)
5. Nakamura, K., Kabashima, Y., Saad, D.: Statistical mechanics of low-density parity check error-correcting codes over galois fields. EPL (Europhysics Letters) 56(4), 610–616 (2001)
6. Stein, W., et al.: Sage Mathematics Software (Version 4.3.5). The Sage Development Team (2010), http://www.sagemath.org
7. Yeredor, A.: ICA in boolean XOR mixtures. In: Davies, M.E., James, C.J., Abdallah, S.A., Plumbley, M.D. (eds.) ICA 2007. LNCS, vol. 4666, pp. 827–835. Springer, Heidelberg (2007)

Author Index

GPSR Compliance

The European Union's (EU) General Product Safety Regulation (GPSR) is a set of rules that requires consumer products to be safe and our obligations to ensure this.

If you have any concerns about our products, you can contact us on ProductSafety@springernature.com

In case Publisher is established outside the EU, the EU authorized representative is:

Springer Nature Customer Service Center GmbH
Europaplatz 3
69115 Heidelberg, Germany

Batch number: 10704691

Printed by Printforce, the Netherlands